CYBERNETICS

IIquIIIII

Edited by
Claus Pias and Joseph Vogl

CYBERNETICS

The Macy Conferences 1946–1953

The Complete Transactions

Edited by Claus Pias

diaphanes

First Printing

ISBN 978-3-03734-598-6

Cover: Thomas Bechinger and Christoph Unger
Layout: Claus Pias
Printed in Germany

www.diaphanes.com

CONTENTS

EDITORIAL NOTE

This edition includes the complete text of the five volume documentation of the Macy Conferences on Cybernetics, edited by Heinz von Foerster.* The pagination of the original document is set in the margin. Notes by the editor are enclosed in square brackets. I would like to thank the *Josiah Macy, Jr. Foundation* and its president June E. Osborn for generously allowing this edition, Tobias Nanz and Katrin Richter for their editorial assistance.

Claus Pias, Berlin 2015

* Heinz von Foerster (Hrsg.): *Cybernetics. Circular Casual and Feedback Mechanisms in Biological and Social Systems, Transactions of the Sixth Conference (March 24–25, 1949)*, New York 1950
Heinz von Foerster/Margaret Mead/Hans Lukas Teuber (Hrsg.): *Cybernetics. Circular Casual and Feedback Mechanisms in Biological and Social Systems, Transactions of the Seventh Conference (March 23–24, 1950)*, New York 1951
— : *Cybernetics. Circular Casual and Feedback Mechanisms in Biological and Social Systems, Transactions of the Eighth Conference (March 15–16, 1951)*, New York 1952
— : *Cybernetics. Circular Casual and Feedback Mechanisms in Biological and Social Systems, Transactions of the Ninth Conference (March 20–21, 1952)*, New York 1953
— : *Cybernetics. Circular Casual and Feedback Mechanisms in Biological and Social Systems, Transactions of the Tenth Conference (April 22–24, 1953)*, New York 1955.

THE AGE OF CYBERNETICS

CLAUS PIAS[1]

Although aspects of cybernetics can be traced back to various points in history,[2] the proceedings of the so-called Macy Conferences, which have been edited for this volume, represent its modern foundational document. Held between 1946 and 1948 under the cumbersome title "Circular Causal and Feedback Mechanisms in Biological and Social Systems," the papers delivered at these conferences were soon thereafter, at least as of 1949, referred to as contributions to *cybernetics*. Sponsored by the Josiah Macy, Jr. Foundation (which was and remains concerned above all with promoting advancements in the medical sciences), organized by Frank Fremont-Smith (who for good reasons was known by the nickname "Mr. Interdisciplinary Conference"), and moderated by Warren S. McCulloch, the Macy Conferences perpetuated the standards of interdisciplinary research groups, as established during the Second World War, into the era of the Cold War. As tempting as it may be to doubt, in its details, the success of this dialog, the possibilities and limitations of which were repeatedly problematized at the conferences themselves (as in the question of group communication, or the relationship between European and American scientific traditions, or by means of "To Whom It May Concern Messages,"[3] for instance), it is nevertheless necessary to acknowledge the systematic intentions and unrelenting efforts of the undertaking, its desire to integrate concepts that had hitherto been kept far apart from one another,[4] and its aim to design overarching orders of knowledge with nothing short of epoch-changing implications (though the results would always turn out to be more suggestive than anything else). Gregory Bateson would later write (with his characteristic ambiguity): "I think that cybernetics is the biggest bite out of the fruit of the Tree of Knowledge that mankind has taken in the last 2000 years."[5] And he was apparently not alone in this opinion.

The meetings themselves were painstakingly organized. As early as the first conference in 1946, it was firmly established what ingredients would be needed to formulate a general theory: the principles of the current computer generation, the latest developments of neurophysiology, and finally a vague "humanistic" combination of psychiatry, anthropology, and sociology.[6] This idea more or less determined the schedule of the talks, which was

1 Support for this research was provided by a fellowship at the University of Konstanz's Institute of Advanced Study, which is part of the university's "Cultural Foundations of Social Integration" Center of Excellence, established under the framework of the German Federal and State Initiative for Excellence.

2 Otto Mayr, *The Origins of Feedback Control* (Cambridge, MA: MIT Press, 1970); Eberhard Lang, *Zur Geschichte des Wortes Kybernetik*, Grundlagenstudien aus Kybernetik und Geisteswissenschaft 9 (Quickborn: Schnelle, 1968); Joseph Vogl, "Regierung und Regelkreis: Historisches Vorspiel," in *Cybernetics/Kybernetik: The Macy Conferences 1946–1953*, ed. Claus Pias, 2 vols. (Zurich: diaphanes, 2003–2004), 2:67–80.

3 Erhard Schüttpelz, "To Whom It May Concern Messages," in *Cybernetics/Kybernetik*, 2:115–30.

4 In his typically aphoristic fashion, Marshall McLuhan once wrote: "[T]he electric age of cybernetics is unifying and integrating." Quoted from Marshall McLuhan, "Cybernation and Culture," in *The Social Impact of Cybernetics*, ed. C. R. Dechert (New York: Simon and Schuster, 1966), 98. In this regard, Hans Lenk has used the term "Experten für das Allgemeine" ["experts in the general"]; see his *Philosophie im technologischen Zeitalter* (Stuttgart: Kohlhammer, 1971), 107.

5 Gregory Bateson, "From Versailles to Cybernetics," in *Steps to an Ecology of Mind: Collected Essays in Anthropology, Psychiatry, Evolution, and Epistemology*, by Bateson, 2nd ed. (London: Jason Aronson, 1987), 475–83, at 481.

6 Letter from Frank Fremont-Smith to Warren McCulloch, 8 February 1946 (APS). Unless otherwise noted, the documents abbreviated (APS) derive from the Warren McCulloch Papers, American Philosophical

based on the dual treatment of themes (the second talk of each panel was referred to as an "exemplification"). Von Neumann, that is, spoke about computing machines, and Lorente de Nó responded with an analogy from biology; Wiener spoke about goal-seeking devices, and Rosenblueth provided a biological analogy; Bateson discussed the need for theory in the social sciences, and Northrop offered comparisons from physics. Finally, certain "problems in psychology and psychiatry" were addressed in light of mathematical game theory.[7] "The agenda[s] speak for themselves."[8]

Although Steve Heims has summarized the Macy Conferences in his commendable work *Constructing a Social Science for Postwar America*, the title of the book is only partially accurate.[9] It is of course true that, from the beginning, there was an interest in cybernetics as a model for social, economic, or political means of control or intervention, i.e., as a model for *machines à gouverner*[10] that promised to fulfill the changing needs of the post-1945 government.[11] That said, interest in the agendas of the Macy Conferences seems to have been dispersed across a number of disciplines beyond the social sciences. In that cybernetic theories, for example, no longer treated technical-material structures but rather logical and mathematical operations as the *tertium comparationis* between the brain and the computer, they gained a degree of flexibility that could accommodate neurologists just as well as engineers. Thus the great majority of publications from this time are concerned with questions of medicine and computer technology. Looking at contemporary reviews, one finds that the Macy Conferences were in fact regarded as being devoted above all to the field of medicine, which is hardly surprising given the prominence of its sponsor.[12] Here one also finds a prevailing sense of irritation: Most of the reviewers simply resorted to the simple adjective "stimulating." It is repeatedly stated that the content of the events could not be summarized, after which a mere list of the individual contributions is provided. "It is impossible to review the content of such work," notes one reviewer, though somehow he is nevertheless able to regard this same work as representing the zenith of "current thinking." Judgments of this sort abound, even though it is likely that none of the reviewers truly understood the comprehensive scope of the cybernetic project. To them, on the contrary, the talks were rather characterized by their "looseness of reasoning" and "lack of precision in thinking."[13]

Apparently it was not (yet) possible to identify the unifying contours of cybernetics.

Society, Philadelphia, B/M 139 (Series I, Box 13 [*Josiah Macy, Jr. Foundation*]; Series II, Box 9 [*Macy-Meeting IX, March 1952; Macy-Meeting X,April 1953*]; Series II, Box 19 [*Josiah Macy, Jr. Foundation 1948–1952; Macy-Meeting I, 1946; Macy-Meeting VII, 1951*]). Documents abbreviated (HvFA) are housed at the Heinz von Foerster Archive, Vienna, DO967 (*Macy Corresp.* und *Macy 49–53*).

7 Agenda for the first conference, March 1946 (APS).

8 Letter from Warren McCulloch to Gregory Bateson and Margaret Mead, 11 February 1946 (APS).

9 Steve J. Heims, *The Cybernetics Group, 1946–1953: Constructing a Social Science for Postwar America* (Cambridge, MA: MIT Press, 1991).

10 This expression is borrowed from Dominique Dubarle's review of Norbert Wiener's book *The Human Use of Human Beings: Cybernetics and Society* (Boston: Houghton Mifflin, 1950); see Dominique Dubarle, "Une nouvelle science: la cybernétique. Vers la machine à gouverner," *Le Monde* (28 December 1948).

11 Though difficult to reconstruct in its details, the verifiable attention paid by the CIA to cybernetics is just one of many noteworthy facts in this regard.

12 The reviews in question can be found in *Archives of Internal Medicine* (May 1952); *Southern Medical Journal* (July 1952); *Bulletin of the Johns Hopkins Hospital* (August 1952); *Journal of the Franklin Institute* (August 1952); *Journal of Medical Education* (September 1952); *International Journal of Psychoanalysis* (October 1952); *The Academy of General Practice* (October 1952); *American Journal of Public Health* (November 1952); *Journal of Applied Psychology* (December 1952); *The Psychiatric Quarterly* (January 1953); *Yale Journal of Biology and Medicine* (February 1953); and *Josiah Macy, Jr. Foundation 1930–1950: A Review of Activities* (New York, 1955).

13 George Rosen, "Book Review: Cybernetics – 8th Conference," *American Journal of Public Health* 42 (1952), 1481

These contours would first be sharply delineated by McCulloch, who arranged the elements of cybernetics as though drawing up a precise blueprint. He created meticulous lists of those who would represent each of the sciences: three mathematicians plus three physiologists plus three psychiatrists plus three sociologists plus three psychologists... and so on.[14] Recommendations of potential participants would only be considered if the parity among the different sciences could be maintained; moreover, he stuck to a rigid policy as regards invitations: The conferences were to remain closed, and requests to participate by interested parties were rejected.[15] In the event, however, that someone would be admitted, memoranda were issued whose precision resembled that of a formal job offer from a government agency. Yet at the same time the hope was to attract "big names" to contribute to this exclusive project, and those who declined to participate shed some light of their own on the founding of cybernetics. Bertrand Russell, for example, conceded that the topic sounded "exceedingly interesting," adding that he would have been inclined to accept the offer had he not experienced considerable fatigue during a recent "jaunt" to America.[16] Albert Einstein declined his invitation with an ironic remark that charmingly deflated the ambitions of the grand theory: Surely, he wrote, cybernetics is just a new branch of "applied mathematics" (not the most reputable of fields, in other words) that will become an "important tool" for its "specialists," but he considered his knowledge of this field to be too "superficial" as to be of any use.[17] Finally, Alan Turing, who ought to have been interested in the opportunity, excused himself from the meeting on account of being "a stay-at-home type", on account of a new semester beginning, and last of all (and somewhat more significantly) because he was doubtful that he "could get permission to be away."[18] It is possible to surmise – and this was a repeated theme at the conferences – that issues of confidentiality were in play.

What subsequently took place behind the closed doors of the glorious Beekman Hotel on New York's Park Avenue is only hinted at by the peculiar textual records associated with the conferences: "We fight in our shirt sleeves, and we do not even publish the proceedings until every man has had a chance to go over what he has said and delete the more objectionable phrases."[19] In order not to air any dirty laundry from the events, the participants' presentations and discussions were heavily redacted for the press. And in this regard alone, the Macy Conferences were a poetic creation and an enormous aesthetic achievement. Among the editors of the volumes, it was only Heinz von Foerster, whose draft for the preface to accompany the published proceedings of the eighth conference has fortunately survived, who wished to acknowledge his pride in this accomplishment.[20] It is not without some irony that his proposed text was revised and returned to him by Hans-Lukas Teuber (and Margaret Mead) along with the question of whether he still might be able to recognize it as "a child of his own" – even though hardly a single word of his draft was preserved.[21] From his text they expunged both an observation concerning the media-technical conditions of the Macy Conferences as well as a cybernetic description of how to describe cybernetics and thus – if one is willing to go so far – von Foerster's venture into what would now be called second-order cybernetics.

14 An overview of the scientific disciplines to be involved and their representatives, 1946 (APS).

15 See, for instance, the letter from Warren S. McCulloch to J. A. Winter dated 13 January 1953 (APS).

16 Letter from Bertrand Russell to Warren S. McCulloch, 4 February 1953 (APS).

17 Letter from Albert Einstein to Warren S. McCulloch, 2 April 1953 (APS).

18 Letter from Alan M. Turing to Warren S. McCulloch, undated [1953] (APS).

19 Letter from Warren S. McCulloch to J. A. Winter, 13 January 1953 (APS).

20 Heinz von Foerster, "A Note by the Editor: Proposal by H.v.F.," 12 December 1951 (HvFA); see also: Heinz von Foerster, "Circular Causality. The Beginnings of an Epistemology of Responsibility", in *Collected Works of Warren S. McCulloch*, ed. by Rook McCulloch, (Salinas, Calif.: Intersystem Publications, 1982), Vol III,, 808–28.

21 Letter from Hans-Lukas Teuber to Heinz von Foerster, 1 February 1952 (HvFA).

Records of the first Macy Conferences were taken stenographically, and Foerster later recalled how difficult it had been, at least initially, to create from this shorthand the poetic reality of a fluent and authentically functional dialog. It was up to the editors, he wrote, "to arrange, to smoothen, to clarify, to condense," and to (re)construct a certain atmosphere in the text: "[F]irst names should be maintained as well as some jokes and acidities as long as they serve the purpose of those delightful enzymes whose presence facilitate otherwise inert reactions." The spontaneity and liveliness of the discussions could not be captured without some degree of narratological ingenuity. Things would have turned out much differently, as John Stroud stressed to the editors after the publication of the first volume, had the event been recorded on tape (he even offered, if necessary, to bring his own device to the next gathering).[22] In this transition from (symbolic) writing to (real) sound, cybernetics should have been to some extent conscious of itself (that is, of its medial difference). Suddenly it became apparent that certain co-authorial forces were at work: "the tone of the voice, the gesture, the smiles, the attention directed by the turn of the head towards one person or another." This was something entirely novel for scientists ("hard scientists," at that) to take into account. For suddenly it was the case that – beyond the predominance of numbers, research results, and argumentation – the "grain" and melody of the scientists' voices attracted attention and gained significance,[23] and it became relevant whom someone was looking at while speaking, what sort of body language was used, and so on. For such reasons the organizers thus found it glaringly necessary to document the details of the conferences all the way down to the order of seating.[24] What is suggested by all of this is a cybernetic interest in cybernetics itself, an interest in the "weak" currents that happen to control those that are stronger. The epistemological shifts of the cyberneticists were governed by details – by blinks of the eye, intonations, and gestures. In fact, it is from such realities that Paul Watzlawick, who himself had taken part in the conferences, would later (and in the best cybernetic tradition) develop his axioms for couples therapy: the fact, for example, that every communication has a content and a relationship aspect such that the latter classifies the former; that every communication, moreover, has a digital modality (with a complex logical syntax but an inadequate semantics for relationships) and an analog modality (with the semantic potential for relationships but with no logical syntax);[25] and finally that, unfortunately, "content-based" or "digital" conversions will be of no help at all in solving a couple's relationship problems.

From this it is possible to draw certain conclusions about the peculiar interdisciplinarity of the Macy Conferences, which were rather heavy on the analog modality of communication. Heinz von Foerster proudly noted (and this, too, was omitted from the final preface), that, despite meeting together over the course of six years, the group had not developed any sort of "in-group" language or jargon. The common component of their language lay elsewhere: not on the "verbal level" but rather "in a sort of ethos within which tones of voice serve as a common currency of communications." What does a sort of interdisciplinarity look like, however, that is based less on an understanding than it is on relationships? At the Macy Conferences – to quote von Foerster yet again – "the thing that is shared is *not* simply a belief that the different disciplines ought to understand each other better, *nor* a body of shared material to which different methods of analysis are brought together, *nor* a single problem towards the solution of which the members are bending their differentiated and united efforts, but rather," he goes on, "an experiment with a set of conceptual models which seem to be useful right across the board and which themselves

22 Letter from John W. Stroud to Heinz von Foerster, 2 September 1949 (HvFA).

23 See Roland Barthes, "The Grain of the Voice," in *Image, Music, Text*, translated by Stephen Heath (London: Fontana Press, 1977), 179–89.

24 A sketch of the seating arrangement for the sixth conference, 24–25 March 1949 (HvFA).

25 Paul Watzlawick, Janet H. Beavin, and Don D. Jackson, *Pragmatics of Human Communication: A Study of Interactional Patterns, Pathologies, and Paradoxes* (New York: Norton, 1967).

provide a medium of communication also – when shared."[26] By sharing models, it was possible to understand even without understanding, and this act of sharing involved an atmosphere – a mood of optimism and expectation – whose content-specific details were and remained personal and particular to everyone in attendance.

*

This "set of models" had three main elements, each of which derived from American research conducted in the early 1940s: first, the logical calculus of Pitts and McCulloch's neuron model; second, Shannon's information theory; and third, the behavioral theory formulated by Wiener, Bigelow, and Rosenblueth.[27] In other words: A universal theory of digital machines, a stochastic theory of the symbolic, and a non-deterministic yet teleological theory of feedback were combined at the Macy Conferences into a *single* theory that could then claim validity for living organisms as well as machines, for economic as well as psychological processes, and for sociological as well as aesthetic phenomena.

Walter Pitts and Warren McCulloch's twenty-page article from 1943, "A Logical Calculus Immanent in the Ideas of Nervous Activity,"[28] begins with the most ambitious goal imaginable, namely that of writing, in McCulloch's words, "a theory in terms so general that the creations of God and men almost exemplify it." With a mixture of notations from Carnap, Russell, and their own invention, the authors sought to devise a sort of logical calculus of immanence: They expressed neuronal interactions as propositional functions that, in turn, could be expressed as neuronal interactions.[29] And this meant, *first*, that to understand a given aspect of the nervous system it would be sufficient to conceive it as an embodiment of Boolean algebra. The material reality of slimy brain matter is at best a sloppy instantiation of pure and elegant switching logic on the (Platonic) "instruments of time." This concept of embodied mathematics implied, *second*, that logical notations could be applied for any number of uses – be it to describe synapses or vacuum tubes, switches or ink on paper. Pitts and McCulloch's ideas were thus able to serve simultaneously as neurophysiological, philosophical, and computer-technical concepts: concepts that operate and function, that can explicate both theoretical and practical entities, and that can be used both to model neuronal structures as well as to construct artifacts – exactly as John von Neumann, with this essay in hand, set out to construct digital computers. And this meant, *third*, that if all neuronal functions could be recorded as embodiments of logical calculus, it would probably have to be admitted that everything that can be known could be known *in* and *by means of* logical calculus. Epistemology merged with psychology; Kant's synthetic *a priori* became a circuit and was thus no longer a purely human matter. Or in other words: For every conceivable thought, a network could be devised that connects to it and is thus able to think, whereby the mind or "spirit" (*"Geist"*) suddenly finds itself on the engineer's desk.[30] "Mind no longer goes more ghostly than a ghost," as McCulloch wrote.

26 Heinz von Foerster, "A Note by the Editor: Proposal by H.v.F.," 12 December 1951 (HvFA). Regarding the inclusion of another person's *Weltentwurf* or conception of the world, see also Niklas Luhmann, *Love as Passion: The Codification of Intimacy*, translated by Jeremy Gaines (Cambridge, MA: Harvard University Press, 1987).

27 On the relationship of cybernetics to the European scientific tradition, see Henning Schmidgen, "Zeit als peripheres Zentrum: Psychologie und Kybernetik," in *Cybernetics/Kybernetik*, 2:131–52.

28 Warren McCulloch und Walter Pitts, "A Logical Calculus Immanent in the Ideas of Nervous Activity," *Bulletin of Mathematical Biophysics* 5 (1943), 115–33.

29 See Lily Kay, "From Logical Neurons to Poetic Embodiments of Mind: Warren McCulloch's Project in Neuroscience," *Science in Context*, 14 (2001), 591–614.

30 The concept of this was an early "expulsion of the spirit from the humanities" (*Austreibung des Geistes aus den Geisteswissenschaften*), as Friedrich Kittler would later call it. See, for instance, Helmar Frank's remarks in his *Kybernetik und Philosophie: Materialien und Grundriß zu einer Philosophie der Kybernetik* (Berlin: Duncker & Humblot, 1966), 103: "It can be supposed that, by the end of our century, [...] the 'humanities', which by then will have been 'modernized', will be characterized by no longer being

With this deconstructive turn, it was thought that the human had become a special sort of information machine, and information machines had become the general concept behind all "communication." The human self appeared to be "computationally constituted" (in McCulloch's words), and this is because the self not only made sense of its experiences through the (conscious) manipulation of symbols; it had also, moreover, ensured that experiences of any sort could only be made possible by means of the (unconscious) manipulation of symbols. Here a reference could be made, for instance, to Lacan's model of the psyche, which was based on the machine.[31]

This new, cognitive "conception of man" in terms of logical circuits possessed not only the elegance of a micro- and macroscopically functioning model of universal symbol manipulation; but for this very reason, it was also perfectly compatible with Claude Shannon's information theory, which was likewise based on the digital.[32] For, *first*, the latter functioned with binary operations to determine the content of information, just as McCulloch's abstract synapses knew only "all-or-nothing" conditions. *Second*, Shannon's theory regarded information as a new category beyond matter and energy – as something that could thus be transmitted without loss regardless of the materiality of its instances, just as McCulloch's circuits could be implemented, without loss, in flesh or metal or silicon. And *third*, it operated with the same statistical event probabilities that McCulloch considered to be neurologically responsible for the possibility of recognizing universals (in the Aristotelian sense).

Finally, logical calculus and information theory also enmesh with the concepts of feedback that Norbert Wiener, Julian Bigelow, and Arturo Rosenblueth had developed around the same time in their article "Behavior, Purpose, and Teleology."[33] For, *first*, the pursuit of various "goals" and "non-deterministic teleology" are based on differences whose (im)probability can be regarded as a need for information. *Second*, such deviations may not be discontinuous but rather have to be dampened and measured in rhythm with discrete intervals of time, just as any type of thinking requires switching time. *Third* and finally, digitally established "living organisms and machines" require the concept of feedback in order to be productive themselves. Memory and phantom pain, stuttering and neuroses, schizophrenia and depression, laughter and pure concepts of understanding (to name just a few themes from early cybernetics) are to be observed within opened black boxes as circuits with cycles in which their signals are incessantly being processed and in which the network itself generates new or supplementary knowledge that, rather than needing additional inputs, simply bends back to its own outputs.

The common precondition of these three foundational concepts of cybernetics – switching (Boolean) algebra, information theory, and feedback – is digitality. It is thus only when humans and machines operate on the same digital basis, when the knowledge of humans and that of machines can be made compatible, that the epistemology of cybernetics is itself able to be productive. It is thus no surprise that all ten of the Macy Conferences repeatedly revolved around the meaning and significance of the concepts of "analog" and "digital" – around their conceptual scope and their empirical "truth," which, even though fundamental to the strategic mechanism or *dispositif* of cybernetics, was not immediately recognized as such. From the very first discussion to the last, each

concerned with the 'spirit' ['*Geist*'] and its derivatives but rather by having fragmented it into multiple components and thus having 'despiritualized' it into systems of information and information processing" (transl. by Valentine Pakis).

31 See Mai Wegener, *Neuronen und Neurosen: Zum psychischen Apparat bei Freud und Lacan* (Munich: Fink, 2004).

32 Claude E. Shannon, "A Mathematical Theory of Communication," *Bell System Technical Journal* 27 (1948), 379–424, 623–57.

33 Norbert Wiener, Julian Bigelow, and Arturo Rosenblueth, "Behavior, Purpose, and Teleology," *Philosophy of Science* 10 (1943), 18–24.

of the many recurring negotiations about the nature of "analog" and "digital" brought new conceptual dichotomies to the surface: entropy versus information, continuous versus discontinuous, linear versus nonlinear, singular event versus repetition, probability versus improbability, the real versus the symbolic, nature versus artifact, and so on.[34] Finally, whenever proponents of the technical efficacy of the digital asserted their views against those who found good reasons to implement analog or hybrid models,[35] what was in fact at stake was the discursive efficacy of cybernetics, which at one point had to be imposed by cutting off one of the speakers.[36] Thus the "Summary" of the last conference draws a point of connection with the first conference's opening lecture: "We considered Turing's universal machine as a 'model' for brains, employing Pitts' and McCulloch's calculus for activity in nervous nets."[37]

⋆

Beyond being yet another affront to anthropology, all of this was also a philosophical and historical challenge of the highest order – a challenge that, as the cybernetic episteme began to spread, prompted a great deal of concern throughout the next decade. Martin Heidegger, for instance, would proclaim the end of philosophy and, when pushed by Rudolf Augstein's questioning, would name cybernetics as its successor.[38] In light of the extent to which cybernetics had frustrated the relationship between the natural and the artificial, Gotthard Günther would claim that a bivalent or polyvalent ontology, in conjunction with a trivalent logic (at least), represented the "finale" of Hegelian reflection metaphysics.[39] For Arnold Gehlen, what would stand out about cybernetics is its objectification of the mind or spirit (*Geist*), which seemed to be synonymous with the perfection of technology and the last technical stage of human history.[40] For his part, Max Bense would be taken by its conciliatory "sphere of technical being," which is "more comprehensive than the sphere of that which was once called nature or spirit (*Geist*). [...] *The human as technical existence* – to me this seems to be the most imposing task of tomorrow's philosophical anthropology."[41] And Pierre Bertaux would predict: "The human beings who are integrated into these apparatuses must necessarily become different human beings. They will no longer fit the previous concept of 'human'. The mutation of humanity is a necessary accompaniment to the rise of apparatuses. [...] There will be a transition into a new and fourth form of material organization – after the mineral, plant, and animal kingdoms will come a kingdom in which the human, though admittedly playing a significant role in this transition, will perhaps only be participating in a phenomenon whose implications and consequences

34 See Claus Pias, "Elektronenhirn und verbotene Zone: Zur kybernetischen Ökonomie des Digitalen," in *Analog/Digital – Opposition oder Kontinuum? Zur Theorie und Geschichte einer Unterscheidung*, ed. by Jens Schröter and Alexander Böhnke (Bielefeld: Transcript, 2004), 295–310.

35 The staunchest defender of the digital was John von Neumann, who held the following position: "I shall consider the living organisms as if they were purely digital automata"; quoted from his article "The General and Logical Theory of Automata," in *Cerebral Mechanisms in Behavior: The Hixon Symposium*, ed. by Lloyd A. Jeffress (New York: Wiley, 1951), 1–31, at 10.

36 It supposedly only happened once that McCulloch felt it necessary to interrupt a discussion with the words "no, not now" (see p. 193 in this volume).

37 See p. 723 in this volume.

38 Martin Heidegger, "'Only a God Can Save Us': *Der Spiegel*'s Interview with Martin Heidegger," in *The Heidegger Controversy: A Critical Reader*, ed. by Richard Wolin (Cambridge, MA: MIT Press, 1993), 91–116 (transl. by Maria P. Alter and John D. Caputo).

39 Gotthard Günther, *Das Bewußtsein der Maschinen: Eine Metaphysik der Kybernetik* (Krefeld: Agis Verlag, 1963).

40 Arnold Gehlen, *Man in the Age of Technology*, translated by Patricia Lipscomb (New York: Columbia University Press, 1980).

41 Max Bense, "Kybernetik oder die Metatechnik einer Maschine," in *Ausgewählte Schriften*, ed. by Elisabeth Walther, 4 vols. (Stuttgart: J. B. Metzler, 1997–98), 2:429–46 (originally published in 1951).

transcend him."[42] It would be easy to add further examples to this list of end games and/ or unifying perspectives.

In any case, Michel Foucault's famous image of man being "erased, like a face drawn in sand at the edge of the sea" has a cybernetic precedent, that is, a scientific-historical basis and a technical-historical *datum*:[43] Kant's "analytics of finitude," according to Foucault's diagnosis, put an end to the question of absolute knowledge and at the same time opened up another question – "What is the human?". The critical project had been attempting to do away with an *illusion* that could not be dispelled. Kant had employed his concept of illusion to define, above all, the functionality of "transcendental illusion," which (unlike "logical illusion," for instance) is unavoidable and "natural" and which maintains the productive force of reason itself.[44] The removal of the "transcendental illusion" could thus only be achieved at the expense of an "anthropological illusion," under whose conditions the human sciences had purportedly been operating ever since.[45] In order to rouse philosophy from its "anthropological sleep," this would require, in Foucault's estimation, an "uprooting of anthropology" – a rediscovery of a "purified ontology or a radical thought of being."[46] The condition for embarking on such a return to the beginning of philosophy was, at any rate, the end of the human, and this would entail the following: no longer treating the human as the starting point for any pursuit of truth and no longer speaking of his dominion or liberation. Rather, it would be necessary to conduct a "counter-science" with which to question the human sciences, to take into account their positivities, to engage in formalizing instead of anthropologizing, to demystify instead of mythologize, and finally "to refuse to think without immediately thinking that it is man who is thinking."[47]

It is not terribly difficult to recognize in this the very points of departure that defined cybernetics only two decades earlier. Cybernetics, moreover, activated the alarm clock of counter-science not on the basis of thinking radically about philosophy but rather by thinking radically about technology. McCulloch's design of neuronal networks beyond the distinction of humans, machines, and symbols; Wiener's common space for "the control and transmission of information in living organisms and machines"; or Shannon's statistically generated language with which to analyze language itself are only the most prominent examples of formalizing, demystifying, and even of designing theories according to which it is no longer necessary to think in terms of "the human."[48] Rather than being a reversion to the classical *episteme*, however, the all-encompassing approach of cybernetics introduced a new epoch altogether, one that altered the order of knowledge so comprehensively and reconstructed the archive in such a profound manner that it enabled, in turn, an entire ensemble of pronouncements to appear within the same functional system and enabled the greatest variety of discourses to formulate, systematically, the objects about which they would make such pronouncements. Whereas before,– in other words – such things as life, language, or work were united in the concept of the human being, they now encountered one another beyond human limits in control circuits of information, switching algebra,

42 Pierre Bertaux, *Maschine – Denkmaschine – Staatsmaschine: Entwicklungstendenzen der modernen Industriegesellschaft (9. Tagung am 25. Februar in Hamburg-Bergedorf* (Hamburg: Hamburg-Bergedorfer Gesprächskreis, 1963), typescript.

43 Michel Foucault, *The Order of Things: An Archaeology of the Human Sciences* (New York: Pantheon, 1970), 386.

44 Immanuel Kant, *The Critique of Pure Reason*, translated by Paul Guyer and Allen W. Wook (New York: Cambridge University Press, 1988), 384–87.

45 Michel Foucault, *Introduction to Kant's Anthropology*, translated by Roberto Nigro and Kate Briggs (Cambridge, MA: MIT Press, 2008), 106–08 (originally published in 1961).

46 Foucault, *The Order of Things*, 341.

47 Ibid., 380, 342.

48 See Stefan Rieger, *Kybernetische Anthropologie: Eine Geschichte der Virtualität* (Frankfurt am Main: Suhrkamp, 2003).

and feedback. And one might ask, following Foucault's way of thinking, how and where this new "critical" project of replacing the anthropological illusion would be achieved at the expense of liberating a cybernetic illusion – something that would also involve a change in the relations of power. At least three conjectures come to mind, and these will form the background of my remarks to follow:

First, the concept of information insists on there being a third quantity that is neither matter nor energy and that thus undermines the dichotomies of form and content, processes and results, subject and predicate. It is simultaneously concrete and abstract, physical and logical; it exists simultaneously in the field of real and ideal relations of being, *tertium datur*. This is indicative of the peculiarity of cybernetic knowledge, which, within the realm of the sciences, is situated in a position that differs entirely from the "precariousness"[49] of the human sciences. It is a theory that also operates, that explicates both theoretical and practical entities, that functions – in a timelessly logical manner and yet in instruments of time – in the case of humans and animals, brain tissue and digital computers, air defense systems and television transmitters.[50] Thus it is perhaps possible to speak of a new "empirico-transcendental duplication"[51] – i.e., something that is known and yet is simultaneously the enabling condition of knowledge itself – that may replace that of the human but is no less problematic in doing so.

Second, basing cybernetics on switching algebra, information, and feedback, which according to McCulloch would henceforth serve as the basis of "all understanding of our world,"[52] is probably just as paradoxical as the universal creation of the human in all places at which it was valid to sense a certain non-basis of knowledge. One would thus have to ask: If the human served to lend unity to disparate histories, which histories are those that the anthropologically constructed human is no longer helping along? And how is the relation configured between "the" cybernetics and the respectively singular cybernetic ensembles, which is no less unjust than that between "the" human and every individual human being?

Third, the anthropological illusion consisted in ignoring the complex of power and knowledge involved with creating the very notion of "the human" and in covering up, by means of self-naturalizing, the fact that this is a product of technologies of power. It would be worthwhile to investigate whether and when a similar theoretical shift took place in cybernetics. For, despite all the talk of forging a singular vision, the Macy Conferences remained somewhat fluid and were more concerned with questions than with certainties. And despite all of the "applied" work conducted by the participants, the conferences seem to have been events concerned not with individual theories and individual apparatuses but rather with the epistemologies within which such things might be instantiated in the first place. This effort to design new orders of knowledge – within which heterogeneous elements could tentatively be arranged and in which the borders could tentatively be eliminated between man and nature, man and machine, subject and object, *psyche* and *techne* – was referred to by McCulloch as an "experimental epistemology." It would be worth asking whether, when, and where such liminal and integrative thinking might have faded and yielded to a trivializing or naturalizing certitude about a universal pattern of explanation and how experiments and instruments might likewise have taken turns to obscure the correlation between power and knowledge.[53]

49 Foucault, *The Order of Things*, 345.

50 See for instance Louis Couffignal's definition of cybernetics – in his book *La Cybernetique* (Paris: Presses Universitaires de France, 1963) – as "the art of preserving the effectiveness of action."

51 Foucault, *The Order of Things*, 374.

52 See p. 719 in this volume.

53 On the exclusion of computer science from the "lofty dreams" of cybernetics see for instance Wolfgang Coy, "Zum Streit der Fakultäten: Kybernetik und Informatik als wissenschaftliche Disziplinen," in *Cybernetics/Kybernetik*, 2: 253–62.

*

It is not difficult to identify, in the wake of cybernetics' early designs, the origins of that diagnosis which seeks to define our present society as one of "knowledge," "information," or "control"[54] – societies that are distinguished by "limitless postponements," that generate "undulatory" existences, and whose "postmodern" forms of communication and interaction can be dated back explicitly to cybernetics. For, according to Lyotard, *postmodern knowledge* is characterized by "problems of communication and cybernetics, modern theories of algebra and information, computers and their languages, problems of translation and the search for compatibility among computer languages, problems of information storage and data banks, telematics and the perfection of intelligent terminals."[55]

In an even broader context, the question of the future itself proved to be one of the greatest challenges posed by a "future world." For, under the cybernetic conditions of a non-deterministic teleology, and as the many popular representations of cybernetics have long assured us, the prevailing relations between temporality and the future can be highly peculiar. The attempt to align the "physical functioning of the living individual" with that of the "newer communication machines" (in Wiener's terms) initiated a set of problems that can be likened to lifting a full glass of water to one's mouth. For whoever (or whatever) might be doing the lifting is no longer the Cartesian active subject with its sequence of willful acts and consequences but is rather a sequence of "real-time" data and calculations concerning a path to the mouth in which case the mouth will always already have been the future of the glass. In order to ensure that this takes place, target-oriented adjustments of motion are required, and for these it is decisive at which distances comparisons are made between present and future values and how drastically the motion needs to be corrected. It is decisive – in short – to know how the "constraints" of the system have to be measured so that it will function. If the feedback is too drastic or frequent, this will lead to "clumsy behavior": The drink will spill precisely on account of the motion that is put in place to prevent the spilling itself, and the whole process will enter a state of oscillation that can otherwise only be observed in experimental conditions with subjects suffering from so-called "intention tremors."[56]

Such "target-oriented" actions become more and more complicated if the target in question does not stand still. A cat that wants to catch a fleeing mouse does not jump to the place where the mouse presently is but rather to where it will next be: It jumps toward the future of the mouse. Whoever wants to shoot down an airplane must be able to read the evasive tactics of his enemy and interpret this data in order to strike it in its future location. The magic word for all of this was of course "prediction," and the powers of prediction improve with increased amounts of data. Just as, on a small scale, a cat jumps into the future of a mouse or a missile is guided into the future of an enemy airplane, so it suddenly seemed conceivable – on the grand scale of societies, economies, and politics – to program "conscious human targets" that, so long as the appropriately oriented mechanisms

54 See Gilles Deleuze, "Postscript on the Societies of Control," in *Cultural Theory: An Anthology*, ed. by Imre Szeman and Timothy Kaposy (Chichester: Wiley-Blackwell, 2011), 139–42; Daniel Bell, *The Coming of Post-Industrial Society: A Venture in Social Forecasting* (New York: Basic Books, 1973); Alain Touraine, *Return of the Actor: Social Theory in Postindustrial Society*, translated by Myrna Godzich (Minneapolis: University of Minnesota Press, 1988); Zbigniew Brzezinski, *Between Two Ages: America's Role in the Technetronic Era* (New York: Viking, 1970). On the concept of the postindustrial age as an anti-communist strategy, see Richard Barbrook, *Imaginary Futures: From Thinking Machines to the Global Village* (London: Pluto, 2007), 137–83.

55 Jean-François Lyotard, *The Postmodern Condition: A Report on Knowledge*, translated by Geoff Bennington and Brian Massumi (Manchester: Manchester University Press, 1984), 3–4.

56 First mentioned in the seventeenth century (by the likes of Franciscus Sylvius and Gerard van Swieten), intention tremors designate involuntary movements that only occur in conjunction with voluntary movements.

of communication and control are in place, would always already have been met.[57] The tense of cybernetics would thus be something like the future perfect: Everything will have been.[58]

The significance of cybernetics (and thus also of the Macy Conferences) for the present perhaps lies in the question of the present itself, that is, in a new and cybernetic order of time. It is perhaps as problematic (if not paradoxical) as it is to reflect about today's digital cultures because it is the cybernetic ensembles of digital media themselves that have fundamentally reconstructed the category of the present. They form the "bias," the distortion, or the schematism that has always cooperated in the concept of the "present." If one cared to identify a signature feature of the electronic world, it would probably be that its present is characterized precisely by an excess of presentness – by a sort of "absolutism of the present" (in Robert Musil's words), by the cybernetic ensembles of digital media that consolidate everything into themselves and in which nothing before our loud present actually counts as the present.[59]

One could dismiss this by observing that modernity, ever since the transitional period around 1800 (Reinhart Koselleck's "saddle age"), has treated the present as the point at which time becomes reflexive and at which there is no longer any separation between observation and motion. The repeatedly proclaimed "futures" of digital cultures would then still belong to the register of this temporal orientation, which regards the present as the decisive point of transition between the past and an (open) future. Conversely, however, one could also ask whether this temporal structure – this chronotope – still possesses any validity or whether it has not (over the course of the development of cybernetic epistemology) been replaced by something entirely different.[60]

Historically, cybernetics gained significance from the fact that its systems of governance and control – regardless of whether drinking glasses, mice, or airplanes are at stake – seemed to be highly scalable at unprecedented levels and according to entirely new standards (its proponents could hardly conceal their optimism about this). In real-time systems with appropriate feedback mechanisms, Norbert Wiener himself believed to have recognized what had been missing from typical critiques of society. A society without feedback is, simply enough, "an ideal held by many Fascists, Strong Men in Business, and Government."[61] The future task of cybernetics would thus be to install such *machines à gouverner* in the realm of politics and to model them according to state-of-the-art technical systems. "Non-deterministic teleology" became a magic expression that led some to believe that they could define goals, introduce a system, and then walk away from it all with the expectation that their desired results would necessarily come to be.[62]

This redefinition of the temporal relations and power principles of control, which cybernetics supposedly adopted (in its own way) from the classical art of statecraft, represented a chance to test the reliability of the latest media technology. "The only way to run the complex society of the second half of the twentieth century," as Robert Theobald noted in 1966, "is to use the computer."[63] And Pierre Bertaux proclaimed in

57 Karl Steinbuch, *Falsch programmiert: Über das Versagen unserer Gesellschaft in der Gegenwart und vor der Zukunft und was eigentlich geschehen müßte* (Stuttgart: DVA, 1968), 151.

58 Peter Bexte, "Uncertainty in Grammar/The Grammar of Uncertainty: Some Remarks on the Future Perfect," in *From Science to Computational Sciences: Studies in the History of Computing and Its Influence on Today's Sciences*, ed. by Gabriele Gramelsberger (Zurich: diaphanes, 2011), 219–26.

59 Wolfgang Hagen, *Gegenwartsvergessenheit: Lazarsfeld, Adorno, Innis, Luhmann* (Berlin: Merve, 2003).

60 Hans-Ulrich Gumbrecht, *Our Broad Present: Time and Contemporary Culture* (New York: Columbia University Press, 2014).

61 Norbert Wiener, *The Human Use of Human Beings*, 15.

62 See Peter Galison, "The Ontology of the Enemy: Norbert Wiener and the Cybernetic Vision," *Critical Inquiry* 21 (1994), 228–66.

63 Robert Theobald, "Cybernetics and the Problems of Social Reorganization," in *The Social Impact of Cybernetics*, ed. by Charles R. Dechert (London: University of Notre Dame Press, 1966), 39–69, at 59.

1963: "Within the span of half a century, the 'tried and tested' methods of government could not avoid two world wars and countless other wars of various scales with more than fifty million casualties. [...] A French expression is appropriate in this regard: *gouverner, c'est prévoir.* The art of governance is the art of prediction. For humans, however – for their organic and cerebral way of thinking, for thinking that involves words – the dimension of the future is difficult to comprehend. This is because it is not possible for the brain to oversee, all at once, the countless elements that influence what is happening. [...] This unfortunate fact can be remedied by machines. I am convinced that the future will belong to those groups of people who are first to recognize clearly that the most profitable investments they can make will be in the 'projections,' forecasts, and technical predictions that can only be realized with the help of government-run thinking machines."[64]

Above all, the lasting legacy of cybernetics is precisely this phantasmatic excess of faith in gaining control over the future by yielding, in a targeted manner, control over some of its aspects. During the Cold War, delegated homeodynamics seemed to be our best chance to survive a threatening future of nuclear war and totalitarianism, overpopulation, pollution, and the depletion of resources. When Jay Forrester, an expert in early-warning systems, designed simulation models for his book *World Dynamics*, the zig-zagging and escalating curves representing future capital investment, population size, pollution, and natural resources seemed only to refer to an "intention tremor" of the political, to the feedback – either too much or too little, too early or too late – involved with the pursuit of a goal, but the mechanism of controlling the future by pursuing goals was itself no longer called into question. The main concern was rather that, according to Forrester, "the human mind is not adapted to interpret how social systems behave."[65] In other words, our minds are not able to recognize the windows of time during which investment and regulation have to be thought about in order for a given goal to be achieved. The result of this lack of theory regarding complex and dynamic systems – systems whose "workload" could not be modeled in advance – was to move on and remove all "intuition, judgement, and argument" from the political, given that the latter are not "reliable guides to the consequences of an intervention into system behavior."[66] There would certainly be no need for a king – off with his head! – if computers could govern things more effectively on their own.

The hope of being able to set goals and yet delegate their manner of achievement, however, was soon confronted by an epistemological problem, that is, by the question of knowledge, its scope, and its ability to be processed – or, in short: by the matter of having to determine which series of data would be necessary for making successful predictions. In the case of Forrester's models of system dynamics, data taken from world almanacs were still considered sufficient, so long as enough intelligence were invested in figuring out their relations and feedback loops. Elsewhere it became clear, even at the time, that the matter was more complicated than it seemed and that more data – and data of different sorts – would be needed. According to the cyberneticist and management theorist Stafford Beer (writing likewise around 1970), electronic governance would have to leave behind the age of statistics, its delays and its aggregated data in order to make a transition into an age of "real-time control," into an age of large-scale disaggregated and real-time data (the term "big data" had not yet been coined).[67] Levels of political representation would simply

64 Bertaux, *Maschine – Denkmaschine – Staatsmaschine*, cited above. Bertaux's long-term hope was in the elimination of national borders and in the development of a "common prediction apparatus for the well-being of mankind."

65 Jay Forrester, "Counterintuitive Behavior of Social Systems," *Technology Review* 73 (1971), 52–68, at 52.

66 Idem, *World Dynamics*, 2nd ed. (Cambridge, MA: Wright-Allen, 1973), 97.

67 Such fantasies from the 1960s are enjoying a revival in today's notion of "big data'; see, for instance: http://www.futurict.ch.

no longer be part of the equation. Thus cybernetic government – and this much was clear from the beginning – would not only allow the notion of statehood to crumble; it would also aim to eliminate the borders of the political by basing the practice of governance on an extensive, undulating registry and on a will to knowledge that does not neglect any area of inquiry and whose interests never stagnate.[68]

Perhaps the end of ideology, as formulated under cybernetic conditions, remains indebted to Carl Schmitt's diagnosis concerning the dissolution of classical distinctions and the "breakdown of all conceptual axes." Whereas Schmitt, however, responded to the decrease of political differentiation with a theoretical defense of sovereign authorities and by underscoring intensive distinctions such as friend and foe, cybernetics endorsed forms of control or government that were based less on individuals, institutions, and legal entities than they were on approaching the question of political power in terms of a diffuse operational field of milieus, scenarios, and feedback and on developing processes of incessant monitoring and assessment.[69] So it is that not only the issues involved with changing the mentality of government but also discussions concerning the issue of the modern temporal order (and thus the issue of the openness of the future) have to be considered in light of the cybernetics of the 1950s and 1960s, the archaeology of which is simultaneously that of our present day.

Just such an attempt at interpretation was undertaken, nearly twenty-five years ago, by Vilém Flusser.[70] If, according to Flusser, there really is a network of cybernetic machines that share feedback with one another, that behave adaptively and process malfunctions independently, and that allow – by means of what we now call "big data" – the data traces of subjects to be conflated with the prediction of forms of subjectivation, then the relation between what is and what ought to be collapses along with the chronotope of modernity and its idea of an open future. Like other thinkers before and after him, Flusser referred to this situation as "post-history." Within this temporal order, according to his diagnosis, there can for logical reasons no longer be any argument, critique, or anything political in the modern sense. Whereas the modern subject is constituted in the present as a transitional point between the past and the future, all that remains according to Flusser, in light of the dominance of a cybernetic temporal order, are various types of functioning within "apparatuses": functionaries, people in despair, technocrats, terrorists, environmentalists (etc.) are social types of a present that is becoming increasingly frayed on account of prediction and feedback.[71]

★

From its beginnings, cybernetics was less a disciplinary science than a general methodology of action. To this can be attributed at least two consequences that have, in turn, deeply inscribed themselves into the history of those theories with which we are now attempting to understand the present – a present that itself has resulted from the resounding success of cybernetics.

On the one hand, stability could no longer be assumed to be the essential core of that which exists (*das Seiende*); it rather had to be understood as a problem, to be solved constantly, of communicating feedback systems. As regards scientific agendas, this meant that causality had to be embedded, in a non-ontological manner, into systemic categories

68 See Eden Medina, *Cybernetic Revolutionaries: Technology and Politics in Allende's Chile* (Cambridge, MA: MIT Press, 2011); Claus Pias, "Der Auftrag: Kybernetik und Revolution in Chile," in *Politiken der Medien*, ed. by Daniel Gethmann and Markus Stauff (Zurich: diaphanes, 2004), 131–54

69 See Michel Foucault, *Society Must Be Defended: Lectures at the Collège de France, 1975–76*, translated by David Macey (New York: Picador, 2003); Jacques Donzelot et al., eds., *Zur Genealogie der Regulation: Anschlüsse an Michel Foucault* (Mainz: Decaton, 1994).

70 Vilém Flusser, *Gestures*, translated by Nancy Ann Roth (Minneapolis: University of Minnesota Press, 1991); originally published as *Gesten: Versuch einer Phänomenologie* (Düsseldorf: Bollmann, 1991).

71 Flusser, *Gestures*, 17–18, and 35–36.

of purpose.[72] Such has been the concern of *systems theory*, which is closely affiliated with cybernetic epistemology. In fact, systems theory can be understood as the result of a division of labor between those who construct systems and those who describe them – as a product of the separation between functioning for the sake of description and the description of functioning, though each is based on the same epistemological foundation.

On the other hand, the possibility emerged of allocating actions to systems, that is, of organizing, delegating, or augmenting them as networks of more or less "intelligent" components of humans and machines. Such has been the concern of *actor-network theory*, which owes its recent prominence to the need to reexamine "things" after the craze of postmodern theory had come to an end. As a "weak" theory, it attempts to gain strength by describing the common origins of knowledge and history through human and non-human components. Although it admittedly manages to reconcile realism and constructivism, it is prone to rehashing aspects of cybernetic epistemology that had already been known (and often far more comprehensively so) by the cyberneticists themselves.

Both of these theoretical traditions could perhaps be understood as a reaction to (or as a result of) a rise of system-oriented sciences in the wake of cybernetics' success in the 1950s and 1960s. Description and design – the analysis and synthesis of systems – have since gone hand in hand, regardless of the fact that, following an early interdisciplinary phase, they have been divided between different university faculties. The study of the behavior of cybernetic systems (also in the sense of McCulloch's "experimental epistemology") thus proceeded with a sort of epistemological modesty that stands in contrast to its claims of universal validity. Abraham Moles, for instance, made the following remarks in 1959: "This principle [of functional analogy] allows cybernetics to be defined as a *science of models*. The nineteenth century endeavored to describe the world as it is. [...] Science of the twentieth century will above all be the science of models. [...] As soon as it is able to construct such a model, cybernetics will be able to answer the question [of what something *is*]."[73]

In this sense, the ongoing trend to create computer simulations can be regarded as one of the most significant scientific-historical legacies of cybernetics, at least to the extent that such simulations aim to imitate, for the purposes of experimentation, the behavior of dynamic systems over time. Joseph C. R. Licklider, a participant in the Macy Conferences, noted as early as 1967 that computer simulation would bring about a new epoch in the history of science, one whose magnitude would equal that achieved by the advent of the printing press. Computer simulations have since become the domain of sciences devoted to the behavior of systems. The latter have not only transformed the scientific and experimental cultures of engineering and the natural sciences; as postmodern or "mode 2" sciences, they have also influenced the "world views" and political frameworks of globalized societies.[74]

The most striking current example of this is probably the climate debate, which over the past few decades has come to occupy the systematic place formerly held by nuclear war (in the production of weapons for which, as is well known, computer simulation had

72 Niklas Luhmann, *Zweckbegriff und Systemrationalität: Über die Funktion von Zwecken in sozialen Systemen* (Tübingen: Mohr, 1968), 107–09.

73 Abraham A. Moles, "Die Kybernetik, eine Revolution in der Stille," in *Epoche Atom und Automation: Enzyklopädie des technischen Zeitalters*, 10 vols. (Frankfurt am Main: Limpert, 1958–60), 7:7–8 (transl. Valentine Pakis).

74 See, for example, Evelyn Fox Keller, "Models, Simulation and 'Computer Experiments'," in *The Philosophy of Scientific Experimentation*, ed. by Hans Radder (Pittsburgh: University of Pittsburgh Press, 2003), 198–215; Paul N. Edwards, *A Vast Machine: Computer Models, Climate Data, and the Politics of Global Warming* (Cambridge, MA: MIT Press, 2010); Peter Galison, "Computer Simulations and the Trading Zone," in *The Disunity of Science: Boundaries, Contexts, and Power*, ed. by Peter Galison and David J. Stump (Stanford: Stanford University Press, 1996), 118–57; Claus Pias, "On the Epistemology of Computer Simulation," *Zeitschrift für Medien- und Kulturforschung* 3 (2011), 29–54; Gramelsberger, ed., *From Science to Computational Sciences*.

enjoyed its first success). The metaphorical portents may have changed; after the thaw of the Cold War, the fear of a cold nuclear winter may have given way to the fear of an increasingly warmer planet. The situations are comparable, however, in that both of them initiated computer-based and transnational efforts to forecast global scenarios, to "think the unthinkable" between science and science fiction,[75] and finally to modify future behavior for the sake of *avoiding* an event.

Climate research is especially noteworthy in this regard because, as I have written elsewhere,[76] hardly any other domain of knowledge is epistemologically so dependent on the historical state of hardware and software: on the observable leaps in quality enabled by sheer computing power but also on a history of software in whose millions of lines of poorly documented or undocumented code have sedimented archaeological layers of scientific thinking that, for good reason, cannot be touched or rewritten but merely expanded and globally standardized and certified.

Thus far the routines of critique have been at a loss to address that which alternative worlds (and not mere prognoses, for instance) might yield to guide our behavior and self-perception. The common reflex of citing the "constructedness" of knowledge achieves little in this regard, for it fails to take into account any activity that takes place in scenarios that are conscious of their own constructivism (with respect to parameterization, for instance). And the criterion of falsifiability espoused by classical scientific ethics is – not merely for reasons of capacity but for systematic reasons as well – simply not practicable in this case because it is impossible to experiment with the climate as an object of science and because the sciences in question are themselves incapable of reconstructing that which is being processed by their software.

Given their epistemological status, potential climate scenarios are thus an example of that which is causing the aforementioned transformation of the modern temporal regime. Out of future possibilities (once regarded by modernity simply as "different futures") arise potential actualities. Neither what was nor what could or ought to happen, but rather more and more exclusively what "is to be expected," serves today as the authoritative basis of knowledge supporting an uninterrupted feedback loop between the past (validation), the present (action), and the future (scenarios). And the problems at stake are made no less real by the fact that they can be related to their media-historical foundation, to their very predictability.[77]

If, in the meantime, serious climatologists have postulated a new cosmology in order to be able to explain our behavior on a global level, then this is simultaneously, as far as knowledge is concerned, a departure from modernity's concept of transparency – from a sort of cybernetic blackboxing that put an end to hermeneutics and the Enlightenment while also enabling such approaches as technical discourse analysis (Friedrich Kittler), deconstruction (Derrida), or systems theory (Niklas Luhmann). On account of its incommensurability, the legitimation strategy of computer-simulated scenarios would be far better suited to the premodern political register of sovereignty; to some extent, it is a new *science royale* that discloses everything (literally: "open source") and yet cannot be betrayed. In the place formerly occupied by the wisdom (or whim) of the ruler, which, as an unbetrayable secret, was protected by a metaphysical limit to knowledge, there is now data processing, which has drawn a new limit to demarcate that which is constitutively evasive on account of being secretive according to its (now highly technical) "nature." The cybernetic epistemology of functional, systemic analogies and its implementation in the digital media of computer simulation have radicalized the fact that modernity always

75 Eva Horn, *Zukunft als Katastrophe* (Frankfurt am Main: S. Fischer, 2014).

76 Claus Pias and Timon Beyes, "Transparenz und Geheimnis," *Zeitschrift für Kulturwissenschaften* 8 (2014), 111–17.

77 Elena Esposito, *Die Fiktion der wahrscheinlichen Realität* (Frankfurt am Main: Suhrkamp, 2007).

operated on the basis of insufficient critical knowledge to a tipping point at which its own project has become dubious.

Ultimately, modernity had rendered the cosmological legitimization of the secret expendable by making it seem as though the (open) future would henceforth be uncertain. It had thus transplanted, in a sense, the unbetrayable secret of sovereignty onto time itself.[78] Yet in light of cybernetic epistemology, as expressed for instance in climate scenarios, we are possible eyewitnesses to a grand reconstruction of the cultural and socio-technical network of this same modernity, whereby the question now looming over us is that of a new *arcanum*, of a functional secret of digital cultures that (like the basis of sovereignty or the notion of the future before it) does not need to be kept secret because it is simply incommensurable.

In any event, the established methods of understanding have clearly reached their limits and now serve merely to indicate even more conspicuously how strongly they have been influenced by the cybernetic technologies that they seek to describe. Systems theory and actor-network theory derive from cybernetic epistemology itself and can thus do little but refer back to their own basis. The media theory of Friedrich Kittler, which taught us to read hardware and to write software, was admittedly determined – as could be expected of an approach influenced by desktop PCs, imperative programming languages, and the hacker ethos of the 1960s and 1970s – to let no secret remain, but at the same time it was melancholy in the knowledge that it could only advance past the outermost gates of digital cultures: "[A] total media link on a digital base will erase the very concept of medium. Instead of wiring people and technologies, absolute knowledge will run as an endless loop."[79]

If it is true, however, that a cybernetic epistemology of real-time, of prediction, of functional blackboxing, and of scenarios has saturated our world – from large-scale political decisions all the way to our microscopic networks of intensities, moods, or emotions – then the question of historical time and thus that of the future itself would yet again be ripe for discussion.[80] In digital cultures, the "space of experience" (that which can be evoked as a memory of one's own or another's knowledge) and the "horizon of expectation" (that which seals us from the future as a forthcoming space of experience) would blend and be reduced to a new form of present – to a chronotope that is fundamentally distinct from the order of time that has prevailed in modern history since the Enlightenment. And perhaps this is ultimately the "cybernetic illusion." To such a present, after all, our thinking would not owe any burden of proof – only curiosity.

78 On the premodern role of secrecy, see Albert Spitznagel, "Einleitung," in *Geheimnis und Geheimhaltung: Erscheinungsformen – Funktionen – Konsequenzen*, ed. by Albert Spitznagel (Göttingen: Verlag für Psychologie, 1998), 19–51; Niklas Luhmann and Peter Fuchs, "Speaking and Silence," translated by Kerstin Behnke, *New German Critique* 61 (1994), 25–37; Aleida Assmann and Jan Assmann, *Schleier und Schwelle*, 3 vols. (Munich: Fink, 1997–99).

79 Friedrich Kittler, *Gramophone, Film, Typewriter*, translated by Geoffrey Winthrop-Young and Michael Wutz (Stanford: Stanford University Press, 1999), 1–2.

80 Reinhart Koselleck, *Futures Past: On the Semantics of Historical Time*, translated by Keith Tribe (New York: Columbia University Press, 2004), 255–75.

CYBERNETICS

CIRCULAR CAUSAL, AND FEEDBACK MECHANISMS

IN BIOLOGICAL AND SOCIAL SYSTEMS

Transactions of the Sixth Conference
March 24 - 25, 1949, New York, N. Y.

Edited by
HEINZ VON FOERSTER
DEPARTMENT OF ELECTRICAL ENGINEERING
UNIVERSITY OF ILLINOIS

JOSIAH MACY, JR. FOUNDATION
565 PARK AVENUE, NEW YORK 21, N. Y.

PARTICIPANTS

WARREN S. McCULLOCH , Chairman
Department of Psychiatry, University of Illinois College of Medicine; Chicago, Ill.

HEINZ VON FOERSTER, Secretary
Department of Electrical Engineering, University of Illinois; Urbana, Ill.

HAROLD A. ABRAMSON
Department of Physio University; New York, N.Y.

GREGORY BATESON
Veterans Administration Hospital; Palo Alto, Calif.

J. H. BIGELOW[1]
Electronic Computer Project, Institute for Advanced Study; Princeton, N.J.

HENRY W. BROSIN
Department of Psychiatry, School of Medicine, University of Chicago; Chicago, Ill.

RAFAEL LORENTE DE NO[1]
Rockefeller Institute for Medical Research; New York, N.Y.

LAWRENCE K. FRANK
Caroline Zachry Institute of Human Development, Inc.; New York, N.Y.

RALPH W. GERARD
Department of Physiology, School of Medicine, University of Chicago; Chicago, Ill.

G. E. HUTCHINSON
Department of Zoology, Yale University School of Medicine; New Haven, Conn.

HEINRICH KLÜVER
Department of Experimental Psychology, Division of Biological Sciences, University of Chicago; Chicago, Ill.

LAWRENCE S. KUBIE
Department of Psychiatry & Mental Hygiene, Yale University School of Medicine; New Haven, Conn.

PAUL F. LAZARSFELD[1]
Department of Sociology, Columbia University; New York, N.Y.

HOWARD S. LIDDELL
Department of Psychology, Cornell University; Ithaca, N.Y.

DONALD B. LINDSLEY
Department of Psychology, Northwestern University; Chicago, Ill.

DAVID LLOYD
Rockefeller Institute for Medical Research; New York, N.Y.

DONALD G. MARQUIS[1]
Department of Psychology, University of Michigan; Ann Arbor, Mich.

MARGARET MEAD
Associate Curator of Ethnology, American Museum of Natural History; New York, N.Y.

FREDERICK A. METTLER
Departments of Anatomy and Neurology, College of Physicians and Surgeons, Columbia University; New York, N.Y.

F. S. C. NORTHROP[1]
Department of Philosophy, Yale University School of Medicine; New Haven, Conn.

WALTER PITTS
Department of Mathematics, Massachusetts Institute of Technology; Cambridge, Mass.

ARTURO S. ROSENBLUETH[1]
Departments of Physiology & Pharmacology, Instituto Nacional de Cardiologia; Mexico City, D. F., Mexico

LEONARD J. SAVAGE
Institute of Radiobiology and Biophysics, University of Chicago; Chicago, Ill.

THEODORE C. SCHNEIRLA
Curator of Animal Behavior, American Museum of Natural History; New York, N.Y.

JOHN STROUD
U. S. Naval Electronic Laboratory; San Diego, Calif.

HANS LUKAS TEUBER
Department of Neurology, New York University College of Medicine; New York, N.Y.

GERHARDT VON BONIN
Department of Anatomy, University of Illinois College of Medicine; Chicago, Ill.

JOHN VON NEUMANN[1]
Department of Mathematics, Institute for Advanced Study; Princeton, N.J.

NORBERT WIENER
Department of Mathematics, Massachusetts Institute of Technology; Cambridge, Mass.

JOSIAH MACY JR. FOUNDATION

FRANK FREMONT-SMITH
Medical Director

JANET FREED
Assistant for Conference Program

1 Absent

INTRODUCTORY DISCUSSION

FREMONT-SMITH: May I open this sixth meeting of the Conference on Circular, Causal and Feedback Mechanisms in Biological and Social Systems and welcome you and tell you how glad I am that you are here. May I explain to those of you who are guests with us for the first time and to the group as a whole the purpose of these conferences. The Foundation has now eight[1] conference groups in action including this one. The others are on Liver Injury, Blood Pressure Regulation, Biological Antioxidants, Infancy and Childhood, Blood Clotting, Metabolic Interrelations and Problems of Aging. As you will see, the topics cover a rather wide range.

The Foundation's interest in these conferences stems from its experience, over some years, with the problem of advancing research and from increasing recognition of the need to break down the walls between the disciplines and get interdisciplinary communication. This failure in communication between disciplines seems to be a major problem in every phase of science. Such communication is particularly difficult between the physical and biological sciences on the one hand, and the psychological and social sciences on the other. The problem of communication is largely a problem of human relations and for its solution requires intensive and comprehensive scientific study of man. In order to study man it is necessary to bring in every one of the physical and biological sciences and every one of the social sciences also. In the concept of psychosomatic medicine, we have the connecting link from the physical and biological through man to the psychiatric, psychological and social sciences. Thus in the study of man we may find eventual unification of all the sciences.

This group, of all our groups is, I like to say, the »wildest« because we spread over the whole range of all the disciplines. I think we have found that communication among us is by no means easy. We really have found a good deal of difficulty and I suspect that we still will find difficulty in today's program. I hope we will, because I believe if we were to find any easy way of talking across this range of disciplines we would be fooling ourselves. I suspect that the most we can accomplish will be to get a feel of, and perhaps occasionally to specify, the nature of the obstructions to our inter | communication. This business of communication across the disciplines is one of the key problems facing the world today, both in going forward in any field and in having any field to go forward in – by which I mean that the physical sciences have developed to such a point and have gotten so far ahead of the social sciences that there is grave possibility that social misuse of the physical sciences may block or greatly delay any further progress in civilization. Professor Wiener in his Introduction in »Cybernetics«[2] points out one aspect of this problem, namely, that the complexity of the computing machine type of mechanism is so great and can be pushed so far now that it potentially threatens individual decision. Is that a fair statement?

WIENER: Brings it at low levels.

FREMONT-SMITH: That is one element of it. One can say also that the physicists have given us the ultimate weapons of hostility. Now perhaps it is important for all of us, including the physicists, and the mathematicians to learn something about the nature of hostility.

WIENER: May I make one little remark here? The physicists have given us the ultimate of hostility and the psychologists have conditioned us to be able to use it.

STROUD: I resent that. It was done by somebody else.

1 Since this conference the Program has been enlarged to include five additional groups on: adrenal cortex, renal function, nerve impulse, levels of consciousness and diseases of connective tissue.

2 John Wiley & Sons, New York, 1948.

MEAD: That is simply not so.

FREMONT-SMITH: The psychologists are pushed, as all of us are, by the impulses which come from group feeling, and one of the things we need to know about is group feeling. One of the things we have right here is group feeling, group tension, and I hope some resolution and some communication will take place. To some extent we can be conscious of the dynamics of our own group relations. Perhaps we should occasionally take a look at ourselves. Mostly we won't be conscious of the group relations. But I want to point out that in all these conferences the Foundation is interested not only in advancing a particular subject, whether it be the liver or feedback mechanisms, but also in setting a frame of reference in which communication across disciplines can take place.

Six of the other conferences have prepared transactions of their discussions for publication and we would like to do the same thing with this group. The discussions across this range of disciplines are too interesting to be wholly lost, so we are having a complete record made by stenotype. In our other groups I have emphasized the fact that [11] the published transactions are only the »tail« and are not | to wag the »dog«. The »dog« is the free discussion, the give and take, and the communication that results when discussion is conducted in a friendly atmosphere. That is the »dog« and the transactions must always remain the »tail«.

It is appropriate to mention at this time the International Congress on Mental Health which met last August in London, because that was perhaps the largest effort at multiprofessional. communication between the social sciences and psychiatry at the international level. I think that that was an interesting experiment in which it was actually possible to get a 20,000-word statement agreed to by twenty-four people from ten nations, representing some eight disciplines in psychiatry and the social sciences. This statement is an actual consensus. As far as I know that never had been attempted previously. In a sense it is the kind of thing that we are all working for in this group.

Now I will turn the meeting over to your Chairman, Dr. McCulloch.

MCCULLOCH: I bring two messages, one at the technical level and one a general warning. At the technical level the team at Princeton is at the present time in possession of a tube designed by an Englishman, Williams, which is going to serve as memory in their machines. These are rather interesting tubes from our standpoint so I will say a word concerning them. The tube, which looks like an ordinary cathode ray tube, has a beam which plays on a screen where the items are to be stored. According to the voltage of that beam you can place a negative charge on the screen with ease and, as you increase the voltage you get to a place where the condition becomes unstable and you might, by knocking off a negative charge there, leave the spot positive. What happens then is that the electrons knocked out of that spot where you overhit, pile up around it so that your positives are surrounded by a little hill of negatives. This spot can be detected from the opposite side of the screen and serves as the trace that there was a mark at that place. The place is simply defined by position of the beam with the voltage to its regular positioning pulse. There are 32×32 points in the present screen. The interesting thing is the way that they found it profitable to use it. The persistence of that screen is not very great. There the memory is of this kind: they store spots and sense the spots all over this screen in assigned positions. But now the beam is made to sweep over this screen in the following rather elaborate fashion: it puts on marks and every alternate time it goes after a point, and if there is a point there it reads it, erases it [12] and re-records it. If no | spot, it takes the same time for it. During the alternate times it is freely under the control of the computing part of the machine and the programming part of the machine, it alters signals as dictated by the computing machine and the programming. So that you have a beam which has two motions over the screen, one, a stepwise motion occurring, let us say, first, third or at all the odd counts. Then its whole business is to preserve the memory intact. The other half of the time you have a completely irregular motion of the beam under the control of the computation which is involved. The second one is the one whose business it is to put things in and take half out, not merely to preserve them. I think that may become of interest when we come later to the problem of human memory. What determines the necessity for it is, namely, that the traces on such a screen fade of themselves and in the course of time would be lost, were they not reproduced.

WIENER: In other words, this is a clarification of the telegraph type of beats which we had spoken of before where we said that the message had to be rewritten. It is a practical way of doing it at high speed. You might permit me to make a comment. One of the great things about this is the small physical dimension of the record. That is our great obstacle in imitating nervous systems, the fact that the record of the nervous system, whatever it was, was certainly on a much smaller scale than on any previously existing machine. This represents a tremendous extension of the machine towards the small.

HUTCHINSON: What are the dimensions?

MCCULLOCH: The screen is approximately 4 × 4 inches.

WIENER: Half-tone dots.

MCCULLOCH: The second is Von Neumann's warning to all of us, that at the present time 10^{10} neurons used as simple relays are utterly inadequate to account for human abilities. He says this becomes very apparent if one goes to lower forms such as the army ant where you have some 300 neurons that are not strictly speaking sensory or strictly speaking motor items, and that the performance of the ordinary army ant is far from complicated than can be computed by 300 yes or no devices at the simple level. He asked me to come here specifically to talk with him about the possibility of items at a lower level.

I would like to tell the story a little bit differently because it is one of those funny things that crop up so often in science. Von Neumann is very insistent that whatever the items are, lying beyond the property of neurons as mere all or none devices, they must still | be quantized, or digital or logical in their structure. You simply cannot in [13] an affair of the size of our brains get away with any analogy devices. It simply is not possible to handle enough information that way. The question is: to what must we look inside the dimensions of a neuron and what is the general order of complexity?

Let me begin the story the other way. In going over the similarity between nerve and muscle, I was very much struck one day with the relation of adenosine triphosphate as the source of energy for both nervous impulse and muscular contraction. The evidence today is pretty clear that both use the same source of energy. What is more, from what is known of the work on acetylcholine and what is known of the times of the appearances of the potentials and the local signs of action, the two processes resemble each other. If acetylcholine plays the significant role in nervous conduction that it plays in the triggering of the reaction in which the energy is derived from adenosine triphosphate in muscle and in the nerve, – very obviously the first thing would be to suppose that nerve, like muscle, had a protein structure, and that the protein by some alteration in shape or internal field, whatever you want to call it, lets go with a bang, even as it does in muscle. If you remember, myosin is normally a kinked fiber lying at right angles to the muscle belly. When it reacts with adenosine triphosphate, it straightens out. When it straightens out it pushes apart, so to speak, the sides of the muscle and pulls the ends together. A similar structure in the lattice of a protein could account for the abrupt front of the nervous impulse, treating it as if it were a mechanical alteration.

A few days later I was invited down to Halstead's seminar, at which he and Katz proposed a theory of memory, placing the burden on the protein molecules, which they supposed to have this property of organizing, – that they formed some kind of a lattice in the membrane, that this lattice at the time of the nervous impulse is temporarily altered, that any one protein molecule serving as a template for the design of other protein molecules –

WIENER: How? Reproduction?

MCCULLOCH: The utilization of this existing protein molecule, that they wanted in the lattice of the membrane to reproduce others of its kind which might key in the whole of a neuron to behave as one piece in this sense, might lead to specificities in the response of one protein membrane to an adjacent protein membrane at the synapse.

The difficulty with the theory from my point of view – and I expressed it even then [14] – was that this leads rather to the soldering | of one neuron to the next, than to a differentiation within it. To my mind, they have it at the wrong place. Nevertheless, it was interesting to see exactly the same demands made for the membrane of the neuron, the protein molecule again changing shape with the passage of the nervous impulses. There is evidence of a shift from gel to sol in the passage of nervous impulses. The evidence is not too powerful yet, but it is there so that even the axoplasm might be involved in things of this sort.

The next day our biochemist, Jim Bain, came in, having come to almost exactly the same conclusion concerning the possibility of proteins that I had as far as their relation to membrane is concerned. He wanted to know what was the best source of surface protein from nerve in order that he might go after it by a procedure which would not denature it so as to tell whether it might not have contractile properties corresponding to muscle. He figured that it would take only an electron transfer the length of the protein molecule – that is a nice short time – to yield the necessary changes in the property of the membrane. He is at the present time thinking of going after the giant axon of squid, extruding the cytoplasm and extracting the surface proteins with a waving blender.

BATESON: Who?

MCCULLOCH: The biochemist. He has to go where there are squid axons.

WIENER: One of our good old dreams has gone to pot. There is no giant axon of the giant squid.

MCCULLOCH: What a shame! The giant squid has small ones.

WIENER: No.

HUTCHINSON: You had one?

WIENER: I asked that question. It may have been that I asked it of Dr. Lloyd here but the answer is that that is a forlorn hope.

MCCULLOCH: The other thing is something which was pointed out by Lettvin, who is working on the theories of synaptic transmission. He has been over every scrap of data that is worthy of the name and has come to the conclusion that while one might hopefully attack some problems of the triggering of one nerve cell by afferent impulses on a somewhat statistical basis, there is at least in the cortex of the cerebellum fairly clear evidence that, the impulses being delivered to a bunch of fanned out dendrites, the timing of the impulses descending those dendrites from the place where they had been kicked requires coincidence at the base of the dendrite in order to trip the cell [15] with almost any theory you can rig. |

In that case it is perfectly clear that the recipient membrane is fanned out in such a way to do your crucial thing by the timing of your impulses of the membrane. On the other hand, where you have a large number of terminal knobs situated on the cell body, the chances of working that out on any such basis are rather small, but the cell possesses a fibrillar structure within it, a banding of the proteins which can be seen in the living cell. This has been successfully photographed in the Bell Laboratories by ultraviolet light.

FREMONT-SMITH: The neuron itself?

MCCULLOCH: The neuron.

WIENER: It ties up with some of the work with the electron microscope at the Massachusetts Institute of Technology.

MCCULLOCH: The possibility exists that the tying together of points on the cell membrane by strands of protein running through its depths may lead to much more complicated requirements for firing quite comparable to the corresponding complexity when you fan out the dendrites and make contact with them in special ways. That is, you then have possibilities of working out requirements of coaction within the cell, structuring the cell from within.

WIENER: You mean the body, not the axon?

MCCULLOCH: The body, internal to the body. The protein connections may turn out to be significant.

WIENER: Yes.

HUTCHINSON: May I make a remark on the internal protein in connection with this? I think it bears in a very general way on the whole of what we have just heard, although it may appear very remote initially. A former student of mine, Dr. V. T. Bowen, has been doing some studies of mineral metabolism of insects, and for reasons we need not go into, he was led to study barium metabolism in hornets. A very remarkable thing turned up; if you add to the food of the hornet a small amount of barium of the order of a few million atoms per hornet, so that no question of the solubility of salts comes in, it is picked up by the gut cells and after about a day there is in a radio-autograph of the gut cells, a definite black mark along the luminal surface. If you then make radio-autographs of sections over a long period of time, the order of twenty days, you find that this black band advances slowly down the cell and reaches the end in about 20 to 25 days. The only possible explanation that we can think of is that these cells, which secrete and absorb and therefore have what one would regard as a rather active cytoplasm, have also a fibrillar protein structure which is parallel to their long | axis or at [16] right angles to the surface of the gut lumen and that there are positions on the protein fibers which hold barium atoms very tightly indeed. Random thermal agitation occasionally will let a barium atom come off and then it makes the next base, and this process occurs in spite of the fact that the cytoplasm is doing all sorts of other things, secreting enzymes and picking out food, etc. But apparently it goes right across that, and quite regularly, so there does seem to be even in the gut cell a perfectly definite structure which is positional, which can be occupied by certain kinds of atoms very tightly and will give a very primitive sort of trace pattern. But it is there.

FREMONT-SMITH: May I just interject that Stetten[1] in a very recent issue of Science has proposed on the basis of data which I cannot reproduce now, a very parallel kind of structuring for the secretion of hydrochloric acid in the stomach acid cell? You probably saw that. It seems to me it is so parallel to what you say.

HUTCHINSON: Yes.

WIENER: Everything that has been said here concerns the cell and not the axon. We don't have to leave it all in any way for the classical axon organ. What we do need is a

1 *Science*, **109**, *256 (1949)*.

process much more complex where the elements are much more complex for what happens at the synapse and in the cell body. There is no argument there against that being the case?

McCulloch: No. After all a neuron is a living cell and the structuring there of the protoplasm – and I assume it is the proteins – seems to be adequate to determine relatively complicated sequential patterns of output in response to complex inputs.

Wiener: I want to say definitely that the work Dr. Rosenblueth and I have done on the normal action of the cell has made it possible to estimate lower bounds for the number of separate impulses that you must have for the action. It is statistical work that we have done. I think that that would be very worthwhile to compare the degree of complexity of Huxley. I want to point out that the degree of complexity of Huxley can go one degree further, and I feel this is going to be of tremendous importance. That is, the observations we make have been essentially observations of the cell under constant conditions. If we are going to work with cells with variable thresholds, which may be a long-time phenomenon, we can get into much higher complication than [17] that. That is the mechanism, I think, which is relevant to what you are | saying. The mechanism of variable threshold is one that has appealed to me very strongly in connection with the problem of memory. I believe we have every indication that various things, chemical and otherwise, in the body do change that: that in other words, we have an evidence from the existence of memory, for the existence of learning of a fine structure in the cell of changeability of thresholds according to specific patterns, which seems to be in no way inconsistent with this sort of a theory.

We should look upon the possibility, and to my mind this remark is relevant, that there is evidence that we cannot consider the neuron to be a constant without a variability of the threshold induced perhaps by chemical substances, an adequate mechanism for learning. Your complexity, the greater complexity, can be just at that point where we need it for learning.

Kubie: I don't think we have thought of the neuron as constant since Lucas. It is only for a set of given conditions.

Wiener: That is perfectly true but the point is for the preliminary survey of the mechanism we have tended to make it.

McCulloch: We have all kept the conditions as constant as we could.

Pitts: It strikes me the point ofVon Neumann's remark that no variation in the gross threshold of the neuron as a whole will be sufficient to account for the fine structure and the fact of simply varying the structure as a whole does not permit us to do what we want. We have to introduce greater structure on the surface of the neuron.

Wiener: Yes.

Pitts: One can easily set upper limits for the specification of the single cell. If it receives inputs from distinct afferents it cannot possibly do otherwise than classify them in two classes, the combination which will fire it, and those which will not fire it, and there are 2^n such combinations possible. Therefore, 2^n, n is the number of distinct afferents which go to it, is the maximum. I don't see how you can get up to 10^{10} in a possible cell.

Gerard: I would like to do a bit more backtracking or dendrite retracting before we get too excited about this. The notion of a specific protein organization in the membrane, able to change under the action of ATP, is very nice. As a matter of fact, two of my own colleagues have worked intensively along those lines. Dr. Libit succeeded in showing in the squid axon that the enzyme splitting ATP is practically, if not entirely, limited to the membrane of the cell; and Dr. Tobias is trying to get myosin-like material from | neurons to see how it is modified with ATPase, ATP, and other relevant [18] materials.

Nonetheless, it seems to me that we are putting ourselves into a position here that is really quite untenable. I have just come from another meeting of men concerned with morphogenetic factors in the nervous system of nerve growth and reconstruction. One thing overwhelmingly impressive in that work is the complete fluidity – and I am talking about structural fluidity – of an adult neuron under normal conditions and its extensive disruption and reconstruction under very slight pathological conditions. In each of our brains at this moment a neuron is not sitting there like a figure on a cardboard diagram, as we ordinarily think of it. Each is giving out pseudopods, retracting its fibers, moving forward and back, swelling and shrinking and moving from side to side. Every time one sees moving pictures of these things, no matter how often, one is impressed by the fact that here is hardly anything more than a thin gel.

Of course, that would not preclude the formation of specific protein molecules with a characteristic organization within themselves, as when structurally specified antibodies float in plasma. But I find it difficult to think of such a molecule as set permanently at a fixed place in the membrane. The membrane is continually forming and being removed; endoplasm changes to ectoplasm, in the moving ameba, and goes back again; slight exposure to alcohol causes the neuron to vacuolate and retract and the axon to pull away long distances. It may then reconstitute in a different position, certainly with different molecules in the membrane.

Further, the individual protein molecules, as we know from all sorts of tracer studies, are changing their constituent atoms, as these electrons and energy states. The spatial orientation is changing. While some kind of structural organization could remain, it would have to be more like that at the immunological level and I think we are pressing our good fortune enormously if we try and get, not 10^{10} but 10^{2} discriminating fragments of the neuron of the cell body alone which would maintain a sufficient temporal integrity and identity and connection to give the sort of thing that you are asking for.

WIENER: May I point out though this variability in time here postulated will do in fact the sort of thing that Von Neumann wants, that is, the variability need not be a fixed variability in space but may actually be a variability in time. I have a suspicion, in other | words, that exactly what has been spoken of may constitute the greater number of [19] degrees of freedom which are asked for.

STROUD: I might suggest it might even be necessary if you look at a very large macro-organism called a destroyer you would see people aboard doing a tremendous number of things having to do with metabolism, chipping paint, painting, etc. As a matter of fact, you would never arrive on the destroyer and find the same organization except in gross structure, but believe me there is never an hour, day or night that you can become playful with the destroyer. It can sting the living tar out of you regardless of the million circumstances which may be existing there at the time. So the essential stable function of this rapid, fighting ship, which is never the same – and incidentally most of these changes you see are her metabolism – does not make it inconsistent that there be this tremendous fluidity, yet with this in the midst of change when men are never at one assignable point it is a fighting ship and can deliver a terrific blow.

GERARD: That is the destroyer as a whole. I am worrying about the fragmentation and fixed sub-units.

STROUD: The fixity is not the parts of the destroyer. The fixity is the fact that the whole destroyer is capable of very complicated fighting actions all the whole time.

ABRAMSON: May I enter into this discussion from the physicochemical view? I agree with Dr. Gerard when one looks at the cell membrane, or a leukocyte – and I have looked at the cells mentioned – one is apt to see a great deal of change and one can show what is known as sol-gel changes or thexotropic changes. However, when one looks at these changes, when one uses optical devices or other devices, one rarely gets anything but an observation which is dependent upon the method of observation. A sol-gel change need not necessarily mean a sol-gel change. I should like to emphasize that. If you observe a red blood cell suspended in a gelatin gel and if the gelatin gel is not too thick, (is not too plastic), and an electric field is applied the red blood cell will migrate through that gel just as if the gel were not present. So, although the gel as a whole is intact, although the fibrillar structure of the gel, or the internal forces of the gel remain intact, the application of an electric field at a given point within the gel will cause the blood cell to move with its usual electric mobility in a sol relative to the gel. In other words, we may have macroscopically a sol-gel change but microscopically there is a very great deal of the organization persisting which may have to do with [20] many of the things which you mentioned. So I don't think | Dr. Gerard's point in regard to macroscopic activity or reorientation of molecules necessarily does away with the idea that the very physico-chemical forces which are operating with changes and have to do with all the activities of the surface membrane or internal structure of the membrane need change the basic lattice framework of that membrane. Because of the very nature of the sol-gel change, a sol-gel change need not necessarily indicate that the structure is lost.

GERARD: That is right.

ABRAMSON: I want to bring up one more point apropos of that. There was another matter in connection with the theory of memory on a protein lattice structure in the membrane. I understand that Dr. McCulloch mentioned that one protein molecule serving as template leading to specificity of an immunological type would sort of solder the molecule in place, is that right, and therefore limit the nature of the forces operating?

MCCULLOCH: The notions were two: first, that protein molecules and only protein molecules seemed to serve as templates for determining like structures to be produced. That is, the trick of reproduction seemed to be the trigger proteins rather than other things.

WIENER: Nuclear.

VON BONIN: To serve as template, appears to be the exclusive property of nucleoproteins, perhaps even only of desoxyribonucleoproteins.

MCCULLOCH: I am giving somebody's thought.

VON BONIN: Quite, long chain nucleic acids can be templates for proteins, particularly since the distance between side chains is the same in both types of molecules.

MCCULLOCH: I am trying to render somebody else's statement here; actually the phraseology was borrowed. The trick of reproduction was the trick of the protein molecules.

PITTS: I would like to ask for some clarification of this. We can still state the characteristics of the neuron in which it catches an afferent impulse at which time it will fire it. What kind of a change does this affair involve in the use of the protein molecule as the template in that characteristic and under what kinds of condition, I don't exactly see how that operates and what it does? Does it operate so that it makes one form inoperative, the synapse inoperative, something of that kind?

MCCULLOCH: That is the item for which Ward and Katz had originally proposed it. That is by no means the thing which I think would result from it. What I think would [21] result from it is something | like this: if within the proteins of the cell there are developed closer ties between points sitting at the same original positions on the surface of the cell, the existence of those protein ties might lead to other combinations of cells being affected than those which were previously affected.

WIENER: May I bring in one thing here? I think it is extremely interesting, namely, that I have been using the notion of threshold. We all tend to use it. The existence of inhibition makes it more complicated, that the real question of the firing of a cell is not what is the threshold. What are the precise combinations in detail of impulses that lead it to fire? These can be and there are many more combinations than there are possible thresholds. I mean that that runs the complication up. It runs it up particularly if the elements coming in are not merely single synapses but something much smaller than what we now consider to be a single synapse. In other words, the variety of behavior we can get I think can be stepped well up by considering the fact that the different inputs of the cell may combine in ways much more complicated than additive in order to determine the output.

MCCULLOCH: Let me put it this way: if we have a cell with a given protein configuration – I say »protein« for the sake of argument – which can now be fired by a certain set of buttons but if among those some other button or buttons are firing or not firing then the firing or not firing of those other buttons at the time these fire may determine alteration in protein structure of that neuron so that thereafter it may be fired by other combinations than those which originally were capable of firing it. Hence possibility of building in alterations of this kind into the cell that Von Neumann wants.

ABRAMSON: Would the probability of those alterations occurring be helped by Rothen's idea that long range forces extend up to about 500 Angstrom units from the surface of protein molecules?

MCCULLOCH: I think it would be hurt rather than helped.

ABRAMSON: Would you mind explaining that?

MCCULLOCH: Because there you want greater specificity.

ABRAMSON: That does not destroy specificity. On the contrary, he claims that immunological specificity extends so that the radius of any action in specificity -

MCCULLOCH: In that sense.

ABRAMSON: Would increase rather than decrease the probability of having long-range forces.

MCCULLOCH: I thought you were thinking of such forces. | [22]

There appeared in the last issue of Nature an article to the effect that plane polarized light shined on starch where it behaves, in splitting the starch to sugar as if it had to be in the right wave lengths.

ABRAMSON: I mean connected with the antigen-antibody reaction.

MCCULLOCH: That would be O.K.

WIENER: I am very suspicious of the idea that these long-range forces should be interpreted as radiative forces rather than fixed forces. Of course, they are not changing the nature of your problem at all.

PITTS: I would like to distinguish three stages of the complication of the relation between the input and output of the cell. The first is of course the simple one where all the synapses act together additively, possibly with different ways, possibly with a different number of buttons coming from the same afferent cell, never add all over the whole cell. Von Neumann objects to this because that makes the cell learn too much.

The second stage, more generally acceptable than the first, is the notion that what is required for the firing of the cell is that a sufficiently close together clump of afferent synapses should fire close enough together. That of course increases the information or the efficiency of possibility of discrimination considerably because what it does in effect is to replace the one cell by a group of cells. You could suppose that there was a fictitious cell corresponding to every clump of such synapses which would be essentially independent. I should wonder whether possibly that might not be enough of a complication.

The third stage would be where we have combinations that are not combined additively in any way which we may select arbitrarily. That of course is carried to its furthest degree when we have a cell which some fixed arbitrary combination of afferent synapses – and only that combination – will fire. That is absolute maximum efficiency information which it transmits.

Then there is a final stage where we can allow careful time in the case of dendrites where impulses may be delivered at different times and have to be precisely relative times of different combinations. I don't know whether the protein molecule hypothesis is supposed to secure the last as well as the next to the last step, but I do wonder whether Von Neumann's requirements actually require more than the second step.

McCulloch: What he wants is something which would give a general level of performance corresponding to a number of neurons [23] | for man that is of the order of at least a thousand times as great as the actual number on the assumption of mere number being enough.

Bateson: I remember in our first or second meeting when Von Neumann was presenting the physiology of computing machines that this 10^{10} figure then appeared to be a rather generous figure and that there was talk in which neuron was treated as the analogue of the tube; but is it the neurone or perhaps the synapse? There was a confused conversation at that time but I thought it was clear than Von Neumann was thinking of the neuron as the analogue of the tube. If the synapse is the analogue of the tube then the figures would be quite different.

Wiener: I agree completely.

Mettler: Do I understand that if the cell body is activated directly on the neurocyte and not through the dendrite, the dendrite itself has a sort of retroactive refractory period as it were, or is it necessary for the impulse to travel through the dendrite in order for the dendrite to have this refractory period? The reason I raise this question is if the former circumstance is the case, then it would be impossible for the sort of thing to happen which was spoken about a moment ago.

Fremont-Smith: It has to travel through all of them.

Wiener: I think the thing is this: can a dendrite be rendered inactive by an impulse coming down the axon which does not go through that particular dendrite?

Pitts: If you initiated an impulse in the cell body it sweeps out over all the dendrites.

Wiener: They are able to carry it?

Pitts: Yes, because they are not active.

Hutchinson: Perhaps this is the wrong place to interject it but there is a system which behaves exactly the way you want it to behave and that is the mating reaction in the paramecium where there are a large number of different races, all immunologically distinguishable. Quite definitely there is propagation of some stimulus from the cilia of one to that of the opposite mating kind that initiates a large number of changes that must be due to protein configuration on the ciliar surface, as far as I can see almost exactly a stationary analogue of the thing you talked about. It is very well established.

Savage: I wanted to say two rather separate things. In the first place all this speculation on how to more complicate the nervous system goes too far. I believe it would be nice to think of why Von Neumann thinks it is necessary. After all, a few weeks ago we [24] | thought 10^{10} was a very generous number. Now we think it is a very stingy number. There are approximately 300 thinking neurons in an ant. Who says that is not enough? How can you tell?

Then the other thing I want to say is really the opposite, namely, if we are going to speculate how to make the thing more complicated I would like to mention then a rather daring speculation along the same line, namely, that what goes along the nerve is basically pretty well described by an electric phenomenon, but might conceivably be just a part of what is occurring. We are looking at the nerve very much as through a ground glass. I suppose it is conceivable at the Buck Rogers level that this impulse is only a very crude part of what happens and that it is also conceivable that there is a specific immunological sort of change running down the nerve which rushes on to the next one so that it may be that the blip which has been studied macroscopically carries under natural conditions, if not under laboratory stimulations, an ultra-fine structure from nerve to nerve. That would give Von Neumann 10 – to several powers. I don't know enough about it to even pretend to take a serious view of such affairs. It is simply a logical possibility which has not been brought up yet.

GERARD: I thoroughly agree with Savage's first point; for the rest of the discussion, it is time to get some orthodox neurophysiology into the picture. If these proteins are to be forced into permanent immunologically-specific postures, by the passage of a nerve impulse, it must be done by electric currents. The nerve impulse will pass over one millimeter gaps in the nerve fiber, where the only thing that could possibly connect the two regions is the electric eddy current; the velocity of the nerve impulse along a fiber can be changed by altering the external resistance, which could only affect eddy currents; and much like evidence makes this position very firm.

Anyone is perfectly welcome to make the assumption that these electrical currents produce structural configurations or malformations in the protein molecules; but then every time the electric currents glow those things have to happen. I think you will find that we have put ourselves again into a straitjacket, which does not help but hinders us.

Incidentally I have been waiting to come back to the point Abramson made. Of course what be said about the apparent fluidity of the gel, the essential structure, remaining when it does not seem to be there, is right. But I don't think that applies to these cases, because there one gets complete structural reorganizations. A nerve | axon [25] may pull back from an ending, round up, even amputate itself, the remainder go forward again; as for the ultramicroscope or electron microscope type of analysis, Frank Schmitt, making just such studies, was emphatic on these points at the meeting of which I spoke.

I don't doubt, Dr. McCulloch, that you can make use of these specific protein configurations again, maybe even with distance forces of the Rothen type. They probably have important influences in determining stickiness of surfaces, even actual movement of surfaces towards or away from each other, as Hutchinson pointed out in the case of paramecia. But these are, at least so far as I can see, an entirely different order of phenomena from the kind you require in connection with this problem.

PITTS: You can lose the information also.

WIENER: I don't think anybody has pointed out – this is with respect to Dr. Savage's remarks – any need for the axon to carry more than a yes or no.

MCCULLOCH: That is right.

WIENER: In other words, at this stage I see no reason why the electrical theory cannot be adequate or the electro-chemical theory. I don't want to distinguish between those. I can't. The point is not that. Even the ant, if you could get a fair degree of complication in the behavior of the neuron, then could have the carrying from one to the other done quite adequately by axons. That I think is very likely at least.

PITTS: I should like to answer Savage's question. I can conceive how one could, using Wiener's information theory, demonstrate that the observed behavior of the ant is incompatible with the supposition that it has only 300 thinking neurons, but I would like to know something more about the details of such computation and exactly how it comes about before requiring any great modification of our customary hypothesis.

VON BONIN: In the first place, I am not quite sure yet what this change in protein structure would do. It would not increase the freedom of the system at any given moment because for any given moment you have a certain behavior of the neuron prescribed. It may alter over time. Could it explain memory? Is that the sort of thing that Von Neumann wants, or the degree of freedom?

MCCULLOCH: I believe so.

VON BONIN: All this time I thought of some observation which Polyak made on the retina. (S. Polyak, The Retina, The University of Chicago Press, Chicago, Illinois, 1941). He showed different| synaptical relationships of the bipolar cell both to cones [26] and to ganglion cells and considered a different function for each variety of synapses. If you could accept that view, you get at any given moment a higher degree of freedom if you take the exact location of the synapse into account. There is in the retina, at least, some hint that the location of synapse makes a difference. In other words, a neuron has probably two, probably several degrees of freedom and you can, if you put that in, get a much higher number of degrees of freedom than by simply counting neurons.

MCCULLOCH: Suppose I try to answer first the question, if we can, as to quantity of information and numbers of trace carriers, etc., required, if we are to understand human memory. I would like to begin by asking Mr. Stroud to give us the picture of the psychological moment as he has studied it. So we will begin from the psychological end now.

KUBIE: Will you make it possible to break down this whole concept of memory, because we are not talking about a theory but we are talking about an abstraction here?

THE PSYCHOLOGICAL MOMENT IN PERCEPTION [27]

JOHN STROUD
U. S. Naval Electronic Laboratory, San Diego, Calif.

Some of you who were at Pasadena know that Dr. McCulloch scared me to death by asking me to speak without warning. I don't remember what I said (1)[1]. Others said I said something. Since then some very reassuring things have occurred. I hope I can now speak with a little more assurance, however, before I speak of my own study I should like to pay tribute. Those of you here who are familiar with McCulloch and Pitts and their work on »How We Know Universals« remember it as an attempt to solve the problems of orders of things that men do on the basis of known facts about neurophysiology and neuroanatomy (2). McCulloch and Pitts came to the conclusion that a very useful, indeed almost a necessary mechanism, would be a sort of scanning process – sequences of computations – at tremendous savings in space and material. That approach of McCulloch and Pitts was one with which I was not familiar until very recently.

To find that this idea of the periodic functioning of a net would be desirable from such a widely separated view, or starting point from my own, was very reassuring. Later too I found that this matter had been approached from another angle by an Englishman whose name was Kenneth Craik (3). He was killed recently which I regret because he was very brilliant and was studying very practical things. As you know, in the firing of guns we have to use human operators to make certain decisions but today we have to fire them very rapidly. We have tried our level best to reduce what the man has to do to an absolute minimum. The guns are much too big to be manhandled, so they are handled by servo mechanisms. These have to be studied very intimately in detail. We know as much as possible about how all the associated gear which brings the information to the tracker operates and bow all the gear from the tracker to the gun operates. So we have the human operator surrounded on both sides | by very precisely [28] known mechanisms and the question comes up »What kind of a machine have we placed in the middle«. Craik came to the conclusion that the human operator was an intermittent servo. The typical servo works, as Dr. Wiener has illustrated in picking up a pencil, »by the amount by which I have not yet succeeded in doing what I intended to do«. This is not the kind of servo system which is involved in the case of the human operator. For many purposes in which short-time changes are not important, we can consider the human operator as a good counterfeit of this model of Dr. Wiener's but the detailed performance of the human operator is of a different sort.

We do use systems which work like Dr. Wiener's model and they do have operators but the operator has only the function of deciding when a target is a target. Once he makes up his mind and makes his decision known to the machine, the machine then works all by itself receiving its information continuously and giving its output continuously. It is at base an error-operated servo system. There are necessarily always mistakes, statistically speaking, for it is only by the nature of their mistakes that such systems can do anything. As I said before the human operator may seem to be able to operate in this way grossly, but when you study him very carefully, he does not operate this way.

WIENER: May I call attention to this: have you ever seen the governor that is used on these two-cycle gasoline engines, the pumping engines?

STROUD: I am sorry I have not.

1 All references in Mr. Stroud's paper are to be found on page 163.

WIENER: It is very interesting. When the thing is going too fast it misses a stroke or two.

STROUD: The type that interrupts the timing?

WIENER: That is the intermittent.

STROUD: What happens is that a little cam which moves continuously finally gets into a position so that it will not let the circuit breaker in the ignition system make contact and thus it prevents the firing of the cylinder.

WIENER: When you speak of intermittent operation, the cycle is intermittent. This is the cycle, very much like the one you are speaking of.

STROUD: There are other things of interest about the human operator. One of them is that he is a curious mechanism in that he can go on doing something which is already right, that is, when he is not making mistakes by which to guide himself. This is at [29] present not a characteristic of most mechanical systems, although | having studied the problem, Craik laid down a design for a beautiful mechanism which could do this.

FREMONT-SMITH: Have we evidence that it is absolute for man, or is it slightly wrong for him also but to a very much less degree of wrongness?

STROUD: If you wish to be that precise, you can point out that it is wrong but it happens to be of a wrongness which can make no difference to the man.

WIENER: I would say it is very much like the pumping governor.

STROUD: I would say it is very possible for a well-trained tracker following simple laws to anticipate motions of the target so that his pointer is not out of line with the target pointer by an amount which he can reliably detect. The fact that I have the instrument that can detect the difference is not important because the man cannot.

FREMONT-SMITH: Consciously or physiologically he cannot detect it?

STROUD: You can present these two points which are misaligned to this man and from now until doomsday he cannot make this distinction.

FREMONT-SMITH: You mean he cannot make the distinction consciously or physiologically?

STROUD: He cannot respond to it in any way we know how to detect. He is just unresponsive to this. There is no information.

WIENER: There is a noise level. The noise level is the width of the line. The problem is then that if the signal is less than a certain amount the confidence with which it can be detected is very low; therefore there is no use in making the correction.

KUBIE: This brings up very important questions and an important source of error because you are leaving out the most important stimulus to the man, the stimulus that has not occurred.

STROUD: Pointing up what Craik did, I was just at one of the Naval laboratories where the boys are still following the British footsteps in the matter. Craik did not make any serious blunders. We find out too that the order of information in this case can vary and be quite complicated. If you consider the spatial displacement of the two pointers as a function of time, the human operator is perfectly capable of being guided by the fact that the two lines are separated from one another but he is also perfectly capable of being guided by the fact that the first derivative of displacement happens to be different from some preassigned value which is right for the situation. Furthermore, [30] the second derivative can be different from some | preassigned value and this can guide the operator. We have a little gadget in which a single knob can control the movement of a pointer. It can control the movement of the pointer by controlling its displacement. By changing an adjustment we can make the knob control the pointer by controlling its velocity. By still another adjustment, we can control the pointer by controlling its acceleration. As the most difficult method, we can adjust to control the motion of the pointer by almost any combination of these three ways. None the less the human operator is capable of handling the situation. Here he has control operations which are to a degree control of displacement, control of the first derivative of displacement with respect to time, and control of the second derivative. With such a control system, the human operator can solve problems even though you did not, and in many cases could not, tell him in advance how the system was set up.

SAVAGE: You give him a pointer to run, you mean, and he does not know?

STROUD: Maybe I can draw it graphically. Picture a track which your pointer is going to follow; in this particular case the indicator and pointer are the same thing. There is a little spot moving horizontally and a vertical line and the operator is supposed to keep the spot on the line. In time the spot is going to wander right and left. There is an assigned irregular track it is going to follow. This is the displacement with respect to time.

SAVAGE: So the spot would move?

STROUD: That is the way the spot would move if the tracker did not have his own controls to bring it back to normal. It happens he can control the movement of this spot either as a displacement, a simple displacement by the movement of the lever, or he can control it in terms of the first derivative of displacement. The velocity at which....

SAVAGE: At his will or the experimenter's will?

STROUD: The experimenter determines what set of parameters there is to be.

SAVAGE: The stick is governing the displacement or governs acceleration?

STROUD: It can be any combination of any, or all three to any degree of sensitivity the experimenter likes. This servo mechanism man is capable of handling the first derivative, the second derivative, the original equation, any mixture of all three and still keep these things together.

SAVAGE: Won't the crudest servo mechanism do that? | [31]

WIENER: The crudest servo mechanism will not because the crudest servo mechanism has a fixed setup, but it is quite possible to build a servo mechanism – I want to say something about that as soon as we get a little further – in which the pattern of its behavior is changed by its own success or failure.

STROUD: Craik pointed that out.

WIENER: We are building one.

STROUD: Such a servo I might say would be very very useful.

WIENER: We are doing the following thing: you know my predictor, the ordinary predictor. There the pattern of prediction is done on paper and it is fed in to the setting of the predictor which then does the prediction job. We are doing now, or we have under contemplation, a predictor which has two scales which we will call the ordinary time scale and the secular time scale. On the secular time scale it will go through the computing motion, compute the auto-correlation and the entire set of patterns for prediction theory, transfer those to the direct predictor which is on a higher time scale and give us the prediction, continually changing its pattern of prediction if there is any change in the statistical pattern of the data. That would be very close to this sort of a thing. In other words, if the particular occasion involved working on the first derivative only, it would work on the first derivative only. If it involved going to the second, it would go to the second, etc. That is a problem which is very interesting and one that we are handling. That is on our schedule for construction.

STROUD: If you will also permit me a slight aside here, ever since I studied mathematics I have been irritated about the things teachers told me. They never told me how you learned. With all the posturing and gestures, writing on the board, they did point out to me and give a name to something that went on in the central nervous system. If the teacher succeeded in doing this I became a mathematician and if he did not I became a flunk, and to be able to demonstrate that the central nervous system can handle the integral and differential calculus so automatically and nicely without any college training is a great pleasure.

WIENER: It is a great pleasure to discover that he found he had been talking prose all his life.

STROUD: So much for the order of complexity of this piece of machinery, man, which we have put in between two already well-known machines. It is of itself interesting to find that man can do such things. It is even more interesting to find that there is a still [32] higher order of activity. Man is a predictor and says »I shall con | tinue to do whatever my last solution predicted will be right so long as no detectable difference arises«. Errors of control thus operate one step above the levels of displacements, velocities and accelerations. I am very curious to know just how far we can push this human operator.

WIENER: The machine I am talking about would literally do this. It would check whether the error was the predicted error. I am not talking about the continuous machine. It could be made discrete quite as easily. Then when the error got out of hand – when the difference between the actual error and the predicted error got beyond a certain percentage – it would repeat itself, re-examine itself statistically.

STROUD: Work up through orders of derivatives?

WIENER: It would not actually do that but it does what is mathematically equivalent to that.

STROUD: The next thing of interest about the human being as a servo mechanism is that he acts regardless of which system of correction is necessary – perhaps I should set the situation up a little better. One way of doing tracking is with the simple hand wheel. The hand wheel itself is directly connected but you will readily observe that this human operator can be concerned about various attributes of this hand wheel's motion, how far did it move, how rapidly and how did it change its motion, so with this simple hand wheel control your typical operator starts off with simple servo relationship. He does his adjustments first in terms, when you analyze his records, of how much did it miss on the simple displacement scale. A little later on it begins to be apparent that the operator is now doing his adjustments partially on a simple system and partially on the basis of how much the velocity is involved. A little later on he may include acceleration. If then you introduce some sudden change you will see that his solution breaks down, he then returns to his original set of adjustments and works through the various orders of derivatives producing successively better and better solutions of the problems. With good luck and a good set of solutions he winds up with the solution that is right rather than wrong, at least not detectably wrong. He does this in a very peculiar way. He does not do it continually. Typically he is a half-cycle corrector for he does corrections every other half second or every third of a second. His corrections are cut and dried. You can demonstrate this. If you blank him off so he cannot see his pointer you discover his next corrective action is just about as good as if [33] he had been able to watch what he was doing. So the interesting thing | is that the information which determines his next corrective action is not obtained during the course of the corrective action itself.

WIENER: There is something quite interesting which we intend to use also and that is a predictor as part of a corrective machine of this sort so that we check not against the actual motion but what the motion is predicted to be. One interesting use we intend to make of this, quite practically, is to see if we cannot use this mechanism to push frequencies up higher in radio sets. We intend to cut by prediction the effective transit time of the molecules in the gas so that we check against a predicted value instead of an actual value. I think we can actually do that. We intend to use that as an actual piece of apparatus for gas tubes.

GERARD: On this last point, I wonder if it is the same phenomenon that one runs into in many studies of the nervous system. If you take a completely untrained subject, sit him in a chair, elicit the knee jerk at regular intervals and then stop the hammer without warning, the leg is very likely to kick for several responses at the »expected« time intervals.

STROUD: I think that is very probably the typical performance of the entire central nervous system.

Suppose we have a task where our operator receives his information by way of his eyes, very large numbers of photons are absorbed at the retina at stable statistical rates so that we may speak of them as being received continuously. This is where the information goes into the human organism, and it goes in, to all intents and purposes, continuously. When we analyze what comes out of the organism, every set of records of sufficient sensitivity, which have thus far been analyzed, has shown low frequency periodicities, frequencies of the order of two or three per second. There is a period of the order of one-tenth of a second during which a corrective action is taken, during which some change is made in the characteristics of the manual output of the man. There follows a period of about two-tenths of a second in which nothing new is done and then another period of about a tenth of a second in which new corrections are made, and so on.

KLÜVER: How many changes have you made in this system? Unless you have tried numerous systems you cannot say, it seems to me, that there is always a discontinuous way of correcting what has been done.

STROUD: I hesitate to say. I am terrible at this sort of thing. I am not very good at rote memory. I sort of edit things I read. There | have been at least 20 sources of experiments going on, just duplications of those of the Englishman (4, 5). [34]

WIENER: May I say why this must be the case? The transmission is essentially discontinuous. Your individual neurons go all or none. It is only by means of sampling that you get anything approaching a continuous input. If you take too short a time your sampling is going to be decidedly bad. Your average of the inputs coming in won't be reached in any precise way. Therefore, in order to get a really significant correction to correct by, you must wait. That I think is one of the chief reasons why you get this discrete performance. If you try to make another correction too soon then the number of nervous impulses, although large, would not be large enough to give an accurate density; to give an accurate density you need a certain amount of time. Therefore the limitations of working with all-or-none transmission force a grosser all-or-none bunching of the corrections.

STROUD: There is a line of argument.

KUBIE: Can I put in a word about this? This is a demonstration of a fact which has an old and venerable history, namely the fact that human beings have variable reaction times. Reaction time was not split up into its component parts when it was first studied, which has an important bearing on the development of modern science. That reaction times vary from one individual to another is known. We also know that an individual's shortest reaction time is not constant. All kinds of emotional influences affect it. Tension, anxiety, over-eagerness, etc. all can make him beat his own time, either accurately or inaccurately; while at other times he lags way behind his basic reaction time.

METTLER: Do I understand that this correction is literally made in total darkness and that he proceeds with as high a degree of accuracy as if he had cues for a period of time, which is what?

STROUD: Let us put it this way: it does not make any difference whether he has cues in the preceding one or two-tenths of a second or whether he has not. What he is to do has been decided by prediction. It may be a very complicated action and it goes through as such and there is nothing he can do about it once it is started. If the information changes in the meantime, he has had it.

METTLER: He does not keep on doing this?

STROUD: Immediately he makes a gross error he stops. The next time he makes no correction. He is not going on indefinitely making corrections on the basis of no information. All I wish to point out | is that the information received during the period of the correction has nothing to do with the correction. [35]

FREMONT-SMITH: It deals with the next correction?

STROUD: It deals with the next one. It is information for the next one. With a sudden interruption like that, a typical one where he may be following a moving spot and the thing just suddenly blinks out and disappears, he makes the next correction in terms where he predicted the spot would be and then stops making corrections altogether because he has no further information.

FREMONT-SMITH: There is a lag between information and action and the reaction is always not to the last information but to the last input?

STROUD: Yes.

MCCULLOCH: The interesting thing to me is that it is quantized in units of the order of one-tenth second for gap and two or three-tenths variable for the rest.

STROUD: Depending upon the complexity of the response the whole cycle itself can take place in a period of a half second, that is, the observation of the results of the last action and the initiation of the new action.

SAVAGE: You talk about this kind of tracking in which he persists – in which there is no error. You don't mean to say that he can follow a complicated contour with that kind of tracking?

STROUD: For quite fair periods of times if he happens to fall on the right solution and he recognizes it as the right solution he does not vary it until he has a reason to. Just for example, one of the tracking problems is a very simple one, three non-harmonically related sinusoidal motions added together. It is possible to stumble on the solution of that. A good tracker can follow that sort of thing for several seconds and never make a mistake, at least not an observable one.

SAVAGE: Several periods?

STROUD: Several.

SAVAGE: Several sinusoidal?

STROUD: Yes.

MCCULLOCH: I want to get Stroud's evidence out on the table. There is one question from Teuber. I would like to have him have a chance to finish the statement so I can get the evidence.

TEUBER: Maybe he should go on first.

STROUD: I did not mean to drag in such a red herring. I know this is interesting. The main thing I want to point out is that here is an entirely different origin of a theory. From data like this Craik | came to the same conclusion that you must necessarily consider the central nervous system as operating on a sort of quantal basis. [36]

MCCULLOCH: On a quantal?

STROUD: He has to work by cycles of operation, very much like an ordinary mechanical computer. He does very many complicated things but there must be a fundamental periodicity of the order of the gaps. Craik suggested some experiments one of which I happened to be doing at the time he was writing. If I may, I will take up my own work (6). I have never been able to explain why I started this, so we will skip that. I would like to explain how I was able to demonstrate this periodicity. My initial experiment was a very simple one, one which all of you may have at one time or another done, or something very similar to it. I sat a man in front of an ordinary cathode ray oscilloscope. On it was a moving spot, moving just ordinarily back and forth, across and returning very quickly twenty times a second. I gave the man a telegraph key. When he punched the key it made a little bip on the line. I said, »Now Joe, sit here and tap the key until you can make the bip stand still tapping just as fast as you can.« With about twenty minutes of practice he could make the bip »stand still« for ten seconds at a time. I found the duplicating bips were falling 100 milliseconds apart, plus or minus 2 ½ milliseconds. What was the stimulus? Why the last bip was. What was the reaction time? Exactly 100 milliseconds plus or minus 2 ½ milliseconds. I asked what he saw and he said that he saw two lines of trace and one bip. The amplifiers caused a slow hunting of the line so that he could see two lines.

»How many bips,« I asked?

»Exactly one.«

What was there to be seen, in the physical sense? Physically there was a little spot which was slightly oval. It was always moving.

He saw two lines of trace, exactly two, one bip, exactly one. The bipper was just a little gadget which was not too well-built. Occasionally it did not give a bip. That was the time he saw two lines and no bips at all.

From this experiment I made certain inferences. I said arbitrarily that experience is quantal in nature. It is up to me to find what the quantum is. When I find it its name is »moment«. From this experiment I inferred very simply that one of the things about my moment was that since this was a 20-cycle trace and each one of these walks across the tube was 50 milliseconds that within this moment there must be represented a sequence of events which gave absolutely no indication as to which came first. Why? Because he | should have somehow or other been able to tell whether that spot was [37] moving from right to left or left to right and that he could not do. He could pull certain tricks and find out, but by just looking at it he could not tell. It is just a line of trace and you cannot tell whether the spot moves right to left or left to right. I am sure some of you have forgotten which was the case on the oscilloscope you were working with and made mistakes and misadjusted, even at low frequencies. I said the moment refers to events which are posted at some place in the central nervous system and that all events represented in this moment carried no information as to relative priority in time. Furthermore, you will notice that the subject never sees two bips. He never sees three lines. He never sees a line and one-half. He sees two lines. Therefore there is a boundary between the content of the moments which is pretty sharp. As nearly as I could determine, information is either in one moment or another moment. That is all. You found out after the fact and you could not necessarily predict which moment the information would be in stimulation and its congealment into moments. This fundamental frequency of the interruption of the train I call the moment frequency. Dr. McCulloch calls it the scanning frequency. So there is scanning and the results of the periods of the scanning are the periods of the moments.

I decided to test this hypothesis in another way. I wanted to find out how a man would add up information about brightness in time. So I set up a test field which was split vertically. One-half of the field was lighted intermittently at a constant intensity, but with variable duration and repetition rates. The other half of the field was set up so that the subject could control a continuous brightness and he was told to match these two for peak brightness. Later on I found the instruction was perfectly silly. He could match them consistently only this way anyway.

The idea behind it was this: you could consider any given flash to be broken up into any arbitrarily small number of individual units of information. If one was not going to be able to tell anything about the difference in time, then these units within any given moment must quite simply add up. If they don't that is information about priority. You see each one of our little sub-flashes refers to exactly the same space in exactly the same proportions and had exactly the same spectral distribution. The only thing which could distinguish one sub-flash from another would be its date. If it were true, as I had postulated, that dates within a single moment were meaningless, then this difference of date could not make any dif|ference. So that if there is to be no information about relative priority, they must add linearly. [38]

My results I could plot as in diagrams A, B, C and D, choosing scales running from zero to 100%.

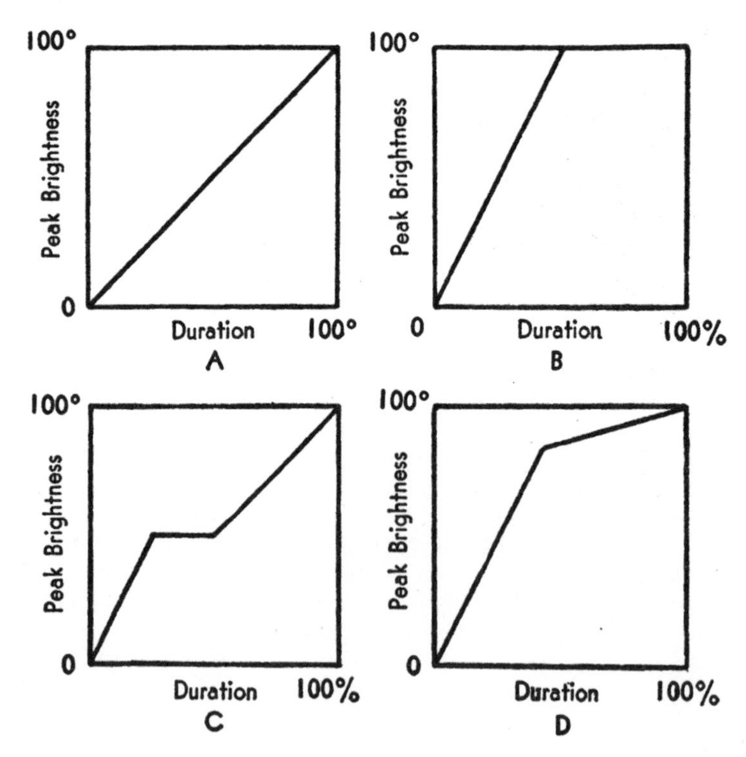

I could plot along the ordinate the subject's estimate of how bright the flash field appeared to him as a percent of his estimate when the field was on all the time. Along the abscissa I could plot the percent of time that the flash at its constant peak of intensity was in fact available. Experimentally for all subjects, all conditions, all frequencies between 8 cycles a second and 20 cycles a second, I got quite nice linear curves running from the zero percent, zero percent corner to the 100 percent, 100 percent comer as in A. It can be shown quite rigorously that these curves are precisely what we should expect from our postulates if there were one or some integral number of flashes per moment. It can be shown that such summation curves would be predicted only if the moment frequency of the viewer and the flash rate of the test field were in harmonic relation to one another. It can still further be shown that the classical assumptions of temporal summation of brightness will not predict the | data obtained. That [39] such harmonic relations should obtain was one of the postulates that was inferred from the earlier tapping experiment. If you remember, the observer saw two and only two lines of trace, one and only one bip, which could be the case only if such harmonic relations obtained.

We could make a prediction for repetition rates which had periods quite long with respect to the periods of moments. Under such circumstances, I could have a chap making summations faster than the flash rate but still in harmonic relation to it. If he were summing at the rate of two moments to one flash, his results should make a curve like that of B in contrast to A. Here he would continue summing linearly until one moment was fully lighted, at which point the light should be as bright as it would ever appear.

Data below 8 cycles per second did not conform to the simple predictions. At 6.4 cycles per second, which was a frequency which I explored, the data very frequently looked like C when plotted. It did not matter which way the subject started. We might start with low values and the curve would start off as if the subject were aiming for some point about 50 percent duration and 100 percent brightness. When one moment was about fully lighted, the subject would get a little confused and erratic in his judgments. As the flashes grew a little longer the curve would finish up as if there were one or some integral number of flashes per moment.

SAVAGE: The variables are brightness?

STROUD: The variables are, along the abscissa, the percent of time the flash of constant brightness is available and, along the ordinate, the subject's estimate of the greatest brightness of the flashing field. The subject would start off as if he were going to have two moments per flash, get confused half way up and have one moment per flash.

SAVAGE: This high end is where the field is almost continuously illuminated?

STROUD: Where the field is almost continually illuminated. At somewhat lower frequencies subjects produce data which conform to the predicted values temporally but with respect to brightness they confounded me. The value did not reach 100 percent where I predicted it should have. It fell somewhat short as in D.

It was amazing the degree of accuracy that these people could achieve. I built a 5 percent instrument and thought it would be good enough. More exactly, I built a good instrument and calibrated it with an accuracy of 5 percent. My observers' data were so internally consistent that they showed up my poor calibrations | very pain- [40] fully. If I ever do this sort of thing again, I am not going to make the mistake of selling the human operator short. If you ask him the right questions, he is capable of giving you beautiful and precise answers.

This is very useful. We can do a great number of things with this postulate system. Thus, for example, we can get rid of the bugaboo of reaction time and its variability. Why is reaction time so indeterminate? You recognize that when you are in a typical reaction time experiment, you present a man with a stimulus, give the man no chance to synchronize stimulation to anything, so you have taken a complete chance on what phase relationship there is to the moment. To do something you must presumably be aware of the cue and so you must be aware of it in some moment. It is very unlikely that you are aware of the content of the moment until all of the initial stimulus excitation which it is going to represent has occurred. There will be some definite time after the cue has occurred before any definite action can be taken. The net result would be, in an experiment where a man operated at a perfectly constant moment frequency, and perfectly accurately, that his reaction times would show a perfectly square distribution of lengths exactly one moment long. Surprisingly enough under many conditions very well-trained individual operators have given distribution of reaction times which approach this limiting case of the square distribution, one moment long and with remarkable accuracy. I remember a very beautiful thing, a curve in Woodworth from some of Jenkins' data that shows this very nicely. Almost everyone of those reactions is included in a little block, one moment long. I had noticed as had others that the distribution of reaction times of single subjects under many conditions are multi-modal. These are statistically significant. I tried an experiment by means of which I hoped to tie the multiple modes of such distributions to this postulate system. I didn't succeed but I discovered something which I have not forgotten. Don't tinker with moment functions in vision with a piece of equipment that gives auditory cues as to what is going on or which has any distracting auditory periodicity. You will tie the scanning to something that you do not want to study and your result will range from confusing to completely indecipherable.

The Talbot-Plateau law which is of purely empirical origin can be predicted simply from these postulates. Above the critical flicker frequency all estimates of apparent brightness become the same; maximum, minimum and average are meaningless in an [41] apparently | flickerless field. The Talbot-Plateau law is simply the prediction which we made for the conditions represented by A in this range of repetition rates. Another way of stating our prediction could have been that if date is to be of no consequence within a moment, the brightness of a varying light is simply that of a light of the same mean energy which is on all of the time.

Some other things I did not anticipate came out of this experiment. Every time I got a curve that indicated he was working on a two moment to a one flash period basis, you had phi in the field. I did not set out to study phi.

McCulloch: Apparent motion. You are looking at a field which is uniformly illuminated, bright or dark. When you get a phi, the illumination seems to travel over the field.

Stroud: The phi's that we got were not at all rigidly determined. This was not a set-up to demonstrate phi. It was set up for experimental brightness summation. Half of the field flashed and half did not. It bad a razor blade separator, If any of you care to duplicate the experiment, have your field of division very sharp because the degree of separation between the two half fields introduces ambiguity of judgment – it does not change the shape of the curves. I know, I tried it with a cardboard separator. It was too wide and I tried it with a razor blade and the results showed much less variability.

Usually the naïve subject would see the periphery of this field collapse at the diameter and swell out again, like half of a balloon.

Klüver: What you just described seem to be gamma movements rather than phi phenomena.

Stroud: You can choose to call it anything. You find an apparent movement. I could have a guillotine. I did all my judgments chopping up and down, whereas most people do not. Whatever kind of apparent motion I saw, it did not make any difference between my curves and other peoples'.

Then we went on experimenting as to how many kinds of contraptions we could imagine could be behind the flashing light as a shutter. If you imagine a little door pivoted at the top and swinging like a pendulum, you could see a little door pivoted at the top and swinging like a pendulum. Apparently any kind of hypothesis which you could self-consciously advance, which could conceivably agree with the kind of information that you had, was verifiable by the apparent motion. Of course there was in [42] the physical sense no motion at any time. |

From this I advanced the postulate that for apparent motion to exist there must be a difference of the spatial representation of information between successive moments. If successive moments are thus differentiated you have the basis for testing hypotheses as to what constitutes the motion which accounts for this difference. At this point I suddenly realized the implications of what we know about the absorption of light in eyes. When a quantum is absorbed in this system it gives directly only the following pieces of information: the approximate date of arrival and the approximate energy. In and of itself a quantum cannot tell you whether it came from a near or a far object, a stationary or a moving object, or even from which direction it came. From the structure of the eye, we can infer the most probable direction. None of this information is about motion directly. Any notion of motion must be an inference, an hypothesis, no matter what kind of an eye is used. It is simply not in the nature of light to give directly any information about motion. You never see motion under any circumstances. You merely have material with which you can test an hypothesis about motion.

FREMONT-SMITH: Isn't that perfectly analogous to the projection of an ink blot in the Rorschach?

STROUD: Except in this case you are time bound.

FREMONT-SMITH: The process is the same, the projection process.

STROUD: For lack of a better term I just call it hypothesis testing. How shall we say, it is of the order of something which does not come from the information itself?

FREMONT-SMITH: Isn't that exactly the same? When you see a bear in the ink blot the bear does not come from the ink blot.

STROUD: In the case of the Rorschach, the hypothesis you can more or less test. It is a timeless one. In the case of successive moment with different structures, the hypothesis is not a timeless one but involves motion. You get a sequence of moments which differ in space value in organization. If there is a space difference then you have a means of testing an hypothesis that something moved.

WIENER: Has any work been done on moving Rorschachs?

FREMONT-SMITH: I don't know.

STROUD: That is right. That is what I am trying to tell you, that the motion that you see, is mediated by the absorption of quanta, quanta arriving at the retina at a place and date. Do they tell you, »We came from a moving object?« I have not heard anybody say that the retina could detect the Doppler effect. Even if it could | detect the [43] Doppler effect we would have to have advance information as to what state of what emitter was involved.

The only thing that seems to be at all binding is that you get a sort of feeling of belief or unbelief. You can try a wrong hypothesis. One that won't fit, just won't fit. Do you get what I mean? At this level if your test hypothesis explains the data of your moments, it works. Please understand I would not for a moment try to propose any mechanism for accounting for it.

SAVAGE: How do you systematize your general hypothesis? You see the guillotine motion is not consistent even in its motions.

STROUD: That I fail to see. In this particular case I can only report things as I observed them. If you want to »do« or »have« guillotines or sliding doors then they are quite apparent. You, Dr. Savage, might have a vested interest in not seeing them and so you might not. I know that my wife could never see a swinging door although she had no trouble at all seeing balloons or guillotines.

FREMONT-SMITH: When you say testing, you mean when it happens to be a happy one?

STROUD: Forgive me for being so naïve. You are happy with your hypothesis of motion or you are not. In a typical life situation – you may be in a train and glancing out the window at the train on the neighboring track you may come first to the hypothesis that your train is moving. Then suddenly you realize that the confirming information about the acceleration of your train is not there and you are then happy with the hypothesis that it is the train on the next track which is moving and not your own. You will remember though while you were still happy with your first hypothesis you were completely convinced that it was your train that was in motion.

FREMONT-SMITH: You had a moment of anxiety.

STROUD: Perhaps you thought it was not time for the train to leave. Perhaps your wife who had come aboard with you but not for the journey, had not yet left, then in the next succeeding moments you come to the more comforting conclusion that it is the other fellow's train that is leaving the station.

SAVAGE: Either hypothesis is compatible with the visual geometry.

STROUD: They are compatible.

FREMONT-SMITH: You don't test at all, except purely subjective reaction.

STROUD: There are moments that do differ in their spatial configuration. Moment A has a field which is wholly lighted. Moment B has a field half of which is not lighted. [44] Now you have two moments | referring to the same space succeeding one another in time. You have the basis of phi.

TEUBER: Of »apparent motion,« not »phi!«

STROUD: »Phi« is a shorter word to say although I am doing injustice to its definition.

TEUBER: Do you assume there are any sensory processes in which you don't have this sort of hypothesis?

STROUD: Personally I don't, but I have every reason to believe, and so has Dr. McCulloch, that this identical mechanism is operating for both vision and hearing and at the same time.

MCCULLOCH: Would you please describe the simple experiment of auditory driving of visual perception?

STROUD: We have a very interesting problem coming up in which we are to empirically or theoretically determine the best way of presenting the same information for both vision and hearing for comparative purposes. That sort of problem is coming up very frequently. In the course of this we got into a discussion of how well we could link these two up. You see my colleagues are accustomed to thinking in my theoretical terms. They were saying, »What we really need is to tie the auditory system to the visual moment function. How are we going to do it?« They were kidding about it by asking in which part of the brain should they screw the spark plug.

It is possible to take the electrical signal which moves the spot of a cathode rate tube horizontally in an ordinary oscilloscope and turn it into an auditory signal. This sounds like a sharp click which has harmonics all over the map. The low harmonics are, however, in the main well below the noise level. What you have here is fundamentally a sharp burst of high frequency.

What happens to a chap with earphones listening to these little clicks as the beam slips by, say around 10 or 12 cycles a second? He can now do interesting things. If he connects a microphone to the vertical amplifiers on the oscilloscope he finds that he can adjust the pitch of his own voice to an exact harmonic of the sweep on the basis of the clicks. If he wants to look at one vowel sound »A« he finds it infinitely simpler to adjust the system by means of his nervous system and its control over his voice than by twiddling the dials of the oscilloscope. He can put in a 270 or 540 pitch so closely synchronized to a 9-cycle sweep that the pattern barely drifts by on the oscilloscope.

McCulloch: It is amazing how accurately you can do that.

Fremont-Smith: Say that again? | [45]

Stroud: Here he is getting a little click every 9th of a second, and with this curious bit of information he is able to synchronize the tone which is many times that frequency, say 270 cycles a second. That is 30 times the click frequency. Now with this click he can tell that in this case there are exactly 30 complete cycles of this tone in each of these moments. He can do this with his eyes closed and when he opens them he will see the corresponding trace spread out before him on the oscilloscope moment by moment, none of it duplicated, none of it lost.

Dr. Arnold M. Small, the head of our division, is an excellent violinist and many of our number are fine musicians. I must confess that I am not. I could only be called a futile flutist. Our favorite bass violinist thought that I was kidding when I first described this strange relationship between a 9-cycle click and a 270-cycle tone. He promptly came over to my laboratory and started listening for this relationship which I tentatively called »beats«. He tried for about 10 minutes and said at that point »It is your vivid imagination«. I had to leave the laboratory for a few minutes and when I returned I found him as excited as I had been. »Now I have it,« he said, »it's a right and wrong feeling.«

Another of our workers who was interested in speech sounds had railed against the fact that he was constantly adjusting and readjusting the oscilloscope without much success at seeing a meaningful trace. He felt that this was a very nice little invention. With the click as a cue, he had as it were only to twiddle the dials of his own nervous system and let the dials of the oscilloscope alone. With a little practice, I found him gaily singing tunes and comparing the vowel sounds of the various pitches with the greatest of ease.

McCulloch: The auditory stimulus is here beating your system to the pace where you can perform visually and vocally in pace with the mechanical system.

Stroud: It is beating out the time for it. Apparently the problem is not how to synchronize moment frequencies. They are always synchronized to something and the problem is to get synchronism with something you want. These things are the sort of things happening in my laboratory today. I speak only with the excitement of the moment and rechecking may change something. I feel very happy about this for if this works out I am not going to have to do anything very desperate to my subjects at all in my special problem. The machinery is going to supply some auditory driving cues related to the visual material. | [46]

McCulloch: Will you describe the experiment with the earphones?

Stroud: Dr. Licklider was connected with this. He came out to our place. He was doing something with binaural beats, not the ordinary kind but another where the maximal occurs in the right ear and the minimal occurs in the left, and it swaps back and forth; otherwise a very ordinary beat situation. This is again a spatial situation, one ear a maximal and one ear a minimal, one ear hearing one thing and the other hearing the other. One can have auditory phi. It is a little bird bouncing between the ears or spinning around the head, from left to right or right to left. I always wondered why it moves only in the horizontal plane. He was using Lissajous figures to count these beats because his beat counting circuit was not finished. My favorite apparent motion for simple Lissajous figures is that of a wedding ring, a wedding ring spinning. While I listened to his beats I watched the ring spin. I found if I increased the frequency of the beats the wedding ring spun faster and faster and finally at a frequency both the apparent motion of the ring on the oscilloscope and the movements of my bird broke down together and became in the case of the bird merely buzzing sound between my two ears, with no definite spatial location. On the other hand the wedding ring phenomenon broke down into a sort of wobbling parallelogram. When you reduced the frequencies you were again able to see the spinning wedding ring and to listen to the little bird. From watching at higher frequencies I had a sort of feeling that the frequency at which the little distorted parallelogram was oscillating was somehow a bit behind the buzzing frequency in my ears. I thought this was interesting so I asked the chap who was operating the equipment – he was most naïve – if he felt any difference in the timing between the buzz and oscillating parallelogram when the simple apparent motion had broken down. He too felt that the oscillating parallelogram ought to hurry up just a little to catch up. This is something else that is interesting to find out, that there is the possibility of measuring the difference in transit time for information from the eardrum and arriving from the retina.

KLÜVER: As a psychologist you are undoubtedly familiar with the Tau effect which appears when three stimuli are successively exposed at equal spatial but unequal temporal intervals. For instance, if the time between the appearance of the first and the second stimulus is shorter than that between the second and the third stimulus, the first distance appears shorter than the second. This effect has been demonstrated in [47] vision, hearing, and touch. |

STROUD: Now we are getting into questions that I want to know more about myself. It is possible in these cases, with the Englishman's data about tracking and mine, with the various and sundry things we have talked about, to have the assumption that the central nervous system is a nice computer with a period of a tenth of a second, as a whole –

METTLER: Did you find that different subjects had different time intervals? The reason for asking this question is because I am influenced by Halstead's data and also some I had myself. Did you find that the individual has his own frequency which varies from individual to individual?

STROUD: I found this. Perhaps I was a little hurried and did not state it. If you ever start to think about it, we do see motion very clearly. In order to be able to account for seeing motion very clearly, as a purely theoretical proposition, I originally advanced the idea that this moment function frequency can be shifted back and forth as a frequency over at least one octave so that it can find and set upon a harmonic relationship to a stimulus field or any relationship to a stimulus field, which gets it the maximum amount of information. I should like to point out that in a normal individual this synchronism is never perfect. There seems to be in normal people a little demon which says, »Never get right on it otherwise you will mistake an intermittent signal for a continuous one« so he rides, always hunts just a little bit. That is important information for apparently normal people. Thus you can see a flicker at 50 cycles per second when you have every reason to believe that there are five flashes per moment, and with the amount of lag and transit time you would expect these to be pretty well blurred but you can still see flickers at 50 cycles per second. But, as Head pointed out, this ability to distinguish between flickering and continuous stimulus field is often destroyed by injury.

METTLER: Do the individuals differ?

STROUD: They differ as individuals and they differ within themselves from time to time. Just for example, I thought I noticed a tendency for individuals to show a lower limit in the evenings. Thus, for instance I might get the broken curve (Diagram C) and apparent motion at a frequency with a certain subject today at 10:00 a.m., and I might get a perfectly straight one for the same subject and the same conditions at six at night. The whole octave had shifted as the result of physiological changes. Whereas he was operating on a range where a straight one moment per flash relation was feasible [48] for this evening, during the morning its general range | of frequencies was somewhat higher. The worst possible case was that of my wife. She suffered unknown but probably severe injuries as the result of rheumatic fever. She has no binocular vision and sometimes her visual field is blanked out by a star-spangled screen with a mere tunnel vision in one eye. Just what it is we do not know. However, her data were, if anything, more precise than any other subject's in this field. She has on occasion been tested when there is this summation period, just one moment to one flash period, which could be as long as 200 milliseconds.

KLÜVER: I hate to bring up the Brücke effect.

STROUD: This enhancement of brightness effect?

KLÜVER: Yes.

STROUD: That was something I expected to find and did not. I later learned from Halstead that it was not to be expected at the levels of illumination that I was working with.(8)

KLÜVER: You were working with very low levels of illumination.

STROUD: Yes.

KLÜVER: You can easily set up the proper conditions for producing the Brücke effect. How would you deal with it?

STROUD: I would like to know. As we all know, the mechanism of adjustment in the eye is by no means fast enough and good enough to account for the kind of seeing that we do. I have assumed that the fellow who made the human being put in A.V. C. circuits[1] in addition to the ordinary ones of chemical adjustments and adjustments of pigment materials, pupil adjustments, and so forth. A.V. C. systems in general have a rather interesting property in that they have some fixed time constants and if you impose upon them signals which have time constants of the same order, the A.V. C. circuits tend to bounce. These are not always completely damped systems and I believe that the enhancement which Brücke observed in some cases as did Bartley, is an upsetting of the A.V. C. circuit.

KLÜVER: What is the brightness in millilamberts?

STROUD: I brought all the original data sheets.

KLÜVER: Offhand?

STROUD: I cannot tell you, I am sorry. I would be able to look it up. They are very low. I was limited by equipment conditions and I could not spend much money. I settled for my light source on a 75 cent Christmas tree fluorescent lamp which happened to have the decay times, linearity, spectrum and so forth required. I wanted to get away from color shifts, so I chopped this off at 700 milli | microns for a number of reasons. [49] One was that it was invisible scotopically. Another one, I wanted to make sure that the subject looked pretty straight at it. At these levels he is going to see a red spot and at these levels we do not get color shift with this sort of light. We either see red or not at all. That put me at a very low level indeed. I was very much disturbed because originally I had not expected this sort of thing at all and I had expected to find the Brücke enhancement (7). I thought this was to remain the fundamental property of seeing. I found out from Dr. Halstead that in cases of low levels of illumination there was no Brücke enhancement effect. But this is something you might perhaps expect. In commercial receivers for example we use delayed A.V. C. That is, the A.V. C. circuit does not start operating until some satisfactory level of signal has already been put in. Here I suspect that the designer of man used the same notion. For low levels of illumination there is no volume control, only when the signal level rises to more than adequate values does the A.V. C. come in. I worked at a level below A.V. C. I did not perturb the

1 Automatic Volume Control circuits.

A.V. C. and Bartley worked above the A.V. C. level. At certain frequencies he found what was expected, that he could perturb it. Dr. Halstead worked on both sides and in his experience that is the way it worked.

SAVAGE: What is this effect you are speaking about?

STROUD: If you have a relatively bright light flashing on half the time and off half the time at the same frequency as normal alpha rhythm, the normal individual will say that the light is as bright as a light of twice the peak intensity which is on all the time. In other words, he will act as if he had received four times as much energy as he did. It is quite a startling effect.

SAVAGE: Exactly four or quite a bit more?

STROUD: It is of that order, not exactly. The degree of enhancement varies as you shift about.

TEUBER: Let me labor this point: the Talbot-Plateau law states that a periodically interrupted light will look just as bright as it would if the same light energy per cycle were equally distributed throughout the entire flash cycle (9, 12)[1]. For instance, if you have a flickering light with a light-to-dark ratio of one (that is, the light is on half of the time in each cycle and off half of the time), then that light will look just as bright as a steady light with a physical intensity which is exactly one-half of that of the intermittent light. [50] | It is as if the visual mechanism were responding to the average intensity in the cycle, and not to the peak intensity of each flash.

However, this Talbot-Plateau law holds only for flash rates sufficiently high to produce the subjective appearance of fusion. The rate of intermittence has to be above those rates at which flicker is still perceived – it has to be above critical flicker frequency, or »cff.« Below »cff«, we have a different relationship: a slowly pulsating light actually looks brighter than it ought to look if the Talbot-Plateau law were to be valid at all frequencies of flicker. The Brücke effect is just that: an enhancement of apparent brightness in slowly flickering light (5). Now Bartley has pointed out that the maximum Brücke effect – the greatest brightness enhancement – always occurs at frequencies of flicker around nine to ten per second (2, 3).

STROUD: The maximal always occurs at the same frequency.

TEUBER: Bartley's argument about the Brücke effect runs like this: he thinks it's more than just a coincidence that the maximum effect appears at nine to ten per second – the usual frequency of the cerebral alpha rhythm. His reasoning is similar to yours, in a way. He believes that whenever a flash rides in on the peak of an alpha wave there is enhancement of apparent brightness; whenever it falls into the trough between alpha waves, you get a depression of apparent brightness (1, 4). In this way, one could perhaps establish a relation between simple electrophysiologic rhythms and rather complex perceptual phenomena, such as visual brightness – except that I don't believe it will work. The coincidence of Brücke maximum and alpha frequency may be fortuitous after all. There is some evidence pointing in that direction.

STROUD: I would like to point out some other things. Remember, I postulated there should be an octave or more of variation in the frequency of moments. I did an experiment which I would not trust too far. I was testing the driving function of the clicks and the oscilloscope trace by means of vertical frequencies harmonically related to the sweep frequency. The gist of the situation was in setting up patterns which could be seen as either a single stationary pattern or as a pattern showing motion depending upon how much time was integrated into a single moment.

1 All Dr. Teuber's references will be found on pages 163.

With myself as the sole subject, on only one occasion was I able to infer from these tests that I *did* have a moment frequency range of about 10, (plus ten, minus five) cps – about the same range as that for photic driving. | [51]

TEUBER: The trouble is you can dissociate Brücke effect and alpha frequency. You know one can accelerate the alpha rhythm by heating a patient (6), or by giving him thyroid; your electroencephalogram shows that his alpha rate has been raised above ten per second. The maximum Brücke effect still occurs at about nine per second even though the alpha rate is higher than that.

STROUD: That I would like to test. Dr. McCulloch wrote to me about that.

TEUBER: I also tried it when I had fever a couple of days ago.

MCCULLOCH: Did you get a shift?

TEUBER: The alpha changes but not the Brücke effect.

MCCULLOCH: I know the Brücke effect does not. The problem is still to examine the moment function. There is an experiment.

TEUBER: I did not know it.

MCCULLOCH: I think some of the group in St. Louis did it.

STROUD: That is interesting to know.

TEUBER: I do want to bring out that it can't be that simple to prove the existence of your general »moment function.«

STROUD: I look at the alpha in this light: here I have a system which works on a fundamental periodicity. I must somewhere have a period generator; regardless of whether or not I am running a computer, I keep the motor running. I am inclined to look upon alpha rhythms as what might be called no-load current, the no-load signals of the system. Immediately you load it, it is like an overmodulated carrier. The carrier practically disappears. One might find that the characteristic requirements of the computer in use will tolerate quite a wide range of operating frequencies whereas the unloaded system may respond only at the frequencies for which it is initially designed to operate.

MCCULLOCH: One more question and then I want to come back to this.

TEUBER: I'm still pointing out that it seems impossible to me to find any simple relationship between the intrinsic alpha rhythm of the cerebrum and those particular perceptual events that we have when we observe flicker in a slowly pulsating light, and fusion of flicker in one that pulsates at a higher rate. There is another argument about this: if you expose a flickering light in different parts of the visual field, and measure the fusion thresholds by manipulating the frequency of the flicker, you get very different values for different parts of the field of vision (13, 14). Generally, fusion thresholds are lower in the periphery – there is a rather steep | gradient between the center of [52] fixation and the outlying parts of the retina. Still, the alpha rhythms are the same all over the occipital lobe, I presume.

STROUD: I would like to sort of turn myself off.

MCCULLOCH: Before you do, may I for the sake of this question concerning the content of memory, come back? We have something now that operates of the order of ten times per second, and we have a nervous system which will have ideas at a rate of about ten per second. Have we any knowledge of the complexity of the items of information which can be stored per tenth of a second? May I say this, my knowledge of this field comes about in an entirely different way. It comes about from the semaphore. A man in World War I could receive 100 unscrambled letters in a period of ten seconds. Immediately following the receipt of those, he was unable to say what he had seen and therefore he could simply shut his eyes and review the whole list, giving back the entire 100. He was tops at that time, but that gives some idea of the amount of information, because every semaphore letter is at least five decisions; you have close to 30 letters.

STROUD: There seems to be considerable complexity. Please understand I am just sort of feeling my way around to guide my own experimental program. We already have, in the case of phi, evidence that the intermomental relationships are part of the process. As to the matter of the operation of the central nervous system, it is a terrible injustice if you just chop it up into a bunch of moments. It is as true of sensory data as of motor action. Perhaps it is the place where the gearing takes place. That does not deny gearing of higher order. We pointed out how we could do differentiation and integration. This must be done by continuous process.

I have been forced to consider the idea of a train of moments after the idea of a train of waves. This train of moments can eventually, as it were, build up a very complicated and highly meaningful terminal moment which I suspect can be filed in toto. After a slight interruption, re-arrangement of the computer system, the old moment can be pulled out and started into a new computation. I can illustrate something like this out of my own behavior. I start dreaming up new circuits. I work on and on; there are a few little interruptions; in the end I have got a few circuits completed, very complicated and difficult to copy down. A friend comes in and interrupts my thoughts. Suddenly I have lost the whole thing, two hours of work, because I had not fixed the ideas in written symbols. | [53]

MCCULLOCH: Any information for computation to be stored would be what one could get per moment. Have you any knowledge from signaling, of other kinds than semaphore, say dot and dash or what not, as to the maximal amount of information that a man is capable of taking in per minute?

STROUD: As a test of my hypothesis I tried to predict Morse code speeds. At the time I did not know what they were and used my ignorance as a protection from bias in the attempt to compute the number of words per minute that a person could take in Morse code.

MCCULLOCH: Nonsense words?

STROUD: On the basis of how many dots and dashes there were in the alphabet, groups of five scrambled alphabet words, I made two sets of assumptions. One of them was that we shall say that a dot shall be merely above the noise level at the moment but a dash shall occupy so much of a moment that it will be recognizable as such and that the interspace shall represent a moment. There would be one moment interspace between a dot and a dot or a dot and dash. Between two characters there would have to be two moments. Between two words there would have to be three. When I got through I found that I had computed the average speed of the average man who gets out of the average Navy code school and can take something like 20 words a minute. Receiving Morse code is about the simplest kind of information that we ordinarily deal with. Morse code is a simple little thread of information which says that I'm either not here or I am here in one of two forms. The forms are »I'm here as a dot« or »I'm here as a dash.«

By making a second set of assumptions I found that it was possible to compute higher possible code speeds. I assumed that the moment with a dot in it could be sufficiently different from the moment with a dash, that you would not have to have any interspaces between dots and dots or dots and dashes as such. This would give the code a sort of undulating tone within a character. Then between characters there would need to be a space of one clear moment, between words, two clear moments. With these new assumptions and further assuming that it would be possible to drive code reception to the same moment limits that were the upper limits of the photic driving alpha, it turned out that something like about 90 or a little over 100 words per minute of normal text should be the upper limit of this particular set of assumptions. I found out that a chap exists who can do this, who can in fact take over 90 words a minute. I discovered thankfully that one of our chaps in my own | department had spent a large [54] part of the war in the code school. He told me some rather interesting things. The average fellow does receive code very much as I originally indicated it. But the man who does it very rapidly does it under conditions in which it sounds like a burbling tone. The »hot« code receiver does hear a burbling tone. He does not hear blank spaces between the dots and dashes of a given character. He told me something that was even more interesting. A typical thing occurs. A man can print and if he prints as fast as he can, he has a top printing speed, as in the case of the teacher, of 19 words a minute. But when he is receiving code, the code is somehow able to drive him until he can print at a speed of nearly 30 words a minute. Yet without the code reception driving his action he never could approach this same limit when printing free text of his own imagining. Apparently the thinking circuit is very much like that of an oscilloscope. When you run your sweep circuit and put more and more synchronizing voltage in it you can push the sweep frequency up to almost double the unsynchronized sweep frequency. And apparently the same principle is involved here. Here was a man who could not print over 19 words per minute, but who could easily be driven to 50% increase in speed.

McCulloch: Auditory drive?

Stroud: He said it is a very common sort of thing. The well-trained operator – driven by the basic frequency of his signal stream can work at speeds above his free running speeds.

McCulloch: These would give us again something less than what we got out of the semaphore. The semaphore is apparently more items of information.

Stroud: In the semaphore you have more dimensions to your moment. In this you have only one, just the dimension of sound amplitude. It is a fixed frequency oscillator. There is no information about frequency. The only information I have is a burst of tone above a noise level or I don't have a burst of tone above noise level.

McCulloch: Would it be fair to everybody's way of thinking, now having compared the linear and two-dimensional affair which only gave us phi items, to say that in all probability one never gets more than about a thousand items per millisecond, per tenth second, that he would ever have to store?

Stroud: I suspect it would probably be smaller.

McCulloch: About the order of 100?

Stroud: I often wondered about that.

McCulloch: Has anybody any evidence on tachistoscopic presentation of data at around alpha frequency? | [55]

Stroud: There has been lots of evidence on the basis of the presentation of material well enough illuminated to give all the stimulation in one moment. If I understand correctly, if you have a prefabricated hypothesis to fit this material into, you can sometimes get quite an amazing span of material; but in the absence of a prefabricated hypothesis, something like three or four categories is all you can ever get out of the thing in any one moment.

McCulloch: You think that we would be safe if we said not over a thousand and probably we would not be safe if we said less than 100 if you are going to keep it up?

Stroud: As far as building things up, I have not any way of finding out but I have an idea that some brilliant men under stress and pretty difficult conditions can build up terminal moments where a thousand items of information would not be high but it might be a great deal higher than that.

McCulloch: Simple computation then will show you the thing which worries Von Neumann. A man lives, we will say, who takes 8 frames a second. The reason for picking these is for simple multiplication. If he takes 8 frames a second, and he does it 64 seconds per minute, 64 minutes per hour, 16 hours per day, 256 days a year, 64 years to his life, he would have something of the order of 2^{33} frames. If the content per frame reaches something of the order of a thousand per maximum he has something of the order of 10^{13} ultimate terms.

Savage: Why take the maximum?

McCulloch: That is what we are troubled with. It is the maximum. I think the reasons for taking the maximum will be fairly clear to you if you remember the kind of evidence you get out under hypnosis – the amazing amount of detail that is actually stored in a man. Something of the order of 10^{13} ought to be enough. The great point is that the quantity stored belongs to an order of magnitude somewhat larger than the total number of neurons. Consequently no simple soldering of neurons together can ever take care of it. I think that is the fundamental thing which is troubling Von Neumann.

Wiener: We are not making the neuron the place of the storage, the synapse?

Pitts: No one ever made the neuron the place of storage, so I think we can make some computations about that which will give us more than 10^{13}. We are of course going to take the maximal case here, otherwise it is difficult to make the estimate show [56] us the kind | of thing we can do. Suppose we start with a given number of neurons, say N, and make the simplest hypothesis. All afferents to a given neuron simply add up without reference to place or where they came from; that is the way it gives us the least information, so we make a minimum assumption there. Each neuron will have a given threshold and there will be, if we care, a pair of neurons A and B. We can have connections between A and B of several different kinds. In the first place in case we consider only excitatory afferents we might hypothesize on the possibility of different proportions of the threshold of B that the fibers from A with various branches could make. Let us suppose that there are N distinct possibilities from A to B, and that is true in fact. If we add inhibitory ones that increases the number. If we consider then some possible connections between A and B and similarly between all pairs, B and C, etc., so as to get a network in general, except possibly for a certain confluence, we should expect that the dynamical behavior of the interconnected network will be different for every possible collection of connections since we suppose they are all meaningful in the threshold. There may be confluences. There are reasons to suppose there are. If we are able to test, with distinct inputs, the supply to all of the separate neurons in the network, we should be able to distinguish which network we actually had by observing the response. Very well, there are N distinct possible sorts of connections between A and B. Then the number of possible networks in essence will be of course N^2N. That will then represent the number of degrees of information on the whole that we should be able to expect, the number of units of information that we should be able to distinguish in the sense of the number of distinguishable networks, and therefore the number of networks which in principle we should expect as a learning process. Even on this hypothesis that we have only N possible kinds of connections it enables us in essence to store pieces of information, not in single neurons. We do not put the decision in one neuron and lock it up. What we are doing is assigning each yes and no decision to each network, so what we have to consider is the number of distinct networks. The simplest will be of the order of N^2N, or $N^{10\text{-}20}$, that will be this number of fibers for the typical one, so there we have 10^{20} even on the supposition that the possible connection between A and B is either a single fiber of an excitatory type or nothing. That is certainly a very conservative supposition. All we can say about it is whether A sends a fiber to B or nothing else. | [57]

WIENER: That involves probability for nearby and remote?

PITTS: Yes, we are trying to make the maximum.

WIENER: I agree with you fully that Von Neumann's argument is completely unconvincing.

MCCULLOCH: You believe the entire amount of information could be stored so to speak if you could grow and discard junctions between cells?

WIENER: Or change levels of threshold.

BATESON: Isn't there a fallacy when we talk about the number of pieces of information? Aren't we in fact talking about something else?

WIENER: No, of decisions.

MCCULLOCH: Number of decisions, items in this sense, in the sense in which the whole alphabet is five decisions.

FREMONT-SMITH: Yes or no.

STROUD: How many items of information are there?

PITTS: The essential point is that even with the supposition of alteration in synapses under all possibilities the unit of information is the network, not the cell.

METTLER: Why do you worry so much about how many neurons you have in the nervous system? You can get rid of the occipital lobe, you can get rid of the temporal and frontal lobes.

WIENER: The factors are not much more than 10^3.

MCCULLOCH: You would have to reduce the nervous system to something like a tenth of its volume before you could begin to effect a change.

METTLER: Certainly!

KLÜVER: I hate to think that the ten thousands of pages of Wundtian psychology have been written in vain. A whole generation of psychologists has been concerned with such problems as the »span of attention« or the »range of consciousness« for simultaneously or successively presented stimuli. If you ask how many elements, items, or objects can be simultaneously apprehended the answer is that no definite number can be given unless you specify numerous conditions in the internal and the external environment. Experiments may show, for instance, that the number of simultaneously seen items is 4-6 if the items consist of certain kinds of stimuli, such as isolated lines, numerals, and letters, and are presented under certain conditions. However, this number may be smaller or far larger if the nature of the stimulus material and the conditions of presentation are altered. It certainly makes some difference whether | I [58] present »isolated« elements or organized »wholes,« meaningful or meaningless, familiar or unfamiliar stimuli. A dot is not the »same« dot if part of a figure, and a letter not the »same« letter if part of a word. The question as to the number of »elements« or »items« seen or responded to cannot be divorced from the question of the organization of the field in which the »elements« are segregated. In fact, a Gestalt psychologist would insist that the whole question is in a way meaningless unless stated with reference to particular field conditions.

KUBIE: It would be a meaningful question for a hypothetical animal who had never had prior experience and who was tested in the first moment of life, and who therefore brought to the test no apperceptive mass of any kind, no organized prior experience with which to charge his new impressions.

WIENER: The experience and what is built in, adding up the experience and the degree by which the animal differs from itself after the experience.

TEUBER: Schumann has a nice illustration for that (11). He used to say he was never able to teach anybody to »see« 20 letters on a single exposure lasting 1/10 of a second. But it is really very easy to make a person read a 25 letter word on the tachistoscope with exposures as short as 1/10 of a second.

WIENER: The decision is much smaller.

TEUBER: Precisely – the question is only why.

KLÜVER: I cannot see how a definite number can be given unless you specify at the same time numerous factors operative in the »inner« and the »outer« field.

SAVAGE: Can a person read any ordinary English word just in a flash? I mean the kind of words that are in satisfactory basic English? I take it that I can read such a word in a flash, yet that represents a vocabulary of thousands of words and it seems to me that that means in the sense of this discussion that I can and do make dozens of dichotomous decisions in a second, or not in a second but a moment.

MCCULLOCH: May I say in answer to you, that Shannon has recently published two papers in the Bell Laboratory journal in which he has picked not merely words by themselves at random, short words, but words and their most probable followers, for two words, two words, and for three words, three. I don't think he has gone beyond that. This is most peculiar in that you form hypotheses inevitably when one word follows another. You do not any longer get through [59] | the words you should. The context picks up first. Say if written by pairs, four or five words seem to make sense to you and you seem to have an idea out of it only to have to disband it, start all over and take some that come later to make sense out of the rest. The result is that you simply cannot read this as you could a set of nonsense syllables. It is utterly impossible to do this. This kind of carry-over is definitely an interference with the simple question as to how many items of information are there per nonsense syllable?

FREMONT-SMITH: Don't you misread, carry on your idea and read words not there?

MCCULLOCH: You do. You find that you can read only at a relatively slow pace because of interference from the hypothesis you set up.

WIENER: Proofreading is done better by mathematical people who don't know anything about the content.

VON BONIN: As soon as you test by reading, even with nonsense syllables, you do not test exactly the amount of information that can be stored but you test somebody who has at least learned the alphabet if he has not learned the language. It seems to me the process is so much more complicated than for instance the phi phenomenon or brightness discrimination, or something like that. It is not on the same level at all.

KLÜVER: In this connection it seems worth-while to recall some old experiments performed in Köhler's laboratory which suggest that the laws of organization governing perception also hold for recall and recognition. In some of these experiments the subjects were asked to learn series consisting of ten elements, represented, for instance, by one number and nine syllables or by one syllable and nine numbers. When retention was tested by recall or recognition it was found that the element that was different in a homogeneous series, for example, the one syllable among nine numbers, was more often correctly remembered than the other elements.

TEUBER: This particular effect, the Restorff effect (10), is very instructive if you try it with visual material. You have a series of homogeneous visual patterns, and among them is one item that stands out – one that is quite different from all the other items. The items are presented to you under low illumination, or with brief tachistoscopic exposures, in a random sequence. Under such conditions you can easily miss the outstanding item when it is first shown to you. You have been seeing homogeneous material for quite some time – the item that is different is overlooked. A moment goes by –

you see more of the homogeneous material, and then you realize | »there was that thing that stands out« and then you remember what it was. The same thing happens quite often in audition – you have heard a word but you didn't listen. A little while later, you suddenly understand what you have heard. Something of this sort is probably going on all the time while we are listening to somebody else's speech. [60]

FREMONT-SMITH: You do the same thing with a mistake you make and suddenly catch it, a few seconds, moments or even longer afterwards.

KLÜVER: Before the Restorff effect saw the light of day I had performed and even published some experiments demonstrating the same effect. These experiments were undertaken chiefly for the purpose of lengthening my scientific life by shortening the time spent in establishing discrimination habits in monkeys. In general, the comparative psychologist presents *two* stimuli whenever he wants to establish differential responses to stimuli differing in brightness, area, color, shape, distance, or some other variable. To be sure, there is no reason why a monkey should be interested in responding to a pentagon instead of a hexagon, to a large instead of a small rectangle, or to blue instead of yellow, but by rewarding or not punishing the animal you will succeed – and often only after a rather long period of training – in attaching a positive response to one of the two stimuli. In practically all training situations of this kind there is nothing in the stimulus situation itself that is effective in producing immediately a positive response to one of the two stimuli. I tried then to determine whether this could be changed by giving up the use of stimulus pairs. Instead of two stimuli, let us say, a square and a circle, I presented four stimuli, namely, one square and three circles. Similarly, I presented one circle and three squares, one yellow and three blue squares or, more generally speaking, I presented in every trial, four stimuli one of which was always different with respect to area, brightness, color, weight, or some other characteristic. I found that no training or practically no training was required to elicit immediately a positive response to the stimulus that was different in a homogeneous series.

MCCULLOCH: I want to get Walter Pitts on the floor for one moment at the blackboard to explain whether neurons can account for learning or whether we have to go to some subneural level before we consider the theory of neurosis.

PITTS: It just occurred to me that I calculated the number of distinguishable networks in rather unrealistic ways. I think I can rectify | that. I still think the same conclusion follows. The point is that if we consider a particular neuron it cannot have an indefinite number of afferents; if we suppose that the number of other neurons from which it can receive afferent fibers is limited, and say this is equal to the average, this will reduce the number of distinct networks very much. Let us see if we can estimate what the number of distinguishable networks will be under these conditions. Let us suppose that each neuron can receive afferents from 100 distinct additional cells. It can pick these cells, we shall suppose, out of all the rest in the brain independently and at random. Since there are 10^{10} other cells minus 1, which is in essence somewhere around 2^{30}, and since it disposes of approximately 30 units of information in picking each one of its afferents, if it picks 100 of those it will dispose of 3,000 units of information, that is, connections to the given neuron constitute approximately 3,000 units of information. There are 10^{10} neurons altogether, so that would give us approximately $3{,}000 \times 10^{10}$ or 3×3^{13}. [61]

SAVAGE: That is just right.

PITTS: That is a much lower figure than we had before but it is of approximately the order that under the most liberal suppositions we assumed to be necessary. It means that Von Neumann does not have a compulsive argument. If he does not have that he does not have any argument. Since he is going to ask us to adopt a radically new hypothesis about the neuron which is to change our view altogether, he is maintaining the positive side of the case. That means he can substantiate his view only if he can prove that the possible multiplicity we can get in our way would be quite decidedly less than needed under the circumstances.

SAVAGE: Might I say in the interest of realism, Von Neumann is not here and we really don't know his position.

MCCULLOCH: I did my best.

SAVAGE: Didn't you say that this part of the argument, this 10^{13}, was your own computation?

MCCULLOCH: I said that he wanted at least a much larger number than even these data will support. He wanted the neurons to be something of the order of complexity with respect to their own quantized components that the nervous system is with respect to its neurons. This would be 10^{10}.

MCCULLOCH: I wanted for the present to get back to the central theme of the conference, namely, feedback. I don't know whether you realize it, but, so far as I can discover in the entire literature [62] | of the nervous system, Kubie was the first ever to propose that there were such things as closed paths in the central nervous system.[1] That was at a time he was working with Sherrington. The reason for proposing them, I gather from the history of that day, is that if they exist, a good deal of the over-behavioristic psychology of the over-behavioristic school falls by the wayside at the neuronal level. John Romano, who is not with us, was the first to show such activity within the central nervous system, – the activity of the closed circuits. To my delight when Larry Kubie proposed his theory of neuroses he again made use of the notion of closed circuits and of a fixity in the behavior of those closed circuits.[2] First, things carried into reverberations, and later the paths of those reverberations becoming in some way so determined that even if the activity were temporarily stopped, it would start again around the old path, giving a reiterative quality to the neurotic process. I was afraid that I might have to propose his theory for him but I see he is set to go.

KUBIE: Let me make one or two minor corrections, just on that historical introduction. I happened to be working at Queens Square at the time, and also was in my own analytical training. Maybe the notion of closed circuits – things going around in circles – resulted in that dichotomy in my own student years. However that may be, I spent quite a lot of time up in Oxford and took my manuscript to Sir Charles Sherrington. If it had not been for his encouragement I would never had the courage to publish it. It was the purest speculation for which I did not have the fundamental equipment to try to test and verify.

One other thing, S. W. Ranson and J. C. Hinsey published a paper in the American journal of Physiology in which they presented a theory of closed circuits in the spinal cord.[3]

This paper was ready about the same time as mine but happened to get into print a little later, so that I never felt that it was more than a happy circumstance that the notion had occurred to me. The others really got their teeth into the notion and gave it some experimental validation.

In my paper I am going to ask you to do something that I know is a little difficult; namely, to begin first at the other end of the scale and try to get a picture of the neu-

1 Kubie, L. S.: A theoretical application to some neurological problems of the properties of excitation waves which move in closed circuits. *Brain*, 53 – Part *2, 166 (1930).*

2 Kubie, L. S.: Repetitive Core of Neuroses. *Psychoanalytic Quarterly*, 10, *23 (1941).*

3 Ranson, S. W. and Hinsey, S. C.[!], Reflexes in the hind limbs of cats after transsection of the spinal cord at various levels, *Am. Jour. Physiol.*, 94, *471 (1930).*

rotic process as we see it clinically | and then end up by bringing it back to where we [63] started with this concept of reverberative circuits.

I talk of a neurotic process and not of a neurosis because I don't believe that the idea of neurosis is helpful. It makes you think of an illness which is implanted on a human being like a cold, whereas Freud shows a neurosis is part of our essential developmental process.

I am going to take a few moments to say, as I have been taught by Allan Gregg to do, what a neurosis is not and a few words about what the neurotic process is, its derivation, what its ingredients are and its relationship to our fundamental process of abstract thought.

[64]

THE NEUROTIC POTENTIAL AND HUMAN ADAPTATION

LAWRENCE S. KUBIE[1]

Department of Psychiatry and Mental Hygiene,
Yale University School of Medicine

We sometimes marvel at the sagacity of nature. We may for instance feel awed by the fact that matter is attracted to matter to make the law of gravity. To me such reverence always seems a bit naive; because if particles of matter repelled one another there could be no universe, but only an infinity of particles flying out in all directions in an infinity of space; and at the rapidly emptying center of this activity no one would remain to make reverential and awe-inspired remarks, not even a doubting Thomas like myself. Somewhat similar considerations apply to the relationship of adaptation to normality on the one hand, and of mal-adaptation to neurosis on the other. Certainly if behavioral normality had nothing to do with the ability to adapt to the world as we find it, then in short order we would go our unadapted ways to destruction; and again no one would be left to wonder how it all happened. This relationship therefore is obligatory, and neither accidental nor teleological: in short, it is what we call a »natural law.« If some such relationship did not exist, we could not survive.

Yet all natural laws may sometimes lead to trouble. That same law of gravity which holds the world together, can also make us either fly or fall downstairs. Which will happen depends upon our understanding and use of all of the forces which are operative under the law. The same principle applies to the relationship of normal behavior to adaptation. The appraisal of normality cannot be made purely in terms of adaptation, but must include an evaluation of the relative roles of all of the mechanisms which energize and shape behavior. Behavior which appears to be well-adapted in a specific situation does not necessarily turn out to be normal when all the forces which have produced it are studied. It is not what we do but why we do it, which in the ultimate analysis determines normality. | [65]

I. Limitations on the Usefulness of Adaptation as an Index of Normality

It is especially important to realize that temporary adaptation is often achieved at the price of neurotic illness. Therefore if it is to be used at all as a criterion of normality, adaptation must be considered in long-run terms and in varying situations. Let me give several illustrations. A writer was free from anxiety and lived happily and worked productively as long as he was in the city, but was paralyzed by unreasoning terror and could not write a word, whenever he had to go to the country. Being an inventive fellow, he would manufacture ingenious excuses to avoid leaving the city. He even developed a system of aesthetics which proved conclusively that the country is homely and that cities are beautiful. Should we call that man normal in the city where he was adapted, and neurotic in the country? Surely he was the same man in both situations; the only difference being that one situation fired off in him a pattern of distress from which he was protected in the other. Or consider an outstanding lawyer who lived on a flat plain with no hills or trees or high buildings. He functioned freely and productively until chance brought him first into a mountainous region and then into a city with tall buildings. Here for the first time he became aware of the fact that he had a

1 Instructor, N.Y. Psychoanalytic Institute.

height phobia. Without his knowing it, his peace of spirit and his freedom in work, in love and in play had been conditioned on his avoiding this special situation. When he faced its challenge he was tumbled into panic, shortness of breath, shaking knees, sweating, trembling; and became almost voiceless. Or take a familiar wartime phenomenon: the psychopathic fighter, who was so well-adapted to war that he won Congressional medals during combat, but was constantly in trouble in times of peace or even in training camp. Another familiar example is known to every educator. This is the youngster who has a neurotic fear of competition. If his terror embraces all forms of rivalry, the youth will be wholly paralyzed; he will have to recognize his neurosis early. If, on the other hand, the terror is attached to just one type of activity, then compensatory drives in other activities will enable him to make a good adaptation for years. Thus the lad who is physically timid often compensates by becoming an outstanding student. His impressive academic record is a compensatory mask for neurotic terror. The concealed forces which lie behind his intellectual attainments may not cause trouble for many years. Ultimately, however, they may rob him of his chance for peace and | happiness through success, or of success itself. Conversely a student with a [66] neurotic inhibition in the intellectual sphere may compensate by a compulsive athletic over-drive. Athletic success will carry him along during his school years, giving the appearance of an excellent social adjustment. Once he is out in the adult world, however, he will begin to pay the price of his neurosis. Thus both the student-grind (so-called) and the athletic-grind (less frequently recognized as such) may appear normal and well-adapted for years; even though these activities are energized by unconscious neurotic forces. There is a group of patients whom I call »campus heroes«, who live well-adjusted lives until the changing circumstances of adult years force their concealed neurotic mechanisms into the open.

These few examples from everyday life should make it evident that neither a good temporary adjustment nor an adjustment which is conditional upon some particular set of circumstances can be used as an enduring test of the normality of a personality.

The same reservation applies to the individual's own awareness of distress. Our unconscious ingenuity is so great that we can fashion situations which allow the neurotic process to flourish without pain to ourselves. Indeed many men who pass as normal have maneuvered their lives adroitly so as to exploit and put to work for them their neuroses. The hypochondriacal invalid suffers no pain; his family pays the price of his neurosis. The rarely recognized compulsive benevolence, and the more familiar compulsive work drives are further examples of painless neuroses. As long as a man can express his neurotic needs and conflicts in ways which are socially acceptable, which meet the demands of his own conscience, and which at the same time feed his self-esteem, he will feel complacent and even happy. What we call the neurotic character is usually an individual who without knowing it has been able unwittingly to establish his neurotic patterns as the standard of the good life. He then goes on to make of them a law for his family. Macfie Campbell once said, »A family is an autocracy ruled by its sickest member.« The same can be true of states. Under these circumstances (and they are far from rare), a man and his neurosis become one. He becomes proud of his very illness, treasuring it as the distinguishing mark of his individuality. In the course of time, with advancing years and changing circumstances, the neurotic adjustment breaks down, and painful psychological or psychosomatic symptoms will appear. Only in rare instances does this fail ultimately to happen; but the day of reckoning may not come for years. | [67]

Thus all around us are seemingly contented individuals, leading socially valuable lives, yet whose very contentment and productivity are energized by concealed neurotic forces. Some examples of this are banal and familiar, others quite strange; yet whether strange or familiar, their significance for human happiness and welfare and their importance as a challenge to science have never been fully appreciated.

Let me illustrate this further. I think of a warm, affectionate, gifted, artistic and musical woman in her late fifties. She had been brought up in a cultured home. Through her attachment to her father, a man of great learning, she developed a spontaneous interest in literature and the arts. During her early years these preoccupied her almost to the exclusion of social life; and in late adolescence she married an older man of similar tastes and interests who had been one of her father's outstanding students. It was a good marriage; and she gave herself to it wholeheartedly and happily. The years went on, however; and in the course of time her husband died, one son was killed in the war, and two of her children had to live on the other side of the world. When the youngest made a happy and suitable marriage, the woman broke down and had to seek help. Retrospectively it became clear that her devotion to literature and the arts and even to her family had served two groups of inner purposes: one healthy, and the other neurotic. Throughout her life, from puberty on, she had suffered from an overwhelming phobic terror of social challenges. Without her knowing it, her studies, her marriage, her home, her children, and her intellectual and artistic interests had served to mask this phobia. Even in her home, she had been a silent and secretly tense hostess. During all of those years she had never been forced to face her neurosis, to acknowledge it, or to seek help for it. When the defense provided by her home and family was removed, she had to endure the unmasked and unresolved terror of her childhood. Fear made her retreat into an unwanted isolation. In her enforced loneliness, she lost all pleasure in the inanimate beauty of music, paintings, or a sunset. She developed profound psychosomatic disturbances, an intractable insomnia, and finally an almost psychotic depression. Eighteen years of happy marriage had served her family and her community well, but had served the patient badly by masking the dynamic residue of an untreated and unresolved childhood neurosis.

Or consider another woman who grew up with an intense and hostile rivalry with her older brother, of which she was totally unconscious. Equally without her realizing [68] it, this had spread to | include all men. Early in life this rivalry had masked itself happily in a socially active, bachelor-girl existence, with talented writing and a vigorous participation in liberal politics and other community affairs. Ultimately, however, this same rivalry with men led her to marry a gifted but weak man who turned out to be impotent. She did not realize that she had been drawn to him by those very traits which rendered him impotent, and which now frustrated her and intensified her hidden feeling that to be a woman was to be unlovable. After two years of this, her unstable adjustment as a woman broke down. None of her previous activities could serve their original unconscious purposes any longer; and she lapsed into a severe neurotic depression, in which she shut out her friends, turned away from all community activities, and could not write.

Perhaps the most dramatic example of the fact that external success is not an infallible indicator of internal health is the frequency with which man reacts to success by going into a depression. One sees this at all levels and in all aspects of work and play. We see it in the tennis player who can never let himself win the important tournament from men he can always defeat in practice. We see it in the business man who gets into a depression when he earns a million dollars; in the writer who commits suicide when his novel becomes a best seller; in the man or woman who reacts to the launching of what should be a happy marriage by deep and destructive gloom. The Bible says that we cannot add a cubit to our stature by taking thought: but without realizing it, an earnest student was trying to build up his biceps by becoming a Greek scholar. For him therefore success could spell nothing but defeat. The world did not have to wait for psychiatry to discover how often success and fame turn to dust and ashes; but psychiatry has given us some understanding of the reasons for this nearly universal human tragedy. It discovered for instance that the long struggle to climb the mountain is energized by a compulsive drive for some unconscious and unattainable goal, and that it is only when the climber nears the summit that he begins vaguely to realize that he has been fooling himself. Then the greater has been his effort, the deeper is his depression; and this even in the face of brilliant external success and valuable social contributions.

Perhaps it is not necessary to give more examples of what I mean. We see it all around us: in the lives of scientists, painters, musicians, writers, businessmen, teachers, clergymen, and housewives. These show that we cannot use the social value of a life as an indicator of freedom from neurosis. Neurotic mechanisms may drive | activities [69] which are useful and creative: and these mechanisms may be as neurotic as those which produce alcoholism, stealing, and other patterns of socially useless or destructive behavior. Man can be neurotically good as well as evil, neurotically constructive as well as destructive; neurotically industrious as well as neurotically lazy; neurotically gregarious as well as neurotically misanthropic; neurotically generous as well as neurotically selfish; neurotically brave as well as neurotically cowardly. If we hope ever to solve the problem of the neurotic component in human nature, we dare not overlook the fact that activities which are intrinsically wholesome and productive may serve two masters within one individual: the one healthy, and the other neurotic; and that even a slight change in the configuration of external situations or the mere passage of the years can shift the controlling influence from one group to the other, thereby tumbling what has seemed to be a well-adapted life into profound illness.

These observable clinical facts of human life challenge us to ask whether human behavior can be explained in terms which will make this seeming paradox understandable. Any attempt to answer this question demands a precise definition of what we mean by normality and by the neurotic component in human life. I will attempt to give such a definition.

II. The Essential Contrast Between Normal and Neurotic

This definition will be in terms solely of the balance between conscious and unconscious psychological processes in the determination of conduct. Such a contrast between normal and neurotic can have nothing to do with the statistical frequency of any act. The fact that 99% of the population has dental caries does not make cavities in the teeth normal. Nor has it to do with the legality of an act or its conformity to social mores or its divergence from them, since one can be good or bad, conformist or rebel for healthy or for neurotic reasons. Even the apparent sensibleness or foolishness, the usefulness or uselessness of an act is not the mark which distinguishes health from neurosis; since one may do foolish things for sensible reasons (for instance as an initiation stunt), and one may do sensible things for very foolish reasons indeed, as for instance out of phobic anxiety … All of this will seem strange only to those who think of neurotic as synonymous with queer or eccentric or foolish or weak or immoral or rare or useless. We must learn instead that there is literally no single thing which a human being can think or feel or do which may not be either normal or | neurotic or, and [70] more often, a mixture of the two; and the degree to which it is the one or the other will depend not upon the nature of the act, but upon the nature of the psychological forces which produce it. This is true of work and play, of selfishness or generosity, of cleanliness or dirtiness, of courage or fear, of a sense of guilt or a sense of virtue, of activity or indolence, of extravagance or penuriousness, of ambition or indifference, of ruthlessness or gentleness, of conformity or rebellion, of playing poker or writing poetry, and even of fidelity or infidelity. Determining all of these there is a continuous, unstable, dynamic equilibrium of psychological forces; and in this flux it is the balance of power -between conscious and unconscious forces which determines the degree of normality or the degree of neuroticism of the act or feeling or trait.

We start with the fact that in every moment of human life our conduct, our behavior, our thoughts and our feelings, our decisions and plans, our hopes and purposes, and our reactions to one another are determined by a complex group of psychological processes. Of some of these psychological processes we are fully conscious, while of others we are wholly unconscious. (The presence of the buried layers can be determined only by special methods of investigation and evaluation, of which psychoanalysis is the pioneer and still the most important).[1] This basic fact, namely, that man operates psychologically on at least two levels, is of more than academic interest. It has a quite practical importance in human affairs; because the consciously and unconsciously organized levels of the personality have different characteristics, and exert quite opposite influences on behavior.

That conduct which is determined by conscious processes is flexible and realistic. Because its motivations are conscious, they can be influenced by conscious appeals to reason and feeling, by argument and exhortation, by success and failure, by rewards and punishments. In short it has the capacity to learn from experience. Therefore normal behavior is in the truest sense of the word free; free, | that is, to learn and to grow in wisdom and understanding. In contrast to this, that behavior which is determined by unconscious processes is rigid and inflexible. It never learns from experience. It cannot be altered by argument or reason or persuasion or exhortation or rewards or punishment, and not even by its own successes and failures. Since by its very nature it can never reach its unacknowledged and unrecognized goals, it is insatiable and endlessly repetitive, repeating its errors as often as and perhaps even more often than its successes, and marching ahead on blindly stereotyped paths. This happens whether the pattern of behavior has brought success or failure, and whether it has been a source of happiness or of unhappiness either to itself or to others. Thus neurotic behavior can learn nothing. It cannot change or develop or grow. It is enslaved.

[71]

It would be a mistake to assume from this that any act or thought or feeling is determined exclusively by conscious or exclusively by unconscious forces. Instead a mixture is always at work; and the modern concept of the neurotic process derives from this fact. Whenever most of the determining psychological forces are conscious, the resulting conduct will merit being called *normal*, because it will be free to learn and capable of adapting flexibly to changing external realities. On the other hand, where unconscious forces dominate, or where conscious and unconscious forces pursue incompatible goals, then the behavior which results will deserve to be called *neurotic*, precisely because it will be a rigid, repetitive, *unadaptive*, ineffectual compromise, serving the needs of neither the conscious nor the unconscious aspirations and motivations.

If these statements are valid, then we may state categorically that if there were no such thing as unconscious psychological processes there would be no neuroses. There

1 The full significance of free associations is generally overlooked. Actually it is the basis of all that is new and scientific in psychoanalytic technique. Conversational thought and speech, questions, arguments, and expositions all require a continuous unconscious screening of what is going on in the mind, an automatic selecting of certain ideas for attention and expression because of their chronological or logical relationships, and the rejecting of others. It is on this that our ability to think clearly and to communicate our thoughts depends. The product of this process, however, is an atypical and weighted sampling of the psychological flux; whereas, by contrast, free associations provide a true random sample of total psychic activity at any moment. It is this virtue of free associations as a method of sampling which makes the psychoanalytic technique more scientific than any previous method of studying psychological processes.

would not be the neuroses which manifest themselves in obvious symptoms and which we encounter in daily practice as the symptomatic psychoneuroses. Nor would there be those masked neuroses which express themselves insidiously in distortions and exaggerations of the customary patterns of living, nor in those quirks which we look upon as the eccentricities of normal people, nor in the neurotic processes which result in delinquency. From this we may conclude further that if the psychological conflicts of infancy and childhood could take place in the full light of consciousness, then the neurotic process would never be launched in human life. | [72]

III. The Derivation and Ingredients of the Neurotic Process

This leads us to the fact that early in the ontogeny of every infant and child a fateful dichotomy occurs, a dichotomy between those psychological processes which develop on a conscious level, and those which evolve on an unconscious level, and which exert their influence on our lives without our knowing of their existence. There is no single chain of events in human affairs which has greater consequences, since it is this dichotomy which makes possible the neurotic process. …

In this process, there are four essential ingredients, each of which requires further investigation. In the first place there is the basic process by which the dichotomy between conscious and unconscious processes occur. Secondly, some if not all of the forces of the repressed and unconscious psychological processes become detached or »dissociated« from their original connections. If this did not occur, unconscious processes would be quite incapable of influencing in any way our conscious behavior, and we would also be wholly unable to penetrate to unconscious levels or to learn anything about them. It is clinically and experimentally demonstrable that a continuous interaction between conscious and unconscious levels is taking place at every moment in life. The third ingredient of the neurotic process is the fact that as a consequence of the dissociation of its forces, the stream of unconscious psychological processes is represented in conscious behavior by a variety of compromises and symbols. These may be discrete symbolic acts or a phobia of heights or of insects, and the like. These are the »neuroses«, so-called, which were looked upon as wholly irrational until we learned to translate them into the language of unconscious conflict. More often, however, the unconscious process is represented by subtle distortions of ordinary behavior, i.e., of the way we live, play, work, eat, sleep, love, hate, etc. These distortions usually pass for normal; yet they consist of mixtures of compulsive exaggerations and phobic inhibitions of normal activities. The fourth essential ingredient of the neurotic process is that which produces its characteristic quality of obligatory repetition. We have already indicated the reason for this, but it may be well to repeat this here. No neurotic impulse can ever attain its goal, since it pursues this goal only symbolically. Therefore it remains forever unsated, and consequently repeats itself endlessly. That is why there is so much repetitive patterning in the petty details of much of our daily living. And that is why the adult, whether he is an artist, a musician, a business man, or a scientist, repeats himself as though he were still | playing one of the stereotyped and exhaustingly repetitive games of infancy and early childhood. [73]

Many questions about the neurotic process remain unanswered: (1) How does the initial dichotomy occur? (2) Is it all to be explained as due to a process which is called »repression«, and if so how does this process operate? (3) When does it start? (4) Does it occur spontaneously and inevitably? (5) To what extent can it be influenced by variations in family structure, in educational mores, in economic processes?[1] (6) Does the dissociation of »energy« from unconscious material occur spontaneously? (7) What forces influence it? (8) Is it comparable to the clinical states of dissociation, and to such experimentally induced dissociated states as hypnosis? (9) Does this dissociation occur when people are in different states of consciousness, and if so, what difference does this make to the nature and fate of the dissociated forces?

IV. The Neurotic Process and Human Progress

This, then, is the process of tragedy in our human comedy: the dichotomy between the conscious and unconscious; the dissociation of unconscious »energies«; their symbolic representation; and the obligatory repetition of neurotic psychological experiences. This is the process which accounts for most of that which is irrational, rigid, inflexible, unlearning, enslaved, and unadapted in human life. Yet these same potentialities of the human spirit also make possible our power of abstract thought and feeling. Thus our neurotic potential and our highest capabilities, (i.e. the ability to communicate through the symbols of language) are linked closely together. Indeed the neurotic potential is as essentially human as is speech itself, with common roots in this initial dichotomy. All hope of human progress depends upon our ability to discover how to [74] pre|serve the creative consequences of this dichotomy, while at the same time limiting and controlling its potentialities for neurotic destruction. We know of no culture in which the basic dichotomy between conscious and unconscious does not occur during the early years. Whether and to what extent we will ever be able to direct this process no one knows. The most that we can say is that we should attempt in every way that is possible to extend the area of conscious motivation and purpose and control in human life; and to shrink and circumscribe the territories of that darker empire which is ruled by unconscious forces. In this lies our hope of shaping human progress towards flexible and truly adapted living.

DISCUSSION:

Kubie: »You will notice that in this paper I have not discussed in detail the application of the feedback principle. This omission was purposeful; because I wanted to make clear the complexity and subtlety of the neurotic process as it is encountered clinically. Without this we are constantly in danger of oversimplyfying the problem so as to scale it down for mathematical treatment.

Nevertheless the principle of reverberating circuits and of feedback relationships comes to mind repeatedly in connection with the clinical problems of neurotic patients. Let me list a few of their possible applications: (1) in the relationships between conscious and unconscious areas of psychological organization; (2) in rela-

1 This is an important area of research for cultural anthropology. It is usually assumed that socio-economic, socio-political, socio-cultural forces all play a significant role in the genesis of the neurotic process. Certainly this assumption would seem to be reasonable, but it has never been critically investigated. This is largely because our concept of the neurotic process has itself been so hazy. Now, however, we may be able to sharpen up our ideas on this whole issue. It can be taken for granted that cultural forces influence all the secondary and tertiary manifestations of the neurotic process: but the influence of cultural forces on the original dichotomy between conscious and unconscious psychological activity remains to be clarified and demonstrated. If, for instance, we review the case of the older of the two women mentioned above, it is obvious that cultural forces had many effects on the fate of her neurosis once it had developed: but it was not so clear that cultural forces had caused or influenced the dichotomy which produced her initial phobic attitudes toward people. The same is true in the history of the second woman; cultural forces are of obvious importance in the secondary and tertiary manifestations of the neurotic process, but their more fundamental influence on the primary dichotomy is obscure. Cultural anthropology must soon turn to the investigation of this basic question.

tionships between various types of emotional processes (i.e. the vicious circle between rage and terror, or between rage and guilt or depression); (3) the possibility of feedback circuits in the explanation of tics. The nervous organization which underlies tics must be close to that which mediates migraine or status epilepticus; (4) similarly related to reverberating circuits are the obstinate manifestations of the complex compulsive behavior patterns or obsessional patterns of thought; and (5) finally, there are the rigid manifestations of psychotic depressions and psychotic elations, which seem to be maintained by re-entering circuits which are almost impervious to psychotherapeutic influences, but which respond to overwhelming electrical stimulation or other profound physiological changes, such as marked hypoglycemia.

Many other phenomena of the underlying neurotic process and of the vast spectrum of neurotic symptomatology seem to be clearly explained by adducing the feedback principle. It seems to me, however, that before attempting to use the hypothesis in further detail we should dissect the clinical neurotic process itself with its funda|mental [75] units, so that we can be sure that we know what it is that we are trying to explain.«

BATESON: If there were no consciousness would there be neuroses?

KUBIE: I don't know.

FRANK: Do cows have neuroses?

KUBIE: I don't know.

WIENER: People speak of contented cows.

STROUD: We have been assured by the Department of Agriculture there are neurotic cows and their neuroses affects the production of their milk.

KUBIE: There is a danger in looking upon the state of emotional discontent to being synonymous with neurosis.

PITTS: You exclude experimental neurosis as ordinarily used under neurosis?

KUBIE: I put a question mark. I just don't know.

LIDDELL: I fully agree with you.

KUBIE: We are not justified in saying what the neurosis is identical with.

LIDDELL: We do not have to make common cause.

KUBIE: To say that that is identical with human neurosis is assuming a great deal of things.

FREMONT-SMITH: Part of the situation here is just as Dr. Kubie has said, namely, that he is making certain assumptions about human neurosis which involve a recognition of unconscious processes and that we cannot really deal with unconscious processes in those terms in animals.

LIDDELL: It is an operational problem, as you will see.

FREMONT-SMITH: It is not possible to answer the question whether they belong in the same category.

LIDDELL: That is an argument that never arises any more. We have our division and all agree on that.

PITTS: You mean that the unconscious goal is always of such a character that there exists no reasonable plan for achieving it or merely by the fact that it is unconscious, a person never actually tries any reasonable procedure? The cases you gave were obviously those.

KUBIE: That brings up a question which is the focal point of research: namely, what are the forces that determine the dichotomy? Is it always due to repression which occurs because you are dealing with unacceptable and/or unachievable drives or can the dichotomy result from other forces as well? | [76]

LIDDELL: The unconscious goals are unrealistic?

KUBIE: Those at least that cause us trouble are.

GERARD: They are unattainable under the circumstances when the neurosis was formed.

LIDDELL: The time for achieving may have passed in the life course?

MEAD: Or not come.

LIDDELL: They are not in touch with circumstances as they exist.

KUBIE: Another example is a man of 45 who lost his mother through her suicide shortly after the birth and then the death of a baby sister. This man was about two when this happened. He did not know what had happened but his father, meaning to soften the blow said to the boy: »Your mother went away to take care of your little sister«. This man became a map expert, and an encyclopedia of time-tables. He knew the trains and maps of every corner of the globe; and ultimately became quite a traveler. He did not realize until his analysis that he had been searching for a dead mother all through his restless and unhappy life.

LIDDELL: What did his father say?

KUBIE: »Mother has gone away to take care of your little sister.« My patient also did not realize that he himself wanted to become that little sister. This created in him a violent unconscious struggle between normal masculine goals and a desire to shed his masculinity and to become his little sister.

BATESON: When psychiatrists talk about neurotic satisfaction or suggest that a person derives unconscious pleasure from such and such actions, does not this contradict what you are now saying – that the unconscious goal is never attainable or never attained?

KUBIE: We are ingenious enough to attain some degree of secondary consolation, even when we do not attain our primary or major goals.

BATESON: The secondary satisfaction but not the primary.

KUBIE: Like the satisfaction of being taken care of if you turn into a chronic invalid.

FREMONT-SMITH: If an individual had a repressed severe hostility towards a sibling might he not actually kill the sibling?

KUBIE: I have a patient who tried twice to run over his little brother. He did not succeed. He won several medals in the war. He was in the position where he was not required to do any fighting; but he would slip away from his official function to arm [77] himself and go out and kill. He received many citations for his bravery. |

PITTS: Would he be cured if he had succeeded?

SAVAGE: Did he really try to kill the little brother?

KUBIE: He really tried to run over him with the car.

SAVAGE: Also did the symbolic thing which would not satisfy anyhow.

KUBIE: When you want to beat up Joe Doakes it does not do much good to beat up Tom Brown. You remain unsatisfied.

FREMONT-SMITH: Yet there is temporary satisfaction.

KUBIE: There is an interesting and famous case in point – in the life of Joe Louis. The reporters who had worked with him commented on the remarkable change that occurred after he beat Schmelling[!] in their second fight. They all remarked on the primal savagery with which he went into that fight. It was for blood in a way that he never fought before. He had always been a cold and determined fighter; but this one made a difference in his relationship to the whole white race. After that fight it was as though he had gotten something off his chest.

PITTS: If your patient actually did try to kill the little brother, was it the unconscious?

KUBIE: In a way. He did not quite know what he was trying to do. He took the brake off the car when it was on a hill and let the car coast down and really tried to pin the child against a wall. Another time he tried to brush the boy off the side of the car against the house.

TEUBER: Unconscious, is that »unreportable« merely?

KUBIE: Certainly, it is unreportable, but not *merely* unreportable.

PITTS: Certainly mostly what goes on inside of our head is unreportable.

KUBIE: That comment brings up the question of the nature of forces which are at work which make it possible to distinguish between the fringes or periphery of consciousness, i.e. irrational processes which are not in the focus of consciousness, but which are accessible to the ordinary processes of self-observation, and for which no special excavating devices are needed: and those unconscious processes which cannot be recaptured without special technical procedures.

METTLER: I presume that you do not mean to imply in saying that these things are unconscious that unconscious things are necessarily evil. As you specifically warned before, unconscious urges can be for good or evil, depending upon how they affect the individual's position and behavior in society. May I ask if it would be possible for man to spend as much time in living as he does without pushing | much into his [78] unconscious? Isn't that inevitable? Must that not necessarily be so?

KUBIE: That is a very helpful question; but it is one to which we do not yet have full answers. In the first place the basic process by which the original dichotomy occurs and the forces which influence it have not been adequately studied. We do not know to what extent variations in family structure, variations in educational processes, variations in the total structure of the culture, or economic factors influence this dichotomy. That particular moment or phase in human life has not been the focus of careful comparative cultural study. We do know that much of the highest potentialities of the human spirit, including our intellectual and artistic creative powers, is closely related to this whole process. As Max Eastman pointed out many years ago, when a man says »Gird up your loins«, we call it poetry: when he says »Pull up your socks«, we call it slang. And when a patient is obsessed with the fear that he is walking around with his trousers falling down, his unconscious symbolic processes have taken over the same symbolic language to express its own masked needs and conflicts. These are three different ways of playing on the same fundamental themes. Specifically, how the creative intellectual and artistic aspects of the human spirit are differentiated from the neurotic process is something that people have talked all around but not a great deal of study has been devoted to it.

FREMONT-SMITH: Is it not true that there is great economy in relegating many acts to automatic or unconscious control? William James in his chapter on habit pointed out that if one had to direct consciously every muscular action of dressing in the morning, the process would take all day and leave one utterly exhausted. Moreover, unless we were able to turn over the majority of our learned functions to automatic or unconscious control and to exclude from consciousness the majority of incoming stimuli, we would be unable to give focussed or sustained attention to any train of thought or action.

KUBIE: Not differences in degree alone. There should be no confusion between *automatic and unconscious*: the automatic functions are always accessible to conscious introspection.

FREMONT-SMITH: Unrecognized if not known by the person while he is doing it?

KUBIE: There are acts to which we do not need to pay attention, and which we handle in completely automatic fashion until a moment arises where we need to attend to them, whereupon they are instantly accessible. This is the difference between the clinically | automatic and the clinically *unconscious*: between the automaticity of our use of [79] our hands in writing, and the unconscious automaticity of an hysterical paralysis or tremor of that same hand.

FREMONT-SMITH: There is a difference between repression and unconscious.

KUBIE: Repression is the name of a forceful process by which psychological processes are rendered unconscious.

WIENER: We have to lose them because we do them so often and the other is an emotional one. There is no emotional strain one way or the other in losing the detail whereas in this there is a tremendous emotional strain either in keeping or losing.

KUBIE: May I add one word to the answer. I want to emphasize that this same capacity that we have to repress and then to represent symbolically enters into all abstract thinking. The capacity to communicate by means of language symbols, and the capacity to become neurotic are very close together. I do not think that the precise link has been worked out. Certainly it would seem that any hope of human progress would depend upon our learning how to preserve the one and prevent or limit the other.

WIENER: I want to speak of a very strong experience I had at one time. I was just in the last stages of a piece of research work when I got pneumonia and was in a feverish condition. There is a peculiar identification between the mathematical difficulty I was struggling with and the discomfort and emotional tension I was under, I transferred and actually used the emotional discomfort, as if it were a term, to represent the incompleteness of the mathematical thought. The two processes fused in a very curious way.

KUBIE: Otto Loewe once described to me a similar dream experience as he was working out the neuro-humeral system. Kekule also described a dream of six snakes in a circle, each with the tail of the other in its mouth, from which he awoke to describe the structure of the benzene ring.

MEAD: It would seem to me to make considerable difference whether you are going to say any activity in which you are 51 percent unconscious and 49 percent conscious is neurotic without regard to culture, without regard to the situation, or whether you are going to say that in any given culture and in situations within that culture when those processes which are normally expected to be in one ratio to another appear in a different ratio with another, this possibly may be very bad.

KUBIE: In this culture we are supposed to work, eat, play, marry, have children, read, [80] and do a lot of things. There is not a single | one of these things which cannot be done neurotically quite as frequently as done normally. It is not correlation with any culture which makes an act or thought either normal or abnormal. It is the balance of inner forces which determine this.

MEAD: What I am saying is that in a given culture, for instance in an Australian tribe, there is an expectation of which pieces of behavior should be flexible, and subject to learning by trial and error or by reward and punishment. These expectations may be different from another culture which expects another area of behavior to be subject to learning; in any given society the particular acts, the particular behavior in which conscious and therefore flexible learned behavior is to play a given role may vary. If you do not include such variations in your definitions, then it will not only be an abstract balance, 51 percent or more unconscious, but it will also be a balance that is inappropriate and unfitted to the cultural experience of that group of individuals and the reality of that Society, which may or may not be prepared to deal with individuals who have that much conscious behavior. It makes a great difference.

KUBIE: Let me give you an example of what I mean. First take a simple homely and everyday learned experience. We expect children at a certain age to become clean in their habits. Up to that time the bathroom functions have been social events; suddenly they became something which you had to make yourself an outcast to do. You must go into the bathroom and close the door. In our culture, in some degree every human child goes through this. Thus every human child learns to look with scorn on his own body, its products and processes. Certainly this cultural distortion has affected everyone of us. Yet there is an amazing difference in the quality of this effect on different individuals, depending on whether it is built up into a conscious system of ideas and fears, or repressed and expressed only in a neurotic development which is unconsciously determined, and which, therefore, is crippling. Now you can say, if you will, that this is a culturally induced neurosis which we impose; yet it seems to me to say that it is a culturally induced distortion which only becomes a neurosis in those who handle it unconsciously.

METTLER: It is not what he does but how much it interferes with him or what his affective reaction is to it.

KUBIE: No. This is not an adequate basis. Many neurotic forces are socially productive. Macfie Campbell's remark, quoted previously about the family applies here because there are individuals | whose neuroses become the law which governs the family, and even in certain instances the law which governs the state. They pay no penalty for their neuroses, leaving this penalty to others to pay. [81]

SAVAGE: It does seem to me that you have not answered Dr. Mead squarely. It is perfectly true that some of us are neurotic, say about our toilet behavior, and some are not, and that all of us in our culture repress it to some degree but as I understand it, Dr. Mead would say the person we call neurotic about it is the one who relegates it to the subconscious to a degree which is unacceptable to this culture. For example, among the Japanese, where I understand repression of these things, is much more intense, it may be the typical Japanese who displays this behavior. I mean that subconscious as well as overt behavior, which for us characterizes a neurotic would not in Japan be considered neurotic because the behavior is normal for a Japanese.

PITTS: I suppose you might say possibly by way of answer to that, that society can prescribe why you do it and not what you do, because the latter is not practical.

SAVAGE: It is fairly practical.

PITTS: How can society find out about your unconscious?

SAVAGE: It finds it singularly difficult to know the anatomy but not difficult to see that a patient is behaving poorly, as Kubie said. We have lots of instinct for it and rather analytic ability, and we think everybody is queer but us, and even we are a little queer. We see all around us, people behaving strangely and we can judge when a patient is behaving strangely.

KUBIE: Let me take up Pitts' society. It prescribes what we do but it cannot make us do it. Every individual act serves many masters, some of whom are represented by society and some by inner purposes of which we are not aware. We do this in work, and we do it in play, indeed in everything we do. Everywhere the quality of normality depends upon the relative roles of conscious and unconscious forces in the act.

MEAD: What I wanted, Dr. Kubie, was to make a statement about cases where the proportion is wrong not quantitative, but leave it qualitative, recognizing that we don't know what that means for sure. Then you get for the particular behavior a definition which is simply based on the disproportionate role of the unconscious.

KUBIE: I would like to put it this way: the degree to which any act is serving conscious purposes has a direct correlation with its essential normality, and the degree to which it is serving unconscious purposes has a direct relationship to its neuroticism. | [82]

PITTS: Normality, you don't mean statistically but what is most common?

KUBIE: Normality has nothing to do with statistics. Let us consider a child with an eating compulsion. We all know that we do not eat only to gratify our biological need for a certain number of calories and a certain amount of water and salt, etc. We eat sometimes because we are scared, sometimes because we are depressed, sometimes when we are mad, sometimes because we are lonely. Eating often serves many kinds of phantasies, both conscious and unconscious. When the unconscious fantasies dominate our eating it becomes an unmodifiable neurotic ritual. If you are eating too much merely because it is fun, and you are advised to diet, you will diet. On the other hand, if you eat too much because of unconscious conflicts and drives, then you cannot stop eating, no matter how urgently necessary it is to do so. Anybody who has dealt with a child with eating compulsion knows this.

LIDDELL: May I ask how you would react to my interpretation of what you have been saying. You mentioned dichotomy. Would this be an acceptable statement? During the development of every child this dichotomy of conscious and unconscious may be due to socially communicable affect versus non-communicable affect.

KUBIE: Maybe. I don't know. We have not really studied this process closely enough.

LIDDELL: It is not consistent with what you have been saying.

MCCULLOCH: May I point out one thing, because I think the way it has been said does not necessarily come to the core of what I think you mean. When you distinguish between a conscious and unconscious act, I don't think the distinction fits because of the things we do, of which we are not aware, but which we have learned to carry on in the course of our day's business, like buttoning up our vest. Let us say we could become aware of those. On the other hand there are an enormous number of things, such as regulation of body temperature or regulation of pulse rate, of which we are not aware; certainly these are not things that we can think about and bring to awareness.

METTLER: Yogis, we are not, but they can get lots of things to awareness.

MCCULLOCH: They can get some. Those are definitely not included in the things that you would include as neurotic.

[83] KUBIE: They are also not on the level of psychological organization. |

PITTS: You are not using the word »unconscious« as the antonym of conscious?

KUBIE: Unconscious means to be *unable* to know the conflicts and drives out of which your behavior arises.

GERARD: May I put in a plea for the rest of Dr. Kubie's exposition?

KUBIE: I think that I have said all that is essentially relevant.

MCCULLOCH: May I put it this way for a moment? Before we go on I want to be sure that I have your idea straight. You disregard as unconscious in your sense those things which otherwise would be conscious processes but which are excluded according to –

KUBIE: The process of exclusion of is an active persistent force which must be overcome by special methods.

FREMONT-SMITH: Which otherwise –

STROUD: Which were conscious at one time and which presumably might again be conscious now and are not.

MCCULLOCH: May I finish my question?

KUBIE: I am not certain whether everything that is repressed and forcibly rendered unconscious was once conscious, or whether some things split off during the developmental years of childhood without ever having become conscious. This is a question which is still unsolved.

McCulloch: Concerning the whole question of cryptesthesia, there are a whole number of things to which we obviously respond, of which we might have been aware but of which in fact we never were, and do not give symbolic expression to them in that sense. Would you include those or would you exclude those?

Kubie: I would like to give you an example. I was studying at my desk one beautiful spring morning in Baltimore, a Sunday morning. The windows were open. A friend of mine with whom I frequently went walking on Sunday morning lived down the street. As frequently happened, he whistled a tune to signal that he was ready to start. I heard and waved to him. I put my books away, went down the street to join him. As I walked along the street I found myself thinking a tune that was not the tune he had whistled. I said, »What the devil is that tune? I can't place it. Dave knows a lot about music. I will ask him.« So the first thing I did as I greeted him was to whistle this tune and then asked him, »What is this tune which has been running through my head? I can't place it.« He said: »I was whistling that to you for half an hour trying to catch your attention; and you paid no attention at all. Then I changed to another tune and you stuck your head out of the window.« Thus I had received it, registered it and reproduced it, all | without knowing consciously that I had heard it. This means that [84] there are processes which we can experience which do not penetrate to the level of organization which we call conscious.

McCulloch: Would those be in your sense unconscious?

Kubie: Not unless they have been actively excluded by obstructing forces.

Fremont-Smith: Not conscious at that time. If we use the unconscious to cover all the things which are not conscious and then use the word »repressed« in various degrees of depth to cover those which are not easily available, would not that simplify the question from a semantic level?

Kubie: It so happens that the words which are in current usage suffer from the fact that they have been translated from other tongues and not chosen with precise attention to their meaning. William James spoke of the fringe of consciousness for which there was no dynamic barrier. Therefore it was not identical with the Unconscious. We speak of *unconscious* only for that which is held beyond dynamic barriers.

Mead: This is essential to your argument. You are dealing with things that are excluded from consciousness and the exclusion is central to this argument?

Kubie: With one reservation, that we really do not know precisely how much is included in this.

McCulloch: It might have been in this pre-conscious rather than something of which you were fully aware.

Hutchinson: The same properties.

Wiener: Isn't there the same thing about a tone, where we speak about a mechanism of pushing away or not? There may be no mechanism already away but associated with something dealing with the emotional tone or effective state.

Kubie: It certainly implies internal conflicts.

Savage: I think the general experience in medicine and related subjects is that it is good to try to achieve non-normative definitions.

It would be wrong in the long run to define infection in a normative way. It is not good because it is fruitful for science to extract from the bad effects of infection generally. In the same spirit, I wonder if your concept of neurosis cannot be separated from the concept of what should be?

You have emphasized the things that are excluded from consciousness, what can »exclusion« mean? It suggested to me things that cannot enter consciousness even when they *should*, and if that is the interpretation of exclusion, the definition is normative. | [85]

KUBIE: I am troubled at that question because my whole point is to get away from a normative definition.

WIENER: There is one very important thing here. There is a strong normative element in what you said but I think you are placing the normative element in the wrong place, namely, the nature of exclusions, an appeal to a process which we ordinarily interpret to be normative. I believe it is the cart before the horse. When we say »excluded« we mean this thing is accompanied by a process which we ordinarily interpret to be normative, not that we say one thing is good and the other is bad, but in dealing with normative matters and in dealing with this exclusion, we are dealing with the same class of phenomena. Normative statements are based on our exclusions, not the other way around.

KUBIE: I am puzzled here because I am not quite sure how this confusion has arisen. Let me retrace my steps for a moment. I began by saying that the differentiation between normality and neurosis which I was suggesting would not rest upon narrative sources, nor legality or usefulness, i.e. none of the qualities which are usually used as basic criteria of neurosis. I pointed out that that is the layman's concept of neurosis; and that there are, however, two more fundamental attitudes of behavior. Normal behavior has a quality of flexibility and of adaptability to external influences. It has this because the predominant determining forces are on a conscious level and therefore they are in contact with the external world of reality. There is another vast area of determining forces which also play a role in everything we do but of which we are unconscious; and when these play the predominant role, then the resultant behavior inevitably develops the qualities of rigidity and insatiability. Those are the natural phenomena with which we have to start, just as you have to start with the natural phenomena of any scientific problem. That in essence is what I have tried to say.

BATESON: When you said the word »reality«, was not that the thin end of the normative wedge?

PITTS: I think your own remark is that you disapproved of it.

KUBIE: Reality does not wait for my approval or disapproval any more than the law of gravity waits for me to approve or disapprove of it. But if I neglect reality I will land myself in a great deal of trouble.

FREMONT-SMITH: Isn't it neurotic when there is a deviation between the pressure of the unconscious and of the conscious, because if the conscious is reinforced by unconscious motivation, I don't think you have necessarily neurotic behavior? On the other hand, it would | [86] be highly neurotic behavior if there was a marked conflict between the direction in which the unconscious pressure pushed you and the direction of your conscious efforts, as for instance, when an obese individual having determined to lose some weight »discovers« that he had eaten two helpings of bread and butter at luncheon.

METTLER: That is not how Dr. Kubie defined the situation. He specifically started by finding out that the neurotic pattern might be reinforced by the conscious, that is, the unconscious neurotic pattern might be reinforced by the conscious elements and might produce a useful effect, but it would still be neurotic.

WIENER: I think that the whole point simply is where the norm comes in. There is no question at all that the unconscious has to do with norms but the norms are not attained by saying that the man does the wrong thing by being neurotic. It is the fact that he interprets a process himself as normative. There must be a normative interpretation on his part or at least an emotional or affective coloring interpretation coming in there which is part of the mechanism which leads to the formation of norms. In other words, I think the norm situation comes in but not at the same point. As Dr. Kubie said, there is something always in the unconscious that deals with a process by an affective coloring. This affective coloring is the raw material out of which norms are made, so it isn't a pathology. It isn't a normal pathology but a pathology of norms that we are interested in in[!] this situation.

KUBIE: I am not sure I understand.

WIENER: You have to know the affective tone coloration. We would scarcely call it unconscious of your sex. The raw material of which norms are made is the emotional affective coloring, the series of experiences.

PITTS: That is, the patient's norms, not Dr. Kubie's norms?

WIENER: That is what I mean.

FREMONT-SMITH: Isn't there an emotional tone to all of this?

PITTS: The emotional tones to Dr. Kubie's are objective effects.

KUBIE: They are not norms which I or anybody else imposes on the human being. It is a question of harmony between conscious and unconscious forces actively operating in a human life.

WIENER: No, normal development would not be unconscious in your sense.

SAVAGE: Your talk has revolved a good deal about the dichotomy between conscious and unconscious in some things and you have said that they interplay, that a given kind of behavior can be so, much unconscious and so much conscious – | [87]

KUBIE: Our behavior is conscious. The determining forces which produce that behavior may be conscious or unconscious or both.

SAVAGE: Is a force clearly conscious or unconscious, or is there any class of objects – I don't know what they are – which fall into this dichotomy of conscious and unconscious or is the process, or the aspects of the process, a certain degree of conscious and unconscious? You say the unconsciously conditioned act is the inflexible act, the act that is not amenable to reason, to experience; but very often we have compulsive acts which are not strongly compulsive and are amenable. I say, »Look here, you've washed your hands three times«, and you say »Why so I have!« and you don't wash your hands a fourth time....

I have asked it as well as I can. Isn't that a case of an act which is close to being completely conditioned by unconscious forces but is not altogether so conditioned?

KUBIE: If I understand you that is precisely what I have said. We always have an admixture of the two working together in varying proportion. We have no instrument by which we can measure with precision the relative role of conscious and unconscious forces. We do, however, have certain effects flowing from the confluence of the two groups.

SAVAGE: I have not said it right. What I should have said is, need we think that it is a blend of two or three extreme things? You see the person who has done this may fairly well understand why he has done it. In other words, there is nothing there that you can say was totally unconscious or totally excluded. There were things which he tended to exclude but you see that is the little difference. Is it that the behavior is a mixture of things which are totally excluded or things that are totally admitted, or is the behaviorism a mixture of things, in each one of which there is some tendency to exclude and some tendency to bring forward?

KUBIE: I would say it is a spectrum.

MEAD: I think what Dr. Savage is feeling for there is the assumption which I believe you are making, that the laws of unconscious behavior are different from the laws or mechanisms of conscious behavior and to a degree you do make a considerable dichotomy between the type of thinking that you would call unconscious, into which these excluded events are put, and the type of thinking which you would call conscious. It is that dichotomy that you are bringing up rather than another sort of dichotomy, isn't it?

Brosin: I will give Dr. Kubie a chance to catch his breath. I want to express my [88] admiration for his presentation to date. There is no | one here who has not read a good deal on the hypothesis about which Dr. Kubie has given an exposition, and some of you probably are as conversant with it in your own way as we are, so anything that I may say is not with a tone of dogmatism, but purely from my own point of view. Of all the teachers I know who have tried to define the functions of the unconscious in relation to the neuroses, I think Dr. Kubie uses the least loaded vocabulary. I am happy to see vigorous nodding. A tribute should be paid to the many years of mastery of vocabulary which has enabled him to present this material. I also want to pay tribute to the skill with which he has reserved judgment and shown restraint about answering some of the challenges which are classic problems, questions which most teachers of analysis, I think, would jump at answering as a matter of course. Some of them would say that that which is now unconscious was once conscious because they would leave out all the subliminal learning possibilities which some psychologists now bring in. He was extraordinary in his restraint, like Freud, in not giving you easy answers even though there are any number of dogmas which would attempt to give you an answer. I think this group is receiving a relatively skillful, unloaded, neutral exposition. The normative problem probably came up because of his avoidance of this tricky issue in order to keep the definitions clear.

Wiener: There is one thing I would really like very much to ask. I would like to suggest something that I cannot call a theory because I don't think I am good enough for it, but as a working model of the sort of problem that comes up here. It also associates with what we said this morning and that is I do not believe for a moment that the neuronal processes about which we talked earlier are all the processes of mental activity and I believe that the moment we begin to orient into the mental activity at least the skeleton of the emotional processes in addition to the neuronal processes we shall see that the whole thing takes on a very definite meaning. This is entirely conjectural. I am giving it for what it is worth rather than something to be believed, pathology if you like. I have a strong suspicion, I said this before in my book, that there are two modes of communication in the human body, the one that belongs to the neuronal system strictly and the »to whom it may concern messages«. I suspect the »to whom it may concern messages« are a) closely associated with emotion, and b) at least partly humorally carried. Now if we have in such a system the possibility that the »to whom it may concern« message may have a differential action on the synaptic or similar [89] mechanisms which are carrying a mes|sage at the time and those which are inactive at the time there is at least a basis for a mechanization of association of learning of conditioned reflexes. In other words, I have a strong suspicion that learning is associated with humoral messages and humoral messages are emotionally released. One cannot separate entirely the nerve part and the humoral part of the message. If that is the case one will have a fairly definite physiological correlate of emotion of affective norm, and which can go wrong together or apart from the other mechanism. In other words, I don't see that this will lead, as it does in the Pavlov case, into something like norms but it does not occur because of the norms. However, I suspect it is a mechanism which we have assigned to norms. In other words, to learn is conditioned by emotion or by affective state. If this is the case this gives a basic mechanism which could be perfectly mechanized to bring in the affective state as an important part of what happens to the nervous system, and would allow a distinction between things that are more conditioned by the affective state and the things more conditioned by the regular nervous mechanism. The mere fact that you can get a working hypothesis like this does indicate to me at least a channel by which the investigation of the unconscious or the, subconscious can be tied to a fairly definite communication theory.

BATESON: I would like to follow that remark. When we were talking about Von Neumann this morning, he was quoted as saying very definitely that the brain could not be an analogic calculating machine but must be a digital one.

Dr. Wiener has now spread our thinking from the brain to the body as a whole with its humoral communicating system. It seems to me we get back to the problem of neurosis very importantly when we see the body as a whole as a possible analogic calculating machine.

WIENER: A machine with an analogic part and a digital.

BATESON: With the analogic part able to contrive analogies with the observed actions of human beings with whom we communicate.

PITTS: These analogic parts are not primarily concerned with storing of information.

WIENER: With the use of it.

BATESON: Not storing but with experiment?

WIENER: Yes.

BATESON: It may be storage too?

PITTS: But the primary purpose is not simply to furnish another battery of memory? | [90]

WIENER: No, the modification of the synapse, if you want to call it that. In other words, the mere fact that a large part of our thinking is done by a digital machine, which we all grant, does not in my opinion exclude the existence as you have said of important parts of an analogic machine.

STROUD: I would like to raise one question. In conjunction with neurotic processes we so often find reported the symptoms that go with it, at least fatigue and very frequently obvious trauma to the body as a whole, of the sort generally mediated by, at least in part, by hormonal systems, following along quite normally as part of the neurotic processes, and these cannot be avoided even though we are quite unconscious of the processes in their other aspects.

WIENER: Roughly speaking, the division is analogous to that between the ordinary central nervous system and the system which is partly nervous and partly humoral, which we would call the autonomic system.

KUBIE: I think again you are getting away from the essential features of the neurosis, and are thinking again in lay terms of neurosis. You are thinking of neurotic people as people emotionally upset. Some of the most neurotic people in the world don't show a trace of emotion.

FREMONT-SMITH: And are not aware of any.

KUBIE: And show none of the physiological concomitants of emotional processes. Furthermore there is more bad work done, more bad science, in the effort to find correlations between somatic disturbances and the neurotic state than in any other field of medicine. So many studies fail to establish more than the coincidence of the two, and provide no illumination on the causal interrelationship between the two. We see patients who have had somatic manifestations of a neurotic process for many years. One man had had 13 separate and complete gastrointestinal studies, and yet after a history of 25 years of gastrointestinal difficulties, he was well in seven months. I am not overly impressed by the vulnerability of the body to the neurotic process merely because of its duration. There must be certain factors other than time which determine structuralization of the neurosis.

STROUD: That is why I said fatigue at least, or often outright trauma.

FREMONT-SMITH: Often there is no fatigue. I think there are neurotic people who are [91] extremely energetic, who are not giving any signs or complaints of fatigue. |

STROUD: That is a difficult thing to say because you have as a rule little by which to compare them. Each man in that case is his own norm.

FREMONT-SMITH: I would like to say that one of the most impossible things to define is fatigue and all the physiological studies – and there have been thousands and thousands of dollars spent on them – end up with one thing, that is, at the physiological level we don't know much about fatigue and that the psychological component is so important that the physiological approach alone is quite inadequate.

STROUD: The man is no longer able to continue doing what he was able to do in the past.

FREMONT-SMITH: That certainly won't apply to neurosis because the neurotic man perhaps persistently continues to do what he was doing in the past and it is the repetitive aspect of what he is doing and the compulsion to repetition which is evidence of his neurosis.

STROUD: I would be quite willing to agree that we shall say that he will continue endlessly doing A, but to continue the test which I would like to apply is to see how he would do tests B and C which we cannot test for because he persists in doing A.

PITTS: When you resolve what you consider to be the original cause of a given way of acting and the symptoms does not disappear, you always have a choice between two hypotheses: first, you have the wrong cause and secondly, the symptoms are self-perpetuating. How do you choose between them?

KUBIE: That is a fair challenge and if we are honest we have to say that we can only be empirical. When dealing with a patient we go ahead on the assumption that since human behavior is a complex business there will usually be more than one storm center in life, and therefore you go hunting around for more. At some point you acquire a conviction or feeling either that you have made an error or else that you have run into one of these processes which has become organically rooted. This is not a satisfactory state of knowledge. At the present, however, we have to say merely that we have no instruments for precise measurements. Perhaps we should ask the clinical psychologists to provide us with such instruments with which to measure the unconscious and conscious aspects in the processes of life. Certainly we need an instrument which does not now exist.

GERARD: It is the reverse thing which worries me more; that is, how do you handle the phenomenon of a sudden resolution of everything with one blinding insight, in a [92] moment so to speak, though it may have persisted for an indefinite number of years | and resisted all sorts of violent treatments and would presumably be structurally engraved as much as anything could be?

KUBIE: I don't know that. In the first place, if you are honest with yourself, you know the therapeutic argument is not very sound. The fact that you get a therapeutic success is not proof that the therapeutic argument is sound. You probably used vitamin D for a very fancy reason, so you cannot argue from the therapeutic result that the theory is sound. You begin to got results from this with a certain persistence.

FREMONT-SMITH: Dr. Kubie, I believe I remember an experiment which you and Richard Brickner performed in which a transient neurotic reaction was induced in a human subject by interfering with his carrying out a post-hypnotic suggestion. Do you remember it?

KUBIE: A report of that experiment was published in the *Psychoanalytic Quarterly* 5, *463 (1936)* under the title of »A Miniature Storm Produced by a Superego Conflict under a Simple Posthypnotic Suggestion.«

FREMONT-SMITH: The point that I would like to make is that this subject was given a post-hypnotic suggestion – I think it was that he was to drink a glass of water so many minutes after he came out of the hypnosis. He struggled and fought to resist it until he really developed quite a degree of acute anxiety. Suddenly he leapt up and rushed into the kitchen, grabbed the glass of water and drank it with tremendous relief.

I think the thing which is important, which has not been touched on here, is that there are phenomena which are reproducible experimentally, not with everybody but with some individuals. They can be told under hypnosis to behave in a certain way at some specified time after the hypnosis is terminated. These subjects will carry out such hypnotic suggestions but with no memory of the hypnosis. They go to great lengths to explain or rationalize their behavior, and if it is interferred with they become emotionally disturbed. An excellent description of hypnotic study is given by J. Eisenbud (*Psychiatric Quart.,* **11**, *592 (1937)*). He describes a patient who suffered from severe headache whenever he repressed his aggressive feelings. It was possible to reproduce the headache experimentally by arousing aggressive feelings under hypnosis and then terminating the hypnosis at the height of his aggressive feeling. The resulting violent headache would disappear promptly when the details of the hypnotic suggestions were recalled to his memory. That was repeated ad lib. | [93]

There are phenomena of that sort. Headache is good because it is subjectively recalled. You can have the same in terms of compulsive behavior or compulsive function such as vomiting. I think it is important to bring these in. Maybe I am wrong, because I have the feeling this whole discussion, which I also think Dr. Kubie handled with extraordinary discrimination, has been mostly at the level of abstractions; we have not had any data. That is why I was hoping that Dr. Kubie would give us a case, some actual behavior of a human being in a neurotic situation because then I think we would be much more able to come to grips with it in our discussion.

GERARD: Let me give you a case and then I will again ask my question, which I don't think quite got across. The only theory involved in my question was neurological not psychiatric.

A man came for treatment with one of these terrific washing compulsions. He was a doctor who had been forced to give up his practice because his hands were always raw because he spent most of his time washing them. He had had the compulsion for many years and had received shock and various other treatments. Here was a permanent, ingrained, structured mechanism if ever there was one. In the course of analytic treatment, which was a long and tedious one, suddenly there came the recollection of a childhood scene in which he saw his mother bleed to death in a washtub while a baby was being born. From that time on the compulsion was gone. How does it go that suddenly? That is what I am asking.

MCCULLOCH: May I have one more word? This concerns a repetitive dream, replete with incidents over and over again which I had for a period of eight to ten years. There is no explanation of the dream or anything else concerned with it. Driving along the road in a part of the country where I had been a dozen or more years before, I suddenly recognized the terrain of the dream and the incidents connected with it and I have never again had the dream. So at any level the problem is the same.

KUBIE: Dr. Gerard, I do not know the answer to your question. Often I am equally astonished both ways, astonished when a symptom disappears after it has been in existence for ages and astonished when much younger symptoms prove to be persistent.

GERARD: How seriously are you offering reverberation, whether it remains dynamic or becomes structuralized, as an explanation of neurosis? This seems to me one stimulating but insufficient attempt to make physiological or morphological sense out of neuroses and related phenomena. I am asking how much *you* believe in it? | [94]

KUBIE: It is not a question of belief. It is to me simply one mechanism by which we can understand how these things can finally become organized on a structural basis in such a way as to persist even after the original conflicts have been disposed of. I cannot say that anybody has ever demonstrated that reverberating circuits are more abundant or significant in neuroses. There are some things about the psychology of depressions and elations which suggest this strongly; but there is no unequivocal proof as yet.

MCCULLOCH: In the case of psychosomatic disorders in man in which there is an out-and-out autonomic outflow, or in the case of sleep, you have a pretty clear evidence of impulses coming out years on end, so there is at least some process active in the organism.

KUBIE: On the other hand, there are extraordinary cases such as the patient who has had a severe ulcerative colitis which clears up, whereupon he develops a severe skin lesion.

LINDSLEY: I wonder if some of the experiences of the individual are not of the sort that cannot be classified with personal and meaningful experience as it exists to date. It seems to me if you look upon it from the therapeutic viewpoint, when it does come into some relationship with the past experience some clarity is apparent and the problem resolves itself. On the other hand, what we are talking about here is the situation which just resists classification in our experience and is thus held in abeyance and is repressed, often with incomprehensible emotional reactions.

FRANK: The point I wanted to raise was this: we have been spending quite a lot of time on the question of the conscious and the unconscious. I hoped we would give a little time to the feedback aspect. Speaking of reverberatory processes as taking place only inside of the organism, I wonder whether we should not take into consideration the organism in environment realizing that the environment in which human beings live is not the actual but the symbolic. The definition learned in the course of life's experience is put into the environment and then that feeds back and provokes the kind of behaviorism symptomatology.

GERARD: That is the »goal«.

FRANK: It seems to me that is the normal process of human living, we don't just live as organisms. We live in a defined world of symbols and meanings which we have put in. Why do we put them in? Because we have learned to think and feel that way. If we can change – I don't know whether that is one way of adequately defining psychotherapy – if you can get the person to redefine the environment in which he lives you [95] interrupt the process of putting | in meaning and the meanings coming back which provoke the neurotic process. Is that compatible with your definition of repetitive process of neurosis? If we don't get out of the organism we are completely organic and not psychologic.

FREMONT-SMITH: Isn't it true that in many cases the neurotic process is kept going by the repetition of a particular kind of a stimulus from the environment, i.e. a particular interpretation of the environmental stimulus?

KUBIE: I think we are becoming confused here between the activation of any specific neurotic symptom and of the underlying neurotic process. Surely the individual who has a phobia of purple cows will not be afraid often because he is not likely often to run into purple cows. The man with a height phobia who lives on a flat plain will not feel fear. This does not mean he is free of the phobia but only that he does not run into the situation in which it becomes operative.

FRANK: I am only following you in this business of the getting away from the model and normal behavior. The process of living means putting meanings into life and then living according to those meanings. You have to create, because each one of us creates an environment in which we live. That is one aspect of the feedback. I want to get in that theoretical consideration because if we are going to get back over from neurosis to feedback, it is more than just reverberation? If it is only that it does not seem that our discussion is getting very far.

HUTCHINSON: Have you any idea of what the time of reverberations is? Is it a question of seconds, days or years?

KUBIE: I cannot possibly answer that.

FREMONT-SMITH: Let me put something in there? You made a distinction between the core of neurosis and a neurotic symptom. I think the only way we can answer your question at all is in terms of behavior and hence in terms of the symptoms, whether the Symptoms are related in the core or are periphery. Therefore, then the thing that Larry Frank spoke about, the feedback aspect of the environment, comes in and the time relationships would be significantly determined by the timing of appropriate stimuli from the environment.

HUTCHINSON: That is why I asked the question. Primarily if it was purely neurological you might expect much greater frequency.

KUBIE: Let me put it in the form of a case. Dr. Fremont-Smith has asked for that anyhow. It poses the question with all of its unsolved problems about as well as this can be done. Here is a | European woman in her late fifties. She has had four children of [96] whom two are dead and two are alive. One son was killed in the war. One daughter was killed by the Nazis together with her husband and my patient's grandchildren. Furthermore for seven or eight years she did not know whether her other sons who were fighting in the underground were alive. Ultimately one of them was killed and one of them survived. She took all of that, as she had taken the loss of her husband years before, with extraordinary fortitude. This was no »neurotic weakling.« (Incidentally most neurotics are not weaklings). Now to go back a moment, she came from a European university family. She had been a very beautiful young woman, and very close to her father. Indeed she married one of her father's outstanding students, a man a little bit older than herself, but not unduly so. She had an exceptionally happy marriage; and she did a beautiful job with her children, with her home and with the community. Then at the age of 57, after surviving all of these terrific emotional strains, her youngest child gets married, and she suddenly breaks into a state of anxiety, depression, insomnia. That is why she came for treatment. Here is a life which has been extraordinarily fruitful and well-adjusted. You cannot point to a single element which has the earmarks of illness if we are going to judge neurosis by symptoms; but with the curious and extraordinary ingenuity with which we manipulate our problems and slither around them without facing them, she had actually spent her entire life evading a severe neurotic problem which had arisen quite early in her childhood. Before she was 8 or 9 years old she had become extremely shy, so shy in fact that she would develop terrific panics in any social situation. She lived actively in and through her home with her children and music (which was her major interest) and, in the setting of her family she could entertain at home. When the last child moved out into the world on her own, she was forced to face the unresolved neurotic problems of her childhood and to pay for these dearly. We see this over and over again. Although she has been a beautiful child and woman, anxiety made an acutely anxious and self-conscious child; and the problems out of which this symptom arose has remained in abeyance in her all through her life. For years these were not manifest except in the path she chose by which to evade the problem. With the passage of years, death and the maturing of her children took away her last defenses. How are we going to formulate this in terms of a nervous system which can keep such problems on ice through so

[97] many years, | problems which today remain as highly charged with »energy« of some kind as they were years ago.

FREMONT-SMITH: Didn't she project meaning? Taking Larry Frank's suggestion, didn't she project on to the marriage of her child a meaning which when the marriage took place reverberated back onto her and precipitated this outbreak of neurosis.

KUBIE: No. The marriage merely pulled the trigger, exploding a charge already present.

FREMONT-SMITH: There was not a symbolic meaning to the marriage of her child?

KUBIE: It meant that she had to meet people unattended and this had a meaning for her, surely.

HUTCHINSON: What was the nature of the original thing?

FREMONT-SMITH: She was unattended during the war.

KUBIE: No, she had two of her youngsters with her, first one and then the other.

MCCULLOCH: I want to come back to this question of energy for a moment. Is it something like this: every repetitive or every process of the neurotic type, whatever you want to call it, while it is going locks up a certain number of our neurons in that path, and the number of neurons remaining for thinking and for acting is thereby reduced? The number of neurons that one has is relatively fixed, and the rate at which they can run is a relatively fixed quantity. The question of psychic energy has always seemed to me better quantified if one thought not in terms of energy, which is certainly wrong for the nervous system, but in terms of the amount of information that can be handled, and is being handled by those circuit elements which are still free to work. Instead of thinking of a certain amount of energy as being locked up in a neurosis, why not think of a certain number of relays being locked up in your process? I think you would keep yourself a lot better off dimensionally. One of our main troubles in psychiatry is that we do not have a decent analysis, a dimensional analysis of »gremlins,« for these are quite comparable to »gremlins.«

BROSIN: Would it help if I restated your proposition in this way, that what you call the power in the relays was, as it were, a detonator to a large unstable system, potentially quickly available, as in an explosive. In the case cited, one does not have to envision gremlins of N dimensions and qualities, and so many ergs, each of which represents one of the child's unresolved problems, but one does see organized systems of symptoms arise when they are appropriately tapped. Most psychiatrists will agree that as we [98] study the cases | more and more we find that there is almost the specificity of a key fitting a lock, as to which precipitating events were crucial in setting off a chain of behavior. Numerous traumata are not always simply additive phenomena as in the metaphor of the straws on the camel's back, but rather one insult is the specific incident which sets off a major reaction. In the case referred to, the desertion of the daughter seems to be the most important determining event. This knowledge may determine what therapies one chooses, for one need not waste time on lesser issues. Reorganization of the patient's energies can be planned in a more economical way, employing in a flexible manner whatever methods seem most appropriate.

MCCULLOCH: It is certainly not the energy of our muscles, glands, etc., that we refer to as psychic energy. It is certainly some property of the organization of the neuron mechanism.

WIENER: I have always that same reaction, wrong dimensions anyway.

MCCULLOCH: It is not the rate of handling information. That is the crucial item.

ABRAMSON: Is it only rate? Dr. Brosin pointed out that not only rate, but also specificity is involved.

BROSIN: To describe the detonator effect upon a system, we can ask the concrete question: why does a woman of 58 in the face of the daughter's desertion suddenly reorganize herself so that she no longer lives harmoniously, and this where the internal consistencies apparently were very good? What concept of power would you envision to describe this detonator effect upon the total system whereby you get new channels for expression?

MCCULLOCH: Wiener will tell you I have the notion we are dealing with gremlins here. They are a curse to us because for the most part we have not yet developed a decent way of thinking about them.

WIENER: I don't think we have. I am going to make a suggestion which is wild. Please remember this, a large part of the pattern of her activity has been released from any normal action. That means as this is released there are lots of things that it can be reinvaded by. There is a tremendous change of her balanced activity. I can easily see situations in which a release of traffic can cause a traffic jam. In other words, what is going to feed into this part of her existence that has been suddenly released for other activities? I can easily conceive of that being a destructive sort of circulatory function. | [99]

MCCULLOCH: May I say one more thing here on the subject of gremlins before I turn it back? One of the familiar gremlins of ordinary radio is an automatic volume control on an instrument with high gain. So long as the instrument has signals of sufficient value the automatic volume control is in operation and the set does not break into oscillation. When the signal is withdrawn the set howls.

WIENER: That is the sort of thing I meant.

MCCULLOCH: This is a familiar type of woe. Again I think if we wanted to use the word »energy« it certainly would be wrong. Power also is the wrong notion to think of. It is a matter of organization and information.

WIENER: Isn't that just what we talked about, volume control can go haywire? Let us take a phonograph scratch remover which depends upon volume control. If there is nothing going over it, message or noise, the whole thing will be forced to act in an abnormal way and you get all sorts of jam coming out.

MCCULLOCH: Man is a host of such devices.

ABRAMSON: Do you think I should put in a word or more than a word? Which do you want?

MCCULLOCH: More than a word.

ABRAMSON: One of the things which disturbed me for some time during my teaching of medical students was the fact that when the medical student left college and got to medical school he proceeded to forget much of what he had learned in college. He had learned physics, he had learned chemistry and he had learned a little algebra. His problem in medical school was to assimilate as rapidly as he could, the dogma and the tradition of the practicing medical man. In other words, he wanted to know enough to pass his examinations and to know what doses to use without applying the basic scientific methods which he had learned at college.

MCCULLOCH: Even arithmetic.

ABRAMSON: Even arithmetic. In a course in physiology I gave at Columbia I ran into the following situation: I put on the blackboard, »What is meant by the term ›work‹?« That was the first question. The second question was, »Discuss the work done by the normal heart?« At once a third of the class raised their hands and they all had the same question. »We have prepared for Question 2. We know how to answer the work done by the normal heart but what do you mean by the first part of the question?« This story is not an exaggeration. | [100]

I at once proceeded to discuss work with them, electrical work, mechanical work, thermal work, surface work, and decided to introduce dimensional analysis into this course in first-year physiology. I think that you all know what dimensional analysis is better than I perhaps, although there have been some dimensional errors made in the last ten minutes.

FREMONT-SMITH: I don't know what dimensional analysis is, so go ahead.

ABRAMSON: Dimensional analysis is a very simple technique used in physics and engineering to get either concepts or test equations. It assumes that fundamental units or dimensions as they are called, of mass, length and time can be used to describe physical events, and that other events are derived from mass, length and time. For example, one of our speakers very casually said you can either spot distance the first derivative or the second derivative. I am certain that some of us did not know what he meant. But the first derivative of length with respect to time is velocity; it is L/T and the second is L/T^2. I believe that is what was meant.

STROUD: Yes.

ABRAMSON: You can handle dimensional analysis. Those words »first derivative« and »second« were not clear at first to me since I was uncertain of what was being measured.

STROUD: I am sorry, I should not have used it.

ABRAMSON: With the support of the Macy Foundation I have compiled for medical students a little booklet called »Dimensional Analysis for Medical Students«. After utilizing the treatment in the Encyclopedia Britannica and Bridgeman, I worked out certain problems about work done by the heart and found in certain current textbooks that the force of the heart, the power of the heart and the work done by the heart were all confused, even in the very best books. It was not surprising, therefore, that my students were also confused. That done, I got more interested in psychodynamics and realized that one of the problems of communication which was possibly more important than any other was the inability of psychiatrists to communicate with physicists who were well-versed in psychodynamics. I did not see how the psychiatrists who were well-versed in motivation, in unconscious motivation, could communicate with the physicists who used an entirely different system of language. That was well brought out I think in our present discussion in which the energy of motivation, the power of motivation, the organization of motivation finally got down to a volume control I [101] think. |

WIENER: This dimensional analysis has come out clearly in our information because Dr. McCulloch has brought out that the thing we have been discussing here has not been energy but information on the one hand or the rate of transmitting information on the other. We have made in these meetings a dimensional analysis of information.

The dimensions of our problem which do not change particularly have been shown to be a negative logarithm of a probability, a zero logarithm. The dimensional analysis is not enough. We have seen the same sort of quantity in a more general sense than dimensions of entropy.

PITTS: There is a factor of proportionality.

ABRAMSON: What are the dimensions of the proportionality factor?

PITTS: Entropy.

WIENER: Entropy can be given.

ABRAMSON: Dimensionally as Q/T.

WIENER: As the logarithm of probability.

ABRAMSON: That would be another way.

WIENER: Using that way of expressing an entropy, probability of information is also the logarithm of it but the sign is reversed. However, that does not change anything analogous. The point I am making is that the real difficulty here lies in the fact that the quantities we are dealing with are essentially dimensionless.

ABRAMSON: I have another way of saying that. Since I have not been through the process which you have been, I have been faced with the idea that it was necessary for the preservation of our culture that very practical methods of communication be established between the pure scientists on the one hand – and the people who understand the weapons of hostility as was mentioned earlier and the people who understand, as Dr. Kubie and Dr. Brosin do, the motivation of hostility. I feel that it is a most urgent and pressing problem, that the practicing psychiatrists, who understand hostility, and the practicing physicists who understand, so to speak, the weapons of hostility, have a common language.

WIENER: The point of dimensional analysis will not save you in this case.

ABRAMSON: I am not so sure you have foreseen what I have in mind. I do believe that one of the ways of establishing communication is to have a simple and clear language. I do not think you would deny that.

WIENER: Not at all. | [102]

ABRAMSON: It may not solve all problems but we have to have a simple method of language. We have to have an alphabet to go on.

WIENER: We do.

ABRAMSON: I feel that dimensional analysis as part of that alphabet, (in my own case someone who has run the gamut from physical chemistry to psychoanalytic theory), aided me when I was not certain what the physicists were talking about thus communication was established in difficult areas of communication.

WIENER: The thing I am saying is that, dimensionally, energy is a very bad idea. By dimensional analysis we know that they are talking in the wrong terms in bringing energy here. It does not correspond to energy in physics.

ABRAMSON: But the psychiatrist uses terms like »motive force«, »motive power«.

WIENER: He is wrong.

ABRAMSON: Telling them won't solve the problem.

WIENER: I have made a definite attempt.

ABRAMSON: But they have to understand.

WIENER: What I am saying is we have come to the conclusion that the notion of entropy, strictly as it occurs in physics, is transferrable to the study of information except for a negative factor Now that does not fit into the ordinary dimensional analysis language. There is a semantic problem here and a difficult one, but the dimensional analysis alone will not handle this particular semantic problem.

ABRAMSON: I do not believe and have not said that dimensional analysis alone would handle it.

WIENER: How would you handle it?

ABRAMSON: Since you pointed out, and I agree with you that we need new methods of communication, that you have provided a new method, merely picking a dimensional method does not mean you are going to solve the problem for the psychiatrist.

WIENER: That is a negative statement. I would like to know what you propose here to take the place of dimensional analysis.

ABRAMSON: For the sake of clarity I should like to repeat and amplify what I have in mind.

During the era of the evolution of psychodynamics, psychiatrists frequently employed physical terms, such as energy, motive force, motive power, dynamics, thermodynamics, etc., to lend physical significance, enhanced meaning through metaphor, and quantitative weight to their ideas. Indeed, they still do feel the need for physical [103] expressions in their technical expositions. It is of interest to note | that at this conference, specifically organized to promote communication between the disciplines, misunderstandings arise between representatives of both the same and of different disciplines because quantities are described in terms either not dimensionally correct or conceptually incapable of being described by terms having specific dimensional meaning. This lack of clarity and precision of meaning leads necessarily both to difficulty of communication, uncertainty and often hostility.

It appears that it is desirable and necessary at this time, that those who are planning to work in psychodynamics learn the exact meanings of physical terms and to use them in a way which would satisfy the strict criteria of those who employ them in their own disciplines – that is, in physics, chemistry, mathematics and related sciences.

Why is this urgent and important? Why is it necessary for a simple technique to be developed so that better communication can be established between the physicist and the psychiatrist so that each can really understand the language of the other? It is especially necessary at this time because it is the pure scientists who now alone understand the use of our new and unpredictable weapon of hostility – the atomic bomb. The psychiatrists as a whole, understand the mechanisms of unconscious motivations of hostility within man himself. Those who control the weapons of hostility and those who understand the psychomotive forces originating hostility must meet on common ground as soon as possible. It appears to be a necessary condition that the pure scientific disciplines and the psychological disciplines, antipodal in their very nature but meeting in the body of man, must merge if man is not to destroy himself. This doctrine of the intimate merging of antipodal disciplines, I have called »sympodism«. That is, sympodism is the doctrine which holds that disciplines directly opposite in their character, must be brought together on the basis of complete mutual understanding. I should like to emphasize that the concept of sympodism is not just team work between sciences so well described by Dr. Wiener in his book, »Cybernetics« (John Wiley & Sons, New York, 1948, 194 pp). In the case of the relationship of psychiatry and the pure sciences, this concept of the team engenders certain difficulties connected with the language of communication itself. Teamwork is impossible if the technique of communication has not been properly established. The field of psychodynamics is complicated by the fact that those who work in it must be thoroughly acquainted with the concept of unconscious motivations. Emotional qualities and [104] uncon|scious factors are in and of themselves not measurable by any frame of reference which we have at present. These psychomotive factors cannot be adequately conceived of or expressed in a simple way by the pure scientist who nearly always deals with measurable quantities. Where should a merger of disciplines begin? The beginning must be, as mentioned before, in language – in communication. Certainly my contacts with physicists and psychiatrists make me feel that there is less communication between these two disciplines than between others, say, theology and psychiatry or mathematics and chemistry. Dimensional analysis is a simple device whereby precise meanings can be given to physical terms, the use of which by psychiatrists can form at least one bridge between the pure scientist and the psychiatrist. The method of dimensional analysis for this purpose can be utilized in its simplest form without more training than high school algebra and physics. That is all that is required in effect the first stage of rapprochement needed to provide a basic language of communication between psychiatrists and the pure scientists. Dimensional analysis takes abstruse physical terms and defines them by means of the dimensions, mass, length and time. These dimensions and their derived functions of energy, power, force, etc., will take on more precise meaning when psychiatrists discuss these terms in expounding psychodynamic theory and fact. The use of dimensional analysis as a primary language of communication and the understanding by the pure scientist of unconscious motivation, will probably show that no amount of purely mathematical reasoning can ever take into consideration the complexity of the emotional factors involved in the communication of one man with another. The problem of communication always involves a common language which includes not only agreement in anticipation of events but also the attitudes of the communicating individuals toward the occurrence of the events themselves. It is these attitudes at present, which defy definition in mathematical terms alone. If the psychiatrist and physicist can communicate, as is only possible at present, through a language which because of uncertainties provokes indifference, misunderstanding and anxiety, the language is no longer just a language but a threatening language. Sympodism, carried out to its logical conclusion will lay the basis for the establishment of channels of communication, of languages more nearly free of threat. Both the language of psychodynamics and the language of the physicist contain inherent threats to members of these divergent disciplines because on the one hand, psychodynamics as mentioned, | deals with psychomotive forces characterized by the property [105] of unmeasurability. The physicist nearly always deals with measurable quantities. Dimensional analysis may serve as a language relatively lacking in threatening qualities to the physicist and may establish a relatively danger-free elementary alphabet of definition so that both disciplines can speak of the same meanings – or at least with meanings not so dissimilar as to constitute a basis for anxiety.

I now believe that it is urgent for those students who plan to specialize in the field of psychiatry, more especially in the field of psychoanalysis, as well as for the present leaders in the field of psychoanalysis itself, to review dimensional analysis and to make certain that a course in dimensional analysis be required in the training of the specialist in psychiatry. This will prepare the psychiatrist to speak in more precise terms and to be prepared to accept with facility and precision, many concepts of modern physics which he oftens[!] uses intuitively, but not with understanding, in his work. In particular, I have in mind the operational concept discussed at length by Bridgman in »The Logic, Modern Physics« (Macmillan Co., N.Y., 1932, 228 pp).

I realize that the idea of sympodism does not either pose or solve all of the questions that will arise. I realize that it is only one path in the channel of communication between psychiatry and the scientific disciplines. But it does, I believe, represent a necessary beginning for the merger of the disciplines. If the notion of sympodism becomes rooted it is believed that the next step will be possible. That is, the forces engendered by the joining of the pure sciences and psychodynamics will lead to the possibility of acquainting the community as a whole with the meaning of psychodynamics and unconscious motivations on a broader basis than is possible at present. Much of the difficulties engendered by the differences in terminology and feeling will thereby be eliminated between the psychiatric and the purely scientific disciplines. The possible effects of a rapport of this type on our educational system and culture as a whole is at present unmeasurable and unpredictable. It will certainly be for the good of mankind. Indeed, I believe that it is necessary for the achievement of world peace.

WIENER: I still say in the discussion of a system and any means of communication, I am including in this hormonal or humeral, there is a very fundamental and measurable idea that I am trying to get over. That is information which is measured in numbers of decisions, between two alternatives, which otherwise were equally | probable. That is [106] a perfectly definite quantity, and we can and do determine it in communicating systems all the time. The trouble is that dimensional analysis does not help there because this quantity, while important, is essentially dimensionless. You cannot give the dimensions of a decision.

VON FOERSTER: I think we can sometimes serve one kind of parameter and that is time. We have a real dimension. We can for instance distinguish numbers per time or only numbers and so I think we can decide if we have a time problem or a static problem, a dynamic problem or a static problem.

WIENER: In other words, there are two quantities that can be distinguished, both psychological and engineering entropy and rate of transfer.

STROUD: May I enter a suggestion here? Isn't it just possible what you are fighting against –

ABRAMSON: I am not fighting against anything. Do you think I am fighting with Dr. Wiener?

STROUD: No, heavens no, I did not mean to imply that. I mean that in the case of the incommunicability of ideas there are difficulties involved. Perhaps you are fighting against the fact that the average medical student remembers enough of his high school physics so that when he hears the use of the word »physics« he gets confused, because he is neither fowl, flesh nor fish?

ABRAMSON: He enters medical school well-equipped.

STROUD: Are you not arguing it might perhaps be better to help the medical student to remember the more rigid terms that he uses when they have a physical meaning thereby making it possible to talk across the boundary?

ABRAMSON: Precisely.

STROUD: Rather than pervert well-defined terms, even to go to the bother on his account of inventing new ones of making the laborious effort to change which is very hard to do, to go counter to the stream.

ABRAMSON: It is especially urgent now for those graduate physicians who are going to be psychoanalysts to be reindoctrinated in these definitions by any devices possible, including even additional dimensional analysis which will ultimately be found not as important as I believe it now is.

WIENER: Dimensional analysis is extremely important in training. The only thing I wanted to say is that in that branch of physics which is closest to psychology the main concept – entropy – is a rate per time. The main concept of entropy happens to be [107] dimension|less. It is a concept which can go over directly from the study of the nervous system to the study of the machine; it is a perfectly good physical notion, and it is a perfectly good biological notion. The notions in that field are the significant notions. Here the point is that it is not the classical physics of energy, that is, the relevant physics, but rather the physics of information. That physics is being developed very rapidly. We have a good language for it and we can state things precisely in it. This is now one of the things we are pushing very hard at Massachusetts Institute of Technology.

ABRAMSON: Do you think it will be digestible by psychologists or people expert in psychodynamics?

METTLER: May I get in a sort of lefthanded comment? This problem of Abramson's is one we are dealing with constantly at Columbia University. The psychoanalytic group does send people for advanced training to our laboratories. I think there is about as little chance of the poor doctor who has to make a living in private practice, learning dimensional analysis as there is of his learning the ins and outs of cybernetics. There will be some individuals gifted by the gods and by fortune like Larry Kubie who will put themselves in favorable places to learn enough to form in themselves a unity between the various disciplines, but I think the real hope comes in the sort of thing that the Macy Foundation is doing, namely, bringing people together so that, without too much stress and strain on the individual to learn the details, it is possible for them to learn the trend and tendency. I think if you are going to force the individual who is getting a particular kind of training into a mold whether it be dimensional analysis or something else, you are going to have a great deal of difficulty in making him digest it. Your poor psychoanalyst and your poor future psychoanalyst now face a period of training which is a very terrific one and very prolonged.

ABRAMSON: May I answer your question before you go ahead, because I happen to be in that very class about which you are talking?

METTLER: He has to learn something about conventional physics; he has to learn something about neurophysiology; he has to learn something about social sciences, and it is largely a question whether there is enough time.

KLÜVER: At least he does not have to learn about experimental psychology. | [108]

ABRAMSON: I would like to discuss this and give you facts. I happen to be attending Dr. Sandor Rado's lectures on psychodynamics at Columbia. He employs the operational concept in teaching psychodynamics. The students planning to be analysts are certain to be confused unless they retained or got a review course in dimensional analysis. I don't share your pessimism on the inadequate background of the first-year medical student because ten years ago my class could use dimensional analysis after brief training. I believe they can learn it very quickly in a couple of weeks, so I don't think it is adding very much to a crowded curriculum. The difficulty is overcoming the dogmatism of those who decide what is supposed to be taught to train good doctors.

KLÜVER: You talked about dimensions in connection with psychomotor forces, is that right?

ABRAMSON: I don't understand.

KLÜVER: Six, seven or more questions ago, you talked about dimensionless psychomotor forces?

ABRAMSON: I used the term psychomotive instead of motive forces to indicate that it is not an ordinary force which is expressed in the usual dimensional terms. Now I understand from Dr. Wiener that perhaps that distinction may not be necessary. However, I think after this discussion it is more necessary than before.

WIENER: The whole idea of force and energy is fundamental to the real problem.

BROSIN: Before we stop at five, could we request Dr. Kubie to summarize his position? For my own pleasure, I would like to hear him discuss the properties of the active dynamic barrier which distinguishes the conscious from the unconscious, and the properties of the unconscious itself, since this has caused many arguments. Is it structured or unstructured?

KUBIE: I cannot do that in so short a time, and I fear that it would just add additional confusion for us at this moment. It is something that we ought to take up, but let me perhaps illustrate the problem from the same patient, and I think that I can do that briefly. To condense it a bit and to clarify one point, her son-in-law who is a doctor had just married the daughter, and asked me to see her. He said, »there is only one thing wrong with her. She is not sleeping. It won't take you very long to help her.« The only trouble that she was aware of, which brought her for treatment, was insomnia. It did not take long to realize that that was just the pinnacle of the iceberg that showed above the water. I sent for him and said, »You are wrong. She is going to go into a serious depres|sion. She has been in a depression without knowing it for a good [109] many years. Furthermore, once she starts to sleep her depression will deepen.« When he doubted this, I determined to put it to an experimental test. I gave her enough sedatives so that for three nights she slept well. At first she thought I was a magician. By the third day, however, she came in in a depression. Only then did her real treatment start. What did all of this really mean? What does it mean in terms of the problem we are trying to deal with here?

I then discovered another thing, namely: for many years when another person would have spoken of feeling blue or depressed, this woman would develop gastrointestinal pain. She had taken this to many of the best internists in Europe and here. As her treatment proceeded her depression became clearer and more intense, whereupon she began to sleep well and intestinal symptoms disappeared entirely. She was depressed about many deeply buried problems; and only as she began to find some way of dealing with them, one by one, did the depression begin to lift.

I want to indicate by this that we are capable of segregating the various processes which go on inside of us. We can segregate experiences so as to make them inaccessible. We can segregate our reactions to those experiences so that we don't actually know what we are feeling. It seems paradoxical to talk about feeling and not being able to feel your own feeling, but that is as accurate a way as we have of describing it operationally. Thus this woman had been depressed for years without knowing it, and had marked this with many physiological and psychological symptoms, and finally through her insomnia. This seems to me to be important both in terms of our effort to understand and formulate normal and neurotic psychological processes and also in terms of our understanding of the whole process of memory. An important theme that runs through this whole concept of neurotic process as an integral part of it, is the fact that memory as we know it is an emotionally determined function. Simply to talk of a physiological or physical trace as in a magnetized wire without reference to emotions is very misleading. Memory as a total human experience has many complicated and complex aspects. There is the physical recording of a trace, and then the whole process of making that trace available either as a direct recapturing of a previous experience or in indirect and translated forms. One patient tried unsuccessfully to remember the telephone number of a girl that he wanted to call up. Finally he went to sleep and dreamt the number, but only after changing the image of the girl in such a way as to [110] make her less terrifying. |

All I am trying to say is that if things are as complicated as that we are not going to make advances by trying to pretend that they are simple.

Klüver: Your last example seems to be of special interest in connection with Silberer's »autosymbolic phenomena.«

It is probably worth stressing that, in trying to understand the neurotic process, the attempt is again and again made to proceed from current neurotic manifestations to events in the past, that is, to trace, let us say, E to D, C, B, and finally to some original experience or some primordial scene or event A. If it were not for therapeutic successes or the fact that E, D, C, and B appear psychologically plausible in the light of what supposedly happened at A, there seems to be no special reason why analyses of such kind should not be extended to the antecedents of A, thus demonstrating even more clearly the *regressus ad infinitum* involved here. This is essentially a genetic approach, an attempt to arrive at an explanation by recourse to some *status quo ante*, and is, therefore, subject to the same criticisms as all genetic approaches in psychology. It may be argued that a psychological analysis, instead of trying to show that E, D, C, and B are determined by A, should concern itself first of all with exhibiting the psychological structure at A or, more generally speaking, with specifying as concretely as possible the nature of factors leading to neurosis-producing or traumatic events and experiences. A man may have seen hundreds of persons or horses die or may have seen or experienced hundreds of injuries, yet an analysis of his neurotic symptoms may conceivably lead to the assumption that all his troubles started on a particular September morning, before breakfast and after polishing his spectacles, when he saw a particular horse, to be sure merely another horse, die. What is the constellation of psychological or other factors that sets off this particular situation from hundreds of other seemingly similar or identical situations? It is at this point, I believe, that we can start discussing real problems of psychology. When I discussed problems of this sort for the last time with Paul Schilder he pointed out to me that psychoanalysis has done very little towards illuminating the psychological structure of supposedly neurosis-producing situations and, what is more serious, has been unable even to outline the kind of research or type of experimental approach most likely to lead to a specification of the psychologically relevant factors in such situations. I am wondering whether you feel differently about this or whether you believe that this state of affairs has changed in the meantime. I am also wondering whether you purposely meant | to stay entirely on [111] the descriptive level in your account of neurotic manifestations.

Kubie: That is why we are here. As soon as we begin to talk in terms of forces we have left the purely descriptive psychological level.

Klüver: Am I right, then, in assuming that you did not want to go beyond a descriptive account and that you did not wish to specify any of the psychological mechanisms involved in the neuroses?

McCulloch: We have to stay on the descriptive level until we get to some kind of perversion of the circuit.

We take up now the consideration of memory, starting from psychological data, applying to it other ways of thinking than those which are common to psychologists, and coming out with a theory which I think may make sense in connection with some of the problems we were going over earlier in the afternoon. It is our hope that after we have done this we may get around to the problems of recall and recognition, and that tomorrow morning we may be able to dovetail these stories and the problems of neuroses together, at which time we ought to go over abnormal circuit action which must underlie any neurotic process.

I am going to ask Heinz von Foerster to start off on the theory of memory.

[112] # QUANTUM MECHANICAL THEORY OF MEMORY[1]

HEINZ VON FOERSTER
Department of Electrical Engineering
University of Illinois

Perhaps the best way to report on this theory of memory will be to split up the whole argument into three steps. The first step may be called a phenomenological step. Here I am only trying to introduce mathematical terms to some psychological facts by mere phenomenological considerations. Learning by the failures of the first oversimplified assumption, one is forced to investigate the psychological facts more in detail. This leads to the second phase of this theory which I call the psychological step. Finally, I would like to show you the possibility of giving the phenomenological assumptions a quantum mechanical explanation. Thus the theory gets a biophysical backbone.

Let us start now with the phenomenological phase. Some time ago I was trying to work out a relation between the physical and the psychological time. Certainly, both these times would be proportional to each other if our memory would work like a tape-recorder: any incoming information would be stored indefinitely. Recall of a certain event would give exactly the same time structure as previously observed. We know, however, that isn't so. As time elapses we lose a certain amount of information by forgetting. Hence I tried to start with a simple theory of forgetting. The principle idea is that any observed event leaves an impression which can be divided into a lot of elementary impressions. I think one is justified in assuming this because the sense organs too are divided into a lot of elementary sensory receptors. Suppose now that any event leads initially to number N_0 of elementary impressions. After a certain time t the number of existing elementary impressions may be called N. What we are looking for now is a function which connects the number N with the number N_0 and the time t. In almost all cases of decay in physics and chemistry one starts with a very reasonable assumption, which I am trying to apply in this case: the rate of change per time unit of the number of existing elementary impressions should be proportional to the number [113] of existing ele|mentary impressions. This assumption can be expressed in mathematical terms as follows:

$$\frac{dN}{dt} = -\lambda N \qquad (1)$$

The minus sign on the right side simply means that this process is a decreasing one. The solution of this equation is well known. It is as follows:

$$N = N_0 e^{-\lambda t} \qquad (2)$$

This function merely means that in the first instant when t is considered to be zero the number of elementary impressions is N_0 and after a very long time the number of these impressions vanishes. The magnitude λ can be called a »forgetting-coefficient«, for it gives the function a steep descent when λ is large, a gradual slope when λ is small.

Before we can compare the form of this function with any measured forgetting process we have to remember that the assumption made before is only applicable to a set of elements – elementary impressions in our case – which are independent of each other. Therefore I was looking for a psychological process which deals with impressions of which the elements are as independent as possible of each other. I think I found it in

1 Heinz Von Foerster: »Das Gedaechtnis«, Deuticke, Wien 1948.

experiments with nonsense syllables. In the following I will use results observed by Ebbinghaus during his study of the forgetting process of nonsense syllables.

But if one compares this function with a measured curve one can see that only in the very beginning both curves run together. (Fig. 1)

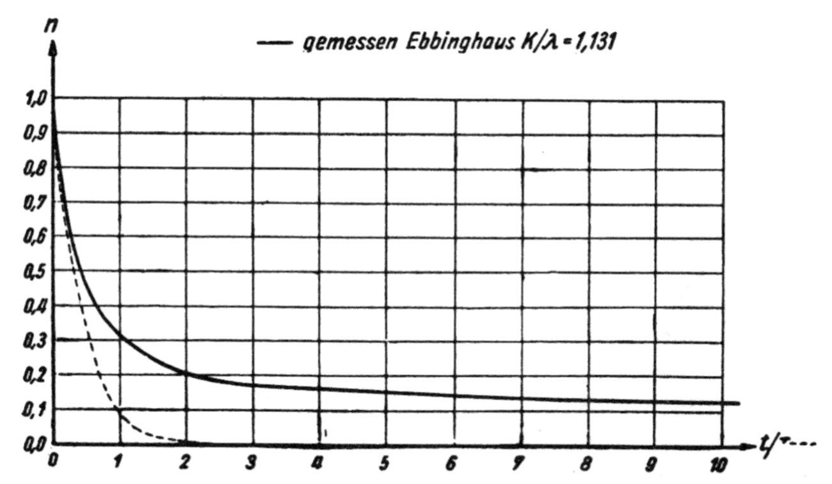

FIGURE 1. Comparison between a measured forgetting-curve according to Ebbinghaus and an *e*-function according to equation (2).

| In this figure the time *t* in days is plotted horizontally; the percentage Z remembered syllables vertically – in our terminology N/N_0. The difference between these two curves becomes more and more obvious with increasing time. The measured forgetting-curve stays on a certain level, the other one computed according to the simple assumption soon approaches zero. One can certainly think that this simple assumption with only one forgetting-coefficient is too weak to describe such a complicated process as the human memory. It is reasonable to assume that besides this single forgetting-coefficient λ other forgetting-coefficients λ_1 λ_2 λ_3 ... λ_i may also be involved in this process. Mathematical formulae can be set up which handle this more complicated case, but one will always obtain functions which after an infinite time approach zero and which are therefore different from the real behavior of the forgetting process. One can go further and can assume that there are not discrete numbers of forgetting-coefficients, but that the elementary impressions are continuously distributed over a continuum of forgetting-coefficients as sketched in the following graph. [Figure 2] [114]

In making this assumption one can try to define such a distribution-function of forgetting-coefficients as a function which fits | optimally to the measured forgetting-curve. In doing so one obtains a very interesting result. The distribution-function of forgetting-coefficients turns out to have a certain number of »negative forgetting-coefficients«. The striped section on the left side of figure 2 indicates this fact. [115]

But what does a negative forgetting-coefficient mean? The answer can only be: learning. How does learning come into this very clean forgetting process? Perhaps we can find an answer to this question by examining how such a forgetting-curve can be measured. It can be done in the following way: the experimenter teaches a group of subjects 100 nonsense syllables until everyone knows these syllables by heart. Then he makes examinations day after day and plots the mean of remembered syllables of all persons on a graph as a function of time. But what happens during such an examination? The subjects are forced to recall these syllables and to pronounce them – a process which is very similar to learning. That means that after such an examination all

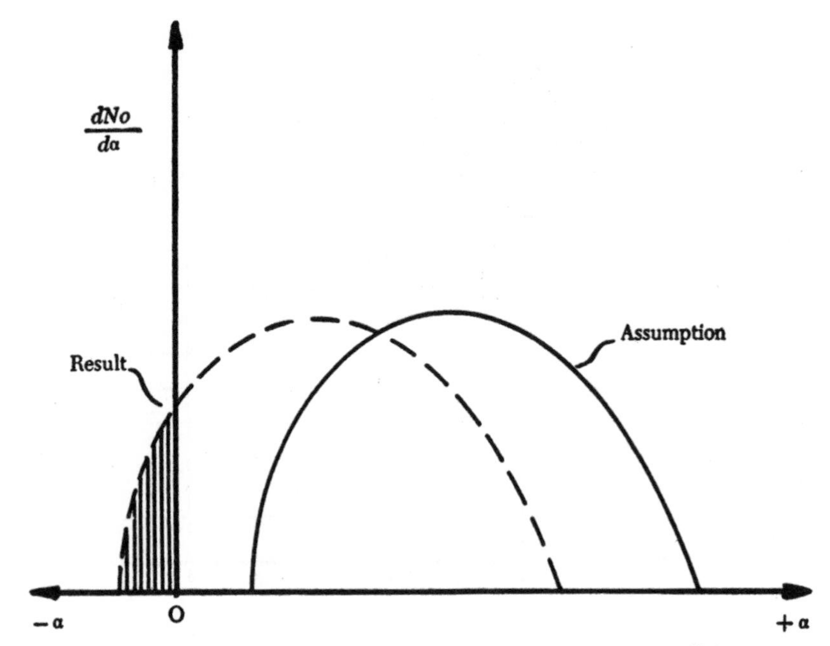

FIGURE 2. Assumed and obtained continuously distributed forgetting-coefficients.

syllables still remembered by the subjects are now fed again into the memory and one has to treat them as if they had just been learned.

The first equation I wrote down doesn't deal with this feedback-procedure. Yet it is necessary to put this process too in mathematical language. But before doing so it is perhaps useful to explain this process a little more in detail, because it is an important part of the theory. This picture that I am now going to develop brings us to the second step which I mentioned above and I called the psychological step. Let us assume that each syllable or part of such a nonsense syllable – whatever we would like to define as [116] an elementary impression – is fixed on a certain carrier, many of which | may be in the brain ready to be impregnated by such an elementary impression. This picture may help you to understand what I mean.

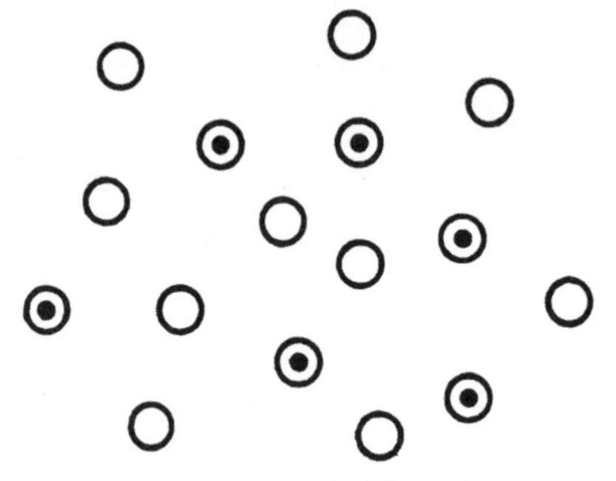

FIGURE 3. Impregnated and free carriers.

The little circles may be the carriers. I will call them »free carriers« if they are not charged with an impression and »impregnated carriers« if an elementary impression is fixed on them. I mark the impregnated carriers with a little dot inside of the circle. What these carriers may be I shall leave open for the moment. In the third step of the theory I will try to give them a physical significance. Let us assume that such a carrier is not able to carry forever its impregnation but only during a certain time and decays after time τ to a free carrier. Every impression impregnates many such carriers, hence they can be treated as an ensemble. In fact, we would obtain exactly the same equation as before except that instead of the forgetting-coefficient λ we have now in the formula the average lifetime τ of such a carrier.

$$N = N_0 \varepsilon^{-\frac{\tau}{\tau}} \tag{3}$$

The relation between λ and τ is simple. The lifetime is the reciprocal of the forgetting-coefficient.

$$\tau = \frac{1}{\lambda} \tag{4}$$

With the help of these carriers it is now easy to picture the examination process I described just now. During such an examination all carriers are scanned and where an impregnated carrier is found its impregnation is transmitted to a free one. Certainly, such a transmission-process does not destroy an impregnated carrier – it only transmits its impression to another one, as figure 4 illustrates by its arrows.

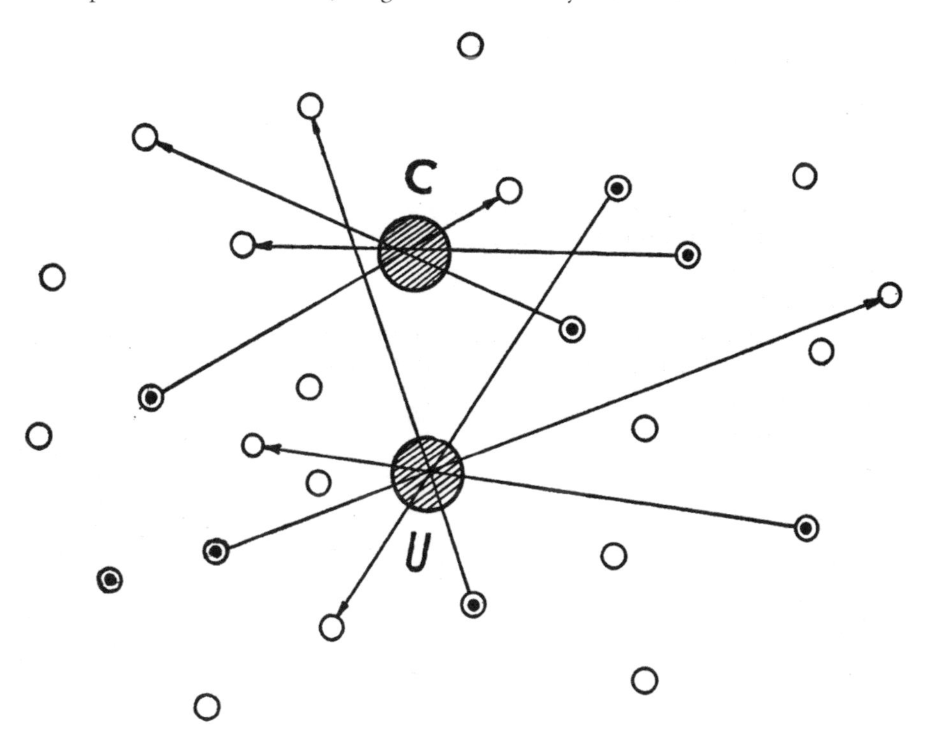

FIGURE 4. Conscious and unconscious transmission of elementary impressions to free carriers.

Such a transmission which takes place during an examination can be made consciously (*C*) or unconsciously (*U*). It is easily seen that such a process would keep some items in the memory, even if each carrier has only a finite lifetime. To formulate this procedure in mathematical terms we have to make at first the most simple assumption we can make and then compare the result with the observed facts. It is certainly obvious that the intensity of such a transmission will be proportional to the number N of impregnated carriers. On the other hand, the transmission can only be successful if a sufficient number of free carriers are available for obtaining the impressions. Therefore, let us assume that the efficiency of this transmission is proportional to the number of [117] free carriers (N_0-N). |

The number of efficient transmissions per time-unit can be defined now as

$$\mu N(N_0 - N) \tag{5}$$

where μ is a certain proportionality-coefficient containing the efficiency-coefficient η of such transmission and ν, the frequency of the scanning-process. The entire process of forgetting and memorizing can now be determined by the following formula

$$\frac{dN}{d\tau} = \lambda N + \mu N(N_0 - N) \tag{6}$$

where the first negative part of the right side describes the forgetting-process and the second positive part the memorization-process. This equation, too, can easily be solved and if one calls $n = N/N_0$ the relative amount of remembered items and $\kappa = N_0\,\mu$ the »memorization«, one obtains the following formula

$$n = \frac{\kappa - \lambda}{\kappa - \lambda e^{-(\kappa - \lambda)\tau}} \tag{7}$$

[118] | It is interesting to see that this formula doesn't become zero, even when the time t goes to infinity. There is always a finite remainder of some items in the memory and it depends only on the ratio of the two magnitudes κ, the memorization, and λ, the forgetting-coefficient – or the »decay-constant«, which would be the proper name for this constant when applied to the hypothetical carriers with their finite lifetime of impregnation – whether the remainder is large, small, or zero. This behavior of the function makes one hopeful about comparing it with a measured curve, e.g., with the forgetting-curve of Ebbinghaus. I computed both constants $(\kappa\lambda)$ involved in the formula above according to the method of Gauss from Ebbinghaus' curve. In figure 5 this function is compared with Ebbinghaus' curve.

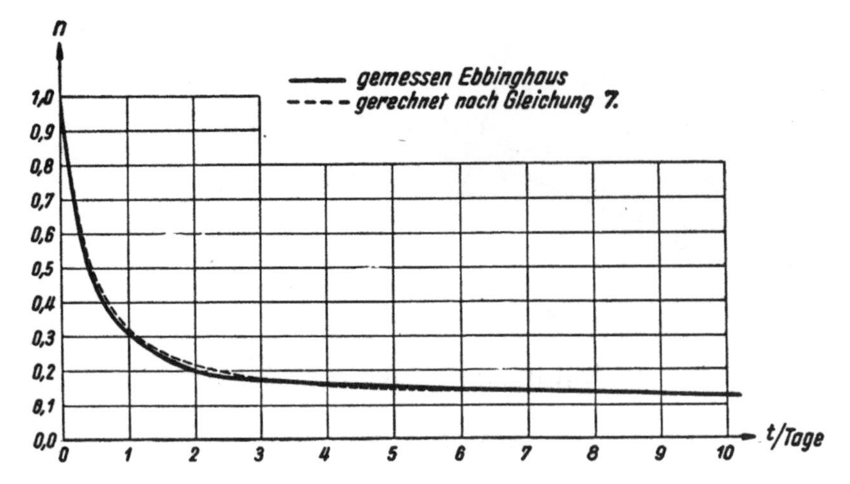

FIGURE 5. Comparison between a measured forgetting-curve according to Ebbinghaus and the »forgetting-function« according to equation (7).

Both run very close to each other, with an average deviation of five percent.

A result which was not obtainable with the forgetting-theory alone – even with six or more parameters – can now be obtained with this simple assumption of transmission and introduction of only two parameters. I would venture to say that the picture is perhaps not absolutely wrong. Perhaps you are interested in the order of magnitude of these two constants; κ the memorization and λ the decay-constant. I found

$$\kappa = 2.755 \text{ per day}$$
$$\lambda = 2.430 \text{ per day.}$$

| In other words the average lifetime of such a carrier is 1/2.43 days that is 0.412 days or about 3.5×10^4 seconds. Later on this figure will be an important clue, when I am going to interpret these carriers by a meaningful physical picture. [119]

May I invite you now to follow me through some psychological consequences we obtain by discussing the theoretical memory-curve. I mentioned before that even for infinitely long waiting times the number of remembered items stays on a certain remainder. This remainder we can easily compute directly from formula (7), letting t go to infinity. One obtains

$$n_\infty = \frac{\kappa - \lambda}{\kappa} = 1 - \frac{1}{\kappa/\lambda} \tag{8}$$

I call n_∞ the »remembrance«, expressed in fractions of the amount of impressions originally received. In the following figure I plotted this remembrance as a function of the ratio of the two constants, defining the whole process, the ratio of the memorization κ and the decay-constant λ.

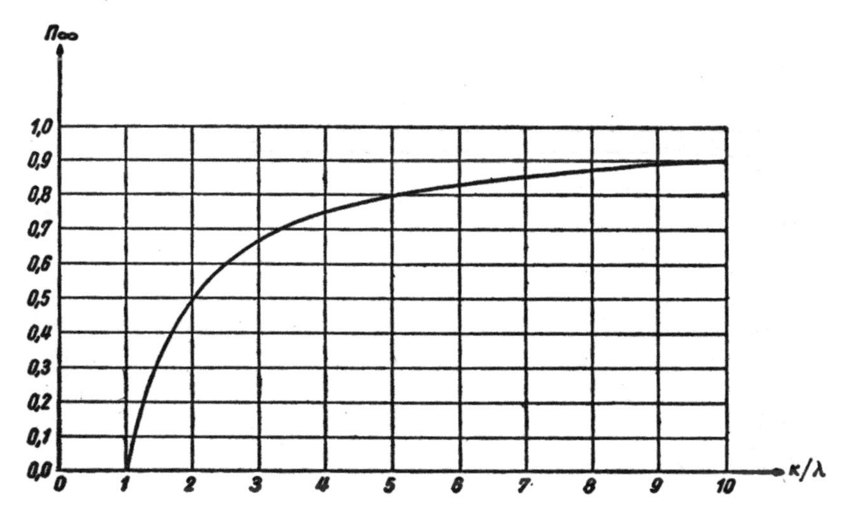

FIGURE 6. Remembrance as a function of the memorization and decay-constant.

As long as the memorization is smaller than the decay-constant ($\kappa/\lambda \leq 1$) we have no remembrance of a certain event. The memorization is too weak. As soon as the memorization becomes more intense than the decay-constant, the remembrance jumps up, but later on with increasing κ/λ becomes flatter and flatter. It never reaches the value one, meaning 100% remembrance of all details of a certain event. The steep start implies that only a slightly increased memori | zation results in a much higher remembrance. Perhaps it explains the large variations in the success of pupils whose attention was only slightly different. A high percentage of remembrance is difficult to obtain. A remembrance of 70-80% implies intense memorization power, four to five times as [120] much as is usual. These facts, directly obtained by the theory, seem to correspond to our feeling for such things.

We can obtain some other psychological effects if we pay attention to the memorization or transmission function alone, the positive member in the differential equation (6):

$$M = \mu N(N_0 - N) \tag{9}$$

If we put into this equation the solution for N (formula 7) and normalize again to the number N_0, we obtain a function representing the number of transmissions per time-unit as a function of time. In figure 7 I plotted this function for different parameters κ/λ.

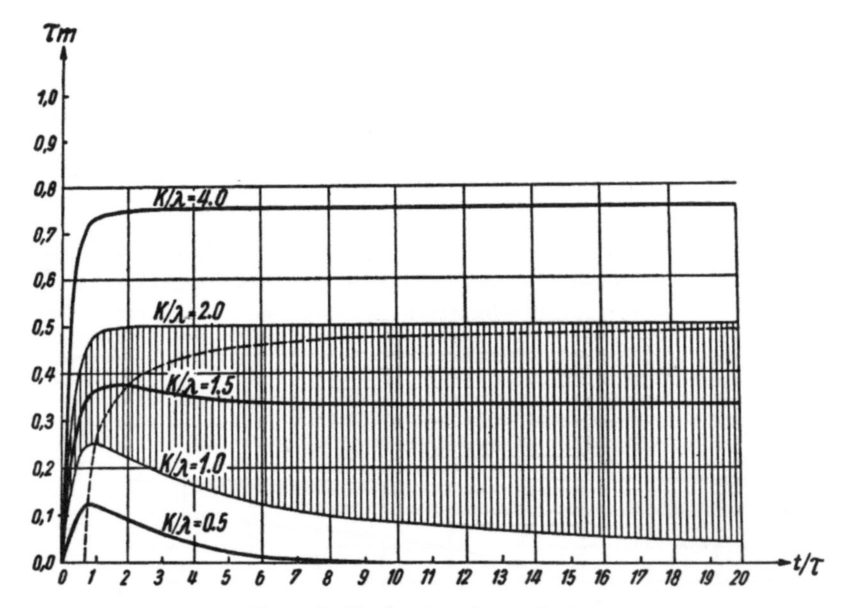

FIGURE 7. The function of memorization

What are these curves now representing psychologically? Let us follow a curve with a relatively small ratio of κ/λ, which means – as we now know – an event with a very small remembrance. Let us choose e.g. the curve where κ/λ is equal to 0.5. This curve starts at zero, runs up at first, reaches its maximum at about 0.8τ, that is after about 7 [121] hours, and decreases afterwards slowly to zero. Let | me give you an example of what happened here. Assume you rode a bus in the morning and had an argument with one of the drivers. He called you names and you called him names. You were a bit upset after leaving the bus, but a short walk to your office helped you to forget about the whole affair. If you look on the curve, you will find that the memorization is still very small. But meanwhile the intensity of memorization of this event inevitably increases in your unconsciousness until suddenly in the afternoon – six or eight hours later – you become aware of this intensive memorization-process. You start worrying and thinking of the very good names that you didn't call him. But after a week or so, you forget the whole event. This effect is psychologically well-known as »aftereffect« and flows directly from this theory without special assumptions.

There are a lot of other applications of these functions to psychological phenomena. But I don't like to bother you with too many details. I would like to show you that this idea of combined decay and transmission is also applicable to the process of observation. It is only necessary to introduce one simple and relatively reasonable assumption; namely, that the sensory receptors can also be treated as carriers. They must be able to transmit their impregnation to the carriers of the memory and must have a much shorter lifetime than their long-term brothers of the memory. This assumption is necessary to keep them always receptive for new incoming impressions.

If we assume the mean lifetime of these short-term carriers to be about 10^{-3} seconds, then, I think, we can explain the process of sensation. Let me call this type of carriers T_1 and the other T_2, with their long lifetime of 3.5×10^4 seconds, belonging to the memory. Then we can describe an experience about as follows: the sense organs impregnate the carriers T_1, which transmit consciously or unconsciously their impregnation immediately to the carriers T_2. Now starts the old game of transmission between carriers of the type T_2 only. We exclude transmissions from T_1, to T_1 as well as reverse transmissions from T_2 to T_1. Figure 8 illustrates what I am trying to explain. | [122]

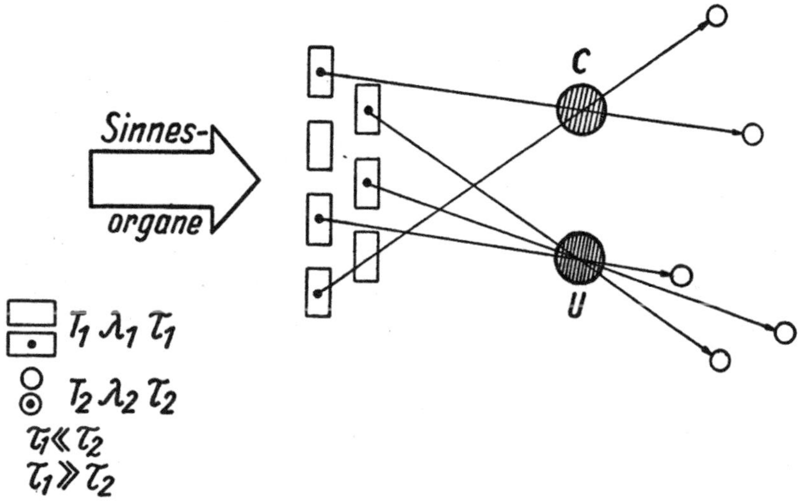

FIGURE 8. Scheme of transmission from impregnated short-term carriers T_1 to long-term carriers T_2 of the memory.

One can make use of these transmissions to describe in a very formal way different phenomena. The conscious observation or the conscious experience of an event may have a transmission formula like this

$$T_1 \rightarrow C \rightarrow T_2 \tag{10a}$$

If the experience is unconscious, we may write

$$T_1 \rightarrow U \rightarrow T_2 \tag{10b}$$

Only as a suggestion I will give you a list of other possible transmissions and their significance.

$T_2 \rightarrow C \rightarrow T_2$	memory experience	(11)
$T_2 \rightarrow U \rightarrow T_2$	unconscious memorization	(12)
$T_2 \rightarrow U \rightarrow T_1 \rightarrow C \rightarrow T_2$	hallucination	(13)
$T_1 \rightarrow U \rightarrow T_2 \rightarrow C \rightarrow T_2$	delay-experience	(14)

If the transmission-chain (14) is produced very quickly, say in one tenth of a second, we have the well-known experience of the »déja-vu« whether a transmission stems from T_1 or T_2.

I developed this more or less phenomenological theory to such an extent that it now becomes necessary to give it a more physical foundation. Let me recall which phenomenological ideas I introduced:

1. The elementary impression
2. The carrier of the elementary impression
3. The decay of these carriers | [123]
4. The memorization as a power
5. The transmission as a process
6. The transmission as a selection.

It would be of some advantage to explain some of these ideas from a physical point of view. Fortunately, it seems possible to explain the peculiar behavior of the carriers in quantum-mechanical terms. Let me therefore describe to you in a very short review those quantum mechanical statements which are necessary to understand the principal ideas used here.

Every microstate of matter – it may be a nucleus of an atom, an excited or ionized molecule, or a molecule-complex – is considered to have a certain energy-level E_1, which guarantees its stability. The most important principle of quantum mechanics states that these energy-levels do not change continuously, but that such a microstate, e.g., a molecule-complex, is only capable of discrete values of energy E_1, E_2, E_3, ... – other energy-values are impossible.

The energy is quantized – that means that a change from one state E_i to the next state E_{i+1} is only possible if a finite amount of energy $\Delta E = E_{i+1} - E_i$ is contributed – and vice versa. Figure 9 shows perhaps this situation.

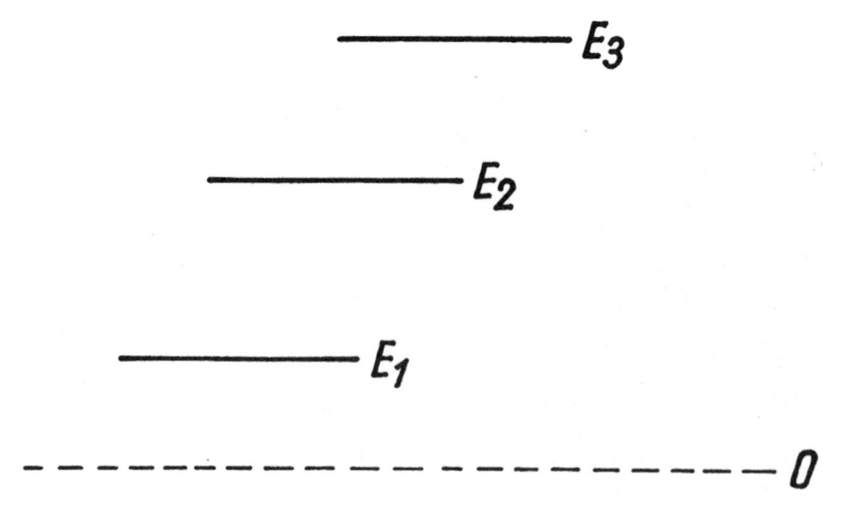

FIGURE 9. Microstates of matter in discrete energy-levels

The quantum theory has to explain why any one of these states does not change automatically to the next lower one, until the lowest state is reached, e.g., a stone on the [124] top of a hill will roll | down until it reaches the bottom of the valley. Quantum theory assumes certain energy-dams between adjacent energy-levels as is shown in figure 10.

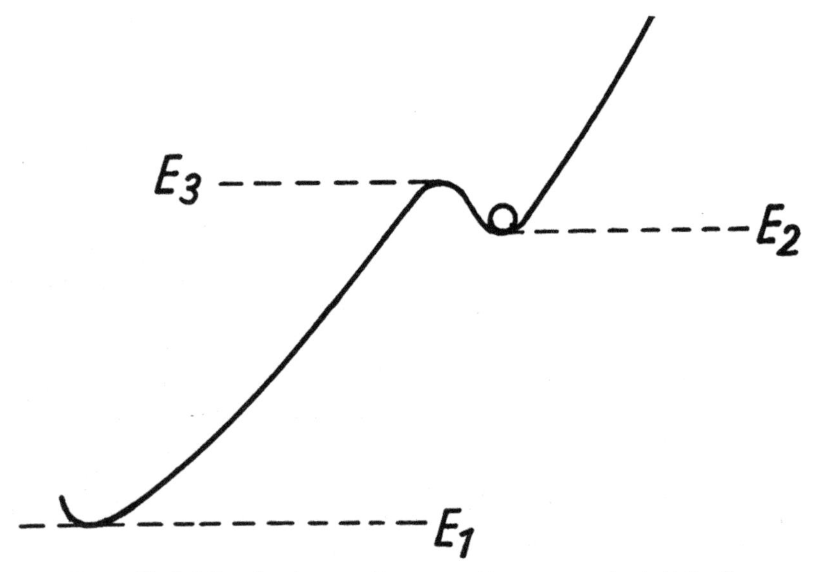

FIGURE 10. Stability of a microstate E_2 guaranteed by an energy-threshold $E_3 - E_2$.

It prevents in this way an automatic breakdown of the energy-level E_2. This picture is similar to that of a stone in one valley, which is unable to fall spontaneously in a deeper adjacent valley. In classical mechanics this stone is absolutely stable. In quantum mechanics no absolutely stable state exists. It is only a matter of waiting time – sometimes very long, sometimes very short – until the lower state is spontaneously reached. This effect is called »tunnel-effect« and describes in this way a very unperceptual process. The radioactive decay is the usual example for these processes.

According to Schroedinger this mean waiting-time, or average lifetime of this state, defines the degree of stability of any microstate. It is strongly dependent upon the altitude of the energy-dam which holds the state in its pseudo-stable position. Is E_z. the altitude of this dam (in Figure 10, $E_z = E_3 - E_2$)? Quantum mechanics gives an expression for the relation between E_z and the lifetime of the corresponding state

$$\tau = \tau_0 e^{\frac{E_z}{kt}} \tag{15}$$

| In this formula τ_0 is an atomic constant of the order of magnitude of 10^{-14} seconds, T [125] the absolute temperature and k Boltzmann's constant ($k = 1.38 \times 10^{-16}$ erg/degree).

An ensemble of N particles, which have the same altitude of these energy-dams, and therefore the same mean lifetime, follows the well-known law of decay

$$N = N_0 e^{-\frac{t}{\tau}} \tag{16}$$

This is exactly the differential equation (3) we used before and describes the negative part of the forgetting-process. This analogy gave me the idea to apply the quantum theory of microstates to the carriers. It remains to be seen whether or not the deductions give a meaningful psychological, biological, and physical picture.

The idea of introducing quantum mechanics into biology is not new. It works very well in genetics, where the genes, the carriers of heredity, are explained as quantized states of complex molecules. The physical picture of the carriers of the elementary impressions would be about the same as that of the genes. Thus, I called the carriers of the memory »*mems*«, to stress the analogy between those particles. What I called »impregnation« of such carriers means now – translated into the language of quantum mechanics – lifting from a lower energy-level to a higher level. Decay is then just the opposite process and means forgetting. What such a molecule does, if it changes from one energy-state to another is not easily stated. According to Schroedinger one has to assume that such a molecule has a very highly organized structure, like an aperiodic crystal. Then such a crystal is stable in different pseudo-isotropic arrangements of its atoms. Each of these arrangements has a different energy-level, the molecule stays most of the time in the most stable one.

Perhaps from another point of view this »all or nothing« principle of quantum mechanics also fits into the neurophysiological picture; the nerve fibres are only capable of all or nothing decisions. The real neurophysiological or physiochemical process I am not able to explain, but with this microphysical picture in mind we can determine some other properties of these carriers. First of all it is possible to compute the energy-threshold E_z of the *mem*, which holds it during the waiting time in its higher energy-level.

One has only to use formula (15), which combines the energy-threshold with the waiting time, for the waiting time is exactly the same as the mean lifetime of the carriers. We know it from our | previous considerations. It was 3.5×10^4 seconds for the long-term carriers of the memory. [126]

The whole process operates at body temperature, that is

$$T = 273.2 + 36.6 = 310° \text{ K}.$$

Now all magnitudes in formula 15 are known and by simple mathematical operations one finds for the energy-threshold E_z.

$$E_z = 1.4 \text{ eV}. \tag{17}$$

But the picture is not as yet complete. If you look at figure 9 you will see that we know the energy-difference $E_3 - E_2$, but not yet the difference $E_3 - E_1$, which plays a very important role. Just as a microstate can jump over an energy-threshold like $E_3 - E_2$, so it is possible for the molecule to jump spontaneously from its lower position over the high threshold $E_3 - E_1$ to its impregnated value E_2; but the probability for such an effect must be much less than the regular one. What is the meaning of such a spontaneous impregnation which does not stem from a regular transmission? Suppose such a self-impregnation is observed from a conscious transmission. It must have significance, although it is not based on any observation or experience at all. It can be interpreted in very different ways.

KUBIE: Isn't that precisely what happens in the dream or hallucination?

VON FOERSTER: It may be – but I am not quite sure yet. I think dreams and hallucinations can be explained by transmissions only. For dreams I suggested formula (12) and for hallucination the formula

$$T_2 \rightarrow U \rightarrow T_1 \rightarrow C \rightarrow T_2 \tag{13}$$

This formula means that an impression stored on carriers of the memory T_2 unconsciously transmitted in the very neighborhood of the sensory receptors T_1 or fed into the transmission-channel of sensory receptors to the memory-carriers, and is then immediately consciously transmitted to the memory again. One has the feeling of an experience of the outer world, though the entire process was a kind of a short circuit. I don't know whether this explanation is right or wrong – I can't decide – but these random self-impregnations are perhaps due to some other phenomena, e.g., ingenious ideas, inventions, entirely new aspects or approaches, and theories. On the other hand, they can lead to disturbances if the person is either too weak to make something out of it or not healthy enough to throw these peculiarities away and forget about them. It depends | entirely on the personality whether the individual listens to these inner [127] observations as a whisper of good or evil demons. I haven't any idea how we could measure the influence of this self-impregnation. The only thing I could do was to make reasonable assumptions about the rarity of such events. Since the waiting time for the regular decay is in the range of one day, I assumed the waiting time for such a self-impregnation to be in the range of ten human lifetimes, that is about 10^{10} seconds. Using formula 15 again, one obtains for the energy-threshold $E_3 - E_1$ about 1.8 electron-volts. Now we are able to draw a complete picture of the energy-levels of the carriers in the range of our interest. The following graph does it. You see, the energy-threshold is still not unreasonably high, in spite of the very small probability of such an event.

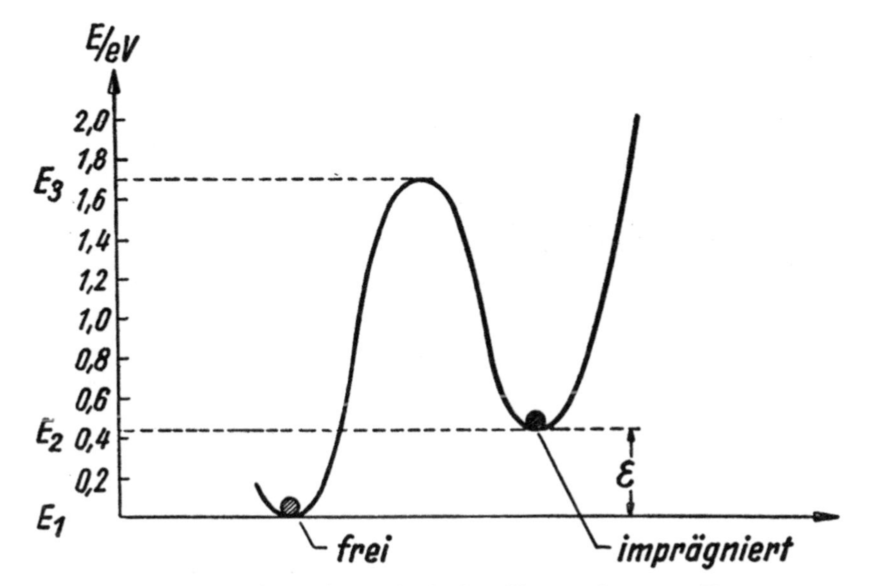

FIGURE 11. Scheme of energy-levels of possible states of a *mem* type T_2.

WIENER: There is something that interests me a good deal here. You know the ordinary classical physics where the quantum quantity comes in by having a certain ground noise level which is not zero and the density per degree of freedom of the system represents a quantum constant. I want to point out that in a system of that sort you also get disjointing, that is, you get jumping because the random motion is enough. There is a chance, even though it is very slight, that the tail of that distribution will enable you to get over any hill if you go out far enough. In the presence of background noise, even without any specific quantum theory, one can with a | certain finite chance get over any hill no matter how big. The probability is there and is summed up in that constant distribution, whereas here we go back to principles of fundamental quantum theory. [128]

VON FOERSTER: That is a very important remark, because it shows that in any system of this sort – even if it is not constructed according to this fundamental quantum theoretical scheme – one can expect some noise-factors. In our case too it is possible to speak about a certain noise-level, if we consider these random self-impregnations. It is simple to compute the probability for such an event. Let us suppose a number of Z *mems* are involved in the game. Z/τ_s *mems* per second become self-impregnated, when τ_s is the waiting-time for self-impregnation. The total number of self-impregnated *mems* Z_s we obtain by multiplying the number of self-impregnations per second with the time of duration of an impregnated state. That is the lifetime τ_2, of the memory-carriers. Thus

$$Z_\sigma = Z\frac{\tau_2}{\tau_\sigma} \tag{18}$$

If we express now the two waiting times T_2, T_s involved in the equation above in terms according to formula (15), one obtains for the total number of self-impregnated *mems*

$$Z_s = Ze^{\frac{E_3 - E_2}{kT} - \frac{E_3 - E_1}{kT}} \tag{19}$$

Since Z is the total number of mems, the probability to meet a self-impregnated *mem* becomes

$$p_s = \frac{Z_s}{Z} = e - \frac{E_2 - E_1}{kT} \tag{20}$$

The energy-difference in the *e*-function is indicated as E in figure 11. It is an important magnitude, showing the amount of energy used or delivered every time a *mem* becomes impregnated or free. With the expression for the probability of a self-impregnation one can go into every communication-theory and can treat it as an expression for the noise-level of an information-source. Perhaps the whole thing describes a kind of »mental noise«.

I would like to draw your attention to the fact that our expressions for waiting times are always strongly dependent on temperature – increasing temperature decreases the lifetime of an impregnation. In other words, increasing temperature increases the [129] decay-constant. |

Since all these processes operate at body temperature, an increase of the body temperature during fever should affect the decay-constant. To check the order of magnitude of this effect, I computed from formula (15) the rate of change of the decay-constant with respect to its original value at regular temperature of 36.6° C. Using all constants describing the properties of the *mem*, I obtained the result shown in figure 12.

In this figure, λ means the decay-constant at any temperature, but λ_1 expresses the regular value at 36.6° C. The first question is: what significance has such an increasing decay-constant for the whole memorization-process, and how does it affect the remembrance? We can find the answer immediately if we go back to that point where we derived the dependence of the remembrance n_∞ of the ratio κ/λ. This function is shown in figure 6. An increase of the decay-constant λ would lower the ratio κ/λ and therefore also lower the remembrance n_∞. That is especially true for relatively small values of κ/λ. Since we assume that the whole remembrance is kept alive by values of κ/λ larger than one, it could easily happen that by an increasing value of the decay-[130] constant this | ratio would become smaller than one and that after a certain time the whole remembrance would vanish. That would be a horribly dangerous effect. One could lose the hard-earned knowledge of innumerable events. But we rarely make such observations. The only reason for the fact that we don't lose our memory during fever, must be that an increasing memorization compensates for an increasing decay-constant. In such a case the ratio of κ/λ would remain constant, however; the decay-constant would increase. If that is true, then the curve in figure 12 gives exactly the growth of the memorization. Intensive memorization can be observed, especially if unconscious memorization does not keep pace with an increased decay-constant. With increased temperature the patient starts with conscious memorization in all parts of the brain: the patient becomes delirious. On the other hand, this connection of the decay-constant with the temperature can sometimes be quite useful. Assume that in the very

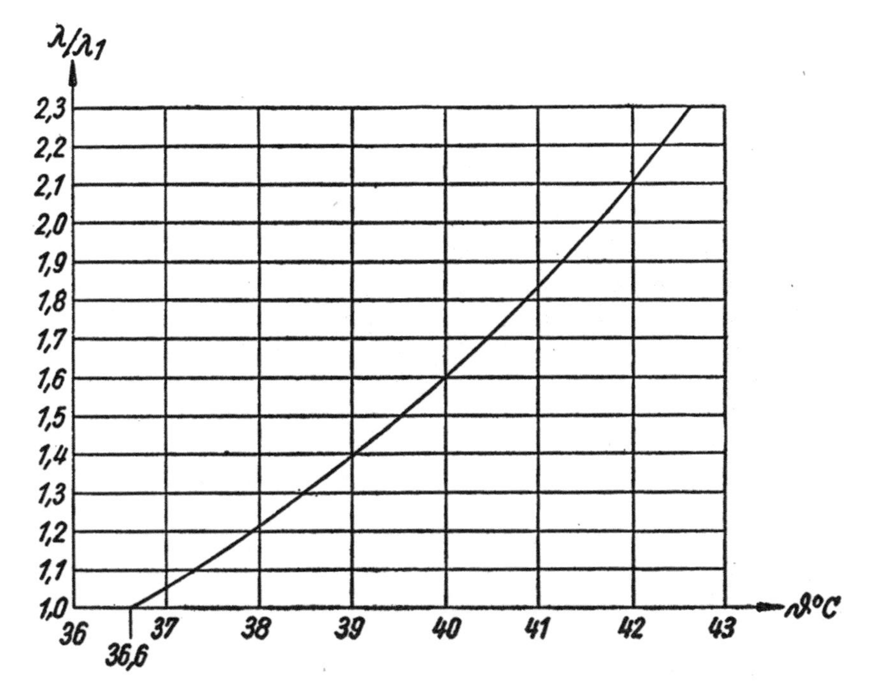

FIGURE 12. Rate of change of the decay-constant with increasing body-temperature ϑ

neighborhood of two regular pseudo-isomorphic states of *mems* there are still other possible energy-states, with only a slightly higher energy-threshold. Then an impregnated molecule remains much longer in this mis-impregnated state than in the regular one. If in a certain area such mis-impregnated molecules are accumulated, then information is stored but not available, because no transmission can take place for the lack of free molecules in this area. The whole area is blocked. A fever-therapy makes the breakdown to the regular energy-level more probable. The blocked area becomes free again for new impregnations and transmissions. The patient is able to remember again. Perhaps a shock-treatment will have the same result.

KUBIE: Could I ask a question? I am troubled by one thing. It seems to me that you are giving quantitative values here not to memory but to *availability* of memory data, which is not the same thing.

VON FOERSTER: I am sorry, I did not understand your question. What do you mean?

KUBIE: You are quantifying the accessibility of stored tracers, not the ability of the central nervous system to record organized impressions; because you are using as your index the rate at which we *forget*, as you called it. The rate of forgetting measures only the availability factor.

VON FOERSTER: You are perfectly right. This theory deals only with available items, because it is then possible to compare theoretical statements with experimental results. I would not hesitate | to extend this theory to the unavailable items also, but I don't [131] see any way to prove it.

KUBIE: Do you want discussion at this point?

MCCULLOCH: I think we better go ahead and finish because I want to get two or three quantities out here before us.

VON FOERSTER: I would like to display some figures, which can be compared with known neurological data.

First of all it is possible to give the ratio between the number of long-term carriers T_2, belonging to the brain and the number of short-term carriers T_1, which are identical with the elementary sensory receptors.

One obtains this ratio directly from the following differential equation:

$$\frac{dZ_2}{dt} = \frac{Z_1}{\tau} - \lambda_2 Z_2 \tag{21}$$

which simply states that the rate of change of the number Z_2 of long-term carriers of the type T_2 is equal to the number of incoming impressions per second transmitted by the sensory receptors T_1, minus the amount of decay of the long-term carriers. In the state of equilibrium the rate of change must be zero. Thus equation (21) becomes zero. Since the decay-constant is the reciprocal of the lifetime, one can write:

$$\frac{Z_2}{Z_1} = \frac{\tau_2}{\tau_1} \tag{22}$$

Both time-constants we know. τ_2 was of the order of 10^4 seconds and τ_1 of the order of 10^{-3}. Therefore the ratio between the two carrier types must be at least of the order of 10^7.

$$\frac{Z_2}{Z_1} = 10^7 \tag{23}$$

McCulloch: Before you go on, may I point out that from what is known of neuroanatomy you can guess your input group Z_1 as of the order of 10^7.

Von Foerster: I didn't know that. But this figure gives then immediately the number of at least 10^{14} memory-carriers.

McCulloch: Do you think that is high?

von Bonin: No, you have a hundred million rods and cones. You have, however, only a million fibres leading off from them.

McCulloch: A million per eye and a million for the rest of the body, 3×10^6.

Von Foerster: This figure of 10^7 sensory receptors interests me a great deal, because [132] it is possible to derive without further assump|tions directly from the theory, a figure which gives the maximal amount of sensory receptors distributed over the surface of the body. To have more receptors would be senseless, because information of those receptors would not be acknowledged. Let me show you that. Remember the first differential equation I wrote down expressing the total memorization process: it was equation number (6). This equation takes on immediately a physical meaning by multiplying the whole equation with ε, the amount of energy used every time to charge a carrier with an impression. Equation number (6) becomes then a power-balance.

$$\frac{d(\varepsilon N)}{dt} = -\lambda(\varepsilon N) + \frac{\mu}{\varepsilon}(\varepsilon N)[(\varepsilon N_0) - (\varepsilon N)] \tag{24}$$

where (εN) is the momentary, and (εN_0) is the initial value of the total energy involved in this process.

Since the structure of this equation is exactly the same as used previously, the result obtained is also the same. What I called κ, the memorization, becomes now a well-defined physical constant:

$$\kappa = \left(\frac{\nu N_0}{\varepsilon}\right)(\eta\varepsilon) \tag{25}$$

In this formula again, ν is the frequency of the scanning process, η the efficiency of transformation, ε the amount of energy necessary for the impregnation of one molecule and N_0 the number of carriers of the type T_2 involved. The first term of the right hand side of the equation above has the dimension of the reciprocal of an action. Since we know that the smallest quantity of an action is given by Planck's universal constant »h«, we can set up an upper limit of the number N_0 according to the following equations:

$$\frac{\varepsilon}{\nu N_0} \geq h \tag{26}$$

therefore

$$N_0 \leq \frac{\varepsilon}{\nu h} \tag{27}$$

The two values ε and h are known. The frequency of the scanning-process I assume to be one cycle per second. With these values

$$\nu \sim 1 \text{ sec}^{-1}$$
$$\varepsilon \sim 10^{-12} \text{ erg}$$
$$h \sim 10^{-27} \text{ erg. sec}$$

one obtains for the maximum value of N_0:

$$N_0 \sim 10^{15}$$

That should be the largest number of carriers type T_2 operating on mutual transmissions. It must be identical with the number | Z_2 we obtained before. Since the ratio of [133] long-term carriers and short-term carriers is given by the reciprocal of their time-constants and is 10^7, the maximal amount of sensory receptors having active connection with the memorization group must then be:

$$Z_1 = Z_2\, 10^{-7} = 10^8 \tag{28}$$

That is not too far away from what you said before, especially if you consider the figure 10^8 as a maximum value.

McCulloch: Would you say just two more things with respect to the energy requirements? First, can the total amount of energy for the process of memorizing be computed?

Von Foerster: Yes, it is possible to do so.

McCulloch: Would you say how that comes out?

Von Foerster: Certainly. The idea is about as follows: Let's go back to equation number (24) which gives the power-balance of the total memorization process. It is possible to extend this equation with a positive term which takes into consideration the constant flow of incoming informations. It is interesting to see that the solution of this new differential equation does not run to infinity, even if the time goes to infinity, but approaches a certain asymptotic value. The power required to keep this state stable is expressed by the following formula:

$$W_2 = A.\varepsilon \frac{Z_2}{\tau_2} \cdot \frac{\tau_2}{\tau_1} = A.\varepsilon \frac{Z_1}{\tau_1} \cdot \frac{\tau_2}{\tau_1} \tag{29}$$

In this equation I used the identity $Z_2/\tau_2 = Z_1/\tau_1$ to obtain more reliable values. The factor A changes the units »electron-volts« in »watt seconds«. If we replace these symbols by their numerical values we obtain:

$$W_2 = 1.5 \times 10^{-19} \times 1 \times \frac{10^7}{10^{-3}} \times \frac{10^4}{10^{-3}} \tag{30}$$

which gives about 10^{-2} watts.

McCulloch: Remembering that the human brain is about 2.4 watts; this is less than half of one percent.

VON FOERSTER: Before I conclude I would like to say something about the way in which the energy may dissipate whenever such a molecule breaks down. According to Debye-Hueckel's theory of molecules it is possible that this energy dissipates either in form of heat or in an electromagnetic radiation. The wave-length of this radiation is automatically given by the fundamental relation

$$\lambda = \frac{ch}{\varepsilon} \tag{31}$$

[134] | from which we obtain for λ about 28,000 angstrom units, which is a large infrared radiation.

MCCULLOCH: These outer ranges in which coupling takes place between enzymes and substrates again fall into the range of our protein fringe.

VON FOERSTER: May I conclude now with a few more words? I started at first with an introduction of six more or less phenomenological ideas:

1. The elementary impression.
2. The carrier of the elementary impression.
3. The decay of these carriers.
4. The memorization as power.
5. The transmission as process.
6. The transmission as selection.

After this quantum mechanical attempt it seems to be possible to give four of them a plausible physical picture. These items expressed in a quantum mechanical language would spell:

1. Lifted energy-level of a *mem*.
2. The *mem*, a complex molecule, stable in close energy-levels. Perhaps a protein-molecule.
3. The breakdown of this lifted energy-level, or »tunnel effect«.
4. The amount of energy necessary to lift up such a molecule from its lower state to the next higher state.

The transmission as a process can only be explained in neurological terms. I am sorry, but my knowledge is too scant to give such a picture. Finally, item 6 – the transmission as selection – cannot be explained in this theory. By this I mean, why do we select from the continuous stream of information exactly that which we want to remember and forget other things we are not interested in? You will understand this when I say that I was only trying to get a picture of the element of this complicated process, not a picture of the process itself. If it is permitted to compare this with mathematics, I was trying here to make a number-theory – but the selection-problem belongs to the theory of functions.

DISCUSSION:

BROSIN: May I state my confusion as clearly as I can, so that the five psychologists and [135] several neurologists present can translate this for me? |

My first point is this. Dr. Gerard has made clear to us that we can deal with behavior on a number of levels: the neuronal, psychological, and social. Dr. Kubie's presentation, admittedly dealing with social and probably relatively with psychological in the true sense, aroused Dr. Klüver to say, »Now we can begin to talk about a psychological problem.« I would really like to reiterate that now we are coming up from the other end of the spectrum. I would like some help on the following: one of the basic assumptions of Dr. Von Foerster's daring hypothesis is that he is dealing with discrete elements. In this group I think we should examine the nature of that. I am not questioning the physical structure, nor the mathematical skills, nor the ingenuity, but simply asking what does this have to do with the experimental fact? Klüver I think, above all people, could give us the exposition, the perception. The elements of perception and of learning certainly do not have the discrete elemental quality which is here postulated. My own youth included graduate seminar work, over a year or so, with V. Henman and experience with Clark Hall on learning nonsense syllables. Since 1924-27 we have come to feel that nonsense syllables are a system of artefacts.

After listening to Dr. Kubie, I think it will be apparent that whatever state of readiness the nervous system is in during the embryonal state, certainly at the moment of birth we can invoke the picture, psychologically as well as physically, which Dr. Gerard gave us of the brain – not as a dead piece of material in the pathological laboratory but a quivering, active metabolizing energy system with a great deal of variability from day one to day two. What I note here is the need for organization versus the discrete element principle. The weakness in the nonsense syllables is that they may represent learning in certain psychological laboratories which has no correspondence to any learning from day one in the human at whatever level of abstraction or concreteness you want to carry it out.

VON FOERSTER: You understand why I took the learning of nonsense syllables for the starting point of my theory. I wanted a group of items with a minimum of interrelation and almost no association with other well-remembered terms. In other words, I wanted something which resembled as much as possible a real »ensemble«. I followed more or less the physical methodology, starting with the simplest assumptions possible, and then developing the theory as far as possible. Therefore I started with the nonsense syllable experiments. | [136]

BROSIN: I appreciate that. A great many decades have been spent utilizing the same technique. Will you please explain to me the relevance of this mathematical derivation to human thinking as I see it in people? It may or may not be possible to explain it. Understand I respect the demonstration but I am asking to be shown.

The next thing I would like to ask about this is: isn't the bringing in of the quantum freedom of the molecule dragging in, as it were, the free will demonstration? I am throwing this in for free. McCulloch I am sure will enjoy it, and I won't try to pursue that.

May I point out that in the examples you used, for instance in the temperature, it has been my fortune for some years not only to use a lot of malaria in the therapy of paretic patients but also to introduce the Kettering hotbox? I ran a couple of them for a whole year and I saw more deliria than I will see for ten lifetimes – than most doctors ever see. I have also used insulin. You see the molecular movement again is not speaking about organization which I think we have to insist on in view of everything we know about memory, at least from the psychopathological examples. That kind of memory, dependent on lowerings, is not lost even under extreme physical-chemical stresses. The apparent accidents and inadvertences of the deliria are just as determined by the past experience of an individual as any other psychological phenomenon. To that extent I am a simple determinist compared to the position which requires the degrees of freedom it seems you are trying to invoke. The man with pneumonia or coming out of the insulin, or the man with the blow on the head, or the fugue, which is the immediate example, all show a high fidelity to organizing principles of some sort, the principles of the central nervous system in a state of becomingness, in states of transition, which are axially coordinated. So again I would ask the neurophysiologists and the psychologists to translate that to me.

MCCULLOCH: You want to take it up Dr. Gerard? I suggest that you take it up for the reason that you, more than anyone else, have thrown the monkey wrench today into the classic picture of the frozen anatomy of the ancient synaptologist.

GERARD: I cannot do anything to clarify Dr. Brosin – he was perfectly clear. I shared the same difficulties with Dr. Von Foerster's exposition. I would like to ask him a question or two for my own understanding of his derivation. This, I would add, struck me as a very ingenious development of a number of concepts, put together in a most stimulating fashion. Right at the start, though, I am not sure I understood some of the [137] rules of your game. Were you | assuming that the half-life of each long-enduring memory point is the same in all cases or that there is a Gaussian distribution of half-lives?

VON FOERSTER: When I was in the first state of this theory I was also almost sure that there must be a Gaussian distribution of the forgetting-coefficients. But the whole thing didn't work. Later I had the idea of the transmission and the quantum mechanical explanation of the behavior of the carriers. Although the half-lifetime or decay-constant of molecule is a well-defined constant, it is itself an expression of the probability of a breakdown. Therefore I started at first with only one constant – and it seemed to be sufficient.

VON BONIN: If it is a probability then it is that of half of the molecules disintegrated.

GERARD: But it could be a different coefficient in each case.

VON FOERSTER: Certainly, that could be. Especially one could be allured to assume much longer decay-constants in deeper brain-regions for deeply engraved impressions. But I think that is not reasonable because in such an area the probability of transmission is almost zero because of the lack of free carriers. And only transmissions are observable. The one parameter which varies, I believe, is the memorization constant κ. The decay-constant is a fixed value, characterizing an important property of these molecules. It differs perhaps only in different animals – or it is a unique biological constant. I don't know – but it is an interesting question.

GERARD: That is what I thought; then you settled on one?

VON FOERSTER: Yes, that's all right. It turned out to be a sufficient assumption.

GERARD: Another point. I think you made the assumption – and I may have misunderstood you – that when you transfer the remembered element from point 1 to point 2 you restore to it by that transfer the full half-time that was initially available. In other words, though you may be near the end of the half-time in point A, by moving it to point B you come back to the original probability of the half-time. I don't see why that should be and it seems to me rather contrary to your energy-level idea.

MCCULLOCH: No, may I put it crudely: all cats die. If no cats had kittens, pretty soon we have no cats. The one that may die at any time would be the cue for the kicking up of another one. It is not the source of energy for the kickup of the second one.

GERARD: How does this transmission come about mechanistically? What is your pic- [138] ture? |

VON FOERSTER: I am sorry, I have not yet a picture. I am not able to give a theory of transmission, because I think it can only be done in neurological terms. My knowledge in these things is too small to explain that.

GERARD: Maybe mine is too great, because I found myself blocking terribly at that point. I just did not see how it could be made to work for any kind of a picture we have of the nervous system.

BATESON: I still don't understand the answer to Dr. Gerard's previous question. If I have 100 molecules with a half-life of a day, and a day goes by, I have 50 left, right? I take those 50 and I make them into another 50, right? These will have the same half-life and at the end of the day I shall expect to have about 25 left.

VON FOERSTER: Oh yes, I see the point! Perhaps I didn't make myself clear enough. Your picture would be perfectly all right if the transmission of an impression from molecule A to molecule B would destroy the impregnation of molecule A. As you said, the transmission of 50 molecules into 50 other ones would never help to escape the steady decay. But my picture of a transmission of molecule A to molecule B does not destroy the impregnation of A. After the transmission I have two impregnated molecules instead of one. They are A *and* B. Take your 50 molecules and transmit them into another 50 then you have 100 again and the game starts as it began. From this picture you can easily see, if the transmission is done a little more frequently than the molecules decay, the amount of impregnated molecules can increase.

STROUD: The process of transmission does not terminate molecule A. It might be better to call it a process of contagion.

MCCULLOCH: Or a process of reproduction.

KLÜVER: In connection with Dr. Brosin's remarks, it is of interest that the learning of nonsense syllables started in Germany. There is no doubt that German lends itself more readily to the construction of series of nonsense syllables than English. Müller and Pilzecker, about fifty years ago, went far beyond Ebbinghaus in trying to construct nonsense syllables free from »associations« and to devise procedures preventing the intrusion of »meanings.« Nevertheless, really pure »nonsense,« that is, material utterly and completely divorced from meanings and associations, seems to have remained an ideal that has never been realized. At best, you start with levels of meaninglessness or, if you wish, levels of meaningfulness. As I understand Dr. Von Foerster, the material he has chosen as a point of departure for his considerations was presumably meaningless and free from cross-connections. | [139]

MCCULLOCH: It is perfectly clear that if there were in such material many items that were very meaningful, and therefore in your sense would have been disproportionately fixed in that memory, then you would not have obtained as good a substantiation as you have. It must be that the items considered were fairly equal in their meaningfulness or non-meaningfulness, or you could not hit the curve within 5 percent on those assumptions.

KLÜVER: It is possible that the various items – or at least some of the items – may have been equal in meaningfulness. In fact, it is conceivable that the meaningfulness of some items may have been on quite a high level.

MCCULLOCH: Whatever they are, they are about equal. That is all he needs for that.

SAVAGE: Are you people really impressed with a fitting curve of five percent? It seems to me that any bent line that starts at one and has the right asymptote to infinity will fit the other one. I don't mean this as an invidious remark but I don't think that this is a remarkable fit. It seems to me that bent lines are much the same; if you allow five percent they all look alike.

VON FOERSTER: I thought the same at first and I was rather sure that a relatively simple, constructed function, monotonously decreasing, would fit Ebbinghaus' forgetting-curve. But I discovered again that mathematics is a wonderful and precise tool. All the functions I tried – and there were a lot – had their own personalities and ran quite apart from the measured curve. Only an increasing amount of parameters brought both curves closer and closer. That is not very significant, however. Suppose you have 10 good measured points and you want to connect them by a mathematical function. You can set up a function with 10 parameters which run exactly through your points and the average error is zero. Since no good and reasonable theory stands behind these parameters, the whole thing doesn't mean very much. Speaking of accuracy we have to distinguish two different questions: one is concerned with the accuracy of the measuring itself. In a physics laboratory an average electrical volt or ammeter has about 3 percent accuracy. A reliable accuracy of 1% takes rather a high effort. Whether or not repeated psychological experiments, like the nonsense syllable learning, give results within a 5 percent deviation, I cannot decide. On the other hand, if I have two theories and their different results run closer together than the average error of the measuring from an experimental point of view there is no possibility of deciding which | theory is the right one. I would prefer the theory, however, having fewer assumptions and needing less parameters.

[140]

In our case where we have two simple and reasonable assumptions, the decay and the memorization, which lead to two parameters, the decay constant _ and the memorization constant κ, and where the computed curve runs close within 5 percent of a measured curve, then I would say it seems that the principal idea is not absolutely wrong.

SAVAGE: It does not show it is wrong.

VON FOERSTER: And it does not show that it is right either but it has perhaps a certain amount of probability; that's all we can say about it.

KLÜVER: When Fechner was still a physicist in Leipzig some of his experiments required observing the movements of a black thread in front of a white scale with black markings and numbers. After carrying on observations of this kind for several hours he found, by closing his eyes or by merely looking at a dark field, that the moving thread as well as the scale with its markings and numbers appeared spontaneously in front of him. The thread and the markings were always clearly distinguishable while the numbers were not quite distinct enough for recognizing their values. Even after 24 hours he was able to see such phenomena whenever he closed his eyes and without any effort on his part to restore the original impression. As you know, the phenomena described by Fechner are commonly referred to as phenomena of *Sinnengedächtnis*. It is worth emphasizing that the conditions under which such phenomena appear can be rather definitely specified and are far better known than for other subjective phenomena.

I should like to know how your theory would deal with the *Sinnengedächtnis*. Would you treat phenomena of this kind in the same way as hallucinations?

VON FOERSTER: In exactly the same way. I wrote down for hallucination the transmission formula

$$T_2 \rightarrow U \rightarrow T_1 \rightarrow C \rightarrow T_2 \qquad (13)$$

which only expresses the rare event where the input-channel is used for a memory transformation. In the case of *Sinnengedächtnis*, where one observes the same process for hours and hours, a relatively small part of the brain will be overloaded with impregnation. In such an overcrowded area transmissions cannot easily be done for the lack of free carriers. In moments of relaxation, where the input-channel is not used as often, the transmission makes unconscious use of this channel – first part of formula (13) – and trans|mits immediately back to other free carriers T_2 in other regions of the brain. Since a transmission $T_1 \rightarrow C \rightarrow T_2$ gives the sensation of an observation, this kind of memorization reveals pictures very similar to observed ones.

[141]

SAVAGE: I would voice the objection everyone else has about organization. A dream, for example, an hallucination is surely not a single quantum act. There is too much to it.

VON FOERSTER: It must not be a single quantum act. It can be a team-play of, say, 10^6 quantum acts. And that is a lot of information.

SAVAGE: Isn't it a hodge-podge of acts?

VON FOERSTER: Not necessarily. If you assume that within a small area associated bits of information are stored, the game becomes quite organized. If it is possible to control, e.g., processes going on in one cubic millimeter of the brain, then it is possible to watch the transmission-game of about 10^9 carriers[.] On the other hand, it seems to me quite reasonable to assume that bits of information which have any relation are also located very close to each other. Before you have the relation you have to go through a hard and painful thinking process to set up the relation for the long-distance transmission one has to carry out. This transmission has to be done in a relatively empty area. But immediately after this transmission, both items are put together and start with transmissions in their neighborhood producing a slowly growing cloud of connected information. I think we have the »same« item stored many times in our memory, always with different connections, e.g., black as the color of ink, black as a spoken sound or a written word, black as the opposite of white – it always is a different »black«, different too from the idea of »black«, which is a logical operator with certain values.

BROSIN: This might be the proper time to bring up the brain localization question. Dr. Lashley's work, with which you are acquainted, speaks against it, and more picturesquely, perhaps, also does the brain of Pasteur, who unfortunately at the height of his powers suffered a cerebral accident. Upon his death an autopsy was done and it was discovered that he had literally one-half a brain, because the other half was atrophied. Pasteur did magnificent work during his later life in the field of abstraction, with all that is implied in that statement, and apparently did not lose the information which he acquired the first forty-some years. The evidence that Dr. Lashley has worked over so extremely well for the last 25 years and the example of Pasteur mitigates against the proposition of finite locale. | [142]

STROUD: I wonder about that.

MCCULLOCH: I am puzzled about that, too.

STROUD: These are enormously fine grain units of information. You could lose say 75 percent of them and still define all the points necessary. To define the perimeter of a circle when you by the dynamics of remembering or imagining, or dreaming, are adding to this system, you are also in the meantime using some elements which have been quiescent some place. You can lose a great percentage of the points on a circle and yet you would be scarcely able to detect the difference between the reproduced and the original because they are enormously fine-grained. An abstraction as complicated or as general as a word must necessarily stand for an enormous number of these little elements.

BROSIN: I know it. I am still trying as I suppose are a lot of others –

STROUD: You can hardly destroy anything.

BROSIN: Is this a classic argument? With five psychologists you ought to get some support. With the loss of frontal lobe function we know something is gone. I have been trying to look at them for years but even now it is difficult to define the nature of the loss.

STROUD: The loss does not have to be any loss of any points of information. You may have lost a few points of information in losing 25 percent of your brain. What might have been more interesting was the loss of the operations that you could perform on these points.

MCCULLOCH: There is a great difficulty here as I see it. One thing is perfectly obvious and that is if one does learn or remember some fact somewhere inside him, something is different from what it was before. Now the question is what sort of a thing are we going to look for? Are we going to look for new synapses? Are we going to look for the disappearance of old synapses? Are we going to look for changes in the protein structure of the individual neurons? Are we going to look for them uniformly distributed through the brain? Obviously not in our sensory input proper. What sort of a thing are we going to look for? So far as I can see the most interesting thing that has come out on memory from all this is a factor somewhere around 10^{13} to 10^{15} for possible storage. If one takes the N terms of items, I don't care where or what one supposes these items to be, in their general number they must be of this order of magnitude. It seems to me it would be a far more profitable thing if we are interested in the problem of what the crucial item of memory, is to ask ourselves a question: what [143] would denaturing | protein do? If one could denature the proteins, one might very easily lose these things en masse. If the proteins of one part of the brain – let us say the mid-brain or the upper end of the mid-brain are denatured as one does with alcohol in Korsakoff's psychosis, even before the third nucleus breaks down, when one gets the encephalitis of Wernicke, there is very obviously a loss of ability to make new traces. I am not at all sure that we have been looking in the right parts of the brain. The number of protein molecules that could be involved in the reticular formation of the mid-brain is enormous. It belongs to this order of magnitude. My feeling is we know no one lesion that wipes out memory unless perhaps it be mid-brain. We know nothing yet as to which structure to observe, – whether we are to look at the coarse grain of the synapse or the fine grains of protein. All we do know is that we have to make an assumption of some structural component which is something like 10^{13} to 10^{15}. I think that is what keeps cropping up in my mind. I think we have two methods of recall, and of recognition and we should keep in mind both of these possibilities when we try to construct any theory.

FREMONT-SMITH: What are your basic assumptions?

MCCULLOCH: Here are the basic assumptions, as to rough sizes of organisms, numbers of inputs, etc. Here are psychological experiments indicative of 10 per second as roughly reasonable value for input, and the number of items may be up to a thousand at any one such time.

GERARD: But nobody for a moment thinks that you remember all that thousand per second for your whole life.

SAVAGE: You can hypnotize them and apparently have them think of them at any moment.

GERARD: Even the hypnotic recall of detail does not demand that.

MCCULLOCH: One could never investigate this because it would take more than a life to express it.

GERARD: That is what I think.

SAVAGE: That is not an argument. You could sample if you wanted to investigate and, if you could get access to places, it might not be beyond the reach of principle.

MCCULLOCH: We get such little glimpses in a man who given at the rate of 10 per second and pause, a series of 100 items in semaphore, every one of which is at least five decisions, is able, after the pause to review the entire series visually and see them all again.

GERARD: Then there is this peculiar phenomenon of the »memory fiend«. He looks [144] quickly through a just-published issue of Time or | a local newspaper and then tells exactly where on any page any word occurs; certainly involving many tens of thousands of memory items. Then he has to make a decision that he will remember or he won't remember. If not careful, he gets his memory permanently cluttered up with these trivia. If he decides he is not going to remember after the performance and then does forget it all, what does that mean neurophysiologically?

MCCULLOCH: The question is: does he forget it or does he ignore and leave it merely kicking around with a retrace.

GERARD: He successfully represses.

McCulloch: That is one of the questions we must go into tomorrow if we are going to make sense out of a neurosis.

von Bonin: It seems that color agnosia is one of the few problems where we know something about the location of these things; it is bound up with the temporal lobe. It has been mentioned a few times here that we can lose large amounts of brain and probably be just as well off as we were before. It seems to me that should very definitely show we cannot assume the individual neuron to be the unit of N, but we have to think of smaller units, because we have to accept the fact that the brain never functions with more than about 20 or 30 percent of its efficiency.

Mettler: Does that mean that the fine grain is distributed over the wide area of cortex?

Stroud: Our notions of long-term memory may have been asking too much of our molecular trace. The trace has to be tiny like a reflector floating around some place suitable, which is available, so if the brain goes into this memory operation, it can be shall we say, »Seen«? When you try to remember something you have seen you are quite self-conscious about the change in operations. Thus, for example, when I want to remember or imagine something I have seen or might see, I cannot see what I am looking at. When I am imagining or remembering, I have to do it in detail. Our memory trace need be nothing more than something available to be »seen« when we are operating our internal »seeing« machinery, as it were, on the past. And our little traces need be nothing more than the little points of reflection at which we look internally. Perhaps we do not need nearly as much machinery for this long-term stuff as we thought. We may need only very tiny little reflectors which somehow or other can become a stimulus pattern which is available for this particular mode of operation of our very ordinary thinking, seeing, and hearing machinery. This particular pattern of reflectors is what I see as it were with my internal eyes just as | what I see when I look [145] at a store window, is a pattern on the retinal mosaic.

von Bonin: The famous example is that of the school boy who does not remember a poem. Give him the first two or three words and he goes rattling off the whole thing. You probably have that sort of thing in mind.

[146]

POSSIBLE MECHANISMS OF RECALL AND RECOGNITION

DISCUSSION:

FREMONT-SMITH: I was particularly thrilled with yesterday's meeting. We are trying to do something about communication across the whole range of the scientific disciplines and in order to do that we have to deal with data from the extreme limits of our divergent or disparate disciplines. Dr. Stroud spoke about the problems that arose in dealing with a situation where a man was the connecting link, if you like, between two machines about which a great deal of information was available. It became important then to ask some pertinent questions: what kind of a machine is this man, or perhaps what kind of a »guy« is this human machine? There may be two ways of saying the thing and perhaps we have to say them both. That in a sense illustrates our problem. We have a very human problem of our own which exists because we, ourselves, who are studying this situation are also examples of this »man-guy« machine who sits, in that instance, between two machines, and in this instance between two other »man-guy« machines. We are here with varying motivations, in which the element of eagerness operates vividly. That very eagerness itself introduces certain problems in communication. We have the eagerness to make statements. This sometimes becomes really limited to making statements *at* the person rather than being conversant with or interested in the receiving apparatus towards which we are throwing these statements. I say »throwing« advisedly because a delicate machine sometimes get injured if you throw things at it or if your aim is not very good your statement may miss its mark; or it may be the kind of machine that knows how to dodge and let the bullet go by. These particular machines have all kinds of distorting lenses in them. Their spectra are not complete, and if we want to really get into them we have to put some attention on the receiving sets and and[!] not merely on the power of our transmitting sets. That is one element of our difficulty. We come here presumably not only with an eagerness to get our statements into those other machines but also perhaps to receive some information. Therefore we ought to give a considerable amount of attention to what the Harvard School of Business Administration calls the »listening technique,« [147] | and really to listen with a little bit of self-consciousness as to what and how much we are hearing.

Yesterday some of the questions that were asked of Dr. Kubie made it perfectly clear that the questioner had a blind spot for some of the very clear and explicit statements which Dr. Kubie had underlined in his remarks. These questioners evidently had not heard them in any conceptual sense. I think we need to try to be aware of our own blind spots in a group which really has as its focus communication.

If you are examining these two machines that Dr. Stroud is interested in, obviously you concern yourself with those people who have had the greatest amount of experience with those machines, both inside and outside, and with the operations of them. Similarly, with the human machine, if you ask who are the people who have had the greatest experience with the human machine, you have to say there is no one comparable to the psychoanalysts. The analysts themselves have certain blind spots to be sure, certain narrow perspectives of their own but these observers of the human machine are in a completely different category from any other scientists as far as I know in the sense that they spend five hours a week for as long as two years observing one person. There isn't anything comparable in terms of observing the behavior of the same individual through time. There is not anything that even approaches it. Therefore, I think I may be wrong – I see Dr. Liddell raises his eyebrows.

LIDDELL: I am thinking of fields of behavior.

FREMONT-SMITH: There are people who have made cross-sectional studies of thousands of machines a little bit. That would not help you very much.

KLÜVER: I hate to mention that individual monkeys have been studied for more than ten years, every day for several hours, in my laboratory.

BROSIN: That is the only exception.

FREMONT-SMITH: I am glad that was brought out.

LIDDELL: Now you see why I raised my eyebrows.

FREMONT-SMITH: Perhaps this was one of my blind spots!

One other point, probably man is never only between the two machines. Certainly he is never only in between two machines when you are studying him because you are the other man who is making an input into the man. You are studying and changing his relationship to the machines by virtue of the fact that you are studying him. You can reduce the effect so that it does not apply | to what you are studying but you have [148] to be quite conscious of your influence on the situation, on the phenomena which are observed. This is a basic in physics, – the observation of a phenomenon modifies the phenomenon. I want to bring that out because it seems to me important if we are going to try to bring the knowledge, extraordinary ingenuity and wisdom, coming from the mathematicians, physicists and the engineers, to bear on the human problem. These scientists must be exposed thoroughly to the data of human behavior and it is not their mathematical and engineering data to which I refer. It is the data on human behavior that the students of human behavior have gathered. That data has its limitations but it has to be looked at by them. Maybe the engineers and mathematicians will later show us how to get better data.

It is also essential, as Dr. Abramson pointed out, that the social sciences have some vocabulary, that is, an alphabet, a common denominator which makes sense to the physicists and engineers. Therefore he has suggested »dimensional analysis,« even though the language of dimensional analysis may not yet serve as a means of communication between physical and social sciences. We all talk English. That is not good enough. We have to learn to talk at some level of common science and probably it ought to be at such a level as dimensional analysis if we are going to start from something about which the social scientists and physicists can say, »Yes, we agree on this. We are talking the same language. How can we build up from there?«

MEAD: May I say just one thing? If we could somehow work it out, the reverse position in this group, so it would be in between the psychologists and the physicists, that would be fine. I think when somebody writes an equation on the board followed by more and more difficult ones, everybody in the room knows when they get left. There are some people like me who get left very soon, and there are some people who never get left. However, almost everybody in the room knows that that point in mathematics is one which I do not understand. I will not understand it in that language. I have to wait until somebody has said it in English, or with a different figure of speech, or has related it to my data before I can understand it. The same thing is equally true of psychology; but I do not see that same response of getting left on the face of everybody in the room when the psychological statement is made. Most people think because they have bodies and nervous systems they know something about psychology. They really know no more than the person, who does not understand mathematics, knows about | mathematics and for purposes of this discussion at least we would get much further [149] systematically if we could assume that.

ABRAMSON: There is one mechanical difference between the person operating in the psychological area and the person operating in the mathematical area. The person in the psychological area may have psychological difficulties which are difficult to explain. The person in the mathematical area it is true may also have psychological difficulties but in general in mathematics it is more of a mechanical difficulty which has to be overcome. As a matter of fact, I would say that mathematics is *the* language of communication. There is no reason that this most precise language of communication cannot be adapted to its primary function. I have never met a good mathematician who could not explain to me what he meant by his own equations. But to explain emotions by means of mathematical concepts is as futile as attempting to describe unweighable quantities by means of a precision balance.

KUBIE: May I throw in one comment, also picking up what you said, Dr. Fremont-Smith? It is true that a human being is never set just between two machines or between two human beings. Let me carry your phantasy a little further. Let us suppose we have the room full of perfect mechanical robots, every one a facsimile of a human being. Theoretically every machine of a certain type and kind is substitutable for another machine. Theoretically if you buy a Ford you can have the motor taken out and put in a new motor. They won't be absolutely identical but within a range of variation they are interchangeable. Theoretically these robots should be interchangeable from one to another; but the important difference is that we endow human beings with special significance partly built on experiences, partly built on phantasies, partly built on strange little associations to minor aspects of those individuals, so that we make relatively interchangeable people non-interchangeable. The problem of communication is not only a problem of communication of ideas across an intangible barrier. This comes into the study of the human being, and is a process which has to be understood in any effort to understand the mechanics of how we operate.

FREMONT-SMITH: It seems to me if we are going to look forward to a time when there is going to be really fruitful communication across this range, all the social scientists will have to receive a minimum of mathematical dimensional analysis training so they can talk freely, at least at a certain level, with the mathematicians and engineers. Engineers and mathematicians must also have a [150] | minimal understanding of psychodynamics so they can talk freely with the psychoanalysts and with investigators working on experimental neuroses. If we don't have that dual level of communication, it is difficult. So I think both sides will need new and additional training. I cannot deal with dimensional analysis and I have, as Dr. Abramson says, a certain psychological block about getting to work on it. I suspect that this operates at the other level too. We have here Dr. Abramson, who has had a really extensive training in physics, physical chemistry and mathematics, and also a training analysis in psychoanalysis. This is a rare combination which should not be so rare and maybe I am overemphasizing the uniqueness.

MCCULLOCH: I would like to pick up the ball myself and start it rolling if I may. In the first place, I would like out of my prejudices to answer Larry Kubie a bit. I am a Scot. I think like most all Scots I fall in love with machines and particular machines, and I am a sailor, and I know that almost every sailor falls in love with a ship, and it becomes as unique as a person, identified in the same manner as our fellow man identifies us. I don't think any greater difficulty rests in the fact that the other machine is a man instead of being made out of wheels or out of canvas.

KUBIE: It is only a degree.

MCCULLOCH: It is not a fundamental difficulty.

STROUD: A machine in my laboratory has a personal label.

MEAD: But the ship does not fall in love with you.

MCCULLOCH: I am not so sure.

WIENER: Tell me this? Is the observer at a greater disadvantage as he observes something like himself? With these resonant phenomena in physics if you had such an instrument you would get in a mess. You cut them to a minimum and try to observe a particle by a light which is not of the frequency of the natural vibrations of the body.

STROUD: May I echo that thoroughly. The human observer is absolutely the most untrustworthy device of my laboratory.

KUBIE: That is one of the reasons for some of the details of analytic technique which make the analyst seem inhuman. He can only be effective by achieving a detached position.

FREMONT-SMITH: May I comment, first that, as Professor Wiener says, in the machine you try to cut reverberations to a minimum, but they can never be wholly eliminated. Similarly with the human machine, such reverberations are crucial and a major cause of unreliability. The psychoanalyst, as a rigid two year part of his | technical training, [151] learns to observe himself, to become conscious of areas within himself which are peculiarly sensitive to such reverberation, and, by making such reverberation conscious, he is able to cut it down to a considerable degree and evaluate the effect of the residual. No other scientist undergoes such self-study as part of his training. Yet in these conferences and at various scientific meetings we see ample evidence of vigorous reverberations which grossly distort the scientist's observations and his communication with other scientists. Such reverberations are rarely seen by the individual in process of reverberation, although they are frequently obvious to the onlooker. A major and unique virtue of the psychoanalyst is his learned capacity to be aware of such interpersonal reverberations and to dampen his own reaction to them. The psychoanalysts, however, too often have meager experience with scientific thinking of the physical sciences, and because of their preoccupation with human phenomena which often cannot be subjected to scientific control, are prone to forget the meaning of such terms as »control« and »basic assumptions«.

I strongly believe that when a larger number of physicists, engineers, chemists and mathematicians have had the self-revealing experience of psychoanalytic training, and when more psychoanalysts continue to participate in scientific work other than analysis, i.e. keep alive their earlier acquaintance with the basic sciences, both the basic disciplines and psychoanalysis will improve, and above all, communication between these now widely divergent sciences will become increasingly possible. The development of effective communication across the scientific disciplines is perhaps the most urgent need of our era. As Dr. Abramson said, until the nuclear physicists, who through their science have developed the ultimate weapons of hostility, can communicate with the psychoanalysts, who through their science have developed the greatest understanding of the nature and control of hostility – until both these groups of scientists can communicate with, i.e. »make sense to« those who are responsible for the administration of human affairs, there will be no hope of applying the principles of science and logic to the problems of social behavior and world peace. I have emphasized nuclear physicists and psychoanalysts as representing perhaps extreme limits in the spectrum of the sciences, but would give equal emphasis to all the other branches of the basic and applied physical and social sciences. This conference group covers nearly the whole range. May I ask a question? Could you | have a radar machine observe? Could you use a microscope to observe a microscope? [152]

STROUD: If you let me shift wide enough apart you can.

FREMONT-SMITH: That is hard to do with the human.

STROUD: The human being is produced on a single production line as it were but with wider variation than one-fourth. Still it is not possible to observe one man by another man and get anything but pure nonsense.

BATESON: I would like to mention that Stroud said yesterday that if you set the experiment up to ask the human being the right question you get very, very close fits to your graphs.

STROUD: If you look at my experiment, I am just standing around in a room watching the machines running. The ideal, which the division chief has okayed, is one on which you worked for months and months, building up the equipment and programming an experiment. In the end you picked the subject up by the collar, put him in and the experimental situation and experiment runs itself off. At the other end results come out with all computations untouched by human hands because in all my processes I dare not use human action because of its tendency to a quantity of uncheckable errors. I can ask the right question but it is my robot friend who must necessarily ask it in practice. I cannot, nor can any other man in my laboratory, observe a human being closely enough or rapidly enough.

FREMONT-SMITH: Not in your laboratory. I think it is fair to say then that you might give some consideration to those people who have spent a lifetime in observing human beings and not trying to interact.

STROUD: I have the greatest respect for a man who takes on a job like that working barehanded.

WIENER: The person's difficulty is there all the time. I have not any doubt that the psychoanalyst as a human being has gone as far as anyone in missing the resonants.

STROUD: Between analysts and theologians, I can scarcely imagine a human instrument being put to better use or to greater advantage. It might be true of them too. I don't know them that well, but of the groups I do know the analysts and the theologians have done about all that it is possible to do by way of setting one human being to operate upon and observe another.

FREMONT-SMITH: You say it is impossible to get a man to observe another man and I [153] am saying that until you try those who have | spent a lifetime in learning how to observe another man you won't know what the limitations are.

STROUD: The tragedy of all of the earliest forms of psychology – I think Dr. Klüver will admit this – that these men had the right problems, they worked desperately hard and yet they failed chiefly because of the limits of their instrumentation, most of all because of the limitations of the human being as an observer. When I got to graduate school in psychology I found that many of these would-be psychologists were of the most sophisticated varieties and were showing their sophistication by making light of the work of all early German psychologists. I resent that. It is possible for a man to *try* with a human observer. As far as psychoanalytic theory is concerned the theory is of tremendous value for a chap like me in what I am trying to do. I am deeply indebted to the analysts for what they have been able to do, but quite sincerely I am not personally man enough to do it. I would much rather build a little mechanical man to do what I cannot do easily.

WIENER: Your problem is such that you would not get a tremendous amount of help in it. The psychoanalyst's method here is of enormous value in getting just what you cannot get, that is, in getting the finer detail of personal arrangements, etc. If we could find a mechanical way of recording which would be as good as a psychoanalyst should and would use, we would give up the mechanical methods. We don't have them to cover the sort of material with no adequate input; we have no adequate transducer for that material, but where we are dealing with limited phenomena of motion you are very highly limited. It is very highly intense at that limit. There the machine has it over the individual every time. I agree with you. In other words, the psychoanalyst cannot abdicate. He must work with what we have already seen to be an extremely noisy transducer. He does it with that because there is not any other transducer available. If there were a choice, any other way for the psychoanalyst to work by using these transducers, he should not.

STROUD: The human being is the most marvelous set of instruments, but like all portable instruments sets the human observer and is noisy and erratic in operation. However, if these are all the instruments you have, you have to work with them until something better comes along.

BROSIN: I want to take that word home. What is a transducer?

WIENER: A transducer is something that indicates an input and output where the relation between the two is reasonably fixed. But | the trouble is that it also has other [154] inputs besides the ones you want and those are called noises.

STROUD: A typical transducer is an earphone. On one side you are putting in energies, pressure systems, and on the other side you are getting out some representation of it but in voltage. What comes out on the other side is supposed to represent what goes in but not necessarily in the same system of energy value, etc., but it has got to be a good representation.

FREMONT-SMITH: It predicts?

STROUD: No, merely it is able to transmit languages.

FREMONT-SMITH: By how much?

BATESON: A codifier.

STROUD: You can have a transducer to predict but it is not necessary for a transducer to predict the thing.

FREMONT-SMITH: But you have to be able to predict.

WIENER: Everybody has a transducer but not everyone predicts.

FREMONT-SMITH: The analyst begins to make predictions. Is that the way he transduces?

STROUD: He can never do anything better than can be done on the basis of the output side of his senses as transducers. Fundamentally he is limited by his transducer output. What he can do by way of computation, on the basis of what is now transformed into usable material, may be enormous in terms of theory construction and so forth. But he is always fundamentally limited. He can never do more than the output of his transducers are permitting him to do, and that is one of the most rigorous and vicious limitations of this portable laboratory known as the human being.

WIENER: Let me point out again this resonance thing. If you want to investigate blue light you don't put the blue light under a microscope that operates in blue light. That is the worst thing you can do.

KUBIE: It may be worth while to say another word about this. Let us go back to the process itself for a moment. If you stop and think, you will realize that what we are doing now is giving a highly selected representation of all the complicated things that are going on in our minds. I am picking out from a chaotic flux certain things for their relevance or apparent relevance. What does the analyst try to do instead? He gets as far away from that as possible and he tries at recurrent intervals to secure true random samples of the total activity. That is what free association means. If you take a sample of the total product of any machine at regular intervals under as controlled and constant conditions as you can | maintain, you maintain your observational platform. If you [155] take a sample of total output of that machine today and the day after, you are going to learn more about that machine than if you simply take a weighted sample of its output. There are many technical difficulties about this in the analytic process. We have not solved all of the emotional difficulties which make the patient do some selecting and which make it hard to secure true random samples. Our problem is to try to eliminate or control the emotion factor in selection. If we can take that one step, then for the first time in human history we will secure systematic observations of a random, unselected product of the human machine, which will be an immensely important scientific step.

WIENER: It is a tremendous step but it is always a step that never goes quite as well as you think. After all, you make a man a Mr. Gallup while doing it.

KUBIE: That is why there is a great deal of interest in the process of automatic recording of random samples.

WIENER: But the fact is that one is working with the same difficulty that the Gallup poll is, and the difficulty with the Gallup Poll is still this resonance difficulty, that it is too hard for a person to jump outside of his skin and sample the population, not as he sees it but as it is.

SAVAGE: It is not terribly hard to do. Gallup did not try. He is not a scientific investigator. People who do that get much better results.

STROUD: It is still a better proposition looking at a brick wall twenty feet high than at a mixed-up population.

WIENER: But your population is not a brick. This is a digression but a very important digression. I have been running into this in connection with small sample work and in a small sample you are using distribution near its center to get information of the tail of the distribution. One of the problems which comes up here is, say, the dam problem. You want to build a dam that will last a century. In order to do that you have got to sample the destructive things that last a century, e.g. storms. And storms are merely extreme cases of a distribution which you thoroughly judge from the center. But when a dam breaks in Massachusetts it is not because of an ordinary storm gone haywire. It is because of a hurricane coming up the Coast and if you want to get a thousand-year break in there it will probably be an earthquake. In other words, you are [156] sampling a different sort of thing. Unless you know a lot more than you do | in most cases, determining what the tail of the distribution will be from the center of the distribution is plain silly.

HUTCHINSON: Isn't that a kind of problem that you are going to use where you get a catastrophe which is running on another law? Most of the techniques that have been invented work for those things which work most of the time and this throws the whole thing out and you begin again.

WIENER: If you are at liberty to.

STROUD: If your ship is clown it is down.

MCCULLOCH: Dr. Gerard wants to know why we are so far off the beam. Is that a fair way of saying that?

GERARD: I just asked what we are supposed to be talking about.

MCCULLOCH: We are supposed to be getting around eventually to recall and recognition, which are certainly after all a sampling problem of their own kind; quite apart from that, I think it is necessary to see the difference between the problems that Stroud is interested in, namely what a statistically normal man can do, whereas the problem that Dr. Kubie is interested in, namely, the problem of a particular man, who may be or is very far from the mean value. I think that one has to recognize that none of our techniques, which are built to discover averages, norms, medians, etc., are going to be too helpful to us when we sit down with a particular man.

WIENER: That is right.

MCCULLOCH: That is the fundamental difficulty that I see.

MEAD: There is one other point that was made here, which I don't think should be allowed to drop, and that is when Larry Kubie does this job, something has first been done to Larry Kubie. It is not a question of minimizing resonance. It is a question of using resonance. It is a question of getting a sufficient training and control of the particular relevant resonant factors so that man becomes an accurate instrument in this interpersonal situation. If we keep thinking, and even take Dr. Kubie's statement without that point, as if anybody could sit and record day after day this sample of the patient's behavior without themselves becoming intricately refined into a specialized instrument, we would lose a great part of the problem.

WIENER: You have to refine your instrument, but the fact is we want to know whether the phenomena we are observing in this particular case are the patient of Dr. Kubie and that is a very difficult thing to do. | [157]

BATESON: I think the first phenomenon we are working on is the phenomenon of resonance. Isn't that transfer?

KUBIE: In a way.

STROUD: I thought that Burrows put that very nicely. The problem in analysis is that of life, in which the analyst and the patient are deeply involved. The thing that is the core of the analytic situation is a very simple one. We want to live and do a good job of it and the analyst perforce becomes a part of the life of the patient and the patient becomes a part of the life of the analyst in a very real sense. You may never be able to do much abstracting in this situation, but if you are lucky you may be able to do a very good job. Psychoanalysis will never, as Burrows pointed out, be a highly favored method. It means if you are going to work with schizophrenics, to a certain extent you have to be a schizophrenic in your life. This is not very attractive as a prospect.

BROSIN: The problem of the human relations between the patient and the therapist I think have been well put by Dr. Mead and Dr. Stroud's position. Of course, they affect each other, and the point that the therapists' position must be more and more refined I think is very well taken. In fact there is greatly increased work on this subject, namely, that of the counter-transference. However, we don't claim the kind of certainty to which this group as a whole aspires. That is why clinicians, or at least I, come here for instruction, to obtain new methods and vocabulary so that I may improve operations in the clinic. One specific way to improve understanding is to implement Dr. Kubie's concept about getting samplings over a longer period. This technique involves much more than simple accumulation of data in the sense that one actually can detect larger patterns of integration of diverse facts with a unified meaning. It is a fact that there is a continuity of expression of conflicts along an axis which becomes extremely revealing. Let us take such specific examples as dependence on a mother, a fight against a father (authority), or sibling rivalry, and work with one of those themes for months in order to reach a clarification which may accelerate both understanding and therapeutic success. To return to the earlier topic of counter-transference, one might suspect that because the analyst perceives the large pattern better it may even have repercussions in his non-professional life. I doubt if anyone claims that the analyst has an integrity which prevents him from being altered at all by the experience gained in therapy. We hope that this experience is an integrating one which will also be helpful to the patient. At best the therapist will now be able to cope better | with his own [158] problems, but intimidation against dealing with the disagreeable aspects of human behavior will be ever present. This intimidation is usually unrecognized by most civilized people because by definition the civilizing process is a repressive one. To be otherwise requires a hardy person, and I think Dr. Stroud makes a real point, that the analyst is a hardy person, if not foolhardy, in choosing a vocation in which he must work with the demoniac, the dreadful, if you like, in man. The naïve criticism of psychiatrists and psychologists is not always altogether wrong, when suspicion is voiced in regard to their ego activities. In a special sense the man who associates with mad men to any large extent is a »mad man«, i. e., daring, in the light of rigid conventional standards, since he works willingly with emotions which other people repress as much as possible. The best single statement that I know on this is Freud's in 1917. He said that one of the difficulties of psychoanalysis, about which he was thoroughly pessimistic, was its growing popularity as a large theory of conduct which apparently could be easily absorbed. He warned his students that this glib acceptance of the hypothesis of unconscious repression, suppression, etc., did not mean that people really would accept or understand it, nor would they undergo the painful process which is necessary in order to get the first-hand information about people which really helps understanding, because of the cost to themselves. A current example of this resistance is in the last issue of a »New Yorker« review, a magazine that is ordinarily pro-analytic. In reviewing Freud's small volume »Introduction to Psychoanalysis« the reviewer says that it is distinguished by a lack of humility. In support of his indictment he quotes from the preface one of the introductory sentences by Freud, where the latter insists that this work has been built up on the observation of thousands of cases and has been well-verified, and that those who aspire to a criticism of the author should at least pay the price of working through these cases to that position. Perhaps this is arrogance, but the reviewer does not give sufficient credence to the fact that unless one has the intimate experience of seeing the data, however crudely, he does not know how much one can learn about the continuum of man's experience.

KLÜVER: I agree with Dr. Wiener that microscopists who wish to study blue specimens are not likely to use blue light for illumination. But if you happen to be interested in the »color vision« of a baboon it seems fortunate indeed that you, a representative of the human primates, possess »color vision.« I doubt that a science of color [159] vision would ever have developed if color vision had been | non-existent in man. The very fact that we possess color vision is of a great heuristic value; it has enabled us to formulate hypotheses which could not easily have been advanced otherwise.

I cannot quite agree with Dr. Abramson's views on the differences he finds between difficulties experienced in the mathematical area and those experienced in the psychological area. It is my opinion that the science of psychology – to express it somewhat paradoxically – has nothing to do with psychology: it has only to do with science.

WIENER: You just mentioned a very interesting case. It is lucky we have color vision. May I point out that the backwardness of the study of the sense of smell is due to the fact that man is relatively anosmic.

KLÜVER: Man may be considered a microsmatic animal.

STROUD: I'm not sure whether it's man to animal or mankind to the society which is anosmic.

WIENER: It makes very little difference. It is very difficult. We are trying and we are studying the dog's smell by determining a residual. We have great difficulty.

KLÜVER: If man had been equipped with a canine olfactory system, his scientific approach to the study of olfactory functions might very well have been entirely different. We appear to be woefully deficient in certain forms of olfactory »resonance.«

WIENER: We have to use what we have. What I have said is not a criticism of psychoanalysis in the sense of being a hostile criticism, it is merely a statement that we are working with tools that are limited. They are the best tools we have, but the problem of working with tools of that sort involves enormous difficulties that are much greater than we have when we are separated from the thing we observe by a big factor of scale or big factor of difference in character.

PITTS: I think one can carry the involvement of the analyst too far. Dr. Kubie and Dr. Brosin have never been allied to demoniac and insane people. They are not mystery mongers. I don't really think that because of the kind of material which you treat it is necessary that you yourself should grow to resemble it, therefore, if it is irrational you have to be irrational.

WIENER: I did not say that.

PITTS: Dr. Stroud said it.

FREMONT-SMITH: One thing Dr. Wiener said is so important, that we have to work with what we have. That is the essence of this whole business. What we have are human beings and you cannot | possibly, Dr. Stroud, eliminate the human being. [160] Therefore what I am saying and trying to emphasize is that, with all their limitations, it might be pertinent for those scientific investigators at the general level, who find to their horror that we have to work with human beings, to make as much use as possible of the insights available as to what human beings are like and how they operate. It was not my thought ever that you could substitute analysts for a machine. However, in considering and trying to set a frame of reference for the human being who has to operate part of your machine and make certain judgments, the analyst might be able to indicate certain ways in which you could do that with less error, less disturbance, or less »noises« than you would get by ignoring the data coming from analysts.

STROUD: I would entirely agree with you that we have to work with what we have, but we don't have to agree to continue to work with no more. I am interested in adding to our kit of tools.

WIENER: That is it.

KUBIE: May I add one word to eliminate one other unnecessary source of confusion? There is a great difference between the analyst working as a therapist and the analyst working as a purely scientific observer. That is true in all aspects of medicine, but particularly in this one. In scientific observations it is possible not to eliminate but to minimize the personal factors arising from the role of the analyst. In the role of therapist you deliberately do not do this, for the specific reason that the analyst is a sample of humanity on whom the patient acts out and lives out a great many of his problems about humans. If you were to eliminate the analyst you would eliminate an important part of the therapeutic experience. In a scientific study of the machine, there are many ways in which the distortion due to the analyst can be eliminated.

WIENER: I will say one more thing about the analyst. I have spoken of the resonance effects for the engineering and mathematical part of the crowd. I would like to put it in the following way. We don't attempt any proper theory of wave filters, we don't hope to, nor can we completely, remove the error that is due to the fact that the noisy message cannot be completely separated. We separate them as best we can. With a filter the separation is at its worst when a large part of the noise and a large part of the message are of the same frequency.

In other words, the very fact of the similarity of the observer and the observed here does not destroy the possibility of an important degree of separation, but it does cut down that degree | enormously. It is quite scientific. I don't want anybody to think [161] that a perfect use of a filter means the limit of what we can do in the predicters and the filters we have been building. This applies to the whole science. We know we are going to get errors. The problem is in some sense to minimize these but we are quite as scientific when we do the best we can with the knowledge that we are not going to be able to get a complete separation.

BATESON: What is on the program for this morning?

McCULLOCH: The program this morning is the possibility of picking up the notions of possible mechanisms of recall and recognition, as the bottlenecks for getting stuff out of memory. Then may be seen to what extent our prejudices, or complexes, or something else, may throw a monkey wrench into these procedures, because I think that is the crucial item we want to clear up when we come back to the notion of a neurosis as something which confines us to a relatively small part of what might do and makes us fixed in our actions.

I would like to point out, if I may, that there is a very different problem between the consideration of mechanical memories and human memory. In all the mechanical memories that I know about, either the time or space, or time and space, coordinates with an item stored as information. In other words, its address is retained and by using this address, one can go back to that specific item that has been stored and find out whether it is a jot, or tittle, or a collection of jots and tittles. This does not hold apparently with respect to human memory. We do not seem in any sense to have an address for the item in that fashion.

I would like to pick it up from our NO 2^{33}, that is about 10^{10} possible frames stored somewhere in our brains as a maximum. If you will imagine that you have in your brain 33 filing clerks – we will place them in the cortex – that each has a punch and that on each of the frames there are 33 tabs and each of your filing clerks decides between a pair of opposites; if it is one of the pair of opposites that leaves the tab on. If it is not, he snips it off. Those 33 filing clerks, each simply exercising one yes or no decision, would label the frame according to the content and you would recall on the basis of similarities of some sort the ideas that had entered into the frames you wanted. Human memory has this kind of approach to it, that one can get back into it on the basis of similarities. This is one of its peculiarities.

The second thing is that the cards of course might be filed anywhere else. There is no reason why they should necessarily be in the cortex, but losing filing clerks for [162] your particular chores, whether | they have gone on a vacation or have died, would leave you without the ability to recall although the frames were still sitting there. It would be the same filing clerks that pushed the button to bring up everything, this tab or lacking that tab.

WIENER: In other words, you need an ability to send »To whom it may concern« messages in order to dig up the hidden storage.

MCCULLOCH: Yes.

WIENER: I want to point out again the importance of »To whom it may concern« messages, and very possibly the importance of the emotional hormonal system which is quite relevant to the psychoanalytical problem in reaching hidden messages. We don't get them back by channel message in all cases, in some cases we have to send a searcher out. That searcher is a »To whom it may concern« message, and I think we can take it as a principle, although we have not really proved it, that the »To whom it may concern« messages are likely to be on an emotional level.

MCCULLOCH: May I come back again this way? If instead of filing these things as filing cards, one drops little cubes over the countryside with one corner cut off and then illuminates them, every little cube sends back its light to that place from which the light was sent. Therefore, if one had his memory scattered all over the lot and sends out a simple flash of light to all of those places, all of those which were of this cubical type would light up. Let us say you have one which works for one wave length or one which works at another wave length, you might easily on the basis of flashing in, come in with the whole distribution of the countryside or the fixed pattern you laid down, quite apart from where the things were stored. In other words, I do not believe that there is a fundamental difference here between a memory which one supposes is structuralized in great detail in fixed places and one which might exist of protein molecules of the right kind scattered hither and yon. One cannot from the overall operation tell which kind of a memory it is but I do think that if we were to go after the problem of recall versus the problem of recognition we might be able to get somewhere with respect to our ability to drag up the things out of the memory we want. I think we want two things. We want some kind of a theory of recall that is workable and I don't think we have it yet.

WIENER: No.

McCULLOCH: Human memory has at least three kinds of things that are distinguishable from the curves whereby one learns. We have a temporary memory, which is probably as characteristically | a reverberation or short-time process as nystagmus is. [163] There is some kind of a process which is involved in skilled acts and is obviously different from the kind of memory that takes snapshots of the world and files them away for future reference, whether or not of any importance at the moment. We do have a memory of the third kind, that is not immediately accessible, that has certain different properties, one from the other. It is this third kind of memory that I think is crucial to us to see how one gets a recall for it. The thing which is carried in reverberation at the moment is already in the works. There is no problem at all. The thing which is a skilled act, or supposed to be, has been laid down by some kind of synaptic structure. All right, it is there. You go to act and the pattern is there already for you. You only key into it and go through it.

WIENER: Playing the piano.

McCULLOCH: But the third kind of memory, which I strongly suspect is more important in neuroses than the rest, I think first needs an examination of the mechanism of recall.

BROSIN: I just want to ask whether you will discuss the possibility of an independent system such as you get in the classic amnesias and related somnambulas and hypnotic dissociations, because I am fascinated with your concept of the third type of memory. I wondered what your thinking of the »independent system« types of recall might be. I say independent systems now in quotes because they are not truly independent.

McCULLOCH: This second type of memory has some very strange peculiarities. If, for example, in the classic phrase you learn to skate in the summer time, the second kind of memory has a peculiarity, if you use it much it is better on test and if left alone it is worse. On the next test it tends to be better again. So it goes away like this when you are testing. Let it alone and pop, it comes again. What this is due to I don't know, but it seems to be rather characteristic where you have a complicated sensory input and a skilled motor act at the end of it. This is the place where this kind of curve crops up. It does not look like the other curves of learning or the other curves of retention.

LIDDELL: Is there any brief way you can characterize it?

McCULLOCH: The first I call simple reverberation. The second I think of as a semi-soldered-in affair, and the third, photographic or snapshot.

WIENER: What is the kind of memory you find most frequently, the arts? | [164]

STROUD: The soldered-in variety.

LIDDELL: What is the example of the first?

McCULLOCH: Nystagmus.

KUBIE: I would ask again that we reconsider the advisability of using the word »memory« here, because »memory« is a total function which includes availability and recall and it carries that connotation in general use. There are permanently stored organized impressions which can be buried 10, 20, 30 years.

McCULLOCH: Which I would call tracers.

LIDDELL: Your third is called »tracers«?

KUBIE: These can be brought to light as vividly as though they happened yesterday. If you use the word »memory« for something, which people speak of differently in popular parlance, you get into semantic difficulties. That is what troubled me last night. Dr. Von Foerster's measure was of availability rather than of memory in the inclusive sense.

McCulloch: Will you say that again? You get into semantic difficulties when you use the word »memory« of that which man cannot recall.

Kubie: Which he cannot recall but goes into certain manipulations and it pops up just as though it happened an hour ago.

Gerard: Are you not worried about the other side when it pops up and cannot be kept down?

Bateson: I think it is very important to take up the point Wiener made about the »To whom it may concern« message being perhaps hormonal and that we should refer that point to the analysts. We were told specifically by Kubie yesterday afternoon to regard the unconscious as memories which face a dynamic barrier which holds them down. I would like to ask how a phrasing would work in your theoretical framework in which the unconscious memories could be evoked conceivably by an affective »To whom it may concern« message, a hormonal state, something of the kind.

Fremont-Smith: May I give an anecdote before you ask the analyst to express it? I think a little data may help and the only kind of data you can get is either a patient or an anecdote. I was sitting alone in my apartment and became aware of the fact that my face was hot. I felt a flush on my face which called to my attention the fact that my face was hot. I became conscious of it and then realized that even though I was all alone, I was acutely embarrassed, thoroughly ashamed and uncomfortable and unaware why. »What in the world is going on«, I ask myself. I have a book in my hands; perhaps I have been reading something that is em|barrassing. [165] I realize as I look at the book that I have not been reading for a little while. I must have been daydreaming. By associating backwards about three associations from where I left off the daydreaming I suddenly came upon a memory of a letter which I should have answered but which I had not and I felt ashamed for not having answered it. The moment I recalled the letter, I said, »Ah, that is it«, and the flush on my face, which by this time was quite uncomfortable, immediately subsided. This is an example of a message which faced a dynamic barrier, with resultant physiological disturbance which lasted until the barrier was broken through.

Bateson: I would like to suggest that we might think about the dynamic barrier in terms of a specific theory: that a particular affective state might be related in some specific way to the »To whom it may concern« message which would evoke the missing snapshot; and that the affective state associated with the snapshot is the thing which for reasons of economy is shirked by the organism.

Frank: Wiener's statement raises this issue which we should recognize explicitly: what kind of selective processes of discrimination, of reception, and of response is operating when a »To whom it may concern« message arrives?

Wiener: That is the point. We have not mechanized that.

Frank: There is one excellent example of this process as revealed by study of the endocrines and the hormones. These various hormones are poured into the blood stream by the so-called ductless glands. In the blood they are freely circulated and each is then selectively absorbed by the organs or tissues or cells when they need it or can use it; each absorbing organ or cell discriminates among all the different hormones circulating in the blood and selects out only those which physiologically *concern* it. The ovaries for example select out and absorb and respond to those pituitary and adrenal and thyroid hormones which are essential to the functional processes of the female cycle.

The endocrinologists have attempted to »mechanize« this process by assuming that each hormone had a »target« organ or tissue at which it was aimed by the gland which produced it. Obviously this analogy of a gun and a target is physiologically impossible since the hormones are freely circulated by the blood in a mixed solution which even the organic chemists and biochemists have difficulty in fractionating.

There is involved in this process, a »To whom it may concern« message, and a selectively discriminatory and responsive organ, both | of which must be conceived as [166] operating in a highly organized manner, but with infinitely varied dimensions of the message, the selective, absorbing operation and the response.

This is the kind of process which occurs in anaphalaxis where the blood stream is sensitized, i.e. made differentially sensitive and selectively responsive to certain substances which it may not again experience. When it does respond, it does so vigorously or violently to their reappearance as do various organ systems which also react selectively to this blood response.

Thus we should recognize certain inherited physiological processes like the endocrine-organ system interactions and also certain processes which are modified, altered or made more or less sensitive to, and selective of, various substances by a learning process.

Now it may be suggested further that the emotional reaction – the overall organic response involving a cumulative reaction of all organ systems and hyperactivity of certain endocrines – may operate to increase the sensitivity and selectivity of the organism to certain »To whom it may concern« messages, internally and also externally. This selectivity, be it noted, is a process of discriminative acceptance or absorption and of response to the message. However, the pattern of the response is usually a function of past experience and therefore may be more *or less* relevant and appropriate to the message as interpreted, accepted and reacted to.

An emotional disturbance may be regarded as similar dynamically to the excitation of an atom when exposed to penetrating radiations so that it becomes radioactive, emitting various charged particles which in turn provoke reactions from other atoms. What is taking place within all atoms, that is the dynamic process of self maintenance, becomes enhanced, under radiation, so that there is a response to the outside of the atom, of an intensity and duration beyond its »normal« behavior. Once radiated, this excited state will persist with cumulative decrease of intensity as the atom »remembers« the episode of being radiated and so reacts thereafter to the present, as it »learned« to respond in the past.

It may be worth noting that when we conceive of these selective activities as organic processes we do not invoke special entities, or *ad hoc* forces, or motives or any of the other heuristic devices of static thinking, so often employed in physiological, psychological and social theory. The processes of the living organism in dynamic relationship with its internal and external environment provide the energetics of the ongoing events. Our task is to develop the capacity for thinking dynamically, which means interpreting all of the activi | ties of the organism-personality in terms of these ongoing [167] living processes and operations and extensive modification, without involving any *ad hoc* agents or forces or imputing additional capacities or special properties, beyond those exhibited by the living organism. These are sufficient to account for personality development and expression if we will search intently and at the same time reject the many survivals of the animistic tradition which are still implied in much of our psychological theories.

This kind of process is, to my mind, inexplicable until we recognize that the message – be it hormone, stimulus, chemical reagent or whatever you choose – has no potency, no power, no force, no dynamics except and until it is selectively accepted, interpreted and responded to by the appropriate agent, organ system, cell or personality who has a concern with that message. The message has no meaning, in other words, except as that meaning is put into the message and interpreted in terms of the selective acceptance and patterned response given to the message. The meaning is defined by the response, for the personality so responding.

Thus we are faced with a feedback situation which is the dynamic process of interaction: the organ system, the organism-personality, has a selective threshold, with discriminating acceptance-rejection and patterned responses, which is continually operating in the interaction of the responding organ or agent with its environment. This I take it is a field situation in which every constituent of the field has a relative dimension, potency, etc. as a function of the field. Hence no one part, action, operation or other constituent of that field can be picked out as the agent or given any specific causal potency as if it were a purely linear relation. This is a circular process – a circular interacting process – which has been evolved by organisms to a high degree and which we have just learned how to construct in machines using the feedback principle.

KUBIE: I have tried to put that in figurative language. We know that every day of our lives we have unfinished business. We have a certain number of experiences which we look back upon with satisfaction, with gratification and a certain number of experiences that are left hanging in the air like an incompleted record, some of them with a definitely unpleasant effective tone, some of them unimportant and we dismiss them. Some of them have a reverberating quality or impact upon us; as we drift off to sleep our minds take those up, the unfinished business of the day, and we start ruminating about them in daydream and phantasy. As we drift off into a hypnagogic state, half-way [168] between sleeping and waking, we lose hold | on reality and they are translated into visual images which are the language of dreams. This happens every day. Anybody who observes himself going to sleep can watch it happen. This unfinished business which we try to seal off is dynamic. It has a lot of juice in it of some kind and can remain highly charged in some way with the power to influence our thoughts and feelings. If you make the experiment of putting people into hypnagogic reveries or light stages of hypnosis, or give mild narcosis of any kind, then the whole associative process goes right back to those storm centers with an immediacy which is astonishing. We saw this in the merchant seamen during the war. As they started off to sleep they would go right back to experiences which had been most traumatic to them, as though they were trying to put a happy ending on something that was unhappy, trying to finish up their unfinished business. That effort gets blocked sometimes as the human being is trying to fall asleep.

We see it taking other forms in the individual who blocks at a certain point so that his memory is completely sealed off. I had a patient around fifty who could remember the outside of a cabin on the Montana ranch of her childhood. Since she was three years old, she could remember everything on the outside. She remembered the shrubbery, the trees, the whole setting, but never could see inside that building until she was put into a mildly narcotized state with certain things that we were working with at the time. Thus, it was as though the door opened and she could move inside and see everything there. This was a highly organized memory, and yet some kind of a dynamic barrier blocked its availability. This barrier can sometimes lift by altering the patient's emotional attitude towards the thing that is concealed. Sometimes you can lift it by narcotizing, i.e., lessening the pain of the memory, like giving a patient a shot of morphine.

BROSIN: One more brief example, which is familiar to all of you, is murder as an acceptable solution in order to avoid some more dreaded difficulty. The classic Oedipal or Hamlet story, and the graphic exposition called »Dark Legend,« by Frederick Wertham, might be used to illustrate this »To whom it may concern« message, which Kubie calls the storm center, with all the charges. It may pick up multiple messages some of which are destructive, but remain more acceptable for overt behavior than others.

FRANK: In discussing the third type of memory we should recognize that there may be recall, not of the events or the facts or any specific content, but only of the meaning, particularly the affective | meaning. We may recall that certain situations and personality reactions thereto are desirable or hazardous but we do not recall the experience, the events, in which we learned that meaning. We may therefore interpret situations and people as having that meaning and respond, to that meaning which we impose upon them, with no awareness of having learned that, indeed no awareness that we are so interpreting events and people. [169]

This kind of recall of meanings without the events is apparently what happens in affective or emotional reactions when the feeling tone or the acute emotional reaction learned in an earlier experience which we cannot recall still operates in our present response to situations which we interpret as having the qualities or properties making an affective or emotional reaction necessary, although no one else in the situation may feel or react that way.

MCCULLOCH: Sometimes it seems to me it comes back with none of the effective meanings and only the detail.

FRANK: The meanings and especially the affective significance may come back more often than the detailed facts in some personalities while the details without much meaning or affective significance may come back in other personalities.

It seems evident that we may recall all the facts and details of a past experience or we may recall the pattern of response-motor and emotional, conceptual, etc. – which was learned in a past experience no details of which can be recalled. The »soldered-in« recall may account for the detailed facts while the other kind of recall may be this patterned readiness to act, especially to carry out what was learned long ago and never achieved or completed, what we may call our personal »unfinished business.«

MCCULLOCH: May I say this, my own chief unfinished business at the end of almost every day of my life is the thing which I have managed not to say when I bit my tongue, when I have smiled, when I have nodded, and what not. Almost invariably as I wake in the morning I put the people back into that situation of the unfinished business and I get off the witticism that I had intended, and I wake up in a gale of laughter. That is my particular kind of unfinished business and it always comes on awakening and never on going to sleep. On going to sleep I try to solve the test problem that I was working on during the day. I get an occupational delirium, or whatever you call it. The day's content churns and churns and on waking I get the thing that seems fairly fixed in me as a pattern of dreaming. | [170]

WIENER: I often have on waking the solution to the problem which I was unable to solve before going to sleep.

MCCULLOCH: That brings me to the question of barrier theories. I have a cousin who was a mathematician, a superb chess player, etc. He received from his friends in Washington – he was then 20 – a complicated set of data where they thought they had enough information to compute something and were not sure of it. If they had enough would he solve it? He put the letter aside and said, »Pshaw!« He went on playing chess and he did that two or three days, sailing during the day and playing chess all evening. We slept next to one another and in the middle of the night he woke up, and with his eyes tight shut, lit the candle, took pencil and paper and sat there with his eyes shut, and wrote down the solution. Mind you he had not looked at the data in three days. He wrote down the numerical answer first, then the arithmetic leading to that answer, then the general method of solving problems of this kind, put the paper under the blotter, blew out the candle and went to bed without opening his eyes. Three days later he said, »Warren, I have the funniest feeling that I solved that problem for the Geodetic« and so on but he was unable to recall any bearing on it. By the end of the fifth day he had become fairly sure he had a solution of it. At that time I asked him to look under the blotter. I have no idea how long it would have taken to reconstruct the solution in the waking state. There was something he had labored at, done, and it was lost completely out of the stream. What sort of a barrier to recall exists in such a situation where the content of the memory was something he intended in his waking state to do and merely set it aside for the pleasure of chess. These are the kinds of barriers.

WIENER: I think there is something very interesting there because I have had similar experiences. Perhaps it was the release of tension. It was disposed when he had solved the problem. There was no longer an emotional pull, even though not finally taken care of.

KLÜVER: The general impression prevails that only emotions are effective in blocking memories and that anything that is emotionally dehydrated, like thinking, does not possess such blocking powers. Nowadays there seems to be very little interest in mechanisms of thinking. A generation of psychologists that has seen to it that 60 million standardized tests have been administered to at least 20 million people has apparently also seen to it that only about two psychologists are left in this country who are really [171] interested in the analysis of thought processes. It seems that a strong interest in | intelligence tests does not imply, and even excludes, an interest in the mechanisms of thinking.

Let us consider »thinking« or, more particularly, »abstract thinking.« In thinking, we may isolate or »abstract« one property or one characteristic from a set of widely different properties. Abstraction implies that heterogeneous events or objects become similar or identical with reference to certain »categories« or »dimensions.« It is by virtue of certain similarities that heterogeneous items are referred to the same »series« or the same »dimension.« Apparently processes of abstraction proceed by constantly shifting »dimensions,« that is, by perpetually destroying similarities and perpetually creating new ones. It is perhaps not surprising that Freud reached the conclusion that the factor of similarity is of paramount importance in the mechanisms of dream formation; it is not surprising since the same factor can be shown to be operative in thinking, memory, perception, and numerous other psychological phenomena. However, William James was right in emphasizing the »lawless revelry of similarity.« Obviously, heterogeneous materials of events may be similar or dissimilar in a thousand and one ways. The recognition of Similarities in heterogeneous materials implies that one aspect or one property or set of properties is »abstracted,« »isolated« or »dissociated« from numerous other properties or even an almost limitless number of other properties. The fact that I abstract only certain kinds of properties and not others may be of decisive importance in determining the directions and turns of my behavior. It may play a great role in recall, recognition, and perception. In fact, such mechanisms as are involved in »abstract thinking« may be considered as powerful as emotions in blocking memories. While there may be a lawless revelry of similarity there is no doubt that the abstraction or isolation of a particular property from heterogeneous materials on the basis of certain similarities is lawfully determined and a function of certain conditions in the internal and the external environment. Unfortunately, our scientific knowledge as to the nature of the conditions accounting for particular abstractions in the presence of particular materials by particular organisms is very meager and one may, therefore, regret that vanishing of a psychology interested in mechanisms.

What I have just said appears to have certain implications for the analysis of memory deficits in brain pathological cases. Suppose a monkey has learned to respond differentially to various stimuli in the external environment; suppose it has learned, for instance, to respond to sensory stimuli in terms of relations, such as »brighter | than,« [172] »nearer than,« »smaller than,« »more angular than,« »heavier than,« etc. We may find after removal of certain portions of the brain that the animal is no longer able to select or respond to the stimulus aspects to which it previously was able to respond. When we compare pre- and postoperative error curves we may find that the previously learned differential responses are seriously disturbed or not retained at all after the brain operation. However, this does not seem to be the whole story. When I analyzed differential responses in which an errorless performance could not be obtained without comparing the stimuli *successively*, as, for instance, when comparing weights or sounds, I found, on the one hand, that various brain operations led first to a complete or almost complete disappearance of the monkey's comparison behavior. On the other hand, that trials, in which comparisons were actually made, were errorless or resulted in correct responses far exceeding chance from the very beginning of the retests. It seems, then, that destruction of cerebral tissue does not necessarily lead to a loss of previously learned responses but may merely produce a temporary loss or a marked disturbance of comparison behavior. We also found that after a cortical injury the tendency to compare stimuli in terms of certain relations may remain intact with respect to certain stimuli, such as weights, and may be seriously disturbed or lost with respect to other stimuli such as visual stimuli. Again, we found that under certain conditions the tendency to compare may be present only with respect to certain kinds of visual stimuli, and may be absent with respect to other visual stimuli. Available data strongly suggest that abilities to compare and relate various items, in brief, the *Beziehungsfunktionen*, can never be entirely destroyed by cerebral lesions. No doubt, particular memory deficits cannot necessarily be understood by reference to the operation of *Beziehungsfunktionen* alone. Whether or not certain stimulus properties exist at all for a given animal or man may be equally significant. Obviously, no comparison behavior can lead, for instance, to the detection of color similarities if »color« as a property of objects no longer exists or never existed for a particular organism.

McCulloch: One of the beauties we ran into when we were testing monkeys first for discrimination of pitch, and second, for audibility, was that after removal of the auditory cortex the values most of the time were just any old place, – at the animal's best he was right back to the old values.

Klüver: Did your experiments involve comparisons of two pitches? | [173]

McCulloch: Comparison of pitches and ultimate threshold for audibility. Those came right back where they were.

Wiener: The animal could not throw itself into the position where it was at its best after the operation but if it happened to be there everything went well.

Klüver: The fact that an animal or a man when confronted with heterogeneous stimuli responds in terms of certain relations only, or is capable of relating items only along certain lines but not along others, implies that numerous stimuli properties are ineffective in determining the response and thus do not play a role in perceiving remembering, acting or reacting. Negatively expressed, relational functions may assume the role of blocking agents.

McCulloch: So far as you are concerned, you would regard the blocking in the case I described as effective so that the man while playing chess does not think about the mathematical problem, and while thinking about the mathematical problem does not tend to think about chess. One would exclude the other?

Klüver: Yes.

Stroud: It is limited to the program?

McCulloch: It is limited to the program.

Stroud: You can put in several computations but not an indefinite number.

Mead: Would it be useful to classify the series of sorts of messages that one could send? There are other than the »To whom it may concern« messages. For example, you embarrass somebody and then you leave him free to freely associate under a controlled situation and he may work up a whole series of other embarrassing incidents or he may forget a name and may say then, »I can get that name back by going back to the situation in which that person occurs. He was a professor at Oxford 25 years ago. Let me say a lot of other names about Oxford«, and you get it back.

Here the search is to create a cluster of related, simple, associative things in the situation, as for instance, where I want to get back a whole culture, and I feel cultures as much as I can in a lump, the language, the names, the geography, so that they don't get in my hair when I am doing other things. I don't want to remember seven South Sea wards from an atoll when I want to go back to the Mincopian culture. If I want to start anywhere in a culture I am going to walk in the village, stop this one, pick this out, pick this up. I can start with the visual motor kinesthesia in that complex and bring it back. Then there is another sort, to bring back the uncompleted task. I cannot bring it [174] back, where the material has been | done and laid aside. There must be as many varieties of storage and of blocks to availability as there are varieties of these searching methods. If we can classify them, we will get your figure of speech, You are using different wave lengths and we are using different notions.

McCulloch: It would not matter to me what the cubes were.

Wiener: I may say this ties up with the suggestion, and again I want to repeat it was only a suggestion, that the mechanism in association is something like a change of permeability which spreads, which a certain effective message or chemical message, whatever it may be, produces by acting on circuits closed at the time, circuits active at the time. Notwithstanding the absence and presence of good physiological evidence for the suggestion that I made about learning, the idea of the association mechanism of hormonal trigger or by the »To whom it may concern« message fits in beautifully with the whole picture.

Hutchinson: There is one thing which seems relevant perhaps to the question of the completed business and that is the rather curious fact that is exemplified, for instance, by the late Lord Raleigh of whom it was said that when you saw him reading something very intently he was reading one of his old papers. We have all had that experience of coming back to something we have read but forgotten. The recall or return to it carries a particular pleasurable feeling about it.

Mead: That we have read?

Hutchinson: That we have written, and have come back to read, having forgotten what is in it. I think it gives one the impression of a circuit that has been facilitated and is no longer useful but is still easy and one gets a certain feeling of virtuosity by being able to run around it.

Kubie: So that there is a very important distinction between a loss of availability, which is not due to the operation of a very strong dynamic barrier and the loss of availability which was due to pain of some kind associated with a particular net.

Hutchinson: Yes.

Wiener: That is a very good point.

Stroud: This discussion – and it is something I don't like – I have a prejudice against Dr. Von Foerster's general class of memory molecules, perhaps because so much has been hung on utterly untestable theory of memory traces. The discussion last night and this morning has taught me something. I don't know whether it means anything to anybody else. You made the distinction between traces and | the memory process. I [175] think it is a pretty good one. I would like to use even the term, just mere marker. The thought occurs to me that always in the past when I had gotten the whole process lumped together, in so doing I might have gotten a better view of the whole forest but I could not see the trees. It occurs to me that when we are looking for the mechanisms of long-term memory we want to look for purely passive elements. I will tell you why. If these markers or tracers were anything other than passive, what a horrible mess they would make out of the program. So that the first thing, if we for example found that Dr. Von Foerster's molecules were meaningful, they do at least have that quality that they are perfectly passive, like the little corner reflectors. They remain where they are, if they are still intact, until you send out an organized search for them. I believe both the Gestalt psychologist and the analysts will tell you that you can utilize many of the same markers, in fact, almost substantially the same markers under different circumstances and come up with a resultant markedly different memory. As it were, we don't go around remembering things, just off the cuff. It would be a terrible occupation. What happens is that we have set before us the problem of what it is we are to remember. Then we begin searching for a pattern of markers and finally we come to the conclusion that either there is such a pattern of markers or there is not such a pattern of markers, and sometimes we are not sure whether there is such a pattern of markers to correspond with the hypothesis we are trying to test by remembering. In a certain sense we know before we start remembering that we know what we are to remember, at least in abstract, and the problem on this long-term thing is to find out if in fact there are in our field enough markers to lay out this pattern for us. We want these little markers to be passive; we don't want them to supply any energy to the system. We merely want them to retain their integrity and their relational properties so that when we send out an organized search pattern, »To whom it may concern« message, etc., we can get back our answer. We are going to supply all the necessary energy from the process of trying to remember itself.

It was suggested to me because where I am we have a terrain surveyed and we have a lot of little corners for radar sets at known points. What do we use them for? We have a lot of mobile units; and instead of asking to remember the terrain we do the opposite thing with them. We ask them to tell us where we are, so that you send out a pattern of search and pick out however many of the little markers needed to identify definitely that these are a pat|ern of markers. With a little simple computation then you know [176] where we are, but the two processes are very closely related. I think the problems of blocking, etc. are curiously enough not related to the shape or character of our passive marker system; they are purely problems of the dynamics of the search. There is no formal organization to a past memory. It is just a collection of markers. We start out with an hypothesis: can we recognize that something which we already know, is what we are trying to remember?

The analyst merely provides the patient with an opportunity to try out a lot of normally forbidden hypotheses, which he should have been trying out a long time before, thereby making available to him not the markers, which, as the analyst points out when he finally gets back a hypothesis which has been so long forbidden, are there and the pattern is very vivid. The problem is whether or not the markers are there. The problem is the here now dynamics of remembering and whether or not you can entertain such a simple hypothesis.

Frank: It is not only barriers to memory that are involved; the individual personality is under the compulsion or the direction of past events (or preferably the meaning of past events) that he can no longer recall as events; but the experience is actively operating in the individual's present conduct and feelings, often to reject and conceal what is not recalled.

»Barriers« is an attractive metaphor but it is static and does not imply the circular feedback process which I believe must be assumed to be operating in all personalities. What we call barriers to memory may be conceived as part of the circular process of selective awareness which ignores what is »behind the barrier,« if we must use that metaphor. The personality imputes meaning to a situation as learned by him in past experiences, then responds to the meaning he puts into the situations or persons, with conduct and feelings usually learned with that meaning, but does not recall the learning experience nor have any awareness that he is putting that specific meaning into the situation – indeed he is usually very certain that he is interpreting the situation in accordance with its »real« meaning or the socially sanctioned meaning, not the idiomatic meaning which he gives the situation and to which he is responding.

Some years ago I suggested that we speak of *time perspectives*[1], as a way of conceiving [177] this process of viewing, or approaching, | a present situation in terms of the dimensions and meanings that are a function of past experience – this may be done »consciously« with awareness that we are drawing upon past experience to interpret the present and to see it in that time perspective; or it may be done »unconsciously«, so that we are not aware that our interpretations and our actions are addressed to the meaning derived from a past situation and so may be irrational and incongruous.

Stroud's suggested analogy of a radar set and of various markers in the environment which are »passive,« until the individual's radar set reaches them and evokes a reflection of the meaning he himself has sent out and now receives back, is a dynamic analogy. It indicates how the individual puts meaning into a universe that is potentially meaningful to all organisms; every personality differentially interprets and selectively utilizes those potentialities which are relevant to him – either by inheritance or by learning.

What we call culture – the traditional meanings interpreting all life situations and the group sanctioned ways of responding to those life situations as thus interpreted – operates to pattern the kind of meanings which members of each cultural group will utilize, but always in an individualized, idiomatic manner. Indeed, it is this idiosyncratic use of cultural patterns which gives rise to the personality process.

The conception of »time perspectives« which we impose upon all situations offers a dynamic approach to this problem implying that every personality, while existing in the immediate present, interprets that present situation either retrospectively in terms of memories »conscious« or »unconscious«, or projectively in terms of expectations which are generated by past experience and operate as anticipations of what will occur after the present, if and when we deal with the present as interpreted by our past experience. At least this suggests how memory of the »past« may be conceived to be operating in the present through the current process of interpreting, giving meaning, to the present according to what we have learned in the past.[2]

The errors we all make, the neurotic interpretations, and the psychotic reactions to situations and people are all evidences of this circular dynamic process, the psychological feedbacks. This process operates in the common public world of current events and relationships by idiomatic selection, acceptance, interpretation and patterned [178] response to whatever meanings the individual puts into | the public world, as he discriminates out of the totality, to which he responds selectively (rejecting and ignoring

1 *Time Perspectives* in *Society as the Patient*, Rutgers University Press, 1948.

2 L. K. Frank – *The Locus of Past Experience* – Journ. of Philo., Psych. and Scientific Methods. Vol. XX, No. 12, 1923, pp. 327-329.

what has no meaning for him). The whole present situation is a complex of innumerable »To whom it may concern« messages and every organism and every personality is a dynamic process operating in its own selectively perceived and individually meaningful »private world«.

This, I take it, is what makes the clinical problem so difficult and perplexing and, may I add, makes it a different problem from the laboratory search for uniformities and regularity of events, linear cause and effect relations and the long familiar problems of scientific investigation which are largely in static terms.

STROUD: I was troubled over that myself and I kept thinking about it. I realized when I thought of approaching it as you do that this is a nice white table cloth, that this process does not just start some place cold. This process of bouncing searching hypothesis off from reflectors and back again is a continuing process. There is no problem in this business of maintaining the hypothesis; the very business of living is constantly presenting us with endless searching hypotheses.

In the blocked case, where are and what are the little markers? The minute the search hypothesis even starts it gets back a forbidden reflection which stops it and the whole thing breaks down.

LIDDELL: May I mention a personal experience? Using the psychoanalytical concept of psychic accent, sometimes a glamorous theory will provide so definite a psychic accent on one's observational framework that one fails to see what is in front of one's eyes. Even though there has been a displacement of emphasis in observation, what has been seen has been stored away. Then later one must go through the labor of reaccenting what has been seen and remembered. For example, Pavlov's leading concepts were conditioned and unconditioned reflex.

According to him the power for operating the conditioning machinery was supplied by the unconditioned reflex. This was his solution of the instinct problem. The prepotency of this concept of the unconditioned reflex was so great in my own case that it has taken me 25 years to realize that this concept was a psychically accented or loaded hypothesis which distorted my observations of conditioned behavior. However, since I have the data at hand in memory, I can bring them back to mind, review my past errors of interpretation, and reorganize the past data – differently accent them. | [179]

MCCULLOCH: I want very much to come around to a good deal more from you in just a minute. Do you want to interpolate something, Dr. Kubie?

KUBIE: These things seem to me most troubling and most perplexing. What is the nature of the barrier? We obviously are in agreement that there is some type of dynamic barrier. What is the nature of it? What physiological explanation is there of it? This barrier is curiously selective because the direct representation of the material is not possible but nonetheless the material exercises an influence.

MCCULLOCH: This is just exactly the point I am most interested in. In the first place, it would be entirely in line with your hypothesis that the initial state of any neurosis, even if afterwards it becomes a matter of fixed traces, is in a state of reverberation. I would like to point out a very simple thing when we come to the mechanism. Our general line of construction is a set of servos, which are matters of inverse feedback, and any feedback system can be made to go regenerative under two sets of circumstances. In the first place, the gain of that system may be simply increased, in which case it will go off even at its own frequency i.e. at that for which it was supposedly inverse. In general there will be frequencies above and below that for which it was inverse, where it gets reversed in phase and there it will resonate.

LIDDELL: With the first increase in gain?

MCCULLOCH: The first increase of gain will do, and the second is a matter of timing.

What I want to look at for a moment is the problem of setting up a dynamic barrier around any process that tends to go regenerative.

Let us look at Gasser's notion of reciprocated innervation:

Let P and Q be afferents playing into the internuncial pool, a, b, c which play on the efferents d and e. And set the threshold $\Theta = 2$ synaptic nobs simultaneously active, but let the threshold rise with rapid firing. Now let Q be active, firing b and c simultaneously and so exciting e. This will continue only so long as the stream of impulses continues over Q. But from e back to Q we are dealing with a reflexive circuit. So the [180] output of e will diminish or stop the stream at Q. |

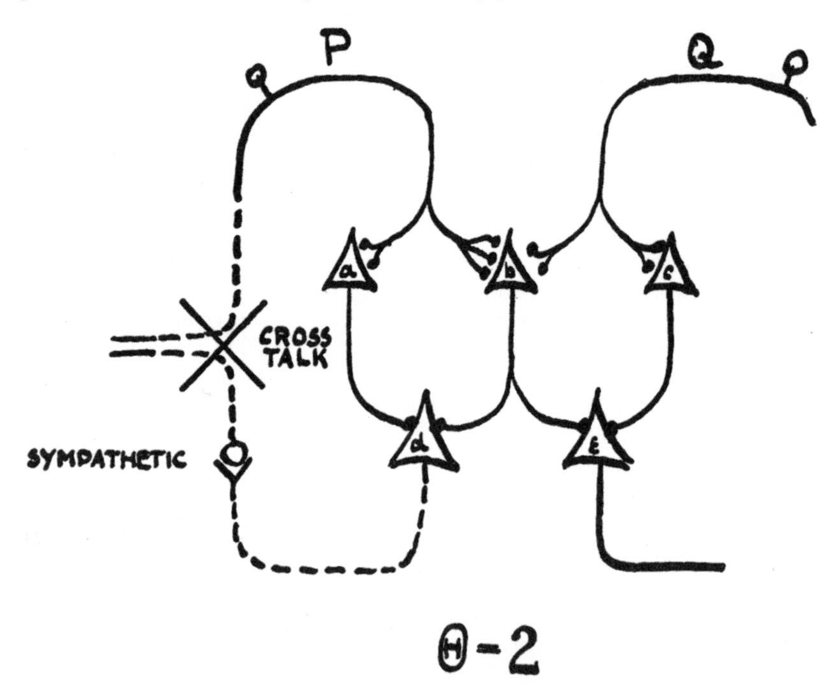

Now consider what will happen when P becomes active. I have given it three nobs on b. It can therefore fire b even when its threshold rises with repetitive firing and so it can steal b from Q and fire it simultaneously with a, converge on d and so excite sympathetic pre- and post-ganglionic fibres. Let us next damage the nerve at X so that cross talk occurs there between sympathetic efferents and fine-fibre afferents conveying burning pain. These will again excite a and b and so d. I have given this circuit through P three nobs on b as a short description to replace drawing other fibres parallel to P, which is what actually happens – for, as more impulses go out over d and cells in parallel with d, more cells in parallel with P are swept into this circuit.

Before the injury these sympathetic efferents would have so altered the circulation through the body or band or foot as to diminish the activity of P. The injury, by cross talk, has effectively reversed the sign of the feedback from negative to positive. It now goes regeneratively on and on and so keeps b in phase with a. In this manner the regenerative circuit surrounds itself with a halo of stolen internuncials that prevent it from being interrupted.

Stroud: May I carry that further with my reflector hypothesis? They are not fixed. [181] They are almost momental in character. If ever | you were to set up any sequence of a finite number of hypotheses succeeding one another, such that A followed B followed

C, and at any point you ever returned to A, then so long as this circuit, this regenerative circular system, is maintained you would be forever stopped from ever remembering anything that did not-belong to that system of hypotheses.

McCulloch: That is right if reverberation goes over the main works but if it goes off on one side so as to steal some of these internuncials there is nothing to prevent the reverberating system from spilling out into the rest. There is everything to prevent the rest from getting hold of the reverberating loop. Even the simplest assumptions are entirely in line with what one knows about neurons and in general it would account for this system once gone regenerative – making itself inaccessible.

Kubie: The influence that this represents is then blocked off. Then the influence would not arise from this but from the regenerating circuit.

McCulloch: This, b, would be the stolen internuncial and the thing kept *spilling*.

Bateson: The blocked material is the spilling circuit?

Stroud: Anything is inaccessible as long as it continues to spin.

McCulloch: The source of information (Q) is shut out by this halo of stolen internuncials (b). The reverberation may spill out. The normal afferent (Q) cannot get in.

Fremont-Smith: Which is available to recall?

Stroud: Only the items available to the circuit.

McCulloch: If running over the recalling mechanism at the time. If you are trying to do your daily business, this reverberant affair is running on its own, it is unavailable to you.

Liddell: It is out of action.

McCulloch: Let me take this as the prototype of all neurosis – namely causalgia, wherein we develop cross talk between outgoing and incoming systems. That cross talk is regenerative and there are the possibilities of blocking it at the input or cutting your sympathetic chain and blocking it at the output, or there is the remaining possibility of getting in enough impulses of a normal kind to sweep some of these cells (b) out of the reverberative circuit and back into the normal affairs. Now in your causalgic limb, any touch of that limb is painful, is appreciated as pain rather than anything else. The idea is to get enough hormonal impulses by to begin sweeping these (b) out of phase with these (a). That | you know was the chief trick that Dr. William K. Livingston [182] used in cleaning up his causalgias. He got by with a minimum amount of surgery, sometimes by doing temporary blocks of peripherical nerves or sympathetic chains, and above all, by getting enough impulses in over other fibers to sweep these cells (b) out of the reverberant circuit.

Bateson: The amputee hitting his stump deliberately.

McCulloch: What you start with is putting the man's arm in the whirlpool and giving him control of the whirlpool so he can, more and more, build up normal stimulation of that hand with a minimum of painful stimulation.

Hutchinson: You have to have asymmetry.

McCulloch: Dr. Liddell, the question then arises how long are these things actually carried in reverberation when one produces a neurosis. I would like to get a word in on your sheep, if you will, first as to the matter of time relations in the development of neuroses, and second, above all on the persistence of the phenomena, the persistence of active reverberation in the works.

Liddell: I would like to make two or three general statements before discussing the actual data which I think bear on Dr. McCulloch's interest here. In the study of animal behavior I think we have, as in many studies, a scale of activities. At one end of the scale we have crude empiricism and at the other a purely theoretical interest. My own bias has inclined me toward empiricism. I became involved in the study of the sheep's behavior through an investigation of structural and functional changes following thyroidectomy in this animal. I persisted for years in trying to think of behavior as physiology but it always turned out to be behavior. I was employing Pavlov's procedures in attempting to show the effect of thyroidectomy upon conditioned reflexes. I reached a stalemate because my interest became focussed on an accidental outcome of an attempt to compare conditioned reflexes in a normal sheep and a thyroidectomized sheep. The normal sheep went into a persistent state of agitation which made him useless for our experiment. Anrep had been in this country lecturing on Pavlov's studies of conditioned salivary reflexes in the dog and described the experimental neurosis. It seemed to us that we had a similar experimental neurosis in our sheep. But what to do about this. The thyroidectomized sheep with its slowed down behavior did not develop the neurosis. In fact, we have never been able to develop experimental neurosis in our thyroidectomized animals. Removing the thyroid in the neurotic animal [183] abolishes the neurosis. Because of the con|tinuous salivation in ruminants we used mild electric shock to the foreleg instead of food as the reinforcing agent. The animal maintains a level of quiet watchfulness in its restraining harness and the reaction to the occasional shock is a precise, deliberate flexion of the stimulated forelimb. With a signal preceding the shock, the animal, sheep or goat, developed neurosis; if the signal lasted too long before the shock arrived to validate the animal's expectation the result was long-delayed conditioned reflex. It also developed neurosis if signals for shock and for no shock were too similar for the animal to discriminate. Then we made a discovery which puzzled us because we were unable to psychologize the developed neurosis if signals for shock and for no shock were too matter. Before discussing it, I want to make the following point. We have discussed the psychoanalytic frame of reference. I must say that the best progress I have made in my investigation has been, in my opinion, through the collaboration and aid of my psychoanalyst friends. They have been of more practical help to me than physiologists or academic psychologists or any other group of scientific people with whom I have been in contact. And this is not accidental, for Thomas French, the first time he visited my laboratory, got down on his knees and scrutinized the sheep exhibiting a neurotic pattern. Then Fremont-Smith, Kubie, and Rado, all took a similar sophisticated interest in our neurotic animals – thinking of them clinically. Those of us who work on, day after day, a session each day, with the same animal or patient, sometimes for years, develop an interest in this animal's or patient's behavior as behavior. While physiological speculation interests us, my psychoanalyst friends have been through the business and it is they who have given me the most practical tips for further experimentation based on the experimenter's intuition.

To revert to our discovery. We finally found the simplest method for developing experimental neurosis in our animals and it brought to mind Dr. Kubie's notion of reverberating circuits – the repetitive core of the neurosis. When a sheep or goat has become accustomed to standing quietly in a simple restraining harness with arrangements for recording head and leg movement, respiration and heart, a telegraph sounder clicks once a second and at the eleventh click a mild shock is delivered to the foreleg. The animal soon comes to expect this shock when it hears the clicking and deliberately flexes the foreleg. In other words the clicking elicits a state of positive expectancy which the brief mild shock validates. After this moment of expectancy of [184] shock always followed by the shock, the | animal relaxes but not completely. It steadily maintains, during the experimental session, a level of watchfulness or vigilance at seemingly increasing physiological cost. Toward the end of an hour's session the breathing is increasingly labored although the situation involves only a multiplication of simple messages or specific ten-second episodes of expectancy. To precipitate an experimental neurosis in this situation it is only necessary to proceed as follows. The telegraph sounder clicks once a second for ten seconds and on the eleventh click a mild shock is delivered to the foreleg. Each day twenty of these ten-second signals are given, each one reinforced by the shock and they are separated by intervals of two minutes. This rigid time schedule is followed day after day. As a consequence of this procedure abnormal behavior insidiously develops as shown by increasing difficulty in flexing the forelimb at the signal for shock. At the signal the foreleg becomes rigidly extended and is raised from the floor by movement from the shoulder. Other evidences of this state of congealed vigilance or expectancy are seen. The free rhythmical movements of respiration gradually give place to a series of gasps separated by long apnoeic pauses. Moreover, unlike normal sheep and goats which never become housebroken but always urinate and defecate freely in the laboratory room, the animal subjected to the rigid two-minute training schedule never urinates or defecates in the laboratory after the abnormal rigidity of the forelimb develops. The bowel and bladder sphincter also come under the grip of hypertonicity. The heart rate remains low and increases little or not at all at the signal for shock. In other words, we have induced an imbalance of autonomic action in favor of the parasympathetic hyperactivity. This chronic distortion of behavior will be promptly reinstated after the animal has been given a long vacation in the pasture – a vacation of many months.

A radically different neurotic pattern will appear in the sheep and goat if the two-minute interval separating the ten-second positive signals is increased to five minutes. Now the animal must wait expectantly five minutes between signals during the daily test. In this case, the neurotic behavior appears abruptly and takes the form of diffuse nervousness. The animals tension in the laboratory (and even in the barn at night) is constantly spilling over into repeated tic-like movements of the trained forelimb, sudden starts, continued head movements, and rapid irregular patterns of respiration accompanied by rapid irregular heart action with many premature beats. In addition, the animal frequently and copiously urinates and | defecates during the tests. Here the [185] picture suggests predominance of sympathetic activity in the overaction of the autonomic.

In these experiments we are maintaining the most elementary form of communication with our animals in the laboratory. We are employing constant ten-second signals consisting of clicks of a telegraph sounder at the rate of sixty per minute and these signals delivered by an automatic clocking device and monotonously repeated at fixed intervals (two or five minutes) each signal followed by a mild shock. We rule ourselves out by our automatic conditioning device and by having different people bring the same animal to the laboratory on different days so that whatever simple transference possibilities exist between animal and experimenter are purposely not exploited.

BATESON: I am not clear on the story, the two-minute sheep. First of all, he recovers after three years in the pasture?

LIDDELL: He does not recover. The two types of neurosis in our sheep and goats seem to be equally persistent.

BATESON: The two-minute sheep did not come back and have the five-minute treatment? It was a different sheep?

LIDDELL: It was a different sheep.

FREMONT-SMITH: How did you get this disorganized behavior shift from the two-minute to the five?

LIDDELL: Two different populations were involved. One group of animals was subjected to the five-minute monotony, the other group to the two-minute monotony. Interestingly, in the goat subjected to the ten-second signals every two minutes the animal may at first go through a brief phase of diffuse nervousness before it develops the typical tonic immobility with rigid forelimb, gasping respiration and so on.

McCulloch: My interest in this problem goes way back to my undergraduate days. At that time the telephone companies were confronted with neuroses developing in switchboard operators.

Stroud: They are still confronted by them.

McCulloch: It turned out to be a very funny one. If we took the typical switchboard operative problems, that is, those which occur in the switchboards of any large city, before dials were installed, and we ran a simple thing like the IQ test on the operators, we found that the operators that broke down were those of high IQ. Actually there was a rather narrow band pass for operators because if they really have a low IQ they cannot handle the work. If they had very high IQ they got out of it. So you were [186] confined to a very narrow band, somewhat subnormal intelligence. Fortunately the | population is large there. The thing that happens to a telephone operator is quite obviously analogous to what is happening to Liddell's sheep, except that the timing is not fixed in that sense but in the other. A telephone operator must be on the alert for the signal. When the signal is received the operator plugs in the line and must listen long enough to make sure that the connection has been established before she cuts herself off. That means she gets a bout of words very significant for her, thrown in at the very moment when she must shut off and attend to the next. You have again, as in Liddell's sheep, a warning, an alerted background; you have a warning signal and you have a signal which is not in this case the simple electric shock but a bout of words thrown in, and that is probably why intelligence makes it tough. In other words, I believe there is in this case a closely parallel human neurosis which is extremely troublesome to the telephone company. This was one of the things which made them spend very large amounts of money on dials, for dials are cheaper than maintaining homes for neurotic operators.

Liddell: May I make one more statement to complete my narrative. It has some theoretical interest in connection with Pavlov's assumption that the unconditioned reflex supplies the power to make the conditioning machinery operate. The shock which we employ in conditioning can have quite different significances. For example, if the animal is standing at the alert and the experimenter accidentally presses the shock key the animal briskly flexes its leg but in a perfunctory manner. Its heart speeds up but soon quiets down – typical of the usual alarm or vigilance reaction. If on the other hand the animal has through gradual training developed a delayed conditioned reflex of a hundred seconds to the telegraph sounder clicking once a second, that animal is accustomed to a good long bout of expectancy. During the hundred seconds of the signal's duration the animal's vigilance will rise before the shock comes. Now the shock has a quite different significance. After the shock the animal relaxes very much as a tense spectator does when his team makes a touchdown. When the day's test is over and the experimenter enters the room to release the animal its heart will pop off just about as it does in response to an unsignalled shock.

Pitts: Would you expect that psychoanalysis would be as helpful in curing one of your »broken down« telephone operators by one of these methods? That is very interesting because in the cause of it there is no obvious unconscious source, or am I mis- [187] taken? |

Fremont-Smith: I would say that the telephone operator who did break down under those circumstances would tie in her experience in the telephone room with a whole lot of personal problems, past and present. The attitude of the supervisor, or the possibility that the supervisor represents the attitude of her mother or a variety of other problems, might come in so that you would get this thing soldered in, if you like, over a time.

BROSIN: For corroboration, I can tell you that I have seen at least six telephone operators who were sufficiently ill clinically to come to the hospital; all had paranoid reactions but they were all middle-aged ladies.

PITTS: Is this a special kind of neurosis which you can distinguish from other kinds?

MCCULLOCH: Not grossly by symptomatology as far as I know, except it is apt to carry a great burden of hostility.

STROUD: My wife was supervisor all the time I was in school and had many operators break down. It is the same thing you had in combat neurosis. The structure of the neurosis was the structure of the weaklings, of the whole individual, and all you had here was a nice neat little hand grenade punch that set things off.

KLÜVER: Do you believe, Dr. Liddell, that the animal during its three-years' rest in the pasture is a changed animal? I know how time-consuming behavior studies are, but I am wondering whether you ran any tests or obtained any experimental data to throw light on the nature of the behavior alterations existing during the three years outside the laboratory?

LIDDELL: Unfortunately we did not test our neurotic animals during the rest interval in the pasture because at that time we thought the neurotic condition might be transitory so we went on to testing other animals. However, to be doubly sure about the durability of the neurosis, when we changed our laboratory from one farm to another we thought we would test our old neurotic animals and discovered them to be neurotic still.

KLÜVER: You believe, then, that the modification of behavior induced by your conditioning procedures is chronic?

LIDDELL: We have reason to believe the neurotic condition to be chronic. Every time the dog got in the pasture it was the neurotic sheep that were killed. They ran off by themselves, apparently, and were thus killed.

KLÜVER: If I recall correctly, there are reports in the literature indicating that discrimination habits acquired, for instance, by | dogs in the laboratory may suddenly no [188] longer exist if the dog is confronted with the »same« stimuli outside the laboratory.

If monkeys are housed in the same cage for several years one of them will sooner or later get the upper hand. Problems of social rank order have, of course, been studied in great detail in a variety of animals. I have noticed that, if two Java monkeys are caged together for a long time, one of them may become more and more lethargic and even develop »pseudo-cataleptic« manifestations. In fact, one of the animals may be found on the floor of the cage in convulsions. It is the »socially inferior« monkey which develops these symptoms, including the convulsions. I have made such observations repeatedly during the years of my laboratory existence and, of course, at first thought of possible organic factors precipitating the convulsions. Being socially inferior in a monkey cage undoubtedly implies getting a diet inferior in quantity and quality to that consumed by the dominant monkey. I doubt, however, that such nutritional deficiencies as may have existed can produce the observed symptoms. I have finally concluded that psychological factors are involved and even play a decisive role. At any rate, as soon as such a monkey is removed to another cage all symptoms immediately disappear. In other words, it seems that we are dealing, not with permanent behavior alterations, but with transitory symptoms produced by one monkey »conditioning« another one.

LIDDELL: I am sure that the neurosis is not transitory. Sutherland's neurotic sow after her neurosis developed became quarrelsome and remained so. Her disposition never changed for the few years we kept her. After an absence of a year Sutherland on returning to the farm was attacked in the pasture by this neurotic sow for whose neurosis he was responsible.

FREMONT-SMITH: I would like to confirm the role of emotions precipitating convulsions – whether or not an observable organic lesion in the brain is present as the underlying or predisposing factor. A convulsion is a behavior pattern which is available to all mammals, and probably goes down below frogs which also can be made to convulse. There is no reason why that behavior pattern should not be brought about by physiological stresses initiated by emotion as well as by physiological stresses initiated by non-emotional stimuli.

MEAD: In the normal submarine situation you get a social situation that is related to what you put your animals up against. The men have to sit there maybe for two or three days with nothing happening, with relatively nothing to do, and with the captain the only one who can see through the periscope. On all the submarines | [189] that we had any data on in World War II, it was exactly the same social pattern – the men ate all the time. That was the one constant feature. The Germans were stuffed with salami and the Americans were stuffed with ice cream. But in each case, they ate and drank continuously. There were plates of chocolate or nuts on the table; every inch of the submarine was stuffed with food.

MCCULLOCH: It was the same story in World War I.

BATESON: There is a nice projection story which I heard the other day about a submarine. A submarine was lying off one of the Japanese ports; it was spotted from the land and the destroyers came out and tried to destroy it. The submarine lay there absolutely passive on the bottom of the water; the Japanese could not spot where it was but kept dropping depth charges in the neighborhood. There was an awful racket; everybody was keeping as still as possible and one of the crew said to the captain, »Gee, aren't we giving them hell«?

I would like more on the sheep and what they can project on their environment and the effects they can have on the environment, picking up Mr. Franks point of feedback interaction with a false premise interjected on environment creating its own truth.

MEAD: When they get the dogs to bite them and when not.

LIDDELL: I think this submarine story illustrates the position I have arrived at. These men obviously were not consciously afraid or consciously angry. They were maintaining a level of vigilance or watchfulness at a physiological cost which provided a background tension, this is what occurs in our animals in the training frame. We build up their residual tension by our monotonously timed episodes of expectancy.

MEAD: We don't know about the physiological cost. One of the problems was to keep them from fighting with each other. I should think that Dr. Bateson's story suggests the handling of the physiological cost.

LIDDELL: There is, of course, the matter of what man can do and the animal cannot.

VON BONIN: I would like to throw something else in. You mentioned that you had a sympathetic overactivity in the first period and that it changed later to parasympathetic?

LIDDELL: The parasympathetic overactivity occurred with the two minute spacing of the signals.

VON BONIN: I am thinking of Hess who pointed out in his work in Zurich that the sympathetic innervation was somehow readying the organism against the outside world and the parasympathetic | [190] was concentrating the animal on itself, its inside. Is that too farfetched?

LIDDELL: Cannon had much the same view as to the parasympathetic system.

von Bonin: They think of the outside world only as the threat that comes later on. They only think about themselves, what will happen to them, something of that sort.

Liddell: I don't think the behavior of our animals is complex enough to make that statement concerning it.

Fremont-Smith: You had one more observation in your pig, which comes down to Dr. Mead's statement about the men playing a low level checkers. As I remember it when you were trying to produce the neurosis in the pig, the question for the pig was whether or not to eat the apple that dropped. The pigs at first avoided the issue, as I recall, by turning the head to the side.

Liddell: That is true and we finally had to hem them in further by electrifying the box cover so that they were punished if they played with it in the absence of the signal for food and also by electrifying the wire fence of the training compartment so that they were shocked if they bit at the fence.

Fremont-Smith: It was only when they accepted the straitjacket of standing perfectly still, that the shifting of your signal was able to throw them into neurotic behavior. Well, that is what Dr. Mead points out, that these men accepted the straitjacket in the submarine but here they were able by eating or by playing checkers at low level, to escape breakdown. When the pig was not allowed even the diversion of rubbing its head against the side of the fence, it broke down.

Liddell: May I give two more illustrations? First, we made the amusing discovery that certain goats which had become pets of the students were almost useless for conditioning experiments. They would not submit to the necessary level of quiet watchfulness. They were always peering around and wanting to fraternize with the experimenter, biting at the wall and exploring everything with their mouths. They were a nuisance.

The most dramatic instance was the following: We decided to reduce expectancy to coincidence. A single click of the telegraph sounder coincided with the shock to the leg and the shock was discontinued occasionally to see if the single click had acquired expectancy value. We were surprised to observe rhythmical behavior which suggested the operation of reverberating circuits. The goat heard the click and at the same instant got the shock but then | continued making stereotyped rhythmically repeated reaching movements of the shocked forelimb. [191]

Fremont-Smith: After a long time?

Liddell: It kept on repeating this behavior day after day.

Fremont-Smith: Not the first time you gave him the signal?

Liddell: Toward the end of the first period in the laboratory after about ten combinations of click and shock it began its mysterious leg waving routine. We took records in the belief that it was some sort of rhythmical spillover. None of the other animals in the experiment behaved this way. We found later that this goat had been raised as a pet and the children in the family had tried to teach him to shake hands.

Frank: Liddell has introduced the term expectancy and the conception of varying states of readiness or vigilance toward what an organism or personality has learned to expect because his experience has made him interpret life situations with that meaning.

Again, I would suggest that the expectations of an individual are the ways he has learned to evaluate present situations in terms of anticipated events or consequences to which he maintains a state of expectation, readiness, vigilance with apprehension or anxiety, or with hopeful euphoria. These expectations are toward the future but are derived from the past experiences from which the individual learned that meaning or way of evaluating present situations.

Certain experiences set up expectations but the individual cannot recall those experiences; he recalls and uses the meanings he learned then and now imputes to situations with no awareness of how, when and what governed his learning. Thus the individual may be coerced by past experience, compelled to endless repetition by the operation of this process of imputing meanings to situations, defining events in terms that make his customary, habitual, repetitive conduct the only possible or seemingly appropriate response. In a sense the individual extrapolates from his past experience into the future since that is his only way of interpreting present situations and the only way he can anticipate the future and prepare for it.

We have learned to deal with the world as seen in »spatial perspectives« correcting for our distortion of it as seen by perspective vision. That is, we respond not to what we see in perspective, but to the corrected, reconstructed interpretation of what we see. Now we are recognizing that we must learn to deal with the world and human relationship as viewed in »time perspectives« by correcting for the distortions of our [192] individual time perspectives, the indi|vidualized meanings and expectations we impute to all situations and people in terms of our personal life experiences.

WIENER: May I say the tools used mathematically for handling a lot of problems analogous to this are time series.

FREMONT-SMITH: Is this related to Korzybsky's »time-binding«?

FRANK: No, he was calling attention, as I understand it, to the way man deals significantly, not with the geographic environment but with the symbolic world of his cultural traditions and does so according to his memory of the past, the general memories of his group which we call traditions and the idiomatic memories of the individual personality. It is these memories which govern his expectations and by so much direct his present conduct.

Thus far in our formulations we have not reached the clarification of how culture-traditions operate dynamically. As we do clarify and gain understanding, we will develop better insight into personality development and expression since as indicated, culture-tradition operates in human beings and is revealed by individual personalities living in that cultural field which they collectively maintain – another example of feedback.

VON BONIN: I think it deserves to be stressed. I missed it yesterday to a large extent. I think we decided that the brain is probably not essentially but certainly in one important regard a forecasting mechanism. I think Dr. Wiener brought that out very well in former lectures. I am amazed we have not used it. I am glad Mr. Frank brought it out now.

STROUD: That forecasting property I would like to emphasize again. It has got to be forecasting because all the computing operations are at a minimum of a tenth of a second behind. You are never going to be able to exercise a control which is effective without forecasting because all your data is ancient by that amount of time.

FREMONT-SMITH: Physiologically there is another. Remember that goes back to George Coghill's work with amblystoma. He was observing the first swimming motions of the embryo amblystoma. He showed that this little organism started swimming motions before any sensory connections from the skin had grown into the spinal cord. Muscular movement was initiated before any sensory connections had been established. As the sensory connections came in, the point of origin of muscular movement migrated up the spinal cord towards the head, always keeping cephalad to the highest level of sensory input. In other words, there is a driving mechanism within the organism which is almost self-starting, i.e. relatively independent of the external [193] environment, which makes it different | from the machine. This driving mechanism, it seems to me, requires forecasting by its very nature.

LIDDELL: May I make a quantitive estimate concerning the animal's forecasting or conditioning machinery. The dog can probably forecast an hour. Pavlov showed that if a dog is fed every half hour he will begin anticipatory salivating pretty accurately on the half hour. With our sheep and goats the limit of expectancy is probably around ten minutes. The outside limits then would be for the dog, say an hour, and for the sheep ten minutes.

STROUD: There is a little chipmunk, he can forecast six months.

WIENER: I would like to make one remark about the sheep. Your sheep must eat continually.

LIDDELL: One should not confuse the unconditioned stimulus, food or shock. It is shock, not food, which the sheep or goat anticipates in our experiments by about ten minutes.

MCCULLOCH: Suppose we got you a camel, the camel can go a long time without a drink and without eating. Would you expect the camel not to differ in the brain from a sheep?

LIDDELL: I would expect that with regard to the episodes of expectancy which we build up under experimentally controlled conditions the camel would fall within the limits of expectancy of the sheep and goat.

WIENER: Another thing that is quite interesting in this is: what would you expect with the elephant?

HUTCHINSON: There are a great many small invertebrates that are conditioned to nocturnal rhythm which requires a 24 hour. It is a 10^3.

KLÜVER: Some arthropods can apparently be trained to expect and seek food at a particular hour during the day. They will then expect to find food at the same hour 24 hours later although illumination and other factors likely to produce periodic fluctuations have been kept constant. They will even look for food at the particular hour as long as 8 days after completion of the training.

LIDDELL: What animal?

KLÜVER: The honeybee.

HUTCHINSON: There is a great opportunity of temperature, for fundamental molecular dynamic effects.

LIDDELL: We can damage this expectancy in the sheep and goat by lowering the metabolism by thyroidectomy.

KLÜVER: Do you consider the sucking activities observed in rabbit fetuses as well as in the human fetus a forecasting of the future?

FREMONT-SMITH: We are talking about human beings and we need human data. So I am going to give you another episode. The ques|tion of memory came up at a tea [194] party when I had said that most memory losses, most inability to remember names, have an emotional content which the individual does not understand, whereupon a man said, »That is perfectly ridiculous. I am always forgetting names and it does not have any emotional context whatsoever. For instance, this morning I could not remember – (and then he turned to his wife) what was the name I could not remember this morning«, and she said »Jane«. He said, »Yes, Jane. There is no reason in the world, no emotional reason why I should not remember that name«.

Let me stop there. This is not the end of the episode. Here there was a recall. There was no evidence of any emotion associated with that recall and he did remember after his wife gave him the name. He definitely had it. Then in my stupid naïvite, I remarked, »But I guess there was a ›Jane‹ once«. All of a sudden he began to blush and it was evident that my remark had recalled to his mind who the forgotten »Jane« was. This suggests that we have two levels of barrier. There was the level which blocked the memory of the name »Jane« and the level which blocked the meaning of »Jane«! Thus even after his wife reminded him and recalled the name »Jane«, he had no recall of the embarrassing significance to him of this name. The name alone evoked no emotion, but when the meaning was recalled the emotion was very disturbing. Here there were two levels of repression, both being active.

STROUD: When you ask a man to recall you present him with a hypothesis as it were, outlined, ready-made.

FREMONT-SMITH: He asked it.

STROUD: That is a different case. That is the case of having A follow B, follow C, etc. back to A. But it seems probable that there are two general classes of this. First where the circle is external and the subject is aware of what is going on much more like typical compulsion. In the second case, in the case of repression, the circle is there nonetheless but it is locked out of awareness by being constantly busy. In this relationship, anything that is included in this set of concealed hypotheses is by virtue of its »busyness« always unavailable to the other half operating at the level of consciousness. As a result the analyst is forced sometimes to break a component before this vicious circle or any hypothesis belonging to the system, or even some of the responses of the reflectors upon which it is playing, becomes available to general consciousness.

FREMONT-SMITH: Are the reflectors passive?

[195] MCCULLOCH: No. |

HUTCHINSON: Is there any possibility of building a passive reflector? From what we know about the protein, even the kind of mechanism that we were hearing about last night, it seems to me difficult in that we know that the protein molecules are being pulled down and built up again. I am also not altogether clear that the quantum type of decay could possibly occur in a cellular system undergoing the kind of rate that we know from our studies is the way the thing is being pulled down. I said yesterday there is a persistent structural pattern in spite of the ameba walking around. There is a specially arranged structure relative to the system but I don't believe that any kind of passive system can exist except perhaps if you built it in the skeleton.

MCCULLOCH: Even the skeleton, the bone is a frame which has the elements. That is passive in that sense.

STROUD: Passive in function, not passive in structure.

MCCULLOCH: The structure must be reproducing itself or pretty soon it would not be there.

STROUD: Mere thermal agitation would put an end to it in time.

MCCULLOCH: It is a matter of impulses running in one place or another. Memory does not have to be.

HUTCHINSON: Neuronal but not neurologically.

STROUD: As far as being able to use this curious little system of reflectors, I think it's quite practical. The dendritic system of some cortical neurons which take 120 milliseconds to act are that way because probably we are doing delicate tasks with them and not stupid ones. Now a system which is of that order could sort out a number of protein molecules of varied structure. Remember this is taking place at a very short distance. For me it is just as easy to think of the notion of being able to see a complex protein molecule a matter of microns away, as it is for me to consider the notion that I may see my hat which is on the hat rack across the hall. The relative scale of factors is not much greater.

GERARD: As to the mechanism, I would like to say something which has been on my mind for some time. We have made some real progress in formulating our problems but so far, in attempting to get at mechanism, we have just been knocking over knock-overs. We have not got the answers to any of them, it is true; but we are even worrying about answers only to the easy parts of the problem and, except for one remark that Larry Frank put in quite early, are paying no attention to the really difficult part of this problem.

There is first the question of storage of memories. There are plenty of places for them. They can be active or passive and certainly a | passive storage is much more efficient. That is a fair assumption to start with. [196]

Secondly, there is some mechanism for recall. This may be generalized hormonal or hypothalamic discharge, whether a chemical in the blood or the discharge of the multiple radiating neuronal system from the thalamus to the cortex, which anatomically and physiologically exists, or it may be more like the scanning beam of an iconoscope. At present, it makes no difference; somehow or other a process can go over the memory elements and contact them in some way. That almost certainly is active; it is involved with energy, with attention itself.

But the real problem – and we have not touched upon it – is that of specificity. When you scan, by whatever mechanism, and contact these memory stores, whatever and wherever they are, how do you pick out this one or that one and bring it into consciousness? That in turn reverts to the question of barrier. Here is the basic problem, and I have not heard any positive suggestions about it. I don't know whether we are in a position yet to give them. Still this remains the critical issue which faces us.

McCulloch: I tried to bring in the notion of the barrier.

Gerard: Your mechanism is fine, something to play with. I don't see any specificity.

McCulloch: You don't see that the process out of it cannot get into it?

Gerard: That is all right. How do you get into it when you say just the right keyword to that subject?

McCulloch: That would be a configurational problem of the kind of triggering off the particular complexes that would manage to give you enough activity piling in to steal back some group of internuncials or what not. That does not seem to be the great difficulty. The point I would come back to is the reverberant circuits in a man with an active neurosis. There is a reason for thinking that this mechanism may be rather crucial even in complex human phenomena. If you treat such a system with carbon dioxide, raise the level of it in general, you shift up both the voltage and the threshold of the neurons. If you make the peripheral nerve less fatigable, if that cell be made less fatigable, there is a good chance that it could be swept out of the orbit of the first system and caught by processes in the rest of the system, so that if one found that he could break up any reasonable fraction neurosis in human beings, he ought to give this kind of notion a fairer break physiologically than it has had. The effect of carbon dioxide on the central nervous | system has never been studied adequately. The fact remains that in our own experiences with 117 miscellaneous neurotics, separated for one exclusion miscellaneous in all other ways, we have been having better than a 50 percent yield of cases, some of them over three years now since they received their last dosage with carbon dioxide where the neuroses have cleared. Other people now have longer series, that is, more cases than we have but the fact remains that you do break up the process and the things that the patients go through when they are coming out from under carbon dioxide are fantastic. If you were to take 30 percent carbon dioxide, you would pass out and you would come to and nothing much would happen. But with the group of neurotic cases you would think they were being raped or in the middle of a fight, or something else, when they are coming out. [197]

The limiting factor in the treatment is this, that in all of the cases anxiety tends to build up. If the anxiety itself is a very severe symptom to begin with you simply cannot use this treatment. The patient comes back once or twice and gets the heeby-jeebies to such an extent that he cannot go on with it. What I would like to know is whether a similar mechanism may not be occurring in the sheep. If we could get runs of the very simple things, the effect of carbon dioxide in the sheep – and we know that it is going to do circuit acts and wherefor we can check it fairly easy in the cord – we might make sense out of it.

That gave you an active barrier, which is the thing I wanted first and foremost. That is a very simple physiologic assumption. Am I clear?

LIDDELL: Yes.

FREMONT-SMITH: The essential thing would be the diminished fatigability which would allow neurons to get in, whereas otherwise they would be in the refractory period.

MCCULLOCH: Sweep them out of phase.

STROUD: Change the constants of the reverberating circuits. Then you can do something about it which you could not do in the normal state.

LIDDELL: We are in a position to check McCulloch's notions about the sheep. We have a metabolism chamber in which a Pavlov frame can be installed.

MCCULLOCH: I know you can do it.

STROUD: It will be an interesting effect to break out these longstanding neuroses.

[198] LIDDELL: We know. |

MCCULLOCH: There is a crazy drug, myanecin, which is a very interesting drug because it knocks out multineuronal reflexes of the spinal cord without affecting, or only slightly increasing, the monosynaptic reflexes. For instance, the reflexes you get on pinching a part are entirely gone whereas the knee jerk is as large or larger than before. That drug in reasonable doses, somewhere around a fifth of the doses that have been tried harmlessly in man, knocks out anxiety, presumably because anxiety goes over somewhat similar mechanisms. We have the first few cases of anxiety neuroses on it. Now it is too late to know to what extent we will be able to block it up. There we have a chance to break up the anxiety and if the anxiety be part of that reverberant circuit, it may be that such a drug will end it.

FREMONT-SMITH: Do you know what this drug is?

MCCULLOCH: It is a benzene ring attached to a glycerine. It has a few oxygens stuck around the benzene. The structure is well-known. We know a whole set of its confreres but again the difficulty seems to be it is the only one of them that is nicely fat soluble which seems to be necessary to get where it is going.

MEAD: This sort of cure would be a situational one? You would pull out a lot of cells that have become involved. You pulled out enough so that the life went on but the core of the neurosis would presumably be left by this method?

MCCULLOCH: What you are knocking out is the reverberation.

MEAD: The ones that belong there as well as the ones that were stolen, is that your assumption?

MCCULLOCH: The ones that are responsible for anxiety as well as the ones that are responsible for your complex reflexes.

BROSIN: The commonplace observation should be made that in the carbon dioxide cases – and presumably in the history of all drug therapy – a great many operations other than the drug-induced physiologic changes are active and that this does not constitute proof. It may well be that there is facilitation or interruption of circuits present, but I know from watching Dr. Meduna work with cases that it is the usual human treatment situation in which suggestion, reassurance and similar operations are inevitable. I would like to bring up Dr. Gerard's excellent question.

McCulloch: May I answer you that Meduna is entirely in agreement with you as far as its being inevitably mixed with the treatment situation of the patient? He does everything he can to minimize it and it is still there. That cannot be helped but, as to the difference between interviews, I don't care how consoling you try to make | them [199] with your patient or how suggestive you try to make them without carbon dioxide and with carbon dioxide, the point is this that you have some tool which is enabling you to get access to things in a hurry in parts of the nervous system.

Brosin: It might well be, but all of us know the cases of schizophrenia that recover from typhoid fever or a blow on the head.

Stroud: These things are capable also of interrupting reverberating circuits.

Brosin: All right. Could we refer Dr. Gerard's question to the case of Dr. Frank Fremont-Smith's »Jane«? In what way did the recollection of »Jane« open and close new circuits?

von Bonin: The question is how »Jane« was recognized. What Dr. Gerard asked is why »Jane« was recognized as the open item that was searched for.

Fremont-Smith: He recognized the name and unrecognized it as someone who was acutely embarrassing to remember at that point. There was a double cross of partial recall and very much further recall and probably even when I asked that I am sure there was a »Jane«; he did not recall all there was of »Jane«.

Stroud: Or necessarily what there was of »Jane«. It would not be essential.

Fremont-Smith: It would not be essential.

von Bonin: This is another example, when you were talking here you fished around for Livingston. Livingston came and that was recognized as the problem. I take it that was the sort of thing.

Gerard: When you make these massive attacks on the nervous system, when you bang a person on the head with an electric current, or drug, or anything else, you do God knows what to God knows what parts of the nervous system. Therefore, it is perfectly possible to draw an unlimited number of logical sequences of particular cause and effect relations. But there is absolutely no assurance at present that the particular sequence anyone can dream up is the right one and the chance of its being the right one at the present I think is vanishingly small. All of which in no way touches the question of this exquisite specificity that one gets in these phenomena.

As Drs. Brosin and von Bonin pointed up with all the stimuli coming into that man's brain and all the times the word »Jane« had been said, and »do you remember this« and »is this in your life«, etc.; why does this particular concatenation succeed in doing something to a particular memory which then floods out over everything and | [200] has enormous repercussions throughout the whole body? I don't think we are anywhere near the gist of that.

Klüver: It should perhaps be stressed that such exquisite specificity, as has been considered here in regard to memories, exists also in the field of perception. We have specificity or selectivity the very moment we perceive objects or events. It may even be argued that different persons never see the »same« thing. Mechanisms of transformation, simplification, articulation, distortion, etc., are immediately operative when we are confronted with sensory stimuli. It is, therefore, not surprising that we later find a selectivity as regards the items that are or are not remembered.

Gerard: It exists at all times?

Klüver: To understand certain specificities in recall it may be profitable to determine first the constellations of factors bringing about specificities in perception. Experimentally, this problem can be more easily attacked than that of a selectivity found in recalling events long past.

BATESON: Could your specificity question be answered with a model something like this: think of a fish net suspended in a frame. If you twang on some of the threads near the edge where they are attached to the frame, according to the combination of threads which you twang a certain wave rhythm is set up in that net which is reverberated or persistent for a period. Now suppose that net or its attachments to be so rigged that the potentiality to have waves of oscillation of a certain pattern is retained and facilitated every time that particular combination of twangs goes into the net. The question which I would like to throw to the engineers to complete that model is: could we have a net of relays characterized by a vast variety of inputs which could be differentiated – each setting up its specific type of wave distribution. Would it be possible to get back the original wave distribution by repeating the original combination of twangs? Could such a system show hysteresis – an increased tendency to repeat wave patterns which it had once experienced?

WIENER: In other words, can you set a net the tapping of which can be changed by these twangs so that with one tapping it will come to one memory system and with another tapping it will go to another system?

STROUD: It is done regularly with our radar systems.

VON BONIN: How do you know the twang? I am fishing around say, for the author of a book; I cannot recall his name. I twang the net. How do I know how to twang it to [201] get it? |

BATESON: You twang by remembering the color of the book; what it looked like. All these things Dr. Mead talks about.

GERARD: Your analogical rather than digital type of system, that is what we keep missing except as you bring it in as a type of hormonal action.

FRANK: Dr. McCulloch also assures me when I want to bring in potential feelings and other things in the field that I am losing information about specificity and I will have to believe him.

MCCULLOCH: May I answer that in a very simple way? It is fairly obvious if one thinks about it in this way; if the world be ultimately made up of small particles which are neither here nor there; not half way between, and if the ultimate units of our universe are going to go analogically we have to deal with it analogically.

WIENER: There is another point.

BATESON: I said can we have a net which can remember different patterns of statistical behavior?

MCCULLOCH: The answer is obviously yes.

WIENER: May I say we are making one. The way we are doing it is this: we are making a predictor which will actually examine its own statistical experience and do its circuit in accordance with the statistical experience if the data changes in character and will change itself to suit the new statistics and data.

PITTS: The brain must be digital, and the heavenly bodies must move in a circle because it would be better that a device should be digital and analogical – that does not seem to me to be connected.

WIENER: It is purely a question of what load it could take. I want to make a distinction between the digital and analogic. The distinction is not sharp. Every digital device is really an analogical device which distinguishes region of attraction rather than by a direct measurement. In other words, a certain time of non-reality pushed far enough will make any device digital.

Supposing I have a block here, supposing I drill conical holes into it, now this could be used as an analogy. I could put things here or there. I distinguish these reasons, however, not by actually giving a map of the region but what the ball will roll into. In other words, as I emphasize as the important measurement, the whole field of attraction of these, the probability that the ball will stand at the edge of regions in balance has become extremely small. I could do this to a degree by introducing not an absolute separation but a quantity which went up faster than the first power. I could get devices intermediate between digital and numerical devices. The important thing of the digital device is the use of non-linearity in | order to amplify the distinction between fields [202] of attraction and that can be done to a greater or lesser degree. I am considering it now from the physical point of view of the human instead of taking the places it rolls into. I introduce the force that went up to the higher power of the distance from the center. My indeterminacy would be something intermediate between what it would be with a pure analogical and a pure digital device. I think it is necessary to consider the physics of digital devices.

[203]

SENSORY PROSTHESISES

NORBERT WIENER
Department of Mathematics,
Massachusetts Institute of Technology

The first thing we have been talking about is hormonal analysis. Now we will talk about a problem of synthesis, using some of the ideas we have concerning messages to do something. We have the problem of seeing what we can do for the totally deaf. The history of this problem is that Dr. Wiesner came to me asking what suggestions I had to make an aid for the totally deaf so they could participate in conversation. The idea that I suggested was the same as the one he had originally had. We found out later on that the Bell Telephone Company owned patents on this but had not actually made the apparatus. We have actually made the apparatus and with Dr. Baslow's aid we are conducting experiments.

The principles that we are using are two: we are trying to educate the deaf-mutes so they can speak decently without the horrible sound they generally use; that is a problem of having the deaf-mute monitor himself. We want a feedback – not a feedback merely in learning but a feedback in continual use. There is the problem of the deaf-mute who only learns occasionally and is in the same position we would be in if we heard people speak only occasionally. The strain would be enormous, like talking into a dead mike where we don't hear our own voice. This is very disagreeable.

We are going to approach the problem through a sense-like touch which is inferior to what we have in hearing. We therefore must limit the amount of information that goes in, the amount of information that is essential, and then build an external cortex to do the job ordinarily done by the cortex after the sound gets in. We have the notion that we do not use the full pattern of the vibration of speech to give us intelligence but, as has been shown by Volk and Coulter, we get a very good intelligible speech if only the relatively slow type pattern for different frequency ranges, perhaps four or five, five we will say, gets through. In the ordinary speech this can be done by reconstructing ordinary speech for the hearing person. This suggests that what we need [to] do is the following: a) to take ordinary speech to a bank of filters, let us say five filters, [204] of the | same length in octave, which means of different length and frequency; b) after we have gone through these relatively crude filters – we want them to overlap to get more discrimination by that rather than less; c) when the sound has gone through each filter we want then to rectify so we get the envelope of the thing. We are only interested in a relatively slow envelope, and not in the phase elements. We then use this to activate a relatively low frequency, perhaps 500 cycles. We carry these vibrations to the fingers of the hand, altho we would like to carry them to some other part of the skin and may do that. The fingers are by far the best bet to start with. We have done this. The result is a man feels vibrations in five different regions. We have carried this far enough already to know that: (a) distinguishable words are recognized as different, if they are to the ordinary sense, in various cases and we have actually learned to get people to recognize a small vocabulary – quite small, ten or fifteen words. (If you can do that you can do a lot more); (b) the same words spoken by different people are recognizable as having the same pattern.

Dr. Baslow is organizing the teaching and training and we are going right ahead with this. How far we will go I cannot say.

We are planning aids for the blind which will be portable and will facilitate his movement from place to place rather than his reading. The reading problem is a different one. That is equivalent to the Vocoda in showing how much needless information we ordinarily carry. In sight, boundaries are far more important than the things they bound. The test of that is the fact that we can read line drawings very easily.

We want a photo cell which can be pointed and we are putting a man to work on this; this photo cell is to be sensitive not to light intensities but to boundaries, which means logarithmic derivatives as you move the things around. This information is to be carried to the ear. But here is one mistake that has been made in a lot of work: it is difficult to try to build inside of a person decent educative channels that will interpret the acoustic matter spatially whereas the blind man automatically interprets kinesthetic sense spatially. It is the best he has. Therefore, we want this apparatus to use the ear only as a monitor to determine when he is on range, and we want to have him get his spatial sense. There will be a little vibration that is put in for picking up the sound. The picking up of the sound is not that which will give him the spatial experience. It is the following of the edge. There is the first space. | [205]

FREMONT-SMITH: How does he follow it?

WIENER: By pointing.

STROUD: A ten foot finger.

WIENER: This is the difficulty of having monocular vision as far as it goes. We can give him two pieces of monocular vision. He holds them in the same hand and affects their convergence by squeezing. These two pieces go to the ear as vibrators of slightly different frequency so that the two sounds sizz and sazz. They are distinguishable. If they come to the edge at the same time he squeezes to indicate the distance. There again the strength of squeeze, which is kinesthetic is the thing to relay distance from and not the time.

TEUBER: There are two German patents that have been described very briefly in the ophthalmological literature of the last year, so that it is hard to get any idea as to what they have been doing. Both are helps in orientation for the blind. One gadget involves the use of ear phones with photoelectric cells (Pallas (8)). The patient is scanning his surroundings and gets tones of varying pitch in his receiver, corresponding to variations of brightness in his environment.

The other gadget involves much more. It is difficult to find out how it is done – there is just a paragraph (Pallas (7)). A simple radar set maps parts of the man's visual surroundings onto his forehead; brightness gradients are translated there into tactile impulses by means of spatially patterned electrical stimulations.

WIENER: Neither the tactile sense nor the hearing sense used for picking up scanning are automatically associated with any good space description, whereas the kinesthetic sense is. Therefore the best aid for the blind man is to use the kinesthetic. While there is less learning, the learning is the pattern of his game. We are giving him a more sensitive and more mobile cane. There are more ways of doing it.

We are led by two motives. The kinesthetic sense ordinarily feeds into the visual cortex, even of the blind man, fairly well. That is the usual way he maps the world, even if he describes it visually, and we want a participating reaction, a feedback rather than a passive operation. That is really the same statement almost over again. | [206]

DISCUSSION:

GERARD: Is that part of the reason why you chose to translate hearing into touch rather than vision?

WIENER: There was another reason. We want the thing to be portable, something that a man can use continually with him.

GERARD: He won't have to keep looking at it.

WIENER: I know even putting it into the hand is bad. You could put it in the gloved left hand or make an aid to squeeze with receivers. You don't want to sacrifice the man's good sense. If you can sacrifice an inferior one.

PITTS: The upper lip is extremely sensitive and you don't use it for anything else.

WIENER: The upper lip might be used. The principles here are fairly clear. They can be used for general sensory replacement, just as I feel for empathy. I would say the principles are, a) a feedback; b) an estimation of what information is important or not. Prefiltering the information and furnishing the inferior sense only with what will get through the body is vitally important to this work, and it is promising.

MCCULLOCH: This is a matter of how hard you squeeze?

KLÜVER: You estimate distances?

WIENER: He estimates how hard he squeezes.

MCCULLOCH: He squeezes harder when he is nearer, less when he is far away?

WIENER: He knows he is on the beam by the coincidence of the bips.

FREMONT-SMITH: He squeezes until they coincide and that tells him how far away he is?

WIENER: He uses things that are very like the things the blind man ordinarily uses, the kinesthetic sense rather than using hearing as kinesthetic. Even touching on the forehead is not a natural way of doing things, asking for a fine discrimination of a sense not ordinarily associated with fine space discrimination, whereas the kinesthetic sense is what the blind man would use anyhow.

BATESON: Thinking about the space discrimination of hearing, one has fairly good localization of objects.

WIENER: Not good enough.

STROUD: This is contingent on a lot of things. The trouble is it is good to you subjec- [207] tively. That is because you are happy with your hypothesis but could I fool you! |

MCCULLOCH: Before and behind are strangely similar.

STROUD: If you had to depend upon your hypothesis you would break your neck. I can kid you and those in the entire room. In there a man is standing, then walking around and talking. What is there? There is nothing but a radio receiver hanging up on the wall and the right technique of recording. I have seen it done.

MCCULLOCH: It goes wrong awfully easy.

REFERENCES

[208]

References (Mr. Stroud)

1. Hixon Symposium on Cerebral Mechanisms, California Institute of Technology, Pasadena, California *(1948)*.
2. McCulloch, W. S. and Pitts, W.: How we know universals. The perception of auditory and visual forms. *Bull. Math. Biophysics,* 9, *127-147 (1947)*.
3. Craik, K. J. W.: The theory of the human operator in control systems I. The operator as an engineering system. *Brit. J. Psychol.,* **38**, *56-61 (1947)*.
4. Searle, L. and Taylor, F.: Studies of tracking behavior. Rate and time characteristics of simple corrective motions. *J. Exper. Psychol.,* **38**, *615 (1948)*.
5. Hick, W. E.: The discontinuous functioning of the human operator in pursuit tasks. Quart., *J. Exper. Psychol.,* **1**, *36 (1948)*.
6. Stroud, J. M.: The Moment Function Hypothesis. M. A. Thesis, Stanford University *(1948)*.
7. Bartley, S. H.: Some factors in brightness discrimination. *Psychol. Rev.,* **46**, *337-358 (1938)*.
8. Halstead, W. C.: A note on the Bartley effect in the estimation of equivalent brightness. *J. Exper. Psychol.,* **28**, *524-528 (1941)*.

References (Dr. Teuber)

1. Bartley, S. H.: Temporal and spatial summation of extrinsic impulses with the intrinsic activity of the cortex. *J. Cell. and Comp. Physiol.,* **8**, *41-62 (1936)*
2. Bartley, S. H.: The neural determination of critical flicker frequency. *J. Exper. Psychol.,* **21**, *678-686 (1937)*.
3. Bartley, S. H.: Subjective brightness in relation to flash rate and the light-dark ratio. *J. Exper. Psychol.,* **23**, *313-319 (1938)*.
4. Bartley, S. H. and Bishop, G. H.: The cortical response to stimulation of the optic nerve in the rabbit. *Amer. J. Physiol.,* **103**, *159-172 (1933)*.
5. Brücke, E. W.: Üeber[!] den Nutzeffect intermittirender Netzhautreizungen. *Frankfürt*[!], *Moleschott's Untersuch.,* **9**, *367-394 (1865)*.
6. Hoagland, H.: Enzyme kinetics and the dynamics of behavior. *J. Comp. & Physiol. Psychol.,* **40**, *107-127 (1947)*. | [209]
7. Pallas, E.: Rückblick auf die technischen Fortschritte der Augenheilkunde der letzten Jahre. DRP. Nr. 745, 339 Klasse 30d, Gruppe 26 (2-12-1943): Orientierungsgerät für Blinde. *Klin. Mbl. Augenheilk,* **112**, *362 (1947)*.
8. Pallas, E.: Rückblick auf die technischen Fortschritte der Augenheilkunde der letzten Jahre. DRP. Nr. 717, 223, Klasse 30d, Gruppe 26 (24-9-1940): Orientierungsgerät für Blinde. *Klin. Mbl. Augenheilk,* **112**, *364 (1947)*.
9. Plateau, M.[!]: Sur un principe de la photométrie. *Bull, Acad. Roy. Soc., Brussels,* **2**, *52-60 (1835)*.
10. Restorff, H. v.: Über die Wirkung von Bereichsbildung im Spurenfeld. *Psychol. Forsch.,* **18**, *299-342 (1933)*.
11. Schumann, F.: Psychologie des Lesens. Ber. üb. d. II. *Kongr. f. Exp. Psychol. in Würzburg. Leipzig,* pp. *153-183 (1907)*.
12. Talbot, W. H. F.: Experiments on light. *Phil. Mag.,* **5**, *321-334 (1834)*.
13. Teuber, H. L. and Bender, M. B.: Critical flicker frequency in defective fields of vision. *Fed. Proc.,* **7**, *123-124 (1948)*.

14. TEUBER, H. L. AND BENDER, M. B.: Changes in visual perception of flicker, apparent motion and real motion after cerebral trauma. *Amer. Psychol.*, 3, *246-247 (1948)*.

CYBERNETICS

CIRCULAR CAUSAL AND FEEDBACK MECHANISMS IN BIOLOGICAL AND SOCIAL SYSTEMS

Transactions of the Seventh Conference
March 23-24, 1950, New York, N.Y.

Edited by

HEINZ VON FOERSTER
DEPARTMENT OF ELECTRICAL ENGINEERING
UNIVERSITY OF ILLINOIS

Assistant Editors

MARGARET MEAD
AMERICAN MUSEUM OF NATURAL HISTORY
NEW YORK, N. Y.

HANS LUKAS TEUBER
DEPARTMENT OF NEUROLOGY
NEW YORK UNIVERSITY COLLEGE OF MEDICINE

JOSIAH MACY, JR. FOUNDATION

565 PARK AVENUE, NEW YORK

PARTICIPANTS

Seventh Conference on Cybernetics

MEMBERS

WARREN S. McCULLOCH, Chairman
Department of Psychiatry, University of Illinois College of Medicine; Chicago, Ill.

HEINZ VON FOERSTER, Secretary
Department of Electrical Engineering, University of Illinois; Urbana, Ill.

GREGORY BATESON
Veterans Administration Hospital; Palo Alto, Calif.

ALEX BAVELAS
Department of Economics and Social Science, Massachusetts Institute of Technology; Cambridge, Mass.

J. H. BIGELOW
Electronic Computer Project, Institute for Advanced Study; Princeton, N. J.

HENRY W. BROSIN
Western Psychiatric Clinic and Hospital; Pittsburgh, Pa.

RAFAEL LORENTE DE NO[1]
Rockefeller Institute for Medical Research; New York, N.Y.

LAWRENCE K. FRANK
72 Perry Street; New York, N.Y.

RALPH W. GERARD
Department of Physiology, School of Medicine, University of Chicago; Chicago, Ill.

G. E. HUTCHINSON
Department of Zoology, Yale University School of Medicine; New Haven, Conn.

HEINRICH KLÜVER
Department of Experimental Psychology, Division of Biological Sciences, University of Chicago; Chicago, Ill.

LAWRENCE S. KUBIE
Department of Psychiatry and Mental Hygiene, Yale University School of Medicine; New Haven, Conn.

DONALD G. MARQUIS
Department of Psychology, University of Michigan; Ann Arbor, Mich.

MARGARET MEAD
American Museum of Natural History; New York, N.Y.

F. S. C. NORTHROP[1]
Department of Philosophy, Yale University School of Medicine; New Haven, Conn.

WALTER PITTS
Department of Mathematics, Massachusetts Institute of Technology; Cambridge, Mass.

ARTURO S. ROSENBLUETH[1]
Departments of Physiology and Pharmacology, Instituto Nacional de Cardiologia; Mexico City, D.F., Mexico

LEONARD J. SAVAGE
Institute of Radiobiology and Biophysics, University of Chicago; Chicago, Ill.

THEODORE C. SCHNEIRLA[1]
American Museum of Natural History; New York, N.Y.

HANS LUKAS TEUBER
Department of Neurology, New York University College of Medicine; New York, N.Y.

GERHARDT VON BONIN[1]
Department of Anatomy, University of Illinois College of Medicine; Chicago, Ill.

JOHN VON NEUMANN
Department of Mathematics, Institute for Advanced Study; Princeton, N.J.

NORBERT WIENER
Department of Mathematics, Massachusetts Institute of Technology; Cambridge, Mass.

GUESTS

J. C. R. LICKLIDER
Psycho-Acoustic Laboratories, Harvard University; Cambridge, Mass.

TURNER McLARDY
Maudsley Hospital; Denmark Hill, England

CLAUDE E. SHANNON
Bell Laboratories; Murray Hill, N.J.

JOHN STROUD
U. S. Naval Electronic Laboratory; San Diego, Calif.

HEINZ WERNER
Department of Psychology, Clark University; Worcester, Mass.

JOSIAH MACY, JR. FOUNDATION

FRANK FREMONT-SMITH
Medical Director

JANET FREED
Assistant for Conference Program

1 Absent

JOSIAH MACY, JR. FOUNDATION CONFERENCE PROGRAM

[7]

FRANK FREMONT-SMITH
Medical Director

I WANT to tell you how happy we are to welcome you to this Eighth[!] Conference on Cybernetics. For the benefit of the guests present may I take a few minutes to explain the nature and purposes of the Foundation's Conference Program.

You have been brought together to exchange ideas and experiences, data, and methods in an effort to further knowledge in this field. However, the Foundation is also interested in investigating the broad aspects of the problem of communication and integration. Experience gained from many research projects presented for consideration has led to the conviction that one of the greatest needs today is a reintegration of science, which at the present time is artificially fragmented by the isolation of the several disciplines or specialties. We feel that the setting up of physiological and – what is probably more important – psychological barriers between the several branches of science is seriously interfering with scientific progress. Although the fertility of the multiprofessional approach is recognized, adequate channels of interprofessional communication do not exist. The Conference Program hopes to encourage this reintegration.

Thirteen conferences are now in operation covering the following fields: aging, adrenal cortex, biological antioxidants, blood clotting and allied problems, connective tissues, cybernetics, factors regulating blood pressure, infancy and childhood, liver injury, metabolic interrelations, nerve impulse, problems of consciousness, and renal function. Each of these conference groups holds annual two-day meetings for a period of five years.

When a new conference is organized fifteen scientists are selected by the Chairman in consultation with the Foundation to be the original members. In this selection every effort is made to include representatives from all pertinent disciplines. For the purposes of promoting full and free participation of all members and guests attendance at any meeting is limited to a total of twenty-five.

In contradistinction to the usual scientific meeting we place | the emphasis upon discussion and not upon the presentation of formal papers. The introductory presentations at our conferences are merely the launching of the ship – the voyage is the important thing! The person opening a discussion is similar to the person who breaks the bottle of champagne over the bow of a new vessel. In other words we feel the heart of these meetings is the discussion. Even though everything said is taken down by the stenotypist you will be given opportunity to edit your remarks or delete any which you do not want to appear in the published transactions. [8]

From our experience with conferences we have learned that if one desires to communicate successfully with another person, one cannot limit oneself merely to making statements *at* him, and to increasing the power of one's transmitting set when he does not understand. Some consideration of the receiving set is needed. One point which should be stressed is that between the disciplines there are real difficulties in communication-partly emotional and partly semantic. Emotionally some investigators accept only data derived from methods or disciplines with which they are familiar. On the semantic level the physical and biological sciences can understand each other without difficulty as can the medical, psychiatric, and social sciences. However, to bridge the gap between the physical and biological sciences on the one hand and the psychological and social sciences on the other is very difficult. Through the Conference Program this Foundation hopes to foster communication and reintegration and in the published transactions to give a clearer reproduction than now appears in the scientific literature of what takes place in the laboratory and what goes on in the minds of scientists.

This program is an experiment and you are part of the experiment. The success of the undertaking is measured entirely by what the participants gain from such an experience. We encourage your critique and hope continuously to improve our conference techniques.

INTRODUCTORY REMARKS

WARREN S. McCULLOCH
Chairman

THERE are two or three things that I wish to say. In the first place, several of you are new to the group. You may have a little difficulty with some of us as we speak. If you do, please do not hesitate to interrupt in order to make sure you understand what the speaker is saying.

Second, to those of you who are new to the group, we ask you to join in the discussion: we want you to do that right away; but don't feel you are going to be called on for a presentation until you know what we are like.

Sometimes we become agitated and interrupt a person too often, or we find that we are asking him questions about the sentence he is going to say next. When we let someone have the floor, we should permit him to have his say at once, interrupting only if we don't understand what he is saying.

Some presenters like to pause at times to ask if there is any discussion at a certain point. We encourage that, but it depends entirely upon the wishes of the man who is making the initial presentation. Others prefer to be interrupted only on the matter of understanding, preferring to discuss actual content after they have completed their statements.

I hope you all received a copy of the proposed agenda for the meeting, and if there are no objections we will start roughly according to that proposal. I don't believe that we will follow the schedule rigidly.

Dr. Gerard, will you open?

SOME OF THE PROBLEMS CONCERNING DIGITAL NOTIONS IN THE CENTRAL NERVOUS SYSTEM [11]

RALPH W. GERARD
Department of Physiology, School of Medicine, University of Chicago

I SHOULD like to begin by saying, especially for the benefit of the newcomers, that this particular group is the most provocative one with which I am associated. I owe more new ideas and viewpoints to the meetings we have had over the past few years than to any other similar experience; our gatherings, therefore, have evoked some insights. The subject and the group have also provoked a tremendous amount of external interest, almost to the extent of a national fad. They have also prompted extensive articles in such well known scientific magazines as *Time*, *News-Week*, and *Life*. Some of these events have, in turn, led me to speak to you this morning.

It seems to me, in looking back over the history of this group, that we started our discussions and sessions in the »as if« spirit. Everyone was delighted to express any idea that came into his mind, whether it seemed silly or certain or merely a stimulating guess that would affect someone else. We explored possibilities for all sorts of »ifs.« Then, rather sharply it seemed to me, we began to talk in an »is« idiom. We were saying much the same things, but now saying them as if they were so. I remembered a definition of pregnancy: »the result of taking seriously something poked at one in fun,« and wondered if we had become pregnant and were in some danger of premature delivery.

Since this group has been the focus and fountainhead of thinking along these lines, we surely have a very real responsibility, both internally and externally. Internally, since we bring expertness in such varied fields, no one can be sure another's statements are facts or guesses unless the speaker is meticulous in labeling suggestions as such. Externally, our responsibility is even greater, since our statements and writings – which may extend beyond an immediate area of competence – should not give a spurious certainty to a credulous audience, be this audience the lay intelligentsia or that precious company of young physical scientists now finding the happy hunting ground in biology. | [12]

The language, experience, and ways of thought, say, of communication engineering, seem to be admirably adapted to make us recognize explicitly that the nerve impulse is not merely some physical-chemical event but a physical-chemical event carrying meaning. It is therefore a sign or a signal, as the case may be; and this is very important in physiological thinking. To use the best mathematical techniques and tools is obviously highly desirable. Everyone here would agree, however, that mathematics, being essentially tautological, cannot put into conceptual schemes something not there in the first place. Moreover, I doubt if anyone in this room believes for a moment that we have made even a majority of the necessary basic biological discoveries of how the nervous system works. We cannot safely build upon presently available biological knowledge rigorous conclusions about the nature of brain action with any confidence in their enduring validity. Overoptimism has appeared before in this very area. In the early 1800's a flood of mathematical articles based upon the teachings of phrenology and exploiting them quantitatively, issued from the best minds of the time. That material is now known only to such encyclopedic minds as that of Heinrich Klüver, who told me about this.

To take what is learned from working with calculating machines and communication systems, and to explore the use of these insights in interpreting the action of the brain, is admirable; but to say, as the public press says, that therefore these machines are brains, and that our brains are nothing but calculating machines, is presumptuous. One might as well say that the telescope is an eye, or that a bulldozer is a muscle.

This brings us to the more immediate problems, particularly that of digital and analogical mechanisms in the brain. We have spent much time discussing these two types of functioning, and probably all here will agree that both types of operation are involved in the brain; but perhaps I disagree with the majority in the relative emphasis put on the two kinds of mechanisms. I personally think that digital functioning is not overwhelmingly the more important of the two, as most of our discussions would seem to imply, and I want to present some evidence for this view.

In the first place, everyone agrees that chemical factors (metabolic, hormonal, and related) which influence the functioning of the brain are analogical, not digital. What is perhaps not fully recognized is the tremendously important role that these play not only in the abnormal but also in the perfectly normal functioning of the nervous system. The influence of carbon dioxide, of acidity, | of the sugar level, of the balance between sodium and potassium, of calcium and a trace of magnesium, and the influence of the thyroid hormone, the ketonic group, which is coming into prominence as influence on the nervous system, and the action of still other factors, such as temperature – these are not only theoretically possible, but, in extensively documented experimental analyses, are demonstrably great. Variation in them can produce or remove convulsions, hallucinations, voluntary control, consciousness itself. [13]

BATESON: I am a little disoriented by the opposition between analogical and digital.

GERARD: I was going to say a few words about that shortly, but instead I shall explain now. The picture that I have of analogical and digital, owing to the expert tutelage that I have received here, primarily from John Von Neumann, is this: an analogical system is one in which one of two variables is continuous on the other, while in a digital system the variable is discontinuous and quantized. The prototype of the analogue is the slide rule, where a number is represented as a distance and there is continuity between greater distance and greater number. The digital system varies number by integers, as in moving from three to four, and the change, however small, is discontinuous. The prototype is the abacus, where the bead on one half of the wire is not counted at all, while that on the other half is counted as a full unit. The rheostat that dims or brightens a light continuously is analogical; the wall switch that snaps it on or off, digital. In the analogical system there are continuity relations; in the digital, discontinuity relations.

To return to the thesis: the chemical aspect of neural functioning is entirely analogical; there are continuities of concentration and consequence.

Second, much of the electrical action of the nervous system is analogical. The brain waves themselves, the spontaneous electrical rhythmic beats of individual neurons, particularly the well known alpha rhythm, are analogical. I am quite satisfied, and I think most neurophysiologists are also, that these represent a continuously variable potential, not the envelope of discontinuous spikes. Further, steady potential fields exist about the nervous system and have been shown by us and others to vary with the physiological state of the brain or, conversely, when varied artificially, to modify the physiological state of the brain. These fields are also analogical. I hope later to say more about the alpha-wave aspect of these. | [14]

Third, remember that the existence of a digital mechanism is of itself no particular guarantee that its digitalness has functional significance. The skeletal muscle fiber, even the whole heart, is as digital as anything in the nervous system, perhaps even more completely so – the all-or-none law, as it is called, applies to all –

Finally, in this group of considerations, I would emphasize that the synapse itself (and with that the nerve impulse) probably does not function digitally in a great many, perhaps in a great majority, of the cases in the central nervous system. This needs elaboration.

The point about digital and analogical and continuous and discontinuous relations can be developed further in this direction: the nerve impulse is digital in character, it has the all-or-none property. That is, if a stimulus is progressively increased in intensity nothing happens, as far as any propagated message down the nerve is concerned, until some further tiny increment in intensity of the stimulus sets off a full-sized nerve message. The response is all or none, a characteristic digital response. Closer examination shows what really is involved: after one region of the nerve fiber has been activated, the excitation which it in turn generates, and which then becomes the effective stimulus to the next region of the nerve fiber, is well above the threshold for the next region. In other words, when region *A* has been activated, by whatever artificially applied stimulus, it itself develops a stimulus intensity which is much greater than the minimal intensity necessary to activate region *B*. That is, in both engineering and physiological terminology, there exists a high factor of safety. The factor of safety in the nerve impulse and nerve metabolism, as several of us have estimated, is five or more; there is about five times as much electrical current generated by the active region of the nerve as is necessary to excite the next region which is to be activated. This region, in turn becoming active, generates five times as much stimulus as is needed for the next; so propagation, once started, is guaranteed. Even relatively large fluctuations in the condition of the nerve, in the response of one region or the threshold of the next, will not disturb this overimpelling drive to go forward.

Now let us examine the situation at synapses. One synapse that has been studied, by Bullock (1), a single giant fiber synapse in the invertebrate squid, has a safety factor of about three. Some vertebrate synapses also have safety factors well above one, for each presynaptic impulse crosses to the postsynaptic fiber with no problem of summation or the like. This is true, for example, for | the synapse from sensory neurons from muscle [15] receptors to sensory paths running up the spinal cord – as reported here last year by Lloyd (2). Aside from such particular cases, the story for central synapses is one of safety factors below unity; and this means analogical functioning.

McCulloch: I am sorry, I did not understand it. Will you say it once more?

Gerard: The safety factor for excitation to cross synapses in the nervous system, in most cases studied (primarily in the spinal reflex group), is less than one. I am going to document that.

Von Neumann: That means?

Pitts: The single afferent will not fire.

McCulloch: O.K. Agreed.

Gerard: First, the general phenomenon you know as subliminal fringe: when one impulse arrives it may do nothing; another impulse, which also does nothing itself, combined with the first one will produce a discharge. This is just the point about which you were asking.

Fremont-Smith: They don't have to be simultaneous?

Gerard: They don't have to be simultaneous but probably they have to be close together.

Wiener: For that reason there is a rather short excitation period.

GERARD: There may be a zero combination period, but when other cells are involved there may be a very extensive combination period. I don't want to develop this line further because other evidence is much more direct, and I shall give three or four samples.

One is the nerve-muscle junction. This also is ordinarily digital in the vertebrate; one nerve impulse elicits one response of the muscle, and it does so even under conditions of fatigue, drug action, and many other disturbances. Yet in some vertebrate junctions, and in all those of many invertebrates, there is not a digital relation with a safety factor of more than one, and a variety of summation effects are necessary before responses occur.

Another case is the squid synapse, already mentioned, with a high safety factor. Even under slight fatigue, nothing more than would probably occur during ordinary physiological activity, the safety factor at that synapse drops to less than one. Repeated incoming impulses are required to fire it and, even more, the response becomes highly variable. Presynaptic impulses, repeated perfectly regularly, sometimes give tetanic outbursts, sometimes nothing at all, and irregular fluctuations between these extremes. [16] The same sort of variability appears in artificial synapses and in | the nerve fiber. It is especially seen in invertebrate fibers, in which a given stimulus may lead either to a full propagated response or to none, but all show gradations of local changes. The stimulus produces local potential oscillations which may die out gradually or quickly or increment gradually or quickly, so that as long as 30 milliseconds after a seemingly ineffective stimulus (a fantastically long period for nerve), a discharge occurs.

The clearest and most important evidence of analogical behavior of synapses is implicit in the work Lloyd (2) reported here last year. You may remember that I was asked to comment on it at the time; but I was not smart enough to see at once some of the more far-reaching implications. Let me remind you of the phenomena: he was dealing with a particular spinal reflex, the muscle-stretch reflex, in which the afferent neuron connects directly with the motor one – a monosynaptic arc. The motor nerve response to sensory nerve stimulation involves transmission across a single synapse. The size of the efferent discharge is, of course, a function of the number of motor neurons that discharge in response to a given afferent stimulus. Two other nerves play upon that reflex center. One of them, when stimulated with or just before the main afferent, will greatly facilitate the motor response. The other will similarly inhibit the motor response. A standard shock to the primary afferent nerve, at a regular interval of a couple of seconds, gave a constant motor nerve response; and the effects of stimulating the other nerves were tested against this stable background. So far, all this is standard neurophysiology.

The important finding was that rapid stimulation (tetanizing) of any one of these impinging nerves tremendously exaggerated the effect of that particular nerve on the reflex arc. The normal afferent nerve, given the standard stimulus a second or more after a brief tetanus to it, would produce a manifold greater reflex response. The facilitative nerve would, similarly, be much more powerfully facilitative after it had been tetanized, and the inhibitory nerve would produce a much more profound inhibition. In each case the changed effect was limited to the particular afferent nerve that had been tetanized, reflex responses to the other nerves being unaltered. Other evidence showed that this effect of tetanization was produced in the incoming nerve fiber, not at the synapse; and the magnitude and timing of the effect was related to the positive afterpotential of the nerve impulse. In other words, the size of the electrical message going along the nerve to reach the synaptic system determined the number of synapses [17] crossed. An increase of *10 per cent* in the electrical intensity of the nerve | impulse reaching the synaptic group led to an increase of ten times in the number of neurons that were effectively engaged and were stimulated to respond.

One interesting implication of this finding is that the mechanism of transmission at the synapse is electrical; but we don't want to go into that. Another one is that inhibition is not essentially different from excitation, as it should be on Eccles's theory; but that also is beside the point. The third implication is that these synapses are not acting digitally. If the situation at the synapse is such that a small variation in the incoming impulse, a 10 per cent fluctuation in one quantity associated with it, can determine whether one or five or ten or no synaptic units fire, the action is more nearly analogical and continuous than digital. The factor of safety is close to one, rather than the high value needed for true all-or-none discontinuity. Small variations in a nerve fiber, well within the physiological range, can determine whether or not a given impulse is effective.

Although it remains true that nerve impulses are atomic in character and that they move or don't move, I think it dangerous to go on from there and conclude that the functioning of the nervous system can be expressed essentially in terms of digital mechanisms of the all-or-none behavior of the units in the system and, particularly, of their connections. That does not for one moment mean that I don't believe digital functions are present, that nerve nets operate, or that the analysis of the properties of such nets is going to be useful. I am certain all of these are very important. I *am* saying that if we focus our attention too exclusively on the atomic aspects of the nervous system, we are likely to leave out an at least equally and perhaps more important aspect of the mechanisms of neural functioning.

I promised not to take over half an hour, therefore I shall stop at this point. Later I may say something about the several difficulties that arise in regarding the alpha rhythm as a scanning device to resolve problems of perception, and the other difficulties in resolving the problems of memory by recourse to reverberating circuits. I should like, however, to ask one question of the group. Do any of you know of definite experimental evidence that the reverberant circuits in the nervous system, which we all accept and use freely in our explanations, do actually exist? At least at the microlevel I can think of none. There is evidence, and it is quite conclusive, of long returning loops from one part of the nervous system to another; but if there is really decisive proof of interneuron circuits running round and round in a small area I hope someone will present it. | [18]

WIENER: May I refer to the *Life* and *Time* articles? I have not been able to prevent these reports, but I have tried to make the publications exercise restraint. I still do not believe that the use of the word »thinking« in them is entirely to be reprehended. I do not maintain for a moment that the detailed operation of the machine is too closely similar to the operation of the nervous system, but I do want to say that I am equally convinced, as I have said formerly and as I say more explicitly this morning, that the action of the nervous system is not purely digital. Processes like learning, and so forth, seem to me to involve what I spoke of last year, at least the possibility of »to whom it may concern« messages, messages that are not strictly channeled, that are probably hormonal. While I spoke of them as very possibly chemical, I don't want to exclude the possibility of their being to some extent nervous. I am definitely sure that they are. Where I think the working of the nervous system is at least digital is exactly where the speaker has said it is least digital; namely, in synaptic thresholds. I believe that the channeling of messages in the nervous system is extremely important. The nervous system is certainly not just a vague means of merely spreading messages in which the channels have nothing important to do. The channels are very important in the nervous system, but I think it is also clear that the determination of the thresholds of the synapse is something that is variable where we have no evidence at all of the principal factor, that is, a channeled factor.

I think that the freedom of constructing machines which are in part digital and in part analogical is a freedom which I profoundly believe to exist in the nervous system, and it represents, on the other hand, with humanly made machines, possibilities which we should take advantage of in the construction of the automaton. I have also felt that the computing machine, which has been an extremely important factor in the study of nervous transmission, is the best machine for the study of that type of behavior at present. It is not the numerical side of these computing machines that is most important for the nervous system, but the logical side of the digital machines. I feel that the machines we build in the future for a great many purposes should take advantage of nondigital ways of modifying the threshold of digital machines. I do not see any reasonable explanation for the learning process which does not take advantage of these things. In other words, I do not feel that there is the sharp antagonism between the [19] different groups which appears on the surface. I believe we | have taken an important existing factor and studied it, but I see absolutely no reason not to believe that these other factors are present.

McCulloch: May I add one thing? I know that Ralph Gerard feels that it is perfectly certain that the alpha rhythm of the cortex is not analyzable into the responses of small components, that is, that it is not analyzable into a distribution of nervous impulses. I don't know that the evidence for his view is clearer than the evidence of a microscopic circuit actually reverberating. There are many cases in which we know of anatomical closed paths. To my mind it is quite conceivable that the alpha rhythm, as we record it, is nothing but an envelope of disturbances proceeding, let us say, over fine axonal ramifications and fine dendritic ramifications. The individual impulses under those circumstances would be below the noise level of our instruments for the most part. I don't see how this question can as yet be settled.

As to whether anyone has recorded the activity of a small reverberating circuit, I think the question can be answered most easily. If you look at various interpretations that have been put on the work of Lorente de No on the oculomotor system, where the question was first proposed, you have either to suppose that reverberation occurs within individual neurons in some way or that it occurs in a closed loop of those neurons. There does not seem to be a third possibility. I don't know if that makes too much difference to the question of whether or not a system is digital. The evidence in question is evidence from microelectrodes, which are only semimicro, placed in the oculomotor system at a time when a nystagmus has started up. There is a sequence of impulses, first from one group, then from another group of neurons, then back from the first, corresponding to the slow motion of the eyes and then a snap back. This persists for minutes in some cases after the end of the excitation. The only question is whether you are dealing with repetitive activity by the individual components or with a circular path going from component to component. You have to suppose that you have a reverberant process either within the individual components or else between them. It does not seem to me in either case that the question of whether or not it is digital is raised.

Von Neumann: I should like to formulate a »caveat.« I certainly agree with the ideas that Professor Gerard expressed, but there seems to me to be a need of circumscribing some of the terms more precisely. It is very plausible, indeed, that the *underlying* mechanism of the nervous system may be best, although some|what loosely, described as an [20] analogical mechanism. An example from a different field which, however, should not be taken as implying too close a comparison, is this: an electrical computing machine is based on an electric current, which is an analogical concept. A detailed analysis of how a responding elementary unit of the machine (a vacuum tube or an electromechanical relay) stimulates another such unit, which is directly connected to it, shows that this transition of stimuli is a continuous transition. Similarly, between the state of the nerve cell with no message in it and the state of the cell with a message in it, there is a transition, which we like to treat conceptually as a sudden snapping; but in reality there are many intermediate shadings of stages between these two states, which exist only transiently and for short times, but which nevertheless exist. Thus, both for the man-made artifact as well as for the natural organ, which are supposed to exercise discrete switching actions, these »discrete actions« are in reality simulated on the background of continuous processes. The decisive property of a switching organ is that it is almost always found in one or the other of its two extreme discrete states, and spends only very little time transiently in the intermediate states that form the connecting continuum. Thus there is a combination of relatively fixed behavior first, then a rapid transition, then again a relatively fixed, though different, behavior. It is the combination and organization of a multiplicity of such organs which then produce digital behavior. To restate: the organs that we call digital are, in reality, continuous, but the main aspects of their behavior are rather indifferent to limited variations of the input stimuli. This requires in all cases some amplifying property in the organ, although the corresponding amplification factor is not always a very great one. All such organs must be suited to be connected to each other in large numbers, pyramided. Thus the question regarding the continuous or digital character relates to the main functional traits of large, reasonably self-contained parts of the entire organ, and it can only be decided by investigating the manner in which the typical functions are performed by larger segments of the organism, and not by analyzing the continuous functioning of parts of a unit or that of a single unit apart from its normal connections and its normal mode of operation.

It seems to me that we do not know at this moment to what extent coded messages are used in the nervous system. It certainly appears that other types of messages are used, too; hormonal messages, which have a »continuum« and not a »coded« character, play an important role and go to all parts of the body. Apart | from individual messages, certain sequences of messages might also have a coded character. It would also seem that the coded messages go through definite specialized pathways, while the hormonal continuous messages are normally messages at large. In any case, there seem to be very intricate interactions between these different systems. The last question that arises in this context is whether any of the coded ways in which messages are sent operate in any manner similar to our digital system. If I understand the evidence correctly, it is nonexistent in this regard. [21]

GERARD: I agree.

VON NEUMANN: For neural messages transmitted by sequences of impulses, as far as we can localize the state of the transmitted information at all, it is encoded in the time rate of these impulses. If this is all there is to it, then it is a very imperfect digital system. As far as I know, however, nobody has so far investigated the next plausible vehicle of information: the correlations and time relationships that may exist between trains of impulses that pass through several neural channels concurrently. Therefore I do not think that one can claim to know anything conclusive about this subject at this moment. In the same sense, all statements regarding reverberating circuits, feedbacks which may be critical and are at or beyond the verge of oscillation under various conditions of observation, and the like, seem to me premature. In addition, even if they were valid, they would only apply to rather small parts of the total system.

WIENER: May I speak of the real distinction between the digital and the analogical situation? This is a comment on what Professor Von Neumann has said. Suppose that we take an ordinary slide rule. In the ordinary slide rule we have to get the precise position of the slider to give us a number. There is nothing to hold the slider in position. However, if we put little granulations in the slide rule and if we push it beyond one, it would have to slip into the next one. The moment we do that, we introduce a digital element. In other words, the digital element lies in the fact that the things to which we are referring are not precise positions but fields of attraction which impinge upon one another so that the field where there is any substantial indetermination as to whether the thing goes to one or the other is as small as possible. I will illustrate that by tossing a coin. Actually, if I toss a coin there is every possible position for the landing of the coin, a certain region where the coin stands on edge and one where it does [22] not. That is the thing which makes the coin essentially a digital possibility. The | dynamic probability of the coin standing on edge is very small. In other words, we convert; in every analogical system we have a certain region that corresponds to a number in one way or another. In the digital systems these are made so that they consist of fields of attraction. We try to make the regions corresponding to the number, corresponding to the fields of attraction with indeterminate regions, as small as possible in between them so that the particle will develop itself in one position or another.

GERARD: May I pick up both of those comments and again say what I think the important point I was making to be? It is not in disagreement, of course, with either of the comments, but deals with the actual character of the synaptic mechanism. This is organized contrary to the assumption we have all been making, that it behaves discontinuously and would land, like the coin, on one side or the other; that the nerve impulse is clearly set up or is clearly not set up. Actually, there are gradations, as in non-Aristotelian logic, where a proposition can have shades of truth and falsehood.

VON NEUMANN: I should like to submit that the following is an acceptable equivalent of what you are saying: There has been a strong temptation to view the neuron as an elementary unit, in the sense in which computing elements, such as electromechanical relays or vacuum tubes, are being used within a computing machine. The entire behavior of a neuron can then be described by a few simple rules regulating the relationship between a moderate number of input and output stimuli. The available evidence, however, is not in favor of this. The individual neuron is probably already a rather complicated subunit, and a complete characterization of its response to stimuli, or, more precisely, to systems of stimuli, is a quite involved affair. There are some indications that one important trait among those that determine this response has rather loose and continuous characteristics; it is something like a general level of excitation. This is quite plausible a priori, especially for neurons which have many thousands of synapses on their surface, that is, many thousands of inputs. However, this does not exclude the possibility that there may be other important relations within the system of input stimuli, which determine other parts of the response, and that can be best described as coded relations between individual stimuli, or between intensity levels of various subgroups of stimuli.

GERARD: There may be coding factors involved.

MCCULLOCH: May I interject some remarks that may help Von Neumann? In this case [23] the motor neuron is the place on which | you get the greatest convergence of dissimilar signals, that is, signals from dissimilar sources. It is the point at which it matters least which neuron fires, because the muscle will add tensions. Consequently, if it is motor neuron A rather than motor neuron B in a given pool, you are all right. All you need to do is to determine roughly the number, and you will determine roughly the amplitude of that contraction of the muscle or the force of the contraction. When you go to input channels or ascending channels, you usually do not find one of these large fieldish types of organizations of termini but a tight grip of one neuron on another, so at the most either one or two contemporaneous impulses will fire it. If you look at the curve, let us say, for facilitation for the cells of the column of Clark, whose axons go up to the cerebellum, you don't find the motor-neuronlike performance but rather something that goes up very rapidly to a totality at a given number of fibers responding for increasing numbers of afferent fibers excited in a given muscle nerve.

GERARD: That is right. That is the case I mentioned with a factor of safety of more than one. But when you get up to the top brain again, we don't know which or how much of each kind of synapse is there.

VON NEUMANN: Isn't this the critical question: Which of these principles of organizations exits in the brain? We know very little about this.

MCCULLOCH: We know much about it in some instances. I don't want to go into it now, but there are cases where the time has to be a matter of approximately 30 microseconds between impulses coming from two ears. That is rather an exact requirement of time that is precomputed before it is sent up to the cortex. Anything else of that sort may be done in a very small region of the brain stem or in the nerves as they come in. From there on, relayed impulses cannot possibly preserve phase relations accurately enough. Suppose you have sound impinging upon your ears. Thirty microseconds' difference between the time of the impulse starting in two ears is sufficient to give you direction. Is 30 microseconds correct?

STROUD: Right. For sharp transients a temporal difference of 30 microseconds is quite sufficient for you to get the bearing of the source of sound. This was a very old experiment performed in the last century. We have done it over and over again, and it always comes out 30 microseconds for a sharp transient and about 70 microseconds for rather smooth tone.

VON NEUMANN: This may nevertheless be analogical. | [24]

WIENER: It is an analogical mechanism that functions.

MCCULLOCH: How is the mechanism going to transmit its information into other portions of the nervous system? A single click will do it, won't it?

STROUD: Yes, but in the sense that the center that is receiving it is receiving it over a very large number of neurons.

MCCULLOCH: Oh, yes.

STROUD: Which have origins which are quite close together. I believe there is some evidence that as the impulse travels up the tympani there is a sort of compensatory lag in the neurons themselves which tends to make all arrive at the central point at about the same time.

LICKLIDER: There is a suggestion.

VON NEUMANN: Many mechanisms exist which will tell you whether the distance in time of two consecutive systems is of the order of tenths of microseconds. You have to transform this into some statement of an intensity. This statement can then be transmitted at leisure.

MCCULLOCH: Right.

FREMONT-SMITH: May I say a word here? It seems to me that there are a couple of points that could be made. Professor Gerard spoke about permanently valid conclusions. Of course I think we all agree that there is none, and that it is the basic system of science that all conclusions are impermanently valid. Similarly, the question of prematurity is relative. All statements are premature, but some of them are very much more so than others. When Dr. Von Neumann spoke about the lack of the atomic nature of the neuron, the lack of complete discreteness, it occurred, to me that that also now enters our concept of the atom. Isn't it true that we have an entirely different concept?

VON NEUMANN: Forgive me.

Fremont-Smith: Am I wrong?

Von Neumann: No.

Fremont-Smith: We have a very different viewpoint of atomicity than we used to have. The nearer we approach knowledge on any topic, the more the concept of relativity has to be considered.

Von Neumann: What I meant was something less sophisticated.

Fremont-Smith: Correct me and put me in place if need be, but isn't it true that we are discussing the question of sameness and differences, and that if you add the words »with respect to,« then part of the difficulty disappears? If you specify in what respect [25] they are the same or are units, and in what respect they are merging – |

Von Neumann: I mean the very practical operating question of whether in attempting to describe the function of the nervous system you reach simple pictures by assuming that the nerves are elementary units which are described simply, or whether it is preferable to assume that they (or that some of them, or the majority of them) are large distribution centers.

Fremont-Smith: With respect to what?

McCulloch: Behavior of the nervous system.

Fremont-Smith: Both might be preferable because behavior of the nervous system is multifold and in some respects it might be preferable to describe it in one way.

Pitts: It is possible to make very relevant statements on this particular question, because I believe that the part of the nervous system that Professor Gerard is talking about is precisely one where I believe there is reason for supposing that the relation between the two possible ways of describing its behavior should differ from the results to be expected from it. That is to say, in the lower level in the spinal cord midbrain, where primarily we are concerned with the mechanisms for maintenance of posture and the carrying on of motion, where we have to deal with continuous dynamic advance, it is necessary for the system to act as if it were analogous in the sense of having its ultimate input continuously variable, or variable as the output, no matter what happens in the lower levels. From what we know, the toes have a wasteful process; namely, the process does not code at any point. Certainly it does not on the simplest reflexes. At least it represents the intensity of muscle stretch or tension on muscle simply by the proportion of the total number of neurons which come from that source and which respond in this particular way. You have, particularly, the inverse phenomena in which it is wise as a simplification to describe a continuous variable by a discrete one by simply classifying its values into two classes. Here it is much more convenient to describe the behavior of a large collection of dyadic variables by simply describing their sum in the sense of giving all the really important information. In this particular case, that describes the nervous system perfectly well. I should consider it extremely unlikely a priori, in all parts of the nervous system that Professor Gerard was describing in particular, and certainly in the spinal cord, that there was any coding in significant degree except in the sense of one-to-one pathways upward.

Von Neumann: Is the evidence really cogent?

Pitts: This is perfectly good evidence. The mechanisms for maintenance of posture [26] and motion, the reflexes, are operated on | an analogical basis which is constructed by summing digital elements.

Von Neumann: What is the evidence?

Gerard: That is a very good point and one well worth our consideration.

Pitts: The only way to get muscle contraction in different degrees is by exciting different portions of the neurons going to this.

Gerard: May I finish? The suggestion that perhaps –

PITTS: That is the last place we should expect to find coding mechanisms.

GERARD: One should look for different kinds of neural mechanisms in the cerebrum. But the fact is that so far our thinking about what goes on in the cerebrum has been predicated overwhelmingly upon the factual knowledge we have gained of other parts of the nervous system; and this does suggest some concrete experimentation. One should go after the cerebrum now and see if some of the behaviors which do hold for the cord are not present in the cerebrum. Such results would be very illuminating.

MCCULLOCH: Hutchinson is next.

HUTCHINSON: I don't know whether I am injecting something frivolous or not, but if one takes the phylogenetic standpoint, starting from unicellular organisms and going upward to the vertebrates, it would seem a very extraordinary thing for the brain to evolve as a purely digital machine. It is likely to be digital on an analogical basis; and I think that where the analogical properties appear to crop out, they are very likely rather primitive. If you want to keep it digital, you must have an intracellular digital setup of the kind that has been suggested.

PITTS: My exact point. I suggested behaving like an analogical division on a digital basis, but it is perfectly true that the intracellular is behaving on a digital basis by analogical means. I think the digital mechanism was introduced later in the phylogenetic series, probably for the purpose of handling larger quantities of information.

BATESON: It would be a good thing to tidy up our vocabulary. We have the word »analogical,« which is opposed to the word »digital.« We also have the word »continuous,« which is opposed to the word »discontinuous.« And there is the word »coding,« which is obscure to me. First of all, as I understand the sense in which »analogical« was introduced to this group by Dr. Von Neumann, a model plane in a wind tunnel would be an »analogical« device for making calculations about a real plane in the wind. Is that correct? | [27]

WIENER: Correct.

VON NEUMANN: It is correct.

BATESON: It seems to me that the analogical model might be continuous or discontinuous in its function.

VON NEUMANN: It is very difficult to give precise definitions of this, although it has been tried repeatedly. Present use of the words »analogical« and »digital« in science is not completely uniform.

MCCULLOCH: That is the trouble. Would you redefine it for him? I want to make that as crystal clear as we can.

VON NEUMANN: The wind tunnel, in attempting to determine forces of a particular kind upon an analogical model airplane, presupposes similarity in almost all details. It is quite otherwise for the differential analyzer, which is supposed to calculate the trajectory of a projectile. The parts of the analyzer look entirely different from any parts of the projectile. It is, nevertheless, analogical because the physical quantities of the true process are represented by continuous variables within the analyzer, for example, by coordinates or by velocity components of various parts, or by electrical potentials or current intensities, and so forth. This is clearly a much more sophisticated connection between the true physical process and its symbolization within the computing machine than the mere »scaling« in wind tunnels. All these devices have, nevertheless, a common trait: certain physical quantities that have continuous motions are represented by similarly continuous processes within the computing machine. Interrelationships are entirely different in a digital model.

To conclude, one must say that in almost all parts of physics the underlying reality is analogical, that is, the true physical variables are in almost all cases continuous, or equivalent to continuous descriptions. The digital procedure is usually a human artifact for the sake of description. Digital models, digital descriptions arise by treating quantities, some of which or all of which are continuous, by combinations of quantities of which each has only a small number of stable (and hence discrete) states – usually two or three – and where one tries to avoid intermediate states.

WIENER: I should like to say something that bears directly on this from the standpoint of the engineering problem; I am considering the question of automatic factoring in the automatic factory. Probably the best internal brains we can use for it will be digital models. I would not say purely digital, but what we now call digital computing machines. To work a chemical factory, for instance, we should have separate organ effectors. These separate organs will involve the stage in which analogical quantities | [28] are converted into digital. This organ will read the thermometer, will have to convert this reading to, say, rotation of a shaft and then into a unit digit, a tenth digit, a hundredth digit, and so forth, for the machine to be able to take it up. Finally, at the end of the machine we will have an effector. This effector will be something that will turn a tap, let us say. This turning of the tap would be done by some machine which will take a series of digits, one of which will determine the place of the thing to within one-tenth, within one-hundredth, and so forth.

The point that I want to make is that the digital machine for analogical purposes is something that we are going to have to contemplate for the engineering applications of this idea. There is no reason to suppose that it does not happen in human-animal applications as well.

MCCULLOCH: Stroud is next. Say a word about following a curve, will you?

STROUD: Dr. Gerard's anxiety about the fact that the general *Time*-reading public wants to change to some single, absolute explanatory principle makes us feel very uncomfortable for ever having said any thing about it. Personally, I refuse to feel guilty about the foolish mistakes that the general public makes in its limited ability to think or in its laziness. I know of no machine which is not both analogical and digital, and I know only two workable ways of dealing with them in my thoughts. I can treat them as analogical devices, and if this is a good approximation I am happy. I can treat them as digital, and if this approximation works I am happy. The devils are generally working somewhere in between, and I cannot understand how they work accurately. I should like to illustrate. This process of going from a digital device to an analogical to a digital device can go on in vertical lattices *ad nauseam*. You begin with the rather highly digital electron, conclude the next step with the rather analogical hard vacuum tube, use it as a »flip-flop,« which is primarily a digital element, and so on. When you have gone through enough stages, what you are finally dealing with depends upon function. Either of the two approximations is confusing. An ordinary amplifier, if you put a signal in it at the right level, is an analogical device. If you use too much signal, it begins to clip off, with two states, a maximum plus value and maximum minus value, and goes from one stage to the other with the greatest rapidity. If you put in too little signal, you get noise from the shot effects, part of which are quantical effects arising out of the motion of individual electrons in the circuit elements. | [29] [Figure 1] | [30]

If you remember, last year I talked about some tracking devices that NRL had by which it was easy to show that the human system was capable of setting up some »guesstimates« that practically involved the idea of being able to solve for displacement, velocity, and acceleration. With much less complication of external machinery, I set up an interesting tracking problem recently. The problem was to track an object with the eyes. If you set up a spot moving horizontally and ask a man to look at it so that he sees it as a spot and not as a streak, you discover that he does not know how the spot is

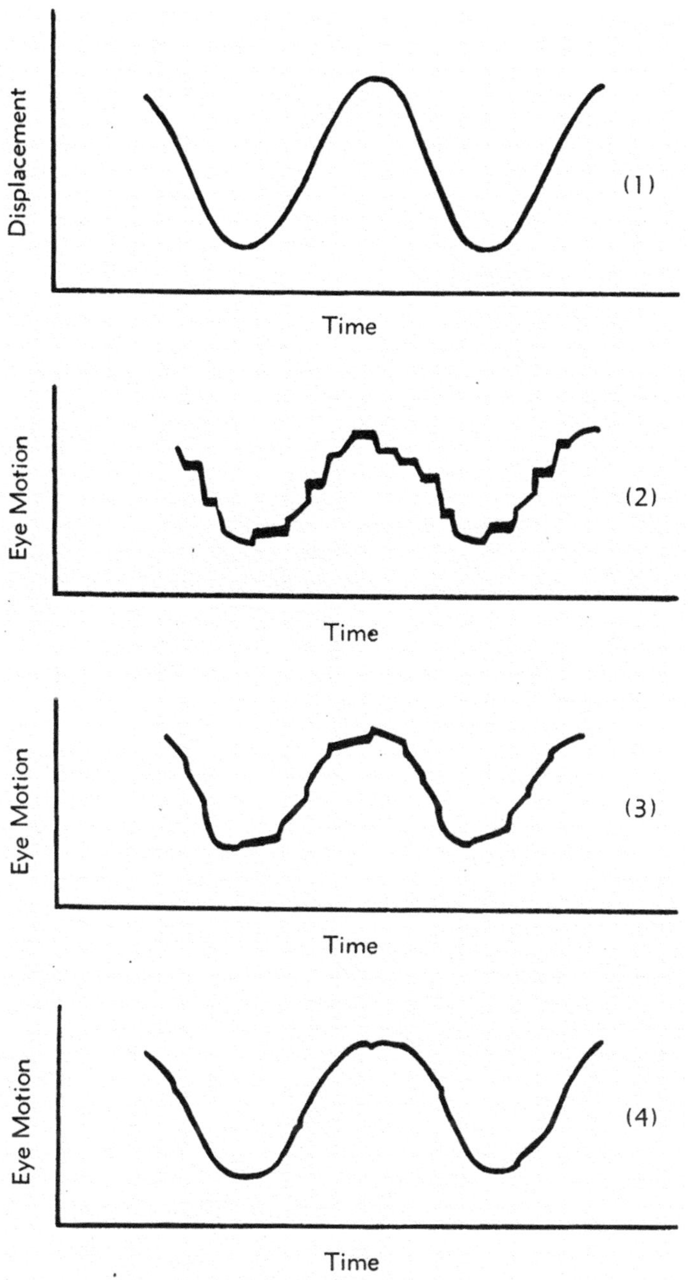

(1) Actual motion of the target.
(2) First solution of tracking problem.
(3) Second (velocity) solution of tracking problem.
(4) Third (acceleration and velocity) solution.

FIGURE 1

going to move. In fact, in the problem it was moving slowly back and forth, with a sinusoidal motion.

First the man followed it in jumps. These jumps happen to fall at the rate of about 4 to 6 per second. You would not confuse that with the simple analogical explanation unless you were permanently biased at the start.

BIGELOW: What is the motion of the eyes as he attempts to follow the spot the motion of which he does not know –

PITTS: What is the analogical explanation?

WIENER: Vertical.

STROUD: The analogical explanation.

PITTS: Is it a digital explanation?

STROUD: I am more inclined to the notion of analogical processes as the statistics of a large number of quantical events. You do not say that the man's eyes move continuously in pursuing a spot which is roughly following a course like that.

PITTS: By analogical and digital do you mean continuous and discontinuous? Many people mean merely that.

STROUD: You treat them as if these transition states did not exist. It is quite a good way of treating them. A little later this chap catches on to the solution of this simple sinusoidal motion and he begins to put in at the same rate little sections of line which are sloped, achieving a much closer approximation of the movement of the object. Here the eye is moving simply and continuously along a sequence of positions. The fits are not good, and you can still see the new little points at which the new constants are taken out and discontinuously posted to a gadget which is now quite obviously working as an analogical device.

BIGELOW: Is the head fixed?

STROUD: The head is fixed. It is simple eye motion taken with a Dodge (eye-movement camera), with a cathode-ray-tube spot as the target. With still more learning he gets to put the curvature in here, until in the final process you have an almost perfect [31] tracing of the ½ cycle curve, with a wobbly noise along it. You have | to use a fine source of light and a good optical system to differentiate the points at which the very minor changes (changes, if you like, in the differential equation describing this sinusoidal motion) have to be posted from time to time. For purely practical purposes I find that the statement that Pitts made is entirely true for all output systems. You consider them as analogical devices, and in this case you find out that the necessary constants that are needed for this operation are the things which are likely to be computed in groups and posted at particular intervals. There are other circuits which are not quite so simple and which involve an eye and a hand, yet showing the same essential characteristics except that their frequencies run in the range from two to three. They are probably a little more complicated.

MCCULLOCH: Two or three per second?

STROUD: Two or three per second. This, by the way, is the fastest set of corrections of which I know.

VON NEUMANN: Did I understand you to say that there is an experimental level of resolution in which motion looks continuous but the derivative looks discontinuous?

STROUD: Yes.

VON NEUMANN: Is it published?

STROUD: No. We want to find out more about it. Besides, I get into trouble every time I try to publish something. I don't know how to do it. I try to get out of it and I also want to pursue the subject in greater detail for purely practical reasons. I must confess that I always keep the Navy's business in mind, and this thing has some immediate practical applications that I am very much interested in.

McCulloch: You are dealing with something that is possible, the correction every fifth of a second or third of a second?

Stroud: That is correct. In this system you can detect the difference because it is a vastly overpowered system and for some purposes can be considered independent of mass; its sharp jumps are very marked and easy to see.

Bigelow: The eye cannot move in infinite acceleration. What is the slope?

Stroud: I wish I had measured them. They are a full decimal order of magnitude greater than any of the slopes I have been imposing upon the eye as motion.

McCulloch: The eye can move far faster.

Stroud: As soon as I set up oscillatory motion which will get anything, even one-tenth or one-hundredth of the slope of the saccadic movements, I find that I have overcrowded the computer | that tries to post the computations and that it breaks [32] down. It simply cannot follow, therefore the eye no longer attempts to solve the problem of motion. The saccadic motions are a whole order of motion faster than any organized effort to follow the motions of a target.

Licklider: I am reverting to Gerard's original problem. I want to say two things. The discussion first showed that there was general analogical continuous substratum to digital processes. Stroud brought out a whole hierarchy of analogues, and so forth. One of the extremely interesting points arrived at, not by getting too many hierarchies but by knowing that many of the things are of great interest to neurophysiologists and psychologists, too, was that there are processes in which we obviously have pretty much all or none of the impulses working as our basic elements. There are so many nerve fibers in a bundle, 25,000 in the auditory, a million in the optic nerve, that it seems inconceivable that each element of detail is important. We have rather the mean behavior of a system, including many parts, each part being perhaps digital. I want to see if I can get the attention back on that level of the problem. I also want to say that I think the time has come when some of us must really know what the distinction between analogical and digital is, besides that of continuous and discrete. Analogue and digit are not words that the ordinary person, even the intelligent person, holds up and says: These are opposite. I can conceive of digital system which is the digital process and the analogue of another digital process, and therefore really analogical. I need clarification. I wonder what the distinction is.

McCulloch: May we start again with the question that was on the floor, that is, the question on continuous and discrete?

Licklider: The question is simply this: We have been using the words »analogical« and »digital« to describe computers. To a lay man analogical and digital are not opposites in any very clear sense. We understand the distinction between continuous and discontinuous or between continuous and discrete. We understand roughly what an analogy is, but we would like explained to us here, to several of us and to many on the outside, in what sense these words are used in reference to the nervous system.

McCulloch: Dr. Pitts, will you tackle it?

Pitts: It strikes me, first of all, that we should speak of physical systems in general, not computers, because a computer is merely a special kind of machine assembled for a special purpose so that we can watch it and derive conclusions from it that we would not be able, perhaps, to find out for ourselves. The physical system in | general is a [33] complex of variables which can be continuous or discrete and connected by various dynamic relations which cause the variables to change as time changes, a complex which can be altered and affected by external inputs. I say the variables can be either discontinuous or continuous. They are usually continuous in the physical systems, with the possible exception of electron spins, and so forth. I am sure we don't have to consider that here. In certain special cases, among which computing machines in the brain come most obviously into review, it is often possible to make a simplification when you try to calculate, account for, or consider the behavior of the system. Suppose that of the variables which compose a physical system and which are bound by various relations to various other variables, one is continuous; suppose that its effect on the change in all the other variables and the subsequent history of the system depends upon a single fact; namely, upon whether its values which may fall within a continuous range belong to one class or another class, or say, are less than B or no greater than B, and that that factor alone makes the difference to the rest of the system with respect to that variable, all you can say about it, all that matters for you is whether that particular variable lies within one range or another range. Well, then, as far as we are concerned, in describing the system we can replace that variable by another one, not continuous at all but only capable of two values, say, zero in one range or zero in another, in any place you please. We can replace it by a discrete value whose knowledge would tell us all that is really important about the knowledge of that second variable from the point of view of the rest of the system. You see, whether it is possible to ignore the actual continuity of a physical variable in that sense depends upon the whole dynamic system and upon the relation between that particular variable and other ones connected with it. It does not depend upon whether it is in its own nature continuous or discrete. Therefore, continuous variables can be so ignored in the sense of affecting their neighbors only by virtue of their exceeding or not exceeding a certain quantity. The simplest example, of course, is the neuron, under certain conditions, at least. Others are the computing machines that are constructed almost wholly out of variables which are capable of finite number of states called digital computers. The ordinary desk computer is primarily an example of that kind and of the *other* kind where we try roughly to make the continuous variables ape the continuous variables of the physical system in which we are really interested. We also try to make the functional relations between them like the true laws of nature, | in which we want to calculate laws in true nature. So the whole system becomes a model in the sense of a true physical system. These are called analogical simply because there is a detailed analogy between the computed system and the computing one. Actually, the notion of digital or analogical has to do with any variable in any physical system in relation to the rest of them, that is, whether or not it may be regarded for practical purposes as a discrete variable. Simplified, I think that is the essence of the distinction.

[34]

McCulloch: Does this cover what you had in mind?

Savage: I think I might add a trifling gloss to it. The word »analogical« may suggest a little too strongly a computing device acting in analogy to the problem situation. Thus, for example, if a multiplication for a laundry bill be computed on a slide rule, the problem is purely digital. Yet the slide rule is properly called analogical precisely because it does not behave in analogy with this digital problem but rather in the essentially continuous fashion typical of an analogical computing device.

Pitts: In such a way continuity is essential.

Savage: That is right.

Wiener: Here is the important point: ordinarily in an analogical machine each digit goes down as we go along in the digits. We are performing a single measurement in which we really are adding the tens digit, the unit, the tens, thousands, and so forth, so that our smallest digit is corrupted by the error in our biggest digit. That is essentially a vicious way of handling things with precision. In the digital machine we make a deliberate effort to have a measurement without any particular degree of error, one which is a Yes or No measurement. In this particular case the probability of error in a particular measurement belongs to that measurement only and is not carried over to one of the others. That probability then can be reduced to a very low quantity. It is like the coin standing on edge. We know that it is either one thing or another. Now the point of the digital machine is that we get our precision of workmanship by extremely close estimates of each digit by itself and not in a situation in which a mismeasurement of one very large digit will corrupt a small digit together with it. To say that a thing is digital is to say that we use this technique of accuracy in the machine instead of the technique of accuracy which consists in extreme precision of the measurement, but in which the error of a big measurement is linearly combined with that of the small measurement and corrupts it.

BIGELOW: I should like to say a few words on a very visceral | level. It is very easy to say things that are true about these words, but it is very difficult to say something which completely conveys the picture. The picture conveyed will be complete in the case of the mathematician, but I think that somebody ought to make the very platitudinous remark that it is impossible to conceive of a digital notion unless you have as a reference the notion of a continuous process by which you are defining your digit; that is to say, the slide rule has continuous length and it has on it numbers which are digital. [35]

WIENER: Certainly.

BIGELOW: The statement that »something is digital« implies that you have as a referent something else which is continuous. The second point I wish to make is that in the actual process of perceiving it is my frequent observation that the continuous property of things as they appear to our sensory organs constitutes an experience between one class of machine or one class of observing device and the outside world. Human beings see light as a continuous phenomenon, although they may be ignorant of its wave or composite nature. Actually, some evidence may indicate that you see a continuous phenomenon, but you may refer this to a digital system which is entirely artificial, but which you yourself produce as a means of interpreting a phenomenon in which you are not otherwise satisfied with your own methods of interpretation.

Thirdly, it does not seem to me enough to describe a digital process as being one in which there are two or more discrete levels in which you are only interested in saying whether you are at level *A* or level *B*. I think it is essential to point out that this involves a forbidden ground in between and an agreement never to assign any value whatsoever to that forbidden ground, with a few caveats on the side.

Finally, I think most people who think about digital machines also have in mind a definition of »coded.« What is meant by the word »coded,« which so far has not been cleared up in this meeting? I don't know that I can define it, but I believe there is this element to it: if you have two or more levels of a quantity, such as a voltage, or if you have two or more periods in time, and if you take the same event and assign to it different numerical values, for example, then you are in some sense coding. You are coding with reference to levels or with reference to segments in time sets. For example, a binary pip, at one moment in time in the computing machine, might equal 7 or 8. At another time it might equal 1. The only way you know 1 in one case, or 7 or 8 in another case, is by referring again to some outside reference system, so that | coding means in effect a technique aimed at gaining increased efficiency by having a simple signal possess different values when referred to a different referent. [36]

WIENER: Yes.

MCCULLOCH: You are next, Dr. Pitts.

PITTS: There is a certain difference between the brain and the computing machine, because in the brain there is not the possibility of variation that there is in the actual construction of computers. In the case of the brain you cannot alter the meaning of the signal coming in along a fiber in the optic nerve. It always means that a certain amount of light has struck a certain place.

BIGELOW: Is that true?

PITTS: It is absolutely true.

WIENER: In combination with another signal it may mean a different thing.

PITTS: Connected with that particular place permanently.

BIGELOW: If I see a green light on the corner of the street and I am driving a car, it certainly means one thing. If the light is at another place, it means something else.

PITTS: If a signal comes along, a given fiber is struck. A light strikes a given point in the retina under given conditions and you cannot contemplate change in the wiring, which is implicit at levels of that sort to increased efficiency.

MCCULLOCH: In foveal regions they are soldered one to one.

SAVAGE: Is Licklider satisfied with these answers to his questions?

LICKLIDER: Is it then true that the word »analogues« applied to the context of the computer's brains, is not a very well-chosen word; that we can do quite well if we stick to the terms »discrete« and »continuous,« and that when we talk about analogy we should use the ordinary word »analogy« to mean that we are trying to get substitution?

BIGELOW: I should object to »substitution.«

LICKLIDER: I mean the object we are trying to compute, using »analogue« in the way it is used by most people and not in the way used by the computing machine.

WIENER: I think perhaps »discretely coded« would be good words for »digital.«

LICKLIDER: I think we could communicate better.

SAVAGE: We have had this dichotomy with us for four or five years, Mr. Licklider. I think the word has worked fairly well on the whole. Most of us have been familiar with its meaning. There would be some friction for most of us in changing it now.

[37] MCCULLOCH: I should be happy to abandon the word except that | I don't see how any simple word like »continuous,« as opposed to »discrete,« would take the place of it. I think one would have to say, as Wiener suggested, »discretely coded« or »continuously coded.« I think that is the chief obstacle.

LICKLIDER: I did not think I would be successful in getting machine computers to use the word »conscious.« That has been around a long time and now has to have a group to find out what that means.

FREMONT-SMITH: And they didn't.

TEUBER: When Dr. Gerard spoke, he seemed to try to take us all the way back to the beginning. To me, he conveyed the idea that in the peripheral nervous system, we do have something that can be described – on some low level at least – in terms of discrete or digital functioning. No matter what we do to the nerve, we either set off a spike or we don't. It has been known of course, all along, that we can do all kinds of things short of setting off the spike – all sorts of things between firing and not firing, but the question is: Does it matter? We can raise the local potential by manipulating the environment of nerves. Still, whether we set off a spike is a matter of yes or no, zero or one.

Now the point is made by Dr. Gerard, if I understood him at all, was that the situation in the central nervous system might be quite different. Apparently, we need more than one afferent in order to set off an efferent, and we have many afferents converging on a single synapse. How many convergent afferents have to be active for transmission to occur depends upon the structure and state of the central nervous system in that region. But this again would not detract from the fact that we can only either set off the synapse or we don't. It may be all right to say that transmission in the central nervous system is not obligatory, but I don't see why this should make it impossible for us to use the »digital« analogy, if I might add to the confusion by coining this term.

To assume digital action is permissible as long as we remember that we are dealing with a model. The only justification for using the model is its heuristic value. It may turn out to be inapplicable to the central nervous system, but by finding out *why* it is inapplicable, we shall have discovered facts about the nervous system which we don't have in our hands at present.

What is it that we know now, precisely, that would make the »digital« model inapplicable? It cannot be the factor of convergence and non obligatory transmission in the central nervous system. I wished Mr. Pitts would have made that point, since this is the reasoning he has used for the past three years. I think I did | not quite understand Dr. Gerard's argument. I would rather have the discussion revert to him. [38]

One more point: In the retina we have an interesting difference in neural structure. In the central region of the retina there is opportunity for one-to-one connections from cones through bipolars to ganglion cells: That does not mean that there are not ample opportunities for cross-talk through collaterals and horizontal cells, even in that central foveal region. Still, transmission can take place in one-to-one fashion. This is quite different from the periphery of the retina, which constitutes the bulk of the structure. There we have a tremendous amount of convergence.

McCulloch: About 200 to 1.

Teuber: Anywhere from 80 to perhaps 200 or more rods for each ganglion cell, again with reciprocal overlap in intermediate layers, and all the structural complexity characteristic of the central nervous system, as Polyak has shown. In that sense, the retina is a piece of cortex pushed out towards the outside world, rather than a peripheral end organ.

Now this anatomical difference between central and peripheral retina is there; the question is, does it make any functional sense? The usual interpretation is that optimal spatial discrimination has to be mediated by the central retina, where there is opportunity for one-to-one connections, and maximal light perception is mediated by the periphery where there is considerable convergence from rods onto ganglia; still, even if the anatomical difference were directly related to the hypothetical difference in function, both regions of the retina could work according to digital principles, as far as the firing-off of individual ganglia in the optic nerve is concerned. In sum, I didn't quite grasp Dr. Gerard's point.

Pitts: I want to chime in there to say exactly what you said, except that I should like to state in the form of a caricature what one of Dr. Gerard's arguments appeared to me to be. It may make clear to him what is disturbing me. His argument about Lloyd appeared to caricature roughly the following: we are trying to prove that the behavior of nerve systems is not completely descriptive of all-or-none impulses because Lloyd's experiment shows that the effect of an all-or-none impulse sent along given paths at the other end is altered by the all-or-none impulses that we may have sent previously along either the same or other fibers. Therefore the nervous system is not describable completely in terms of all-or-none impulses. That is the way the argument appeared to me, why it appeared to me not to have relevance. I think I must have mistaken its application. | Because presumably continuous processes are intervening at the other end, whereas all the all-or-none impulses sent along the tracts can affect the results of new ones we send in, therefore we cannot describe the behavior of the nervous system in terms of all-or-none impulses. But this does not seem to me to be cogent. [39]

Savage: I take it your point is that digital machines behave that way, and that their response to digital stimuli does depend upon the past history of digital stimuli?

Pitts: The question whether or not there is an intervening variable makes little difference and is of no consequence.

McCulloch: I want Dr. Gerard's opinion at the conclusion of this.

Kubie: I want to consider the digital and analogical concepts at another level. Plain words lead to confusion when we do not know what we are trying to use them for. I cannot conceive of any measuring device, whether a machine nor not, that is not ultimately digital. If you measure, you count. If you are going to count, you must be able to recognize identical discrete units. But in science we often try to measure where we cannot even identify the units. Here we have to work by analogy.

Consider the clinical thermometer. Is a clinical thermometer a digital measuring machine or is it analogical? It is both; because if you think simply in terms of the temperature of the aperture into which you insert it, it is digital. If you think one step beyond that, in terms of the internal processes about which you are going to make some deductions on the basis of estimated quantitative changes in *un*isolated units, then it is analogical. Therefore, whether a machine is digital or analogical depends on the use to which the machine is put. As a measuring device, however, a machine must always be digital.

The reason this is so important to me is that in all our theories of human behavior the word »dynamic« implies a capacity to measure in an area where we cannot make the distinction between analogical and digital at all. I will have occasion to return to this later.

Klüver: I wonder what Dr. Gerard would have said if he had discussed these problems, not as a physiologist, but as a biochemist. For instance, does the Krebs carboxylic acid cycle involve analogical or digital mechanisms? In fact, is it particularly fruitful to consider it in terms of such a dichotomy? That is one point.

There is another point which might be stressed in connection with Dr. Teuber's remarks on the retina. Again and again the attempt has been made to determine the [40] functional meaning of | the high degree of anatomical differentiation in the visual system or the functional significance of the point by point representation of the peripheral sensory surface in the cerebral cortex. Lashley once suggested that the difference between anatomical systems with little or no subordinate localization and systems with a high degree of internal specificity is related to a difference between nervous mechanisms regulating intensity of response and mechanisms involved in the regulation of spatial orientation [Lashley, K. S.: Functional determinants of cerebral localization. *Arch. Neurol. and Psychiat.*, **38**, *371* (1937)]. Many investigators have tried to relate the facts of spatial differentiation in the anatomy of the visual system to facts of visual functioning, for example, to the spatial differences of visual stimuli. More recently, in an anatomical study of thalamocortical connections, Lashley found that the differentiation within the anterior thalamic nuclei and their cortical fields was as precise as in other sensory systems. Since olfactory experience is lacking in spatial character, he found it difficult even to imagine an attribute of odor represented by an accurate detailed spatial reproduction of the surface of the olfactory bulb on the cortex [Lashley, K. S.: Thalamocortical connections of rat's brain. *J. Comp. Neurol.*, **75**, *67* (1941)]. If such a topographical arrangement is functionally meaningless in the olfactory system – so runs Lashley's argument – there is no reason to give it a functional interpretation in the visual or other systems unless that is done on nonanatomical grounds. Topographical arrangements may very well be »accidents« of the mechanism of embryonic development. To be sure, the situation is even more complex, since voices have recently been raised that the anterior nuclei of the thalamus are not concerned at all in mediating olfactory function and that they represent structures in an anatomical circuit concerned with emotions.

To return to the visual system, it may be argued that the eye and the subcortical and cortical structures of the visual sector of the central nervous system represent more than merely delightful opportunities for anatomical and electrophysiological researches. I may be forgiven for mentioning the old-fashioned but nowadays apparently somewhat radical idea that these structures have something to do with *seeing*. You would think that any investigator who really tries to relate brain mechanisms to visual behavior would be seriously concerned with the question of how nervous mechanisms as they present themselves on the basis of electrophysiological data are related to mechanisms as they are uncovered in investigating actual processes of seeing. Unfortunately, the visual | sector of the central nervous system appears to be an excellent device for [41] achieving a remarkable degree of independence from the intensity and energy fluctuations on the retina. It is apparently for this reason that some years ago I was unable to find any evidence for electrophysiological correlates even of »primitive« visual functions operative in actual seeing [Klüver, H.: Functional significance of geniculo-striate system. *Biol. Symposia.*, 7, *253* (1942)].

This brings me to a third point. It seems to me that the discussion of the problem whether nervous system activity involves digital or analogical functioning or both has been chiefly concerned with a nervous system constructed by electrophysiologists and anatomists. I happen to be interested in the nervous system that is actually operative in behavior, let us say, in seeing. I find then, for instance, such facts as that a given line suddenly appears perceptually longer and that this increase in perceptual length is – and in another situation is not – associated with an increase in objective length. Or I find that increasing the length of a certain line transforms what appeared to me as chaos into the face of a devil or, if you wish, into five devils. I do not know how you propose to deal with the reality of polymorphic phenomena on a digital or an analogical basis. In fact, I do not even know how the factors governing the appearance or disintegration of even simple visual *Gestalten* are related to analogical or digital functioning or to what an extent, if any, an experimental analysis of such factors may benefit by digital or analogical models.

Teuber: May I correct one point? When I quoted Polyak's evidence on the retina, especially on the fovea, I did not mean to imply that I accepted Polyak's own interpretation of the anatomical arrangement. However, the fact remains that in the fovea we have this opportunity for discrete transmission in one-to-one fashion through the retinal layers, and from then on up, all the way up to the cortical retina, where there is orderly projection of the macula. Lashley has often made the point – and I quoted him here, about two years ago, I think – that the orderly anatomical arrangement in the visual system might be quite fortuitous, an embryological accident, so to speak. In the olfactory system there is also orderly projection; if one assumes that the anterior thalamic nuclei project olfactory activity to the cortex, then one has to be puzzled by the perfectly orderly spatial projection of these nuclei onto the cortex: What corresponds to space in olfactory experience? Lashley used this sort of argument to discredit the notion that spatial projection may have anything to do with experienced | space, and [42] in that respect he is probably correct. Yet, orderly spatial projection might have a definite functional significance, all the same. For the olfactory system, matters begin to look different now. Since we first discussed this point, Adrian has picked up electrical activity from the exposed olfactory bulb in the rabbit, while the rabbit was stimulated with different odoriferous substances. Depending on the substance used you get an audibly different »orchestration« of the electrical activity, as picked up from the olfactory bulb and fed into a loudspeaker. These differences may depend on differences in diffusion gradients, different substances diffusing in different patterns over the entire expanse of the olfactory membrane. That means that some spatial ordering of activity may be of importance in the olfactory system; diffusion patterns may have to be mapped centrally in some fashion, even though in our own experience the result would not correspond to anything like visual space perception. Study of the structures will not lead us to the ultimate of experience; still, models constructed on the basis of present knowledge of neural structures suggest concrete hypotheses as to how function comes about. As long as the hypotheses are testable, and as long as we keep the obvious model-character of our notions in mind, these models are worth retaining till we have better ones.

PITTS: You mean position on the olfactory cortex essentially refers to position of the odoriferous substance on the olfactory membrane rather than on the quality of the odor?

TEUBER: Intensity, and possibly quality, too. Rapidity and extent of diffusion over the olfactory membrane might give intensity of any one odor, but possibly different odors might give characteristically different diffusion patterns as such, or might be selectively absorbed. Different olfactory stimulations would lead to characteristically different space-time patterns of neural activity in the olfactory system.

PITTS: Space pattern?

MCCULLOCH: Space coded information of some sort.

TEUBER: Even though it does not give us a space experience.

MCCULLOCH: That is right.

WIENER: I should expect a good deal of discrepancy between the possible coding of information in the olfactory system and actual coding, inasmuch as we are animals who are already on the downgrade as far as smell goes. Our smell sense is unquestionably largely a residual sense and cannot be expected to give us a true picture. You are lucky to find anatomical fossils and the connections there that no longer have the same [43] physiological meaning that they had with euosmite animals. |

PITTS: An animal with thirty light spots has not a good conception of visual space.

LICKLIDER: I know which machines are called analogical and which are called digital. I don't think those terms make sufficient distinction and that is all there is to it. It won't help me to talk about differential analyzers. I know all about that.

FREMONT-SMITH: Good with respect to what?

LICKLIDER: It confuses us in communication here. These names confuse people. They are bad names, and if other names communicate ideas they are good names.

FREMONT-SMITH: They are not good means of communication in this context.

MEAD: It would help if we knew when this distinction was made in describing the machines, that is, if we knew the historical use of the term »analogical.«

MCCULLOCH: I don't know how old it is.

WIENER: I would put it at about 1940, when Bush's machine was already developed and when the rival machine, the differential analyzer, and the machines which were working on the principle of the desk machine electronically were being developed. That began to be an acute issue about 1940, and I doubt if you will find any clear distinction older than that.

MCCULLOCH: They used to be called logical machines or analogical machines before the word »digital« appeared.

HUTCHINSON: Analogical, if I may use the word, the difference between the natural and real numbers, is hiding in the background all the time, but you must go back to the Greek mathematicians.

WIENER: If you want to say that in one case you are dealing with counting and in the other, with measuring, the concept of the machine goes back to the Greeks.

HUTCHINSON: That is a neat way of putting it.

WIENER: Yes.

LICKLIDER: Continuous and discrete.

GERARD: May I speak now?

MCCULLOCH: No, not now. I want to make one point here. There is no question in my mind that there are many continuous variables affecting the response of neurons. The question in my mind is different. If you will, for the moment accept as a distinction between analogical and digital, the question whether information be continuously coded or discretely coded. My question is whether these continuous variables, which are undoubtedly present in the nervous system, are conveying information or not; that | is, they may not be coding of any sort of information any more than the [44] voltage which battery or power pack is delivering to your set, which may vary.

PITTS: Or they may be simply representing the effect of past all-or-none actions?

MCCULLOCH: Yes.

FREMONT-SMITH: Or future, setting conditions for the future.

PITTS: Or they may have intervening actions precede the discrete and following one.

BATESON: There is a historic point that perhaps should be brought up; namely, that the continuous-discontinuous variable has appeared in many other places. I spent my childhood in an atmosphere of genetics in which to believe in »continuous« variations was immoral. I think there is a loading of affect around this dichotomy which is worth our considering. There was strong feeling in this room the night when Koehler talked to us and we had the battle about whether the central nervous system works discontinuously or, as Koehler maintained, by leakage between axons. The present argument seems to me to be the same battle.

PITTS: Whether better or worse, if insulating partitions were put between separate synapses.

SAVAGE: The battle is whether that distinction is worth making or not. The most important question is that which Dr. Licklider said: Is the nomenclature confusing us, or is the nomenclature a promising one? That has always been a central issue here.

FREMONT-SMITH: With respect to what? It might be confusing in certain contexts and very helpful with others.

SAVAGE: To be sure.

WIENER: That same issue did come up in the matter of genetics end; the continuous variables which were driven out with the pitchfork at one door came back. Are not characterizations as observed simply certain factors which combine with certain variables to give us the characterizations as observed? That is precisely the situation that has occurred here.

HUTCHINSON: However, you always use such situations in the mathematical handling of pure genetics.

WIENER: Use them both ways. The moment you begin to consider survival factors, you have to consider how these affect it.

HUTCHINSON: The continuous situation.

LICKLIDER: I am afraid we disturbed the course of things by talking about this too much. We really ought to get back to Gerard's original problem. We will use the words as best we can. | [45]

MCCULLOCH: All right, we have got about a half-hour before we stop for lunch.

GERARD: I shall try to discharge my duty to the group. Perhaps there will be some other points that will occupy us for another half-hour. Most of the comments made in the last few minutes seem to me very clearly to point up the critical issue, but I should like to go back just a bit and explain what I think has happened. Incidentally, I think this whole discussion has been extremely worth while to us even at this late date in our history.

Stroud used the example of the electron tube working on its characteristic or off its characteristic; on the characteristic one has continuous relations or an analogical behavior; off the characteristic it »flip-flops,« Yes or No. Several others have pointed out that there are always these analogical factors in the functioning of any machine or system the ultimate output of which is effectively digital. Now most of the physiological evidence I gave was to the effect that in the nervous system, in the body of afferent receptors and efferent effectors, in the synapses, and perhaps even in the nerve fibers, operations are much more on the characteristic of the tube than we have usually assumed in our physiological thinking. It is operating, to use Pitts's term, in the forbidden region of the system. Now, just to the extent that that is where the system is operating, it is, not only by definition but also in terms of the interpretations of what it does and of the significance attributed to what it does, operating analogically.

Dr. Pitts made another critical point which was picked up by Dr. McCulloch and some of the others. He said that the essence of the digital machine was that, whatever happens inside, so far as the significance of its operation is concerned, all that matters is whether something is below or above a certain number in this category or that category. Whatever is happening as the wheels stir or the electrons shoot, what counts in terms of the functioning of the machine is its Yes or No answer. Dr. McCulloch pointed that up nicely by saying, »The question is, is coding done digitally or analogically?« My point, in emphasizing the actual functioning of the nervous system in the forbidden continuous region, is that much thinking about the nervous system and much of the theoretical interpretation of memory, learning, and many other things has been based upon its *not* functioning in this region, upon its working digitally, upon the all-or-none behavior of firing or not firing an impulse. What I am suggesting is that, [46] although it is certainly true that impulses either do or do not fire through large | parts of the nervous system, this may not be the critical mechanism in its effective functioning. Just as has been said, in order to operate with a continuum one breaks it down into units with which one can work; but that is just an incidental procedure, which could perhaps be avoided, and might have little relation to the ways the whole functions. So, first, we do have many continuous mechanisms operating in the nervous system and my feeling is, although I confess at once this is not established except by collateral evidence, that some of those continuous mechanisms have coding value and are critical to the functioning of the nervous system. I am further suggesting that, even though we find digital operation in the nervous system, this may not be the essential mechanism accounting for its behavior but may be incidental, to pick Teuber's nice term, to the orchestration.

I fear to use the word »*Gestalt*« at this point, let me use the term »envelope.« Perhaps all through the brain, as is certainly the case in the periphery, the fact that discrete nerve impulses travel is not important. What is important is the total pattern of time intensity. No variation of the impulse, and whether messages go by discrete impulses or by some other mechanism which is not discrete, would essentially alter the total performance of the nervous system. That is stated as an extreme, and I ask how nearly it is valid. I have no doubt myself that in some cases in the nervous system the discreteness of the impulses, the digital behavior, is critical to effective functioning of the nervous system. I am also suggesting, however, that, to a much larger extent than has appeared in most of our discussions and indeed in the thinking of most people in the field, the fact that nerve impulses run discretely may be accidental. In the final determination of function, it is the total accumulation of impulses, here, there, and in the other place, which is critical. That is the initial difficulty.

SAVAGE: One of the things Gerard's discussion has brought out is a concept which has not been given much emphasis in this group, largely because of our perhaps untoward tendency to classify all computers as digital or analogical. I refer to computational procedures in which random devices – gambling apparatus, to speak figuratively – play an essential part. As a matter of fact, such methods are currently quite fashionable in applied mathematical circles, where they are romantically referred to as the Monte Carlo method. In the nervous system there is such a multiplicity of similar elements that one can imagine that some computing is done by »games of chance« in which some of these elements are drawn upon at random, relying on statistical averages | for [47] the appropriate effects. We have already come across this concept in our sessions, for example, in the treatment of clonus by Rosenblueth and Wiener. Statistical averages played an essential part in their hypothetical mechanism and in some of the phenomena Gerard has talked about today.

WIENER: May I say we have an example which plays right into your hands here. We are using shunt effect for analysis of the electric circuits in exactly that way. Excuse me! I want to pass that along because that was apropos.

BIGELOW: A few minor points that Gerard mentioned: the possible existence of operation in the forbidden zone, needless to say, is a contradiction in somebody's terms. If a device operates in an in-between zone and if that is a meaningful behavior it seems to me one either has to throw out the term »forbidden« and admit that the zone is an acceptable one having a value, or else assume that there are as many values as you please and therefore as many zones as you please, and that therefore there is a continuum of zones, in which case the digital property really has vanished and you are talking about analogical concepts.

GERARD: Forbidden for one type of functioning.

BIGELOW: It seems to me that most people who are approaching this Conference from a mathematical or machine side, as I do, would be happy to throw the following thing up in the air: What we mean by neurons are not cells as they are described in somebody's book on cell structure; we mean that the neural cell is exactly that part of the system which has the property of carrying out processes like computation, that is, the property of carrying out operations which are in fact digital. I think that actually the physicist would be willing to use this as a definition of what the nervous system is, all and everything that the system is, calling all else another system.

GERARD: I think the physiologists would be likely to say that that is just like a physicist.

BIGELOW: There is one more point. As Dr. Savage has been saying so many times today, the useful aspect of the idea of the computing analogy and the computing aspect of the nervous system depends upon whether or not by exploring such analogies you can come to any new insight into what goes on, that is whether these notions contain useful, descriptive properties of what goes on. It is clearly true that sooner or later – and perhaps we are there now – we will reach a state where the business of describing some other point where computational properties of the nervous system are not like the model and will be advantageous and probably the best way we can explore it. | [48]

GERARD: That is a very good statement.

PITTS: There is a third between the two, because they are not opposite. The digital and analogical sorts of devices have been defined quite independently and are not logical opposites. We called them analogical because we think we meant roughly this: when we want to solve equations and construct a device to do it, often the way to do it is to construct a system. You can map variables in which we can make a one-to-one representation between the variables in the computer and the variables in the system obeying the equation, a representation such that the connections between them are the same equations that we want to solve. Beside that simple sort of analogy and direct relationship, we might also consider the possibility of a true sort of continuous coding where the mathematical corresponds between the machine and the original system which obeys given laws, although it is one-to-one, and therefore from the results of the computing machine we can calculate, we can find our answer. It nevertheless does not have such simple topology. We might not be able to map variable A in the first system into variable A_1 in the second, B in the first and B_1 in the second, and so forth, and the connections between A and B and the connections between A_1 and B_1. It might be topological where we map A and B together on A prime. If we were not interested so much in the accuracy of the measurement, something of that sort would really approach much more closely to true continuous coding in the strict sense of coding than what is ordinarily thought of in the case of analogical devices, where the coding is simply a strict one-to-one correspondence of a very simple kind of an order, where order is preserved and practicability and continuity are preserved, and so forth. You would really almost need that in the brain if it were to operate and do all the things it does with the truly confused mode of behavior, because the dynamical laws are not at your disposal, whether continuous or discrete.

SAVAGE: Would you describe the sort of coding once more?

PITTS: Where the correspondence between the variables in the machine and the variables in the problem is not a one-to-one topological correspondence, it does not have simple continuity properties.

WIENER: That is easy to do.

PITTS: That is necessary in any brain that would operate on continuous principles simply because the equations that govern the relations between neurons or pools of [49] neurons that are near, one to each other, are fixed and cannot be fixed around those | laws. They cannot be changed to suit the convenience of the problem.

WIENER: I am encountering something of this sort in the work I am doing in connection with prediction. If you are dealing with a discrete time series, the all-or-none sort of coding is a natural thing to do. However, in the presence of noise and a continuous time series, it is still possible to introduce functions in such a way that there is a hidden coding. The coding isn't given directly by the values. At one time it is a quasi. It has the appearance of being continuous, and yet you have discrete lumps of information which come up after a certain time but not all at once. There arises, for example, in this work that I am doing on nonlinear prediction –

PITTS: What a single value on the one side represents depends upon the whole course of values on the other side, so that the transformation is not a simple functional one?

WIENER: Yes.

BATESON: On the question whether or not the sorts of logic involved in an analogical computer would be essentially different from the sorts of logic involved in a digital computer, I don't know with what rigor Whitehead makes the point that the shift from arithmetical to algebra is the introduction of the »any« concept.

WIENER: Quantification.

BATESON: Arithmetic is quantification.

WIENER: In the logical sense.

MCCULLOCH: Quantification in the logical sense.

SAVAGE: It means any.

PITTS: It means every.

McCulloch: Introduction of pronouns.

Bateson: Introduction of pronouns in a sense. Is there difference of that order?

McCulloch: No. It is much more as if you shifted a problem, let us say, from the calculus of propositions to the calculus of relations – something of that sort – and it is a much greater shift.

Frank: May I ask a question that follows that? Is there any light or any understanding of how the transition or transformation from discrete to continuous takes place? Are they two utterly opposed processes? The second question is, is it conceivable that organisms which have had a very prolonged evolutionary history have developed a capacity for making that transformation from discrete to continuous that we are not yet capable of conceptualizing in language? That is a very important point. I get the impression that we are dealing with processes that we can approach | from the concept [50] of discreteness or the concept of continuum and that it depends upon the way we phrase our problem which will appear to be more significant. I wonder if we have the same situation as that pointed out years ago by Eddington when he said that physics was classical on Monday, Wednesday, and Friday and quantum on Tuesday, Thursday, and Saturday. We are not confronted with irreconcilably opposed viewpoints when we realize that there are two ways of recording events which exhibit both discreteness and continuity.

McCulloch: Let us put it this way: as long as the probability of a state between our permitted states is great and has to be taken into account, we have still a flavor of the continuous. When the probability of the *Zwischen* state is zero or negligible, we think chiefly in other terms. That is, I think, purely a matter of practicality.

Wiener: I think it is entirely a matter of practicality whether we approximate a situation which neither corresponds to an absolute number of theoretical lumps nor to a complete continuum with all the derivatives and extremes by either means. That arises all the time in mathematics; it is the correct procedure and annihilates no theories.

Wiener: You simply do whichever is convenient.

Bigelow: We don't have to settle that question here, do we?

Wiener: No.

Frank: I hope not.

Wiener: I say that the whole habit of our thinking is to use the continuous where that is easiest and to use the discrete where the discrete is the easiest. Both of them represent abstractions that do not completely fit the situation as we see it. One thing that we cannot do is to take the full complexity of the world without simplification of methods. It is simply too complicated for us to grasp.

Fremont-Smith: Isn't it true of neurology today that the *Zwischen* zone is becoming more and more pertinent and that we really have to reexamine the all or noneness of the all or none?

McCulloch: A very much more peculiar thing has happened: we have begun to find parts of the nervous system in which a sufficient number of digital processes are lumped so that one can treat them as if they were continuous.

Wiener: Yes.

McCulloch: Mock continuous; that would occur, let us say, in such a thing as the spinal reflex.

Fremont-Smith: But if you go back to your neuron, it seems | to me one can and [51] should – I brought this up in the Nerve Impulse Conference – challenge the use of the words »all or none.« When I brought that up I got a violent reaction, which was what I expected from everybody present. It was as if I had suggested something unholy. It was said to me afterward that the challenging of the all or none was something that would be quite important about ten years from now. I thought that was an interesting comment. It seems to me that the center of our problem is the fact that we are basing the neuron on the all or none, while actually we have only relatively all or none.

PITTS: I think I neglected to make my point a moment ago. It was that there can be devices which are computing machines which are continuous without being analogous in the sense that the engineer assumes that the parts of the problem are analogous, the parts of the machine as well as the whole of the machine being analogous to the whole of the machine. If one tries – and it is worth while doing – to see how far one can endeavor to understand the nervous system on that basis, that is the way in which you would have to do it. I think that part of our difficulty is that we have been using terms as opposite which apparently are not logical opposites. We use them only because they are in the properties of the two.

FRANK: Years ago Ned Huntington talked about the continuum in terms of the dense, discrete and continuous. Have we dropped that concept between discrete and continuous?

PITTS: I don't think we have. It is the very point. The nervous system treats the continuous by averaging many of the discretes.

FRANK: As I suggested, there may be a biological process which we cannot conceptualize by our present-day concepts and language.

FREMONT-SMITH: Capillary flow is continuous and the heartbeat is intermittent; it seems we have a perfect example right there. You cannot take any point and decide when the shift from the intermittency of the heartbeat to the continuity of the capillary flow takes place.

PITTS: I think that is a very good point. I should like to go back to eye movements a bit, which appear to be discrete about following the sine curve in which the grade of continuity appears to be increased at successive steps. That really seems to be remarkable. The remarkable thing is, it is true in the best grade of following of the sine curve that the original nodes showed where there was discontinuity from the steps.

[52] STROUD: I have been able, using a fine source of light, the con|centrated arc lamp, to distinguish the discontinuities up to a fair amount of practice, but now I am confounded by the problem of increasing the resolution of my records. It is very easy to follow in the early stages of learning.

PITTS: Always in the same places?

STROUD: Roughly speaking, they are. The steps are nearly the same length. The frequencies are quite constant. I have a subject who can never get beyond the first step that is staged, yet others give smooth transitions from the start.

WIENER: The stepping is unchanged but the mechanism isn't.

STROUD: As though you had a set of one computer working with a good approximation of a continuous equation, which had a good constant and did not supply its own constants. These were changed at intervals to a better fitting set of constants.

PITTS: Those points in the first approximation where the step changes, are they always the same in the same person, are they always the same in number, or do they vary from time to time, or case to case?

STROUD: This I have not explored sufficiently. I can only say that over a forty-five-second period of record on a typical subject reasonably well settled to his task, to a first order of approximation they are quite smoothly repetitive at rates that fall in the range of four to six corrections per second.

WIENER: Trapezoid form, in simple zones, and so forth. I think that will convey what is happening.

TEUBER: I don't understand the physical situation. Does that vary with the velocity of the target?

STROUD: I can tell you this: smooth following motions that do a good copy job even with a lot of practice are not possible at very high rates of periodicity. Half-cycle a second was the speed I chose to do most of the work at here. I got good copy. I can assure that at four cycles per second the whole system breaks up.

MARQUIS: It could also be too slow.

STROUD: It could, I am sure, be too slow for good following motions. I have not found out how slow it could be.

WIENER: This could be useful in detecting forged handwriting.

STROUD: Believe me, this is one of those lucky accidents where in getting into an argument and seeking to prove a point I fell on my feet and we got perfectly readable records the very first time I tried it. But the pressure of business has kept me from collecting more than something over twenty records on about ten or so individuals. It is a very preliminary experiment which, fortunately for me, is quite unambiguous from the start. It is just a fluke, if | you like. Incidentally, it is a bit late, but speaking of the [53] mechanism which Dr. Savage suggested, I have been guilty of promulgating a theory of color sensitivity of cones which requires a somewhat similar sort of thing to take place in a neurological net. If you remember, in color vision for any subjective psychological color there is effective infinity of spectral distribution for each color. This is perhaps not too far afield from what Dr. Pitts was remarking of various kinds of mapping where you do not get the exact topological equivalents because for every point on the psychological color scale there is an infinity of spectra. In an attempt to imagine some reasonable neurological mechanism for color vision – some of the details of this will be published for those of you who have access to NRC Armed Forces Vision Committee proceedings – I used some theory of dielectric rod antennae. To come out with the psychological color, it was necessary to assume that some such process as Dr. Savage suggested was taking place, perhaps in the lateral connections that Dr. Teuber pointed out existed, as well as the direct one-to-one connections, in order to explain color vision. I don't know if it is of any use to any of you, but I have a holier-than-thou feeling because I escape all of this argument by considering these mechanisms not to be properties of some, to me, quite imaginary thing. There are reliable ways in which I may think about what I know, and therefore I find no difficulties if I can find a particular way of thinking about what I know that works. One of them is digital. One of them is analogical, and I suddenly realize that I was very liberal in using quite another one, the probabilistic that I spoke about. If I can think about what I know successfully I leave the rest of you to argue about the essentialness of these various mechanisms for the imaginary.

KLÜVER: I assume that your new color theory is based on investigations with spectral or, to use the psychological expression, »film colors.«

STROUD: There are an infinite number of spectra for each psychological color.

KLÜVER: You are not talking about »surface colors,« that is, the kind of colors that are seen when a surface reflects light in the presence of other surfaces.

STROUD: This is simply in the coding, the psychological experience of, for example, a lighted source.

WIENER: Have you been following the work that has been done by Professor Hardy of the Massachusetts Institute of Technology for printers with reproduction of colored pictures in printing? | Much of the work there is extremely relevant to this sort of [54] thing.

STROUD: The entire practical system of specifying color for the printing and dyeing trade, the so-called »tristimulus« system, is based upon the statistics of large numbers of people in their responses to these various spectra and is a purely mechanical method of reclassifying physically measured spectra in the color equivalent with huge success.

WIENER: There is a probable developmental problem that is actually used to solve it.

STROUD: Modern photoradiometers are computers that are all thimblerigged to come out with the psychologically equivalent color.

KLÜVER: Deane B. Judd, at the National Bureau of Standards, has pointed out that practically all existing color theories refer to film or aperture colors with a dark surrounding field, but not to surface or object colors in an illuminated space [Judd, D. B.: Hue saturation and lightness of surface colors with chromatic illumination. *J. Opt. Soc. Am.*, **30**, *2* (1940)]. According to him, a surface color requires at least six variables in contrast to the three variables of a film color (hue, saturation, and brightness).

STROUD: This is a purely restricted notion of color.

KLÜVER: Practically all the colors we encounter in our environment are, of course, surface colors. In general we do not go around peeping into holes or looking at the clear sky or inspecting objects through Katz's reduction screen.

WIENER: This goes further than that.

STROUD: These things will report the analyzed spectral distribution and convert into acceptable equivalent color so that two entirely different spectra having the same equivalent representation will be psychologically indistinguishable.

PITTS: Let's get back to what we started with: the sine curves. What disturbs me is the secant, not the tangent, and there are only four or six in the whole business. So the secant differs considerably from the tangent at every point. Now how in the beginning do you project where you are, and in the next second, the second? How can you project to move along the secant in the sine wave rather than along the tangent?

STROUD: I can only assure you that I don't think the eye does anything of the sort. At first it merely moves to a new position in this possible row of positions. I drew a hump here in which time is drawn along this way and lateral motion is drawn vertically.

[55] PITTS: That is what I supposed. |

STROUD: The eye moves to a new position and stops in the first attempt to solve. It finds it is in a wrong position and moves to another position. It is still a wrong position; these saccadic movements are extremely fast. The eye is a vastly overpowered system. To a first approximation you can neglect its inertia. It soon discovers what it really wants is to have a moving point of view. So it sets up a moving point of view, starting with a rate, with a starting point of two constants.

PITTS: Our question is how at every node – let us so call the point where there is a chance of behavior nodes – it decides the velocity, knows where it is, and simply changes the velocity. It now knows it is going to proceed for the next period of time with a constant. How is it changed? It set[s] the velocity apparently by the secant and not the tangent, not by the contemporary.

STROUD: It is a predictive cycle.

PITTS: Which apparently –

STROUD: The results of predictions come in discontinuously, but the thing acting is acting smoothly.

PITTS: It is very difficult to tell. It says nothing about the mechanism, because nothing ever changes and because you merely use the sine wave. Nothing ever changes, so you have no idea on what information it relies. It could rely upon information from very far back, since the sine wave was going on for a long time or simply relying on information in the last couple of cycles, or possibly on information in the last sixth of the second.

STROUD: Those are the things I want to find out.

WIENER: I have encountered precisely this in connection with prediction. You may have apparatus which works well on the sine curve but is going to show indecision when it comes to the angle. I think this should be studied.

STROUD: I can give you a little information about how long a simple sinusoidal motion has to go on before the eye gets good, since I happened to have data on the eye-hand circuit which was obtained elsewhere. Under these circumstances, if you have more than one cycle of harmonics the average subject will quickly enough get very excellent approximation of this in one or two preceding cycles.

BIGELOW: What sort of screen persistence are you using on this?

STROUD: I was using the 11, which is a very short one, and I plan on using a 15 to remove any faint doubt that there is any possible persistence at all.

PITTS: If it operates as a linear prediction on that long-time base with things moving in an irregular way, it is going to do very badly. | [56]

STROUD: You have to remember that the third step comes in and begins to introduce acceleration.

PITTS: One or two cycles? You say that is enough to do it. That would be approximately a sixth or half a second?

STROUD: You are asking me about things I have not measured myself.

PITTS: What is the order of magnitude?

STROUD: The order of magnitude of 20 moments, something like that, for a good quick eye.

PITTS: That is about two seconds.

STROUD: I am guessing from another fellow's data using another circuit.

PITTS: I wanted the order of magnitude; linear prediction acting on a two-second time base would be very bad for the eye in general.

LICKLIDER: Not affected time base.

STROUD: It has been shown in other tests that once such a solution breaks down, it does not break down slowly. It breaks down suddenly. For instance, if you are tracking a thing manually, all of a sudden the chap introduces a step function, you make the next prediction, and then your whole system of predictions of acceleration and velocity breaks up and you start off again more or less with the simple position in an attempt to get the following motion.

PITTS: What happens if the frequency drifts so slowly that it does not break up?

STROUD: These are things I hope to find out more about. Remember, I said these are preliminary experiments in which I fell on my feet.

TEUBER: You know of Rademaker's and ter Braak's work on nystagmus? [Rademaker, G. G. J., and ter Braak, J. W. G., On the central mechanism of some optic reactions. *Brain*, **71**, *48* (1948)]. They had a rabbit look with one eye at a moving drum with black and white stripes. The eye that was looking at the striped drum was immobilized. Eye movements were recorded from the other eye, which was free to move, but was completely covered by an eggshell. The covered eye moved in the same direction as the stripes, which were rotated at constant velocity around the rabbit; that means the eye followed, even though it didn't know how well it followed, it couldn't judge.

PITTS: The first eye is immobilized … | [57]

TEUBER: The first eye is completely immobilized. The surprising thing is that the rabbit does follow with the eye that gets no light. The other eye can see the light, but can't move. Rademaker and ter Braak have shown earlier that animals can get nystagmus when a single light moves in a totally dark room. That, too, does not fit into our theories of following movements and nystagmus. However, in the case of the rabbit in which one eye was covered, that covered eye moved much faster than under normal conditions. They state that it moved about sixteen times faster than the stripes which were rotating at constant angular velocity. There was following without tracking, without knowledge of whether you were on the target or not, but evidently with considerable overshooting. Still, you must assume some central mechanism for such optokinetic responses independent of any specific feedback. I think such mechanism should manifest itself in other tracking situations, even in the human. I'm bothered by the fact that you did not get any other oscillations than the one you described.

Stroud: That is one of those things I am very much interested in doing. I want to present, for example, a problem, watch it become a completely solved predictive problem, change the problem, and watch the breakup of the old solution. We are very much interested in intricate details. They have implications which I am not at liberty to discuss. Believe me, we will work this thing to the very bone before we are through with it.

McCulloch: Dr. Gerard, will you summarize briefly?

Savage: Then we go backward.

Frank: Let us continue.

McCulloch: Are you sure questions won't come up again? I think Von Foerster might like to quantize nervous activity at the level of the electron, the basic physical level. I know I should like to quantize at the level of neurons. I know Stroud had to quantize it at the level of the moment, something of the order of a second in order to match our data. We are next going to tackle speech efficiency. We are going to begin with Licklider, and I am going to ask you to show him the same courtesy that you showed to Ralph Gerard. Let us allow him to proceed without interruption except for questions of plain understanding.

THE MANNER IN WHICH AND EXTENT TO WHICH SPEECH CAN BE DISTORTED AND REMAIN INTELLIGIBLE

[58]

J. C. R. LICKLIDER
Psycho-Acoustic Laboratories,
Harvard University

I SHOULD like the blackboard, lantern slides, and permission to walk around a little bit, because I'll get stagnant if I stand in one place too long.

The discussion this morning makes it almost unnecessary to say what I have to say this afternoon, since my remarks will be directed toward the point that – even on the rather microscopic level on which I want to consider it – speech is highly redundant. You can talk a long time and say very little.

I should prefer to consider the problem of speech communication on a single level. It will be necessary, however, since I am interested in effects of noise and distortion (which are on one level) upon intelligibility (which is on another level), to consider the problem on two different levels. Noise is most easily thought of as some sort of a function of time or of frequency. Distortion is most easily thought of as a deformation of a function of time or of frequency. In describing noise and distortion, therefore, we will work with functions of time and frequency. In order to discuss intelligibility, on the other hand, it will be necessary to work with elements: phonemes, syllables, words, or sentences. Therefore we shall have to engage in an exercise in shuttling back and forth between the level of functions and the level of elements.

To facilitate our getting together on the kinds of noise and distortion I want to talk about, let us set up a simple paradigm. Let us write $f(t)$ for the speech wave, a temporal segment or sample of someone's speech, say the pressure variation in the air at the end of this piece of chalk or at the face of that microphone over there. Of course, f is a very complicated function of time t, | but we can think of it easily enough, and we can schematize it and represent it by a jiggly line like this: [Figure 2] [59]

We shall use $f(t)$ as short for the speech wave form. Or, alternatively, we shall use $F(\omega)$, the Fourier transform of $f(t)$, which will be more convenient when we want to operate in the domain of frequency ω. [The complete spectrum $F(\omega)$ gives the amplitudes and phase angles of all the sinusoidal components into which the speech wave form can be analyzed, and therefore represents the same thing as $f(t)$.]

In addition to $f(t)$ and $F(\omega)$, it will be convenient to think of an arbitrary time function $g(t)$, different from $f(t)$, and the Fourier transform $G(\omega)$ of $g(t)$. Noise is represented by $g(t)$ and $G(\omega)$. By assuming various forms for these functions, we can run through the gamut of noises. When we inquire about effects of noises upon speech intelligibility, we are simply asking for a comparison of a listener's reactions to $f(t)$ and $f(t) + g(t)$, or, what is the same thing, to $F(\omega)$, and $F(\omega)$ superposed upon $G(\omega)$. | [60]

In order to handle distortion, we add to our notation by introducing the versatile operator h, which does whatever we want it to do to whatever we want to operate upon. Starting again with our $f(t)$ and $F(\omega)$, we can write:

$$f[h(t)] \qquad F[h(\omega)]$$
$$h[f(t)] \qquad h[F(\omega)]$$

That covers the domain of the distortions in which we shall be interested. $f[h(t)]$ indicates that we distort the time scale. For example, we play a record back too fast or too slow; h is then simply a constant multiplier of t. We might play the record on a poor playback machine that introduces »wow« or »flutter.« Or we might play it back back-

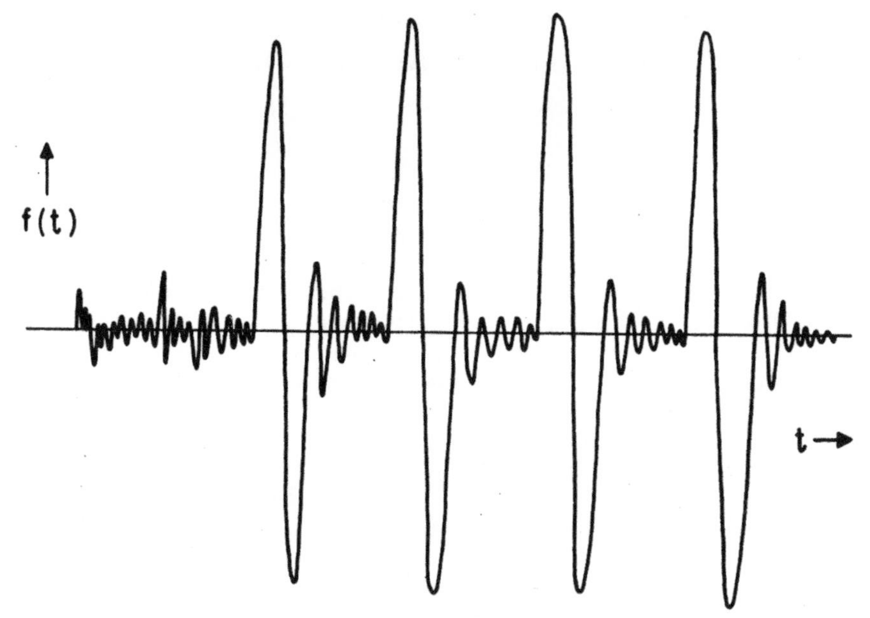

FIGURE 2. Schematic Representation of speech wave, showing sound pressure as a function of time.

ward. These distortions of the time scale are represented by $f[h(t)]$ with simple interpretations of h. In $h[f(t)]$ we distort the time function instead of the time scale. In $F[h(\omega)]$ and $h[F(\omega)]$ we operate upon the frequency scale and upon the frequency function, respectively. There are of course infinitely many possible operations. We shall have to select a few of the more interesting ones, and even those we can examine only briefly.

Now, suppose that we have decided upon a noise or a distortion and we want to determine its effect upon intelligibility. You are probably all at least somewhat familiar with articulation or intelligibility tests. If you are not, it's immaterial; they are ridiculously simple. You could count the number of repetitions it takes to get a message across to a listener. You could determine how long it takes to transmit a certain amount of information. But in the great majority of intelligibility tests, at any rate in this country, you simply *(a)* have a talker read syllables, words, or sentences to a listener, *(b)* have the listener write down what he hears, and *(c)* after the test is over, find out what percentage of the words that were read were written down correctly (that is, correspond with the words on the talker's list). That percentage is called »per cent syllable (or word, or sentence) articulation (or intelligibility).« The relations among syllable, word, and sentence intelligibility are fairly well worked out. If you know what percentage of the syllables gets across from the talker to the listener, you can predict what percentage of words or sentences would get across.

Usually, of course, you conduct the test with a number of talkers and listeners (though often not a very large number, because the test is expensive and time-consuming). The mean or median [61] | of the individual percentages is your measure of intelligibility.

Let us then start with noise. Of the many noises that have been studied, the most detrimental to intelligibility are: first, the babble of a number, say six or eight, of other people, all talking, laughing, and shouting at once. It is somewhat important that they be laughing and shouting as well as talking. That is one. Second, random noise (fluctuation noise) that has had its spectrum tailored to resemble the spectrum of speech. And third, what you get if you apply random noise to a vibrator, a Sonovox (one of those things they use to make musical instruments talk in the commercials you hear on the radio), hold the vibrator to your throat, move your lips and tongue a little, and then amplify the sound that comes out of your mouth. This third noise is very much like a whisper, but it has the advantage (if you are careful never entirely to close your lips) of not having any gaps in it.

The noises that are least detrimental to intelligibility are almost entirely nondetrimental. A sinusoid of 4,000 cycles, for example, can be made so intense that you cannot bear listening to it, yet it will not interfere with the understanding of even fairly weak speech. So, when we examine the effects of noise upon intelligibility, we must keep in mind what sort of noise it is that we are working with.

The favorite noise in the experimental laboratory is wide-band, continuous random noise. It is like the noise of Brownian motion, like the noise inherent in a resistance, but in practice it is usually derived from the fluctuations of current in a gas tube. This noise is uniform and continuous both in time and in frequency. These characteristics simplify the interpretation of its effects. It sounds like »ss-sh-hh-hh-hh.«

If we add together some speech, $f(t)$, and some noise, $n(t)$, and measure the intelligibility of $f(t) + n(t)$, we find that the result depends most critically upon the intensity of the speech relative to the intensity of the noise. That (over a wide range of intensities) it is the ratio of the speech intensity to the noise intensity and not the absolute value of either is a simple but in my opinion very important fact. It means that the listener identifies the sounds of speech on the basis of characteristics that remain invariant despite wide variations of average intensity. Think how much wider our vocabulary of speech sounds would be if »ah« at one intensity level meant something different from »ah« a little louder. On the other hand, think how confusing it would be if the meaning changed abruptly whenever the acoustics of the room or the location of the talker changed slightly. | [62]

With the speech and random noise at the same intensity level, this level being measured in the frequency band 0 to 5,000 cycles per second, we get about 50 per cent word articulation. Varying the speech-to-noise ratio, we find that intelligibility changes in approximately the way shown by this curve: [Figure 3]

The curve slopes about four words per decibel in the middle of the range and tails off a little at the lower end and quite a lot at the upper end. Now, if we fix upon a speech-to-noise ratio and | vary the levels of the speech and noise simultaneously, we [63] find that intelligibility varies like this: [Figure 4]

The broad plateau covers the range in which only the ratio of signal to noise is important. If the speech gets too weak, some of the speech sounds become inaudible. If the speech gets too intense, the listeners either can't or won't listen (I think it is in large measure the latter).

So much for a rough map of the relations between intelligibility and the intensities of speech and random noise. We can now generalize a bit by bringing into the picture noises that are not uniform in frequency (noises we can produce by passing random noise through various frequency-selective networks), and we can uncover something a bit more fundamental about the listener's reaction to speech in the presence of noise. The easiest way to do these things is to describe a procedure for computing the intelligibility of speech in the presence of noise. The procedure is based on simplifying assumptions, of course, but it works rather well. And, to the extent that it yields correct results, it displays in convenient form some of the functional characteristics of speech perception.

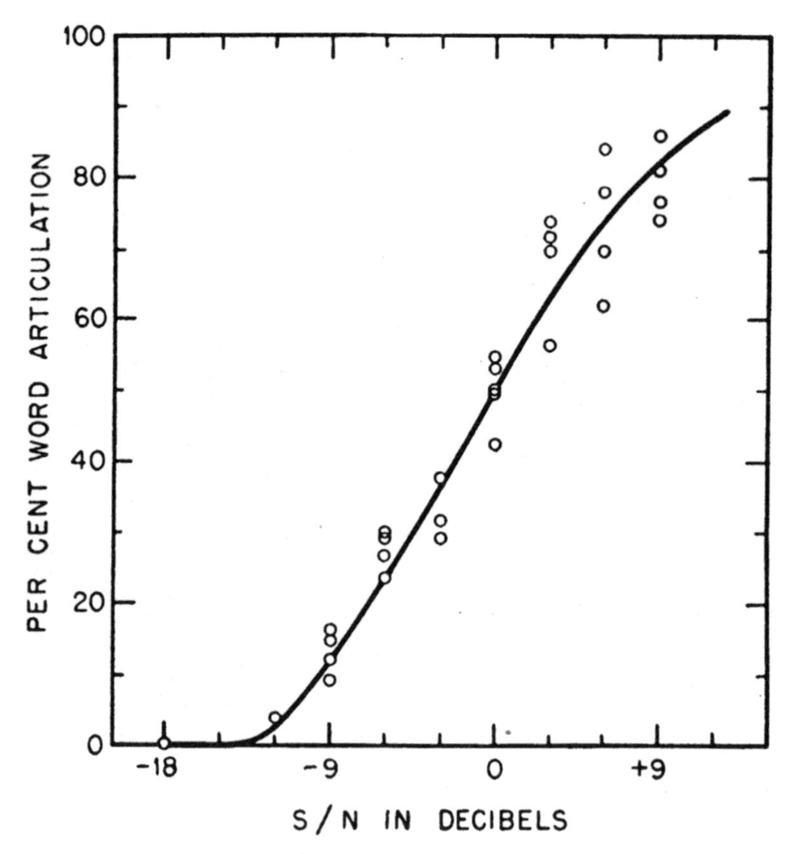

FIGURE 3. Word articulation for speech heard in the presence of noise, plotted as a function of the speech-to-noise ratio in decibels. The average speech level was held constant at approximately 90 db re 0.0002 microbar [MILLER, G. A., and LICKLIDER, J. C. R.: *J. Acous. Soc. Am.*, 22, *167, Fig. 9* (1950)]. Grateful acknowledgment is made to the Journal of the Acoustical Society of America for their permission to reproduce this figure.

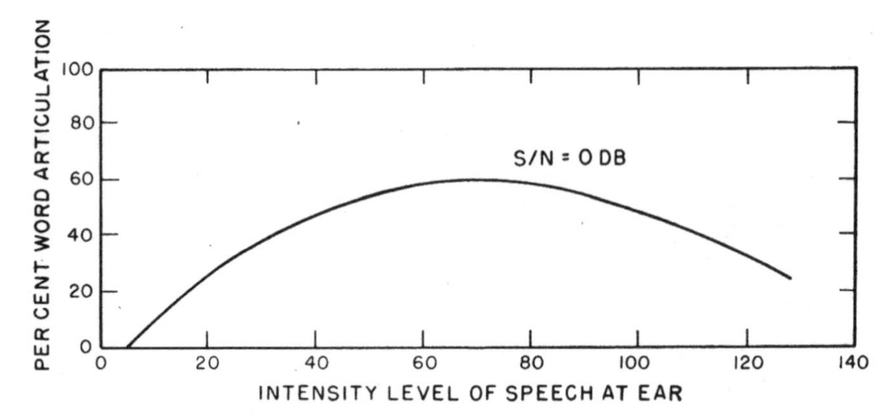

FIGURE 4. Word articulation at constant speech-to-noise ratio, plotted as a function of the intensity of the speech at the listener's ear. This figure is based in part upon the work of K. D. Kryter and in part upon a number of other observations.

The basic assumption of the procedure is that there exists an index, an index of articulation or intelligibility, that *(a)* bears a simple relation to syllable, word, or sentence intelligibility and | *(b)* is a linear sum of a number of parts, each of which is [64] dependent upon the speech-to-noise ratio in a narrow interval of the frequency scale. I shall say no more about *(a)* than that the relation between the index and, say, syllable intelligibility is supposed to be of the same general sort as the relation between syllable and word or between word and sentence intelligibility. If you know one, you can look up the other in tables or curves based on empirical observations. Part *(b)* of the basic assumption, on the other hand, deserves some comment. If you examine the auditory system in any detail at all, you are impressed by its intricacy and by the nonlinearity of its components. If you examine speech carefully, you are convinced that intelligibility is carried by patterns or relations in which various parts of the spectrum are interdependent. These facts make it inconceivable that the assumption of linear combination of contributions from narrow frequency bands can be correct in detail. The computational procedure is only supposed, however, to yield results that are approximately correct in a long-run statistical sense. It is with that fact in mind that part *(b)* of the assumption should be interpreted.

Let us sketch roughly that part of variation of $F(\omega)$ that describes the distribution of speech power in frequency. This is $S(\omega)$, the power spectrum of the speech. On the same plot let us indicate $N(\omega)$, the power spectrum of the noise. It will simplify matters to assume without discussion that both the speech and the noise have continous[!] spectra, and to define S and N in terms of power per cycle, i.e., power in a frequency band 1 cps wide, centered at $\omega/2n$ cps. Now the essence of the computational procedure is that the articulation index I is proportional to the integral of the logarithm of S/N over a *distorted* frequency scale. It will be necessary to say something about restrictions on the range of S/N and about the nature of the distortion of the frequency scale, but the main idea is, as the last sentence said, to add up contributions proportional to the ratio of speech to noise in each of a number of narrow frequency bands.

Making the speech more than 1,000 times as strong as the noise in any narrow frequency band adds little or nothing to intelligibility, so we set 1,000 as an arbitrary upper limit for S/N. If the speech is weaker than the noise, in any narrow frequency band, it contributes essentially nothing to intelligibility, so we set unity as an arbitrary lower limit for S/N. Furthermore we must restrict the absolute value of S by imposing an upper limit beyond which the auditory mechanism would be overloaded, and we must | restrict the absolute value of N by introducing a weak fictional noise to play [65] the role of the absolute threshold of hearing. The basic function, then, is S'/N' where the primes indicate that we have made the necessary restrictions in range. S'/N' is shown as a dotted line in the preceding figure.

The very low frequencies and the very high frequencies are unimportant for intelligibility; frequencies between 1,000 and 2,500 cps do a disproportionate amount of the work of carrying intelligibility. Differences in importance of the various parts of the frequency scale are taken into account by distorting the scale before integrating S'/N'. The distortion can be represented by $p(\omega)$. It is a highly remarkable fact, or so it seems to a number of workers in the field of speech and hearing, that very nearly the same function $p(\omega)$ appears in four different contexts: $p'(\omega) = dp/d\omega$ gives the importance for intelligibility of each elementary interval of the frequency scale; it is proportional to the just audible increment in the frequency of a pure tone; it is proportional to the interval of frequency over which random noise is effective in masking a pure tone; and $p(\omega)$ is proportional to the subjective pitch of a sinusoid of angular frequency ω. Evidently, $p(\omega)$ is a reflection of something basic in the frequency-resolving mechanism of the auditory system.

The articulation index, then, is

$$A = \int \log[S'(p)/N'(p)]dp\,,$$

where k is the constant that makes $A = 1$ when the signal-to-noise ratio is always at least 1,000, or 30 db.

This formulation tells us a number of things about the perception of speech. It incorporates the fact, upon which we have remarked, that it is the ratio of speech intensity to noise intensity, measured in a narrow frequency band, that is the important quantity. It says that that quantity makes itself felt according to a logarithmic rule. And it tells us that the contributions of various frequency bands to over-all intelligibility are weighted in a way that is governed by the frequency-selective characteristic of the auditory system. It is noteworthy that characteristics of the auditory mechanism appear to be much more important than those of the speech signal in determining the weighting.

It will be of interest to many of you to note the parallel between the formula for [66] articulation index and the formula for | amount of information,

$$H = WT\log_2[(S+N)/N]\,.$$

It is possible that the index formulation may be improved by substituting $S+N$ for S (thus following the information formula) and dispensing with the restriction that S'/N' must not fall below unity. In any event, the parallel is sufficiently good to suggest that the listener is indeed receiving information when he understands speech.

PITTS: I have one question about the pitch-frequency relation. Is that such that it distorts the length of intervals very significantly at the extremes?

LICKLIDER: It distorts them considerably. With respect to frequency, the pitch scale is neither linear nor logarithmic; it is halfway between.

PITTS: However, from the linear –

LICKLIDER: Halfway from the linear to the logarithmic.

PITTS: In the range that is important?

LICKLIDER: Yes, say from 100 cycles to 4,000.

WIENER: What does halfway mean? X plus S over N?

LICKLIDER: Mathematicians are always doing that: taking me up on inexact statements. Subjectively, it would be about halfway between. I can quote you the terminals of the frequency bands which contribute equally to intelligibility, or better, which are judged equal in terms of subjective pitch. I can quote them pretty closely. We will see how many there are. Then we can see how many per cent. First, up to 160 cps. Then from 160 to about 400. Then 400 to 670, 670 to 1,000, 1,000 to 1,420, 1,420 to 1,900, 1,900 to 2,450, 2,450 to 3,100, 3,100 to 4,000, 4,000 to 5,100. You can see at about what rate we are going up in frequency. These bands are equally wide, according to the best available evidence, in terms of subjective pitch. In addition, they contribute approximately equally to intelligibility.

PITTS: The reason this is important is that it multiplies the S plus N over N factor in the quantity if you expressed it in frequency, and so expresses the difference from the behavior of linearality. I wondered how much difference was made by that. Here it was going to the difference, but then you rubbed it out.

LICKLIDER: 160, about 400, 670, 1,000, 1,420, 1,900, 2,450, 3,100, 4,000, 5,100. I think, I am not sure, that from there it goes pretty far up, about 9,000 cycles. The relation is roughly linear with frequency to 1,000 cps, logarithmic beyond 1,000 cps. | [67]

BIGELOW: This ought to do.

LICKLIDER: I am sorry; I did not see what you were doing.

This, then, gives a rough orientation concerning effects of noise that is continuous both in time and in frequency. Perhaps I should not say »continuous« before so many mathematicians; it appears to be dangerous to say that a random function is continuous. (Pitts, here, says that random noise is continuous, all right, but not differentiable.) Anyway, the noise we have been talking about »covers« fairly uniformly both in time and in frequency. Let us see what happens when the noise is not continuous in time.

If we interrupt a noise, that is, turn it on and off intermittently, we can introduce temporal gaps. The possibility then exists that we can hear snatches of speech in the intervals when there is no noise. Let me draw for you, here, a graph:

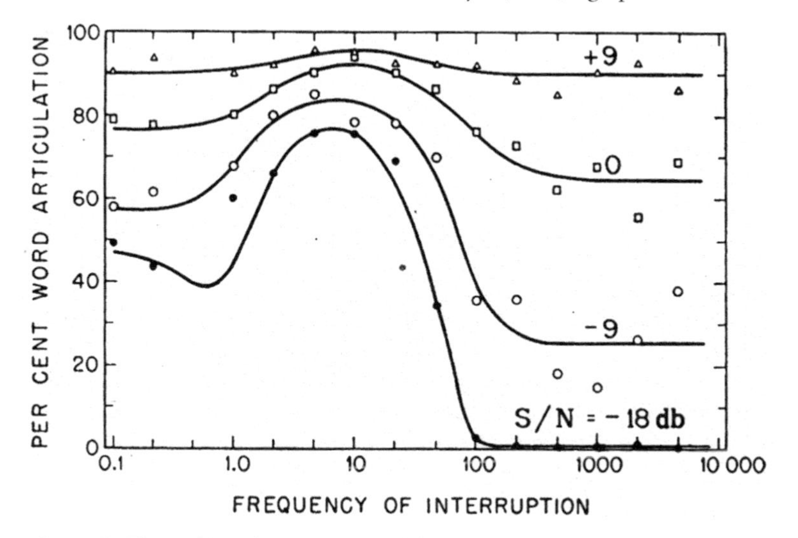

FIGURE 5. The masking of continuous speech by interrupted noise. Word articulation is plotted against the frequency of interruption of the noise, with the speech-to-noise ratio as the parameter. Noise-time fraction, 0.5 [MILLER, G. A., and LICKLIDER, J. C. R.: *J. Acous. Soc. Am.*, 22, *167, Fig. 8* (1950)]. Grateful acknowledgment is made to the Journal of the Acoustical Society of America for their permission to reproduce this figure.

The ordinate is word intelligibility. The abscissa is the frequency of interruption. Let us examine first the curve for $S/N = -18$ db. (In this notation the intensity of the noise is the intensity | measured with the noise turned on steadily.) Now with a speech-to-noise ratio of -15 db (measured in the frequency band 0 to 5,000 cps), in which the noise is something less than 100 times as intense as the speech, you can understand just about no thing when the noise is on all the time. When it is on half the time and off half the time, it has a wave form like this: [Figure 6] and the average intensity of the interrupted noise is only one-half the average intensity of the noise before interruption. The longtime average speech-to-noise ratio is therefore 15 db. If the interruptions occur at a very high frequency, the interrupted noise sounds just as though it were a continuous noise of the same average intensity, and we find that the intelligibility score is about what we should expect with steady noise and a speech-to-noise ratio of -15 db: practically no words are heard. [68]

The graph shows what happens when the frequency of interruption is varied. Above 100 or 200 interruptions per second, the noise masks speech as effectively as if it were continuous. The ear cannot hear through gaps that are less than 1/200 or 1/400 second wide. John Stroud is looking at the peak at 10 interruptions per second, but let us first go to the other end of the graph.

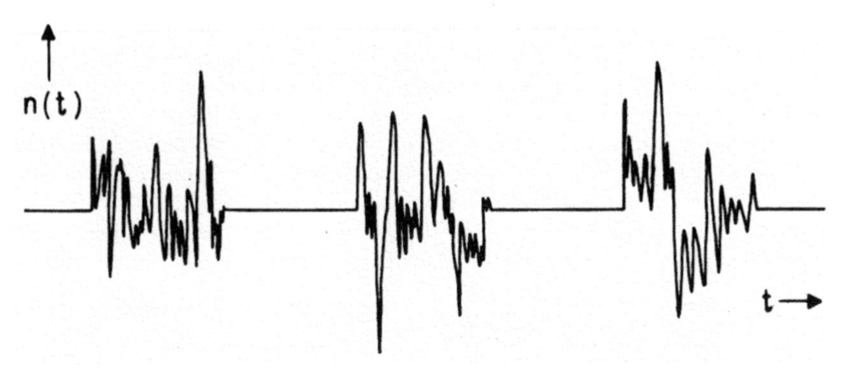

FIGURE 6. Schematic illustration of the wave form of interrupted noise.

We see that if the interruptions are very infrequent, you blank out whole series of words, then pass whole series of words, and you come out with a score that is about one-half the score for speech in the absence of noise. Here that is about half of 100, or 50 per cent.

Now if we were using a weaker noise, intelligibility would follow a roughly parallel [69] but higher curve (see graph). But we | already have the important thing, I think: the noise must be there effectively all the time if it is to mask speech. Or, conversely, the mechanism of speech perception is agile; it can piece together little bits of speech that get through between the bursts of noise.

BIGELOW: Were these tests made on random words or on a text?

LICKLIDER: These are random words.

BIGELOW: No content?

LICKLIDER: No.

KUBIE: Random words or nonsense syllables?

LICKLIDER: A vocabulary of 1,000 monosyllabic words was used for the tests. They were drawn at random from that vocabulary to form lists of 50 words each, and these were read off.

MCCULLOCH: That peak is at 10 per second?

LICKLIDER: At ten.

MCCULLOCH: The peak in intelligibility?

LICKLIDER: Almost identical results were obtained, in so far as the location of the peak is concerned, by simply turning the speech on and off at various rates.

But I want to hurry on. There are many things to cover, and I will probably have to subside before I am really quite willing to.

We have just seen what happens when there are gaps in time. When there are gaps in frequency, we get an entirely comparable effect. It can be explained with the aid of a very rough model of the frequency-resolving mechanism of the auditory system. We can think of the auditory system as containing a large number of overlapping frequency channels. The channel widths vary from a minimum of about 40 cycles to a maximum of perhaps 2,000. The channels handling frequences below 2,000 cps are all less than 100 cps in band width. The curve of band width versus center frequency is [70] shown in Figure 7. |

You may recognize this as the reciprocal of the function $p'(\omega)$ we talked about a while ago. Now if the speech components in one of the channels are to be masked, there must be noise in the channel. Therefore, any noise that has wide gaps in its spectrum will let some of the speech get through unmasked. This is the primary conclu-

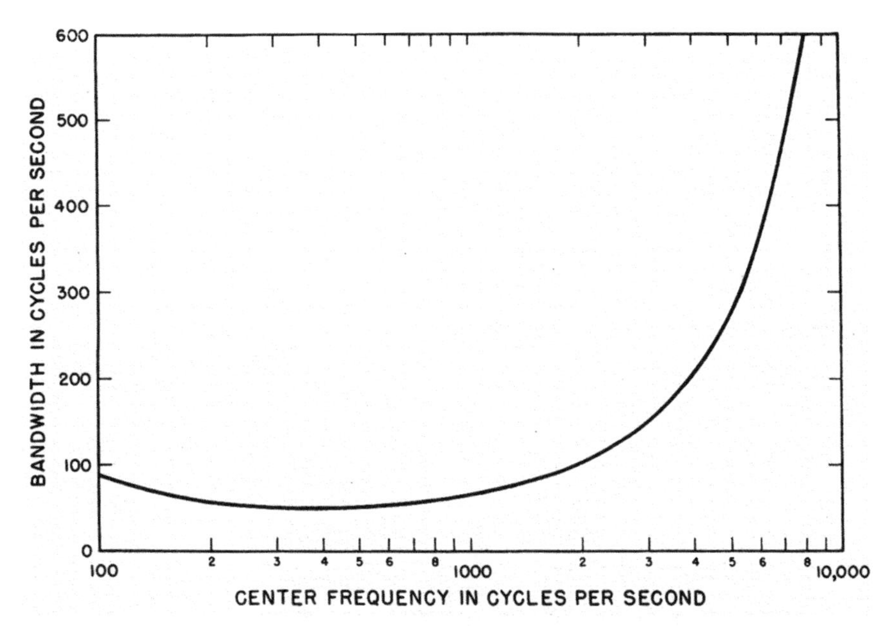

FIGURE 7. Band widths of the critical frequency bands as a function of their center frequencies. Based on data of Fletcher and of Hawkins and Stevens.

sion. I shall not spend time on the secondary one, which is the qualification that low-frequency noises do to some extent mask high-frequency signals.

An interesting situation arises when the interference consists of delta pulses – it doesn't matter whether I draw them in time or in frequency – spaced at equal intervals. | [71]

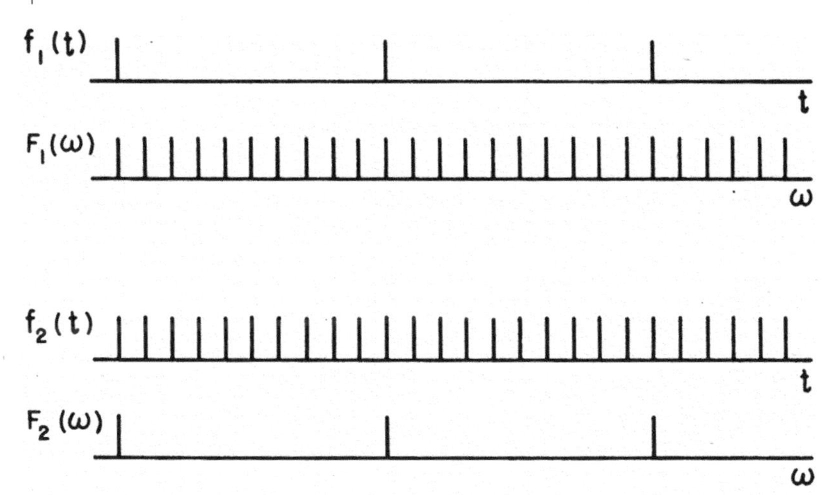

FIGURE 8. Schematic illustration of reciprocal relation between wave form and spectrum of a train of sharp pulses. When the gaps in the time function are closed, the gaps in the frequency function widen.

If the pulses are spaced at intervals of 1 second in time, they are spaced at intervals of 1 cps in frequency. If we decrease the time interval, we increase the frequency interval, and vice versa. If we want to mask speech with the pulses, therefore, we must make a compromise. We cannot have close spacing both in time and in frequency, as we should like; we have to sacrifice in one domain to gain in the other. The question, then, and I think it is a rather cute one, is what is the compromise that minimizes intelligibility?

Well, remember the time-gap graph. The rate of interruption had to be below 100 per second before much speech got through the time gaps. And remember the band widths of the auditory frequency channels. They were a little less than 100 cps wide in the important part of the frequency scale. These two facts suggest that 100 pulses per second should be about the best rate for masking speech. As the following graph indicates, that is approximately correct. | [72]

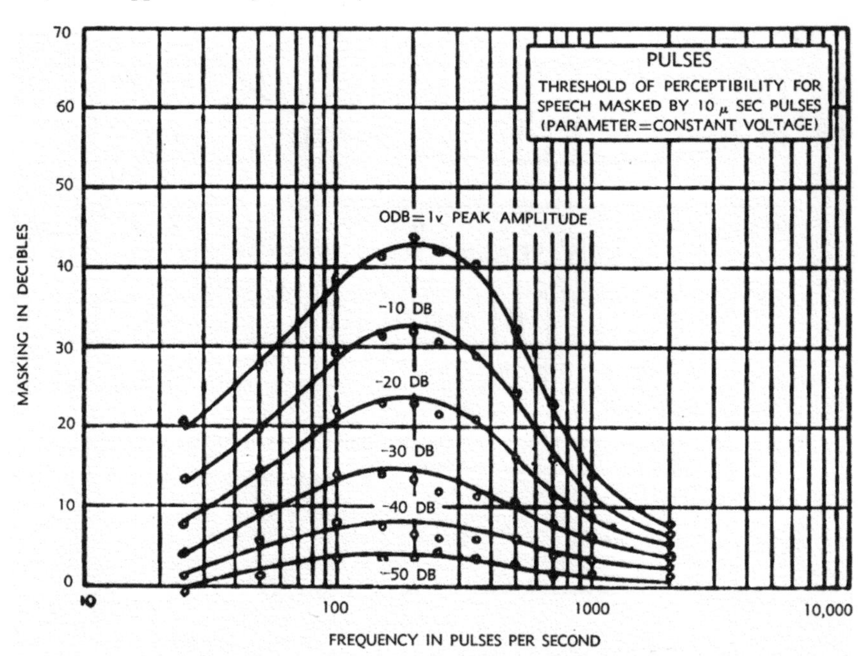

FIGURE 9. The masking of speech by trains of pulses having various repetition frequencies. [STEVENS, S. S., MILLER, G. A., and TRUSCOTT, I.: *J. Acous. Soc. Am.*, 18, *418, Fig. 6* (1946)]. Grateful acknowledgment is made to the Journal of the Acoustical Society of America for their permission to reproduce this figure.

In a way, I am sorry that the critical figure comes out 100 instead of 10 per second, but, John [Stroud], I cannot help you here.

With that, so much for noise. Distortion is for me more interesting to talk about.

KUBIE: May I ask one question? Do you encounter any subjects whom you have to eliminate because on scoring their reactions they deviate very sharply from your general group?

LICKLIDER: No. I think I can give a categorical answer to that. But perhaps not: there are probably some aphasiacs, even bad ones, that might get through high school or college. It would not matter much if the subjects were a little deaf, really. Where there is noise, conductive deafness is not a handicap because the noise is attenuated along with the speech. You would have to except nerve-deafened people. Some people with nerve deafness are said to be completely confused by noise. Other exceptions, no. | [73]

KUBIE: I was thinking In terms of auditory and visual types.

LICKLIDER: I would say it is rather hard to find any »audives« or »visuals.« There is no relationship there –

KUBIE: Have you made any comparisons using random nonsense syllables rather than words?

LICKLIDER: Yes, I have. In fact, many such comparisons have been made. You can find special sorts of stress that will break the rule down, but by and large the rule is approximately this: if we plot the articulation index I have mentioned along the abscissa and other measures of intelligibility along the ordinate, we have a way of transforming from any measure to any other.

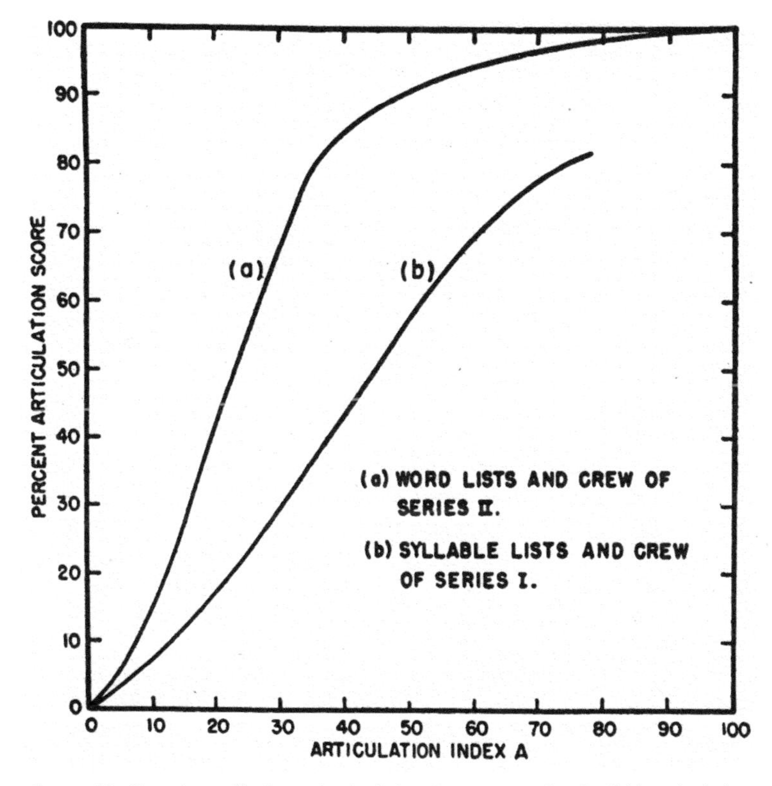

FIGURE 10. Experimentally determined relations between word and syllable articulation and the articulation index. [BERANEK, L. L..: *Proc. Instit. Radio Eng.*, 35, *880, Fig. 11* (1947)]. Grateful acknowledgment is made to the Proceedings of the Institute of Radio Engineers for their permission to reproduce this figure.

| The Bell Telephone Laboratories, who were the pioneers in this work, used nonsense syllables rather consistently. Others who have followed found it difficult to train listeners in the phonetic notation necessary for working with nonsense syllables, and they have worked largely with words removed from context. [74]

BIGELOW: What was the repetition rate of the words you used?

LICKLIDER: As long as the period is 3 seconds or more, it does not matter very much. It is usually around 3, 4, or 5 seconds.

BIGELOW: Between words?

LICKLIDER: Between words. I did not elucidate the method of articulation testing at all. The words are imbedded in a carrier sentence to improve the dynamics – to make the talker read them more naturally. The talker says: »You will write beans now,« or something of that sort. The key word, beans, is imbedded in the carrier sentence.

BIGELOW: All this is for key words in sentences?

LICKLIDER: It turns out that the carrier sentence makes practically no difference. The carrier sentence adds no meaning to the key word. It just gets the talker going.

BIGELOW: It paces him in some sentences?

LICKLIDER: Yes. We have tried the key words without any carrier sentences at all, and we get the same scores.

HUTCHINSON: If it is going at a very high noise level, how can he tell if the key word is part of the carrier sentence? How can he distinguish the key word in the carrier sentence?

LICKLIDER: If you can hear anything at all, you always hear the carrier sentence. You expect it. If the same carrier sentence is repeated with each test word, as is the usual practice: »Now you will write this. Now you will write that. Now you will write the other,« the carrier impresses you as being entirely intelligible, even though if it were presented only once it would not be at all intelligible. Your expectation is built up. If the theory ever develops to the point of trying to explain expectation, it is going to be very interesting. The threshold of agreement or disagreement with expectation is very low. If you diminish the vocabulary, so that instead of working with 1,000 words you are only working with 5 0, you find that the scores go up. There is more to it than getting some words right by chance.

We can distinguish between noise and distortion in this way: Both noise and distortion alter the form of the signal. In the case of noise, the discrepancy between the transmitted signal $f(t)$ and the received signal $f(t) + g(t)$ is simply $g(t)$, and $g(t)$ is in general unrelated to $f(t)$. In the case of distortion, on the other [75] | hand, the discrepancy between the transmitted signal $f(t)$ and the received signal (say) $f[h(t)]$ is a function $H(t)$ that is in some respects like noise, but $H(t)$ is in general quite closely related to $f(t)$ because the operator h defines a definite operation. Noise and distortion are thus both deformations of the signal, but only in the case of distortion does the deformation depend upon the signal.

To take first the case schematized by $f[h(t)]$, where h is a constant, let me show you a slide. [Figure 11]

This shows what happens if you play a record too fast or too slow. In interpreting the graph, we can use a rough rule of thumb: if syllable articulation is over 50 per cent, a listener can get the gist of connected discourse. Applying this rule, we see that the auditory system has a fair tolerance for speeding up or slowing down the time scale.

STROUD: A little over one octave.

LICKLIDER: If we consider our h to mean »add a constant« instead of »multiply by a constant,« we get what we can call [76] | the heterodyne effect. The influence on intelligibility is shown here: [Figure 12]

This distortion makes speech sound bad because it makes the harmonics of speech inharmonic. Nevertheless, we can tolerate a shift of perhaps 100 cps downward in frequency, and even a greater shift upward.

SAVAGE: You slide the frequency spectrum along horizontally?

LICKLIDER: I am sorry. I have gotten ahead of myself. I started talking about $f[h(t)]$, then switched over to $F[h(\omega)]$ without taking you in on the secret. Let me try to set things right.

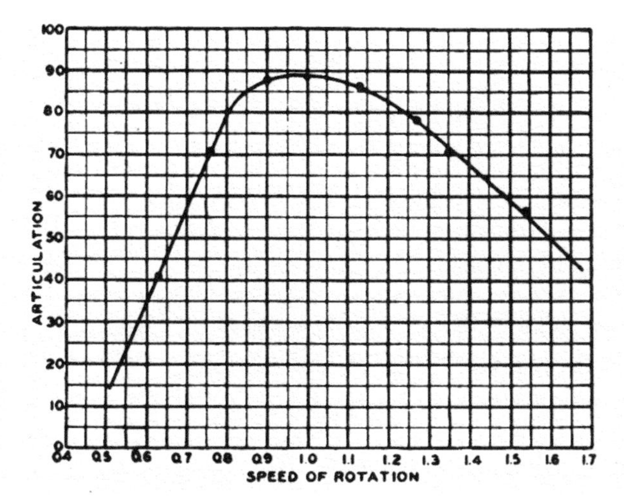

FIGURE 11. Effect upon articulation of speeding up and slowing down the passage of time or, what is the same thing, of multiplying all the component frequencies by a common factor. [FLETCHER, H.: *Speech and Hearing.* Bell Telephone Laboratories, Inc., New York, D.Van Nostrand Co., Inc., *Fig. 145* (1929)]. Grateful acknowledgment is made to the D.Van Nostrand Co., Inc. for their permission to reproduce this figure.

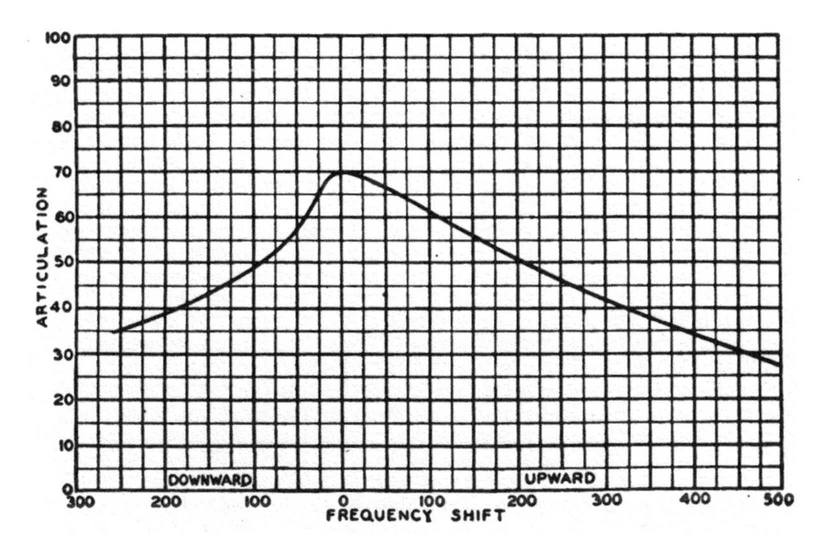

FIGURE 12. Effect upon articulation of shifting the frequencies of all the sinusoidal components a constant distance along the frequency scale. [FLETCHER, H.: *Speech and Hearing.* Bell Telephone Laboratories, Inc., New York, D.Van Nostrand Co., Inc., *Fig. 146* (1929)]. Grateful acknowledgment is made to the D.Van Nostrand Co., Inc. for their permission to reproduce this figure.

We examined the effect of multiplying the time scale by a constant. This was the case $f[h(t)]$, h constant. I should have mentioned at that point that, of course, playing the record too slow by the factor h, which we did to stretch the time scale, had the effect of multiplying all the frequencies of the spectrum by $1/h$. The one set of observations therefore served for both $f[h(t)]$ and $F[h'(\omega)]$, $h = \text{constant} = 1/h'$. Then I should have said that if h means »add a constant,« $f[h(t)]$ is simply $f(t)$ advanced or delayed in time. [77] Advance and delay, of course, | do not affect the intelligibility of a single speech sample, though they might upset two-way conversation. The heterodyne effect that I introduced out of turn and without warning is the result of adding a constant to ω; it is in the domain of frequency the parallel of advancing or delaying the wave form in time. It involves shifting the spectrum along the frequency scale. As we have seen, the listener can tolerate a considerable amount of that kind of distortion, also. The speech sounds terrible, but the listener can understand what is said.

Going on to more complicated *h*'s, we come to sinusoidal modulation of the time scale. This is the wow or flutter mentioned earlier. Actually, to get wow, I guess you have to modulate, not time *t*, but the rate at which time flows – but things are complicated enough as it is, and I shall simply report that work is going on now on effects of wow and flutter, and that although they annoy the listener they do not prevent his understanding the speech – not, at least, until they are so severe that they make time flow backward part of the time. Over in the frequency domain, complicated *h*'s correspond to the inversions and garblings that are used in privacy and secrecy systems.

The effect appears when the transmitter and the receiver of a single side-band, suppressed-carrier, amplitude-modulation system are not tuned to the same frequency. The mistuning has the effect of sliding the spectrum along the frequency scale.

While we have the slide projector running, let us move along to distortions of the time function. In these distortions *h* operates upon $f(t)$. If *h* is simply a constant multiplier, it gives us linear amplification or attenuation. We have already noted that intelligibility is invariant over a wide intensity range. If *h* is a more general linear operator, a linear operator upon the past of $f(t)$, it gives us the whole class of frequency and phase distortions.

Perhaps the most interesting of the linear distortions is reverberation. To a first approximation, reverberation is due to echoes. Reflection and scattering from surfaces and objects set up a large number of more or less faithful replicas of the original sound, and these are all superposed at the ear of the listener. Different echoes are delayed by different amounts, so the listener actually receives a large number of inexact copies of the talker's speech, placed one on top of another in inexact register.

The effect of reverberation upon intelligibility is shown here as a function of the [78] reverberation time. | [Figure 13]

The reverberation time is the length of time it takes the signal to decay 60 db when the source is turned off suddenly. For a reverberation time of 6 seconds – I think I am right about this – you have to have a tile room; in it, you get about 50 per cent syllable articulation. It is amazing to me that any mechanism can tolerate so much distortion, but it appears that the auditory system is able to pick out the first of a number of successive, overlapping presentations of a sound, to pay special attention to the first, and to ignore the rest. This primacy or precedence effect, as it is called, is important in sound localization as well as in speech perception.

Bateson: For some reason it seems to be able to do that with real speech in a room and not able to do it when listening to the same speech recorded on wire.

Stroud: If you stick your finger in one ear you won't be able to do it in the room.

Licklider: If we now leave the lantern slides for a minute and get the lights back on, we see that we have gotten into frequency and phase distortion by letting a linear [79] operator (e.g., a reverberant room) work on $f(t)$. In reverberation, we have frequency | and phase distortion in combination. Now it is possible, with the exercise of considerable ingenuity, to get frequency distortion without phase distortion and vice versa, but very little work along that line has been done. Some very important work has been

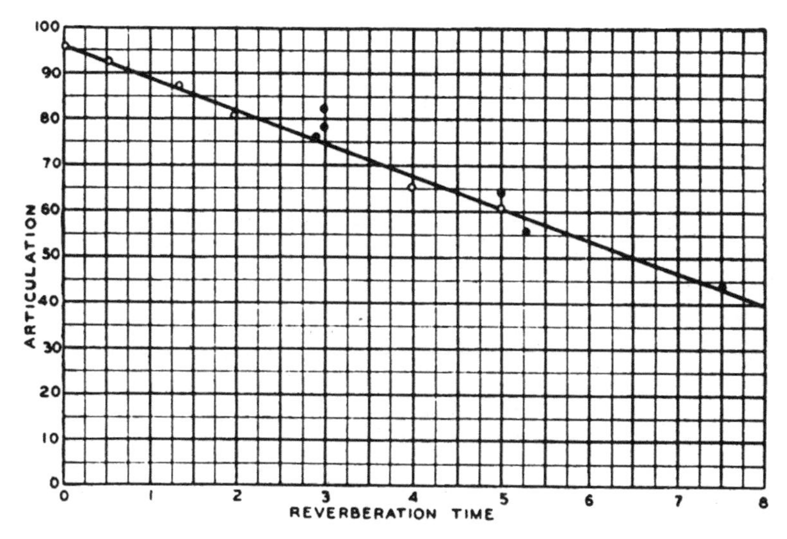

FIGURE 13. Effect of reverberation time upon articulation. [FLETCHER, H.: *Speech and Hearing.* Bell Telephone Laboratories, Inc., New York, D.Van Nostrand Co., Inc., *Fig. 147* (1929)]. Grateful acknowledgment is made to the D.Van Nostrand Co., Inc. for their permission to reproduce this figure.

done, however, in which it is reasonably clear that frequency distortion was the principal variable, and other work in which the same can be said about phase.

In one set of experiments, filters with sharp cutoffs were used *(a)* to pass the low-frequency components of speech while eliminating the high-frequency components, and *(b)* to pass the high-frequency components while eliminating the low-frequency components. The intelligibility of the components that were passed was measured as a function of the cutoff frequency. The results are shown here: [Figure 14] | [80]

The curve labeled »low pan« shows how intelligibility grows as we increase the cutoff frequency of a low-pass filter. The curve labeled »high pass« shows how it grows as we decrease the cutoff frequency of a high-pass filter. Note that the curves cross at about 66 per cent syllable articulation. This means that either the upper or the lower half of the speech spectrum is adequate for intelligible communication of connected discourse.

If the filter is a band-pass filter, eliminating the upper and the lower ends of the spectrum and passing the middle, the impairment of intelligibility is on the whole even still less marked. The speech components in the single octave between 1,000 and 2,000 cps will let you hear everything I have to say. Preliminary results obtained with a number of very narrow bands indicate that reasonable intelligibility can be obtained with considerably less than 1,000 cycles total band width if the bands are located in strategic places along the frequency scale. The general conclusion of the work with sharp-cutoff filters, then, is that a large part of the spectrum can be eliminated without destroying intelligibility. This conclusion points backward to my opening remark about the redundancy of speech, and toward the results obtained in the presence of noise, and forward toward similar effects I want to describe shortly.

Next, however, let us look for a moment at phase distortion, or, what is the same thing, time-delay distortion. Some frequency components are delayed more than others, with the result that the speech wave form is altered. The work on this type of distortion is not yet definitive. It is found that delaying parts of the spectrum is no worse

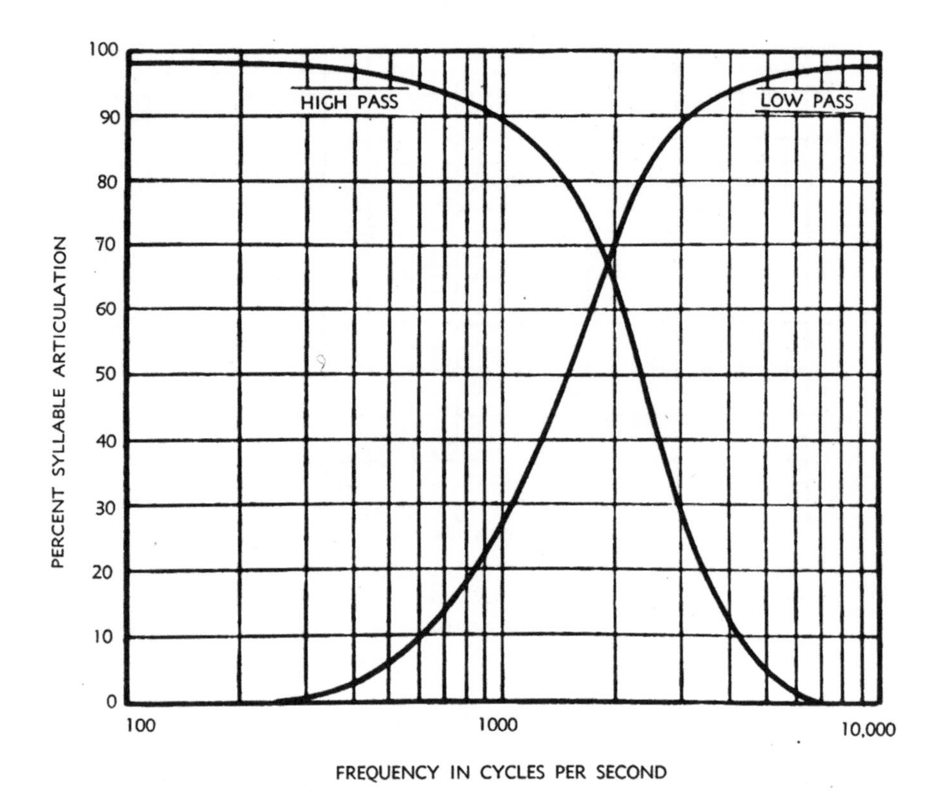

FIGURE 14. Functions relating syllable articulation to the cut-off frequencies of high-pass and low-pass filters. [MILLER, G. A., WIENER, F. M., and STEVENS, S. S.: *Transmission and Reception of Sounds Under Combat Conditions.* Washington, D. C., Office of Scientific Research and Development, *Fig. 18,* Chap. 7 (1946)]. Grateful acknowledgment is made to the Office of Scientific Research and Development for their permission to reproduce this figure.

than eliminating them, and we have seen that eliminating them had, on the whole, less detrimental effect than one might have supposed. Actually, the statement has been made for many years that »the ear does not perceive phase.« That statement is shown to be false, however, by the simple experiment of playing a recording backward. Playing it backward leaves the pattern of amplitude versus frequency entirely unchanged; only the phase angles are changed. It is not –

SAVAGE: That is a tour de force. Suppose the phase shift is only a few hundred degrees, does that really matter?

LICKLIDER: It is quite true that you can shift any of the components up to a full cycle at its frequency and not hear the difference, provided the phase shift changes smoothly with frequency. That experiment was made by Van der Pol, who worked out a very neat phase shifter. If he did not vary the phases too fast (which is another matter), he [81] could not tell that anything was | happening to the speech.

WIENER: That is not true of one, but it is true of quite independent shifts.

LICKLIDER: What you say is true. The problem of phase has not been worked out definitively. Yet it is known – and there is the accumulation of many of incidental observations, and one experiment in which whole bands of speech were delayed –

well, it is not at all certain that you could not work out a set of interrelated phase shifts that would have a marked effect upon speech intelligibility.

BIGELOW: If you take frequencies close together and knock every alternative one by 180 degrees, the results are drastic. You can get cancellation, omega, omega plus epsilon. You knock them at 180 degrees and you get cancellation. There is no physical apparatus to do that.

LICKLIDER: That is why I am wary of a negative conclusion about detrimental effects of phase distortion.

WIENER: I agree, but may I say that the following proves so? In vocoder work, if I understand rightly, you can divide a sound into frequency bands.

LICKLIDER: How about holding that discussion for a minute? I should like to save for last the discussion of abstracting transformations of the speech wave.

WIENER: That will turn out to be essentially unrelated phase shifts. For unrelated reasons they are rather worse. They apply unrelated phase shifts, for different reasons, of fairly broad waves.

LICKLIDER: One rule about the problem of phase: look at the time delays, not the phase shifts but the time delays.

WIENER: I agree.

LICKLIDER: If the time delays are under 1/100 of a second, if you can keep them under 1/100 of a second, then you are really in no trouble.

At last, now, we have exhausted the linear operators, or at least we have talked about all the linear *h*'s I want to talk about. I should like to consider, then, arbitrary, nonlinear *h*'s, operating on *f(t)*. These give us the so-called »nonlinear« distortions, which have the effect of introducing frequency components that were not present in the original signal. Since I have talked too long already, I shall mention only the nonlinear distortion that is my particular favorite.

If we start off with the speech-wave *f(t)* at the bottom of the diagram, and operate upon it by the *h* that is defined by one of the input-output characteristics (Figure 15), we get the wave | form shown at the right. It is the same as *f(t)* throughout the middle range of amplitude, but it is limited, as it were, by a ceiling above and a floor below through which it cannot pass. [82]

WIENER: Rectangularized speech.

LICKLIDER: Just peak-clipped speech. We have clipped the peaks off, both the upward peaks and the downward peaks. The rule is that if you do not reduce the level so that it gets below the noise, this clipping has no serious effect upon intelligibility. The easy way to perform the operation is to turn the gain up every time you reduce the amplitude by clipping. | [83]

WIENER: An interesting thing they have been doing at M.I.T. recently is to make the consonants of ordinary speech more intelligible.

LICKLIDER: This shows how well acquainted with one another's work people are at M.I.T. I am doing work at M.I.T.

WIENER: One of the other boys told me of it on the basis of your work.

SAVAGE: Have you heard the story?

LICKLIDER: I am willing to listen.

WIENER: I think I got that from Levine [Lewin].

TEUBER: There is a certain paper by Licklider from a certain acoustic laboratory.

WIENER: I am sorry.

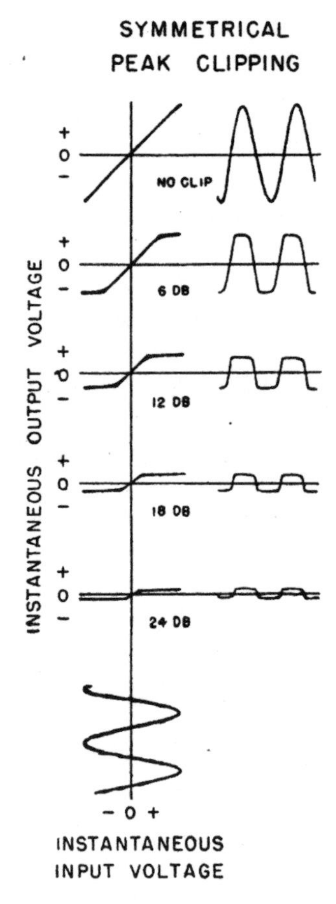

FIGURE 15. Cathode-ray oscillograms, showing instantaneous input-output characteristics of a symmetrical peak clipper. The several characteristics illustrate various amounts of clipping. Below the input-output characteristics is an input wave, and to the right of each characteristic is the corresponding output wave. [LICKLIDER, J. C. R.: *J. Acous. Soc. Am.*, 18, *429, Fig. 1* (1946)]. Grateful acknowledgment is made to the Journal of the Acoustical Society of America for their permission to reproduce this figure.

TEUBER: It was in that, and he will tell us in a moment that clipped speech can be just as intelligible as ordinary speech, or am I misquoting the person who did the experiment?

LICKLIDER: If we continue to lower the ceiling and to raise the floor, we get, in the limit, a sort of speech wave that has, as Professor Wiener implied when he said »rectangular speech,« only two amplitude values. It is dichotomized.

Going from no clipping to infinite clipping, we find the effects upon intelligibility [84] that are shown in this graph: [Figure 16] |

Almost nothing happens to intelligibility until we reach 20 db peak clipping, which reduces the amplitude to 1/10 of its original value. Then the function drops gradually to about 70 per cent word articulation, and there it remains indefinitely. Calling the effect at the extreme right-hand side of the graph infinite peak clipping, we see that one understands about 70 per cent of discrete monosyllabic words in quiet. That is quite enough for intelligible conversation.

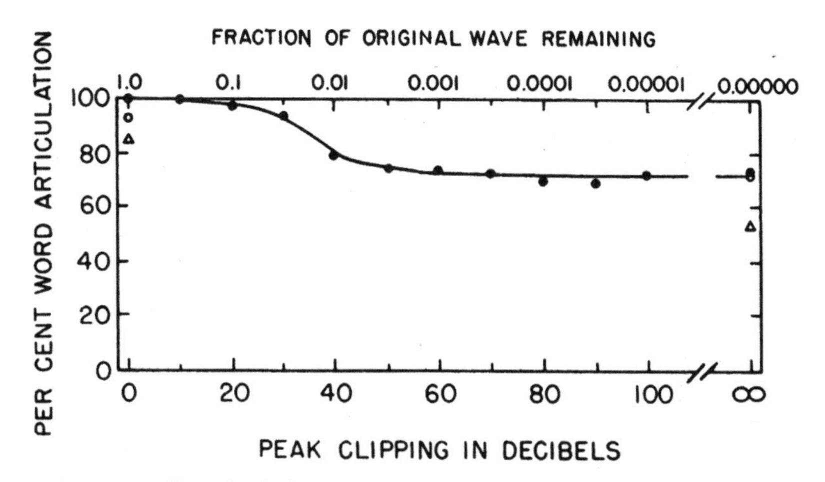

FIGURE 16. Effects of peak clipping upon the intelligibility of speech in quiet. The lower scale shows the amount of peak clipping in decibels; the upper scale indicates the fraction of the peak-to-peak amplitude of the original wave that remained after clipping and before reamplification. [LICKLIDER, J. C. R., BINDRA, D., and POLLACK, I.: *Am. J. Psychol.*, 61, *1, Fig. 3* (1948)]. Grateful acknowledgment is made to the American Journal of Psychology for their permission to reproduce this figure.

Now what Teuber suggested was this. If we are transmitting the speech over an amplitude -modulation radio system, or over any system that can handle amplitudes only up to a certain limiting level, then in the presence of noise we can obtain higher intelligibility by reducing the speech wave to a squarish form by peak clipping than by adjusting the gain to let the highest peaks of the original speech wave get through the system.

Perhaps I should mention that you do not have to hit the exact center of the speech wave. You can take any swath that is reasonably near the middle. There are in fact, an infinite number of different swaths or layers of the wave form that are reasonably intelligible, but this is hardly surprising, since there are an infinite number that are almost exactly like the one in the center. The main thing, as a practical matter, is not to miss any of the consonants. The weak consonants like *f* are easily missed if the clipper gets biased off the center axis.

WIENER: If you could displace the center of the thing, it would be roughly in the center of the energies of it. You would probably avoid most of the trouble with a sort of moving center.

LICKLIDER: You need not go to the trouble of getting a moving center if you have a stationary one that is right along the time axis. All the sounds oscillate around the zero axis. The important thing is not to get away from that.

SAVAGE: What does the stuff sound like? Can you describe it at all?

LICKLIDER: You can certainly tell that something has happened to the speech. As we increase the amount of peak clipping, it goes about like this: | [Figure 17] [85]

We clip half of the peak-to-peak amplitude away (6 db clipping); usually the listener detects no change. It may perhaps sound a little weaker. We turn the gain up and clip another 6 db (which makes 12 db peak clipping and eliminates three-quarters of the original waves); now it sounds as though there were sand in the apparatus. Then 24 db peak clipping; gravel. Infinite peak clipping makes the speech sound extremely bad, principally because the wave always has the same root-mean-square amplitude – just as

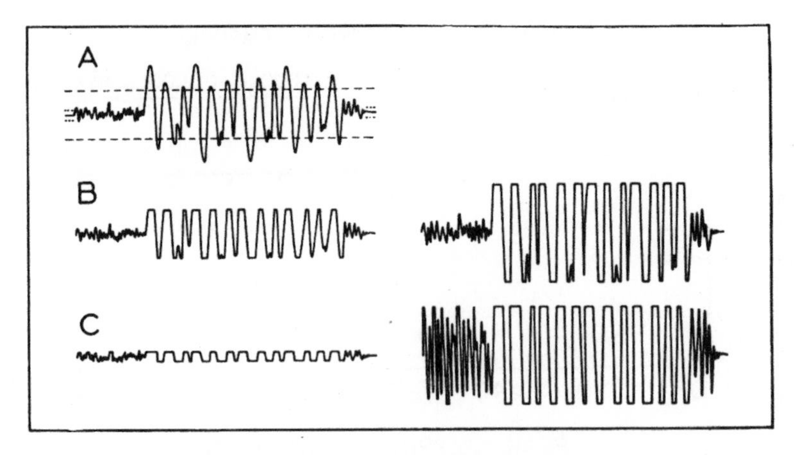

FIGURE 17. Diagram illustrating peak clipping. The waves of the word *Joe* are shown schematically at *A*. *B* shows what is left after 6 db peak clipping, i.e., after reduction to one-tenth the original amplitude. At the right, the waves of *B* and *C* are shown reamplified. [LICKLIDER, J. C. R., BINDRA, D., and POLLACK, I.: *Am. J. Psychol.*, **61**, *1*, *Fig. 2* (1948)]. Grateful acknowledgment is made to the American Journal of Psychology for their permission to reproduce this figure.

high when you are not talking as when you are. The irregular rectangular waves between words sound like noise, and the noise is as strong as the speech.

WIENER: May I say one thing in this connection? They think of using the device for prosthesis. I just caught the constancy of the amplitudes, because there we are going to use electrical skin stimulation on different regions. Since we are going to work with constant voltage, with constant current density, it is very good that we don't have too variable a signal. This may actually be better from the standpoint of our prosthesis for the skin in ordinary sounds.

LICKLIDER: Clipping assures that the speech comes through at a high level. A combination of clipping and what is known as compression – compression also smooths out [86] variations of speech level – works well in hearing aids. |

Let me mention that you do not want to do the opposite of peak clipping and introduce distortion like this: [Figure 18]

This form of distortion, called center clipping, since the center of the wave is clipped out, is very detrimental to intelligibility, even in small doses.

As a last item, let us consider a combination of linear and non-linear operations. A [87] nice linear operation is differentiation with | respect to time. It turns out that the derivative of the speech wave form is just as intelligible as the original speech. Differentiate the derivative to get the second derivative, and now all the consonants are greatly emphasized. You can understand such speech; you can, in fact, understand it clearly and without unusual effort. Going in the opposite direction, we can take the time integral of speech, and then the integral of the integral. The integral sounds boomy but is quite intelligible. The double integral sounds muffled, and close attention is required to follow it. Evidently, however, intelligibility is highly resistant to differentiation and integration with respect to time.

The thing I am most interested in describing is what happens to intelligibility when you introduce both differentiation or integration and infinite peak clipping. The answer to that question is quite simple. If you integrate first and then clip, you lose intelligibility; the result is almost entirely unintelligible. But if you differentiate first

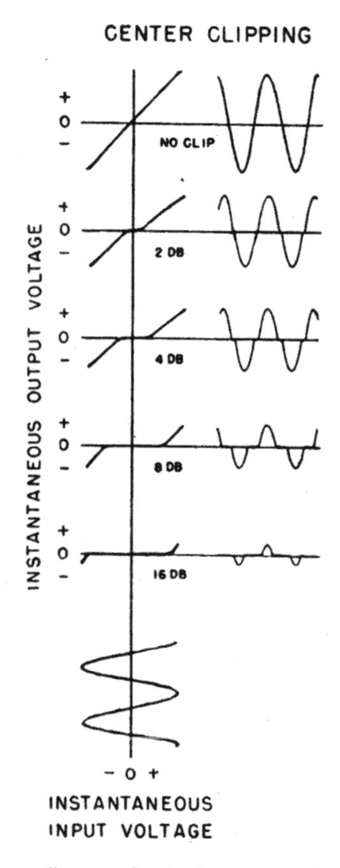

FIGURE 18. Cathode-ray oscillograms, showing instantaneous input-output characteristics of a center clipper. The plots are analogous to those of Figure 14 for the peak clipper. [LICKLIDER, J. C. R.: *J. Acous. Soc. Am.*, 18, *429, Fig. 2* (1946)]. Grateful acknowledgment is made to the Journal of the Acoustical Society of America for their permission to reproduce this figure.

and then clip, you come out with a square wave, as before, but the square wave gives you word articulation scores of about 90 per cent. It sounds terrible, but you can understand it clearly. Finally, if you introduce an integrator in cascade with the differentiator and clipper, so that you differentiate, then clip, then integrate, you obtain a wave that looks like the one in the second-to-last line of the figure, that sounds not at all as though it had at one point been an irregular series of rectangular waves, and that is no less intelligible than the result of differentiating and clipping without integrating. These observations are summarized in the following two figures: | [Figure 19] [88] | [Figure 20] [89]

WIENER: Were you going to talk also about the separation of speech bands?

LICKLIDER: I think that I must subside. In closing, let me say this: The next category of distortions that we should cover (or should have covered) is the category of abstracting transformations, transformations that intentionally throw away information, | but [90] not the information that underlies intelligibility. Peak clipping is obviously such a transformation, but there are others. Perhaps the most interesting are based on the notion of retaining the general features of the distribution of speech energy in time

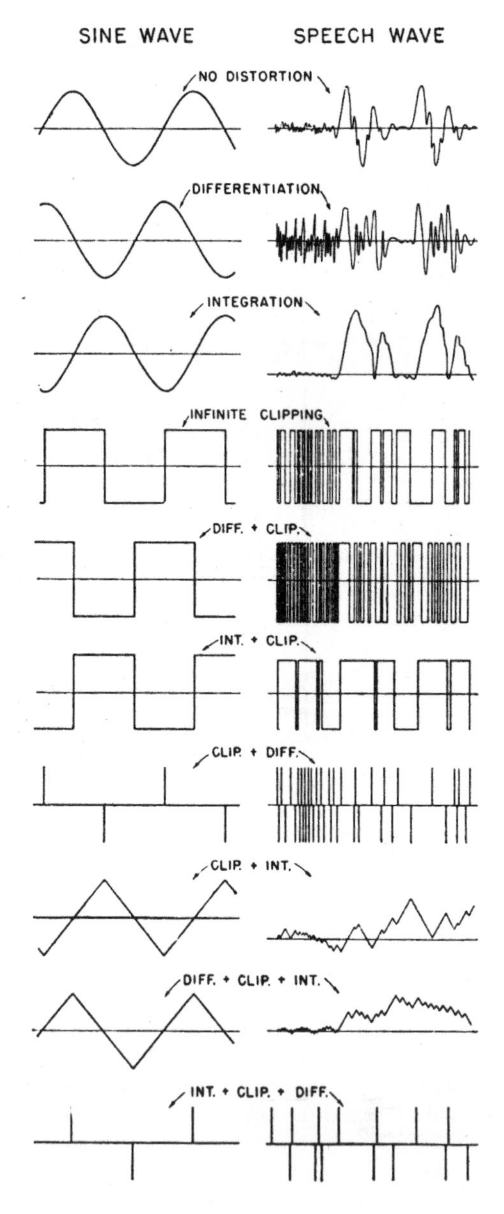

FIGURE 19. Schematic illustration of the effects of the distortions upon sine wave and upon speech waves. [LICKLIDER, J. C. R., and POLLACK, I.: *J. Acous. Soc. Am.*, 20, *42, Fig. 3* (1948)]. Grateful acknowledgment is made to the Journal of the Acoustical Society of America for their permission to reproduce this figure.

and frequency, that is, in retaining the general features of the running spectrum. The vocoder is that sort of thing. Another possibility is to preserve the general features of what we might call the running autocorrelation function of the speech wave. A third possibility is to preserve the moments of the distribution of energy in frequency, that is, the mean, the variance, the skewness, the kurtosis, and so forth, of the running

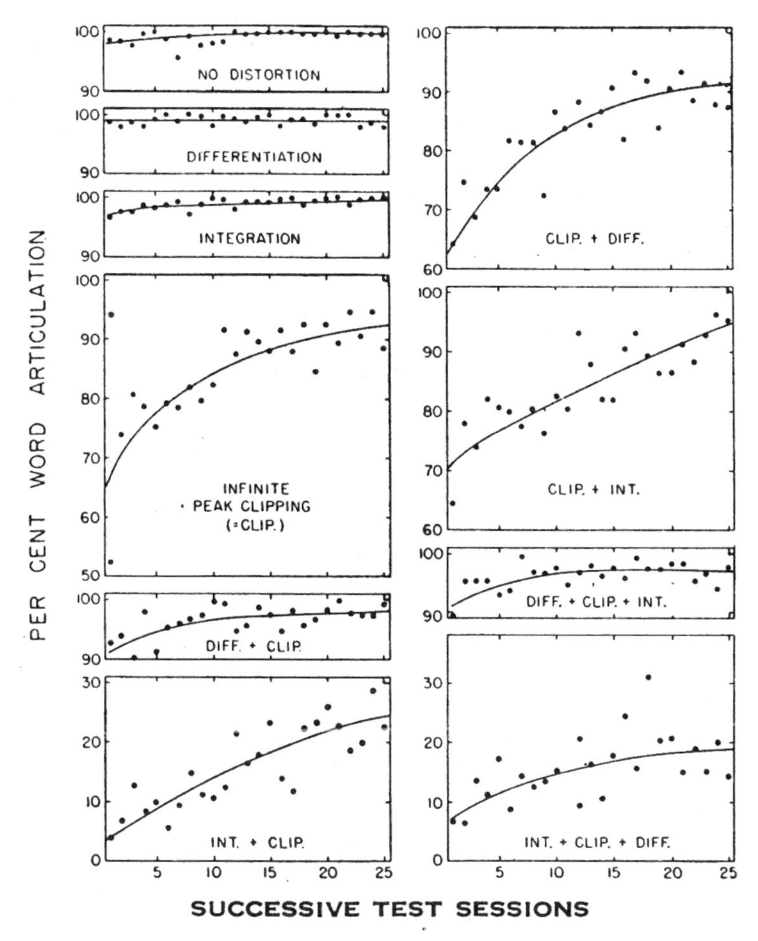

FIGURE 20. Showing the articulation scores for each of the 10 arrangements (see Figure 18) of the distorting circuits. [LICKLIDER, J. C. R., and POLLACK, I.: *J. Acous. Soc. Am.*, 20, *42, Fig. 4* (1948)]. Grateful acknowledgment is made to the Journal of the Acoustical Society of America for their permission to reproduce this figure.

power spectrum of the speech wave. There are a number of such schemes afoot, most of them arising, I think, in the fertile mind of my friend Bill Huggins. I think it might be interesting if, in our discussion, some of you are particularly familiar with such schemes, especially with the vocoder, would talk about them further.

WIENER: I should like to say in this connection what the relation of phase to the vocoder work is. I think that is very important here; that is, you can interpret a lot of the vocoder work as a rather loose phase transformation. If you take your speech band and divide it into a number of bands, rectify, and then use it to modulate an oscillator somewhere near the middle of the band, you will see that what you are doing is to take the general envelope pattern and keep it the same; actually, you are doing a great violence to the phase. You are doing one thing with the phase in one band and another thing with the phase in another band. The only thing that you are doing is to work rather consistently with the phase in each band as a whole. Now that is what is actually being done with the vocoder, if I am not mistaken. It leads to admirable speech.

LICKLIDER: It goes even further than that. A train of pulses is put into all the band-pass filters. These filters simply operate upon this impulse train to make the voiced or line-spectrum sounds.

WIENER: In other words, it is only an analogue that is preserved from each region. Everything else is mutilated terribly; you figured out how much of the information should be in from the frequency range, and the amplitude range preserved by this is less than one-tenth in many cases.

LICKLIDER: I think Shannon has figures. It takes about 250 cycles to transmit vocoder signals, does it not?

SHANNON: That can be done with 250, even less.

LICKLIDER: It is interesting to look at it this way: the band width of normal speech is [91] 7,000 or 8,000 cycles – even more than | that, if you want to include everything. It is possible, however, by picking the pass band from the center of the speech spectrum, to obtain intelligible communication with less than 1,000 cycles band width. Intelligible. It would not sound good, it would not be satisfactory for many purposes, but it would be intelligible with 1,000 cycles and no fancier transformation than that provided by a band pass filter. The vocoder brings it down to 250 cycles or less. But if you built a voice-operated typewriter, that is, an electronic stenographer, you could communicate through a band only 5 cycles wide, or probably even less. (This is assuming a reasonable signal-to-noise ratio, say 20 db.)

WIENER: This is just the reason why this hearing aid we are working on is so promising. Of course that is because obviously it would be impossible to do much on the skip if we had to get all the information of 7,000 cycles through. When you come to 250 cycles, there is nothing at all ridiculous in having, we will say, ten or even five recorders on different parts of the skin, each getting about 50 cycles' worth of through. Fifty cycles isn't an awful lot. The skin can discriminate at that rate fairly well, and the result is that it is perfectly possible to build an automatic skin receiver for vibrations that one can be trained to read.

LICKLIDER: Have you any recent data on the vocabularies of the people?

WIENER: The best thing we have done with that was a vocabulary of 12 words, a run of 80 with six errors. That is obviously not enough. But that simply shows there is enough basis for discrimination to warrant going much further.

There is one thing that I want to say about this from the general point of view of communication engineering, and that is that the measurements that Mr. Shannon and I have of an amount of information going over a line really need a certain comment. In view of this we have been measuring the maximum amount of information going over the line which is recoverable by any sort of terminal apparatus. Now that is very relevant if we have our terminal apparatus at our disposal. In the kind of situation we have here, concerning error with the skin receiver or anything else of the sort, the relevant amount of information is the amount that will get through both the transmission line and through what we may call the phonetic translation, translating oscillation into phonemes or their equivalent for the skin or any other sense. This information will give us no more information than we had in the beginning. There is still the maximum amount of information recoverable from this. In general, from the second law of [92] dynamics, it would give us less information that we put in. |

LICKLIDER: It is probably of interest to set down some figures here. If you figure a speech-band width of 5,000 cycles and signal-to-noise ratio of 33 db (that is, 2,000 in terms of power), and this is at least tried for in commercial radio, then I think it appears that 100,000 bits of information can be transmitted through such a communication channel.

McCulloch: Per second?

Licklider: 100,000 per second. Now, if you speak at the rate of 10 phonemes per second and use a vocabulary of, say (to make it easy), 64 phonemes – we actually distinguish only 40 or 50, but 64 is 2^6 – you generate 60 bits of information per second. This assumes that the phonemes are all equally probable –

Wiener: Yes.

Licklider: And, of course, they are not.

Wiener: Less than that.

Licklider: So the 60 is an upper limit. To get into round numbers, let us call it 50. From 100,000 down to 50. You see what sort of economy of communication facilities is possible in theory.

Wiener: There is also another translation.

Shannon: I have worked out the amount of compression possible with speech. If you made use of all statistics available, as far as I could estimate them, speech could be transmitted over a 20 db channel that is only about 2.5 cycles wide. You can compare with 20 db channels 10,000, if you have a 5,000 to 1 reduction.

Wiener: In the translation from the phonemes to semantic understanding, the words not only have to be heard but also understood, and again there is some further reduction. We really have another stage of translation that may either replace the one I have mentioned or be put on top of it, particularly when we observe animals or observe people, that is, the behavioral translation. In other words, we have what we consider the amount of information, the original theory being that we have the maximum amount of information; but as we have just said, the amount of information received and usable by the particular receiving apparatus is terribly pruned. The interesting thing is that this same sort of pruning not only comes up in speech; it also occurs with language in machines. There are many cases where we talk to a machine and listen to a machine. I am considering, for example, the remote-control substations for hydraulic power, where the power dispatcher has to get messages to the machine and where the machine has to inform the power dispatcher about significant facts. Now there again you have the same translation problem. You must distinguish the language as it goes over the line with the | amount of information going over the line, with the language [93] as it exerts itself as behavior. There again you have the second law of thermodynamics cutting the information that you received; of course it is not good design to have it clipped too heavily, because it means a bad use of the line; but in most cases there is still probably a very appreciable amount of particulate information.

Licklider: Since we have talked about the vocoder, I think it might be of interest to report some very preliminary results of observations being made at M.I.T. now with a vocoder. It is actually an 8-channel arrangement that divides the speech spectrum into 7 channels and uses the eighth to keep track of the fundamental period of the voice. The apparatus is so set up that you can record the 8 signals (7 envelopes and 1 voice-pitch control signal) and play them back too fast or too slow. I believe that the observations are interesting in relation to Stroud's notion about how fast the brain can operate. The device lets you speed up speech without having all the frequencies go up, or down without having them drop. It is beginning to look as though you cannot increase the rate by more than a factor or two, or possibly three, over the rate of ordinary conversation, without –

Stroud: My best high is 2.5. You would be driving a man mercilessly. Two would be the maximum.

Licklider: You start to lose track of so many of the words that you cannot piece together –

WIENER: There is not time enough to get the memory working.

LICKLIDER: You just don't run that fast.

PITTS: When you were discussing the effect on intelligibility of a sharp filter that cuts off below or above a certain amount, it appeared as if the upper curve were simply the lower curve subtracted from one?

LICKLIDER: Oh, no! I mean to indicate that the two curves, low-pass and high-pass, intersect at the level of 70 per cent syllable articulation.

PITTS: Is it not also true that the other one is subtracted from the first, that is, in the sense in which you might then consider that the intelligibility is distributed additively in several ranges? You suggested that.

LICKLIDER: On the logarithmic frequency scale the curves are, in the narrower sense, monotonic functions. I think, though, that you can find a scale (a distorted scale of frequency) on which the two curves would look symmetrical.

PITTS: Not if you plot them both in the same way.

[94] LICKLIDER: Perhaps within the limits of accuracy with which the | curves are determined. I am pretty sure you can.

PITTS: No, I mean the question is whether or not you can regard the intelligibility as being made up of additive components of different bands, suggesting roughly that that was nearly true when you had bands that were distributed over the spectrum in an irregular way, each one being narrow and spread out.

LICKLIDER: It breaks down. Let us reduce the situation to a ridiculous extreme. Say that you take alternate one-cycle frequency bands; take all the odd cycles and compare their intelligibility with that of all the even cycles. Of course you cannot realize filters of that sort, but when you approach the situation represented by that extreme case, you get intelligibility that is too high, higher than the theory based on linear additivity of contributions from individual frequency bands says it should be.

WIENER: That whole subject is a fascinating one, and it explains why there is so much hope of working for the deaf. Has anyone tried to determine what the limits are for similar limits of compression for the eye for television and how much real information we use for intelligibility there?

LICKLIDER: It probably should be said here that we have talked only about intelligibility. While there doubtless is in vision something comparable to intelligibility, it is not what people look for in television sets.

WIENER: I often wonder why people try to look at television.

LICKLIDER: That defies analysis. One of the things that should be kept in mind while discussing effects (and especially non-effects) upon intelligibility is that all these distortions (I should perhaps except some phase distortions) impair speech »quality« considerably. It is amazing how sensitive the ear is in detecting slight changes in spectrum, wave form, or any other aspect of speech. The phase variation is the only one that can be carried to any considerable degree without producing a detectable effect. Intelligibility is unaffected, but differences can be heard. There are two separate questions: How finely can we discriminate, that is, how small are the changes we can detect? And, on the other hand, how strongly does our perception resist restructuring; how large are the changes that we can tolerate without losing the identity of the speech sounds and the meaning of the messages they convey?

MCCULLOCH: Two or three want to speak here. Marquis wants to say something.

MARQUIS: I want to ask whether the so-called »electronic pencil« represents the same kind of compression of visual information when you have five scanning beams going [95] across a line of print translated into electrical on-off signals in five channels? |

WIENER: I think it does.

LICKLIDER: Even better is the Telautograph. Since you are interested principally in contours, the Telautograph draws contours. It is not forced to cover the whole area of the paper. (The Telautograph is the machine they use in hotels, between the telephone switchboard and the desk, for example. You write at one end with a pencil, and another pencil at the other end of the line copies what you have written.) The Telautograph signals can be transmitted – I think William Tuller gave some precise figures once – in a fantastically small band width.

WIENER: There is one interesting thing in all this distortion. Have you ever tried to see whether you can completely destroy the listeners ability to hear music? Does anything in the way of music get through in these violent distortions?

LICKLIDER: There are not any very good results on that. Most of the things are just casual observations.

WIENER: What are the observations?

LICKLIDER: Ocarina music is vastly improved by clipping.

TEUBER: But not violins or flutes, I suppose. Traditionally psycho-acousticians used to be very musical-people like Hornbost[e]l, Stumpf. I don't know whether that tradition continues in New England. A musical person might take violent exception to your transformations. If we think of mere information value, whatever we get is redundant; we can safely reduce it to some iron ration.

LICKLIDER: For understanding.

TEUBER: Of a sort. It depends very much on your attitude. There are many situations where we might hear something, but we don't listen. All we notice, and care to notice, is whether someone is still talking or whether he has stopped. Then again, we can switch our attitudes and listen specifically to inflections, slight variations in pitch, overtones, or what not. Very fine differences can be detected; any of these transformations you mentioned can be detected by the listener, I presume, even though they might not grossly affect intelligibility. When you first published your observations, you put the emphasis in a different place. You spoke of equivalence of stimuli. You pointed out that the intelligibility of such distorted speech demonstrates the extent to which the nervous system is capable of disregarding differences, differences that are irrelevant to the organism at that moment. Certainly your data show how far we can go in extracting one and the same general pattern from vastly different physical configurations. To emphasize equivalence of stimuli would be more in keeping with my own bias. If you state the problem in terms of redundancy of informa|tion, I'm afraid an [96] important aspect of your work is overlooked.

GERARD: That points up something I have been very eager to ask. I don't know if you can answer it from all these modifications that you make. Have you come out with any fairly simple statement of just what the coding is for intelligible speech? What are the elements of matter that must be there? Are they few and simple?

LICKLIDER: In so far as I have come to an opinion, I have come to one that has been held by several others. It is that, if not *the* most important thing, one of the most important things is the distribution of the energy in frequency and in time, the frequency and time analyses being made simultaneously. One way of saying it is this: look at the resonances of the speech spectrum (the places along the frequency scale where there are maxima of the power or energy spectrum), and follow the resonances along in time. Here, I know how to describe this. You probably are all familiar with the Bell Telephone Laboratories' work on visible speech. I brought along a slide showing one of the patterns.

PITTS: You think the results would be the same if you used Arabic words instead of English words?

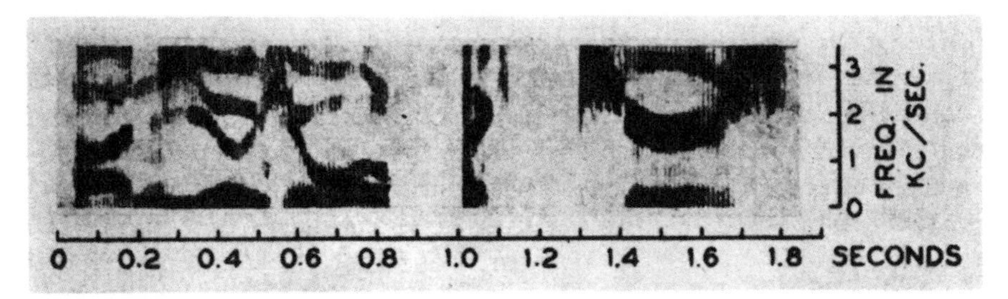

FIGURE 21. An intensity-frequency-time pattern. This visible speech pattern shows how the resonances or concentrations of energy along the frequency scale vary as the talker, saying »Unusual pictures,« progresses from one speech sound to another. [POTTER, R. K., and PETERSON, G. E.: *J. Acous. Soc. Am.*, 20, *528, Fig. 1* (1948)]. Grateful acknowledgment is made to the Journal of the Acoustical Society of America for their permission to reproduce this figure.

LICKLIDER: Why?

PITTS: Arabic has many more gutteral consonants and sounds back in the throat. I wonder if it is due to phonemes.

LICKLIDER: The figure presents a representation of a sample of speech, »Unusual pictures.« The plot is of this sort: the vertical scale is frequency plotted linearly. The horizontal scale is time, also plotted linearly. The darkness of the plot at a given spot represents the envelope amplitude of the output, at a particular moment, of a band-pass filter centered at a particular frequency. Figure 21. As you see, each of the sounds has a [97] characteristic pat|tern. In the vowel patterns there are concentrations of energy in frequency; these are called formants. You can see in the figure how the formants move along, sometimes swinging up in frequency, sometimes down. Now, one strongly supported opinion is that the important thing for intelligibility, the basic clue to the identity of a speech sound, is the pattern in frequency and time taken (in so far as frequency and time can be so taken) simultaneously.

I have the notion that, for the hearing of consonants, especially, there are other quite important features. The durations of consonant sounds, quite aside from considerations of frequency, are important, perhaps crucial in the case of some consonants. It is an interesting possibility that the auditory system either works simultaneously with two or more compromises for resolution in time and frequency, one part of the system resolving accurately in time and another resolving accurately in frequency, or that it skips nimbly from one compromise to another. I think that something of the sort is a definite possibility, but, as so often is the case, there are no really crucial data.

STROUD: If you make up a sound recording which is the same as listening to sounds in one ear, such as you get from this speech recorder, and it includes a minimum number of the multiple paths of transmission, it is possible to get a sense of the relative depth of the source's radial distance, or range, from a single-ear perception, whether you hear it over one or both ears, whereas it takes two ears to get the bearing. One ear will give you the range. Unless you have some precedence computer, it is difficult to imagine how you would get that information that *A* is in front of or behind *B*. That is not all you get. You get subjective enhancement of the loudness, although the physical intensity of the stimulus is unchanged between a single channel recording of a sound and a recording which includes the same sound pattern transmitted with the slight time delay over several paths. The physical volumes can be equated, but the one with the multiple paths which also gives you the sense of radial range is subjectively louder, although it may be equated to physical equivalent power.

McCulloch: Will someone say something about click loudness? I think that is one of the peculiar things that should be described.

Stroud: Last summer Dr. Garner at Johns Hopkins took several subjects. We have a little device which rather brutally clips out a short section of a sinusoid. If you clip it out very brutally, of course, you get not only the frequency from which you clip but many other frequencies. Under these circumstances Garner, clip|ping brutally with [98] naive subjects, found that a sample, perhaps a few milliseconds long, 2, 3, or 4, was quite a good sample, whereby a subject could estimate the loudness of the continuing sound, although he is using, in my terms, only about 2 per cent of the available information. A very short sample, 3 or 4 milliseconds long, was good for an estimate within a decible[!] or two of the amplitude. How loud would it be if I heard it all the time? You have to be quite self-conscious about it. A little later I was working by a rather more difficult method, using filters which restricted the frequency components quite sharply, which meant that I could not make a nice square bundle of sounds, but that I had an envelope which rose and fell in almost probability shape distribution for sounds. However, I found that once you had the idea of what you were trying to discriminate, in this case a sample in which 2 or 3 of the waves passing through this filter reached their maximum amplitude at a thousand cycles, it was again a sufficient sample to estimate the amplitude of a constant signal. The method of estimation, by the way, was to set this same signal at a constant amplitude to the equilavent[!] loudness.

McCulloch: Can you imitate the change in sound at all?

Stroud: Subjectively the sounds at short levels go *tick-tick-tick*, and you have a difficult time assigning any particular frequency to them. You can tell one tick is a little higher or lower than another tick, but to match for frequency is difficult. You miss by an enormous amount. As the samples grow a little longer without getting appreciably »click« louder, they go *tock-tock-tock*, then *beep-beep*, then *bleep-bleep*, and finally *e-e-e-e* and you have practically an exact replica of the sound. Apparently, whenever you fill up about 100 milliseconds of the sample, you have got about as good an estimate of the pitch as you will ever get, but you need to fill up only 3 or 4 milliseconds of it to get an estimate of how loud it is in the click sense.

Licklider: I think the repetition rate of the pips is an important parameter.

Stroud: You do your best in estimating the famous ten pips.

Teuber: Can you do it with single clicks?

Stroud: I suspect you can do single clicks with a great deal of practice. All you would do would be to smear your statistics out. That is all. I may be wrong there. Certainly if repeated, say ten times a second on these very short little bursts, it is not very difficult to make a good estimate with a sample a few seconds in length.

Werner: That brings out one point concerning the practice effect. Have you made any studies on training methods?| [99]

Licklider: Listeners or talkers?

Werner: Listeners. The practical problem involved here would be to see whether there are certain methods of training which are better than others and which would bring about a quicker adjustment to this type of verbal stimulation. As a psychologist I am particularly interested in two things: the practice effect dependent on training methods and the individual differences in the rate and kind of adjustment.

Licklider: During the war, especially when the military was picking people for fairly crucial jobs involving listening – the military had the notion, as you probably all know, that everything must go through many links of a chainlike communication system before it becomes really official – they were interested in the problem of training listeners. Quite a bit of practical cut-and-try work was done to find out how to train listeners, and talkers, too. As far as listening is concerned, the conclusion was this: the way to teach a person to listen is to provide high motivation (you can do that in a number of ways), to provide lots of speech for him to listen to, and to provide knowledge of results. (The listener must know whether he heard it correctly or not.) It is important, also, to verisimilitude. The test situation must be quite like the situation for which you are training. As a matter of fact, in articulation testing work it has been found that after much experience with a particular kind of speech and a particular communication system, the crew will become very proficient. The scores will be uniform. Then, when a somewhat new form of distortion is introduced, or when the speech material is changed a little, the scores drop and a new learning period begins. There is a considerable amount of transfer, of course, but the degree to which the learning is specific to a given communication system or to a given type of distortion is nevertheless impressive.

WERNER: You are not speaking of testing a specific method.

LICKLIDER: A new crew improves throughout a period of two or three months before it levels off on a plateau. Then the specific secondary improvements extend over intervals of perhaps a week or two.

WERNER: You are obviously not concerned with the development and the testing of training methods. You present a person with a great variety of distortions and measure his ability to perceive.

LICKLIDER: There has not been enough work done for me to say anything specific about that question. Actually, I was never personally active in the work on training; [100] colleagues in the laboratory were doing it. |

MCCULLOCH: Have you any idea what human variability is at the level they reach at the end of training?

LICKLIDER: I know less about that than about their initial levels. Let me quantify variability in terms of percentage of words recorded correctly. I should say that if you run enough tests to get stable scores for individuals in a group of 20, there might very well be one person getting 25 per cent correct and another getting 75 per cent correct. The figures presuppose enough tests to rule out fluctuations of measurement; the range of individual differences is that great.

MCCULLOCH: If you keep training that group, do they pull very close together as time goes on?

LICKLIDER: What happens is this: as the scores get higher, they bunch closer together; but that fact is not significant, because the better listeners run into the ceiling of 100 per cent articulation. The thing to do is to introduce stress (in the form of noise or distortion) and thereby reduce the scores to the initial level. If you did that, I do not think that the spread of trained listeners would be much less great than the spread of untrained listeners. It might, but I do not think the effect would be very impressive.

MCCULLOCH: They might? May I ask one more question? Is there any variation between those who can pick up syllables easily, those who can pick up words easily, or those who can pick up phrases or sentences?

LICKLIDER: I don't know. I wish I did.

MCCULLOCH: Does anybody have a measure?

LICKLIDER: Perhaps the Bell Telephone Laboratories have.

MARQUIS: Didn't you find a lack of relationship between individual differences in this ability and intelligence?

LICKLIDER: Very low, if any, correlation between the intelligibility test performance and performance in receiving code. There was an extremely low correlation, negative, perhaps -0.17.

PITTS: Significant.

LICKLIDER: Just about.

MARQUIS: Code?

LICKLIDER: Spoken words and telegraphic code.

WIENER: I think it might be.

MEAD: If you talk about another kind of information, if you are trying to communicate the fact that somebody is angry, what order of distortion might be introduced to take the anger out of a message that otherwise will carry exactly the same words?

LICKLIDER: I do not know. Peak limiting, peak clipping makes speech monotonous, and I would strongly suspect that some of the | clues to anger would be eliminated by [101] clipping. But one thing seems to be quite true: even these emotional things are represented in a rather redundant fashion. The clues appear in amplitude and in time. The pace of the speech and its intensity are affected together. I think it very unlikely that you would lose the angry quality of speech by washing out, say, the amplitude variations (variations in the envelope of the speech wave).

PITTS: It is difficult to express resignation if you have to shout at someone.

WIENER: It would also be understandable in dead-pan tone. I do not think that this is good. I mean the tendency to convey emotion by limiting the time might be a very serious one.

PITTS: Time amplitude and possibly general drift in pitch from the beginning to the end of the phrase.

MEAD: Perhaps the most essential point would be sincerity. If we are going further into this continual broadcasting between countries, or use of broadcasting in various sorts of general communication, what does and does not sound sincere might be something that you could analyze. Our commercial broadcasters are purposely insincere. They must do something. The question is whether that would be manipulatable.

SAVAGE: You mean nobody could be that insincere without special training.

WIENER: I don't think so.

PITTS: It depends upon the person receiving. I imagine that to a Midwesterner anyone who speaks with an English accent sounds insincere automatically.

MEAD: That is another kind of effect.

LICKLIDER: The word »insincere« is used differently by different people.

KLÜVER: I should like to bring up a point which has fortunately or unfortunately no bearing on the international situation. In listening to talking parrots, you may discover that the speech of at least some parrots appears indistinguishable from that of the human teacher. My own observations are chiefly based on parakeets. After hearing some talking behind my back, I have again and again been obliged to turn around to determine whether the words or the sentence had been uttered by the parakeet or its human teacher. I am wondering whether this identity of auditory impressions merely represents some form of equivalence or whether the spectograms of the parakeet's and the teacher's speeches would be more or less the same.

LICKLIDER: My guess is that if it sounds the same, the spectrogram | is going to be [102] quite the same, but I do not know for certain.

KLÜVER: I think it is important to know whether such a striking »identity« of auditory impressions could be associated with widely different spectra.

I should like to return for a moment to the question of motivation mentioned by Dr. Licklider. Perhaps Dr. Werner will remember the old observations by Kalischer on the aphasia of parrots [Kalischer, O.: Das Grosshirn der Papageien in anatomischer und physiologischer Beziehung. *Abh. d. K. Preuss. Akad. d. Wiss.* (1905)]. Kalischer performed brain operations on talking parrots and discovered that he could produce speech disturbances by injuring the mesostriatum in the neighborhood of the Sylvian fissure. He made some very interesting observations on the restitution of speech in these aphasic parrots. After the operation, the parrots regained their vocabulary by themselves (first the rhythm, then the vowels, and finally the consonants) without ever being taught again. The birds behaved as if they were aware of errors and imperfections in their pronunciation and gave the impression of purposely trying to eliminate such defects by assiduously training themselves. At any rate, they were heard practicing the same words or even parts of words for hours at a time until their pronunciation was flawless again. As pointed out by Poetzl (Vienna), this is in striking contrast to the behavior of human aphasic patients. Apparently the »motivation« of an aphasic parrot in trying to regain normal speech is far stronger than that of a human aphasic.

LICKLIDER: That is lack of cerebral cortex.

KLÜVER: I believe one of the more interesting things a psychoacoustic laboratory might do is to analyze the speech of normal and aphasic parrots.

MCCULLOCH: If you wanted to make a Fourth of July oration, which of the modes of distorting speech would you necessarily eliminate?

LICKLIDER: Certainly you would want to eliminate clipping. You might introduce a certain amount of »expansion.« You can heighten the effect by increasing the dynamic range of speech artificially. I think you would want to avoid flutter. Flutter at a rate of 7 to 10 per second introduces a quality that suggests uncertainty on the part of the talker. Your bold, declamatory statement would fade into a question.

PITTS: Because the frequency changed also.

LICKLIDER: It is rather hard to say. I think you want to keep the dynamics, that is, the [103] rates of change of intensity, as pronounced | and definite as they are in the original speech.

MCCULLOCH: Has anyone done this trick of taking a speech said perfectly flatly and then putting in these things that give dynamics to it, in order to see what happens?

LICKLIDER: We tried blowing clipped speech back up by modulating the clipped wave with the envelope of the unclipped wave. That made it sound considerably more natural.

BIGELOW: Does it change the intelligibility?

LICKLIDER: Not much.

FREMONT-SMITH: In the sense of having more effect on it?

LICKLIDER: Yes.

FREMONT-SMITH: Is it possible to make a friendly speech hostile, or vice versa, by building it up in a particular way after it has been clipped?

LICKLIDER: That is a very interesting possibility. You might try, and this can be realized easily with the aid of electronics, having one person modulate another's words.

GERARD: The voice of Jacob in the words of Esau.

FREMONT-SMITH: Kukla, Fran, and Ollie on television. One character in that simply makes the sounds »Tui, tui, tui.« Yet he carries on a conversation by modulations of »tui, tui, tui.«

LICKLIDER: One of the things we have been interested in doing but have not yet done is to start with random noise and see what we can put into it by such methods as imposing upon it the envelope of a speech wave.

MEAD: When you say you can take the modulation of one person's speech, do you mean, for example, that you can take Hitler's speech, analyze it, and take Roosevelt's speech and analyze it, and then, in effect, have Roosevelt give Hitler's speech?

PITTS: That would be difficult because of language differences.

LICKLIDER: Yes.

MCCULLOCH: Not having listened to these test devices, I should like to ask a question about something that puzzles me. It is fairly clear that you can practically destroy everything that I have thought necessary to speech and still have speech that is speech. To what extent is that translation and to what extent is that still speech?

WIENER: It is still speech, sir, for this reason: it is understandable without special new training or with a minimum of new training.

LICKLIDER: That can be qualified. I can present some figures. If we clip speech until all that is left is the series of rectangular waves we have mentioned, and present this to an articulation crew with no practice, we get scores of 60 words in 100. After perhaps a | month, they go up very nearly to 90 words in 100. [104]

WIENER: But the point is that 60 is a conspicuous contribution.

LICKLIDER: It is high enough for connected speech. As soon as you talk in sentences, say, »Pick up the yellow pencil,« or, »Get me the pliers,« it is possible to have the command obeyed immediately, that is, if the listener has any intention of obeying.

WIENER: Isn't the training that you need the kind that a man needs who is deaf, that is, the present-day hearing-aid kind?

LICKLIDER: Perhaps less training than that involves.

PITTS: I wonder more and more how much the results of the degree in variance of given kinds of distortion depend upon the statistical distribution of the phonemes which are present in the thousand words which are possibilities and upon the degree to which different phonemes are fairly likely to be confused or present among them. I wonder if you might not get very different results if you added collections of words with quite different statistical distribution of phonemes in which there might be more or less confusing ones. As to the relative freedom from variants upon a given source of distortion, say, in language of different types, comparing Arabic with –

WIENER: Hawaiian would be the extreme.

PITTS: – with the consonants of almost any other language, would the results be different?

TEUBER: When you put speech through transformations to make it unintelligible, I suppose you want to be sure that you can make your transformations in reverse and get the speech sample intelligible again. This is the principle of a secrecy system: you want to preserve potential intelligibility. But during the war, there also was the much simpler problem of »jamming« – destroying intelligibility by superimposition of some extraneous sounds. Maybe you are not free to tell us how it was really done, but I heard that one of the most effective ways of jamming speech was to superimpose on it some other speech – a jabber of voices – preferably in the same language. This sort of jamming is an experience we have here at times. Now I wonder does it have to be American speech for optimal jamming of American speech, or is it equally effective to use Chinese to jam American speech?

LICKLIDER: I think that Chinese would interfere with American almost as much as American would.

WIENER: May I say there are really two jamming problems. In one you want to jam the other person from getting anything. If you take a transformation of speech that is meant to be gotten back from the apparatus, am I right that after a while anything | [105] that you do to the speech to make it unintelligible where it is supposed to be recoverable by proper apparatus sooner or later is understood by the ordinary ear if the timing is not too badly changed?

LICKLIDER: I am not so sure that that is true.

FREMONT-SMITH: Without changing from time to time, keeping one system constant?

LICKLIDER: I cannot answer that. You could make the system complicated enough so that it would be unlearnable, I think. For example, you might take the speech spectrum, separate it into parts with the aid of filters, invert these parts, heterodyne them to other parts of the frequency scale, and then put it back together; and you might also change the recipe for the mixture at frequent intervals. Then I don't think anyone would learn to understand it.

WIENER: To go further, that is what one would have suspected in the beginning.

SHANNON: During the war some of the secrecy systems that we developed used much scrambling of frequency bands. However, they were directly or indirectly understandable; therefore, it is obvious that you have to do a lot more than you think you do.

PITTS: Possibly anything you do that does not involve changing over too great time intervals would be penetrable or nearly so?

LICKLIDER: One scheme that permits complete recovery of the speech is this: you send two signals in adjacent bands. One of the signals is speech entirely masked by noise. The other is the noise, the same noise that is used to mask the speech alone. At the receiving end you simply subtract one signal from the other, and out comes speech. There is no possibility of learning to listen through that, because the speech can be masked so thoroughly that it is dead and burled.

WIENER: You ferret it out, in other words?

FREMONT-SMITH: You cannot learn selectively to obliterate by virtue of attention?

LICKLIDER: You cannot learn to hear speech through noise that is so strong that you miss by 10 decibels being able to hear the speech through it.

GERARD: This is a good chance for me to ask a question that has been on my mind. With your fully clipped speech, people at first get 60 per cent and improve to 90 per cent. Have you any evidence whether comments that would be pleasant to them are recognized more easily than those that are unpleasant; that is, is there any kind of [106] affective factor in giving meaning? |

LICKLIDER: I don't know. There is, however, an observation that seems closely relevant. When speech is so badly distorted or masked that you cannot hear the words, then Freudian influences appear to exert themselves and to determine what is written down.

GERARD: That is, perhaps, another story.

WIENER: It is well known that people who are so deaf that they cannot follow conversation are not to be trusted not to hear things you don't want them to.

LICKLIDER: That would make it very difficult to measure the thing you suggest.

GERARD: You still have right and wrong answers as your objective measure.

FREMONT-SMITH: Unpredictable, desirable commands.

GERARD: One so often hears the statement referred to that the deaf person hears what he wants to and does not hear what he doesn't want to. That would be a way of seeing whether it is so objectively.

SAVAGE: That pertains to the deaf or the psychology of the people around the deaf?

GERARD: The psychology somewhere.

FREMONT-SMITH: It would be very important to do a hypnotic experiment on the problem of unhearable sound. I am not at all sure that you can say when sound is unhearable unless you do a hypnotic experiment. We do know that under hypnosis in the waking state, in the normal state, one actually perceives and records a mass of material which is unrecoverable except under hypnosis. It seems to me that that might lead to interesting experiments.

LICKLIDER: We once had a student who was interested in doing just that. He made the mistake, however, of doing the converse experiment first. I think he measured the masking of a tone by a hallucinated noise. The student was not supposed to know the literature, or enough about hearing to know how the experiment would come out if the noise were real, but the result he obtained with hallucinated noise was almost exactly what a real noise would have yielded. So we lost confidence in the whole thing.

MEAD: I should like to get back to the question: Is this translation or isn't it? What is translation? If you take a very literate Slav who thinks there is something called *Slavic* and who speaks three of four Slavic languages, thinks of them as Slavic and believes firmly that there is something called Slavic, and that of course anybody who is erudite can speak them all, is he translating from one to the other or has he made a central point of reference | toward which his attitude varied? In contrast, if you took the average American who had never learned a foreign language, you could teach him two absolutely separate Slavic languages and he probably never would discover they had any stems in common. If you took two groups of people who had learned to handle these squared sound effects and told one of them that they were going to learn a completely new system of communicating English words which would be systematically transformed in a particular way involving translation, and if you told the other group of people that they were going to experiment with a lot of words which would involve a slight difficulty in hearing those same words, in one case would it be translation and in the other not? Can you define translation objectively instead of subjectively? [107]

WIENER: I don't think that is possible. The experience you have spoken about is very interesting. I noticed when I was in Mexico, for example, that understanding a foreign language which is close to one's own depends much upon the degree of sophistication. The ability to understand a foreigner's mispronunciation varies greatly between an uneducated person and an educated one.

PITTS: Isn't that what the psychologist called the question of definition of stimulation?

WERNER: Equivalence of stimuli.

MEAD: Having questioned if this is translation, are we talking about stimulus?

PITTS: Every time we hear English we are in a certain sense translating. What we hear is somewhat different versions of the same word. The recognition of them as the same I suppose you could describe as an act of translation into the common exemplary, which is the Platonic idea we hold in our heads. Everybody's education is slightly different.

WIENER: A Spaniard who has never studied Italian can still follow Italian without studying it, but it is quite difficult for him to speak it if he has no idea that they might be related.

STROUD: I had a very curious experience in England. England has more varieties of English than we have. I had been there a week. I was rather naïve because I thought that *Pygmalion* was a wonderful play and I was sorry I was going to miss it. After I had been there about a week, I encountered two soldiers in a railroad station. One of them was trying to tell the other when the next train went to London. (Later, after many months of education, I finally realized that one was from Ayrshire and the other from Devonshire.) I had little difficulty in listening to either of them and I was wonderfully fascinated by the thick English | dialects. Brave Boy Scout that I was, I thought there was difficulty in hearing and began translating for them. Both listened to me. In no time at all we had the whole affair straight as to who was going on what train and where and when. A little later I discovered that all Englishmen understand two variet- [108]

ies of English: BBC and American movies. I, being an American, was perfectly listenable to by anybody in the British Isles. How do you sort that one out? Here were two people in need of a translation, yet I was not translating either form of English. It was still English to me.

McCulloch: That is just the point.

Wiener: My sister had a similar experience in a train between Bavaria and Hamburg. A woman from Rosenheim – she pronounced it »Roosenhame,« and my sister had never studied Platt – had to translate between the two German women who could not understand each other.

Klüver: How far have you gone in analyzing the speech disturbances of neurological patients? For example, have you been interested in sensory aphasia?

Licklider: I have done nothing of that sort at all. There has been relatively little work of that kind, at least by people interested primarily in intelligibility.

McCulloch: May I put it a little differently? What puzzles me is what happens *when* something stops being stimulus equivalent and actually goes over to a process of translation. I don't know any objective way of testing it. I know, for example, my own difficulty in the Midwest was with the word »Umurukum.« It was rather surprising. It was all right with the word »American,« but there are many words where of I simply didn't know which of two or three words a man meant. I used to try a literal process of translation in those cases, where as ordinarily I don't. Having lost the distinction between the vowels, I would suddenly land at the word I didn't know and I would have to translate it. Does it mean this or does it mean that?

Licklider: This stimulus –

McCulloch: The name of the railroad station is one of the worst examples, that is, when they call off the name of the railroad station.

Hutchinson: I had a very curious experience three days ago. I was giving a lecture and I was talking about what I naïvely call »plants.« One of the students was terribly worried and came to another member of the faculty and said that I was talking about something called »p-l-a-u-n-t-s,« and that they seemed to be related to the vegetable kingdom. | [109]

Fremont-Smith: What was it?

Hutchinson: He got all around it but was just not able to hit it on the head, yet it isn't a very great transformation.

Fremont-Smith: Almost a selective blind spot.

Hutchinson: Perhaps he simply did not like the English way of doing things.

Mead: It is perhaps a question of framing. Some people frame all European languages together, so that the idea of learning any Indo-European language is regarded as just another language to be learned; the idea of learning Hawaiian or Japanese, however, would be regarded as something quite different. What is translation for one person is not translation for another.

Licklider: The distinction presupposes that you get inside the black box and see how the circuit works. If you set up a little theory about how one recognizes phonemes or how one understands words and sentences, it might run something like this: in the sensory channels of the nervous system, the process set up by stimulation takes the form of a multiple time series. This multiple time series is compared against each one of a large number of other multiple series which are stored in the nervous system. The incoming time series is »recognized« as being the same phoneme or word or phrase as the stored series with which it is most highly correlated. Doubtless the nervous system correlates series corresponding to words, phrases, or even sentences. But if you want to build an automatic, electromechanical recognizer of speech, you must work with phonemes. You bring in the phonemes, see which standard patterns give the best fits (after taking into account conditional probabilities that govern the expectation on the basis of what has already been received), and put each phoneme in the indicated category. That process we would not call translating. If then we put in distorted speech, and its phonemes fell into the proper categories, we would not call them translating, either.

McCulloch: Two procedures: let me make this perfectly clear. I am riding in a railroad train and the guard comes in and says »Fara.« I have not the faintest idea what I am to look for except a »Fa« at the beginning and a »ra« at the end. I look out the window and see the name on the station. After I have done this two or three times, I don't translate. I just wait until I get the station and sound together. I don't translate and I can go out and say »Fara.«

Werner: I believe Dr. McCulloch does not use the term »translation« in the conventional way.

McCulloch: I mean by it that I look around for a word among | words which is to [110] stand for the same things that this stands for. The other way is to wait till you find the physical object or something outside language to tie it up with.

Werner: Let me make clear, by a visual example, the distinction between »translating« in the conventional sense and something which I would call »configurating.« Suppose you are looking through a frosted window at an object. If you use the reduced cues and arrive at the perception of that object, for example, a tree, you are »configurating« and not »translating.« On the other hand, a primitive man may be presented with a chair, though he has never seen one. No cues would enable him to understand the meaning of a chair. That can be achieved by showing him an object familiar to him which serves an analogous function, a low stool for instance. By equating the chair to something which already exists in his own system; namely, the low stool, he understands the chair: this is »translation.« Verbal translation, then, is based on the finding of a word in my system equivalent to a word in another; configurating a verbal pattern involves the construction of words within the confines of one system.

Pitts: All the people in Licklider's experiment are exactly in the condition that you are in. The first time that the guard comes in and says, »Fara,« they have not looked out the window. They have never had any opportunity of comparing what they hear with the original, and therefore it cannot possibly be translation because you cannot learn to translate unless you have opportunity for comparison; and never in any circumstances of any experiment do they have an opportunity for comparing what is truly said, what actually has been said, with what they hear. If they did, they might learn to translate, and everything would be wrong.

Gerard: The question of giving meaning initially.

Licklider: One qualification in all of these tests is that working with words of a limited set (the listener does not know the set at first; he gets a fair idea of what words are in it after prolonged testing), the listener tends to record only words he knows are in the set. If he hears what seems to be a word not in the test vocabulary, he tends, if he is not pressed for time in recording the words, to hold it in mind, think it over, and come up with the closest word in the set that he can think of.

McCulloch: He has to pick it from a thousand words?

Licklider: Limited context. If you reduce the number of words in the test vocabulary, the listener's ability to hit the right word increases considerably.

Gerard: I think this is a good time to do that little experiment. | [111]

McCulloch: Oh, imperception, standard jest phrase. Most of them know it.

In pine tar is
In oak none is
In mud eels are
In clay none are.

GERARD: Before you do anything else. That is a series of sounds and, with the rarest exception, people get no meaning at all out of it. If I give you a certain set, you will understand every word he says, and he will say it in the same way.

MCCULLOCH: In pine tar is
In oak none is
In mud eels are
In clay none are.

GERARD: Now suppose you know that a man was sent on a mission to find a supply of food and fuel; on his return he repeats:

MCCULLOCH: In pine tar is
In oak none is
In mud eels are
In clay none are.

GERARD: It is a remarkable example of the tremendous gap between any perception and the final giving to it of meaning. Here is a whole complex sentence.

LICKLIDER: One thing is clear: until I got the clue, I could not remember the sounds. You cannot remember sounds you cannot put together into words, and you cannot try out various arrangements of the sounds you cannot remember.

MCCULLOCH: There is a very great difference among people.

FREMONT-SMITH: That is true.

MCCULLOCH: My eldest daughter is the only one in the family that has it. She can hear anything anyone says, and could do it as a small child. She could repeat the whole sequence of sounds with no understanding of what it was at all and babble it right back again.

BIGELOW: I should like to raise a question concerning the effective utilization of information reaching an observer, say, the amount of information contained in a speech which is a little vague. It is clear that the amount of information which is conveyed in ordinary speech is far less than that of the channel used to communicate it, both the middle channel and others. The same thing is true also to some extent of visual communication. It is obvious if the channel is fully used, and if we remember all [112] we | have heard, our memories would become loaded down in no time at all. It seems that there is a biological safety valve of some sort here.

Does anybody know that there are phenomena of recall in speech having to do with the unused odds of information similar to that of familiar recall of vision having to do with unused parts of the picture?

FREMONT-SMITH: Under hypnosis you can get them.

FRANK: It is what you see in the »double take« in the movies, where the listener agrees, then repeats, then suddenly realizes the meaning of what he has heard and unthinkingly agreed to. This always excites laughter in the audience.

May I make this comment? The question raised about the difference between psychological stimulation and translation, it seems to me, might be illuminated by thinking about this: when we listen to words that we are accustomed to, we don't really listen. I think that occurs because most people don't hear what is said. What we do is to put a certain set of meanings into the sounds immediately; as long as we continue to put meanings in we can accept a wide range of stimulus equivalents. In the matter of translation, I wonder whether it is not a matter there of first identifying each word and then putting meaning into it. In that procedure there is an interpolated step.

It is very interesting to learn to read another language. I think most of you know that when you start, at first you translate each word very slowly and painfully. After you learn to read without translating, you do the same thing. You don't look at individual words. Isn't that part of the difference between the psychological equivalent and translation? It is the meaning, the point that Dr. Gerard and Dr. Werner were pointing out. That is important. We put the meaning into the situation almost without listening. In the other case we have to listen to the word, translate it and put the meaning into it.

LICKLIDER I think we really start to put the meaning in before we hear the word. That makes it possible to get the word.

FRANK: Precisely!

LICKLIDER: In other words, you would be able to understand nonsense sequences of words.

MCCULLOCH: Apparently not greatly less intelligible. That was at the forty-fifth degree, and monosyllabic words pulled a little above it. It was only when you came to sentences that you began to fill up space in intelligibility.

LICKLIDER: The words were considerably above. I may not have | done justice to the [113] situation in my rough sketch.

MCCULLOCH: May I ask one more thing with respect to the unintentional auditory Rorschach, or whatever you intend to call it when you get unintelligible speech. Which is worse, mere noise, or the noise which begins to fall into the band width of speech? Which produces the most violent reaction?

LICKLIDER: I think the thing that makes the effect show up most markedly is masking noise that almost but not quite masks the speech. Perhaps one of the formants is just audible, the others not at all. This structures the perception to the extent that the listener knows a particular vowel, or perhaps one of two or three possible vowels, is in the word, but he doesn't know at all what the word is.

WIENER: There is very much of that effect in my understanding the words in singing, particularly in a foreign language, if it is a language I know well. There is just enough masking there to require a very definite effort on my part to get them to come through.

LICKLIDER: I was amazed how much better I could understand French when it was slowed down to half-speed on the speech stretcher. A person who speaks French fluently, who has no difficulty at all with full-speed French, thought that the effect of slowing it down to half-speed was horrible.

FREMONT-SMITH: It is very well known that when people are at international meetings, understanding is enormously increased if speech is slowed approximately to half-speed.

MCCULLOCH: Yes.

KLÜVER: I believe you get the most perfect auditory Rorschach when listening to a parakeet imitating voices on the radio, and at the same time the voices of six or seven people talking, shouting and laughing in the same room. When I listened for the first time to such a performance – only the bird and I were present – I thought at first that I was merely listening to »bird talk« until I found myself again and again on the verge of recognizing meaningful words. It finally became clear to me that I was listening to something like an acoustic recording of a babble.

BATESON: Does it accurately reproduce a slice of babble or make babble?

KLÜVER: My guess is that such a bird occasionally reproduces a slice of babble accurately. In comparison with parakeets, we are apparently at a great disadvantage in listening to babble. For us acoustic events are practically always carriers of »meanings.« A [114] parakeet, however, does not seem to be plagued by the thousands | of meanings that keep us from »hearing« babble or hearing a sequence of acoustic events devoid of meaning.

FREMONT-SMITH: We cannot hear it when the parrot gives it back.

KLÜVER: One of the reasons that we cannot vocally reproduce babble or similar acoustic events is undoubtedly that we cannot even »hear« babble perfectly.

FREMONT-SMITH: How do we know?

SAVAGE: Do we train the parrot to do that or hear a lot?

KLÜVER: The parakeets I am talking about have not been particularly trained to imitate babble, but they have been trained to imitate spoken words and sentences. Such birds must be isolated at the age of five weeks and kept from seeing or hearing other birds.

SAVAGE: You must suppose that it does not need to hear the same sample to teach the same word.

KLÜVER: I am sure that the speech performances of »talking birds« and the stimuli eliciting such performances should be analyzed by objective methods. In the meantime, I consider a parakeet's imitation of babble as interesting stimulus material for an auditory Rorschach.

TEUBER: Among people with brain injuries we sometimes run into individuals who have trouble in perceiving complex, or ambiguous configurations. Such a person may study a picture for quite some time before he can recognize it as you can. His retardation can be demonstrated by flashing the image on a screen. The exposure has to be abnormally long before he can recognize what it is. But there are always exceptions, and they are embarrassing for any simple theory as to why the man's perceptions seem to be slowed down after his brain injury.

During the war we had a prize case, a man who was injured by shrapnel which had lodged in his third ventricle. Of course, the shrapnel did some damage before it got there. He was very slow on the tachistoscope; when we projected pictures on a screen, they had to be exposed for as long as two seconds before he would attempt to identify them, and then he was often wrong. This was true for all sorts of material, even simple line patterns. Then, one day, we flashed a kodachrome showing his ward medical officer, a man about whom he held certain rather strong opinions at that time. On the picture, the officer was not even in uniform, but in civilian clothes holding one of his babies on his arm. We flashed the picture for 1/50 of a second, and immediately our [115] patient called out the name of the officer with a long string of unprintable | Marine imprecations. The next moment he was quite slow again – as soon as we showed him different material.

GERARD: It bears on that question.

TEUBER: Of course, it did not take him below the normal limit; his speed of recognition had just become temporarily normal, but not better than normal.

LICKLIDER: There has recently been much experimental work involving tachistoscopic presentation of figures, work oriented toward the question you had in mind. The results first published suggested that things with high emotional content were perceived faster than neutral items. In more recent work the results have been analyzed in terms of the frequency of occurrence of the objects (words) presented. The conclusion is that the items with high probability are the ones that are rapidly perceived.

GERARD: I am not sure that you mean the words have emotional charge in one's past experience.

WIENER: With the image staying for a fraction of a second, the actual time of flashing is not the time of scanning, and it is possible to recognize things in the image during the period after which it is no longer there but their afterimage is. The point image, the seeing time is not the flashing time.

WERNER: That is correct, though the problem of what one would call seeing time is a very intricate one. One of the reasons is that of qualitative differences in the level of »recognition.« Flashed words may register »organically« without specific visual recognition. A number of experiments have shown that emotionally toned words flashed tachitoscopically may arouse a specific bodily reaction, as indicated by a psychogalvanic response; nevertheless the subject is unable to indicate the word flashed [McClary, R. A., and Lazarus, R. S.: Autonomic discrimination without awareness. *J. Person.* **18**, *171* (1949)].

This situation is probably very similar to auditory speech. If you want to analyze the way people hear word patterns with reduced cues, we might take into account different levels of hearing. This might have a bearing on training to hear distorted speech.

LICKLIDER: Who were these people who responded to words they could not understand?

WERNER: These are normal subjects and the words used are, for instance, nasty words.

LICKLIDER: Why could they not understand them?

WERNER: They would not consciously recognize these words.

LICKLIDER: Normal people? | [116]

WERNER: Normal people.

PITTS: Why not?

LICKLIDER: Why could they not recognize the words?

WERNER: Because they flashed on too quickly. We are talking about the tachistoscopic method.

LICKLIDER: We are worried a little about the word »cannot« in that connection. Did not?

WERNER: »Did« not recognize under these conditions.

KLÜVER: Concerning the time factor in emotions, some interesting data may be obtained by studying the reactions of monkeys to motion-picture films. I once used this technique in investigating the emotional behavior of monkeys belonging to different species [Klüver, H.: Behavior Mechanisms in Monkeys. Chicago: Univ. of Chicago Press (1933)]. The results were particularly enlightening in the case of a female Cebus monkey which responded in a variety of ways and always sat on my lap while looking at the films. The appearance of a python in one of the pictures always led to signs of extreme fear: she uttered various sounds, she defecated, urinated, and quickly disappeared into a corner of the laboratory from which the picture could not be seen. However, these fear reactions occurred only at certain speeds of projection. In other words, this technique may be used to demonstrate that emotional behavior is a function of a certain number of frames per second.

STROUD: Perhaps I am a little presumptuous here. Much of this discussion carries the implication that somehow, of necessity, a meaning comes from being presented with any stimulus pattern, and that somehow or other some necessary meaning should follow from it. The thing I wish to suggest is that any meaning is of the order of a hypothesis which can be tested on the basis of the available information. The length of time required to verify or reject or to examine a family of hypotheses is primarily a function of the number of alternatives, their degree or organization, and everything else. Thus, as Dr. Licklider pointed out, in a thousand-word vocabulary the probability, the amount of information required to accept or reject the hypothesis that this is a particular word out of the thousand, is considerably greater than if the number of alternatives is only fifty. I know that some time ago, and I am sure Dr. Klüver and others could verify it, a great deal of experimentation was done on the tachistoscope presentations of words. You were merely asked if it was a certain word or not. Under these circumstances very minimum amounts of information are required to accept or reject [117] the typical. |

Another experiment that comes to mind is the case of a very low-level presentation of the outline drawing of a banana.

WIENER: Oh, yes.

STROUD: If you present this to the subject, he does not see anything; but if you ask him in which orientation the banana is, his responses as to whether it is horizontal or vertical are well beyond the chance level. So in all of these considerations I prefer to look at the problem as one of the amount of information required to accept some hypothesis, as belonging to some hypothesis system, and to discard altogether the notion that any response in the central nervous system of the man necessarily follows from any particular stimulus configuration.

FRANK: May we make one amendment? What I was suggesting was not a meaning as conveyed by the stimulus.

What I was trying to suggest for consideration was that the individual listening or hearing puts meaning into those situations out of his own past experience, which is a little different process.

STROUD: Let us put it this way: He will never test hypotheses, ordinarily, of which he is not here and now the master. To that extent you can say he puts the meaning there.

FRANK: Yes.

STROUD: He will attempt to test and verify hypotheses of which he is the master on the basis of the configuration of the sensory input he has available, but nothing necessarily follows from any sensory input.

FRANK: What I wanted to reemphasize was that the individual himself puts the meaning into those sensory inputs and that therefore we get a clue to the great diversity of what people hear out of the same speech.

HUTCHINSON: If he has many hypotheses, he will jam a lot more than if he is the master of only one or two.

STROUD: Yes, in a sense. There are other things about hypotheses, the flexibility of this indexing. A typical sort of situation in a training program is one in which you take some chap who is going to see a class of information that he never saw before, presented in a way he never saw before, and from which he is to draw conclusions he never thought of drawing. This is a very typical sort of Navy problem. Before he realizes it, we cram down his throat all the possible hypotheses he is expected to test. It is relatively easy to show he can test, but he tests at a slow pace because he is badly organized. As he becomes better organized, he can, with neither more nor less information, [118] test with greater speed. When I say | »neither more nor less,« I am neglecting the time dimension. He has less by the amount of lesser time that is required for him to test all these factors. It is necessary, then, to consider both the hypotheses and the familial organizations and their accessibility, and perhaps, in another sense, if difficult hypotheses, the length of time and the span of bits that he can hold while he is searching for the right hypothesis.

FRANK: Does that agree with the conception that the individual receives every experience with a readiness or certain expectations?

STROUD: It is impossible for me to believe, now that you ask me, that at any given time you are not entertaining an expectation. This expectation may not be what one might call highly concrete and specific. It may be merely the expectation that I will continue to speak in English and not suddenly change into pidgin English and Chinese. If I did change, I dare say even if you understood Chinese, I would become unintelligible until you started taking down my speech in the new set of hypotheses.

WIENER: I have seen that happen very frequently, the change from a person who speaks perfectly correctly, to a person to whom even his own language would throw him completely off.

KLÜVER: Titchener insisted that it is the business of a psychologist not to commit »stimulus-errors.« On the other hand, it may be argued that it is the business of psychology to determine all »stimulus-errors« that a psychologist or nonpsychologist possibly can commit in dealing with a certain sensory input. In many situations our chief interest is really in the hypotheses, categories or principles in terms of which a man or an animal reacts to certain stimuli or certain stimulus constellations. Heterogeneous stimuli may often be equivalent in eliciting the »same« response, and it is the task of the psychologist to specify the factors responsible for such an equivalence [Klüver, H.: The study of personality and the method of equivalent and nonequivalent stimuli. *Character and Personality,* 5, *91* (1936)]. Obviously, sets of stimuli may be equivalent or nonequivalent for a thousand and one reasons. This does not relieve a psychologist of the obligation to determine, in a given situation, the particular factor or factors underlying particular forms of equivalence and nonequivalence.

STROUD: This gets to be quite practical. Somebody asks what is the man's acuity under certain conditions referring to visual, auditory, or other factors. I frankly come back at him, »You cannot say it that way.« Given a man with a certain set of hypotheses indexed in a certain way, what is the minimum amount of | information he requires [119] for a decision among these alternatives?

MCCULLOCH: I should like to try one very simple experiment. This is a poem. Will anyone who recognizes it speak up?

Til mi no on mounful monfer
Fif is put an enty dreen
Fur they sow as dred as slummer
Wings ha na wha ha seen.

Only one word is from the original, the tenth word.

WIENER: The general rhythm is the same.

PITTS: The degree to which the stimulus is equivalent is largely conventional in terms of the language. You know in a foreign language how the field of stimulus expands visibly when you first begin to understand someone's speech. You usually can understand a small number of people whose pronunciation happens to be particularly clear. For one reason or another, then, the number of people whom you can understand gradually grows roughly according to the similarity of their articulation. The ways in which a given word may be mispronounced apparently are quite different, depending upon the word and depending upon the language; that is conventional, not phonetic. That, I imagine, is why, when apparently a parrot speaks a sentence imperfectly, it has such a sound, whereas when a man does, it is not the same field of equivalent with respect to the sentence.

WIENER: A parrot has no phonemes.

PITTS: He deforms words in a manner different from that of any possible speaker of the language.

WIENER: He has no phonemes.

LICKLIDER: I should like to report just a couple of facts based on articulation-testing experience that bear out some of the things we have said. One of them is that when you have a test crew at work, the crew likes to entertain itself. The listeners will read magazines or books or even play cards if you let them, and they will be unhappy if you don't. We once tried to find out whether it mattered. Their scores were no lower when they read books and wrote the words down nonchalantly (getting several items behind, then catching up rapidly) than they were when we forced the listeners to devote full attention to the job. You can argue, of course, that they were irritated at us in the latter instance and therefore did not do as well as they could have done, but their performance was better than I could achieve when I worked as hard as I could.

WIENER: It has something to do with the similar situation of children studying with [120] the radio on. |

LICKLIDER: I agree with Stroud about the hypotheses. I should emphasize, however, that the hypotheses are not on the level of awareness; they are tested automatically. After listening to recorded tests in which listeners heard the same sequence several times, they were tested to see how well they could predict what was coming. We presented four words in the repeatedly heard sequence, then blanked out the fifth. The listeners tried to guess or remember the word that followed the four they had heard. In this, they had almost no success. Next we mixed the words of the test lists to which the listeners had been exposed with an equal set of words of the same type, in fact, with other words drawn at random from the same source. The listeners could tell with high accuracy whether a word was from the old or from the new set. The listener learns the vocabulary beautifully, but not the sequence in which the words appear.

WERNER: May I ask one very general question? Distorted language may be of two kinds: one in which there is a systematic change of which the subject, bye and bye, may become aware and to which he may become adapted, and another where there is no systematic change. Is the distortion you are dealing with of such kind that the subject will, so to speak, get adapted to the »key,« as it were. The speed and the amount of recovery plausibly would depend on such an adaptation. I am interested in this angle because some time ago I studied the effect of systematic changes in melodic relationships on the appreciation of such relationships. I built what I called a micromelodic system where the physical intervals between the tones were reduced to about one-fifth, or less, of their conventional size, so that a complete scale spanning an octave would come within the range of one whole tone on the ordinary scale [Musical microscales and micromelodies. *J. Psychol.*, **10**, *149* (1940)].

We were able to train our subjects to recognize melodies within another range because we maintained a melodic system whereby the relationship between the intervals, though reduced, was identical with our ordinary diatonic scale. The question actually is whether you have any evidence that your subjects accepted the distortion as mere unsystematic changes or were able to adapt to a new frame of reference. In the latter case, psychologically, we are no longer dealing with distortion.

LICKLIDER: As I defined »distortion,« there had to be some relation preserved. I think it is probably true, as you suggest, that the distortions that retain the most systematic [121] relations are the least detrimental distortions, but I am not sure that that is entirely | the reason. I cannot say. If you work with instantaneous sound pressures, any monotonic transformation is going to leave you with intelligible speech. But that isn't at all amazing when you consider that you can dichotomize the scale of instantaneous pressure and still have fair intelligibility.

MEAD: Isn't it related to learning? In the different cases is it true that the more systematic the order of distortion, the more rapidly your test crew will learn to distinguish.

LICKLIDER: Let me quote the result of an experiment, a preliminary result. You can measure the memory span for letters read in sequence from a book. The memory span for these letters is very great. The subject remembers the words the letters spell, of course, and the sentences the words form, so he remembers a lot of letters. Now a friend of mine, George Miller, thought it would be fun to test the memory span for letters with precisely the same conditional probability structure, but letters that did not make familiar words and sentences. He substituted one letter of the alphabet for another, choosing substitutes at random but always using the same substitute for any given letter. For the substitute sequences, the memory span was just about 6 or 8 letters. Mrs. Miller served as a subject. She practiced for about a month but could never improve her memory span for the substitute letters.

MEAD: She never got the system. She never got the clue to the system.

LICKLIDER: She was given the system. He told her what the substitutions were. It simply did not help. She could not go through the translation process fast enough to let the meaning of the translated symbols help.

GERARD: Do it in reverse. People may have intellectual information about a system, yet may have difficulty entering into it perceptually.

PITTS: It takes very little time to learn to read ordinary material in any particular substitution group. Suppose you make such a systematic transformation, it takes very little practice to be able to read it as if it were plain text. Certainly if you invert the alphabet, A to Z, and so forth, you learn that the regular combinations represent the common words, and after a while you read it very well.

HUTCHINSON: In the melodic system do the complements of the melody go over?

WERNER: The melodic components go over but not the harmonic.

HUTCHINSON: I can see that. I wonder how people feel about | these strange new [122] tunes. Do they feel that they are tossed in or that they are delightful?

WERNER: They feel they are very delightful. I can tell you from my own experience.

PITTS: If a man who learned the system were presented with a form in the micromelodic system of a melodic ditty which you had some reason to suspect he would greatly like if it were presented in the ordinary system, would he find it similarly attractive there?

WERNER: Yes. But I might add that maintaining the micromelodic system requires quite a bit of effort. Compared with our ordinary system the micromelodic system is more or less unstable.

PITTS: It is surprising that there should be any question.

WERNER: The micromelodic system is unstable because it is in competition with the ordinary system with which we have been acquainted since childhood.

PITTS: Because we commonly assume that a number of relationships are essential.

MCCULLOCH: For this evening's discussion Claude Shannon has the floor.

[123] # THE REDUNDANCY OF ENGLISH

CLAUDE E. SHANNON
Bell Laboratories,
Murray Hill, N. J.

THE CHIEF subject I should like to discuss is a recently developed method of estimating the amount of redundancy in printed English. Before doing so, I wish to review briefly what we mean by redundancy. In communication engineering we regard information perhaps a little differently than some of the rest of you do. In particular, we are not at all interested in semantics or the meaning implications of information. Information for the communication engineer is something he transmits from one point to another as it is given to him, and it may not have any meaning at all. It might, for example, be a random sequence of digits, or it might be information for a guided missile or a television signal.

Carrying this idea along, we can idealize a communication system, from our point of view, as a series of boxes, as in Figure 22, of which I want to talk mainly about the first two. The first box is the information source. It is the thing which produces the messages to be transmitted. For communication work we abstract all properties of the messages except the statistical properties which turn out to be very important. The communication engineer can visualize his job as the transmission of the particular messages chosen by the information source to be sent to the receiving point. What the message means is of no importance to him; the thing that does have importance is the set of statistics with which it was chosen, the probabilities of various messages. In general, we are usually interested in messages that consist of a sequence of discrete symbols or symbols that at least can be reduced to that form by suitable approximation.

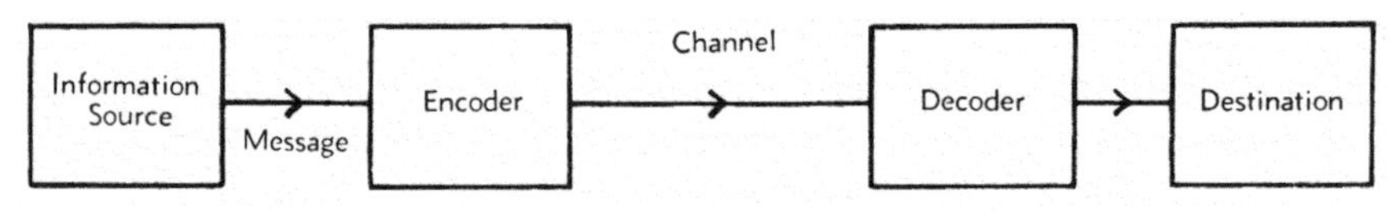

FIGURE 22

[124] | The second box is a coding device which translates the message into a form suitable for transmission to the receiving point, and the third box has the function of decoding it into its original form. Those two boxes are very important, because it is there that the communication engineer can make a saving by the choice of an efficient code. During the last few years a theory has been developed to solve the problem of finding efficient codes for various types of communication systems.

The redundancy is related to the extent to which it is possible to compress the language. I think I can explain that simply. A telegraph company uses commercial codes consisting of a few letters or numbers for common words and phrases. By translating the message into these codes you get an average compression. The encoded message is shorter, on the average, than the original. Although this is not the best way to compress, it is a start in the right direction. The redundancy is the measure of the extent to which it is possible to compress if the best possible code is used. It is assumed that you stay in the same alphabet, translating English into a twenty-six-letter alphabet. The amount that you shorten it, expressed as a percentage, is then the redundancy. If it is possible, by proper encoding, to reduce the length of English text 40 per cent, English then is 40 per cent redundant. The redundancy can be calculated in terms of probabilities associated with the language; the probabilities of the different letters, pairs of letters; probabilities of words, pairs of words; and so on. The formula for this calculation is related to the formula of entropy, as no doubt has appeared in these meetings before. Actually, to perform this calculation is quite a task. I was interested in calculating the redundancy of printed English. I started in by calculating it from the entropy formulas. What is actually done is to obtain the redundancy of artificial languages which are approximations to English. I pointed out that we represent an information source as a statistical process. In order to see what is involved in that representation, I constructed some approximations to English in which the statistics of English are introduced by easy stages. The following are examples of these approximations:

1. XFOML RXKHRJFFJUJ ZLPWCFWKCYJ FFJEYVKCQSGHYD
2. OCRO HLI RGWR NMIELWIS EU LL NBNESEBYA TH EEI
3. ON IE ANTSOUTINYS ARE T INCTORE ST BE S DEAMY ACHIN D ILONASIVE TUCOOWE AT TEASONARE FUSO | [125]
4. IN NO IST LAT WHEY CRATICT FROURE BIRS GROCID PONDENOME OF DEMONSTURES OF THE RETAGIN IS REGIACTIONA OF CRE.
5. REPRESENTING AND SPEEDILY IS AN GOOD APT OR COME CAN DIFFERENT NATURAL HERE HE THE A IN CAME THE TO OF TO EXPERT GRAY COME TO FURNISHES THE LINE MESSAGE HAD BE THESE.
6. THE HEAD AND IN FRONTAL ATTACK ON AN ENGLISH WRITER THAT THE CHARACTER OF THIS POINT IS THEREFORE ANOTHER METHOD FOR THE LETTERS THAT THE TIME OF WHO EVER TOLD THE PROBLEM FOR AN UNEXPECTED.

In the first approximation we don't introduce any statistics at all. The letters are chosen completely at random. The only property of English used lies in the fact that the letters are the twenty-seven letters of English, counting the space as an additional letter. Of course this produces merely a meaningless sequence of letters. The next step (2) is to introduce the probabilities of various letters. This is constructed essentially by putting all the letters of the alphabet in a hat, with more *E*'s than *Z*'s in proportion to their relative frequency, and then drawing out letters at random. If you introduce probabilities for pairs of letters, you get something like Approximation 3. It looks a bit more like English, since the vowel consonant alternation is beginning to appear. In Approximation 3 we begin to have a few words produced from the statistical process. Approximation 4 introduces the statistics of trigrams, that is, triplets of letters, and is again somewhat closer to English. Approximation 5 is based on choosing words according to their probabilities in normal English, and Approximation 6 introduces the transition probabilities between pairs of words. It is evident that Approximation 6 is quite close to normal English. The text makes sense over rather long stretches. These samples show that it is perhaps reasonable to represent English text as a time series produced by an involved stochastic process.

The redundancies of the languages 1 to 5 have been calculated. The first sample (1) is a random sequence and has zero redundancy. The second, involving letter frequencies only, has a redundancy of 15 per cent. This language could be compressed 15 per cent by the use of suitable coding schemes. The next approximation (3), based on diagram[!] structure, gives a redundancy of 29 per cent. Approximation 4, based on trigram structure, gives a redundancy of 36 per cent. These are all the tables that are available on the basis of letter frequencies. Although cryptographers | have tabulated [126] frequencies of letters, digrams, and trigrams, so far as I know no one has obtained a complete table of quadrugram frequencies. However, there are tables of word frequencies in English which are quite extensive, and it is possible to calculate from them the amount of redundancy (Approximation 5) due to unequal probability of words. This came out to be 54 per cent, making a few incidental approximations; the tables were not complete and it was necessary to extrapolate them.

In this case the language is treated as though each word were a letter in a more elaborate language, and the redundancy is computed by the same formula. To compare with the other figures, it is reduced to the letter basis by dividing by the average number of letters in a word.

PITTS: Do other languages have the same frequency or the same degree of redundancy?

SHANNON: I have not calculated them; but according to the work of Zipf,[1] who has calculated the frequency of words in various languages, for a large number of them the falling off of frequency against rank order of the word, plotted on log-log coordinates, is essentially a straight line. The probability of the nth-most probable word is essentially a constant over n for quite a large range of n:

$$pn = \frac{k}{n}.$$

PITTS: But presumably the constants are different for different languages?

SHANNON: That I don't know. They are not vastly different in the examples which Zipf gave in his books, but the difference in constants would make some difference in the calculated redundancy. The equation $pn = \frac{k}{n}$ cannot hold indefinitely. It sums to infinity. If you go out to infinite words, it must tail off. That was one of the approximations involved here which makes the figure somewhat uncertain.

TEUBER: Your probabilities are based on predicting from one letter to the next, or one word to the next?

MCCULLOCH: No, upon the word.

SHANNON: In the particular case (4) this refers to a language in which words are chosen independently of each other, but each [127] | has the probability that it has in English. »The« has a probability of .07 in the English language. So we have seven of them in a hat of 100.

TEUBER: As you led up to it you said, first of all, you take the probability that any one letter will occur in that particular system, and then, given that letter, and the next one that occurs, you predict the one that follows immediately after that?

SHANNON: Yes.

TEUBER: Do you go further than that and say that any one letter will follow the one you took beforehand, regardless of what letter was in between?

SHANNON: That is true in the calculation from 3. This is based on probabilities of groups of three letters. Number 4 goes on a new tack and starts over with the word as a new unit. The words are independently chosen. It is a better approximation to English than 3, since each group of letters forms a word, but the words don't hang together in sentences. At this point there seemed to be no way to go any further, because no one had tabulated any frequencies for pairs of words. Of course, such a table would be impractically large because of the enormous number of possible pairs of words. However, the thought occurs that every one of us who speaks a language has implicitly an enormous statistical knowledge of the structure of the language if we could only get it out. That is to say, we know what words follow other words; we know the standard clichés, grammar, and syntax of English. If it were possible, for example, to translate that information into some form adapted to numerical analysis,

1 ZIPF, G. K.: *Human Behavior and the Principle of Least Effort. An Introduction to Human Ecology.* Cambridge, Mass.: Addison Wesley Press, Inc. (1949)

we could get a further estimate of the redundancy. It turns out that there is a way to do this.

The method is based on a prediction experiment. What you do is shown by the following typical experiment. Take a sample of English text:

(A) T H E R E I S N O R E V E R S E O N A M O T O R C Y C L E ...
(B) 1 1 1 5 1 1 2 1 1 2 1 1 15 1 17 1 1 1 2 1 3 2 1 2 2 7 1 1 1 1 4 1 1 1 1 1 ...

and it goes on from there. Take a subject who does not know what the text is, and ask him to guess the text letter by letter. As soon as he arrives at the correct letter, he is told so and goes on to guess the next letter of the text. In this case he guessed *T* as the first letter, which was right. We put down 1 because he guessed right on the first guess. He was also right on the first guess for *H* and *E*. For the letter *R* he guessed four wrong letters and | finally got it on the fifth. In general the numbers in the lower row [128] represent the guesses at which he finally obtained the right answer. The figures may look surprising, starting off with three right guesses. It is actually very reasonable, since the most common initial letter is the *T*. The most common letter to follow that is *H* and the most common trigram is »the«. In this particular sample, which went on for a total of 102 letters, the score obtained by this subject was as follows:

Right on guess	1	2	3	4	5		> 5
Occurrences	79	8	3	2	2		8
						Total,	102

The number of ones out of 102 letters is 79. He was right 79 times on first guess, that is, about 78 per cent of the time. He was right on the second guess eight times, three times on the third guess, and on four and five, twice each. He required more than five guesses only eight times out of 102. This is clearly a good score. It is more or less typical for literary English. The scores vary. With newspaper English scores are poorer, mainly because of the large number of proper names, which are rather unpredictable.

I should like to point out that in a certain sense we can consider the second line of such an experiment to be a »translation« of the first line into a new »language.« The second line contains the same information as the first line. We have operated on the first line with a device, our predicting subject, and obtained the second. Now the crucial question is: Could we, knowing the second line, obtain the first by a suitable operation? I would say that this property is actually the central characteristic of a translation: it is possible to go from *A* to *B* and from *B* back to *A*, and nothing is lost either way. In the case at hand, it is possible to go from *B* back to *A*, at least conceptually, if we have an identical twin of the person who made the first record. When I say »identical,« I mean a mathematically identical twin who will respond in exactly the same way in any given situation. Having the same information, he will make the same response. If we have available the line *B*, we ask this twin what the first letter is. He guesses it correctly because he guesses as the original subject guessed; and we know that is correct because it is the first guess he made. This process is continued, working through the text. At the letter *R*, | for example, we ask him to guess five times, and at [129] the fifth guess we say that that is right.

Of course, we don't have available mathematically identical twins, but we do have mathematically identical computing machines. If you could mechanize a reasonably good predicting process in a computing machine, you could mechanize it a second time and have the second machine perform precisely the same prediction. It would be possible to construct a communication system based on this principle in which you sent as the signal the second line B from one point to another. This would be calculated by the computing machine which is doing the predicting. At the second point the second computer recovers the original text.

From the data in the second line B, it is possible to set upper and lower bounds for the entropy of English. There is a theorem on stochastic processes that the redundancy of a translation of a language is identical with that of the original, if it is a reversible translating process going from the first to the second. Consequently, an estimation of the redundancy of the line B gives an estimate of the redundancy of the original text, that is, of English. Line B is much easier to estimate than line A, since the probabilities are more concentrated. The symbol 1 has a very high individual probability, and the symbols from 6 to 27 have very small probabilities.

WIENER: This is very interesting to me. In my prediction theory I start from a series with a correlation and I build a series which is equivalent; each is the past linear function of the past of the other. In addition, however, in my new series the choices are completely independent, whereas in my earlier series the series are partially dependent. In fact, the way I built the linear prediction theory was by the reduction of my dependent choices to independent choices. My method has some parallelisms to this. Excuse me for interrupting.

SHANNON: That is perfectly cogent, I think. I was going to say in connection with this that the successive symbols in line B are not yet statistically independent, but they are much closer to independence than they are in the original text. It is approaching the sort of thing we are talking about, an uncorrelated time series.

WIENER: Yes.

SHANNON: To continue the analysis: it is possible to estimate an upper bound and lower bound for the amount of redundancy for time series B. The two bounds for redundancy are based upon the frequencies of these various numbers in the second line. If we | carry out this experiment with a very long sample of text, we will obtain a good estimate of the frequencies of ones, two, threes, and so forth. Let these be q_1, q_2, q_3, respectively.

[130]

The lower bound for the redundancy is given by

$$R \geq \log_2 27 + \sum_{i=1}^{27} q_i \log_2 q_i \, .$$

This follows from the fact that if the line B were an uncorrelated time series, its redundancy would be that given by the right-hand member. Any correlation present increases the redundancy. Since the redundancy of line B equals that of line A, the result follows.

There is also an upper bound of a sort, but it is not quite as secure. It is given by

$$\log_2 27 - \sum_{i=1}^{27} (q_{i+1} - q_i) \log_2 i \geq R \, .$$

This is a provable upper bound if we assume the prediction to be ideal; that is, that the subject guesses first the most probable next letter, second the next most probable, and so forth. Actually, of course, human subjects will not guess quite this accurately, although in the actual experiments I believe they were close to ideal prediction. They were supplied with various statistical tables concerning English to aid in the prediction. All in all it is probably safe to use the upper bound given above.

SAVAGE: Does the theory say that if a subject is an ideal predicter, his pattern of integers will be uncorrelated?

PITTS: Will he exhibit the maximum correlation?

SAVAGE: No, I suspect not. Suppose, for example, that the subject guesses the first letter immediately; it takes him two guesses to think up the second letter and five to think up the third. Does that information imply anything about how many guesses the next letter is going to take him?

SHANNON: Well, normally we don't start from the beginning of a sentence. Most of the actual experiments were done by giving a subject *N* letters of text, asking him to guess the next letter. This | was done 100 times with each value of *N* for 16 different [131] values of *N*.

I am not sure I have really answered your question. At any point the subject knows the text up to this point, or at least he knows *N* letters of it. That will influence his next guess, because he will try to continue the test.

SAVAGE: Will it influence how quickly he will? That will certainly influence the probability that his next guess will be right; there will be certain circumstances on which the guess is sure to be right.

BIGELOW: Like in your previous example. After the fellow got the letter C in the word »motorcycle« he knew at that moment the whole word and got therefore every letter after one successful guess.

SAVAGE: However, if you look simply at a sequence of integers generated by such a performance, do they have implications in probability for the next integer to be generated?

SHANNON: I think they might if you were clever enough.

SAVAGE: I see. There is no general theoretical reason why they might not?

SHANNON: No, there isn't. There are cases in which you rather expect a series of right ones. By an analysis of this, you could say quite a bit, probability-wise, about the next numbers.

WIENER: There should be ways, I think, of sorting this out. That would be more absolute, like codings. I think there would be a reduction of the choices to completely independent choices. That could be worked out.

SHANNON: Well, there certainly is in principle, if you allow the encoding of long sections of text into long sections of uncoded text, but as a practical matter I don't know how to get at it.

WIENER: To do it is difficult, because you would need more complete tables.

PITTS: Very extensive.

WIENER: Very extensive tables.

SHANNON: When I evaluated the upper and lower bounds for redundancy from the experiment, the following results were obtained:

N	8	10	15	100
Lower bound for *Dn*	74%	75%	75%	93%
Upper bound for *Dn*	50%	57%	60%	72%

Dn is the redundancy owing to the statistical structure of English | extending over *N* [132] letters of text.

SAVAGE: Is this only some lower bound or the greatest lower bound?

SHANNON: The only provable bound in case it were an ideal prediction but as I say, I suspect it still does bound the actual value because I think these people were close enough to ideal prediction too, so that the other things involved in this lower bound in this discrepancy more than compensate.

SAVAGE: What mathematical properties of the new language do you use to introduce these bounds?

SHANNON: The lower bound is rather trivial. It follows from the fact that the least redundancy possible with a certain set of letter frequencies would occur if they were independent. The upper bound is more difficult to prove. It involves showing how the ideal predictor would predict. An ideal predictor lines up the conditional probabilities in the order of decreasing magnitudes. You line those up and find the worst set of those that could occur. This will produce the highest possible value of entropy. Then you calculate this value in terms of the q_i.

Of course, these values are not only subject to the conditions given but also to statistical fluctuation, since the samples involve only 100 trials.

PITTS: That is, even where words are considered as being made of letters, where they are not treated ideographically?

SHANNON: Everything is reduced to a letter basis.

PITTS: It might be quite different if one simply carried the parallel through with respect to ideographic words much as you could imagine carrying out the same sort of translation and the same sort of estimation?

STROUD: Didn't some joker do that? He exposed one-, two- and three-word samples and asked the subjects to guess the next word. He did it the other way around. He didn't ask them to predict the text. He simply reported the texts that were created by this method, placed certain other restrictions on it as to subject matter, and then gave you the samples as merely samples.

MCCULLOCH: Miller. His name was Miller?

LICKLIDER: He did do that sort of experiment.

PITTS: Have you carried through exactly the same procedure for words as a whole that you have for letters, and do you have bounds?

SHANNON: In the first place, difficulties arise concerning the number of words there are in the language.

[133] PITTS: You can ask a man for the first guess, second guess. |

SHANNON: Prediction of words as a whole is certainly possible, but it would take a long time to obtain a reasonable sample.

PITTS: There would be so many more words and letters that it would take much longer to guess the right one.

LICKLIDER: There is a problem that bothers me. I am sure you have taken care of it, but I don't see quite how, in having the prediction made, compact prose is always generated.

SHANNON: This is taken from standard text chosen at random out of a book chosen at random. It represents what a literary man would write. When he chooses an improbable letter, the subject usually must guess many times before he gets it right.

BROSIN: I don't know the Zipf evidence. It is actually astonishing how the different languages, the different types of prose, literary, and so forth, follow the graph, the straight line.

MCCULLOCH: Isn't the Zipf evidence for the newspaper the same as that for James Joyce's work?

BROSIN: Yes.

STROUD: Both for James Joyce's work and for newspaper print. Joyce had a fantastic vocabulary, something like 15,000 words.

McCulloch: It seems the law holds approximately for Joyce as for newspapers.

Hutchinson: The number of moths of a given species plotted against the number of species containing that number of individuals, in the catch of a moth trap, or any similar statistics, behaves in very much the same way [Fisher, R. A., Corbet, A. S., and Williams, C. B.: The relationship between the number of species and the number of individuals in a random sample of an animal population. *J. Animal Ecology*, 12, *42* (1943).] The constant can be used as an index of diversity of the population.

Licklider: My criticism was against the other method of doing it, not the one you used at all; it was against having the sample generated by presenting *N* items to a person and having him give you the *N*+1st.

Shannon: I don't think that a statistical sample of statistical letters is true.

McCulloch: May I ask whether it would be possible for you to look up, or to recognize among printed letters, queues to phonemes, and see what the frequency for phonemes, more constricted than the sequence of letters, really is?

Stroud: That is a very important question because of the multiplicity of representation.

McCulloch: Yes.

Shannon: I think it would be quite easy to do most of these | experiments with phonemes in place of letters. They go through rather rapidly after you get into the swing of it. I don't know that choosing phonemes would come as naturally, though, to the experimental subject as choosing letters. [134]

McCulloch: I was thinking that one might link it with the intelligibility of speech of sentences as opposed to that of words, of that of words as opposed to that of nonsense syllables, and so forth. If you knew the sequence of phonemes represented by these letters, you might be able to link it to the intelligibility or to the increased intelligibility.

Licklider: Miller has completed some work on the learnability of speech that ties in with this analysis. The difficulty of learning a sample of synthetic prose is roughly proportional to its information content.

Teuber: By this logic, would baby talk be more predictable or less predictable than the talk of an adult?

Shannon: I think more predictable, if you are familiar with the baby.

Teuber: Do you know that linguists claim that baby talk is highly similar, that is, similarly patterned from baby to baby in different language systems, in terms of its phoneme structure? Wouldn't it be easier anyway to go from one phoneme to the next in predicting speech?

Stroud: In some of the latest orations from my youngest there is only one pair of phonemes.

Gerard: If you had taken simplified spelling, would you have decreased redundancy?

Shannon: Yes, that is right.

Licklider: Conversely, if you used the International Phonetic Alphabet to do the phonetics, you would find high redundancy.

Shannon: There is one other experiment that we performed in connection with this work which did not have much bearing on anything but which proved to be rather interesting. We asked a couple of subjects to predict in reverse, that is, to start in at the end of the sentence and to guess backward letter by letter. It turned out that the scores were almost as good as in prediction in the forward direction. The problem was much more difficult, however, from the psychological point of view. The subject was really tired out after he had worked through a sentence in reverse, whereas in going forward it is quite easy.

GERARD: As a matter of curiosity, how did the guesses go? Looking at those actual guess numbers, the first one is perfectly clear. Then you get to the *I*, and I suppose the [135] first guess was an *I*. | The subject decided it was not *A*, so he guessed *I*. Now how did he get the *N* on the second guess? That I don't see.

STROUD: »There is a,« »there is the,« »there is no,« are some of the most common progressions.

BIGELOW: Assertionary denials.

STROUD: Either followed by the negative or by the article.

SHANNON: I think *A*, *T*, and *N* would be the first three guesses; since »the« sounds a little strange in that construction, he probably guessed *A* and then *N*.

SAVAGE: Then he has *O* for sure. He got space without the *T*.

STROUD: No, most probably, and not the most probable.

PITTS: Or *L* less.

SAVAGE: No, reverse *EHT* there.

HUTCHINSON: Isn't five rather high?

SHANNON: The *R* took 15 guesses, and one for the *E*.

STROUD: How does that appear compared to the probability of *R* as a first letter of a new word?

SHANNON: I think this particular subject was not using tables and probably guessed his initial letters improperly. In later experiments, the ones I got these estimates with, the people were supplied with all the tables we had on the statistics of English and used them in any way in which they saw fit to aid their guessing.

SAVAGE: How did you pick a book? At random?

SHANNON: I just walked over to the shelf and chose one.

SAVAGE: I would not call that random, would you?

GERARD: Unless you were blindfolded.

SAVAGE: There is the danger that the book might be about engineering.

BAVELAS: The book would be.

KLÜVER: I wonder whether Sievers' *Schallanalyse* is in some way related to the problems discussed here. The *Schallanalyse* or sound analysis of Sievers was concerned with »translating« auditory sequences as found in human speech into motor sequences [SIEVERS, E.: Ziele und Wege der Schallanalyse. Heidelberg: Carl Winter's Universitätsbuchhandlung (1924), pp. 65-111. Cf. also Vol. 35 of the *K. Sächs. Gesellschaft der Wiss., philol.-histor. Klasse.*] Sievers called attention to the fact that speech, no matter whether we are dealing with poetry or prose, tends to be accompanied by certain movements, postures, and tonus regulations. He held that the same is true for any written text, since all texts represent potential speech. He published numerous »curves« (Becking curves, time curves, and signal curves) supposedly involved in any [136] form of auditory reproduction. These motor »curves,« let us say, | a circle or the figure eight, may be produced by movements of the hands and arms while reciting, for instance, a poem. Sievers also made use of optical signals, such as brass figures lying before the speaker. These figures were supposed to influence auditory reproduction if viewed by the speaker while talking. However, the chief contention of the *Schallanalyse* was that out of all possible motor curves it is always only one particular curve that »goes with« a particular auditory sequence. It was said that the voice becomes inhibited if any attempt is made to produce curves that do not go with the poem or text in question.

WERNER: The *Schallanalyse* is an extremely interesting but somewhat controversial method. Before Sievers, Rutz worked out a system of tones of melodic rhythm patterns.

KLÜVER: As I remember the story, Sievers really never succeeded in teaching his methods to others. The exception seems to have been a student who flunked all examinations, although he was extraordinarily gifted for *Schallanalyse* in Sievers' sense.

PITTS: I suspect philologists denied the acceptability of the method except for reasons in which discontinuity of the authorship came in.

SAVAGE: G. U. Yule's efforts were bona fide and not mysterious. They were primarily based on English, and as I recall, in one or two cases on the Latin language, which he knows and is therefore able to deal with. He examines the frequency of occurrence of various obvious sorts of things, computes the average length of word, makes other like calculations, and finally makes a judgment based on standard statistical principles and his own very extensive statistical experience of whether the two works in question do or do not have style similar enough to be attributed to a common author.

KLÜVER: As far as Sievers' work is concerned, it rests on certain correlations between motor-kinesthetic and auditory phenomena. Sievers apparently was very gifted in »expressing« auditory sequences in kinesthetic patterns. He always insisted that only one type of motor curve was adequate whenever he recited or read a certain text. If he sensed that certain movements representing a particular motor curve were no longer adequate, his voice simply gave out until he replaced the old curve by one that »fitted.« He was able in such a way to assign a particular motor curve or a sequence of different motor curves to a given auditory reproduction.

WERNER: That's about it as I recall. Looking at a poetic line, Sievers transforms it into a kinesthetic pattern; if various other | lines are produced by the same author they will [137] also fit the pattern. If then he comes to a part which does not go with the kinesthetic pattern, he infers that a foreign element has been introduced.

PITTS: What sort of kinesthetic patterns?

WERNER: Though I do not remember very clearly, he contended that there were a restricted number of patterns. These patterns could be objectified by visual symbolic representations; a triangular pattern was one of them; a pendulum pattern was another.

PITTS: A particular phenomenon would be translated into the triangular?

WERNER: Not quite. The visual representation is an aid for *Schallanalyse*. A poem has a certain rhythm which is repeated and which is represented by a certain visual signal, such as a triangle.

PITTS: How do you carry out the translation? It is not obvious how you would carry out the translation of the poem into the kinesthetic pattern.

KLÜVER: It looks as if the »translation« of a poem into specific kinesthetic patterns has remained a secret of Sievers.

PITTS: It is not an objective method?

WERNER: No. Statistics have been applied, though, and they contend that these statistics bear them out.

KLÜVER: It may be argued that Sievers' »curves« were more than subjective motor-kinesthetic patterns. There seems to be some evidence that they represented objective indicators. If I recall correctly, Sievers was tested by some experimental psychologists who showed him, for example, a text he had not previously seen. This text had been written by two different authors. He read the text while at the same time producing the movements and curves that, according to him, »went with« the text, but then suddenly insisted that he could not go on reading. His voice gave out and he could continue only after having discovered the right kind of motor curve. The point thus located by Sievers was the point that actually separated the texts of two authors as previously determined by other philological methods.

Pitts: Nobody ever ascertained the rule by which he obtained this?

Klüver: William Stern used to say that he would give a doctor's degree in psychology to anybody who could throw light on the psychological mechanisms involved in Sievers' performances. Incidentally, I am sure that Sievers' own »explanations« do not suffice or are wrong. C. K. Ogden of Basic English fame told me just before the war [138] that he had used the methods of *Schallanalyse* | to settle the old question of how the Romans really did pronounce Latin. In fact, I spent part of my last night in Europe making a gramophone record of Ogden's recitation of a Latin poem. And I took this application of Sievers' principles, that is, the gramophone record, to New York.

Pitts: Does this agree, on philological evidence, with the most powerful of the schools with respect to the actual pronunciation?

Klüver: I understand that the pronunciation of Latin as practiced by the Romans remains a controversial matter.

Werner: I may add that Sievers was one of the outstanding German philologists who wrote classic works on old German grammar and texts. *Schallanalyse* was obviously a hobby for him at first which he later extended and included among his methods for textual analysis. Later he felt the *Schallanalyse* was much more satisfactory than the traditional philological methods.

Pitts: Philologists have explored more eccentric methods of argumentation that most people have.

Stroud: Every individual shows a reluctance to repeat himself in the use of a single word. Sometimes there is reluctance to place verbs too close to the noun, and there are other idiosyncrasies of sentence structure which are to some degree characteristic. I wonder if such landmarks could not be guessed at reasonably well with a fair length of text.

Pitts: The only trouble is that the serious questions are practically never concerned with two texts of any considerable length, whether by the same author or not. Usually both texts are very short.

Brosin: I don't know how relevant it is, but the Masserman-Bolk chromatic analyses of parts of speech in the Murray Thematic Apperception Tests with relevance to psychiatric reactions are an effort in this direction. I don't know how seriously you would consider this.

Pitts: You might decide the Shakespeare controversy in this fashion but probably very few others.

Brosin: I don't know how useful it is, but obsessed people use words in certain orders and quantities. Surely the scope of word patterns and, let us say, the kinesthetic rhythms of one type of schizophrenic, say a hebephrenic, will surely have distinctive patterns. Whether these are sufficient to be computed for absolute values, I don't know.

Mead: Milton Eric[k]son has a whole series of texts taken down from different diagnostic psychiatric types. In these the clear formal properties of the language and the [139] types of balance that recur can be distinguished. It would be such patterns, I think, | that you would have to deal with. You would have to use a good many abstractions, such as balance and types of repetition and inversion, in such an analysis.

Frank: Eliot Chapple, who has done the same thing at the Massachusetts General Hospital, has developed a machine for recording the pattern of speech.

Mead: But his records contain the interaction with another person within it. That would be the difference here, would it not?

FRANK: Probably.

MEAD: If you introduce two people into the picture, then Eliot Chapple's chronograph gives a very diagnostic speech pattern when one individual's responses to another's are analyzed.

MCCULLOCH: May I ask one clumsy question? I don't see quite how to put it yet. Most of the things you have been going at, Mr. Shannon, have been on a small scale. Can you work out anything that would handle affairs as large as your ordinary grammatical units, phrases, clauses, sentences, and so on?

SHANNON: I think the 100-letter approximation is beginning to bite into phrases. The person with 100 letters has a fraction of a sentence, perhaps a full sentence, to work on. He makes use of all that information, perhaps finds a key word 100 letters previous. More in the spirit of what you are asking, I feel there is a point at which this statistical approach is going to break down. It is very questionable to me that the very long-range structure of language can be represented by a statistical process, even that there is any meaning of speaking of the probabilities of sequences which are so rare that they never have occurred. Certainly the frequency concept of probability begins to weaken at some point. Also, when you are considering very long-range structure there are questions of whether the stochastic process is stationary or not. The process may not have the stationary properties that are implied by most of this analysis.

TEUBER: Isn't it true – for long passages at least – that it is easier to make predictions after some preliminary acquaintance with the idiosyncrasies of a particular author, or even of a particular group of people with whom you have been in contact?

PITTS: Probably after the passages exceed a certain length, and if you are concerned with the question of deciding whether or not some particular person or anybody at all could possibly have said that. Your chances would be much better if you were to analyze on the basis of whether or not it is a kind of notion or idea that could have been expressed at that time, rather than on the basis of the statistics of a series of words in it. | [140]

TEUBER: That could be based on a single experience. Just say the word »Gestalt« in this group, and you can just about predict what will happen and who will say what. You make your prediction in terms of past experience, but one previous exposure to that interchange of points of view will suffice. This is of course irrelevant to what Mr. Shannon is trying to do. For him, the important thing is to get rid of all idiosyncrasies.

PITTS: There you would find probabilities of notions of the man rather than of the symbols used to denote them. I don't see why this should be necessarily impossible a priori.

BATESON: I was thinking about the extraordinary difficulty of reconstructing stenotypic transcript. If a word like »ratio« becomes »ration,« as it cannot do on the stenotype but may on the typewriter, it may be very hard to get back to the original alternative. Meaningful distortion has crept in where there has been a deformation.

PITTS: You can correct it astoundingly well, probably much better than you think on the basis of statistics about series of words. If you tried correcting it solely on the basis of the statistics of a possible series of words, and extended it, say, to ten-word sequences, even you would probably do it much worse.

MCCULLOCH: If you can find anybody who knows the speech of the person in question and can imitate it, and if you make him read aloud the stenotypic notes, you can reconstruct the speech again and again.

BATESON: Have you ever tried it with a long text in which 10 per cent of the letters, say, are deleted at random? What percentage of letters do you have to delete before unintelligibility sets in?

SHANNON: I tried a few experiments with a 27-letter alphabet, again with the space a letter, and found that you can reconstruct, say, about 70 per cent of the text when about 50 per cent of the letters are deleted at random. The trouble is that the random letters, the deletions, pile up in certain places. If you delete every other letter, you can do quite well with almost 50 per cent missing. You can delete all the vowels in a passage and have no difficulty in reconstructing it. Only very infrequently will you miss a word. Vowels constitute 40 per cent of the letters. You can also delete the spaces, which is another 20 per cent.

FRANK: Is it correct to infer from your remarks that the tables you build up are predicated upon knowledge of the uniformities of English language on the part of your ideal informants, who have then interpreted the text with these deletions according to [141] | the structure of English language and ideas? Is that the way it operates in your procedures?

SHANNON: I would say that all of this work simplifies the complexity of English a great deal in that we say there is one kind of English and there is one set of statistics for English. Actually, English is really many different languages, each with different statistics. If a person knew who wrote the text he is predicting and was familiar with the author's habit patterns, he could certainly do better than if he were just taking it as blind English.

HUTCHINSON: Isn't it like the library catalogue which provides you an English book full of information by giving a single number that can be provided from Washington or New York?

SHANNON: Provided you wanted to send that book; but suppose I write a book. That book is not in the catalogue, yet I want to send that; but you don't have any number for it.

HUTCHINSON: That is true.

SHANNON: You should have numbers for all books and the one that might be written by this information source.

MCCULLOCH: What is known about the frequency of the various parts of speech in the ordinary grammatical sense? To what extent can you guess what is coming next?

SHANNON: I have not seen any tables on that, but it is possible to guess surprisingly well in this kind of experiment. When subjects obtain scores like 60 per cent right on the first guess, they must have known what word was coming.

LICKLIDER: There is obviously a close relation between this and what Rudolph Flesch says in his little book, *The Art of Plain Talk.* The latter is on a less precise and more intuitive level, but I think it could be translated into terms of information and redundancy. Flesch says that if you really want to communicate with someone, you have to make your speech (and more especially your writing) even more redundant than it naturally is. You have to repeat two or three times, then say the same thing in different words. This gives us, perhaps, a very dismal outlook for verbal communication. I suppose it does not make the outlook for conferences like this very hopeful. They have to be long.

MARQUIS: There might be another implication of the same thing: for example, the best communication occurs if you say what the listener expects you to say.

LICKLIDER: There should be an optimum degree of correlation between the talker and the listener. If the correlation is zero, the listener has no expectations and understands [142] nothing. If the correlation is unity, he doesn't need to listen. |

MEAD: If you take a form of verbal communication like the one Harold Laswell uses, involving a vocabulary from about six different disciplines simultaneously in places where such usage is not expected, most people, while they may know all six vocabularies, will find following him exceedingly difficult. He has also adopted the device in his usual communications of interlarding endless redundant clauses, such as, »It is unnecessary to specify.« One can, of course, learn to recognize these interpolations.

PITTS: You mean this is deliberate?

MEAD: He does it all the time. I tried the experiment once of asking him to cut them out. After I had asked him to cut them out, I found it about impossible to listen.

PITTS: On the book page you can find the significant part of the sentence because it is padded with these things.

MEAD: If you are used to the style, you knock these phrases out, and the pauses give you a chance to adjust to the shifts in vocabulary. If you get *him* to knock them out when you are trying to listen, then you realize how difficult it is to make these shifts all the time. This is the opposite of the point you were making about the condition in which a person is saying exactly what is expected. As you move away from expectancy, even in type of vocabulary, you need this padding which is not necessarily saying the thing over but just permitting a slight shift from one frame to another.

PITTS: He could probably speak more slowly and do nearly as well.

GERARD: Friends rarely finish sentences with each other, since one knows what is coming and picks it up.

MCCULLOCH: Mr. Shannon, will you say a word about the assurance of getting a message across as it affects the redundancy? Have you any actual evidence on that?

SHANNON: No, I don't have any numerical evidence. Do you mean into a human being or through a noisy communication channel?

MCCULLOCH: Getting information through noise.

PITTS: Then the redundancy reduces automatically. By the way you calculate the redundancy it reduces if you have to put it through noise. I mean, the way you calculate information means that in case you put it through noise then repetition becomes less redundant than it would otherwise be.

MCCULLOCH: That is right. Have you any quantitative work on it at all?

SHANNON: We have some work, for example, done by Rice at the Bell Laboratories on White thermal type of noise and various | methods of encoding for transmission [143] through it. Rice has developed formulae which show roughly how much delay is required in the encoding operation to introduce redundancy properly so as to overcome the effects of noise and enable correction of errors. It appears that a rather large delay is usually required if you wish to approach the ideal encoding with, say, one error in a hundred transmitted symbols.

VON FOERSTER: Is there knowledge of redundancy of different languages, or only of English?

SHANNON: The only work I know of is in English.

VON FOERSTER: What do you expect for the other languages, the same figure or different ones?

SHANNON: The Zipf curves suggest that the redundancy for other Indo-European languages may be of the same order as that of English.

VON FOERSTER: This has certainly something to do with the closely related grammars among the different Indo-European languages. The grammar of a language is probably more or less an expression of its structure. With respect to the redundancy of a language, it is certainly true that the more freedom of choice the grammar leaves the less redundant the language becomes. On the other hand, a language with an extremely highly developed grammar would be a language with a big redundancy. As for instance, mathematics or symbolic logic are languages with 100 per cent redundancy. I see here two tendencies operating against each other to develop the optimum of a language. The one tendency tries to decrease the redundancy in order to transmit as much information as possible; the other one tends to increase the redundancy by establishing a highly structural order within the language. That means we have to expect certain values for the redundancy in an optimized language. Perhaps the numbers of letters – or perhaps the number of phonemes – play a very important role in optimizing a language. I would like to remind at this point that the first attempts in writing are usually solved by an idiographic system which does not know any letter but has a symbol for every word. I am thinking of the old Maya texts, the hieroglyphics of the Egyptians or the Sumerian tables of the first period. During the development of writing it takes some considerable amount of time – or an accident – to recognize that a language can be split in smaller units than words, e.g., syllables or letters.

I have the feeling that there is a feedback between writing and speaking. After writing freed itself from the archaic rigidity of [144] | idiographs and became more fluent and versatile due to the elastic letter system, I would expect a certain adaptation of the possibility of making words and making sentences. In other words, I believe there should be a connection between the redundancy of a language with respect to its word and with respects to its single letters. These figures must give a certain knowledge about the structure of a language.

SHANNON: I am sure it does. I believe there are a large number of compromises in constructing a language, one of them being that the language ought to be pronounceable on reading it. This requires certain constraints about how the vowels and consonants separate each other. This already implies a certain amount of redundancy. I believe that there are many other desirable features that you require of a language which force the redundancy to be fairly high in order to satisfy these requirements.

WERNER: Of course there is also the phonological aspect of language, which has been studied so extensively by phonologists and which enters into the problem of redundancy. Every language uses a limited number of diacritical phonic signs; certain combinations of sounds exist in one language but do not exist in another. The phonic unit *kn* does not occur in English but occurs in German, and so forth. Because languages differ in phonological structure, there is a qualitative and quantitative difference in redundancy between languages.

PITTS: In simple ways. You know, for example, that one language has common words on the average much longer than another. Then it will be more redundant in that sense. Thus, translation from English into Latin increases the length of the book. As a matter of fact, that is very clear in the Bible, in which the English translation is the shortest, except for the Chinese one. That, I think, almost certainly implies that the redundancy of English is less, on a syllabic basis, than that of most other languages.

If you are interested, however, in the capacity of language, in your sense you probably mean to a greater extent the redundancy on the basis of entire words taken as units.

If you were to take the article »the« and add four symbols to it, in one sense you may increase the redundancy. Another language that has articles may suffer more redundancy than a language that has none at all.

TEUBER: Mightn't that possibly give us a definition of a »primitive« language? Some of the American Indian languages seem to us to »overqualify« all the time. They have to keep going for a [145] | long time before they get where we would get in a few words. At the same time, would there be greater predictability of phonemes, on the basis of their sequence, that is, to be able to predict which phoneme follows which? I raised the question about predictability in baby talk for that same reason. If you have a very redundant language, you could argue that in that language one makes the same noises for a longer time, and one could soon tell what noise tends to follow what noise, so that there is greater predictability. I don't know whether that is so, and whether it would lead to a precise definition of »primitiveness« of a language.

MEAD: I don't think you could do that. With a great knowledge of phonetic structure you might be able to construct an ideal, such as working on your ideal redundancy point, an ideal language of an ideal degree of primitiveness, especially using the child as a model. But with actual primitive languages you cannot do anything of the sort. You get extraordinary variations; therefore you would not be able to make any kind of sequence.

MCCULLOCH: Some are extremely redundant and some are not?

MEAD: Yes. Some are not.

VON FOERSTER: This situation might be illustrated by the problem of translation, where the same thought has to be expressed in different languages. It very often happens that a certain phrase, a poem or a thought can be beautifully expressed in one language and sounds impossible in another one.

For instance, the beauty and clearness of Aristotle in its own language becomes in German tedious and clumsy, whereas in English Aristotle regains his sharpness and conciseness. On the other hand, to read Goethe's *Faust* in French is almost ridiculous. But these two examples don't say anything against German in the first case or French in the second one. It seems to me that different languages are able to express things with different results. Isn't it so, that a language is a symbol for an intellectual world like any other human expression, architecture for instance? I am considering the Greek temples and the Gothic domes. Each of them is perfect in itself, serving the same purpose – and how different they are. I think word-redundancy is not a sufficient key to judge the value of a language.

KLÜVER: Dr. Von Foerster, it has been said that a German can understand Kant's *Critique of Pure Reason* only if he reads it in English. Obviously, an English translation makes it necessary to transform the long Kantian sentences into simple, short sentences.

VON FOERSTER: Yes, certainly. According to Pitts' statement about the Bible, Kant should have written his Critique in Chinese. | [146]

SAVAGE: It should be emphasized again that Shannon has talked about redundancy at the presemantical level. Redundancy of printed speech really refers exactly neither to the spoken language nor to the very difficult problem of semantical redundancy. I think it would require special and difficult experimentation to measure the semantical redundancy of the language, though the experiment last reported by Shannon does bear on the subject to some extent in that the guesser knows English and is utilizing that knowledge in his guesses. But still, to isolate the semantical redundancy, to separate it from the phonemes would probably be very difficult. Yet it is the thing you are all talking about now. It is the thing that refers perhaps to what you would call the spirit or essence of the language.

STROUD: I wonder if you would care to consider such highly artificial languages, if you like, as symbolic logic, or to consider mathematical notations as being perhaps among the least redundant symbols that we have. My reason for bringing this up is really very simple. I planned at one time to use a sample of quite readable text by simply reading off the rules by which you extract the roots of the cubic equation. This can be read in English; and yet only those of you who are mathematicians would have recognized it for what it was, I am sure, and could have reproduced the equation from the instructions. A majority of us might have recognized that it was the process, but we could not have reproduced the process; some of us might not even have recognized it for what it was. I had thought of using this as an example that was perfectly pronounceable at the good standard elocution rate of one phoneme per tenth second, though insufficiently redundant to be absorbed by the listener. I hoped thereby to indicate that in this case that amount of information which I was trying to convey to the subject in question, lacking in many cases the full knowledge of the statistics involved, was too much for our abilities to handle.

PITTS: No, I don't think you can say in general that the artificial languages of symbolic logic express a minimum of redundancy. I suspect that anyone would agree with that who had read Bell's book on mathematical methods in biology in which he explains the elementary principles, or the principia mathematica.

SAVAGE: I would say they are highly redundant. They risk nothing.

PITTS: Exactly.

SAVAGE: Freedom from redundancy is a desideratum of mathematical and, to a somewhat lesser extent, of logical notation. It [147] | is not, however, foremost among the desiderata. To appreciate this it is important to recall that redundancy means here the sort of statistical redundancy which has been defined, that is, the frequent use of long or otherwise awkward expressions. Thus, redundancy includes more than what we call redundancy in ordinary usage; namely, actually saying the same thing twice or using expressions that might well be dropped. The business of trying to make the frequent symbols simple, though more consciously pursued in mathematics than in ordinary speech, has been going on for a much shorter time.

STROUD: I shall have difficulty in believing this until I have seen fair samples of fairly long expositions in symbolic logic subject to this sort of analysis of the diagram, trigram type in which the knowledge of the person reading the material of the rules whereby these were established was ruled out. If you include the rules in a complete knowledge of them, hypothetically, at least, unless the person has been deliberately redundant, you should be able to reproduce the entire text once you got one-half of an equation of it.

LICKLIDER: I think there is a possibility of studying the semantic aspects of language profitably in simple languages that grow up in special situations. I have some friends in the Human Resources Research Laboratories in Washington. They are very much interested in what they call Airplanese. This is the language of the control towers that get the airplanes down at airports, that control takeoffs, and so forth. They have kept track of the words and messages that pass between the towers and the planes. I remember this preliminary result: at an early stage they had 10,000 tokens, 10,000 words recorded; of these, 5,000 were either numerals or place names, such as Washington, Bolling, Andrews. The remaining 5,000 tokens included just 400 types. By any way of figuring, that is quite redundant. The most frequently occurring actions are landings and takeoffs, requests for wind directions, and so forth. The physical problems determine the messages. This suggests, since the semantic aspect of language has to do with the relation between the signs and their referents, that the signs are autocorrelated and redundant in large part because the world we live in is. When we get into routines, we find stretches of language that are extremely redundant. They are redundant because they are trying to describe actual situations that are redundant.

PITTS: Every natural law expresses a great redundancy in nature.

LICKLIDER: That may be pursued farther in that direction. I think it is important to bring the referents into the picture. For example, the rate at which information flows over a Ground Con [148] | trol or Approach link is dependent upon the type of plane that is being controlled. When a jet plane comes in, the rate of direction goes up. When a training plane comes in, the rate goes down. The idea, then, is that you may be able to trace much of the structure of language to the structure of the actual situations in which language is used.

McCulloch: What comes in, then, is mere padding or actual repetitions of directions.

Licklider: In the G.C.A. example.

McCulloch: It sounds like an American radio advertisement. Is that the kind?

Licklider: The G.C.A. operator never goes off the air for any length of time during a talk-down. He talks, keeps talking; the idea is not to let the pilot get the notion that the radio link is dead. The pilot would be much distressed, flying blind, coming into the ground with nobody talking. If the operator is talking to a slow plane, he has to keep on saying the same thing over and over. It is extremely redundant. If the plane is fast, the situation changes rapidly enough for the operator to make every second or third phrase a new instruction.

Savage: Does it suffice to maintain some inanimate acoustical contact or is actual talking preferable?

Licklider: I have heard that the pilots don't like the operators to be overchatty. They would just as soon not have that.

Pitts: He might get a lick in for security to make perfectly sure he has been perfectly understood, since he has to –

Wiener: The direction might be repeated half a dozen times.

Gerard: Such as, »Keep coming.«

Licklider: I think the G.C.A. operator does not talk about the wind direction. It is more like: »You are on course, you are on course, you are doing fine, you are on course; now two degrees left, two degrees left; the heading is so-and-so, the heading is so-and-so; hold that heading, hold that heading.« It goes on at that rate; the number of words said is purposely high. Actually, the whole problem of coding in military communication comes into our picture. There are two rival philosophies. One says that you want to set up a restricted set of messages and enforce the restriction, hoping that an emergency does not arise for which it will not be adequate –

Stroud: And »hope« is the right word.

Licklider: The other says that you want people to use their heads about the phrasing of messages. I don't think there has been a decision between the two yet. | [149]

Bavelas: What may seem redundant from one point of view may not be redundant at all in terms of what one might call second-order information. I am reminded of a little study, done several years ago, in which college students were asked to tell what kinds of things people like themselves could do to help the war effort but which would very probably invite criticism from their neighbors. Also, they were asked to tell what people like themselves could do to help the war effort which would evoke praise from their neighbors. On the basis of that information, two leaflets were prepared. Both leaflets were purported to be statements by a public official urging college students to help the war effort and suggesting what they might do. One of them suggested only those activities which the interviewed group said would be criticized; the other leaflet suggested only those things which the interviewed group said would be praised. These leaflets were distributed to entirely new samples of college students. When these students were asked what they thought about them, it was clear that they attributed to the »public official« favorable or unfavorable characteristics, depending upon the extent to which the suggested behaviors were of the one kind or the other. In other words, the text not only bore information with respect to the activities in which the college student might take part but also information about the author.

Bateson: Communication about relationship between you and the other person.

PITTS: You could code them and send them all at the beginning very quickly.

SAVAGE: I say you could not.

PITTS: Not communication of information. That is the important point, even of second-order information.

FRANK: That is the difference between machine and man.

PITTS: If I could say you are alive in half an hour and continue to say it –

SAVAGE: When you marry, tell your wife on the wedding morning, »I love you, darling; I love you eternally, no matter what I say or do from now on. I love you eternally, remember.« Then you tell her again. If you never refer to the subject again, see what happens.

PITTS: That shows exactly that it is not proper to speak of it as a communication of second-order information in the same sense in which primary information is communicated, because the assumption is reduced to absurdity by exactly that remark.

[150] SAVAGE: She can't decode that. |

STROUD: This concept of redundance abstracts the information from the date. If for any reason the date of origin is part of information, there isn't any such thing as a redundant signal, as near as I can make out.

SAVAGE: I sympathize with Mr. Pitts. It is something of a tour de force to call these things information. The obvious differences, which are brought out by these examples as reassurance, emotional contact, and so forth, though there is communication in them, are connections between individuals which are not just transmission of statements of fact.

BAVELAS: I think they are. If we could agree to define as information anything which changes probabilities or reduces uncertainties, such examples of changes in emotional security could be seen quite easily in this light. A change in emotional security could be defined as a change in the individual's subjective probabilities that he is or is not a certain kind of person or that he is or is not »loved.«

MCCULLOCH: Verily, verily I say unto you.

BAVELAS: What I am saying is this: I want to avoid technical language, but if a man walks in and pats you on the back or winks at you across the table, this may be information in the very same sense that any other message is information if it reduces your uncertainty as to your present state among a possible number of states, your position in the group.

FRANK: Can't you refine that by saying that this may consist of information if you want to call it such, but that primarily the communicated sign or signal or gesture keeps the recipient of your communication tuned to the meaning of the information that you want to convey? In a sense it is getting the person ready with the right expectation so that the information you want to convey will be accepted, received, and interpreted in the terms you want it to be.

BAVELAS: That is one function, but there is also the function of informing his present state in, for instance, a group relationship.

FRANK: Yes.

BATESON: In the straight telegraphic situation you have a whole series of conventional signs for, »Please repeat; I received the last word; talk louder,« and so on. Now those are very simplified analogues of the thing you are talking about, aren't they?

BAVELAS: Of other things too.

BATESON: I mean giving commands.

BAVELAS: Let us suppose that I am a stranger in this group. As I sit here, a gentleman [151] whom I don't know but whom I assume is | one of the leading members of the group smiles across the table at me. When I make a comment, he nods in agreement. Now it seems to me that his behavior bears information in the sense that it makes it possible for me to select certain ones from all the possible relations that I might conceive myself as having to this group.

PITTS: Certainly there is always an informative component in these emotional communications. I think our point was mostly that that is not all; very often it is not the essential part of it, nor is it the reason why people spend so much time at it as they do.

FRANK: There is another aspect of the situation.

PITTS: It is the birthday telegram with us.

FRANK: Often it is a reaffirmation, saying, »I mean,« and then repeating in other words because the facial expression and response that you are watching on the other individual indicates that you must make another attempt at communication because you see you are not getting through. Therefore you restate it, reaffirm it, put it in another way while watching that individual, until you believe that you perhaps have made a communication. So I think that is another aspect, face-to-face conversation.

GERARD: It is about time to tell a story that has been on my mind for a while. Speaking about the stranger in the group makes it relevant. A guest spoke to a group that was intimately knit. One who preceded him said a few words and ended with, »72.« Everybody roared. Another person said a few words, then, »29,« and everybody roared.

The guest asked, »What is this number business?«

His neighbor said, »We have many jokes but we have told them so often that now we just use a number to tell a joke.«

The guest thought he'd try it, and after a few words said, »63.« The response was feeble. »What's the matter, isn't this a joke?«

»Oh, yes, that is one of our very best jokes, but you did not tell it well.«

PITTS: Of course, in a certain sense there is much more information in this kind of communication than one would suppose. If a man tells his wife every morning for thirty years that he loves her, the declaration may convey very little information if she is in no real doubt. If he omits it, however, the omission gives her considerable information. You must consider the possibility of communication not occurring if it is not a message at all.

STROUD: Something that bothers me is the way in which we make wisecracks about this thing we call noise. For many very practical considerations it seems often very wise to consider the whole message as a message, and the noise, if it has any meaning at | all, merely as that portion of the message about which you do not wish to be [152] informed. This has some very practical applications in that you have two ears: there is nothing to prevent you from hearing my voice as it comes to you from several paths, each of which repeats substantially the same pressure pattern, but with a slight time delay. From one point of view, if you say merely that it is sufficient for you to hear one version of this time-pressure pattern and not the other time-pressure patterns, then hearing with two ears is a highly redundant affair. If, however, you begin to find out that because people have two ears they hear these same pressure patterns in several different versions with slight modifications of amplitude and phase relations, you discover that they are able to take the range and bearing of the speaker with a considerable degree of accuracy. The information, then, was not redundant at all. It is perfectly true that they did not use the information to verify that I said »is« when I said »is,« but they did use it for the very useful purpose of informing themselves about where the joker was who was talking to them.

LICKLIDER: And what sort of room he is in.

STROUD: And where he is in the room, such questions as these.

SAVAGE: None the less, it is highly redundant to answer these questions a thousand times over.

STROUD: They are highly redundant in the sense that your past experience tells you that you will not move without having some other source of information, or that the room will not change without some other comparison of information; but since in these matters one requires a very high degree of security, they are not over redundant if you want to attain high orders of probability in not making a mistake.

There is another case: Suppose I want to take an electrocardiogram on a man who is sick in bed in his own home. I try it, but what I get out on the trace is practically pure 60-cycle hum. This was not the information that I came for. I already knew that the local power company put out at that rate. However, it is a very practical stunt to be quite well informed about this, and, incidentally, independently informed. This I do by putting on an aerial that picks up this same information about the power system. In a certain sense this is rather stupid, but I can use this as a minus message in the same sense that Dr. Licklider meant when he was talking about transmitting the noise on one channel and the mixed message and noise on the other, then using the pure noise as a minus on the mixed signal, and coming out with nothing but the pure informa- [153] tion. I use the information about the same self| power supply as a minus message to the confusion I get from the patient and conclude with a fairly respectable electrocardiogram. So I often suspect that it might be considerably more profitable, at least in many contexts, not to be too quick to define what is the signal and what is the noise. I know that I am of ten very amply rewarded by stopping to find out what the stuff I call noise in a message is. Sometimes it turns out to be more valuable than I had implied when I said it was noise, when I said that the »not« signal I was interested in was noise.

SAVAGE: Your mistake was in defining noise as part of the signal you were interested in. The distortion of speech that comes to you from the walls of the room is not in itself noise. It is distortion, that is to say, recoding.

STROUD: How, for example? I am sorry.

LICKLIDER: Then you'd never know about it if there were really some 60-cycle stuff in the cardiogram.

STROUD: That is the difficulty. Perhaps the fellow with past experience with cardiographs would expect me to pick out the 60-cycle.

I study the presence of a message. This message is defined on an a priori basis in the presence of white noise. If this is realized in the experimental situation, do you know what my white-noise generators tell me? They tell me about random movement, charges in the gas tube, in the magnetic field. I must admit that in the vast majority of cases this information is nothing for me to be particularly concerned about, but I would remind you that this restriction upon what is a signal in a message, and what is a »not« signal is often highly arbitrary. Very frequently it is not stated, much to the detriment of the discussion that follows, even in your treatment of it.

FRANK: May we go on from what you said earlier about noise to consider the situation that man and his mammalian ancestors grew up in a world of communication that consisted chiefly of noise? Everything was going on at once and producing a myriad of to-whom-it-may-concern messages. Man gradually evolved the ability to select and orient himself, as you say, with the two ears. Thus we get a conception of language as a codification of events which we have learned to pay attention to and to interpret in specific ways. Each culture has picked out what it pays attention to and what it will ignore among the innumerable messages from events. We think and speak in terms of selective awareness patterned by our cultural traditions, the *eidos* and *ethos* of each cul- [154] ture. Out of the noise patterns it is not *the* message that is received | but rather *a* message that our selective awareness, readiness, or expectation filters out and gives meaning to as a communication which may or may not be authentic or valid. In other words, we may be said to create the communication, each in his own image.

STROUD: I have a suspicion that all of our sensory inputs are capable of supplying us with a tremendous number of items of information, by the billion in the case of the eye, in very short periods of time. The limiting factor in the thing is the computer, and into that we are only able to insert a relatively small number.

PITTS: We discussed at length last time how the optic nerve is greatly reduced as compared to the retina. As a matter of fact there is an immense loss of information.

STROUD: We speak of these as losses, but I suspect they are not simple losses. They are well planned, programmed.

PITTS: It is done by the aid of natural laws generally.

STROUD: The fact that we are able to attend only to a limited number of bits per second in our abstraction leads us to quite unconscious and implicit statements as to what the signal and the »not« signal are in any given set of sensory inputs. It also leads to a similar arbitrariness in handling a set of information which we get over an instrumental, extrasense system, which most of these gadgets are.

PITTS: I don't think, in the sense of the exact theory of information which we have been discussing this evening, that there is any vagueness or ambiguity at all in what is meant by the message, in what is meant by the noise. It is perfectly true that in loose everyday use of the term there may be.

SAVAGE: There is arbitrariness.

STROUD: Yes.

SAVAGE: Consider a telephone. It seems to me that from the engineering point of view it is an arbitrary question whether the telephone is to communicate, say, the meaning of English spoken into it, or the affect – or whatever the psychological word is – of the English as well, whether it is to communicate not only the English spoken into it but also other sounds which happen to be occurring in the room. Practical considerations might dictate the discussion of a telephone from any of these points of view. It is therefore these practical considerations which would determine what is to be considered signal and what, noise.

I think Shannon ought to say a word now. His ideas have been discussed for half an hour, therefore I suppose he is interested in | expressing his own views about the discussion. [155]

SHANNON: There are several comments I want to make on the last point. I never have any trouble distinguishing signals from noise because I say, as a mathematician, that this is signal and that is noise. But there are, it seems to me, ambiguities that come in at the psychological level. If a person receives something over a telephone, part of which is useful to him and part of which is not, and you want to call the useful part the signal, that is hardly a mathematical problem. It involves too many psychological elements. There are very common cases in which there is a great mass of information going together. One part is information for A and another part is information for B. The information for A is noise for B, and conversely. In fact, this is the case in a radio system in which one person is listening to WOR and another to WNBC. You can also have situations in which there is joint information, something of this general nature. You can have a device with information going in at one point A, part of it coming out at B, and part of it coming out at C. It is possible to set up the device in such a way that it is not possible to transmit any information whatever from A to B alone or to C alone, but if two of these people get together and combine their information then you can transmit information from A to the pair of them. This shows that information is not always additive. In this case the information at C is essentially a key for the information at *B*, and vice versa. Neither is sufficient by itself. If two of them get together, they can combine and find out exactly what the input was.

STROUD: The ideal minus message case. I didn't mean to cast any shadows or doubts. I merely wish to make the point that Mr. Shannon is perfectly justified in being as arbitrary as he wishes. We who listen to him must always keep in mind that he has done so. Nothing that comes out of rigorous argument will be uncontaminated by the particular set of decisions that were made by him at the beginning, and it is rather dangerous at times to generalize. If we at any time relax our awareness of the way in which we originally defined the signal, we thereby automatically call all of the remainder of the received message the »not« signal and noise. This has many practical applications.

LICKLIDER: It is probably dangerous to use this theory of information in fields for which it was not designed, but I think the danger will not keep people from using it. In psychology, at least in the psychology of communication, it seems to fit with a fair approximation. When it occurs that the learnability of material is roughly proportional [156] to the information content calculated | by the theory, I think it looks interesting. There may have to be modifications, of course. For example, I think that the human receiver of information gets more out of a message that is encoded into a broad vocabulary (an extensive set of symbols) and presented at a slow pace, than from a message, equal in information content, that is encoded into a restricted set of symbols and presented at a faster pace. Nevertheless, the elementary parts of the theory appear to be very useful. I say it may be dangerous to use them, but I don't think the danger will scare us off.

MCCULLOCH: May I ask a question out of ignorance? I meant to ask it earlier in the day. Has any work been done on the number of simultaneous speeches that one can hear with noise, without noise, with distortion or without distortion?

LICKLIDER: The only work I know is a preliminary effort, made a couple of years ago, to compare the intelligibility of two talkers talking at once with the intelligibilities of the same talkers talking separately. In one test the two signals were simply superposed. In another they were alternated at various rates. The word lists were read slowly enough so that the listeners had sufficient time to get down both words when there were two talkers going at once. None of the listeners was able to do as well, of course, in that case, but it came out between two-thirds and three-quarters as well. Thus, even though the talkers were trying to enunciate the test words simultaneously, there was only moderate interference. It turned out that one of the two talkers had a big advantage over the other: I was the one with the advantage, and I held it over a friend with a much better voice. It just happened that he had a slightly clipped manner of speech (New York State), and my Midwestern words began sooner and ended later than his words did. So the listeners heard me first and last, and presumably therefore better. We tried putting one talker's signal into one of the listener's ears, the other talker's signal into the other. The isolation thus provided did not help much.

MCCULLOCH: How about alternation? You clipped and alternated?

LICKLIDER: Not clip, but blank. An electronic switch turned on first one talker, then the other.

MCCULLOCH: Regardless of rate of clipping?

LICKLIDER: We tried only a few rates, ones that looked interesting. None of the ones we tried proved useful in separating the two talkers.

TEUBER: In aphasics that have gotten practically well, one of the last symptoms of [157] aphasia you can detect is a difficulty, on the | patient's part, to follow a dialogue between two people in the room, neither of whom is directly addressing the patient. He may grasp pretty complicated things if you talk to him directly.

LICKLIDER: When you figure what the pattern of a pair of superposed vowel sounds must be like, it is really a little puzzling how the auditory system ever sorts the things out, especially if the two talkers are talking at almost the same pitch. There are, of course, a number of possible clues.

MCCULLOCH: Have you tried different pitches, and so on?

LICKLIDER: As I said, this was not a very elaborate enterprise. We don't know any more about it.

MCCULLOCH: Has anyone any questions bearing directly on Shannon's theme?

LICKLIDER: I have one. At Christmas time, here in New York, Shannon defined the concept of information – not just amount of information, but information itself. When I heard him, I said to myself: »That is wonderful. Why didn't I think of that?« It was simple and very clear. When I got back to Cambridge, however, I got a hint about why I hadn't thought of it. I couldn't even reproduce it. So I'd like to hear it again, and I think it would be of interest to all of us.

SHANNON: Yes. The general idea is that we will have effectually defined »information« if we know when two information sources produce the same information. This is a common mathematical dodge and amounts to defining a concept by giving a group of operations which leave the concept to be defined invariantly. If we have a message, it is natural to say that any translation of the message, say into Morse code or into another language, contains the same information provided it is possible to translate uniquely each way. In general, then, we can define the information of a stochastic process to be that which is invariant under all reversible encoding or translating operations that may be applied to the messages produced by the process. In other words, we define the information as the equivalence class of all such translations obtained from a particular stochastic process. Physically we can think of a transducer which operates on the message to produce a translation of the message. If the transducer is reversible, its output contains the same information as the input. When information is defined in this way, you are led to consider information theory as an application of lattice theory.

PITTS: Can the transducer wait for infinite time before commencing its translation?

SHANNON: There are actually two theories, depending on wheth|er delays are allowed [158] or not. The more general type of equivalence allows delays approaching infinity, while the restricted type demands that the translation and its inverse occur instantaneously, with no delay. Either type leads to a set of translations of a given information source, each containing the same information.

LICKLIDER: The information is the group that is generated?

SHANNON: Yes. Put it another way: It is that which is common to all elements of the group.

MCCULLOCH: With your permission I am going to omit the presentation of semantics by Walter Pitts and hold him as a whip over our heads to keep us lined up semantically from that time on. He and I have agreed that this is probably the most effective way to make use of him. If we approach the semantic problem before we go into the problems of learning languages, he and I both feel it would be somewhat contentless, whereas after we get the problems of learning languages in the open I think it will be extremely useful to us. We shall continue then, as follows: we shall first ask Margaret Mead to give us a picture of how one learns languages if he does not know the languages of that family or the culture of the people, languages for which there is no dictionary. That is a situation in which an adult consciously attempts to break a code.

References

1. SHANNON, C.E., and WEAVER, W.: *The Mathematical Theory of Communication*. Urbana: The University of Illinois Press 1949.
2. SHANNON, C. E.: Prediction and entropy of printed English. *Bell System Technical Journal*, 30, *50* (1951).

EXPERIENCE IN LEARNING PRIMITIVE LANGUAGES THROUGH THE USE OF LEARNING HIGH LEVEL LINGUISTIC ABSTRACTIONS

[159]

MARGARET MEAD
American Museum of Natural History

I AM GOING to handle this primarily as a case history because work of this sort is so idiosyncratic that it is impossible to make general statements of how anthropologists work. From this general field of experience, I think that what would be most useful to this group is a description of what happens when an anthropologist who has had linguistic training, such as I had with Boaz, goes in equipped with the widest possible frame of reference to face a new language. We had systematically been broken of our Indo-European categories, though not with complete success. Boaz used to try to illustrate how bad it was to use Indo-European categories in taking a grammar of a non-Indo-European language, such as the early Spanish treatments of Aztec. But we learned Aztec faster than anything else, so the demonstration was not as complete as he meant it to be. Nevertheless we had, as students, systematic exposure to a variety of languages, and particularly American Indian languages, which broke down our Indo-European expectations about the way in which language is put together and widened our range of expectation. A general picture of the type of expectation given the student body may be got from Sapir (1).

What we learned was that any type of statement may be made in almost any way. Tense may be embodied in inflection; it may be handled merely as a particle. Almost any syntactic device may carry almost any type of information, or the same information may be simply relegated to vocabulary in one form or another. So you are given an enormous expectation to work within as you approach a new language. You also are given a very wide range of possible mechanisms. You know about suffixing, infixing, and prefixing. You know there are varieties of gender. I think perhaps the easiest illustration is recognition of what we call technically »formal gender,« that gender can be divorced from sex, and that you may expect to find a language that will classify | [160] round things, long things, sharp things, in groups, and handle them as we handle gender in Indo-European, with differentiation of pronouns and numerals. There are a whole series of expectations of that sort that one has. One also has some phonetic equipment. I, myself, have an abominable ear. I am really very bad at the phonetics, so that I will not be able to say anything very significant phonetically. I do not use phonetic notations when I am learning to speak a language. I use approximations out of the English alphabet, sometimes writing four or five letters one on top of another. I found that was an easier way to learn. I was concerned with learning to use the language. It is important to emphasize that my concern was to learn the language. I was not going in as a linguist to get a perfect record of the language, which would add to our knowledge of language, as such. I was going in as an anthropologist. I had wanted to work with the culture, and the native language was the only possible way of working in these cultures. There would sometimes be only four or five reasonably deculturated young adult males who had gone away to work for Europeans on plantations. With them I could communicate to some extent in pidgin English, but they were not the people I wanted to talk to; so while they could be used as a device to learn the language, it was still necessary to learn the language.

Another point that I think is significant is the type of ego involvement of this task. It is very different from the situation that arises when an educated European comes to the United States and wants to discuss things with his scientific colleagues.

I am going to talk primarily about New Guinea because it gives the most vivid picture of an area where we know nothing about what kinds of languages would be encountered in the interior. The European or American who speaks even a fraction of one of these unrecorded languages is regarded as a miraculous person by the native for the reason that no European ever has taken the trouble to learn them before. You do not want to discuss modern scientific thought with the native. The less you talk with the natives and the more you let them talk, the better it is. It is primarily teaching oneself to use a special kind of tool. You do have to establish some points of high specificity. You have to learn to ask questions perfectly. I spend a great deal of time learning to ask questions, mastering the appropriate interrogatory phrases for different things. You have to give commands accurately for your own self-preservation, so you get your dinner, things of that sort. You have to learn to make appropriate remarks; you have [161] | to learn to give the signs of grief, pleasure, and other emotions which are verbalized in the society, the little or partially verbalized aspirations; you have to know when to sigh, when to give a sharp exclamation. If one goes into a strange society and can do these three things, ask a question accurately, give a command accurately, and gloom and exclaim and enthuse at the proper moments, most of the rest of what you have to do is to listen (2).

WIENER: Never use a period but a question mark or exclamation mark?

MEAD: It does not matter so much in interaction. You actually don't, as a rule, turn to a native and say, »You have five clams.« There is really no point in a declarative sentence. What is appropriate from you is an interrogative sentence, »How many clams have you?« a command, »Tell me more about that fishing trip,« or an exclamation, »My, weren't you wonderful!« Those are the three positions in interaction.

PITTS: Position is very much more like the child learning the language in our own culture.

MEAD: Yes. Dr. Wiener, one does make some declarative statements. One says: »I have no more yams.« »I want no yams.«

WIENER: Compare with other statements, with exclamation point or interrogation.

MEAD: They are not primarily the important statements. They are the incidental statements. You work through the lingua franca. I have only once had the experience of having to learn the lingua franca at the same time that I was learning the native language. In other cases I have always had some lingua franca or English that I could use. In learning the Admiralty Islands Manus language, I had a schoolboy who understood a little English and spoke pidgin, which I did not yet know; using him as an interpreter, I had to work out the grammar of the local language. That isn't quite as bad as having to work out the lingua franca completely oneself. Pidgin English is a language which has Melanesian grammar with an extraordinary polyg[l]ot vocabulary from all over the world, and you have to learn how to use a very large number of circumlocutions. You can get on with about three hundred words if you know how to circumlocute accurately. Therefore it is virtually an ideal language to acquire rapidly as a medium working with a real language. A relevant point that I didn't make yesterday is that pidgin is probably the most redundant language in the world. But its redundancy is functional and it is a substitute for the time that it would take any of the people speaking it to learn the other people's languages. You can describe [162] | anything in the world in pidgin.

PITTS: Is this true of all pidgins?

MEAD: I am talking about the pidgin English of the mandated territory (3). It is true of most pidgins that I know about, with the possible exception of Chinook, which I don't think we ought to discuss here.

PITTS: It might be a good case for vocabulary survey of philological elements necessary to say what a language of human beings must have. Of course logical languages are exactly all the same way.

MEAD: On redundancy?

PITTS: That same principle: as few basic elements as possible and build out everything from them in philological combination.

MEAD: Let us consider that one begins by knowing what to do. One wants to be able to listen and one wants to be able to interpolate often enough to see that the material that one listens to is what one wants to hear. One wants to be able to ask: »Who was that? Where were they going? Why were they there? What relationship were they to X?« One very rapidly begins to master the types of questions that it is possible to ask in the society. We wasted much time, for instance, in the village of Bajoeng Gede, trying to get people to answer the question: »Why did so-and-so take an offering to a particular temple?« because we thought they could answer it. It turned out to be a question they could not answer. The quicker you learn which areas are not articulate, the better. Ideally, the ideal method to do that –

SAVAGE: When you say »could not answer,« do you mean they could not conceive the answer?

MEAD: They did not articulate to themselves the reasons why they did these things. Therefore the question was meaningless.

PITTS: You asked why they did?

MEAD: Yes.

PITTS: You should have asked whether.

MEAD: Why did somebody not do something? People in some cultures will give an explanation. The Samoan will tend to give an explanation in terms of physical defects. »Why did somebody not make a speech?« »Because he has a bent toe.« The explanation has no reference to the event, but the explanation will be given (4). So one of the things that you want to avoid, if possible, is meaningless questions.

WERNER: Is that because they want to give an answer, or is that because they are just confused?

MEAD: Because they, the Samoans, have a general tendency to | refer aberrant behavior to obvious and trivial physical conditions. [163]

FREMONT-SMITH: Rationalization in simplest form.

MEAD: It is a statement about other people. It is not an attempt to explain away a motive.

FREMONT-SMITH: It is about other people as part of them.

MEAD: It is an easy social way of doing it.

KUBIE: We do the same thing. Why does somebody do something? The average person says, »Well, everybody does it.« They think of that as an explanation.

MEAD: This is another type. In learning a language, you make specific explorations of the interrogative structure of the language, rapidly, in as many kinds of interrogative statements as you can think of. This also gives you a picture of what it is going to be possible to ask, the things you get into trouble with if you try to discuss them.

To shift for a minute to what you actually do, there is one other point I learned when I first started to work on languages – to learn from children. This may look like a contradiction of going in at a very high level of abstraction based on a knowledge of possibilities derived from the linguistic structures of the world and at the same time learning from children, but it is much the quickest way for me to learn a language. I have done both, learned from adults and from children, in different circumstances. Children will repeat and repeat and repeat; adults will not. It is very often difficult to get an adult to say the same thing twice unless you have trained him to be a linguistic informant. He said it once, and there is not a very good reason why he should say it again. If you ask what he said, he says something else, whereas children are accustomed, in every society that I have worked in to practicing the language. They expect to practice and a child is able to comprehend that you are learning a language. In many of these societies the majority of the adults really do not comprehend the process of learning in another adult. You don't know the language or you do know the language. The minute that you learn to say six things correctly, they are apt to think that you can speak; they make very imperfect allowances for lack of knowledge. This, of course, is not true of someone whom you specifically train as a linguistic informant. But if you make mistakes with children that a child a year younger than they would make, they will criticize and correct it. They still have the frame of reference that says there are people who don't know how to talk their language correctly. I think that is the principle reason I learned from children, except that in any case I usually study [164] | children. Little boys of eight to ten who are teaching other little boys how to behave are probably the brightest things you can find in most societies in the world. They seem to be the freest from other preoccupations. In learning from children, it is a matter of practice. You don't get your major points from children. You practice with children. To construct the grammar of a language is a question of getting a native who speaks a little pidgin. I am speaking here specifically for New Guinea. I never worked in an area of New Guinea where there was no pidgin. You get a native who speaks a little pidgin and you put him on a box. This is quite an important part of learning. You live in a house with chairs on which natives don't sit. There is the highly stylized European-white-versus-native situation in the home; that has to be maintained. You must take a linguistic informant from the many people sitting on the floor. You place one person on a box, which changes his relationship to you. He is then your vis-à-vis in a tutorial situation, in a context where the native is accustomed to being treated more or less as a subservient person. Then you simply start out with very simple questions, such as, »I see a dog.«

SAVAGE: What kind of a question is that?

MEAD: You say: »Suppose you talk place. All right now, how's that you talk! ›Me lookim one fellow dog‹.« If I bowdlerize pidgin, it is close enough for you all to follow, isn't it? That would make it a useful device here. The native gives you a statement in unfamiliar phonetics; he may not say, »I see a dog.« He may say any number of other things instead, because he has not gotten the idea yet. You have to allow for that. We assume that about 50 per cent of the time he does say what you asked him to say. Is that fair, Dr. Bateson?

BATESON: I think that 75 per cent of the time he would say, »You saw a dog.«

MEAD: Exactly. All those varieties enter; then you put in what he said with as accurate phonetic rendering as possible. That is an approximation. That is the place where your ear has to come in while you are trying to isolate the word for »dog.« Once you have isolated it, you can mispronounce it within limit for the rest of your life with safety, but not before you have isolated it.

PITTS: You begin with questions about where, of the form where, when, and so forth?

MEAD: You begin to give simple declarative statements that are variable, such as: »I see a dog; a woman; a dog goes; a woman goes«; but you don't know what tenses are going to be like.

PITTS: What do you ask: »Did you see a dog?« | [165]

MEAD: No. You say, »You talk place.« That the pidgin English for »talk your own language.«

PITTS: I see.

MEAD: You say, »Suppose you talk place.« »Place« is the phrase used in interpretation for »our village« in the European New Guinea world and the native's »You talk place. Now you talk: ›Me lookim one fellow dog‹.« Now perhaps he puts »dog« in that sentence. Perhaps he does not. Perhaps he says, »You said ›dog‹.« Perhaps he puts the pronoun in instead of the word »dog,« and says the equivalent of »I see it,« but you write down what he said.

GERARD: You would not use the ostensive definition of the word by pointing to a dog and saying, »Say that word?«

MEAD: I usually go at the word and crack it.

BATESON: Finger.

MEAD: If you pointed to the dog and asked for the word, as likely as not he would give you the word for finger. You have got to deal with whole statements. You could go around forever trying to get the word for house, dog, woman. You never would know whether you got woman or that particular woman's name, house or the word for the men's house. When we work with a culture, we start with wholes, then get the main pattern, and then work down toward details. You have got to get basic grammatical structure to know what the details are.

Let's assume that he did say, »I saw a dog«; you then ask him and you make a guess, an approximate guess of what in that sentence is »saw« or »looked at«; what is »dog.« Then you work those through. You ask for »I saw a dog, a house, a man,« all sorts of things.

STROUD: Twenty questions?

MEAD: The principle of excluding possibilities. You may have missed something. In that language you may look at a dog with one verb and a woman with another. You may keep on getting replies in which there are still too many variations for you to tell which is the verb or the noun. That is just a question of excluding until you get it. Then you make a quick stab at all the possibilities.

I work on tense as soon as I get a notion of the nature of a verbal statement and a set of nouns that are identifiable, so that I am reasonably certain that this word is »woman« and this is »dog« in the discussion. Then, »Me lookim today.« Pidgin handles tense entirely by words like »today,« »yesterday,« »tomorrow.« »Me lookim yesterday, today, bye-and-bye me lookim,« running through the tenses for the verb »to see,« to get the notion bow tense is going to work out. You start working on, »Me lookim, | you [166] lookim, emi lookim,« You will get »me« for »you« and »you« for »me,« and so forth. But you aim in a very few days to have your basic grammar well enough to know how tense is handled, whether you have got inflection or whether you haven't, whether you are going to have to deal with modifiers that are related to the adjective or embodied in the noun, whether verbal ideas are expressed in the verb or by particles or by a complete change of stem, whether you are dealing with a language in which you have to deal with infixing, and so forth. I have never had to deal with a language that made one of these changes that has to be treated as a series of parentheses. There are languages in which the occurrence of one sound alters the sound pattern in subsequent words; it is like a change of sign with a series of parentheses, as in a mathematical statement. I have never had to crack one of those languages myself. We know that it is possible to do so, and that it is not too difficult to crack it, but it takes a little longer and it depends upon your ear more than some of the other languages do: You don't expect in the first two or three days to discover the refinements; also, you may be trapped in a point of change. For instance, I worked with a multiple-gender language which was changing from a multiple-gender language to a two-gender language. It was changing from a language in which there were a large number of phonetically differentiated genders, differentiated in the singular and plural, in which the pronouns, the adjectives, the numerals all fit the phonetic pattern of two-gender language usage, animate and inanimate, breaking the phonetic congruences. Because the change was being made slowly, people still used both the old forms and the new forms. That sort of difficulty will catch you, and it may take you quite a while to solve it.

WIENER: General languages like Mantu.

MEAD: Sometimes such genders are not meaningful, sometimes they are completely phonetically crystallized. You may have a pronominal pattern that fits with noun classes, but you no longer decide by analysis what that noun class meant. Sometimes you can.

WIENER: Would you call, by the way, the Chinese use of qualifiers a gender extension, *egoshim*, *ekul*, *dogya*, different words for the one piece?

BATESON: They are still related to genders.

MEAD: We call them piece words in the analysis of Malay. They are related. In these New Guinea languages occasionally you have very complex structure. I worked with [167] one that had thirteen | genders (5). Each gender has one to four phonetic peculiarities, and pronouns reflect these phonetic patterns in nominative, objective, possessive forms, and so forth. Now one form of Arapesh was once learned by a German missionary who never discovered there were genders. He had a perfectly good category to apply to the gender. He called it euphony. You have a word for dog which is *nubot*. *T* is the ending. When you say, »I saw a dog,« you say, *Ya terut*, with *t* at the end of it. When a dog is the subject, the pronoun is *ta*. This phonetic correspondence runs all the way through the language and can be called euphony, so these missionaries learned to communicate with the natives without any recognition whatever that there was gender there. But I imagine it took them a great deal longer to communicate than it did me, because I had the possibility of multiple gender to work with. So you have your problem of cracking your essential grammar and then building vocabulary and building it fast enough. I build vocabulary entirely in terms of what I expect to need. Ideally, one would have thousands of pages of text that are to be taken down first. A linguist would first have made a study of the language. One would then know what vocabulary was relevant to the culture and what was not, without wasting time organizing the vocabulary along our lines. That is not possible when you have money for seven months and food for only five and you don't know whether the carriers will take any more food in.

TEUBER: How important would it be in this situation to know something about possible differences between their tenses and ours? There is little chance of finding a one-to-one correspondence between their time structure and the three major tenses of our language.

MEAD: You have got to know how tense is formed, otherwise you would be in great difficulty in no time.

TEUBER: Suppose they have a half a dozen tenses or more, do you find that out in the first week?

MEAD: Oh, yes, in the first week. You have got to get the mechanism for tense, for number, and so forth. You have got to know whether or not you are dealing with a system of pronouns which distinguishes sex.

TEUBER: How do you do it? Do you do it on the basis of some general linguistics, looking for different ways of slicing up the time scale?

PITTS: You learn how to say things yourself, not how to understand them, apparently. You learn the translation of our tenses, even if the translation apparently involves just the use of adverbs | and not tenses. This is not learning to understand arbitrary sen- [168] tences of theirs but being able to express arbitrary sentences that you wish to express in this form of question.

MEAD: I learn to listen. I ask, »How do you say, ›I go; I went yesterday‹.« But you don't start with the intransitive, like that. You start with the transitive because you get the »dogs« and the »cats« in easier. But after you have got a little bit done and you are really sure this is, »I go,« you can have, »I went yesterday, I went the day before yesterday, I went a long time ago, I was going but I didn't go,« and so forth. You can range through the possibilities of tense and mood and conditional statements; you find something you didn't know was there at all.

WIENER: How general do you find true tenses and how general do you find aspects?

MEAD: I don't know what that statement means.

WIENER: Ordinary tenses are placed in time with respect to the present. The aspects are placed in time with respect to other parts of the assertion.

PITTS: Roughly you have to try the statement with all possible adverbial conditions, don't you, in order to see which ones collect together on the common grammatical form?

MEAD: You need to know, in Melanesian, for instance, whether there is a distinction between the incomplete act and the completed. That is essential.

WIENER: That is aspect.

MEAD: That is exceedingly important. Suppose, for example, that the first day you are in a new village, your pidgin-speaking boy announces, »Mary (wife) belong Paleoa, emi die.« You rush over expecting to find a corpse, but you find Paleoa's wife sitting up and talking. You turn to your boy and demand what in the world he meant, and he replies, »Em i no die finish, i die that's all.« As you explore it, you discover new points. When you ask what is commoner, I don't think we know whether it is common to say, »My position in the past or my position in the future,« or to work with aspects; you get all varieties. You don't know what you are going to get.

WIENER: Chinese has no tenses, only aspects.

TEUBER: You check against sequences of actions observed in that particular context?

MEAD: Yes.

TEUBER: Then you try, not only to get what you can in the give and take of the language situation, but you try constantly to fit what you have learned, tentatively, to particular sequences of | action that you observe? [169]

MEAD: Then you start taking tests. Almost at once you start, just as soon as you can hear the language at all. Now Dr. Bateson would start with the text much earlier than I would. He has a much better ear. You would start with text almost immediately.

BATESON: Yes, but I start with substantives.

PITTS: Proper names?

BATESON: No, general substantives. »Man he fight him dog.« Man hits dog. With a little vocabulary I then start telling a story which my informant is giving me back in native language. I tell the story in pidgin, phrase by phrase, and he gives it back in native language.

PITTS: You get into difficulty with the special tenses used for foreigners when you try to use the language in telling the story.

BATESON: The difficulty lies in special languages for foreigners or for castes.

PITTS: Then you might possibly encounter difficulty.

BATESON: And you are outside the caste system – an outcast.

PITTS: You would not learn how anything is said at all.

MEAD: In Bali, Malay is the only appropriate language in which to talk to the foreigner. A Balinese can talk to you a long time in Balinese about the fact that neither of you can discuss anything because you don't speak Malay. You can get in that sort of situation, but there are always such possibilities. In Balinese we distinguished seventeen different degrees of gradations in the language, depending upon who is speaking to whom, or about whom. Such distinctions are a dreadful trap because sometimes you have no opportunity ever to hear the »rough« language spoken to you. There is no way in which you can get into a position that is low enough so that you can be spoken to as a low-caste child. Then you have to use an interpreter and get somebody on a lower level than the interpreter; you say something to the interpreter and he always answers you in high language, but he turns around and speaks to the other person in the low. Then you hear it. It is much harder to move in and speak a language that you never hear spoken to you. You always have to give an answer in a form different from the way the question was addressed to you. That is another complication.

WIENER: I am rather curious about that. To a minimum extent that does exist in the European languages and in Chinese. On the other hand, these different phrases of courtesy are being simplified, not only in most European languages but also in Chinese. Do you find a tendency for that sort of social up-and-down expression | to be fluid and to be changing while you are looking at it, as it were? [170]

MEAD: It is a changeable system; and if Europeans are entering the system, or if there is a breakdown of the social system, that is a part of the language that responds very rapidly. In Balinese »I« is one of the very crucial points; there are whole sequences of »I.« In our mountain village there was an »I« you only used if you were born in the mountain village; it was also very low, familiar, and rough. There was a slightly higher »I« which was not quite as rough. You could use it in low-caste conversation to low caste if you were not born in the village. Then it went up the scale. The Balinese have been trying to get out of saying »I« by every known device for a long time.

WIENER: We are trying to get out of saying »I« or »you.«

MEAD: We are. »I« or »you« is a crucial point in the caste languages. Our Balinese interpreter, who was Dutch educated, used the Dutch »I« to his brother, who used it to him in order to avoid the situation; but he would also use the Malay »I,« which is literally »your slave,« but he did not know that. That was a way of avoiding the hierarchical relationship.

PITTS: Why is it that they avoid using pronouns only when the culture is changing? I suppose if there were a fixed system you would always know what part of the system to use.

MEAD: There is Dr. Wiener's question: Do you find this a particularly fluid point? My answer was that when the culture is changing social organization, when new people are coming in, then it becomes exceedingly unstable.

PITTS: If you left the Balinese alone, you would not have trouble.

MEAD: The Balinese have a composite culture, a caste system imposed on a large group of casteless persons. Sometimes a low caste tries to have the kind of cremation tower appropriate for a higher caste, and so forth. I imagine they have been playing with pronouns for thousands of years.

WIENER: It is almost impossible to guess a Swede's position from his outward dress and courtly manner.

MEAD: In Bali you have a regular conversational form which is used until you find out the caste and status of each speaker, a form of courtesy between equals. You use this to people until you know who they are. You ask where they sit and they tell you. Then you know how to address them. Until you know that, you cannot really talk with them. If you are going in as an anthropologist, you also learn the vocabulary of technical areas rapidly. Kinship terms, for instance, you learn almost at once, because you cannot | operate in the society without a knowledge of the kinship structure; also, it [171] may overlap the pronoun structure. You may, for instance, refer to your mother-in-law in the third person, as »they three,« analogous to the royal »we.« You may call a pregnant woman by a dual and a married couple without children »they three.« Unless you learn those things very rapidly, you cannot understand what is happening. You have to enter those interpersonal situations for information before you can understand the next situation.

PITTS: At the stage we are at, do you translate, do you try to translate a folk tale which you have in your bag in the evening?

BATESON: I do.

MEAD: Thanks, Dr. Bateson!

PITTS: You have no special tense for experiences?

MEAD: A special tense for folk tales?

WIENER: Once upon a time.

MEAD: Yes, once upon a time. In Samoan, for example, it is *ona* ... *lea* with the predicative statement inserted. This form is also used for cooking recipes.

PITTS: That would not be useful for practical purposes.

SAVAGE: You emphasize in the beginning, as anyone who talks about linguistics does, that one has to break the stereotypes of his own language – even of his whole language stock. What has never been clear to me is whether linguistics has discovered enough generalizations and principles so that you are not left with the feeling that anything whatsoever can happen. In approaching a grammatical question about a new language, can you say to yourself, »Well, only these few hundred things can happen, and by systematically dichotomizing, I will find the right one,« or do you have to allow for the possibility that there will be a special tense referring to the future actions of men whose eyebrows happen to be of a certain color?

MEAD: You still have to expect that unpredictable categories may occur.

PITTS: It is fixed. At least one man in the culture has learned pidgin English.

MEAD: This much is fixed, that as far as we know, all human languages are learnable by other human beings. So far no has found a language that could not be learned, although there are languages which present great difficulties. Tchambuli is such a language. Only about 600 people could speak Tchambuli in 1933, and the peoples around did not attempt to speak it. This was a language in which the grammatical structure was in process of | simplification. Some of the boys who went away from [172] Tchambuli to work refused to speak it when they returned; they insisted on speaking a neighboring language to their own parents. But Tchambuli is not as difficult as many American Indian languages. But most American Indian languages were expected to be difficult, whereas in New Guinea most languages are easy enough so that the average native expects to »hear« the languages around him to some extent. They make a distinction between »speaking« and »hearing.« You »hear« the languages that are adjacent to you, and in border villages people sometimes speak two languages without knowing which one is primary. From the standpoint of other villages, this will be a village of which they say in pidgin, »they turn the talk,« but from the standpoint of the people who are born and live there and speak both languages, they are not quite clear what they speak.

Now, Dr. Pitts, to get back where we were. You very quickly try to get a mass of material of some sort. Dr. Bateson gets it by telling a story and getting them to tell it back; it is also possible to get them to tell you a story or to get them to dictate a description of a quarrel that had just happened. I do this very early, if possible, because in this way I have some control of what has actually taken place in front of my eyes. Folk tales I hate. They are likely to be filled with archaic forms, pickled clichés not used in real life. They are not nearly as useful in getting on with life. Folk tale texts are the traditional linguistic method, but they are awfully poor, as a basis for field work. It is as if you were to try to talk to people entirely in terms of *The Idylls of the King* or the King James Version of the Bible. You meet natives who have learned English through the King James Version of the Bible. Unless you know the King James Version type of English well, it is pretty hard to understand. So I stay away from the archaic and turn to the living stuff. I turn it into interpersonal stuff, rapidly building vocabulary. I learn about two hundred words a day of a language and in about six weeks I have a workable vocabulary. How workable depends upon the culture. However, I can get around. One may know every word that relates to marriage and be helpless when it comes to funerals because the vocabulary is so specialized. On the other hand, it may be a language where all ceremonies have about the same form and are treated the same way, so by the time one can verbalize about people getting married one is also able to bury them.

[173] BROSIN: How do you motivate a translator to give you that amount? |

MEAD: You have to do two things. You pay them. This becomes a context in which it is correct for Europeans to pay natives. He becomes a work boy of sorts. And you have to get somebody who is interested. You have to move fast at all times. A linguistic informant has to be an intellectual. The intellectuals seek you out reasonably fast, too. They are bored, and the chance really to talk to somebody is offered them probably for the first time in their lives. They are usually natives who have never had a chance to talk enough, that is, interestingly enough. The first boy that you seat on a box may just sit there; he may get restless and start to wriggle and not stay very long. By that time what you are doing gets around, and other people come and stand about and listen. Somebody else comes into the picture, and finally you get your linguistic informant and he loves it. The main motivation is virtuosity.

For example, in Arapesh there are two ways of absorbing foreign words in the language. One is by looking at its structure roughly. You take the English word »butter,« for instance. If the native has heard the word »butter« pronounced by an American, it then goes into the gender that ends with an intermediate »rl.« The plural would then become *butigu*. If, on the other hand, he heard it pronounced with an Oxford accent and it becomes *buttah*, he can put it into another gender approximating this phonetic ending or he can say that the ending he hears is not like anything in his language. He can say, in effect, »Let us treat it like those words that are indeterminate.« There is a gender which includes creatures of indeterminate sex. The word for »child« is in that gender, objects of indeterminate form where form is specified elsewhere. The plural forms of this gender are used for men-and-women, people, all the different things on this table, several kinds of fruit, and so forth. You can put a foreign word in this gender. We had two linguistic informants (6). One of them had a very formal mind and insisted that all foreign words belonged in the gender for foreign words, even though you could make them fit phonetically into other genders. But our principal linguistic informant had a lovely fluid phonetic flexibility. He would follow the sound around, work around it until he could fit the sound into the gender structure. Both were possible, and it was wonderful to argue about. So your linguistic informant is an intellectual who gets fascinated and to whom you teach the grammar as you learn it. Now sometimes you have to teach more than at other times; you teach such concepts as that of a *word*. You decide how you are going to write the language. If you are going | to [174] write it, you are going to have to see something of the gender that ends in *t* and this *u*. *T* is an objective pronoun; you decide whether to handle it as a suffix or as a separate word. Once you have made up your mind on the subject, you then use the idea of a word, and if necessary of suffixes and prefixes. The Balinese when he wrote in his traditional script did not break it into words at all. But words are convenient for use and easier to think in. So, in the end, if you are using the concept of a word, you teach your linguistic informant in what size units you want him to answer. Sometimes there is a word for names of things; pidgin has the categories »big name,« »lik-lik name«; »big name« is the class name and »lik-lik name« is the name for a particular type of object within a class. So the »big name« for birds, for instance, is »pidgeon.« There will also be the name for a particular kind of bird, and then you may find another type of bird that is rather like a pidgeon but unnamed. You ask, »Now what name lik-lik name belong im,« but there won't be a lik-lik name; instead it will be described as, »This fellow pidgeon i all the same brother belong.« So you have categories, large name, small name, and »something that is related to something that is named, but we have not got a name for it.« Now you work with that sort of loose net and gradually teach your informant how to give you rapidly what you need to crack a new sentence or a new piece of material that you have taken down.

GERARD: You indicated that you learned some 10,000 words in the first few weeks.

MEAD: A good many of which I forget.

GERARD: How many words do you find it necessary on the average to know in order to break it up? That is a tremendous number of particular words.

MEAD: That is entirely a question of the language.

GERARD: The number you need for the field.

MEAD: Manus is a language completely analytical; there are no figures of speech, no sex gender ever, just the third-person pronoun, and very few adjectives. It is a cold, accurate, precise language of people that can count to 400,000 without any form of notation whatsoever and do something with it in their heads. One figure of speech was all that we collected in nine months among the Manus. Such a language you learn very fast and can work with a very small vocabulary. It is analytical, not overconcrete. They don't have one name for the end of an object and another name for the end of another object and another name for the end point, middle piece, and things of that sort. Balinese, on | the other hand, has practically no abstract general words. If a boy is [175] in front of you with a loaf of bread and a knife, whose task has been to cut bread with that knife every night for two years, and if you don't use the verb that means »to cut in even regular slices across the loaf,« but instead use any of the other thirty or so words for cut, such as »to cut in irregularly,« »to cut in asymmetric pieces sideways,« he will look at you with absolute amazement and have no idea whatsoever what you are saying. Balinese is a language in which you never learn to give enough commands correctly. You are dependent upon building a new vocabulary for every event. When we were there, we fortunately had an informant who could read and write. It is a most complicated society. He had to take the vocabulary down for each event. We could never have mastered the vocabulary in time. The range is from a small vocabulary to an enormous vocabulary that requires special techniques for building. If you have to have a complete vocabulary for canoe building, you don't learn that until you are ready to work on canoes; on the day you are ready to work on canoes, you sit down with somebody and work out the two hundred words you need for canoes. You draw diagrams, ask questions, get your models, write the words down, and then you forget them as fast as you can, because you cannot stand that much memorizing and still get on with the work.

What I do is to build myself a very large vocabulary of what I think I will need, which is not surrendering to the culture initially. I think I am going to need words that indicate size, for instance, or I want to know whether or not there is any way of indicating a comparative right away. That is an important thing to know. It is necessary to know what measures there are almost at once, a series of things of that sort. I need to know all the medical terminology I can get my hands on at once because the natives come in and ask for help.

KUBIE: One question, Dr. Mead: It strikes me that this multiplication of names for the same object, depending upon the total relationship of the object, must mean that if that object appears in a myth or a legend or a dream, it in a sense represents all of those potential names. So that if you have two hundred names for a canoe and a Balinese dreams about a canoe, you have a hundred or two hundred starting points for his associations to the unit which he has visualized in his dream. Remember that only the visual image occurs in the dream.

MEAD: I cannot get the visual image. When he tells me the dream, he has already [176] made a selection. Now granting that that is secondary elaboration – |

PITTS: Presumably it is as if we can use the wording to include canoe or possibly rowboat; you might say in a dream that you had dreamed about a canoe or dreamed about a small boat.

MEAD: You work through these connections all the time. You are learning the language and the rest of the culture simultaneously (7). For instance, in Bali there is a word for what the witch does when she throws her hands up and back (demonstrating), *kapar*, which is also the word for a child who is startled by a fear of falling or for what a man does when he falls out of a tree – a gesture representative of fear. Every time you get a new word in the language, you follow it through. You don't simply ask what is the word for »good.« You ask in pidgin and write down »good.« But then you say, »Now what is the word ›good dog‹ or ›bad dog,‹ ›good man,‹ ›good woman,‹ and so forth,« and then you try it on a whole set of connections to see. At first you try to fit things together; you listen; and the first time you hear that word in the new connection, you know. So you are continually operating between the categories you impose out of your need and the categories you hear out of what they do. They can do a lot of exasperatingly difficult things. As I said, the anthropologist has no ego involvement in talking to colleagues in polished scientific speech. But occasionally there is ego involvement of a different sort. Among the Iatmul, if a man has a quarrel with his wife, one of the things his sister's son can do is to stand outside the house and take the parts of the husband and wife, speaking first as one, then as the other – and this is a pretty problem for the recording anthropologist to unravel.

SAVAGE: The man has a quarrel with his wife and the sister's son can do this?

MEAD: The sister's son, who is in a very special kinship relationship to his mother's brothers, comes along, stands in front of the house, and starts speaking, first as the husband and then as the wife, from the extreme positions where he says that she is a something with whom he is never going to have anything to do again and he wishes she would take every article she possesses and leave, but that if she does not he is going to break them all, and so forth. Then he represents the wife answering in about the same tone of voice. This is all dramatized by the sister's son. Gradually, the speakers seem to approach each other; the quarrel gets milder. A few issues are dropped; others are intruded, and he reconciles them at the top of his voice, taking the positions of each. The first time you hear a thing like that, it is difficult to recognize in a strange language. You don't know what is happening unless you | have had the luck, before [177] you hear one of those speeches, to have had it described to you by a native as a possibility. But without the knowledge of the possibility, if you suddenly hear a man speaking as a woman at the top of his voice outside – and it depends upon how much masculinity and femininity there is in the language – it is rather astounding. Iatmul has a feminine »thou,« so you get a clue from that particular point. But different peculiar events of this kind occur simultaneously with your learning, and it is very much a question of luck how fast you learn. You may be stuck for quite a long time because you have not got the point.

There is just one other point that I want to make. I think it might go with child learning here. Cultures differ enormously in the extent to which their members think they have a language, and think about language, and therefore the learning of the child differs; that is one of the things that I experienced when I tried learning as a child. I once published the fact that I make the same mistakes that children do, which was true at the point that I published it; but I don't think it would be true of learning all languages. I make the same mistakes that children do in two different ways. I may make the mistakes of ignorance, of course, as they do. In Arapesh, every time that a child acquires a new noun it is acquired in a way that stresses the gender forms. So the child says: »What is this?« The word for *x* itself may end in a *t*, which would mean that the correct form for »one *x*« is *atut x*, and for »many *x*« would be *minahigi*. But the adult will answer by initially using the forms for »one« and for »many« which belong to the indeterminate gender which I discussed above, but using the correct plural form for the noun itself. So the first time the child hears the placing of the noun, by its plural, this is done in a way in which the usual gender correspondence is violated, and so the abstraction is conveyed to the child.

SAVAGE: The parent does not teach the gender properly but the child knows automatically?

MEAD: No, the child says, »What is this?« and the parent says the word for pencil in its accurate form but uses first the word for »one« which is limited to the nondefinitive gender. Then he says, »*Minihisi*« (many). The »one« and the »many« remain constant in the teaching situation and give the child the specialized statement.

SAVAGE: The child is not confused by this irregularity? He knows this is the irregularity for teaching words?

MEAD: The child learns the genders. The child learns that there are both the special forms and this *minihisi* form, which is any | kind of plural, just as we say to the child, [178] »This is the singular, this is the plural.« But we don't say, »This is the singular, this is the plural,« to the child, as a rule, until it is five or six.

WIENER: There is a very interesting thing: the difference in teaching Indo-European languages with the gender where only in the teaching situation do you give the article to a great many words.

MEAD: Yes, but another sort of parallel thing can happen, as in Manus, which is a very analytical, clear language on the whole. There are three particles in Manus indicating direction, up and away, down and toward, and something that can be defined roughly as »along.« Children make terrific mistakes about whether now I am here or Warren is there. They do not know whether to say, »I go along to Warren,« or, »I go away from myself toward Warren,« or, »Warren comes toward to me,« or, »Warren comes along.« There are three possibilities. Children make mistakes all the time, and they are concretely corrected, you see. So I would say, »I go up and away from myself toward Warren.« They would say, »No, you go along to Warren.« They wouldn't give any reason whatsoever. The child gets that correction. The next day however, if I am working and I am in a position like this, I may say something involving direction and again I may be corrected, but I will not know why. The children keep making mistakes, but by the time they are grown up they don't make the mistakes. The Manus don't know that they use these words because their conceptions are based upon a geographical model that is not conscious, that is, that the world is shaped like a saucer or like a platter. The village in which one lives is conceived of as lying along a long axis, as if it were at the bottom of a platter. If you are going along the bottom of the platter, you use the word »along.« Otherwise you use the words »away from me up.« There is no way for a child to learn the abstraction. When two people are going together, you use »toward me down« or »away from me up.« In this case the child's incorrect usage is because the teaching is arbitrary, because the form has not been articulated.

I think that is about enough, Dr. McCulloch.

McCulloch: I think that gives sufficient idea.

Wiener: I was going to say that that difficulty appears in many European languages, in Spanish, *agui*, *aqui*, and *aqua*, the three different forms of place. The confusion is the same as that in Dr. Mead's example, the near there and the far there.

[179] Mead: Or you may have »near me.« It is very common in | some of the Oceanic languages to specify, »near me, near you, and near him« in combination with »coming to your house« or »going to your house.« In some languages these forms are articulate and teachable, or are implicit in a form of teaching which actually articularizes the form of the language. In others the adults merely correct, and the correction does not contain the generalization.

Kubie: We have similar uncertainties and overlapping forms, all of which carry slightly different emotional connotations, as, for example, the difference between saying, »Come to visit me,« or »Drop by some time.« This is quite like what you are describing.

Mead: Some of these expressions we tell children not to use, some we tell them to use in a different way. Sometimes we tell them a form is »common«; sometimes we tell them it is incorrect; sometimes we tell them it is bad grammar. We have different degrees of articulating class positions or formal position, and it makes a great difference to the manner in which the child learns the language. As far as we know at present, however, from the data we have, children in all cultures learn to speak at about the same time.

Pitts: We correct children by different methods in grammar or manners. In this case we would correct the child where it uses the grammar badly.

Mead: Calling it »grammar« when class is meant, or a class phrasing when grammar is meant, does not help much.

McCulloch: As to the learning of languages by children, you say it does not make much difference whether they teach or do not teach, particularly; that the child learns the language at about the same time?

Mead: Yes. Undoubtedly it learns it very differently, its sense of facility, its sense of being able to move around within linguistic forms. The Balinese don't really think that language is ever easy to use; they conceptualize with great difficulty. They think there are endless situations which cannot be tackled. The child learns in Balinese by hearing the parent put words in its mouth all the time. The mother speaks for a three-months-old baby and makes a speech to other people, thanks them, and uses the very best language she can use. The child always has a model presented for it by someone who is in fear of making a mistake. The child learns to talk, but it also learns this fright, this sense of paralysis if it doesn't know who the other person is, if it does not know who it is in respect to them. Even if it knows, it may make a mistake. Balinese varies every part of speech by caste or position, not only nouns and verbs but also participles. This sense of partial | paralysis survives all their life. You learn to talk but you learn to talk with a [180] sense of near paralysis, you learn to talk with endless ifs, whereases, lests, and so forth.

Among the Mundugumor of New Guinea, children, little boys of three, four, and five years, are sent away as hostages to another tribe while a joint head-hunting party is being planned. The child knows that the people among whom he is a hostage will later be his enemies again, and one of his tasks is to learn the language and learn the roads. If he learns the language, later on this knowledge will be very valuable. Here the tone of voice is entirely hostile, and little children master the tone of voice of their host.

McCulloch: Is there any difference between the linguistic structure in those languages which are taught and those where the child just picks it up as best he can?

Mead: Not that I know of. We don't have the details of the learning of American Indian languages. We have not enough material on the learning of these languages to know to what extent the children learn in lumps where the adults virtually never analyze and simply deal in whole predicative statements. That is something which could be studied. It ought to be very valuable.

McCulloch: May I ask one more question? There is distinguishable in our civilization a prattling among children who cannot form words sufficiently to speak. The whole speech is opposed to the learning of particular words.

Mead: I would say there are two elements: individual differences in children – and some children will prattle under any circumstances – and the manner in which the society attempts to pin you down. The *Manus* teach language in the sense of getting the names of things and getting them very early. Older children teach younger children all the time. The minute the child can form words at all, nobody is interested in his making an imitation of a speech in nonsense syllables. They want him to be able to talk and to talk accurately. The culture can inhibit prattling by patterning the language teaching, especially if older children teach younger children. Manus children sit by the hour making other children count, making other children name their fingers and toes, going over and over the simplest form of learning. This has a very inhibiting effect on the prattler. In a society where nobody is very much interested in what the child is doing and does not care awfully when they learn, the prattler has a little more free range.

Teuber: Prattling and babbling are not the same. You do not | know how to determine which is prattling and which is babbling? [181]

McCulloch: A Fourth of July address.

Teuber: Which comes later?

McCulloch: Which comes a little later.

Stroud: The kinds we invent in sixes and sevens, special private worlds.

Wiener: They cannot speak at all.

Stroud: My one-year-old daughter, with a very appreciative audience, will deliver an oration which consists of the syllable »da,« but I would not for a moment hazard the faintest guess what it is about.

McCulloch: That is what I was referring to. That is the real prattler.

Teuber: I see.

Pitts: Actually you do have to make the assumption at the very beginning that there will be on the whole, and with reservations in different sentences about small common movable objects, a common sequence of the elements that you can isolate?

Mead: You do not have to make the assumption that people will isolate.

Pitts: If you asked for a collection of sentences, »I see a dog; you see a dog,« and so on, and wrote them all down, there would be absolutely nothing in common with the sequences in phonemes; you would be absolutely at a loss.

Mead: You might get a good many sentences which from our point of view ought to have common isolatable elements, but they do not.

Pitts: Suppose you distinguished ten varieties of dogs but never combined them, you would find ten classes but it would be a small finite number.

Mead: It will be finite, we assume systematically. So you can practice. You find a common name for a dog or the name for a bitch, the name for a young dog, an old dog, a dog without teeth. You would assume at every point that it is crackable, that it is a code that is learnable. If you have a large enough amount of recorded material to cross-compare, plus a living person, as of course this sort of method requires, you can crack it.

Wiener: The thing that seems to me apparent is that the languages which belong to the professions and the trades will show an extraordinary specificity of words which are not learned by children but learned by adults. I am thinking of the old-fashioned Maritime varieties of the European languages where everything on shipboard has a [182] name, but where the word »rope« is only used, | I believe, for the bucket rope, or something of that sort.

McCulloch: Tiller rope.

Wiener: You have »staff.«

Bateson: That means attached rope.

McCulloch: Rope on shipboard only refers to the tiller rope.

Wiener: The rope with which you let a bucket down.

Teuber: That is a line.

Wiener: That is a line?

Stroud: Fasten the line to bucket and lower it over the side, is that about somewhere near it? Even the modern American Navy has a terrific hangover. I get confused aboard a big ship, in spite of the fact it looks much more like an electronics factory than like a ship.

Wiener: It is the very interesting case where you have the extraordinary variety of words for specific purposes, something like what you said about the Balinese. This is language only learned by adults, but mistakes are heavily reprimanded.

Mead: You also have children's languages in which every term is highly specialized, like our adolescent jargon. Mistakes are heavily penalized, and later the vocabulary is forgotten. I don't think it is a question of adult language but of the specialized languages which carry both the techniques and the ethos of the particular subject or aspect of a culture. What we try to do when we learn a language is to isolate only those elements we are going to need. We don't learn special vocabularies otherwise, and therefore we don't get confused by overspecificity that cuts oneself off from communication within the group.

Pitts: From the logical point of view, there is considerable restriction on the supposition that the language is based upon all names of objects in a sense.

McCulloch: All are.

Mead: All languages are, all known languages.

Pitts: Mathematical is not defined in functions of space. As a matter of fact, to express the notion of physical objects in physics is practically impossible.

Mead: We had a terrific argument with the Balinese informant on the word »container,« about whether it was abstract or concrete.

PITTS: There is always the assumption that there exists a class of things which are invariant from sentence to sentence, which are names of classes constructed in some way, physical objects, people, and so forth.

LICKLIDER: It would appear that the invariant is closely related | to perceptible constancy. [183]

PITTS: There is a very simple way in which you can imagine any language based upon coordinates and properties of coordinates of, say, visual space.

LICKLIDER: This is an excellent example. It could happen logically but does not happen naturally, and the reason is, as I say, that all human beings have in common a brain that works in a particular way.

WIENER: In modern physics the noun »ether« has been replaced by the verb »undulate.«

PITTS: As a matter of fact, it would be very difficult to express in terms of physics a statement involving persons acting on objects, such as, a man kicking a dog, or something of that sort.

MEAD: There is no known language that does not contain that order of statement, no known natural language.

PITTS: The notion of a person acting upon an object. Well, I am sure that is not the notion of a personal action.

MEAD: It occurs in all languages.

PITTS: As something that a man does?

MEAD: The general predictive statement in which you have a subject, an action, and specific circumstances – which may include an object – occurs in all known languages.

PITTS: Certain relations are not universal. To be able to say that so-and-so acts in a certain way on something else, where the form of acting of the two parties are entities of any kind that you like, is apparently always true in all natural languages.

MEAD: It seems to be true in the real world.

PITTS: It is difficult to say. Suppose you have a movie film; you don't know if any person involved in the object has any idea, series of shapes, or not. You see how the shapes move. You don't know whether any action is involved. You see what I mean? You have colored shapes on the film. You see that the colors, the shapes changed. You can imagine all as amoebae in various ways. You don't know whether any action is involved.

MEAD: You can say that too. You just said it. This does not alter the fact that all languages do deal with predictive statements.

PITTS: That is a very difficult concept.

MEAD: Arbitrary.

PITTS: Suppose you have three patches of amoeboid color on a movie film in such a way that they changed shapes and relative positions, possibly their colors, too, in a more or less arbitrary sort of way, but more or less continually. Then say if all three of them were circular and one bounced from one to the other, you | would say an action [184] had taken place, or that one had pushed the second toward the third. If they were all amoeboid, you would not say that action had taken place?

WERNER: The direct perception of causal relations by means of moving balls is exactly what has been studied by Michotte in his work, »La Perception de la causalité,« *Etudes psychologiques,* VI, Louvain (1946).

PITTS: That seems to be a fundamental notion.

WERNER: In his set-up Michotte had, for instance, presented two balls, one moving toward the other. Ball 1 may reach Ball 2, Ball 2 may move away from Ball 1. This simple visual presentation gives rise to the perception of Ball 1 acting on Ball 2, which is an immediate and unverbalized perception of causality. Michotte had also studied the threshold of such a causality perception: if the movement is too slow, such action is not perceived. I think Michotte makes the statement that the experience of causal relationships is not just a problem of verbal conceptualization but occurs on a much more fundamental level – a primary perception.

PITTS: Fundamental ideas which are apparently common to all languages without being necessary in any sense.

BATESON: I think it is necessary to say that. Well, for example, all people, as far as I know, have the notion of a person (which is not, on the whole, a starting-point notion) that is probably used for a very great deal of differentiation of objects, the notion of action, and so forth. You start from that level rather than from what is logically necessary in an abstract way.

PITTS: It is not a question of what you start from but of what is absolutely necessary. We are not concerned with what you start from in one sense of building out but with what we must have when we finish our construction.

BATESON: Those are two different questions.

PITTS: We can generally derive one from the other, making one more fundamental or the other more fundamental, but the important thing is that we have to end up with [what] is our minimum start.

BATESON: Those two questions become confused.

PITTS: I am sure we should keep them separate.

MCCULLOCH: I think we ought to start with the next presentation. Is there anything that has cropped up in the earlier period that we should get out of the way before we start?

FRANK: Licklider and Shannon yesterday gave us some very significant aspects of the symbolic process – we didn't recognize it as such – wherein the individual invests those [185] symbols with certain meanings and responds to them as symbols with meanings | he puts into them. Both papers, to my mind, significantly show the way in which symbols can be attenuated with the individual still responding appropriately. That is essentially the symbolic process, which is a circular process, as in all learning situations where we put meanings in the situation and then respond to what we have selectively perceived. Symbols then can become progressively attenuated.

BATESON: To illustrate: when the white man came to this country the Indians, in order to make war on him, seeing he had script, invented an ideographic script. It had a time period of about thirty to forty years, starting with beautiful drawings, say, of a bird with feathers. That script symbol degenerated to two lines representing a bird. The point, I think, is that as the recipient of communication learns the codification system in which the messages are being sent, it becomes possible to trim what appears to be a redundancy to a much simpler message form.

WIENER: That is almost universal, in alphabets and other scripts. Although the Chinese have done tremendous things in this system, the best example is the Egyptian hieroglyphics. The hieratic and the demotic script represent successive stages that can be followed in the degeneration of the script until we have something approximating confluent writing, the writing of the least words.

MCCULLOCH: The next thing that we covered, in substance, was the manner in which an adult, fully equipped, with the highest level of abstraction, manages to learn a new language.

What we wish to hear next, and what we should hear next, I think, is the genetic approach. Will you begin, Dr. Werner?

ON THE DEVELOPMENT OF WORD MEANINGS [187]

HEINZ WERNER
Department of Psychology,
Clark University

I HAVE been assigned to the task of discussing the problem of ontogenetic development of language, as well as its breakdown in psychopathological stages such as schizophrenia. As to the latter, I am not expert enough to discuss it further than it pertains to genetic problems; and as to the former, I shall limit myself to discussing developmental problems of semantics within the framework of an experiment I have conducted with a group of children.

Most of you are accustomed to analyzing language in terms of normal adult language. In my discussion, however, I shall use as a starting point certain characteristics of nonnormal verbalized thinking, that of schizophrenics, which was originally described by Domarus (1) and rather recently reformulated by Arieti (2). To show the essential difference in thinking between the normal and schizophrenic, Arieti points to syllogistic thought structure, for instance, presented in the *modus barbara*: All men are mortal; Socrates is a man; Socrates is mortal. In other words, the reasoning presented by this syllogism is in terms of identifying the subject, Socrates, with the object, men, and drawing conclusions from this identity. This is the type of reasoning which characterizes normal logical thought. Arieti defines it as the principle of identical subjects. In contradistinction to this principle he speaks about the »paleological« principle of primitive, schizophrenic thinking. »Whereas the normal person accepts identity only upon the basis of identical subjects, the paleologician accepts identity based on identical predicates.« Suppose that the following information is given to a schizophrenic: certain Indians are swift; certain stags are swift. From this the schizophrenic may conclude that Indians are stags. This conclusion, which to a normal person appears as a delusion, is reached because of the identity of the predicates in the two premises. This is a very interesting analysis; but I am doubtful whether the interpretation given here, that is, the interpretation in terms of predicative thinking, is final. | [188]

It seems to me that the analysis can be carried a step further. Arieti's statements seem to imply that logic is entirely independent of language. Before drawing any conclusions, however, as to what constitutes schizophrenic thought, we first have to inquire into the nature of the language used, the character of words and their meanings. We have now enough evidence that the connotations of words as the schizophrenic uses them are much less »lexicalized,« much more »holophrastic,« as I use the term (3). By holophrastic expression is understood the fact that words carry meanings far beyond the conventional connotation; they have an associative fringe, and therefore to state adequately what the meaning of a single word may be, one might have to use a whole sentence or even a paragraph. Coming back to the example of Arieti, illustrating so-called »predicative identity,« the question arises whether or not this analysis is not too much in terms of that of the normal adult rather than that of the schizophrenic, whether it is adequate to talk about subject and predicate as if a sentence were analyzable into grammatically distinctive functioning words. I would rather assume that holophrastic grammar does not necessarily distinguish between a word and the linguistic context in which it stands. Consequently in the premise, »Certain stags are swift,« »swift« in itself carries the connotation of the whole sentence, that is, »swift« is identified with staglike; and coming back to the premise, »Certain Indians are swift,« it is readily conceived as »Certain Indians are staglike.«

We can probably state with Arieti that the type of logics presented by his schizophrenics is a logic which is primitive with respect to the logics in the sense of syllogistic thinking. That would mean that in the development of childlike thought one should meet similar forms of thinking. Since we mentioned that at the basis of this type of schizophrenic thinking is probably a primordial word formation which we have termed holophrasis, one would expect to find these primordial thought processes directly related to the development of word meanings in the child. Instead of presenting to you a survey of this actual development, I offer results of a test that by its nature should throw light on the development of word meaning. The test consisted of a presentation of twelve sets of six sentences each. The child's task was to find the meaning of an artificial word which appeared in the six different verbal contexts; for example, the artificial word in the first set was »corplum,« for which the correct translation would be »stick« or »piece of wood.« This is the test (4): | [189]

WORD CONTEXT TEST

I. *corplum*
1. A corplum may be used for support.
2. Corplums may be used to close off an open place.
3. A corplum may be long or short, thick or thin, strong or weak.
4. A wet corplum does not burn.
5. You can make a corplum smooth with sandpaper.
6. The painter used a corplum to mix his paints.

II. *hudray*
1. If you eat well and sleep well, you will hudray.
2. Mrs. Smith wanted to hudray her family.
3. Jane had to hudray the cloth so that the dress would fit Mary.
4. You hudray what you know by reading and studying.
5. To hudray the number of children in the class, there must be enough chairs.
6. You must have enough space in the bookcase to hudray your library.

III. *contavish*
1. You can't fill anything with a contavish.
2. The more you take out of a contavish, the larger it gets.
3. Before the house is finished, the walls must have contavishes.
4. You cannot feel or touch a contavish.
5. A bottle has only one contavish.
6. John fell into a contavish in the road.

IV. *protema*
1. To protema a job, you must have patience.
2. If a job is hard, Harry does not protema it.
3. Philip asked John to help him protema his homework.
4. John cannot protema the problem because he does not understand it.
5. A child should try to protema his homework when it is only half done.
6. The painter could not protema the room because his brush broke.

V. *ashder*
1. A lazy man stops working when there is an ashder.
2. An ashder keeps you from doing what you want to do.
3. Mr. Brown said to Mr. Smith, »I don't think we should start with this work because there are ashders.«
4. The way is clear if there are no ashders.
5. Before finishing the task, he had to get rid of a few ashders.
6. Jane had to turn back because there was an ashder in the path.

VI. *soldeve*
1. The dinner was good, but the fruit we ate was soldeve.
2. When we were driving in the evening, we did not feel safe because things on the road seemed to soldeve.
3. The older you get, the sooner you will begin to soldeve.

4. Most people like a blossoming plant better than a plant that is soldeve.
5. Putting the dress on the sunny lawn made the color of the cloth soldeve.
6. Because the windshield was frozen, things looked soldeve. | [190]

VII. *sackoy*
1. We all admire people who have much sackoy.
2. Have sackoy when you start to do a hard job.
3. When you have done something wrong and you are not afraid to tell the truth, you have sackoy.
4. A person who saves a baby from drowning has much sackoy.
5. Soldiers must have sackoy when they are on the battlefield.
6. You need sackoy to fight with a boy bigger than you.

VIII. *prignatus*
1. Boys sometimes prignatus their parents.
2. Mary did not know that Jane used to prignatus.
3. Mother said, »Jimmy you should never prignatus your own mother.«
4. You may prignatus someone but you will not get away with it often.
5. A good man who tells the truth will never prignatus you.
6. If Bob prignatus somebody, he makes sure they don't find out.

IX. *bordick*
1. People with bordicks are often unhappy.
2. A person who has many bordicks is not well liked.
3. The plan to build a house was a bordick because it cost too much.
4. People talk about the bordicks of others and don't like to talk about their own.
5. A person has many bordicks because he doesn't listen to wise men.
6. Your work will not have a bordick if you are smart and work hard.

X. *lidber*
1. All the children will lidber at Mary's party.
2. The police did not allow the people to lidber on the street.
3. The people lidbered about the speaker when he finished his talk.
4. People lidber quickly when there is an accident.
5. The more flowers you lidber, the more you will have.
6. Jimmy lidbered stamps from all countries.

XI. *poskon*
1. You should try to give poskon to other people.
2. If you believe in poskon, you are a good person.
3. The children will like that teacher because she believes in poskon.
4. People will always be afraid when there is no poskon.
5. Some bad people do not like poskon because they don't want to be punished.
6. There is no poskon when a thief is not punished.

XII. *ontrave*
1. Ontrave sometimes keeps us from being unhappy.
2. If you ontrave a good mark, you must also work for it.
3. We ontrave good things to happen to us.
4. It is silly to ontrave things that are not possible.
5. Johnny ontraved that Mary would like him.
6. According to what the doctor said, the children could not ontrave that their mother would get well.

The children taking the test were between eight and one-half years and thirteen and one-half years. The way the child, going from one context to the other, forms an overall meaning reveals | processes which in many ways resemble the semantic processes [191] which one finds in early childhood, under natural conditions, when the child begins to build up word meanings in varying contexts. What strikes us most – and this is something I shall discuss a little later – is the fact that the word meanings, conceived particularly by the younger children, appear to be more or less holophrastic, that is, in many cases the child does not distinguish between the meaning of the word and the meaning of the sentence. Though in other cases the child may distinguish between the word and the sentence, the child may still conceive of the word as containing a meaning which is so broad that in adult language one would have to use a whole phrase or a whole paragraph to define that word adequately. We may turn for a moment back to the early development of speech in the child. Here we find, and this has been stated over and over again, that the first meaningful utterances are extremely difficult to describe in terms of our grammar, that is, we cannot state the grammatical nature of the first meaningful sound patterns in terms of adult language structure. Psychologists have mentioned that such »early words« are not really words but represent sentences, and have called them »word sentences.« I don't think this an adequate term. I prefer to use the term »holophrasis.« If a child says, for instance, »chair,« and means by that, »I want to be put on the chair,« I should call this a holophrastic speech form, that is, a protoform out of which, later on, true words and true sentences emerge.

I may give you one example to show that there is an intimate relationship between the development of words as grammatical units and true sentence formation. I am choosing one of the examples presented by the French psychologist Guillaume. A child two years old understands the command, »Brush mama« (I use a brush and brush the mother); he also understands the command, »Brush papa.« However, the command, »Brush hat,« is not understood; the reason is that hat has a holophrastic meaning signifying, »Put the hat on the head.« We see from this example that in order to develop sentences that articulate into words, the word has to have some circumscribed connotation which makes it possible to relate it to other such units. The development from holophrasis to lexical units is an important problem of speech development, and it is this problem which we are particularly studying in our test.

Coming back to our test, then, we find that the responses of the younger children [192] abound with holophrastic expressions. For | instance, in series IX, sentence 4 reads, »People talk about the bordicks of others and don't like to talk about their own.« The correct translation of »bordick« would be »faults.« One child responds, »Well, ›bordick‹ means here that people talk about others and don't talk about themselves.« The child conceives of the meaning of the artificial word as being identical with the sentence. We can test our interpretation by seeing how the child fits the meaning of »bordick« into sentence 1, »People with bordicks are often unhappy.« The child fits his concept of »bordick« into that sentence in the following way, »People talk about others and don't talk about themselves and are often unhappy.« To the question, »How does this fit?« the child has this to say, »Say this lady hears that another lady is talking about her, so she will get mad at her and that lady will get very unhappy.« You can infer from this example that the child was identifying the meaning of »bordick« with sentence 2 and was pretty definite about it.

I should like to point out that this is but one form of conceptualization which indicates a lack of differentiation between word and sentence meaning. What I have called holophrastic gradient is another instance. Here a word is tied up so closely with its sentential surroundings that the latter are fused with the word meaning. For instance, series X, sentence 6, reads, »Jimmy lidbered stamps from all countries,« which the child completes as follows: »Jimmy collected stamps from all the countries.« »Collect stamps« becomes now an indivisible unit and is henceforth used as the meaning for the word »lidber.« For sentence 2 the child states, »The police did not allow the people to collect stamps on the street.«

WIENER: May I ask one question? I don't think this is altogether fair, because »collect« transitive and »collect« intransitive are treated as one word. A peculiar characteristic of the English language or of a limited group of languages is that there is this double use of a word as a transitive and intransitive. I do not believe there is any simple definition of a word which does not depend on a much more mature knowledge of the language than the child may be supposed to have, particularly in using the same word in a transitive and intransitive sense. Here »libder«[!] does not represent a notion; it represents an actual word in the English language, and there is no good logical definition of the word.

PITTS: The word is used in English transitively and intransitively. I thought the word was »gather,« which has the same general meaning. Transitive verbs may be used intransitively with very similar meaning in English. | [193]

WIENER: In English; but not every word can be so used. There are cases where transitive and intransitive words are different, »fall« and »fell.« This is not sufficiently universal to be done without some sophistication.

BAVELAS: Are you suggesting that the use of the word »lidber« in this case would be the same as the use of the word »pit,« meaning run, in two sentences, one meaning to run down the street and one meaning a run in a nylon stocking?

WIENER: The same degree.

PITTS: I think most children will have learned by a fairly early age that a very large number of common English words can be used in an intransitive sense and in a transitive sense that is the causative of the other.

WIENER: Yes.

WERNER: This is certainly true. I remind you that we are dealing here with children from eight and one-half years to thirteen and one-half years.

PITTS: As a matter of fact, it is a very common thing for children to do just that.

SAVAGE: Is the synthetic word supposed to correspond exactly to the English word?

WERNER: No, not necessarily. We have some examples here where it is almost impossible to do that.

SAVAGE: Your answer does a lot to restore my self-respect.

WERNER: In most cases, however, the meaning of the artificial word corresponds pretty closely to the connotation of an English word. We find that when the test is administered to adults they all come to practically identical solutions. During the development toward adult performance, there is a remarkable sudden change from word meanings used holophrastically to those used circumscriptively; I shall come back to this later on. As far as the particular example is concerned, and the criticism raised by Dr. Wiener, the criticism might be applied to that particular example but certainly not to the process, which I simply tried to illustrate by such an example. We have innumerable examples of the gradient process which do not involve the linkage of a verb to an object, for example, nouns combined with adjectives, or almost any two terms that stand together in a sentence.

WIENER: Let me point out that if you were to use a word such as »assemble« instead of »collect,« the first uses here would all be correct; but »assemble« in its intransitive meaning is the same as »collect.« The transitive meaning is different. In other words, you are getting the identity of a particular English word here | rather than the unit [194] which belongs to a thing, as it were, as your criterion.

WERNER: Again I should like to point out that an adult has no difficulty in abstracting a verb from its object, whereas a child has. Since time is short, I should like to go on to demonstrate another aspect of the close connection of the verb with the sentential context in which it stands. This feature I have called »embeddedness in sentence.« Let me first give you an example.

The child will, for instance, signify in series VIII, sentence 1, »Boys sometimes hit back at their parents.« In sentence 2, »Mary does not know that Jane used to lie.« In sentence 3, »Mother said, ›Jimmy, you should never holler at your mother.‹« In other words, the child signifies in various sentences by using different words which have some common denominator. Now sometimes you will find that the child will express this common denominator in very general terms. If you press a child for what »prignatus« in this particular instance really means, the child says, »Not to be nice.«

KUBIE: Does this tendency to throw up a different explanation for each sentence arise more in certain series than in others?

WERNER: On the whole, no. Of course the example just given is but one of many. There were quite a few children who responded similarly to this series.

KUBIE: Don't you think that the fact that the word »prignatus« is pregnant with meaning for every child between the ages of eight to fifteen has something to do with it? You can't possibly get a child to react with a stimulus of that kind without mobilizing every process in him.

WERNER: That might be so, but I don't feel qualified to elaborate on this question.

MEAD: »Prignatus« is a wonderful word.

KUBIE: Look at that context: »Boys sometimes prignatus their parents; Mary did not know that Jane used to prignatus; you should never prignatus your own mother.«

STROUD: I wonder how that series happened to be constructed.

WIENER: There seems to be a good deal of constructive Rorschach in this.

WERNER: There is undoubtedly a good deal of projection. We did not study the projective aspect in particular because we were more interested in the formal aspects. I should like to add that other words, such as »poskon« in series XI and »sackoy« in series VII, yielded a great many instances of this word embeddedness that I am trying to discuss. | [195]

I shall characterize the essential feature of this embeddedness. In trying to find a concept which would fit into the various sentences, the child is hindered because for each individual sentence the word meaning that he has offered is specialized to such a degree that it cannot be placed in another sentence. Though all these individual meanings have a common denominator, this common denominator cannot be placed as such into the individual sentences. This feature thus possesses two characteristics: 1. it is a concept which in a vague way is common to the specific word meanings used for the various test sentences, and 2. the individual word meanings are so specifically fitted into each of the sentences, and the final concept is so general, that the latter is incapable of replacing the single solutions. Therefore the common concept is, as I call it, »pluralized« in order to fit into the various sentences. We can speak here of »pluralization« as a protoprocess of true generalization.

I shall go on to talk of another aspect of the test, that is, the way the sentences are apprehended. We find here a lack of semantic articulation. As McCarthy (5) has shown, in early childhood sentences may be semantically, though not grammatically, complete. There is a tremendous development between one and a half years to nine years in the grammatical articulation and completeness of sentences, that is, from 80 per cent incomplete to 33 per cent in standardized situations.

MCCULLOCH: Perhaps I have misunderstood. Do you mean that the child uses a grammatically incomplete sentence?

WERNER: That is correct.

MCCULLOCH: You do not mean that what he says is not for him a whole?

WERNER: On the contrary. The study by McCarthy shows that children will use sentences in a structurally incomplete way, though as far as the semantic aspect of it is concerned it is complete. The child actually makes a semantically complete statement.

PITTS: What proportion of the statements made by adult human beings are grammatical sentences?

STROUD: I dare say if my own transcript was a sample, I must be very childish. I don't think I hit the ninth-year level as a rule.

WERNER: You are modest.

STROUD: You did not see the transcript. It was really terrible.

MEAD: It seems to me that the statement, »Children don't complete sentences,« is a statement about written language, and not about adults speaking. Any stenotype record of almost anyone's speech, except those few individuals who are accustomed to mak | ing formal presentations, perfectly rounded propositions, shows a very large proportion of unfinished sentences. Isn't much of this work that has been done on children's languages the comparing of children's spoken language with adult written language? [196]

WIENER: That is true.

STROUD: It is in most of the cases. I had the occasion to dig up –

WERNER: Unfortunately, as I said, we have no study on adults. I would very much like to have studies made on adults under comparable conditions.

MEAD: Something of this sort was done by Theodora Abel, who recorded conversations of adults, comparing them with the results of Piaget's tests on children. The difference between children and adults virtually disappeared; she found incomplete sentences, imperfect sentences, conversation without reference to what the other person said. A large amount of the statements about children that have been made by Piaget were found to be just as applicable to adults.

WERNER: I think Miss Abel used not highly educated adults.

MEAD: Educated high school students working with paper work.

MCCULLOCH: May I say that in the stenotype notes of our conferences Von Neumann almost never finishes a sentence and Wiener almost always does. I don't think you can hang much on the difference of the educational level.

BATESON: Except that Von Neumann is speaking a foreign language.

WIENER: Do you know if there was any substantial difference in the completeness of the sentences now and at the age of nine?

WERNER: It would be excellent to study the speech habits of individuals during growth. There is a tendency among child psychologists to include more and more longitudinal studies that will give us a much more adequate picture of child growth. Only a few places have started such an ambitious enterprise; California, for instance, has.

It is a question of expense and time. Cross-sectional studies that we use today are, of course, crude in relation to what one might obtain from longitudinal records. Controversies, such as those Dr. Mead just mentioned between Abel and Piaget, will probably be ironed out when results from longitudinal studies become available. Everybody will agree that it would be absurd to assume mental development if there were no differences between children and adults.

WIENER: I should like also to call attention to the fact that the | finishing of sentences is not an easily definable thing. When a man with the New England habit of talk says, »Looks like rain today,« is that a sentence or not? There is no »it.« He would not say »it.« »Think it is going to rain,« things of that sort. [197]

MEAD: They are sentences.

PITTS: It is a sentence, certainly.

WERNER: I should like to say that there is a definite method of approaching differences between adults and children. We will never find differences if we select atypical adult responses for comparison with the typical responses of children. In our language there exist many expressions which have the character of holophrasis or even of one-word sentences. But these don't occur typically in those standard situations such as were used in our comparative study between children and adults. An expression such as, »Oh, my God,« said by adults in an emotional situation should not be used for comparison with the productions of children under circumstances of a relatively nonemotional nature.

WIENER: I also want to say – and I think this is along your lines, Dr. Mead – that the questions of what is a word, what is a sentence, differ by language. What is an isolated phrase?

MEAD: The idea of sentence is a construction of our own particular linguistic form.

WIENER: Culture to word, too.

MEAD: I shall give you an example that I think will show it up quite sharply even in our own society. One of the things that Americans who have studied in England say characteristically is, »I wish I could go back to Oxford and/or Cambridge and finish my sentences.« I heard that a great many times, and I thought I would find out what it meant. I then listened to conversations among Cambridge men, and it seemed as if they never by any chance finished a sentence. A good conversation between two or three people who have gone to Cambridge together consists of one person saying three words, the next putting in another bit, the next putting in another bit, with nobody finishing his own sentence. A characteristic, on the other hand, of an American form of conversation, even among people who are accustomed to working together, is that of not listening to what the other person says except to find an opening. You wait for an acceptable end to what they are saying and then shoot your sentence in. The truth of the matter is that what those Americans meant was, »I wish I could go back to Cambridge, where people listened to what I said and where what they said felt like the end of my sentence.«

[198] Now under those two circumstances what is a sentence? A | sentence is an extraordinarily formal unit which we have devised in our handling of language. It is related, of course, to written language, to punctuation, and I don't think the question: What is a sentence? is at all meaningful except within a specific cultural context; it is not even meaningful between Englishmen and Americans, you might say, except in terms of written language. It is a formal but not very useful category.

PITTS: You might say a sentence is something that is capable of being true or false.

MEAD: That is a predictive statement. A sentence is a device or form for writing language and punctuation.

I should like to go back, in just that connection, to the point you made a little earlier about the use of the word »collect.« It seems to me that this is perfectly fair if what we are studying is the way a child learns English, because it is of the nature of English to have words both transitive and intransitive that have four or five meanings. Therefore, these combinations with the exception of two – may I please ask you what »hudray« is?

SAVAGE: Increase.

MEAD: You cannot use increase that way.

KUBIE: Enlarge.

MEAD: This is not a single word.

SAVAGE: Not a single word?

STROUD: Increase, one grows.

WERNER: It doesn't make any difference as long as the child uses words which are in the appropriate meaning sphere.

MEAD: It does not need to be a single word?

WERNER: We are only interested in determining if it is in the appropriate sphere; »Increase,« »enlarge« »get bigger,« and so forth, would be all right.

MEAD: You don't mind if it is not a single word?

WERNER: I don't mind whether the child presents a solution in a word or in a phrase so long as we determine whether the meaning sphere is correctly reached or not. However, the question of a single word representing a sentence rather than a circumscribed meaning is something else again.

MEAD: So you see, I think we can ask the question: What do children do in our society where we have an abstraction called a sentence and where we teach abstractions called sentences, where we teach punctuation in relation to sentences? We say: that is not a complete sentence. We send things back from the editorial offices because every sentence must have a verb unless it is an exclamatory sentence. Then it is a relevant inquiry to find out | at what age children complete sentences or at what high level [199] people give up completing sentences. I think that is highly cultural. In Chinese the problem would probably be entirely different.

WERNER: This brings up the extremely interesting problem of speech development in cultures where the language is of an entirely different type than ours. I wish some comparative studies would be made which would make it possible to generalize findings concerning speech development of the child.

WIENER: I also want to say something about the word »word.« In Chinese you have an entirely different meaning. Take, for example, the word *shia*, which may mean »down,« which may mean »beneath,« which may mean »to lower,« and so on. In Latin *faratultatuum* is one word, because it has been a local tradition in the Latin language to use pieces of different words for one word.

KUBIE: We must also bear in mind the fact that in childhood there is a *pre*verbal language, and that the child acts out a great deal of his speech, only gradually substituting words in a sentence for activity. He does this only as he begins to inhibit activity, which then must be put into words. If the child takes a piece of bread and walks toward you with it, saying »Bread,« he is saying to you, »I want to give you this piece of bread«; but he only *said* one word out of that sentence. There is a sentence-function in what he is doing, if you take in the totality of his behavior and language.

PITTS: Everything a man does is a language, anything he does is. That robs the concept of any usefulness.

KUBIE: As we develop, we substitute for behavior more and more complex and subtle indices and codes, which become the fully adult elaborated language.

WIENER: Many things we would call words, I think, in Chinese, according to our way of thought, will be combinations of characters in Chinese.

PITTS: You can certainly say it is necessary for the concept of language. It should be a substitute for something else. I am not sure exactly what else.

MCCULLOCH: Please be brief, because I should like to get back to the main presentation.

TEUBER: I want to get back to what Dr. Werner said; I was bothered when we began to discuss the question whether »sentences« or »words« could have the same meaning in different linguistic contexts. If I understood Dr. Werner correctly, he | wanted to [200] say that early sentences in the individual development of language are incomplete rather than unfinished. In the language of the psychopathologist, you are talking of some sort of agrammatism, or paragrammatism. I thought that was the thing you wanted to convey.

WERNER: It is exactly this problem of lexicalization and grammatization as a developmental process that I had in mind.

TEUBER: So it is not a question of an unfinished sentence left dangling in mid-air, a sentence that can be completed by anyone who listens, but rather a question of having parts of a sentence differently put together, and some parts left out.

WERNER: The problem of incompleteness has some bearing on Mr. Kubie's question concerning the relation between gesture and language. He correctly points to the developmental feature which I would call increase in verbalization, that is, the fact that communication is increasingly based on verbal expression rather than on gesture, action signals, and so forth.

I'd like to go on now with my general presentation and say something about grammatization, which ties up with what we've been talking about. We find that the younger subjects in our test do not take the sentence as a closed and stable semantic unit. They may combine parts of one sentence meaning with another sentence meaning, a linguistic feature which I call sentence contamination. For instance, in the case of the word »hudray,« one child did respond to the first sentence with »healthy«: »If you eat well and sleep well, you'll be healthy.« For the second sentence the child stated, »Mrs. Smith wanted her family to be healthy.« In the fourth sentence the child stated, »You read about health and you know by reading and studying about health.« Coming back to the second sentence, the child made this statement, »Mrs. Smith wanted her family to be healthy and to read books that teach you to be healthy. The book says, ›Brush your teeth three times a day,‹ and if you brush them only once you won't be healthy.« The child finally comes to the conclusion that all these sentences deal with »health books.« This illustrates well the semantic instability of sentences for the child. This makes possible the meaning assimilation of various sentences.

One of our very remarkable findings is the rather sudden drop of a great many more immature features of language behavior between the years ten and a half and eleven and a half, such as word-sentence fusion, sentence contamination, and the like. That is, after eleven and a half, very few of these grossly immature language characteristics were found. | [201]

MEAD: Is it the same for boys and girls?

WERNER: The same for boys and girls. The sudden drop holds only for the most immature features; there are other immature features where the drop is more gradual. In other words, mature lexicalization and grammatization of speech behavior on the whole show sudden emergence.

BATESON: Is it at that age that the child learns to separate the notion of lexiconization from other notions?

WERNER: There are practically no incidents after that age level.

MEAD: Do they all go to the same school or are they distributed?

WERNER: Most of our subjects came from Brooklyn. We also studied children in Ann Arbor public schools.

Though we had some older children who still showed grossly immature characteristics indicating lack of lexicalization, their number was very small, and those children who did show these characteristics did so very infrequently. At the thirteen-year to fourteen-year age level, only 2 per cent of the responses showed these signs of immaturity compared with 33 per cent at the level of eight and a half to nine and a half years. At this higher level there was practically no lack of differentiation between word and sentence; there were very few of those word gradients that I mentioned; there was little of the instability of sentence meaning and sentence structure which leads to contamination of a part of one sentence with a part of another, and so forth. In other words, at ten and a half to eleven and a half years, something crucial in mental development must occur which is reflected in the sudden change in language behavior. I'm inclined to believe that this change is due basically to the understanding of the abstract nature of the test sentences or, generally speaking, to the acquisition of an abstract symbolic attitude. In order to arrive at correct solutions with a minimum of immature processes, one has to assume a hypothetical attitude removed from the concrete world of action. In speech behavior involved in everyday situations, such an attitude may not be necessary. The child may be able to get along without his lack of this attitude being brought out. But this test, since it demands the abstract symbolic attitude, demonstrates by what it evokes from the child, the lack of this attitude. As is well known, the lack of abstract symbolic behavior is a feature occuring in certain psychopathological states, and I feel that a test like the present one would be valuable in testing immaturity or symbolic behavior in aphasiacs, mentally retarded cases, schizophrenics, and the like. | [202]

PITTS: It sounds very much as if the child had learned what a mental test was in between.

WERNER: In order to solve the problem you have to know –

PITTS: What the riddle is.

WERNER: Yes, you have to know that test sentences are removed from everyday speech.

MEAD: Yes.

WERNER: As I mentioned before, they are to be conceived as hypothetical statements. I should like to give you an example showing what a child might say. In series II, sentences 2 and 6 were considered by a child. Sentence 2: »Mrs. Smith wanted to feed (hudray) her family.« Sentence 6: »You must have enough space to fix (hudray) your library.« Coming back to sentence 2, the experimenter asked, »How does ›fix‹ fit here?« The child said: »No, it doesn't fit. Mrs. Smith wanted to feed her family. She couldn't fix her library at the same time while she is feeding her family.« We see here that the child viewed these sentences as concrete parts of a realistic story. With such an attitude the child cannot solve a problem of this sort. You are, of course, right, Mr. Pitts; the child does not understand the nature of the test. However, it is the job of the psychologist to analyze further and find out the reasons causing this lack of understanding. Doing so, I believe we arrive at the conclusion that this lack is due to a concrete attitude toward language rather than a hypothetical one.

WIENER: Take a German child, or a child in any country where a good deal of teaching of foreign language is instilled, where familiarity with the use of grammar, and with test sentences for grammar, begins, say, at six. I cannot imagine the child not having an idea of what is meant by this.

MEAD: I don't think it interferes if we keep it clear that these are American children in an American school system who are learning what the system will do to them, becoming test-wise. What we are studying here is at what age American children faced with this situation suddenly begin to adapt to it, and it seems to me that we may make a major generalization. What the children really say to themselves is, »This is something like a test in which a meaningless word has been inserted in a series of sentences, and I have got to find out what it means.« At that point they stop doing all these other things which Professor Werner has described. They now correctly define the situation. If you took the same child with a high I.Q. at six and gave it the same test and said: »Somebody has made up a word. They have made up a word and have put it in many different sentences in such a way that you | have to look at each sentence, but do not guess what the word is until you have looked at them all,« the child could then do it earlier. [203]

WERNER: Again I should say that without any doubt the mental age level or I.Q. has its bearing upon solving such a test adequately. But the developmental psychologist, as long as he does not act as a psychometrician, is not satisfied with stating such a relationship. Without going further into that problem, I should like to point out that mental age is only superficially defined by the overt achievement. Basically we have to define it by the underlying patterns of functions, attitudes, and so forth. We also have to deal with individual differences, influences of home environment, and again these influences are translated into the previously mentioned patterns of function; it is then the job of the developmental psychologist to determine the nature of these functions by adequate methods.

PITTS: I think that by eight and one-half the child has almost learned language completely. The language habits are practically the same as those of the adult – I mean the really important things he has always learned about the structure of language.

WERNER: This may be true in a limited sense. As far as language as a tool for everyday communication is concerned, you are possibly correct to assume that there is little difference between the child and the adult; at least there is no evidence one way or the other. However, what we do learn from a test like this is that with respect to situations demanding abstract symbolic behavior, there is a fundamental difference between children at eight and a half and children at thirteen and a half.

SAVAGE: We have to find out what »soldeve« is.

WERNER: That is one of the meanings which is not a conventional English word.

I should like now to come back again to the problem with which I started; namely, the developmental nature of the word and its relation to forming concepts and logical relations between concepts. In other words, the problem here is the relationship existing between child language and child logics. As you recall, I started with Arieti's supposition that logics can be determined independent of language. Such an assumption is unwarranted. I should like to show you that if words are conceived holophrastically, the relationship between concepts, the logical inferences, and so on, will be basically influenced by it. I should like to illustrate this briefly by an example which reminds us of Arieti's examples of the paleologic inference of schizophrenics. In series III, sentence | 3, a child had developed the concept »plaster« for »contavish,« stating, »Before the house is finished the walls must have plaster« (contavishes). According to the child »plaster« also fits in sentence 5 («A bottle has only one contavish«), in spite of his response, »A bottle has only one label.« [204]

BATESON: Good enough.

WERNER: That fits very well, as the child explains it, because »plaster« is used to put on the »label.« This is exactly what Arieti and Domarus have called predictive identification. However, if we examine more closely the reasons for this identification we find it is based on holophrastic word usage rather than on identification of circumscribed terms. The meaning of »contavish« is actually a global situation, »plaster + label,« and in a particular sentence one or the other may be more emphasized but never really isolated.

MCCULLOCH: I should like, in the next moment or two, to get a little more of a picture of how children learn language; I have asked Mr. Stroud if he would aid us. Then I shall ask Dr. Kubie to go on with the next topic.

THE DEVELOPMENT OF LANGUAGE IN EARLY CHILDHOOD

[205]

JOHN STROUD
U. S. Naval Electronic Laboratory,
San Diego, Cal.

IN ORDER to give my talk as rapidly as possible, I show you at first a chart (Figure 23) with some data on it. This is a calendar for the approximate ages at which development in language takes place.

I had to develop an hypothesis on which to hang the descriptive facts. I was plain foolhardy. The notion begins very simply. If the child expresses anything symbolically, it must be its own experience. The development of the symbol action is language in its broadest sense; it must therefore reflect the developing organization of his own experience. In one point of view, what you see developing in the language of the child is perhaps like the developing leaves on the end of a tree. It is impossible to set up a rigorous scale to show how this goes on in the baby until the point about which Dr. Werner was speaking is reached. The reason for this is that what you are looking at is the outside of a considerable underlying development. The general notion was that it seemed reasonable to infer (I won't say from what, for time does not permit) that the newborn infant is roughly capable of distinguishing between *things* and *not things*, and has feelings about –

WERNER: What are »not things?«

STROUD: Not things and things arise very simply. In the central nervous system the addresses of sensory endings are not entirely lost. Thus if a pattern is impressed upon some region, the bounds of the pattern roughly remain as bounds in their representation centrally. So that a *not thing* is that which is on the other side of the gradient. Not things for a child do not have any very elaborate or highly differentiated meaning. They are the raw given of his own experience, such as one might infer existed when he was born.

WERNER: You mean what some would call background experience?

STROUD: What you call background experience might very likely | be included in the not thing, but I would certainly hesitate to say that the not thing of the child is at any such sophisticated level as that. [206]

FREMONT-SMITH: Would you give an example for an adult of a not thing?

MCCULLOCH: Grope your way in the harbor in a fog and you find things.

STROUD: If you wander in a fog at night, there are vague things and vague not things, and you are not even sure of your ability to define. It regresses, shall we say, to your childhood levels.

KUBIE: The experience of darkness.

MEAD: Aloneness.

STROUD: Aloneness; the experience of something which might almost be a light in a very dark night, a light area in a dark area.

FREMONT-SMITH: One over a thing.

STROUD: Roughly one over a thing. It is rather difficult to believe, that is, I find it difficult to infer, that the newborn has any space. All that he has are things and not things;

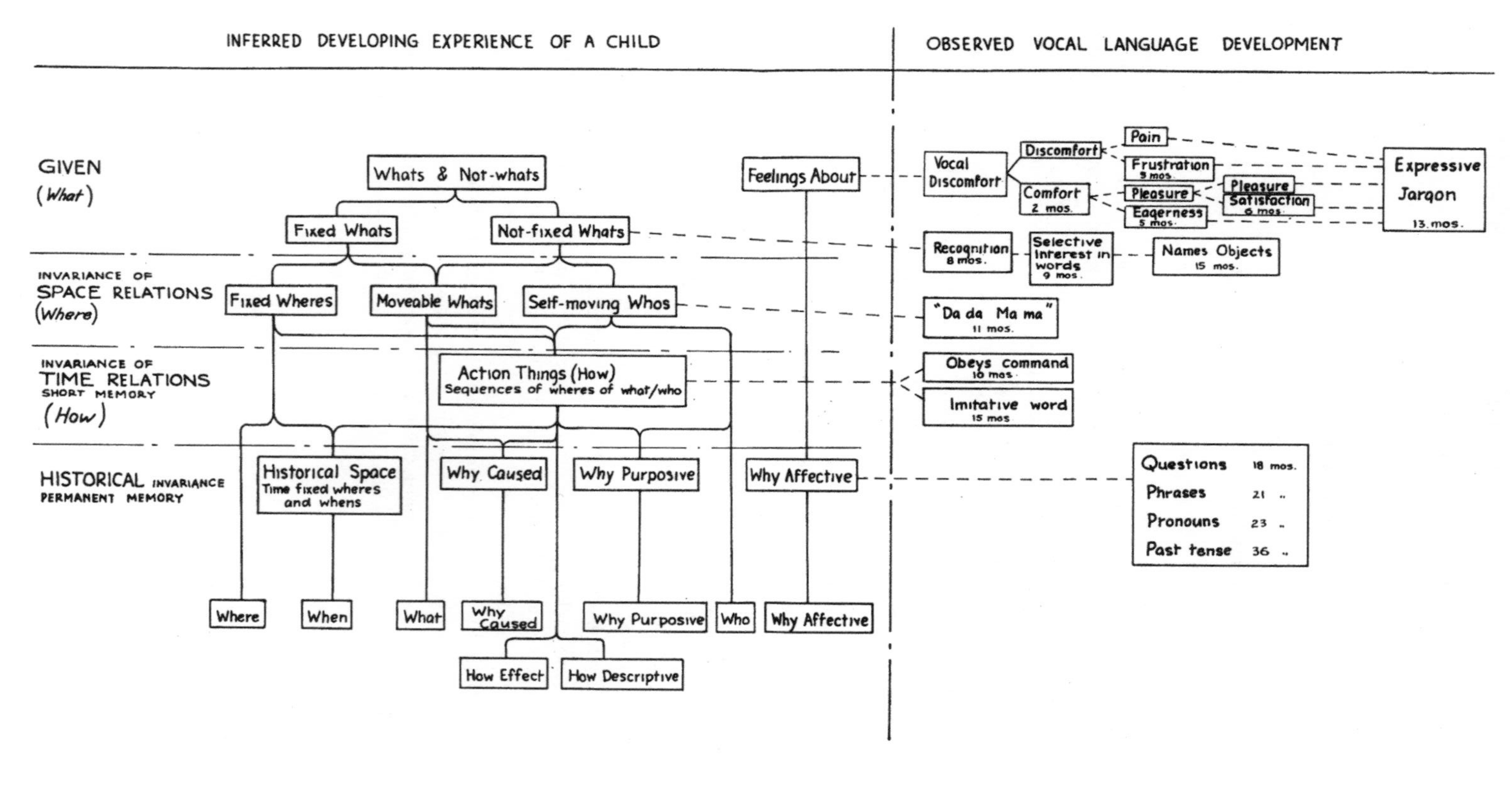
INFERRED DEVELOPING EXPERIENCE OF A CHILD
OBSERVED VOCAL LANGUAGE DEVELOPMENT
GIVEN (What)
INVARIANCE OF SPACE RELATIONS (Where)
INVARIANCE OF TIME RELATIONS SHORT MEMORY (How)
HISTORICAL INVARIANCE PERMANENT MEMORY
Whats & Not-whats
Fixed Whats
Not-fixed Whats
Feelings About
Fixed Wheres
Moveable Whats
Self-moving Whos
Action Things (How) Sequences of wheres of what/who
Historical Space Time fixed wheres and whens
Why Caused
Why Purposive
Why Affective
Where
When
What
Why Caused
Why Purposive
Who
Why Affective
How Effect
How Descriptive
Vocal Discomfort
Discomfort
Comfort 2 mos.
Pain
Frustration 3 mos.
Pleasure
Eagerness 5 mos.
Pleasure
Satisfaction 6 mos.
Expressive Jargon 13 mos.
Recognition 8 mos.
Selective interest in words 9 mos.
Names Objects 15 mos.
"Da da Ma ma" 11 mos.
Obeys command 10 mos.
Imitative word 15 mos
Questions 18 mos.
Phrases 21 "
Pronouns 23 "
Past tense 36 "

FIGURE 23

and looking on, we would say they change, but I doubt that they even change for a child right away – I mean, for a newborn infant. The only thing that you are reasonably sure about is that it can vocalize a simple dichotomy. It yells when it hurts vaguely and sleeps when it does not. There is vocal behavior at birth, and this particular vocal behavior you can follow, if you will note, on the chart. That is one behavior I did follow out well.

The first major transformation which gives any kind of invariant in this experience of the child occurs when it learns how to transform its experience so that it achieves the invariances of space; its former things and not things can now become fixed things and not fixed things, because after a child has learned the trick of transforming his central nervous system input to give invariance in space, there are still moving objects which will not assume fixed position in his recently and primitively created space of perception. This can again develop in such a way that he learns to move his body and to manipulate; the very fixed things become very primitive wheres; the manipulatable things become movable things. There is still a broad class of objects which do not remain fixed but are not manipulated; they are self-moving whos in the most primitive sense. Having managed this first-order transformation giving him his first set of invariants, the invariant relationships in space among unmoved things and fixed wheres, he has the possibility, with his developing short-term memory, to hold the sequences of wheres, of a movable who or what, and he develops | »action things,« like go, run, hit. [207] From such notions of a fixed sequence of wheres, which can be classified or described, from the notion of such a fixed sequence of words of a what, the most primitive notion of time develops. I would like to point out that whether you look at the wheres of the what clock hands or whether you concern yourself with the wheres of the maxima electric vector, in some very fancy oscillatory circuit, which will keep times in seven decimal places, it is still the means by which we define time.

At this stage the infant has action things. With the development of a long-term memory he can congeal these action things and be fixed wheres and the sequences of wheres associated with whats into a four-space framework within which the adult of stored facts can be established, that every fact was what, where, when. This, as I see it, is a very crude but practical primitive notion of the development of the child's handling of his own central nervous system. I don't say he is responsible for it. It is just a description of a process. It would take Solomon to tell whether his primitive action is effective, modifying his position in his world, or operates through the central nervous system of his parents and is thus symbolic. Out of these many actions some remain perhaps mere practice acts. Others are directly effective, and still others are effective through the medium of another life form or through himself at a later date, and these are symbolic acts. As the child develops adequate expressive jargon, he also develops what someone very nicely called whole body language, that is, actions which do not themselves modify the circumstance but can work through the medium of someone else. These become refined to a very effective gesture system at a time when a child has not reached the sophistication of developing more than perhaps a very few names for whats. His verbal language lags behind his developing ability to use his nervous system by many months. In between lies an intervening language which I regret to say has to my knowledge never been seriously studied. It had not as of 1946, after very considerable efforts of McCarthy to write her chapter in Carmichael. I have heard of none since.

As the child continues with his jargon, babble, inflections, gestures, and a few word names, these are soon reinforced by a few words for action things. It is dangerous to generalize from the data of one culture to another. I tried to abstract data that would be as general as possible. These words, as apart from the system of feeling expression, are added to, slowly, and never entirely supplant the original gesture language of the [208] child. |

The approximate developmental levels from mere things and not things and feelings to quite nice sorts of differentiations, as I tried to show, are by no means complete, but in spite of the relatively enormous number of factors which can influence the behavior of the child at any of the intervening stages, statistically this framework is still very largely held. The last major element of logical thinking, as we like to think about it, the manageability of tense, seems to appear at about three years. From this point on, in all the functional fundamentals, if not in all of the abstractions, a child has already acquired his first and fundamental verbal language, incomplete, inaccurate, and badly organized though it may be.

McCulloch: Let me say one more thing. If you read the older attempts to study the speech of children, you will see that we tend to attribute to a child's »why« our meaning of »why,« to a child's »no« the adult meaning of »no,« and so forth. What happens apparently is this – and I give the word »no« as an example – the word »no« has to be developed apparently rather early. A child can merely accept that which it wants, but it has to have some general form of negation. A general form of negation may at first simply take the form of shoving away; afterward it becomes »no.« That goes on to develop later into both the »no« of logic and the »false« of logic. A child cries, »Mama« which has potency over mama so that mama comes; if a child says the word »Mama« and mama does not come, the child will say »No mama,« which is the equivalent of the word »mama« being false. This is linguistically and semantically a higher level of problem than the »no« of not that object. It is already referring to the futility of the language. This kind of thing is altogether in the child in a state in which it does not submit to the interpretation which we later put on the word. The word »why« is the perfect example, because even in the adult the »why« is so confused again and again that you don't know what the adult means by it most of the time.

McCulloch: Dr. Kubie, will you speak next?

THE RELATIONSHIP OF SYMBOLIC FUNCTION IN LANGUAGE FORMATION AND IN NEUROSIS [209]

LAWRENCE S. KUBIE
Department of Psychiatry and Mental Hygiene,
Yale University School of Medicine

WHAT I have to say troubles me, partly because it renders what is already complicated even more complicated. Yet it must be in the picture as part of the totality that we are discussing and have been discussing for these last two days.

I want first to backtrack, so as to add to what has just been said. In 1934 I published a short paper, which this meeting led me to resurrect (1). It was called »Body Symbolization in the Development of Language« [*Psychoanalyt. Quart.*, **3**, *430* (1934)], and with your permission I will submit it to our chairman to be reproduced as part of our proceedings, since much of what I want to say is said there more concisely than I could possibly say it if I were to extemporize. The paper begins by pointing to Pavlov's evidence that new connections (conditionings) can be formed in the central nervous system only when the animal is not in a satiated state, and that consequently all processes of learning depend in some measure upon the existence of a state of craving. Then I pointed out that in various ways (which I will not go into here) analysis proves essentially the same point, that is, that it is only in the state of craving (of deprivation) that one establishes new connections and breaks up old ones. Next I point out that cravings arise in infancy and childhood in body tensions. Consequently it is inevitable that the child's first conceptual structure should begin around his own body, and that his first concepts must deal with the products and the apertures of his body, the needs and the feelings of his body.

Therefore to understand the growth of language we must begin by observing closely what the child wants, what parts of the body become involved in the process of wanting, and ultimately how he learns to represent in symbols these needs and the different parts of the body which are involved and associated with these needs. I emphasize this because my only modification of your | scheme is that it begins *too late*. It begins with [210] the recognition of the phenomenology of the outside world. That cannot come until the infant has already realized a good deal of the internal experiences with which his life experience actually begins.

STROUD: I did not wish to indicate that at the thing-nothing level there was any inside or outside. You cannot get inside or outside until you get to the space transform. It is a meaningless thing.

KUBIE: Let me quote from the paper I have referred to:

> Since the child's world begins inevitably with his own body and since the force which instigates the child to expand his knowledge is always the pressure of internal needs and tensions, since every new fact of experience which enters into our psychic life can make its entrance only by relating itself to that which is already present, then it follows that every new fact which is apperceived by the infant and child must somehow relate itself to bodily things. Schematically you can represent this by a series of steps. If this represents with any fidelity the process of expanding knowledge, it must also represent the process of expanding speech, expanding communicating measures, the meanings. There is at first a long period in which concepts are vague, broad, overlapping. With advancing years these concepts become discrete and distinct one from another. At some point during the early period they are represented by indefinite but meaningful gestures and movements and expressions. As time goes on they are represented by that symbolic code which we know as language.

That is all that I am going to quote. In the paper itself I give many illustrations of something which we all recognize in verse and story, and which we recognize in all the games we play with infants and children in which the parts of the body are named for external objects, animals, and so forth.

Now let me come back to what I had originally planned to say and which is going to make things seem to you even more complicated than they do already. We have to keep in mind the fact that the human organism has two symbolic functions and not one. One is language. The other is the neurosis.

To clarify my meaning, let me quote a few pages from an article now in press for the *United States Armed Forces Med[i]cal Journal* and entitled »The Neurotic Potential, the Neurotic Process, and the Neurotic State«:

I want to make it clear that when we say that there are psychological processes of which we are not aware, these may mean either of two things. There are two kinds of unconscious processes, one of which is readily accessible to conscious self-inspection, the other not. Even the simplest activities of life, such as breathing, sucking, excreting, moving and crying, must be learned through repetition. All of these activities were originally random and explosive. Through repetition, however, they become economically organized toward goals. As they become fully learned, [211] the act can be initiated simply by the contemplation of | the goal and we gradually become unaware of the intermediate steps which make up the act. This is the great economy that is achieved in learning by repetition. It is in this way that we learn to be able to walk without pondering each step, to talk without working out the movements by which we enunciate each word. It is in this way that the violinist and the juggler and the athlete learn complex chains of synergic movements. It is in this way that our thinking processes acquire Seven League Boots: i.e., the ability to leap over many intervening steps as we perform complex arithmetical processes, and the art of intuitive thinking. In each case the intermediate steps drop into the background and disappear from consciousness. Yet they remain accessible to conscious selfexamination. They are in what William James would have called the »fringe of consciousness,« or what Freud called the *pre*conscious (PCS) or the descriptive *sub*conscious, as contrasted with the *dynamic unconscious* (UCS).

The dynamic unconscious, then, is not merely a descriptive concept: it is an area of force or rather of whole constellations of forces in psychic life. Unconscious processes are constantly at work in our lives; but we cannot become aware of them by ordinary self -observation because they are hidden from us by active opposing forces within ourselves. Throughout life these UCS processes exercise a powerful influence on human behavior; and it is out of their influence that everything that is neurotic in human affairs has its origin. In this sense everything that we say and do and think and feel serves multiple functions and becomes a symbolic representation of both Conscious and Unconscious levels of psychological organization. From this we may go further and conclude that if the psychological conflicts of infancy and childhood could take place in the full light of consciousness, then the neurotic process would never be launched in human life.

Why we render psychological processes unconscious in this way is a question with which I will not try to deal on this occasion, except to say that it happens whenever we are unable to discharge internal tension because of opposing feelings of guilt and fear. This is when we render our conflict-borne tensions unconscious and then express them in the masking symptoms of the *Neurosis*.

THE SYMBOLIC PROCESS IN THE NEUROTIC POTENTIAL

The symbolic process which I have in mind includes the formal symbolism of dreams as a minor example, but is of far broader significance. The human being is capable of two related but differing types of symbolic processes. One gives him the ability to make abstract concepts of his experience, to represent those abstractions in symbols and thus to express and communicate his purposes, needs, thoughts, and feelings through behavior, gestures, sounds, words, and their written symbols. The other symbolic process is the one by which man expresses in disguised forms those psychological tensions which he is unable either to discharge or to face. The first is the symbolic process of self-expression through language; the second is the symbolic process of self-deception.

In the developing infant and child these two symbolic processes have a common origin; and the ability to represent internal experiences in symbolic activity is the *sine qua non* equally of the neurotic process and of speech. Whether either symbolic process is possible to a significant degree among animals lower than man is far from clear: which is why it is not certain that the so called »experimental neurosis« in animals (actually an emotional disturbance which also occurs in human neuroses), is identical with the neurosis itself.

Between the two forms of symbolic representation there is a difference about | which we can be quite specific. The difference between the representational process in communication and the representational process in the neurosis is primarily the difference between using a symbol for an internal experience of which we are or can be aware, and using a symbol to express an internal experience of which we are unable to become conscious. The capacity to create and use symbols (or »transforms,« as the engineers would say) is identical in both and is essential to both. The difference resides solely in the fact that the relationship of the symbol to the underlying psychological process is conscious in language, and unconscious in the neurosis. [212]

Consequently, the roles of the symbolic representative of these two types of »unconscious« experience differ. In speech the symbol is like the salesman who represents a firm that is doing a legitimate business. Its salesmen save the heads of the firm much time and energy, because they do not have to visit every customer themselves; yet these principals are known to the customers and can always be reached by them. On the other hand, there is another kind of »salesman,« i.e., the representative of a gang of criminals, or the secret agent of a foreign country. Even if he is captured, and even if it is known that he is the agent of criminals, he will not divulge their identity or whereabouts except under pressure, if at all. In the neurosis the relationship of the symbol to the inaccessible unconscious processes which it represents is of this nature. We must repeat, therefore, that if human beings were not able in the first place to abstract their psychological processes and in the second place to represent these abstractions symbolically, and if in the third place they were not able to render certain unacceptable psychological processes inaccessible to conscious introspection, there could be no such thing as a neurosis. Together, then, these three human capacities constitute the Neurotic Potential.

Thus human vulnerability to the Neurosis (i.e., the Neurotic Potential) arises out of our capacity for symbolic psychological function, without which there could be neither a neurosis nor a thinking process, but merely dreamlike sensory imagery, passive echoes of previous perceptions. Like the neurosis, planful action and speech require symbolic processes by means of which sensory imagery is taken apart and reassembled in new combinations.

Out of this matrix the *Neurotic Process* emerges gradually and progressively. It begins the first time that some complex psychological experience is hidden in such a way that all that is accessible to conscious introspection and all that shows to the world is the symbolic behavior which represents it. This representative or symbol or transform may be simple at first; but with the passage of time the same symbol can come to represent many hidden psychological states; and in turn there can be representatives of representatives, symbols of symbols of symbols; so that the complexity of such a linked chain of symbolic representatives can become very great indeed.

In speech, then, we use a symbol for something of which we are conscious. In the neurosis we use a symbol for something that we are unaware of; and the distinction between language and the neurosis is the difference between being conscious or unconscious of that which we are representing symbolically. The capacity to represent inner experience symbolically is necessary to both. But at some point a dichotomy occurs between the symbolic representation of that of which we are aware and the symbolic representa | tion of those things of which we are unaware. There are many unanswered questions about this: for example, must everything that we represent symbolically at one time have been conscious? As far as I know, this is one of those issues about which people have biases but little evidence. It is certainly true, however, that everything that a child does, including his preverbal activities, can be used in either way; whether this is rocking, crawling, walking, running, eating, excreting, grunting, spitting, or throwing. All of these can be used in infancy and childhood, both as a [213] direct expression of a conscious purpose or as an indirect expression of things of which the child is unaware.

How do we recognize the difference? Again in a simple and pragmatic fashion: by the automatic repetitiveness of the behavior when it is serving a purpose of which the child is unaware.

If we think of our own experiences with infants and children, we see this in the repetitive play of the child, and in his handling of his instinctual functions as well. Thus eating serves not merely the instinctual or biological need of the organism but also a vast superstructure of unconscious needs. This occurs both in the area which we call normal and in the pathological exaggerations such as compulsive and phobic eating; indeed, it occurs in all of the distortions of instinctual function that arise in childhood.

FREMONT-SMITH: Would you not agree that at a very early stage a given symbol may be used simultaneously for both conscious and unconscious?

KUBIE: Not »may be« but *always* is. There is nothing that we do that does not serve both conscious and unconscious masters.

PITTS: Is not your conception of unconscious like the vermiform appendix, in that it performs no function and becomes diseased with extreme ease? It is there for no discernible purpose.

KUBIE: I would not compare it to a vermiform appendix at all. Unconscious processes perform many functions, some of them constructive and useful, some of them destructive. Sometimes they are constructive, yet at the same time cost us a high price.

PITTS: That is the point. What are the constructive ones that are performed all the time? Roughly, what is the function?

KUBIE: I am not a teleologist. I do not begin with that question. I begin with the question whether it exists and how it operates. That it exists is experimentally demonstrable. I will give you a demonstration in a moment. In fact, I will show you that it has very powerful influence in human life. That is also demonstrable. Let me give you [214] two very simple examples. Both, so help me, are true stories. |

One summer night I arrived in the Grand Central Station from the country and got in a taxicab and asked the driver to drive me home. My home is on Eighty-first Street between Fifth and Madison avenues. For those of you who are not entirely familiar with New York I will make a diagram. I am going up Madison Avenue this way. Here is Eighty-first Street. Eighty-first Street is a street in which traffic is bound this way (indicating), but down here, in the sixties, the driver starts to go over here (indicating), which would bring him into Eighty-first Street against the traffic. As he started to swing in this way, I said: »Hey, where are you going. I said Eighty-first Street. You will enter it against the traffic.«

He said: »Right. I heard you say Eighty-first Street but here I have been thinking Eighty-second all the time. Now why do you think I was doing that?«

I said: »I don't know. Maybe you don't like the odd numbers.«

He stopped the cab in the middle of the street, turned around with his eyes bulging out of his head, and said, »Jeezuz, how did you know? I have been betting on them all afternoon and lost my shirt.« Then he proceeded and told me that he had been betting on the odd positions at the post in a bookie shop all afternoon, and that he had lost steadily; until now he could not go into Eighty-first Street and had had to go to Eighty-second instead.

This is interesting for many reasons. In the first place, there was no barrier, no resistance to insight. I only had to say that perhaps he did not like the odd numbers; he filled in the rest.

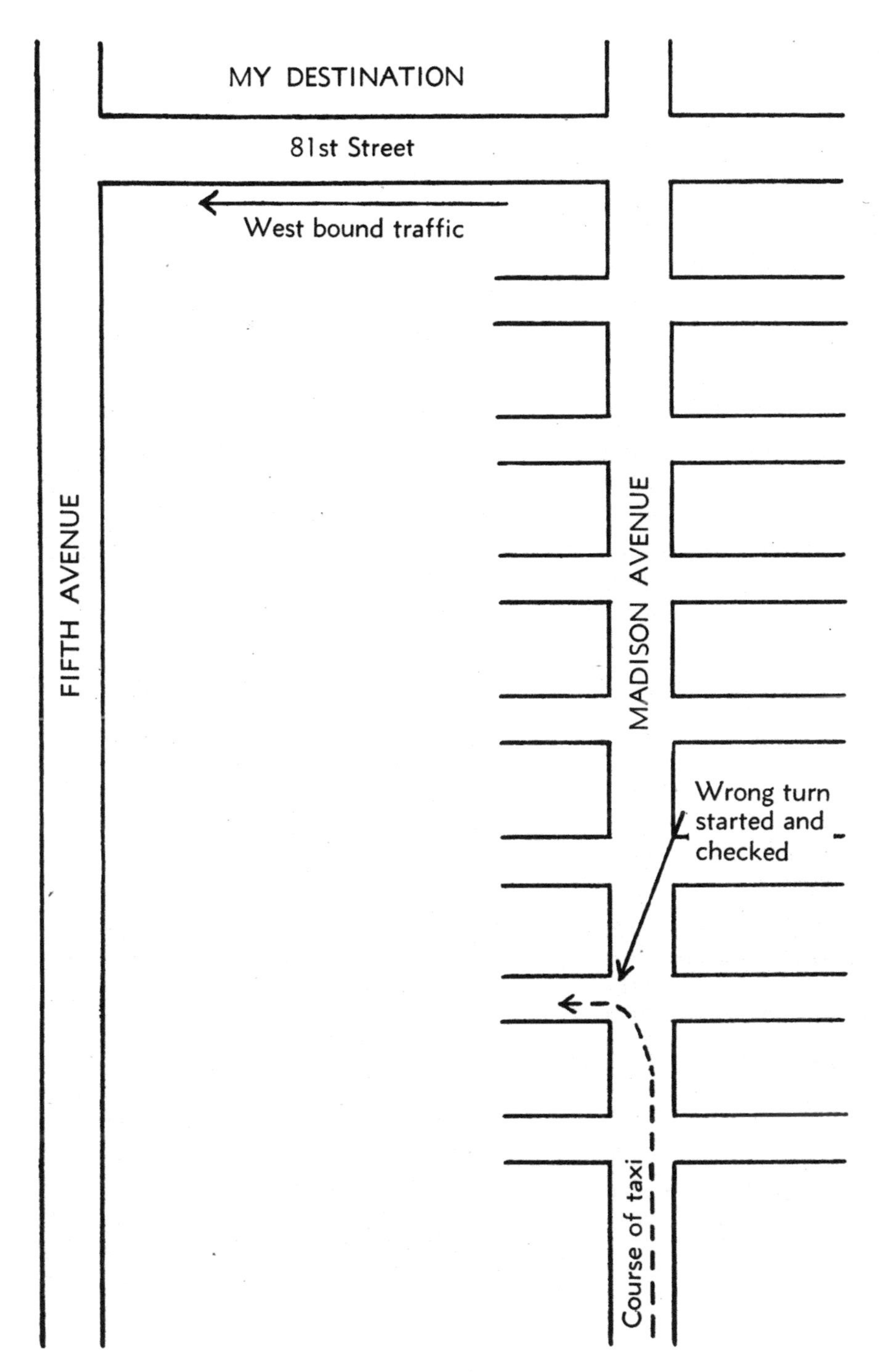
MY DESTINATION
81st Street
West bound traffic
FIFTH AVENUE
MADISON AVENUE
Wrong turn started and checked
Course of taxi

FIGURE 24

My second story concerns the airplane dream described in the paper to which I have alluded and which is appended to these remarks. Therefore I need not repeat it here.

PITTS: The whole style – the thing still does have, it seems to me, the peculiar flavor of the vermiform special organs which exist to be infected and have no function.

FREMONT-SMITH: Dr. Kubie did not explain what function it was performing. It performs a real and very important function.

KUBIE: It was performing an important immediate function in the behavior of the taxi driver. He was ashamed to go home because his wife would ride him like the devil for playing the ponies. A whole human drama was behind this episode, yet not deeply buried. The other example led back to the girl's most traumatic experiences, and to the deepest problems of her life. She did not want to face these at all. She had had to bury the experiences, yet at the same time they had to be represented indirectly in [215] some form that was going to get by. | [Figure 24] |

[216] | I would like to supplement these comments by presenting for your consideration a very brief statement on the role of *Emotions* among the feedback mechanisms. The statement is made in relation to the problem of the evaluation of so called »psychosurgery,« because the quotation is from my contribution to a recent U.S. Public Health Service conference on that topic. The specific references to the surgical problem does not, however, lessen its general implications:

THE ROLE OF EMOTIONS IN THE FEEDBACK MECHANISMS

Let us begin by considering the basic significance of emotions in human life. Although emotional states are themselves products of complex psychological processes, they are also *causal*, in that they exercise a vitally important *feedback* influence on psychic processes. In this circular or feedback function they are like the governor on a machine. Indeed, this is the major key to an understanding of the role of emotions in psychic life. Thus under normal circumstances one group of emotions (i.e., anger and elation) lends a quality to any psychological experience which makes us want to experience it again. In general, elation tends to have this effect consciously, and anger relatively unconsciously. By contrast, and still within normal limits, depression and fear give a quality to any psychological experience which makes us want to avoid its repetition: depression exercising this influence consciously, and fear tending to exercise its dampening, feedback influence relatively unconsciously.

As illustrated in the accompanying diagram (Figure 24), all emotional states can be grouped under these four major categories, each of which varies qualitatively within itself without losing its fundamental quality, and each of which can be combined in various more complex emotional states with one or more of the others. Certain combinations occur more frequently than others; but there is no combination even of seemingly opposite pairs, which is unknown to us both clinically and in daily life. In the diagram, anger and elation are above an arbitrary dividing line, fear and depression below the line, with crossrelations indicated by dotted lines. Psychosurgery frequently releases or facilitates the expression and/or the intrapsychic influence of those emotional states which lie above the line: i.e., of those emotions whose circular influence on behavior is reenforcing. This does not always occur, however; nor can we make any generalizations as to why this sometimes happens and sometimes fails. It must be borne in mind that the effects on the expression of an emotion and on its circular or feedback influence need not be identical. [217] [Figure 25]

| On the other hand, psychosurgery tends more regularly to decrease the influence and/or the expression of fear and depression (i.e., of those emotional processes whose circular influence on behavior is inhibitory). Again this generalization about psychosurgery is not an invariable rule.

From these facts we may draw one simple conclusion. If psychosurgery were attempted in a »normal« person (for instance, in an individual suffering from intractable pain), the most fundamental index of the influence of the operation on the role of emotions in psychic life would be to compare before and after operation the effectiveness of anger and elation in causing repetitions of experience, and conversely the effectiveness of depression and fear in causing us to avoid

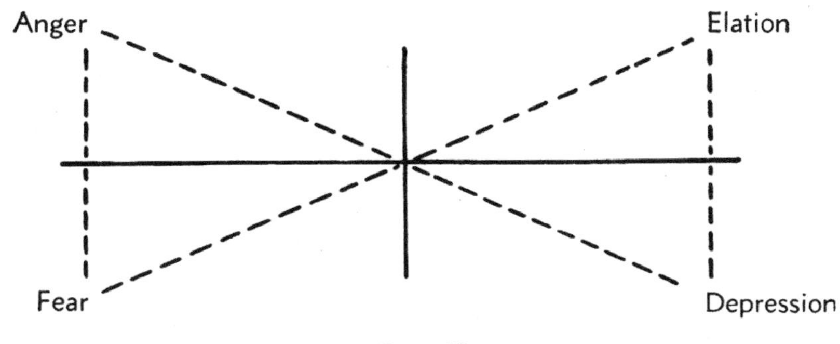

FIGURE 25

such repetitions. For example, we would expect that after psychosurgery the subject even in the face of fear or depression would be relatively complacent about repeating experiences which because of this fear and/or depression he would previously have avoided. This is the basic change to look for, and not merely the influence of the operation on the overt display of emotional fireworks.

PITTS: Suppose one did not have one, what would happen?

FREMONT-SMITH: That is like what happens to the rectangle when you remove the width. What kind of a rectangle have you got?

PITTS: It is not obvious to me why it is not conceivable to have the human being without one of these objects called the unconscious. How would he act and what would he do wrong?

FREMONT-SMITH: Let me give you an immediate answer. It would be impossible to focus attention on anything for even an instant. The only reason that any of us now focuses attention on anything at all is that unconsciously he excludes from his stream of attention all the inflow of impulses excepting the one to which he is directing his attention. That is a very positive activity on his part, and if it were removed he would be blurred and overwhelmed by the inflow of impulses, both from inside his body and from the external environment.

MEAD: Let me add that it is possible to conceive of a culture so built that it would be possible to take the human being at birth, equip him to handle and symbolize and deal simultaneously with far more of these impulses than we do now, conceivably even with all of them. There are some differences among the people in this room. Some in this room can handle six sets of ideas at once and some can handle only one. The people who can handle one would say they cannot concentrate if five other sets are introduced. Those who can handle five or six would still have to eliminate something else. We have not yet invented a society which makes it possible so to equip an individual that it is not necessary to have an unconscious in Dr. Fremont-Smith's sense. I don't think it is quite our | sense. For your sake, Dr. Pitts, I could invent one. [218]

FREMONT-SMITH: They merge. I think one could show, as Dr. Kubie said, that in the taxi driver there was very little that was buried. Everything was on the surface. The same thing is true, at the moment, of incoming impulses, for instance, the sound of the stenotypist's fingers, which you probably did not hear, but which, at this moment, you now can hear. That was not buried and could have been brought to your attention at any moment, and yet some active process excluded them so that you could listen or focus.

PITTS: The existence of that gives me no difficulty at all. It was that I had detected a difference between his concept and yours; it was about his that I had a queasy feeling, not about yours. I think that is inescapable.

FREMONT-SMITH: I think that you will find no dividing line between the two.

PITTS: He thinks there is.

FREMONT-SMITH: Let us ask.

KUBIE: What do you think I think? Tell me what I think.

PITTS: The reason I think that you happen to believe this is that we talked about something more or less similar last time and I think you ended in the position that there was a substantial difference between your notion of the unconscious and that of the other people who had one roughly of that kind.

KUBIE: What kind?

PITTS: In the sense that yours was somewhat more like an isolated box than theirs.

KUBIE: Unfortunately the unconscious can become isolated indeed. If it did not, there could be no such thing as a neurosis; but unfortunately neuroses are ubiquitous. What I am trying to say is that along with the development of the capacity to represent things symbolically (which is necessary for all of our highest human functions), there has arisen for reasons which are not entirely clear, this dichotomy between things which are represented and of which we are conscious, things which are represented and which are on the fringes of consciousness (such as the taxi-driver example), and things which are so deeply buried that we cannot get at them without some special method of excavation.

WIENER: The point is, then, that the second case, that of the woman, was between the two. You did not dig for it, but it was suppressed.

KUBIE: I dug for that by various indirect methods.

[219] WIENER: Yes. |

FREMONT-SMITH: But it came up partly spontaneously when the right time came, when life experience had changed.

KUBIE: When her capacity to face certain problems had been enlarged. When the guilt feelings about those problems had been diminished.

MCCULLOCH: Forgive me if I pull a little closer to the main theme. I am by no means clear in what sense a neurosis is part of the development of language in the sense of symbolism? Is it not possible for a neurosis to exist quite apart from symbolism? Is it of the essence of a neurosis that it is necessarily symbolic? Even if it is necessarily symbolic, what is the connection with the form of symbolism that we call language?

MARQUIS: May I add my question to that please? Has the communication that is added to the neurosis the same connection that the neurosis has?

GERARD: Could you have, conceivably, a social organization in which the neurosis did not develop because the subject consciously became aware of everything that you attempted to symbolize?

KUBIE: We do not know. It is possible for me to conceive of a cultural attitude toward infancy and childhood which would attempt continuously and repeatedly to bring into the full light of consciousness all of those internal ethical struggles which under our present cultural pressure become unconscious. I can conceive of entirely different ways of bringing up children, the goal of which would be to minimize this dichotomy. I am not sure that it is conceivable to me that there would ever be a culture which could eliminate it completely. I repeat, however, that we do not know because it has never been tried.

Related to this is the question of when this dichotomy first arises. Here again it is difficult, to be sure. One sees a great deal of meaningful automatic activity in the preverbal life of infants which seems to evolve into early panics, nightmares, eating disturbances, bowel disturbances, and primitive neurotic episodes of childhood. Certainly it must begin early, yet I cannot answer when with precision. This is one of the more important questions of cultural development: What can we do to control, minimize, shape, eliminate this dichotomy?

McCulloch: Take them backward in time. You answered Dr. Gerard's question.

Mead: Communication.

Marquis: Does the neurotic symptom have the same relations as communication in general?

Kubie: This really relates to Dr. McCulloch's question. Again it | is hard to answer it, [220] because certainly the functions are different. Every neurotic symptom serves multiple functions, among which is always some gesture toward the outside world, whether the symptom is a height phobia, a handwashing compulsion, or a conversion symptom. Whatever it may be, it is a way of telling the world that there are things I will do and things I will not do. Yet this is not its primary function. Its primary function is the effort to deal with certain internal tension states arising out of unconscious conflicts.

Marquis: There is one kind of language which is sometimes thought of as expressive, just emotional expression, which does not always have a communication function. It occurs when you are alone, and so forth. Is the neurosis that kind of behavior?

Kubie: That reverts to what I said before about emotional disturbances which can be induced experimentally in animals. In their fundamental nature and dynamics these are quite different from the disturbances of neuroses.

Klüver: I wonder whether all animal psychologists would agree with your statement that conditioning or learning is always associated with states of craving and occurs only in the absence of satiation. A research physicist once came to me with the idea of doing some work in the field of psychology. I suggested that he should pick a problem in which he could utilize his knowledge of physics. After a few days he came back with an indictment of all objective studies in the field of animal behavior. He felt that the studies done so far had merely produced a one-sided and misleading picture of the psychology of animals, since experimenters always had used hunger or the avoidance of punishment as motivating factors. When I finally inquired into the type of problem he wanted to work on, I discovered that he was interested in such problems as the psychology of a chimpanzee retiring with a full stomach for the night to its tree nest and taking a last look at its surroundings. He was wondering whether such a chimpanzee was capable of gaining new insights by attentively inspecting the landscape or even capable of enjoying the real beauty of an African sunset. I am not going to advocate the use of chimpanzee subject for studies in experimental aesthetics; I merely wish to point out that there is some evidence to the effect that a full stomach does not necessarily prevent conditioning or learning processes in animals. To be sure, even a determination of the properties of visual *Gestalten* involves more than an analysis of perceptual factors. It was Kurt Lewin who used to point out that you cannot analyze or specify such properties – no matter how | objective they are – if you turn your back to [221] them. An objective psychology of visual *Gestalten* still depends on your having to turn around and taking a look at the *Gestalten*. Similarly, the fact that learning may involve nonperceptual or noncognitive factors does not necessarily mean that these factors must be identified with cravings, deprivation or an empty stomach.

In connection with Dr. Marquis' question, I should like to recall one of Buytendijk's studies. Incidentally, Buytendijk, who for many years was in charge of the largest physiological institute in Europe, has now become a psychologist and in a recent paper has analyzed »the first smile of the child« [Buytendijk, F. J. J.: Das erste Lächeln des

Kindes. *Psyche*, 2, *57*]. Since you strongly emphasized the role of internal needs and body tensions in development, I wonder, Dr. Kubie, how you would interpret the first smile of an infant. Is the first smile an epiphenomenon of bodily states or is it a smile at something or somebody?

KUBIE: I am not sure that I quite understand the question.

KLÜVER: Suppose the infant is lying on its back perfectly relaxed and with its physiological needs attended to. Now it smiles for the first time. The question is whether this smile is merely a concomitant or function of bodily conditions or whether it is a phenomenon elicited by, or directed toward, some stimulus or some stimulus aspect of the external world.

KUBIE: I should think there is a considerable probability that the first smile of the child is part of general random activity but that it becomes a method of communication with the grownups that tend to it. It is clear to anyone who has ever brought up children that it becomes a method of language very early.

FREMONT-SMITH: Identification comes in early there.

SAVAGE: Dr. Kubie's interpretation, if taken literally, is testable anthropologically. If it is correct, there should be cultures – many of them – where a smile isn't a sign of pleasure, where it just means something else, or even more frequently, means nothing at all.

PITTS: It should not be very common, should not mean anything. Possibly babies always smile when happy.

MEAD: They smile everywhere, but I don't believe you have to state it that way. If you have as random behavior in response to states, internally recognized states, either crying or smiling, the culture can change the meaning of either. A good many cultures exist that handle tears in a much more positive way than we do, and tears can be turned into expressions of pleasure later; smiles certainly can be turned into expressions of displeasure. In Bali there are a good many situations in which an adult's smiling is [222] | regarded as virtually insane behavior. The person who smiles too much in Bali and the person who smiles at a stranger are considered insane. The expressive forms are modifiable by culture. I still think the data we have at present suggest that a smile occurs initially in states which are diffusely pleasant in some way, such as a full stomach. The smile seems to be a recognized phase, as suggested by Spitz's data on the child smiling at the masked human face [Spitz, R.: Emotional growth in the first year. *Child Study*, **14**, *68* (1947)]. That is probably some of the best material we have.

WIENER: A smile is quite different from what is expected on certain occasions of hostility.

KUBIE: There is much hostile smiling in our culture, too.

BATESON: I suspect that the answer to Dr. Klüver's question, whether the smile is a symbolic statement about the outside or the inside, is that it very rapidly becomes both. And the fact that it becomes both is one of the roots of the unconscious component of language. What Dr. Kubie is really trying to say is that language is a double coding: both a statement about the outside and a statement about the inside. It is that doubleness which gives this conscious-unconscious quality to it.

MCCULLOCH: May I come back to my question, because I think it is very closely related to Heinrich Klüver's? Is it true of the essence of a neurosis that the process is symbolic?

KUBIE: Yes, it is of the essence of neurosis that the process is symbolic and that the subject does not know what it is symbolizing.

TEUBER: To whom?

VON FOERSTER: To himself.

FREMONT-SMITH: That is part of this business. It is that the language is not only communication to the outside world but also to one's self; the neurosis is an interruption of communication to one's self as well as to the outside world.

PITTS: As a corollary, an animal neurosis, so called, and a human neurosis have nothing in common.

MEAD: Dr. Kubie gave a definition of neurosis at the first or second meeting of this group that ought to be appropriate here. You said a neurosis was characterized by an inappropriate method of seeking a goal.

KUBIE: I have never said that.

MEAD: You have.

FREMONT-SMITH: You have neurotic amnesia for it.

PITTS: An irrelevant repetitive process. | [223]

KUBIE: If, without knowing it, we seek unattainable goals, under any mask, then our conduct becomes both repetitive and inappropriate, but it is the unattainability of the unconscious goal that is essential. The repetitiveness and the inappropriateness are secondary attributes.

PITTS: He said irrelevantly repetitive process.

MEAD: That he added also. That was the essence of it.

PITTS: He must know.

MCCULLOCH: You say it is of its essence symbolic. Is that symbolism necessarily of the kind we call language?

KUBIE: Well, Dr. McCulloch, in the neurosis the symbol has many forms. In obsessional states language plays an important role. In compulsive states action plays that role. In the phobias it takes a still different role. Actually, the use of language in the neurotic symbol is relatively unimportant. It is relatively unimportant in which form the neurosis presents itself. What I am saying is that the symbolic process which is used in language, and which is the *sine qua non* of language, is also the *sine qua non* of the neurosis, with the one additional factor that in language we do know by and large what we are trying to represent, whereas in the neurosis we do not.

BROSIN: At the risk of oversimplifying this question, in direct answer to Dr. Marquis, I find the simplest solution or model is to regard the individual with the neurosis as being a complex, integrated series of operations in a matrix of other series of operations. The inappropriate ego functions may be directed toward two kinds of outer spheres of operations. One of these may not be physically present but may represent an inappropriate series of symbols of the other (inner) sphere.

FREMONT-SMITH: Or both at once.

BROSIN: Or both at once.

MEAD: You would include –

BROSIN: In the language of the ego, id[!], superego model, the ego operations concern environmental operations or the instinctual-physiologic, or internalized, conscience. The ego has a galaxy of goals in the inner sphere of operations which may or may not be contiguous and relevant to the outer series.

MCCULLOCH: Are those relations necessarily symbolic, or may they be direct?

BROSIN: Let us recall the work both of semanticists and of philosophers on this question of the specific meaning of such operations. Until we define the detailed processes more clearly, I cannot answer the question. If you would set up examples and a | common notation that would make the question concrete in operational terms, I could venture an answer. We need a notation which is sufficiently neutral and free from [224] emotion or older accrued meanings to permit formulating an experimental question which would elicit a meaningful answer.

GERARD: Perhaps the problem lies in what one means by symbolization.

STROUD: I would suggest that acts are the key. Acts, whether symbolic, irrelevant, or effective, that our subject is impelled to act. He is in a difficult situation. If he acts in such a way as to get out of it, we have no neurosis and no problem. If, however, he acts ...

FREMONT-SMITH: How do you define »getting out of it,« because there is the crux?

STROUD: That is precisely it. Getting out of the situation is to my way of thinking readily identified by the absence of the repetitiveness. If you got out of it, you would not repeat it. You can only say a man repeats it in this sense, because he is still in it now.

KUBIE: I think it is important for us to recognize how closely related is the potentiality for neurosis to our highest symbolic functions. That is why I cling to this.

FREMONT-SMITH: This is the heart of the whole problem of this conference group right now. This is the thing I have been waiting for since I started this conference group: that we who think in mathematical, physical, and engineering terms would come to grips in a genuine way with the people who think and talk in symbolic unconscious terms. There is a tendency to intolerance on both sides which should be avoided. I suggested to our Chairman earlier that for the next meeting we should have the raw data from a hypnotic experience presented to the whole group for discussion. We are going to get that next time, because there are crucial phenomena dealing with human behavior which operate in the thinking of mathematicians; there are interactions between mathematicians and psychoanalysts with which we must come to grips if we are going to understand either the mathematical or emotional end of human behavior.

GERARD: Hear! Hear!

PITTS: It appeared to me that he was separating the unconscious from all normal processes. It seemed to me that if the unconscious was a valid process, it was related.

[225] FREMONT-SMITH: I am sure you misunderstand him, because that is not his concept. |

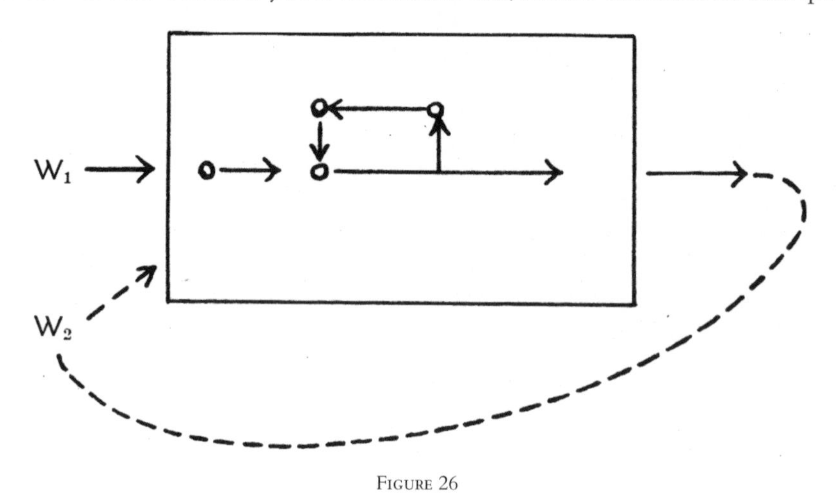

FIGURE 26

MCCULLOCH: I should like to draw on the blackboard a diagram which I think will help to get rid of the difficulty which is standing in our way quite gratuitously. Consider an organism with receptors hooked through sundry circuits to effectors. Whether signals go straight through or reverberate within it, they evidently come out by way effectors into that same world where it has its receptors. If that is all that we suppose happens, I see no reason for calling the process symbolic, but when we consider two parts of the world, say W_1 and W_2, detected by this organism, and the central activities for these two associated so that for that organism W_1 comes to stand for W_2 at some later date, I would call it symbolic. If W_1 and W_2 have any common properties, W_1 can be used to signify W_2 to someone else or –

STROUD: Or to itself at a later date.

MCCULLOCH: That is oversimplification. Every one of these is a distortion.

FREMONT-SMITH: You never have W_1, always W_2.

STROUD: Substitute W_1 for W_2. If I eat a melon, does the empty rind stand for the symbol of eating the melon?

MCCULLOCH: For the basic notion of eating.

STROUD: I don't treat empty melon rind as melon. I do treat in many respects the word »melon« as a melon.

PITTS: It is a perfectly real, difficult stimulus equivalent, to begin with.

WERNER: What about substitution? | [226]

PITTS: We must take him to have meant not stimulus directly but equivalent classes of stimuli.

MCCULLOCH: Surely.

WERNER: Still you have the difficulty of distinguishing between stimuli substituting for others, as in true conditioning, in contradistinction to symbolization. I would not identify that type of substitution with symbolization.

PITTS: No. W_1 and W_2 must be taken to be wholly equivalent of symbol. Since W_1 and W_2 are not class equivalent, but W_1 may stand for any one of a whole class of equivalent stimuli, and similarly W_2, either one may be replaced by a class of such equivalent stimuli.

WERNER: The difficulty is to define the symbol in terms of what this type of substitution is, because we have so many types of substitution. You have to be quite clear about the particular type of substitution which you call symbol.

PITTS: Substitution which does not occur according to classical perspection, which is dealt with in Helmholtz's groups.

FREMONT-SMITH: Right.

KUBIE: Let me give another example of what I mean. Perhaps this will make it clear. I think of an adult who was an X-ray man and an able and outstanding scientist. He got into serious trouble with the law because he was also a Peeping Tom. This brought him to treatment. As a child, he had had a severe eye-blinking tic. It was easy to work out that the eye-blinking tic had represented a defense against his own childish visual curiosity, which had been translated into a career as an X-ray man. Yet the childish needs which had become unconscious remained unsatisfied and finally broke through in the form of a Peeping Tom perversion that finally landed him in the hands of the law. That is a simple example of what I mean when I say that all of these things are essentially symbolic functions.

MCCULLOCH: Would that be any more so than a goose that nurses a bright tin pan all summer by himself and makes love to it, and so forth?

KLÜVER: Since I have no definite opinions about neuroses in infants and children, I should like to speak about a baby monkey. I happen to know something about bringing up baby monkeys. The baby I am now referring to had a rather remarkable mother: namely, a Java monkey in which the connections between the temporal and the frontal lobes had been severed bilaterally without removing any cerebral tissue. Three years after the operation this monkey delivered a normal baby and within the [227] next two | years had two more pregnancies at term. In all three instances the baby was found lying on the floor of the cage. The mother made no attempt to pick it up and definitely refused it even when offered. She showed no excitement when the baby was finally removed. There was, therefore, a complete absence of maternal behavior. It was one of the three babies of this monkey that I took to my apartment, where it spent the first months of its life. Its bedding was a pad of white cotton on which it used to lie contentedly, with its tail grasped in both hands, whenever its bodily needs had been attended to. One day it was taken back to the »hostile« world of the laboratory, where it exhibited at once marked fear reactions. Although it held on to its own tail it did not cease crying until it noticed a pad of white cotton. It ran to it immediately, squatted on it, and then produced what I believe to be the simian equivalent of a smile. In the world of this monkey baby, two things, grasping its own tail and a piece of white cotton, appeared to be essential for producing »social security« and peace. In fact, the same result could be achieved by reducing the pad of cotton to a few square inches or even less than one square inch of white material. The absence or removal of the white cotton always resulted in great anxiety. Unfortunately, I could not go on providing cotton through the following years. I wonder, Dr. Kubie, whether behavior deviations developing in such a monkey would have finally brought the animal in conflict with the »laws« of a respectable simian society.

KUBIE: It may be. I am not denying the possibility of real neurosis in animals. I am just saying that I do not know. It is an even more difficult thing to appraise in animals than it is in humans.

KLÜVER: I have talked about cotton and Dr. Stroud has talked about »things« and »not things.« We want to know the psychological structure of these things and not things in a child's world. These things are certainly not the things of our adult world: they have an entirely different »physiognomy.« It is fortunate that we have Dr. Werner with us today, who has been concerned for many years with the psychological problems involved in »physiognomic characters.« Some child psychologists have claimed that, long before an infant becomes aware that the mouth of the mother is red, her eyes blue or brown, and her hair blonde, it can recognize the face of the mother as being friendly or unfriendly. In fact, it has been claimed that our thresholds for recognizing or differentiating physiognomic characters such as friendly, gay, sad, melancholy, [228] threatening, tempting, and so forth, are lower than | our visual acuity thresholds for brightnesses, colors, shapes, et cetera. The problem is, therefore, how a world divided into things and not things is related to a world differentiated on the basis of physiognomic characters. In the world of the baby monkey the white cotton was undoubtedly more than merely some material with a certain brightness, size, shape, and texture: it was imbued with certain physiognomic characters. And it seems to be the difficult business of psychology to deal with these ubiquitous and powerful determinants of behavior. I wish Dr. Werner, who is an expert in these matters, would comment on this.

WERNER: I do not have very much to add to what Dr. Klüver said. The recognition of physical qualities, or what I have called physiognomic perception, is a primordial way of perceiving, definitely not restricted to human experience. Now some psychologists have termed a primordial physiognomic perception, such as the expression of friendliness in the behavior of an animal, as symbolic. I am a little bit puzzled by Dr. Kubie's use of the term »symbolic« without defining it, and particularly by his setting it off against expressive perception and behavior in the primordial sense. If a statement is made that symbolic behavior is essentially human, and therefore that neurosis, which presupposes symbolic behavior, is essentially human, such a statement hinges on a clear definition of what a symbol is.

McCulloch: Licklider, will you try to define what is meant by »symbol« in such a context?

Licklider: I think that, since one of the important things is to try to talk about neurosis in terms of language and symbols, in terms that are perhaps admissible both by the psychologist and by the mathematician, it might be all right for me to try – since I fall somewhere inbetween – to define »symbol.« To do that, I should like to start off with the notion that any physical process, if broken up into small enough parts, can be represented by an arrangement of Dr. Shannon's black boxes, and that some of these processes have the property (he referred to it last night) that they lose very little, or no, information. In other words, you can reconstruct the predecessor if you can get your hands on the successor. What I am saying is that all processes can be described in terms of transformations, sometimes information-abstracting or information-losing transformations, sometimes information-preserving transformations. Symbols, I think, are the products of information-preserving transformations, products that have the special property of not looking like the things of which they are transforms. | [229]

Kubie: Would you accept one modification that there are two orders: one information preserving and one information burying?

Licklider: Would you elaborate »information burying« a little?

Kubie: The difference between the symbolic process in speech and in neurosis would lie precisely there.

Licklider: Let me say this: the predecessor can be recovered if you put your hands on the successor and if you know the rule of transformation. In the case of information-burying transformations, nobody knows the rule of transformation, unless it is the psychiatrist, who thinks he knows the rule of transformation. Of course, the rule may include the psychiatrist. I have said all I think I should say.

Fremont-Smith: When not conscious it is not easily accessible to translation and when it becomes conscious it becomes translatable. There are some things which are accessible to translation but it is not convenient to translate them, and there are other things which are inaccessible. Speaking at a different level, it might be convenient to translate, but you cannot do it well. It is at that level that we are making the distinction of the neurosis: even though it would be convenient to translate it you cannot; you have to behave because you cannot conveniently translate it when it would be convenient to do so. You are behaving in a neurotic way. Is that close?

Kubie: In that sense, therefore, the neurotic transform for communication cannot be used in the same way that the transforms of speech can be.

Licklider: It may unwittingly communicate to someone who knows the rule.

Teuber: That is the point I was trying to make with regard to symbols, by asking: »To whom?« I cannot think of a symbol without thinking of somebody.

Stroud: The neurotic family is ruled by the most neurotic member. That is the social communication function of these neurotic symbols. That is the next point I wanted to bring up.

Teuber: There is still another point that has been brought up during the last 48 hours, and that is that a symbol does not have to look like whatever it signifies. It is obvious that on Shannon's level it does not look like anything at all. Sometimes it is just a dot or a click or a pulse. It is not an eidolon as the ancients thought, an image which must resemble what it refers to, along some very obvious dimensions of similarity. In that sense I think we can find at least some precarious common ground if we are all agreed that a symbol does not have to look like the thing it symbolizes. | [230]

BROSIN: I don't think the meeting ground is at all precarious once you explain it as well as you have. To reiterate an earlier statement about the importance of the object »to whom« the meaning of the symbol is directed, it may be useful to stress the fact that ego functions include a wide range both of direct and of more remote significance. An act such as throwing a plate during a quarrel may have multiple elements of both. Symbolic ego functions (receptive, integrating, executive, or motor) may be directed to objects in the outer environment or toward the more inaccessible areas in the unconscious. The reinforcement from the older emotional patterns may then furnish the impetus to produce or distort externalized symbol formation for a long time. The neurotic is solving his problems inadequately by symbols which are inappropriate, but these may often be easily understood by the layman in such everyday actions as apprehensive laughter, hostile smiling, ambivalent muscular gestures.

MCCULLOCH: I want to ask Larry Kubie one more question. Then Larry Frank and Frank Fremont-Smith will speak. Is it of the essence of a neurosis that it is symbolic or is it of the essence of a neurosis that some information, whether it be symbolic or not, is in some manner inaccessible? The inaccessibility is the crucial item.

KUBIE: Dr. McCulloch, sometimes in the course of argument you misstate my position a bit. What I have always said about it is that *if there were no unconscious processes, there could be no neuroses.* That is another way of saying what you just said: *if the information were accessible to the symbol, there would be no neurosis.* What is common to the neurosis and language is the transforming process, that is, the symbolic process. What is peculiar to the neurosis is the inaccessibility of the psychological state represented by the symbol.

MCCULLOCH: Then my question is very much in order. I see no reason why the concept should in any sense be linguistic.

MEAD: No.

TEUBER: No.

MCCULLOCH: If the symbol is no more symbolic than the nervous impulse is of the stimulus that evoked it.

STROUD: It is just possible that these symbol functions in neuroses operate for the person who looks at the neurotic but not for the neurotic himself.

MCCULLOCH: Yes, that is the question: To whom is it accessible?

FRANK: That reinforces a point I want to raise. It is too late now to discuss it, but we [231] must give more attention to inner | speech. One gets the notion that inner speech is the way the neurotic maintains his idiomatic speech codes and that therapy consists of bringing the patient's inner speech into line with the actual world as directed by the therapist, so he won't go on giving a neurotic interpretation to events. Isn't that right?

KUBIE: Right.

FRANK: We have neglected in this discussion the important role of inner speech. We know very little about it and we should try to explore its operation so that we can discover more about the symbolic process and especially about language.

FREMONT-SMITH: That brings up the point I was going to make about »to whom,« which is the crucial question here. It is not only to the individual or to the outside world, but also to which phase of the individual. I think that it is essential to our understanding of what we are talking about to accept the assumption that the individual has two or more phases, multiple phases, and that a thing may be accessible to one phase and not accessible to the other phase, or to two or three phases. There are dramatic examples of that in raw data, which I think can be demonstrated if necessary. The moment you bring that in, you have inner speech because inner speech »to whom« is involved. There is good evidence to show that inner speech, which can even be vocal, is to the other phase or to the other person, as we sometimes say. There may be two or more persons, so to speak, or phases of the person in which communication may be going in one direction but not freely in the other; and in order to understand what we are talking about I think we should realize that.

BATESON: I wonder whether we can clear up something about the question of »accessible to whom?« One of T. S. Eliot's poems, »Sweeney Among the Nightingales,« makes a great deal of sense if you read it as a poem, but you cannot say what sense it makes. That poem has been analyzed by Collingwood in his *Principles of Art*, and he reconstructs, probably correctly, the objective situation which the poem is about. But this information does not add in any way to the impact of the poem.

PITTS: Excuse me! What objective situation?

BATESON: The context. I can't recall the poem in detail nor can I remember the analysis of it. I want to get the reference into the record. It is a situation in which Eliot is in the presence of certain people; there is a convent next door, and the life in the convent is visible from where he is, and so on.

PITTS: It is the authorization.

BATESON: The authorization? | [232]

MCCULLOCH: It has nothing to do with the understandability of the poem.

BATESON: That is the case.

PITTS: It is possible, as Teuber said, that symbols need have nothing to do in these. They need not be at all like what they refer to. It is quite different with respect to complexes of them; that is, complex symbols which are made of symbols as parts will have a definite reference; they will have a certain kind of isomorphism to relations among the things for which they stand that will be a parallelism and a similarity. I think that occurs only when we get to the level of complex symbols. That is, a statement about a man and a dog has a correlation; namely, if you take the statement, »The man beat the dog,« there is the man, a transitive relation, and the dog; there are three symbols following one another, one with the relation in the middle, an actual event to which it refers. You can identify the objects and relation, whereas single symbols may have no direct content of their own. The complex symbols may have content of their own, although elementary components may be arbitrary, like letters or simple digits.

LICKLIDER: This takes us back to the conversation of yesterday morning. There may be isomorphism where there is not the perceptually obvious sort of relation that leads one to say, »Here is a clear parallel.«

TEUBER: It need not be point for point. It could turn out to be multiple.

PITTS: Symbol alone, as a matter of fact, conveys no information unless you suppose the implied statement that in such place there is the object designated by the symbol or that something like the information can be conveyed only by the whole statement, which is a complex of symbols.

TEUBER: Given the system to which your symbol refers.

KUBIE: Precisely. The little piece of white cotton could represent the security in your home for the little monkey.

KLÜVER: I wonder if your statement that the human organism has only two symbolic functions should perhaps be qualified. There are symbols, I believe, which have to do neither with language nor with the neuroses.

KUBIE: The language of treatment and of the hypnagogic reverie which Silberer describes is the raw material out of which the dream comes. The language process of the dream is almost entirely in visual imagery (2).

KLÜVER: »Autosymbolic phenomena« of the type Silberer de|scribed may also occur [233] in dreams. At a time when I worked every day with Java monkeys and in spare moments thought about the principles of Kantian ph ilosophy, I dreamed one night that I entered a room of my laboratory, in one corner of which I found a gunny sack standing against the wall. When I opened the sack, I saw that it was filled with Idaho potatoes. While I still looked at these potatoes, every potato turned into a Java monkey. It was clear to me in my dream at this moment that »the synthetical unity of the manifold in all possible intuitions« was directly in front of my eyes.

PITTS: How?

KLÜVER: In phenomena of this kind and in those described by , there seems to be a direct symbolic representation of thoughts.

PITTS: How would that go? I cannot see that.

KUBIE: Kekule's dream in which he evolved the problem of the benzene ring.

KLÜVER: Kekule's own account of how he arrived at his cyclic formula has become very famous.

KUBIE: Loewi had a similar experience.

KLÜVER: But since Kekule was a chemist and not a psychologist, I have wondered sometimes whether his own account of his discovery of the benzene ring should be accepted [Schultz, G.: Bericht über die Feier der Deutschen Chemischen Gesellschaft zu Ehren August Kekulé's. Deutsche Chemische Gesellschaft Berichte, **23**, *1265A*. Berlin (1890); Japp, F. R.: Kekulé Memorial Lecture. *J. Chem. Soc.*, **73**, *97* (1898)] One of the designs frequently observed in entoptic phenomena and in hypnagogic hallucinations is the »honeycomb« design with its hexagonal elements [Klüver, H.: Mechanisms of Hallucinations. Chapter X in *Studies in Personality.* New York: McGraw Hill (1942) p. 175]. This hexagonal pattern was seen entoptically by Purkinje, König, and many other observers. I have seen it on several occasions, not with closed eyes, but on the ceiling after awakening. It is highly probable that Kekule, who was used to watching »repeated visions« and configurations gamboling before his eyes, also saw hexagonal patterns either entoptically or in hypnagogic hallucinations. Unfortunately, my hypothesis that the benzene ring was psychologically derived from the hexagons of this well-known honeycomb design can never be checked. I should mention that hexagonal patterns also appear in mescaline hallucinations, in the visual phenomena of [234] insulin hypoglycemia and when viewing flickering fields under certain conditions. |

STROUD: How do you solve problems in sophomore calculus that way? I did it half one summer that way when I was asleep.

FREMONT-SMITH: Einstein got the formula while improvising at the piano and thinking about nothing whatsoever.

MEAD: I think at some later stage the question Pitts raised will come back, that we might do something with the grammar of dreams, and do a great deal more with the primary process than what we have done so far. Kubie has given us several discussions of neurosis as we went along, and we had a little bit of Rorschach, but we have never discussed in any detail whatsoever the type of grammar, the logic of the primary process, and the way in which that type of grammar and logic are involved in the thinking in visual images and what kinds of speed can be made with visual symbols when they interact with the sort of thinking that I think Pitts calls normal.

PITTS: Rather the rhetoric of dreams.

MEAD: You can call it that if you like.

PITTS: The rules of rhetoric of dreams.

MEAD: Of artistic construction. We have had bits once in a while by Von Neumann. Wiener has given a little bit of dream but we never systematically discussed it. At some future time we will want to put those two kinds of thinking together much more than we have done so far.

FREMONT-SMITH: We will touch on it at the next meeting, when we have the one-to-one relationship of human beings and discuss it in the hypnotic phenomena. As a matter of fact, it is just in that area that Brenman is quite superb, I think, in discussing it. I do feel that this time we have come closer to a discussion in which there was a common denominator for every discipline here. I don't think we ever came as close to approaching that as we did this afternoon. And I think that that is the goal. We ought to have many common denominators for all the disciplines if we are going to reach our goal, but at least we had a common denominator this afternoon, and I think it is the first time we have had it. I feel very pleased. I must say I have been quite impatient for it, and others have been impatient for it. I really think we got closer to it today than we ever have before. I don't know if everybody else had any agreement or disagreement.

SAVAGE: Just how do you mean that?

FREMONT-SMITH: I think we were all talking about what Larry Kubie brought up in a way that seemed to me to be intelligible simultaneously to most of the disciplines concerned. I don't mean to say wholly intelligible or wholly agreeable, but we are coming to grips with it. | [235]

I should like to remind you that in the earlier conferences these topics came up again and again, but in the earlier conferences they were effectively avoided. So much anxiety was aroused in the group when we began to talk about unconscious phenomena that we really never seriously brought our intellects to bear on the problem. This time I felt relatively little anxiety in the group; we tolerated each other and tolerated the subject before us, however difficult and evanescent it was.

MCCULLOCH: I think we ought to close the meeting with a vote of thanks to the Macy Foundation.

FREMONT-SMITH: A vote of thanks to you and the rest.

MCCULLOCH: And above all to Frank Fremont-Smith.

[237]

APPENDIX I
BODY SYMBOLIZATION AND DEVELOPMENT OF LANGUAGE[1]

LAWRENCE S. KUBIE
Department of Psychiatry and Mental Hygiene
Yale University, School of Medicine

PAVLOV HAS proved that no new reflex can be conditioned in a satiated animal, and that therefore all processes of learning depend upon the existence of a state of craving. Approaching the problem from another angle and by another method, psychoanalysis made the same discovery, proving that the acquisition of new knowledge depends upon the existence of a state of instinctual tension and deprivation. Since in infancy and childhood cravings arise in body tensions, it is inevitable that the child's thought world should begin with his body, and that his first concepts must deal with the parts, the products, the needs and the feelings of the body. In order to understand the growth of language one must observe closely what the child wants, what parts of the body become involved in the process of wanting, and ultimately how he learns to speak and think of the different parts of the body and of the desires and feelings associated with them.

These considerations may seem banal; yet they are far-reaching in their significance. Since the child's world begins inevitably with his body, and since the force which instigates the child to expand his knowledge is always the pressure of bodily desires, and since every new fact of experience which enters into psychic life can make its entrance only by relating itself to that which is already present, it follows that every new fact apperceived by the child must somehow relate itself to bodily things. Schematically the process can be represented as follows:

A....................	A'....................	A''.................... etc.
(Body concepts)	(New data of the first order, related to A directly.)	(New data of the second order, related to A' directly, but to A only through the mediation of A'.)

[238] | If this represents with any degree of fidelity the process of expanding knowledge, it must also represent the process of expanding speech. It means that there is at first a long period in which concepts are vague, broad and overlapping; and that with advancing years these concepts become discrete and distinct. Schematically again one might represent the situation as in Figure 1. Therefore it is not surprising to find that in sleep, in a [Figure 1] state of semidozing, and in delirium, we drop back from our topmost level of development, I, at which all concepts are completely separate one from another, to lower levels of imagery such as II, III, or IV, in which ideas and their related feelings fuse and interact. It is also clear then that some of the energy infusing speech derives not from level I alone, but from deeper, broader, more inclusive meanings.

In this paper I shall present a group of naïve and spontaneous examples of this type of »symbolic« language from children and from patients. There are several reasons for [239] seeking examples of this process in the speech and behavior of very young children. | In the first place, under the conditions of modern education, the young child is

1 [Reprinted from *The Psychoanalytic Quarterly*, 3, *430* (1934)]

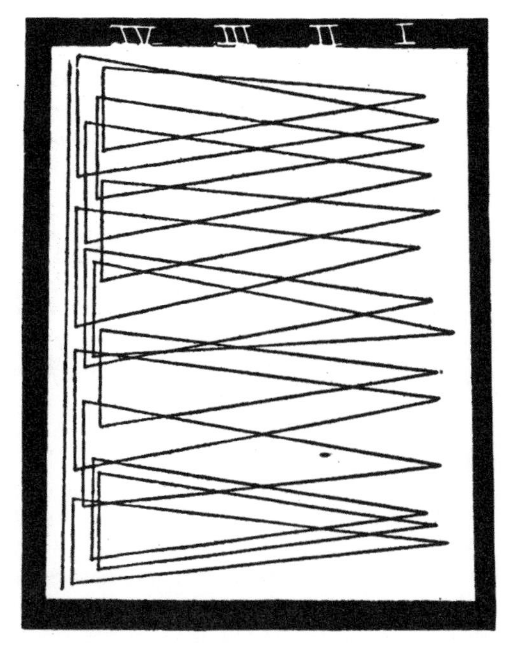

FIGURE 1

becoming more articulate about his fantasies than was true in the past. We hesitate to check a child's naive fantastic production, we are less heedless of what he is trying to express through these fantasies, and we no longer feel impelled to correct him at once with considerations of reality before making an effort to understand him. (It may be, of course, that modern educational methods err by going too far in this direction. Experience alone will settle that; but in the meantime one may profit by current errors to the extent of studying these childhood fantasies.) There is one further advantage to be derived from beginning with the spontaneous and naive productions of childhood, namely, that by so doing one subjects to a fairly critical test many of the psychoanalytic interpretations of adult speech, dreams, and symptoms. If in earliest childhood one finds examples of the naive use of similar symbolizations, where the interpretation is self-evident, and where the possibility of adult suggestion is excluded, it has the value of evidence in support of the psychoanalytic position.

In the following examples we shall meet one surprising fact. It will appear that children, at least in some phases, seem to develop certain individual and characteristic kinds of body-language. One child, for instance, for a time consistently used articles of furniture, familiar household shapes, buildings, etc., as representatives of the body; another child used clothing; a third child used animals. Still a fourth used machinery. We are not yet in a position to say whether this represents different stages of instinctual development, different age periods, the influence of sex, the effects of special interests on the part of the parents, or whether it correlates with different types of personality development, or with different neurotic structures. We can only record the observation tentatively and pose the problem for further investigation.

Let us turn to a very simple example: A little boy of seven, on a picnic in the woods, has a bowel movement which stands up straight on end in the underbrush. Pointing to it with evident amusement and satisfaction, he says, »Look, I made the Chrysler Building.« One can share in his little joke without feeling that the analogy is too »farfetched« or surprising. But then one asks whether such a joke uttered by a naive child provides any hint as to what may be meant when a man in soberer and politer years dreams of making a tall building. Of course, it must not be overlooked that this is only [240] *one* of the possible meanings of such a | dream element. The same little boy, three years before, had made drawings to represent each day of the week. One day would be a »black slide,« another day would be an oven, and so on – each day being represented by some form or shape, mostly suggesting familiar household objects. This is a very complex type of fantasy, evidently related to its source only by a series of intervening steps which in the absence of careful analysis could not be made clear.

Another more complex, and more surprising, bit of symbolic speech and behavior was found in a girl of three and a half, who consistently used little animals as her idiom. The child had been reporting to her nurse and mother repeated dreams of a little mouse that got into her bed and tried to nip her. There had been no evidence of anxiety during the night, no undue restlessness, and no waking in terror as from a nightmare. The child's report, however, was uneasy, as though given half in jest and half in pursuit of reassurance. The little girl had frequently seen her older brother in his bath and had shown a direct interest in his body and in the difference between his genitals and her own. One morning, however, she burst into her parents' bedroom, just as her father emerged naked from the tub. She marched directly up to him and stared at his genital with unembarrassed attention for at least a minute. She then heaved a little sigh, smiled, and looked off thoughtfully into the distance with an arresting expression – which made her father ask, »What are you thinking of?« To this the child replied airily, »Oh, I was just looking at a little mouse.« Here it seems an inescapable conclusion that the child was using both the image and the word-symbol for a mouse interchangeably with the word-symbol for a penis (a word which, incidentally, she knew quite well). Of course, the substitution of a little, furry, biting animal for a part of the body has its special significance.

A further and even more surprising example of the naïve use of an animal idiom is given in the following story. The same little girl, at about the same age, is standing on her father's shoulders looking down at him as she faces him and as he looks up at her. Suddenly she crouches, and quite deliberately presses her genital region against his face, so that he has to draw his head back in order not to participate in her little seduction. As he smiles up at her silently she suddenly says, »Would you like to see me naked?« The father parries with the counter question, »Why? Would you like me to?« to which the child replies, »Yes, but not when I'm asleep.« Then to the amazement of [241] her father, | she slides swiftly down to the ground and says, »*Come, let's play you're a snake.*« Here, perhaps, one is on more debatable ground; and yet the quick sequence of ideas, speech, and action almost makes one suspect that this three-year-old infant had been reading the text books. First there is the deliberate genital approach to the father; then the fantasy of exhibiting her body; then the faint flurry of anxiety as to what dangerous things might occur if this happened in her sleep; and the final resolution in a game in which the paternal partner of all of this play of instinctual trend is to assume the classical role of a snake.

To these anxiety laden fantasies one finds a sharp contrast in those of a little girl of five, whose habitual idiom was to a rather striking extent a language of clothing. It was characteristic of this child, for instance, that in her effort to solve the problem of the penis, she would put little pieces of chalk inside the trousers of her dolls in order to convert them into boys, and insisted on pinning a safety-pin to the front of her little shirt as a »pretend« substitute for the lacking organ. She called the pin her penis. (At her home psychoanalysis was held in highly critical doubt and any active suggestions to the child were scrupulously avoided.) This child had broken a shoe lace, and took a new lace and a shoe to her father so that he could replace the broken one. The repair completed, she moved reflectively into a room where her mother sat, and said to her, »Daddy took the shoe lace and put it in and out of my little hole, in and out, in and out.« Since this was a term which the child ordinarily used to describe her urinary orifice, her mother looked up sharply and said, »What?« To which the child replied, indulgently, »Oh, I meant in and out of the little holes in my shoe.«[1] Then she was silent for a moment and breathed a deep sigh, »I wish I were married.« Here again one faces a perplexing phenomenon, the use by the child of such »far-fetched« objects as a shoe, a shoe lace, and the holes in the shoe to represent her concept of genital reproductive functions, and a thinly disguised fantasy of intercourse with her father. Again one must confess that how such symbolic representations are laid down, is, up to the present time, a mystery. That it cannot rest, to any very large extent, upon the basis of racial imprints of old experience, is proved by the fact that so often the objects used are shoes, automobiles, airplanes and the like, whose racial history can hardly be said to be a lengthy one. | [242]

The fourth child cited above is the only son of an artist. In the atmosphere and daily life of his home machinery plays a negligible role; yet in the child's fantasies machinery is used rather strikingly to represent the body and its functions. At four this child had never had contact either with newborn babies or newborn animals. He had, however, asked the usual questions as to their origin. He was playing with an automobile jack one day, busily unscrewing all the bolts and taking it apart. His father came upon the scene and remonstrated, saying, »I wouldn't do that if I were you. You'll spoil it.« Whereupon the child became quite excited, and protested, »No, Dad, no. I've got to. You see, if I can get way down into this hole, I will find a baby.«

A year later the child was in a drawing class, in which the teacher was wont to call on the children for ideas before they sat down to draw. The youngster shouted out, »I've got an idea, Miss X. – It's night, a black, black night. There are two engines – on the same track – with great *big* headlights. They rush at each other and there's a wreck. I am going to draw a train wreck.« »All right, Johnnie,« said the teacher, »Go ahead.« A few minutes later she strolled over to see the picture of the railroad wreck in the black night, and found that the child had drawn *a man and a woman*.

Food can also become for the child a representative of his bodily problems. There is, for instance, the case of the youngster of four who came to discuss a certain matter with her mother. She said that after she had touched her anal orifice, her finger smelled like chocolate. Here it is less surprising that the child had made a connection in her mind between feces and chocolate, on the basis of general appearance, color, suspected consistency and the like, than that the child distorted her real sensory impression and converted the actual odor from the anal orifice to conform to her preconceived expectation, namely, the smell of chocolate. She repressed, or denied to herself what she actually smelled, and in a sense hallucinated the expected smell. On this basis it is not difficult to understand how that child, when a trifle older, a little wiser in the ways of the world and a little more sensitive to the world's disgust, might reverse the process and refuse to eat chocolate because of the relationship which she had built up in her mind between chocolate and feces; and this even if, in the meantime, the connection had been rendered unconscious.

A colleague of mine actually had an opportunity to watch the development of a transient food-phobia in a boy of six. The birth of a baby had led this child to ask insistent questions, in answer | to which he had been told that a child was carried within [243]

1 Obscene jokes often use this method of simple emphasis upon an innocent word, thus attending to its sexual significance.

the mother's body, and that it grew there as a result of the implantation of a »seed.« Not very long after this the child was eating fresh peas, to which were attached unusually long points of implantation. After examining these with interest, the child turned to its nurse and said, »What are these?« The nurse replied, »They are seeds.« The child looked startled, gulped once or twice, pushed his plate away, and for several months thereafter could not be induced to eat peas at all. An amusing contrast to this is a three-and-a-half-year-old girl, who ate grapes, and with each mouthful said proudly, »I eat the seeds, I like them.« After repeating this boastfully several times, the child grew rather reflective, then suddenly turned to her father on whose knee she was sitting and said, »Aren't these the kind of seeds that Aunt Jane uses to make babies?« Here again one faces the child's persistent notion that conception occurs by the ingestion of something through the mouth, although in response to direct questions this had already been specifically denied. Nor again is it difficult to picture this child at a later date refusing food on the basis of conscious or unconscious fears of pregnancy, just as she, at this early and more fearless age, dramatized her desire to have a baby by proudly swallowing the grapeseeds.

These examples may stand alone as evidence that the indirect representation of those parts of the body which are connected with our emotional and vegetative functions occurs exceedingly early in the formation of language in the growing child. It would seem that the sharp focus and definition of concepts and of words is something which develops only later in life; that in the early years concepts, images, words and feelings overlap to an extraordinary extent, and become separated into independent entities only with advancing years. In the schematic diagram already shown (Figure 1), the bases of the triangles represent overlapping conceptual and verbal units with their concurrent emotional charges. With advancing years these units become more and more clearly defined; but in sleep and under stress in the neuroses and psychoses it is easy to see how the less isolated conceptual units of childhood are brought into play. Furthermore they stand as a fringe and background tonus behind all conscious adult thought and feeling – and it is into that well that one dips with the analytic technique of free association.

It is worth while pointing out that from nonanalytical sources somewhat similar conclusions have been forced upon most critical and objective observers. For instance, [244] Piaget, on page 127 of his | book, *The Language and Thought of the Child* (1), describes a phenomenon which he calls »verbal syncretism,« and in his description recognizes the use of symbolism in the speech of a child, that is, the use of apparently arbitrary imagery and loose analogy.

Occasionally, it falls to the lot of some observer to be presented with a ready-made experiment. In the observations on very young children this opportunity comes not infrequently. As the years go on, however, with the sharper definition of thought, feeling and concept, the opportunities become rare, and the example which proves the case becomes correspondingly more valuable. Such an opportunity occurred during the illness of a patient in the Henry Phipps Psychiatric Clinic of the Johns Hopkins Hospital in 1921. The patient was a gifted and attractive young woman of eighteen, in some ways unusually naïve, and formally educated according to quite old-fashioned ideas. She had literally never heard the name of Freud. Despite her conventional background, however, she had been subjected to certain very disturbing influences through the fact that her father, an alcoholic, on rare occasions had made erotic advances to his wife in the presence of the children, and through the fact that her brother had manic-depressive spells, in which he was sexually exhibitionistic. The patient's illness was a mild depression, which had been precipitated by her first proposal of marriage. Not long after her admission to the hospital she had a nightmare which was as disturbing to her as it was interesting to the physician. The dream was that she was walking along a narrow street whose high, grey buildings converged at one end. Suddenly an airplane appeared overhead and began to shower her with bombs. She repeated twice that it was a »funny looking airplane.« In the airplane was a man whom she later recognized as her brother. In terror she ran up the narrowing street, and finally hid in a box at the end of it – a box which she again characterized as a »funny looking box.« The emphasis which she laid upon the peculiar appearance of the airplane and the box led her physician to ask her to draw them. She began by making a long oblong shaft from the upper right-hand corner of the page. This, she said, was the fuselage of the airplane. Then as she looked at it, she said, »Oh, I know what was funny about it – the wings were round,« and she proceeded to draw two circles at the upper end of this oblong. Then she added, »Oh, yes, the propeller was here in the rear,« and she made a blur of scribbled lines at the rear of the plane. When completed, the drawing was an unmistakable phallus with testicles and pubic hair. The patient's repressive mecha|nism was  so strong, however, that her own drawing excited no comments from her and no recognition. Nothing was said to her except to ask her to draw the box. This she did in the lower left hand corner of the page, in the direction towards which the plane was pointing. She drew a triangular box, then hesitated a moment, and said, »Oh yes, and up here at the base was a funny little bit of a cover that didn't cover the whole box.« Again she failed to realize that she had drawn the vaginal orifice with a clitoris. The completed drawing is shown in Figure 2. It is worth stressing that although when she

Figure 2

had finished these drawings she still did not realize their nature. When shown the same drawings several months later, without any intervening interpretation, she recognized them at once. Not only had she dreamed of sexual objects in this form, but of sexual practices, as an attack, a showering with bombs (semen?), from which she retreats by going back inside a vagina – the box. Beside this dream one might place that of another patient who dreamed that he was about to take his first flight in an airplane; but just as the machine was about to leave the ground he awoke having a spontaneous seminal emission.

A third patient, a woman physician, while talking in the analytic session, suddenly had an image before her eyes of a penis and testicles – but as she spoke of this, they [246] turned first into a | cannon on wheels, and then into a large scissors – i.e., the phallus became first a shooting and then a cutting weapon.

The next example concerns a young woman of seventeen, who was first seen in a mild elation, which necessitated her being sent to Bloomingdale Hospital. Before going to the hospital during one of several visits to the office, she rehearsed with rapid, flighty excitement and resentment a whole series of minor injuries to which she had been subjected in childhood – cuts, sprains and bruises, an accident in which her brothers accidentally broke her collar-bone at five, and finally a tonsillectomy at eight or nine. In a final burst of exasperation, she said, »*They took my tonsils out in the worst* PLACE *possible,*« and then, correcting herself, said, »I mean in the worst *way* possible.« Several months later, after her discharge from the hospital, recovered from her elation but in a mild depressive swing, she returned to the problem of early injuries, referring slowly and thoughtfully to the same series of accidents and operations. This time the meaning of her slip, and the significance which somehow or other had become attached to this tonsillectomy was made clear by the sequence of her thoughts over a series of days of analytic work. In the first place, her earliest years had been filled with an unhappy craving to have been born a boy like her two older brothers. Boys, she said, were her ideals; and as she said this, she described an image of an airplane taking off into the air; and a moment later she visualized a sudden image of a tiny ear of corn. The next day she brought an array of memories of childhood games of playing doctor, the game consisting always of a series of operations with knives, followed by the placing of bandages on the body, the chest and the arms. Two days later she gave a still fuller account of the tonsillectomy at the age of eight, with details of the inadequate warning and preparation which had been given her, of her final paroxysm of blind terror, and her sense of betrayal by her father. In the succeeding days of analysis she provided further data on her sense of injury in childhood, and the feeling that women in general are an injured lot. There were vivid memories of a wounded toad, and of a doddering old man who was an invalid. Then, finally, after another interval, the entire story of the tonsillectomy in its final form came through. It seems that she either had the actual experience or a dreamlike fantasy of seeing her tonsils immediately after the operation. She asked if they were not like little round red balls, about the size of olives, and if they were not like the material which is *underneath* the tongue. She added that [247] she thought she had seen the hand of a man in a white coat holding | these round, red balls like olives, or like scooped-out pieces of watermelon, on a piece of gauze. Linked to it was a memory of her uncle, who was a physician, of how she kicked, of someone holding her legs down, and of terror almost to the point of desperation. Suddenly as she recounted these thoughts she broke off and said, »I just had a very disgusting thought – a disgusting name for a man's organs is balls. I thought of these, too, but I didn't want to mention them.« We see, therefore, a long series of minor childhood injuries and accidents built into a fantasy of the damage that women suffer at the hands of men, and culminating in a tonsillectomy which in turn is linked directly to the idea of literal castration. Furthermore it is significant that this idea of castration contained the picture of two round red balls which were thought of as testicles and yet were made from material from *underneath* the tongue, so that the tongue takes its classical place as a representative of the phallus.

Again it is needless to say that one faces a perplexing phenomenon when one tries to understand how in this young woman's mind, totally unschooled or undirected into psychoanalytic ways of thinking or interpretation, such a spontaneous production of castration imagery in response to an experience of a tonsillectomy can have occurred.

And yet, that it occurred is evident. (It should be stated that the young woman was undergoing analysis in a mild depressive swing in the hope that the analytic experience might forestall further episodes of elation or depression. The material just presented appeared spontaneously in her free associations during the course of the first two weeks of analysis, before interpretations of any kind had been offered her.)

Not long ago an opportunity occurred to observe a bodily dramatization of the same problem, again in a manner which left no room for doubt as to the interpretation of the phenomenon, although in this case not a word was spoken. It was necessary to perform a lumbar puncture on a young man of twenty-six. Some years earlier this young man had been subjected to the same procedure, and at the time had suffered intense pain. As a result, he was excessively apprehensive, and consented to the lumbar puncture only on condition that it be performed under a general anæsthetic. He was admitted to the Neurological Institute and the lumbar puncture was made under nitrous oxide-oxygen anæsthesia. As the equipment was wheeled into his room his terror mounted visibly, and he began to do a rather peculiar thing: although normally modest, despite the presence of the nurses he made quick impulsive gestures which would repeatedly expose his | genitals. Then, to our amazement, as he began to be [248] affected by the anæsthetic, but before he was completely relaxed, he did exactly the reverse: that is, he reached out for all the bedclothes that he could grasp, for his bathrobe, for a pillow, for anything which his groping hands chanced to touch, and piled a protecting mountain of clothes and bedding over his genitals.

Again the question arises why an attack directed against one part of the body is deflected in the patient's mind in such a way that it apparently represents to him an attack upon his genitalia. Such a deflection seems extraordinarily uneconomical from the point of view of psychological tension and peace of mind. Without entering into a discussion of this problem, it is worth stressing that its solution may lie close to the heart of the problem of anxiety and its genesis.

The phenomenon of »displacement upward from below« is only a special instance of this phenomenon; and it might be better to speak of centrifugal and centripetal displacements, that is, of displacements away from or towards the instinctual zones. That displacement can occur in both directions is clear. The displacement centrifugally is readily explained, because it can be employed so readily to lessen tension and anxiety; but the occurrence of centripetal displacement, with its inevitable increase of anxiety, is a perplexing phenomenon.

SUMMARY

I. The basis for the symbolic representation of the body: – 1. The growth of knowledge and the growth of language depend upon states of instinctual tension in the infant and child. – 2. This tension is a body function. – 3. The first learning therefore concerns itself almost entirely with bodily things, the child learns the parts, the products, the needs and feelings of the body, and so on. – 4. All new knowledge must relate itself automatically to that already known. – 5. Therefore all new knowledge must have special points of reference to bodily things. – 6. Therefore as the outside world is apperceived, each new unit comes to have special significance with relation to various parts of the body, – i.e, representing parts of the body by analogy, at first consciously and later unconsciously. – 7. This »body language« is used freely in early childhood, but later in life occurs chiefly in dreams, in dozing states, in delirious reactions, and in symptom productions.

II. The examples given indicate that the indirect or symbolic representation of the body can be classified into two general | types. There is one large group of representa- [249]

tive objects which are drawn from the outside world: household objects and buildings, animals, machinery of various kinds, clothing, food, and so on. There is a suggestion that the type of symbol used may be correlated with different types of personality development, or with different neurotic structures. At any rate it is clear that this representation of the body by external objects is linked to the process of projection, and to the externalization of internal problems in the psychoses and psychoneuroses.

The second main form of representation is that in which one part of the body is substituted for another part of the body; and this can occur in either of two directions. Under certain conditions those parts of the body which are relatively slightly involved in any direct expression of instinctual yearnings can be substituted for parts of the body which are more intimately connected with emotional drives. And contrariwise, the opposite can occur; that is, those parts of the body which are directly involved in instinctual need and expression, may be substituted for the more indifferent zones of the body. In other words, the translation can be made either towards the instinctual zone or away from it. It is clear that in this translation of experience and feeling from one part of the body to another, one approaches closely the problem of anxiety, hysteria, and hypochondriasis.

BIBLIOGRAPHY

1. Piaget, J.: *The Language and Thought of the Child.* London: Kegan Paul, Trench, Trubner & Co., Ltd., 1926.

APPENDIX II
REFERENCES

From Dr. Gerard

1. Bullock, T. H.: Properties of a single synapse in the stellate ganglion of the squid. *J. Neurophysiol.*, **11**, *343* (1948).
2. Lloyd, D. P. C.: Post-tetanic potentiation of response in monosynaptic reflex pathways of the spinal cord. *J. Gen. Physiol.*, **33**, *147* (1949).

From Dr. Mead

1. Sapir, E.: *Language, an Introduction to the Study of Speech.* New York: Harcourt Brace (1921).
2. Mead, M.: Native languages as field work tools. *Amer. Anthro.*, **41**, *189* (1939).
3. Bateson, G.: Pidgin English and cross-cultural communications. *Tr. New York Acad. of Sciences*, Ser. 11, **6**, *137* (1944).
4. Mead, M.: »Coming of age in Samoa« in *From the South Seas.* New York: Morrow (1939), p.122.
5. Fortune, R. F.: *Arapesh, a Study of Papuan Language.* Publications of Amer. Ethnol. Soc., Vol. XIX, Locust Valley, New York: J. J. Augustin Co., Inc. (1942).
6. Mead, M.: The Mountain Arapesh, American Museum of Natural History, *Anthro. Papers*, **40**, *237* (1947).
7. Bateson, G., and Mead, M.: Balinese Character, New York Academy of Sciences (1942), plates 60-62.

From Dr. Werner

1. Von Domarus, E.: The specific laws of logic in schizophrenia, in Kasanin, J. S., *Language and Thought in Schizophrenia.* Berkeley: University of California Press (1944), p.104.
2. Arieti, S.: Special logic of schizophrenic and other types of autistic thought. *Psychiatry*, **11**, *325* (1948).
3. Werner, H.: *Comparative Psychology of Mental Development.* rev. ed., Chicago: Follett (1948).
4. Werner, H.: The test procedure and the results will be published in detail in the forthcoming monograph: Acquisition of Word Meanings. Mono. Soc. Res. Child Devel., Washington, D. C.
5. McCarthy, D.: Language development in children, in Carmichael, L., *Manual of Child Psychology*, Boston: Wiley, ch. X (1946).

From Dr. Kubie

1. Kubie, L. S.: Body symbolization and development of language. *Psychoanalyt. Quart.*, **3**, *430* (1934).
2. Silberer, H.: Berichte uber[!] eine Methode gewisse symbolische Halluzinations-Erscheinungen hervorzurufen und zu beobachten. *Jahrb. f. psychoanal. u. psychopathol. Forsch.*, **1**, *513* (1909).

CYBERNETICS

CIRCULAR CAUSAL AND FEEDBACK MECHANISMS IN BIOLOGICAL AND SOCIAL SYSTEMS

Transactions of the Eighth Conference
March 15-16, 1951, New York, N. Y.

Edited by

HEINZ VON FOERSTER
DEPARTMENT OF ELECTRICAL ENGINEERING
UNIVERSITY OF ILLINOIS
CHAMPAIGN, ILLINOIS

Assistant Editors

MARGARET MEAD
AMERICAN MUSEUM OF NATURAL HISTORY
NEW YORK, N. Y.

HANS LUKAS TEUBER
DEPARTMENT OF NEUROLOGY
NEW YORK UNIVERSITY COLLEGE OF MEDICINE
NEW YORK, N. Y.

Sponsored by the

JOSIAH MACY, JR. FOUNDATION
565 PARK AVENUE, NEW YORK 21, N. Y.

PARTICIPANTS

Eighth Conference on Cybernetics

MEMBERS

WARREN S. McCULLOCH, Chairman
Department of Psychiatry, University of Illinois College of Medicine; Chicago, Ill.

HEINZ VON FOERSTER, Secretary
Department of Electrical Engineering, University of Illinois; Champaign, Ill.

GREGORY BATESON[1]
Veterans Administration Hospital; Palo Alto, Calif.

ALEX BAVELAS
Department of Economics and Social Science, Massachusetts Institute of Technology; Cambridge, Mass.

JULIAN H. BIGELOW
Electronic Computer Project, Institute for Advanced Study; Princeton, N.J.

HENRY W. BROSIN
Western Psychiatric Institute and Clinics; Pittsburgh, Pa.

LAWRENCE K. FRANK
72 Perry Street; New York, N.Y.

RALPH W. GERARD
Department of Physiology, School of Medicine, University of Chicago; Chicago, Ill.

GEORGE EVELYN HUTCHINSON
Department of Zoology, Yale University; New Haven, Conn.

HEINRICH KLÜVER
Division of the Biological Sciences, University of Chicago; Chicago, Ill.

LAWRENCE S. KUBIE
Department of Psychiatry and Mental Hygiene, Yale University School of Medicine; New Haven, Conn.

RAFAEL LORENTE DE NO[1]
Rockefeller Institute for Medical Research; New York, N.Y.

DONALD G. MARQUIS[1]
Department of Psychology, University of Michigan; Ann Arbor, Mich.

MARGARET MEAD
American Museum of Natural History; New York, N.Y.

F. S. C. NORTHROP[1]
Department of Philosophy, Yale University; New Haven, Conn.

WALTER PITTS
Department of Mathematics, Massachusetts Institute of Technology; Cambridge, Mass.

ARTURO S. ROSENBLUETH
Departments of Physiology and Pharmacology, Instituto Nacional de Cardiologia; Mexico City, D. F., Mexico

LEONARD J. SAVAGE
Committee on Statistics, University of Chicago; Chicago, Ill.

THEODORE C. SCHNEIRLA[1]
American Museum of Natural History; New York, N.Y.

HANS LUKAS TEUBER
Department of Neurology, New York University College of Medicine; New York, N.Y.

GERHARDT VON BONIN
Department of Anatomy, University of Illinois College of Medicine; Chicago, Ill.

JOHN VON NEUMANN[1]
Department of Mathematics, Institute for Advanced Study; Princeton, N. J.

NORBERT WIENER[1]
Department of Mathematics, Massachusetts Institute of Technology; Cambridge, Mass.

GUESTS

HERBERT G. BIRCH
Department of Psychology, City College of New York; New York, N.Y.

JOHN R. BOWMAN
Department of Physical Chemistry, Mellon Institute of Industrial Research; University of Pittsburgh; Pittsburgh, Pa.

DONALD M. MACKAY
Department of Physics, King's College, University of London; London, England

IVOR A. RICHARDS
Graduate School of Education, Harvard University; Cambridge, Mass.

DAVID McKENZIE RIOCH
Division of Neuropsychiatry, Army Medical Service Graduate School; Army Medical Center; Washington, D.C.

CLAUDE SHANNON
Murray Hill Laboratory, Bell Telephone Laboratories, Inc.; Murray Hill, N. J.

JOSIAH MACY, JR. FOUNDATION

FRANK FREMONT-SMITH
Medical Director

JANET FREED
Assistant for the Conference Program

1 Absent

JOSIAH MACY, JR. FOUNDATION CONFERENCE PROGRAM

As an introduction to these Transactions of the Eighth Conference on Cybernetics I should like to outline what it is that the Foundation hopes to accomplish in its Conference Program. We are interested, first of all, in furthering knowledge about cybernetics, and to this end the participants were brought together to exchange ideas, experiences, data, and methods. In addition to this particular goal, however, there is a further, and perhaps more fundamental, aim which is shared by all our conference groups: the promotion of meaningful communication between scientific disciplines.

The problem of communication between disciplines we feel to be a very real and very urgent one, the most effective advancement of the whole of science being to a large extent dependent upon it. Because of the accelerating rate at which new knowledge is accumulating and because discoveries in one field so often result from information gained in quite another, channels must be established for the most relevant dissemination of this knowledge.

The increasing realization that nature itself recognizes no boundaries makes it evident also that the continued isolation of the several branches of science is a serious obstacle to scientific progress. Particularly is it so in medicine that the limited view through the lens of one discipline is no longer enough. For example, today medicine must be well versed in nuclear physics because of the tracer techniques and the injury which can result from radiation. At the other extreme, medicine is certainly a social science and, through mental health, must be concerned with economic and social questions. The answer, then, is not further fragmentation into increasingly isolated specialities, disciplines, and departments, but the integration of science and scientific knowledge for the enrichment of all branches. This integration, we feel, can be encouraged by providing opportunities for a multiprofessional approach to given topics.

Although the fertility of the multidiscipline approach is recognized, adequate provision is not made for it by our universities, scientific societies, and journals. And perhaps the presence of other hindering factors must be admitted. Partly semantic in nature, they may also to some degree be psychological. Admittedly, it is oftentimes difficult to accept data derived from methods with which one is unfamiliar. By making | [viii] free and informal discussion the central core of our meetings, we hope to achieve an atmosphere which minimizes as much as possible these emotional barriers.

Thus, our meetings are in contrast to the usual scientific gatherings. They are not designed to present neat solutions to tidy problems but to elicit provocative discussion of the difficulties which are being encountered in research and practice. For this reason, we ask that the presentations be relatively brief and that emphasis be placed on discussion as the heart of the meeting. Our hope is that the participants will come prepared not to defend a single point of view but to take advantage of the meeting as an opportunity to speak with representatives of other disciplines in much the same way that they would talk with their own colleagues in their own laboratories.

We have, now, thirteen groups functioning under the Conference Program. The following topics are covered: adrenal cortex, aging, blood clotting, cold injury, connective tissues, consciousness, cybernetics, infancy and childhood, liver injury, metabolic interrelations, nerve impulse, renal function, and shock and circulatory homeostasis.

When a new conference is organized, the Chairman, in consultation with the Foundation, selects fifteen scientists to be the nucleus of the group, and every effort is made to include representatives from all pertinent disciplines. From time to time new members are added by the group to fill gaps in viewpoint or technique. A limited number of guests are invited to attend each meeting, but, for the purpose of promoting full participation by all members and guests, attendance at any meeting is limited to twenty-five. It is inevitable that in no topic can we possibly include more than a small fraction of the key investigators in the field, and one of the difficulties in forming a group like this is that it is necessary to leave out so many people whom we would like to include.

The transactions of these meetings are recorded and published. This is done because the Foundation wishes to make current thinking in a field available to all those working in it, and because it believes that conveying to those in other fields who are concerned with science, for example, government officials, administrators, etc., the essential nature of scientific research is also an important problem in communication. Logic is a vital aspect of science, but equally essential is the intuitive or creative aspect. Research is as creative as the painting of a portrait or the composing of a symphony. Although logic is, of course, necessary in order to rearrange, to test, and to validate, research thrives on creativity which has its source in unconscious, nonrational processes. | [ix]

Unfortunately, however, in the finished products which are presented to the world through research reports this integral part of scientific endeavor is shriveled by the cold, white light of logic. By preserving the informality of our conferences in the published transactions, we hope to give a truer picture of what actually goes on in the minds of scientists and of the role which creativity plays.

FRANK FREMONT-SMITH, M.D.
Medical Director

A NOTE BY THE EDITORS

To the reader of this somewhat unusual document, a few words of explanation, and caution. This is not a book in the usual sense, nor the well-rounded transcript of a symposium. These pages should rather be received as the partial account of conversations within a group, a group whose interchange actually extends beyond the confines of the two day meeting reported herein. This account attempts to capture a fragment of the group interchange in all its evanescence, because it represents to us one of the few concerted efforts at interdisciplinary communication.

The members of this group share the belief that one can and must attempt communication across the boundaries, and often chasms, which separate the various sciences. The participants have come from many fields; they are physicists, mathematicians, electrical engineers, physiologists, neurologists, experimental psychologists, psychiatrists, sociologists, and cultural anthropologists. That such a gathering failed to produce the Babylonian confusion that might have been expected is probably the most remarkable result of this meeting and of those which preceded it.

This ability to remain in touch with each other, to sustain the dialogue across departmental boundaries and, in particular, across the gulf between natural and social sciences is due to the unifying effect of certain key problems with which all members are concerned: the problems of communication and of self-integrating mechanisms. Revolving around these concepts, the discussion was communication about communication, necessarily obscure in places and for more than one reason. Yet the actual outcome was far more intelligible than one might think, so that the editors felt enjoined to reproduce the transcript as faithfully as possible.

The social process, of which these transactions are an incomplete residue, was not a sequence of formal »papers« read by individual participants and punctuated by prepared discussion. With few exceptions people spoke freely and without notes. Unavoidably some speakers produced inaccurate memories of their own facts, or of those of others, and trends of thought were often left incomplete. The printed record preserves the essential nature of this interchange in which partial associations were permitted on the assumption that closure would take place, at some other time, producing new ideas or reinforcing those that were thought of in passing.

Stimulation, for many scientists, comes from such partial, and sometimes even inaccurate, reproductions of material from widely separated fields, fields which seem dissimilar except for the logical structure of | their central problem. If the reader wishes a format statement of the work and point of view of individual participants, he will have to consult other sources. This can be done with ease since most of the contributors have provided references to previously published material. [xii]

The reader should be warned that the presentations and discussion tend to be responsive to previous meetings of the group. Some statements were designed to answer questions asked months or years before, or designed to evoke some long-anticipated answer from a fellow member. Radical changes in the manuscript would have been necessary had we attempted to rid the group of its history. Such changes would have been distortions, and would prevent the reader from noticing the unfinished state of the group's affairs.

Our editorial procedure, nevertheless, involved some revision of the transcript prior to publication. A verbatim record based on the stenotyped protocol, even if perfect, would in fact have been an incomplete and misleading account. It would have given the verbal content, but the tones of voice, the gestures, the attention directed by the turn of head toward one person or another would all be missing. For this reason, we adopted a more traditional procedure. Each participant was supplied with a mimeographed copy of his transcribed remarks, and was given the chance to revise his material for the sake of clarity and coherence. Not all of the participants availed themselves of this opportunity so that our working copy represented a mixture of revised and unrevised contributions. In the unrevised passages the editors corrected only those statements which seemed to them obvious errors in recording. For the rest, they confined their censorial activities to occasional deletions of overlapping or repetitive passages and of a few all-too-cryptic digressions. Most of the asides, such as jokes or acidities, were preserved as long as they seemed intelligible to people outside the group.

The editors were eager to retain the participants' first names in the printed record when they had been used during the discussion, but this would have been an unnecessary handicap to the reader. However, the reader should realize that most speakers addressed each other informally as a consequence of acquaintance outside the framework of these particular meetings. The occasional shifts to more formal modes of addressing each other was therefore indicative of distance and sometimes of disagreement. The use of last names also underscored the special role of the invited guests and of the subtle differences in pace and tone which some of them introduced into the meeting.

It is noteworthy that these invited guests cannot be identified by any obvious differences between their vocabulary and that of the regular members. One of the most surprising features of the group is the [xiii] | almost complete absence of an idiosyncratic vocabulary. In spite of their six years of association, these twenty-five people have not developed any rigid, in-group language of their own. Our idioms are limited to a handful of terms borrowed from each other: analogical and digital devices, feedback and servomechanisms, and circular causal processes. Even these terms are used only with diffidence by most of the members, and a philologist given to word-frequency counts might discover that the originators of »cybernetics« use less of its lingo than do their more recent followers. The scarcity of jargon may perhaps be a sign of genuine effort to learn the language of other disciplines, or it may be that the common point of view provided sufficient basis for group coherence.

This common ground covered more than the mere belief in the worthwhileness of interdisciplinary discussion. All of the members have an interest in certain conceptual models which they consider potentially applicable to problems in many sciences. The concepts suggest a similar approach in widely diverse situations; by agreeing on the usefulness of these models, we get glimpses of a new *lingua franca* of science, fragments of a common tongue likely to counteract some of the confusion and complexity of our language.

Chief among these conceptual models are those supplied by the theory of information.[1, 2, 3] This theory has arisen under the pressure of engineering needs; the efficient design of electronic communication devices (telephone, radio, radar, and television) depended on achieving favorable »signal-to-noise ratios.« Application of mathematical tools to these problems had to wait for an adequate formulation of »Information« as contrasted to »noise.«

If noise is defined as random activity, then information can be considered as order wrenched from disorder, as improbable structure in contrast to the greater probability of randomness. With the concept of entropy, classical thermodynamics expressed the

1 Shannon, C. E.: A mathematical theory of communication. *Bell System. Tech. J.* 27, 379-423 and 623-656 (1948).

2 Shannon, C. E., and Weaver, W.: *The Mathematical Theory of Communication.* Urbana: University of Illinois Press, 1949 (p. 116).

3 Wiener, N.: *Cybernetics or Control and Communication in the Animal and the Machine,* New York: Wiley, 1948 (p. 194).

universal trend toward more probable states: any physiochemical change tending to produce a more nearly random distribution of particles. Information can thus be formulated as negative entropy, and a precise measure of certain classes of information can be found by referring to degrees of improbability of a state.

The improbable distribution of slots in a slotted card, or the improbable arrangement of nucleic acids in the highly specific pattern of a | gene both can be considered [xiv] as »coded« information, the one decoded in the course of a technical (cultural) process, the other in the course of embryogeny. In both instances, that of the slotted card and that of the gene, we are faced not only with carriers of information but with powerful mechanisms of control: the slotted card can control long series of processes in a plant (without itself furnishing any of the requisite energy); the gene, as an organic template, somehow provides for its own reproduction and governs the building of a multicellular organism from a single cell. In the latter case, mere rearrangement of submicroscopic particles can apparently lead to mutations, improving or corrupting the organism's plans as the case may be. Such rearrangement may indeed be similar to the difference brought about by the transposing of digits in numbers, 724 to 472, or by transposing letters in words such as art and rat.[4]

Extension of information theory to problems of language structure has been furthered by psychologists and statisticians.[5, 6, 7, 8, 9] There are unexploited opportunities for additional applications of the theory in comparative linguistics, and more particularly in studies of the pathology of language. Yet available work is sufficient to show how communication considered from this standpoint can be investigated in mechanical systems, in organisms, in social groups;[10] and the logical and mathematical problems that go into the construction of modern automata, in particular the large electronic computers,[11] have at least partial application to our theorizing about nervous systems and social interactions.

A second concept, now closely allied to information theory, is the notion of circular causal processes. A state reproducing itself, like an organism, or a social system in equilibrium, or a physiochemical-aggregate in a steady-state, defied analysis until the simple notion of one-dimensional cause-and-effect chains was replaced by the bidimensional notion of a circular process. The need for such reasoning was clear to L. J. Henderson, the physiologist, when he applied the logic | of Gibbsian physicochemical [xv] systems[12] to the steady-states of human blood,[13] and to integration in social groups, down to miniature social systems.[14] Quite independently, social scientists had been tending in the same direction, as witnessed by the work of the functional anthropologists Radcliffe-Brown[15] and Bateson.[16, 17] In ecology, the concept of circular causal

4 GERARD, R. W.: *Unresting Cells.* New York: Harper, 1940.

5 FRICK, F. C., and MILLER, G. A.: Statistical behavioristics and sequences of responses. *Psychol. Rev.* **56**, *311* (1949).

6 MILLER, G. A.: *Language and Communication.* New York: McGraw-Hill, 1951.

7 MILLER, G. A., and SELFRIDGE, J. A.: Verbal context and the recall of meaningful material. *Am. J. Psychol.* **63**, *176* (1950).

8 NEWMAN, E. B.: Computational methods useful in analyzing series of binary data. *Am. Psychol.* **64**, *252* (1951).

9 —: The pattern of vowels and consonants in various languages. *Ibid.*, 369.

10 BAVELAS, A.: A mathematical model for group structures. *Appl. Anthropol.* **7**, (part 3), *16* (1948).

11 VON NEUMANN, J.: The general and logical theory of automata. *Cerebral Mechanisms in Behavior (The Hixon Symposium).* Jeffress, L. A., editor. New York: Wiley, 1951 (pp. 1-41).

12 GIBBS, J. W.: On the equilibrium of heterogeneous substances. *The Collected Works. Vol. 1. Thermodynamics.* New York: Longmans, 1928 (pp. 55-371).

13 HENDERSON, L. J.: *Blood: A Study in General Physiology.* New Haven: Yale University Press, 1928 (p. 390).

14 —: Physican and patient as a social system. *New England J. Med.* **212**, *819* (1935).

systems has been employed by Hutchinson,[18] and further applications in statistical biology and genetics can be expected.

The remarkable constancy in the concentration of certain substances in the fluid matrix of the body led Claude Bernard originally to posit the fixity of the »milieu interieur« as one of the elementary conditions of life.[19] Cannon[20] designated as »homeostasis« those functions that restore a disturbed equilibrium in the internal environment – the complex self-regulatory processes which guarantee a relative constancy of blood sugar level, of osmotic pressure, of hydronium ion concentration, or of body temperature. Many of these processes are at least partially understood, but, as Klüver[21] has pointed out, we know next to nothing of the physiological functions which underlie our perceptual »constancies.«

Normal perception is reaction to relations, to »universals« such as size, shape, and color. Perceived objects tend to remain invariant in their size while distance from the observer varies; perceived shapes and colors retain subjective identity in varying positions and under varying illumination. This crucial problem for the physiological psychology of perception is rarely faced[22] and the neural correlates for our reaction to universals are still *sub judice.*

[xvi] Recent attempts at identifying a possible neural basis for our re|actions to universals[23] have adduced hypothetical sustained activity in neuronal circuits as one of the prerequisites for the central processes which guarantee perceptual constancies. Persistent circular activity in nervous nets had been postulated on theoretical grounds by Kubie[24] over twenty years ago, thereby anticipating the subsequent empirical demonstration of such reverberating circuits by Lorente de Nó.[25] The importance of Lorente de Nó's disclosures for neurological theory lies in the fact that, earlier in the century, many investigators considered the central nervous system as a mere reflex-organ; the mode of action of this organ, despite all the evidence to the contrary, was viewed as limited to the relating of input to output, stimulus and response corresponding to cause and effect. The possibility of self-sustained central activity in the nervous system was overlooked. Thence the denial of memory-images in early behaviorism, the emphasis on chain reflexes in attempts at explaining coordinated action.

To this day, many psychologists tend to see the prototype of all learning in elementary conditioned reflexes, a tendency which cannot be understood unless one assumes, with Lashley,[26] that these psychologists are still handicapped by »peripheralistic«

15 Radcliffe-Brown, A. R.: *The Andaman Islanders. A Study in Social Anthropology.* Cambridge, England: Cambridge University Press, 1922 (pp. XIV and 504).

16 Bateson, G.: *Naven: A Survey of the Problems Suggested by a Composite Picture of the Culture of a New Guinea Tribe, Drawn from Three Points of View.* Cambridge: Cambridge University Press, 1936 (p. 286).

17 —: Bali: the value system of a steady state. In: *Social Structure. Studies Presented to A. R. Radcliffe-Brown.* Oxford University Press, 1949 (p. 35).

18 Hutchinson, G. E.: Circular causal systems in ecology. *Ann. New York Acad. Sc.* **50**, *221* (1948).

19 Bernard, C.: *Leçons sur les phénomènes de la vie communes aux animaux et aux végétaux.* Two volumes. Paris: J. B. Baillière, 1878-79.

20 Cannon, W. B.: *The Wisdom of the Body.* New York: Norton, 1932 (p. XV and 312).

21 Klüver, H.: *Behavior Mechanisms in Monkeys.* Chicago: University of Chicago Press, 1933 (p. XVIII and 387).

22 —: Functional significance of the geniculo-striate system. *Biol. Symposia* **7**, *253* (1942).

23 Pitts, W., and McCulloch, W. S.: How we know universals; the perception of auditory and visual forms. *Bull. Math. Biophys.* **9**, *127* (1947).

24 Kubie, L. S.: A theoretical application to some neurological problems of the properties of excitation waves which move in closed circuits. *Brain* **53**, *166* (1930).

25 Lorente de Nó, R.: Analysis of the activity of the chains of internuncial neurons. *J. Neurophysiol.* **1**, *207* (1938).

notions, the unwarranted idea that central nervous activity cannot endure in the absence of continuing specific input from the periphery. Undoubtedly, the action of reverberating circuits can be overgeneralized, and has been abused. Long-range memory may need more permanent neural changes, but the notion of such circuits has suggested models of neural activity which are potentially testable and therefore of value.[27, 28]

Activity in closed central loops thus has to be carefully distinguished from the older and simpler notions of neural circuits, circuits which join periphery and centers, as in the classical conception of postural reflexes. Sir Charles Bell[29] spoke of a »nervous circle which connects the voluntary muscles with the brain.« Through such circuits, muscles in a limb maintain a given tension, as long as motor impulses flow into the muscles according to the sensory signals which issue from these same muscles. The idea antedated the discovery of the sense organs (muscle | spindles) which monitor a state of [xvii] tension in the muscle and, by increasing their rate of centripetal firing, set off centrifugal volleys which shorten the muscle in response to imposed stretch.

Numerous analogues for such recalibrating mechanisms can be found in those modern electronic devices in which output is regulated by constant comparison with input. The automatic volume control circuit of a radio receiver prevents »blasting« by decreasing the volume as the signal is increased and counteracts »fading« by increasing the volume. A speed control unit slows a motor down when its revolutions exceed a desired value and speeds it up when revolutions fall below this value. Such »feedback« or »servomechanisms«[30] are man-made models of homeostatic processes. They are not exclusively found among electronic devices. In the days of the thermal engine, Maxwell[31] developed the theory of the mechanical »governor« of steam engines. Small versions of this governor are still found today in old-fashioned phonograph turntables. Two massive metal spheres are suspended by movable links from a vertical shaft which spins with the main shaft of the machine. As speed increases, centrifugal force drives the metal spheres apart and increases the drag on the shaft; the machine slows down. Again, with the spheres sinking low, the drag on the shaft is decreased and the machine speeds up. Such a governor insures approximate constancy of speed in the engine, by the simplest mechanical means, and, in contrast to many more complicated devices, the mechanical governor shows little likelihood of going into uncontrollable oscillations.

Recent complex electronic devices are not only »error-controlled« (like a mechanical governor), but can be so built as to »seek« a certain state, like »goal-seeking« missiles which predict the future position of a moving target (at time of impact) by extrapolation from its earlier positions during pursuit. Such devices embody electronic computing circuits, and the appearance of »purpose« in their behavior (a feedback over

26 Lashley, K. S.: Discussion. In: *Cerebral Mechanisms in Behavior (The Hixon Symposium).* Jeffress, L. A., editor. New York: Wiley, 1951 (p. 82).

27 McCulloch, W. S.: A heterarchy of values determined by the topology of nerve nets. *Bull. Math. Biophys.* 7, *89* (1945).

28 Hebb, D. O.: *Organization of Behavior: A Neuropsychological Theory.* New York: Wiley, 1949 (p. XIX and 335).

29 Bell, C.: On the nervous circle which connects the voluntary muscles with the brain. *Proc. Roy. Soc.* 2, *266* (1826).

30 MacColl, L. R.: *Fundamental Theory of Servo-Mechanisms. New* York: Van Nostrand, 1945.

31 Maxwell, C.: On governors. *Proc. Roy. Soc.* 16, *270* (1868).

the target!) has intrigued the theorists[32, 33] and prompted the construction of such likeable robots as Shannon's electronic rat described in this volume.

The fascination of watching Shannon's innocent rat negotiate its maze does not derive from any obvious similarity between the machine and a real rat; they are, in fact, rather dissimilar. The mechanism, however, is strikingly similar to the *notions* held by certain learning theorists about rats and about organisms in general. Shannon's con-

[xviii] struc | tion serves to bring these notions into bold relief.

Recent emphasis on giant electronic computers as analogues of the human brain should perhaps be considered in the same light. The logical and mathematical theories demanded by the construction of these computers raise problems similar to those faced on considering certain aspects of the nervous system or of social structures.[11] It is no accident that John Von Neumann, a mathematician who is currently concerned with the theory of computers, should be more generally known for his analysis of human interaction in games and economic behavior,[34] and that Norbert Wiener, after working on computers and guided missiles, turned to the consideration of the social significance of these mechanisms.[35] Brief consideration of computers may therefore be in order.

Computers are constructed on either of two principles: they may be digital or analogical. In an analogical device, numbers are represented by a continuous variation of some physical quantity, a voltage, say, or a distance on a disc. A digital device, however, represents numbers as discrete units which may or may not be present, e.g., a circuit that may be open or closed, and the basic alphabet of the machine may be a simple yes or no, zero or one.

Peripheral neurons act on an all-or-none principle, and synapses in the central nervous system are frequently considered to act similarly. Theories of central nervous activity have consequently often paralleled those required for digital rather than analogue machines (cf., Pitts and McCulloch[23]). The applicability of digital notions to the actions of the central nervous system has been questioned,[36] but the calculus worked out for handling them is certainly applicable to electronic digital computers,[37] and the very fact that testable theories of nerve action have been proposed is due to the availability of the electronic models.

We all know that we ought to study the organism, and not the computers, if we wish to understand the organism. Differences in levels of organization may be more than quantitative.[38] But the computing robot provides us with analogues that are helpful as far as they seem to hold, and no less helpful whenever they break down. To find

[xix] out in what ways a nervous system (or a social group) differs from our man-made | analogues requires experiment. These experiments would not have been considered if the analogue had not been proposed, and new observations on biological and social

32 Rosenblueth, A., Wiener, N., and Bigelow, J.: Behavior, purpose and teleology. *Philos. of Sc.* **10**, *18* (1943).

33 Northrop, F. S. C.: The neurological and behavioristic psychological basis of the ordering of society by means of ideas. *Science* **107**, *411* (1948).

34 Von Neumann, J., and Morgenstern, O.: *Theory of Games and Economic Behavior.* Princeton: Princeton University Press, 1944.

35 Wiener, N.: *The Human Use of Human Beings.* Boston: Houghton Mifflin, 1950.

36 Gerard, R. W.: Some of the problems concerning digital notions in the central nervous system. *Cybernetics.* von Foerster, H., editor. Trans. Seventh Conf. New York: Josiah Macy, Jr. Foundation, 1950 (p. 11).

37 McCulloch, W. S., and Pitts, W.: A logical calculus of the ideas immanent in nervous activity. *Bull. Math. Biophys.* **5**, *115* (1943).

38 Schneirla, T. C.: Problems in the biopsychology of social organization. *J. Abn.. Soc. Psychol.* **41**, *385* (1946).

systems result from an empirical demonstration of the shortcomings of our models. It is characteristic that we tend to think of the intricacies of living systems in terms of non-living models which are obviously less intricate. Still, the reader will admit that, in some respects, these models are rather convincing facsimiles of organismic or social processes – not of the organism or social group as a whole, but of significant parts.

How this way of thinking emerged in the group is difficult to reconstruct. From the outset, John Von Neumann and Norbert Wiener furnished the mathematical and logical tools. Warren McCulloch, as the group's »chronic chairman« infused it with enthusiasm and insisted on not respecting any of the boundaries between disciplines. The Josiah Macy, Jr. Foundation, through Dr. Frank Fremont-Smith, provided the physical setting but actually much more than that: the social sanction for so unorthodox an undertaking. The confidence of the Foundation and of Dr. Frank Fremont-Smith made it possible to obtain a type of cross-fertilization which has proved rewarding over a period of six years.

The gradual growth of the principles can be recognized from dates and titles of successive Conferences. A nucleus of the current Cybernetics Conference seems to have been formed in May 1942 at the Macy Foundation Conference on »Cerebral Inhibition.« Among the participants were: Gregory Bateson, Lawrence K. Frank, Frank Fremont-Smith, Lawrence Kubie, Warren McCulloch, Margaret Mead, and Arturo Rosenblueth. All of these later became members of the continuing group devoted to the discussion of »cybernetics.« The publication of the article on »Behavior, Purpose and Teleology« by Rosenblueth, Wiener, and Bigelow in 1943[32] focused attention on several of the problems which led to the organization of the first Macy Foundation Conference in March 1946 devoted to »Feedback Mechanisms and Circular Causal Systems in Biological and Social Systems.«

The fall of 1946 found the group very active. Two meetings sponsored by the Macy Foundation followed rapidly upon each other: first, in September, a special meeting on »Teleological Mechanisms in Society«; then, in October, the second Conference on Teleological Mechanisms and Circular Causal Systems. Next, the group formed the nucleus of a formal symposium on »Teleological Mechanisms« held under the auspices of the New York Academy of Sciences.[39]

In the following year, 1947, the third Macy Foundation conference was held, retaining the title of the second meeting; the fourth and | fifth conferences, in 1948, [xx] were entitled: Conference on Circular Causal and Feedback Mechanisms in Biological and Social Systems. The fifth conference concerned itself particularly with considerations of the structure of language.

With the publication of Norbert Wiener's book *Cybernetics,* a term appeared which was unanimously chosen as title for the sixth conference in the spring of 1949. The title *Cybernetics* was maintained for the seventh and the present eighth conference 1950, 1951 with the subtitle *Circular Causal and Feedback Mechanisms in Biological and Social Systems.*

Through the fifth conference, no transactions were published. With the sixth conference (1949) our program of publication began, so that the two preceding conferences, the sixth[40] and seventh,[41] are available in print. The reader might suspect that this urge to fix the group process in printed form is the beginning of fossilization. He may be right, but we prefer to think of it in terms more favorable to the group.

39 Frank, L. K., *et al.*: Teleological mechanisms. *Ann. New York Acad. Sc.* **50**, *189* (1948).

40 *Cybernetics.* Von Foerster, H., editor. Trans. Sixth Conf. New York: Josiah Macy, Jr. Foundation, 1949.

41 *Cybernetics.* Von Foerster, H., editor. Trans. Seventh Conf. New York: Josiah Macy, Jr. Foundation, 1950.

For well over two thousand years, the »symposium« has been a setting for the matching and sharpening of ideas. Evolved from the Attic stage, the literary form created by Plato has persisted through the Middle Ages and the Renaissance. Until recently, it was unrestrained by stenotypists and tape recorders. Few of the classical symposia were anything but prose poems of one man's making. They brought a simple message, stated in contrapuntal fashion. Now we have our modern devices for the recording and storing of information. Communication transmitted has been infinitely multiplied in volume, but the thinking of simplifying ideas has not kept pace.

Whether the meetings here recorded contain such simplifying ideas, the editors would not presume to say. Some of us believe we can see such ideas here and there. For this reason, we preserved the record, and exhibit it to others for their judgment.

Pressure of time made it impossible in the last two years to sum up the historical background and to formulate an editorial policy. The published records of the sixth and seventh conferences, in 1949 and 1950, therefore appeared without any introduction. We hope that this note will serve as a preface to the earlier reports, as well as to the present publication.

HEINZ VON FOERSTER
MARGARET MEAD
HANS LUKAS TEUBER

New York, N.Y.
January 29, 1952

COMMUNICATION PATTERNS IN PROBLEM-SOLVING GROUPS

[1]

ALEX BAVELAS
Department of Economics and Social Science
Massachusetts Institute of Technology

I AM GOING to describe two very simple experiments which are related and which I have chosen because I think they give a good picture of the way we are trying to get acquainted with our problem. I believe they give an idea of the motivation and the spirit of the work. They are not elegant experiments, and the work is in an exploratory stage. We are not striving for niceness of design but, rather, for a maximum of interplay between the experimenter and his material.

I should like to state the problem, describe the experiments, and then tell you about some of the notions from which we are trying to build a theoretical framework for understanding what is happening. The problem is this: If a task is of such a nature that it must be performed by a group rather than by a single individual, communication is usually necessary; but does it make any difference who may communicate with whom? In other words, if I draw circles to indicate people, and draw arrows between people (Figure 1) showing that a message may go from one to the other, does it make any difference, in a group situation, where and in what direction these arrows go? The experiments I shall describe deal entirely with a special set of such patterns which we call »connected groups.« By »connected group« we mean such a group that, taking any pair of individuals, it is possible, over some route, for the individuals that make that pair to exchange messages, i.e., the group shown in Figure 2 would not be a connected group; but the groups shown in Figures 3, 4, and 5 would be connected groups.

Of course, as long as there is an experimenter who imposes a task and who looks at the results later, I suppose we ought to indicate that he is present as a sink and a source. We feel, however, that the experiments were so conducted that so far as the subjects were concerned, they were operating without the experimenter in a sink-source role for them.

If you were a subject in one of the experiments that I shall describe, the first thing that would happen would be that I or some other member of the research group would appear in class about five minutes after the class had begun. I would give you a souped-up story about a | [Figure 1-5] | research experiment that was being carried on and I would explain that it was supported by the Air Force, that it was quite important, that it had to do with communication, and that we would like some volunteers. [2] [3]

When you arrived at the laboratory, you would be asked to sit down and you would be presented with five cards. On each of those cards would appear five symbols – perhaps a cross, an asterisk, a square, a circle, and a triangle. Each set would be different. They would be so arranged that each symbol would appear on four of the five cards, but one symbol would appear on all five cards. You would be told to look at those five cards and tell as quickly as you could which symbol was common to the five cards. You would be asked to do this for several sets of cards.

Then we would say: »What you have just done is the experimental task. Instead of doing it yourself by being able to see all five cards, the task will be done by five people. You will be separated from the others, and each of you will have one of the five cards.

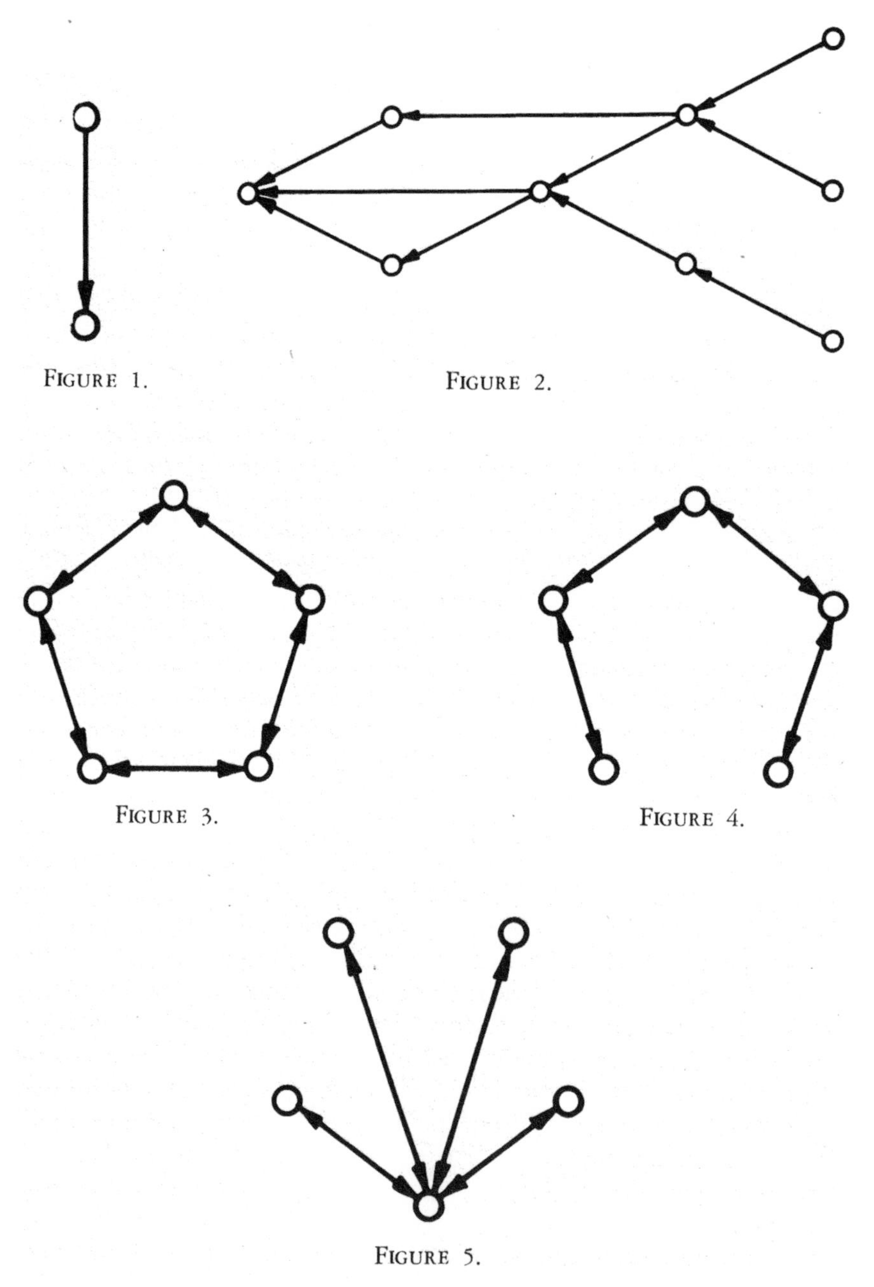

Figures 1 through 5. Reprinted by permission from
BAVELAS, A.: *J. Acoustical So. Am.* 22, 725 (1950).

By writing notes to the others, you will determine as quickly as possible what the common symbol is.« Then you would be led to another room where you and the other four subjects would sit around a table. The table would have partitions built on it so that you couldn't see the others. Each subject would have a little box full of cards of the same color as the cubicle in which he was sitting. The cards would be numbered in sequence. If you wanted to send a message you would write it on a card and put it into the slot that had the color of its destination. You would be instructed to keep all the messages you received; anything you sent out would have to be on cards of the color assigned to you.

There would be no restriction as to what you might write on the cards. You could write anything you pleased. As soon as you knew the common symbol, you would turn to a little box with six switches on it, each switch labeled with one of the six symbols, and you would press the appropriate switch. You could send the answer out to the others, too. These switches would operate a lightboard in another room on which the responses were counted. A single trial would end as soon as each man had a light on; in other words, it would end when each man had pressed a switch.

SAVAGE: Does each individual know by looking at his own setup whom he can reach, and does he also know, presumably by having it explained to him, whom the other participants can reach, or does he have to learn through participation this kind of information?

BAVELAS: We began these experiments with the notion that we would show each group what the network was. After they had sat at their places, they would look at a little diagram which would show the net|work; but we found that we couldn't discover a way of doing that without biasing the performance of the group. We decided finally to tell the subjects nothing about the network, so that when each individual sat down at the table all he knew was that there were two slots that were open, that one was colored green and the other blue, and that when he sent a message in that slot it went to somebody designated by blue; but he didn't know where that person was sitting, whether it was next to him or not. [4]

FREMONT-SMITH: He has only two slots?

BAVELAS: He may have one, two, three, or four, depending on the net, but each man knows what he has. Gradually, they discover what the network is, in some patterns. In other patterns they never quite know.

KUBIE: Some men will have one or two slots open, and others will have three or four in the same trial?

BAVELAS: They might; the net is imposed in the beginning.

KUBIE: But they don't all have the same number of open slots?

BAVELAS: They might or might not.

VON BONIN: He doesn't know who is Red? He doesn't know with whom Blue can communicate?

BAVELAS: Not unless Blue tells him. Blue, of course, may tell him. This is a very interesting thing to watch.

SAVAGE: How can Blue tell, if he doesn't know where he is himself?

BAVELAS: What frequently happens is that a man's first message to the next man is, »I have a square, and I can send messages only to you.«

BIGELOW: Does the curtain go up on everybody at the same moment?

BAVELAS: You mean when we »uncover« the table?

BIGELOW: Yes.

BAVELAS: Yes. After they have taken their places, we uncover the table and say, »Are you ready for trial No. 1?« They know what the task is, and we have described the table with pictures, and so on, to them, so they know what to do. They go right to work. There are rarely any mixups.

HUTCHINSON: Is there any effect due to psychological associations with the color names?

BAVELAS: We don't know of any.

SAVAGE: Are the individuals permitted, or even encouraged, to keep notes and diagrams of their own? Does each have a blank sheet of paper?

BAVELAS: They have blank paper and a number of pencils.

We chose three patterns of groups to experiment with; I shall report only on these three, although we have done work with some others. We picked one pattern, which [5] is shown in Figure 3, because the topo|logical properties in each place are the same; that is, any differences with respect to some of the things we were measuring and hoped to locate in pattern relations should be zero here. We hoped that it would give us some notion of how big a difference was a real difference. Another pattern chosen is the one shown in Figure 4. The last one is shown in Figure 5.

In discussing the results, I am not going to present numbers; rather, with words, I shall give you a general notion of what seemed to happen, and I shall mention only those things for which the numbers give us some reason to believe that the differences are significant.

First, of all, there is the matter of speed. We measured speed by counting the number of seconds between the beginning of the trial and the end of the trial. The pattern shown in Figure 3 turns out to be quite slow; the pattern shown in Figure 5 is quite fast; and the one shown in Figure 4 falls somewhere in between.

FREMONT-SMITH: Do you want to give us a rough idea of the duration of a trial?

BAVELAS: Yes. Each group does fifteen trials. By the time they have done six, the picture with respect to time is fairly steady; that is, it doesn't change much after that; in fact, it hardly changes at all. After they have reached this steadiness, the pattern shown in Figure 3 is doing a trial at about 60 to 75 seconds.

PITTS: By the sixth trial they would be assumed to know the network. Is that correct? Or does it change? The fifteen trials are all run with the same connectivity?

BAVELAS: Yes, and the group is never used twice.

SAVAGE: So one group faces only one connectivity?

BAVELAS: Yes.

SAVAGE: And then they are presumed to learn the connectivity in the early trials, so that in the later ones there are typically no messages saying, »I can speak to So-and-So«?

BAVELAS: That's right.

SAVAGE: And is it possible, at least legally possible, for them to form a conspiracy, as it were; that is, to agree on a way of getting the thing over with quickly?

BAVELAS: Oh, yes, and of course, they do. They are working against time. They are told that other groups are doing this, that we are trying to find the best group, and that they are to be compared with those other groups.

PITTS: Do you find it takes longer to learn one connectivity than another, in the sense of leveling off?

BAVELAS: Some are never really well understood. In general, an individual in a net [6] does not know what the net looks like, beyond two | transmission links; along the shortest path, of course.

MEAD: There is no illegality in giving people information?

BAVELAS: No.

MEAD: That is important to point out, I think.

BAVELAS: There is no information that is barred.

SAVAGE: When you say some are not learned at all, do you mean that, even among these three patterns, there are some that are not learned at all?

BAVELAS: Yes. In the pattern shown in Figure 4, the individual at one end doesn't usually know what is at the other end.

BIGELOW: Whereas in the pattern shown in Figure 3 he would know?

BAVELAS: Yes, and in the pattern shown in Figure 5 he would know. We have tried other nets, but by and large it looks as though more than two links away they really don't know; they guess. If you plot the things they tell you, it is very hard to get a clear picture.

The pattern shown in Figure 5 averages 20 to 40 seconds per trial. The pattern shown in Figure 4 lies between the other two.

Another thing we tried to measure was errors. If we look at the errors made and average them for the nineteen groups and make a graph, we find that they, too, become remarkably stable after the fifth or sixth trial. The relative differences are the same between the patterns. The pattern shown in Figure 3 makes very many errors. It stabilizes at about 15 per cent. The pattern shown in Figure 5 stabilizes below 1 per cent. It makes very few errors. And the pattern shown in Figure 4 lies between 4 and 5 per cent when it settles down.

SAVAGE: The group is never informed whether its last trials were correct, so it can't learn by experience to avoid errors?

BAVELAS: No. The group may infer that it has made an error, but the experimenter never says, »You three made an error,« or something of that sort.

SAVAGE: Does the experimenter ever say, »You five made an error«?

BAVELAS: No, he never does. But it is possible for them to know that they have made an error from evidence inside the net.

SAVAGE: How could that happen? It would mean talking in the next trial about the last trial, which they presumably don't have time to do.

BAVELAS: Well, it happens because an individual may throw his switch but not be able to correct it in time. However, he gets conclusive evidence just a moment later that it was wrong.

Another thing in which we were interested was what happens with respect to the emergence of organization. We have looked at that in two ways. First of all, if you plot the frequency with which messages go from one place to another, do you get stability at all? In other words, does an operational pattern emerge? Do you find, for instance, | [7] in the pattern shown in Figure 5, that after a few trials messages just go into the center and the answer comes back out? Looking at it that way, the pattern shown in Figure 3 never acquires anything you could say was a stable pattern for sending the messages around.

BIGELOW: You mean stable within a single group or stable within group to group?

BAVELAS: Within a single group. In other words, if you look at any one of the groups which operated in the »circle« pattern, you can't tell from one trial to the next who is likely to send, in which of the two directions he may send, or where the answer will occur first, and so on. What the individuals in these groups tend to do is to send the information they have in both directions as fast as they can, and sooner or later somebody gets all of it.

After the experiment is over, the subjects are interviewed, and one question that is asked them is, »Did your group have a leader?« Now, perhaps one of the most interesting findings as a result of this question is that nobody ever asks, »What do you mean by a leader?« Figure 6 gives the percentage of responses which indicate a leader by position in the pattern.

FREMONT-SMITH: No special attachment to color?

BAVELAS: No.

SAVAGE: It is relative to ego. Is that right? In the first pattern it is really perfectly symmetrical; and when you say there is such-and-such relative per cent designated as leaders, they are so designated relative to ego, which is some place on the diagram?

BAVELAS: Yes.

KUBIE: When you say 8, 12, 4, and 6, you are starting from some one individual viewpoint. You are starting from the viewpoint of 4, 12, 8, or 6?

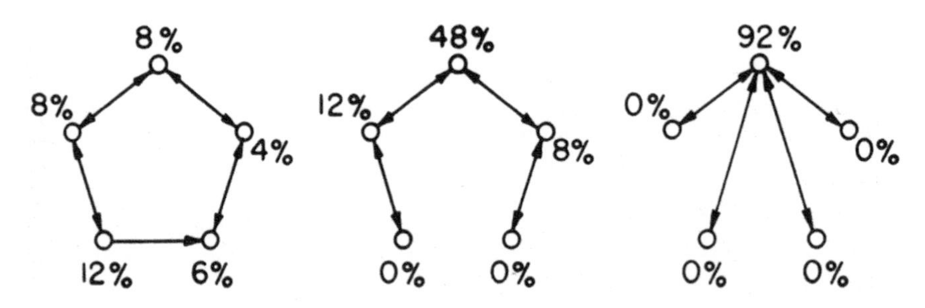

FIGURES 6. Emergence of recognized leaders. Reprinted by permission from BAVELAS, A.: *J. Acoustical So. Am.* 22, *725* (1950).

[8] | BAVELAS: Well, each person who said there was a leader was asked who it was, and he said, »Red.« We tabulate those.

SAVAGE: Oh, you tabulate the color, not the position?

BAVELAS: The color, yes.

FREMONT-SMITH: What happens if he says his own color?

BAVELAS: We tabulate that color.

FREMONT-SMITH: But you make no distinction?

BAVELAS: No.

SAVAGE: Should you not, in dealing with that pattern, recognize that it is really perfectly symmetrical (except for such connotations as may be attached to color names), and that therefore the position of the presumed leader relative to the individual who is speaking – that is, the one I have been calling ego – is of central importance? This general idea seems to me of primary sociological and psychological interest here, particularly in connection with highly symmetrical patterns.

BAVELAS: You mean, we would like to know whether an individual in a given position is more likely to indicate one of the adjoining men as a leader?

SAVAGE: Yes.

FREMONT-SMITH: Or himself.

SAVAGE: Yes, or himself.

FREMONT-SMITH: It seems to me it would have a particular significance when he does indicate himself.

KUBIE: But he doesn't know who is adjoining.

BAVELAS: Of course he doesn't know who is adjoining him.

FREMONT-SMITH: But he does know himself.

BAVELAS: Yes, he does know himself.

FREMONT-SMITH: Therefore, the man who chooses himself does something –

BAVELAS: Oh, I can tell you that nobody chose himself.

FREMONT-SMITH: Nobody did? Ah!

BAVELAS: Nobody did here, no.

FREMONT-SMITH: That is very interesting.

SAVAGE: But he knows with whom he can communicate, that is, to whom he is, in a topological sense, adjacent.

KUBIE: This is a spaceless experiment from the point of view of this subject.

BAVELAS: You have a wire but you don't know where it goes.

SAVAGE: But there is adjacency, in a sense. Those with whom you can immediately communicate may properly be called adjacent to you. Does he pick those as his leaders or the next two or –

BAVELAS: I am sorry but I can't give you that analysis. I am sure I could dig it out very quickly. | [9]

SAVAGE: I suggest that it may be important, especially in so symmetrical a situation.

BAVELAS: Yes, I agree with you.

PITTS: Then the percentages you have depend wholly upon color, upon whatever associations a man may have with colors?

BAVELAS: Well, I don't know that color really enters into the choice.

MEAD: Oh, no.

BAVELAS: I regard the differences among the positions in the pattern in Figure 3 as pretty much zero differences – in the light of the differences one gets in the other patterns.

PITTS: You don't consider the differences of statistical significance at all?

BAVELAS: I wouldn't say that that is important.

GERARD: But those aren't fixed figures. You have just given a random sample there.

BAVELAS: They are pretty close to right, I think, for nineteen groups.

ROSENBLUETH: If they are right, they can only mean color selection. I don't see what else they could mean.

SAVAGE: They could be just at random.

BAVELAS: There is a question as to whether this difference really means anything.

GERARD: Put it this way: Are the twelve always in the same color? That is the point.

BAVELAS: Of course not. I'm sorry. In tabulating the data, you have this difficulty, that the colors in position don't have any very good reference point unless you start with the ego point for reference. I am sorry I don't have that analysis. Color and position were deliberately –

BIGELOW: Permuted?

BAVELAS: Yes, permuted.

SAVAGE: But, again, all that can be tabulated. The tabulations should be made with reference to position; thus you would be able to find out such things as that the man in the center frequently refuses to name himself, that those on the periphery invariably name the center. Does the man on one end of the linelike configuration see himself as a leader, or does he see no leader? Is it only the man in the center of that configuration who sees leaders, and so on?

BAVELAS: I think the full answer to your question can be obtained. I am sorry I don't have it.

HUTCHINSON: Is there ever an individual present who is more interested in analyzing than in doing a job for you, who shows curiosity about this system and is a deviant from the standpoint of your experiment?

BAVELAS: Well, after the experiment is over and the subjects have | been interviewed, [10] we let them sit around and talk if they wish. There is a great deal of curiosity and many ideas as to how the job could be done better.

Hutchinson: But nobody ever holds up the experiment by getting too interested in the formal properties?

Bavelas: This was never true when speed was the main incentive. When we ran other experiments and said, »We don't care about speed, but you must try to do the job in the fewest number of messages possible,« the group spent a lot of time in the first two or three trials studying the thing out, and then performed much faster than the groups which were working against time.

To go on, we asked some questions in the interview with respect to things like this: »How well did you like the job you were doing? How good a group do you think this is? What kind of things in your opinion prevented the group from doing better?« In general, based on the answers to these questions, it is quite clear that the »circle« groups were quite happy with the way things were going. They were slow and they made a lot of errors, but they were quite satisfied. They would volunteer to come back and do many more trials. They would often hang around the laboratory to discuss the experiment and to tell us they could do better if they could try it again. In the pattern shown in Figure 5, except for the man in the center, the individuals were either quite aggressive or apathetic about the whole business. The evidence of this is very striking if you look at the messages they were writing. If you take the messages and give them to a number of individuals and say to them, »Can you pick out of this group of messages those which seem to you aggressive or which say derogatory things about other people, or those that are intended to stop the operation of the group?« they have no trouble picking them out. One individual, for instance, sends out information but sends it in Spanish; another tries to get his neighbor to play tick-tack-toe instead of doing the job; another tears up the messages he receives instead of saving them.

Bigelow: Do you ever get any smart-aleck situations? For example, as you were speaking, it occurred to me that if the subjects decided that the idea was to effect the quickest termination of the experiment, they could easily set it up to produce a trivial solution along this line: Send one message to every member of the group, telling everybody to take the first item on the card, then pull the lever for that; then take the second item on the card and pull the lever for that; take the third item on the card and pull the lever for that, and when you get a coincidence, you will ring the bell.

Bavelas: But you can have only one lever.

[11] Bigelow: What I mean is this: Suppose everybody does follow such | a procedure. You are going to tell them indirectly when they are right, so that they would be using you for a test device in the experiment.

Savage: But the rules are such that if each participant has thrown one lever, and one has thrown an incorrect lever, the group loses; that is, it is scored »incorrect« on that particular trial.

Bigelow: How about stopping the experiment? You said you stopped the experiment.

Bavelas: Yes, when the trial was over.

Bigelow: Don't they know that the trial is over when they get a correct answer?

Bavelas: No, when five answers have been given, the trial is over. You get the picture: those with more errors are happy, and those with fewer errors are quite unhappy. When you ask them how good a group this is, the individual in the center feels the group wasn't quite as good as it should have been, but the others think it was all right.

Bigelow: Do you get a great deal more statistical variation in these centrally arranged groups than you do in the behavior of the ring?

BAVELAS: That is a very interesting question. All of the error in the pattern shown in Figure 5 was contributed by two groups, which became confused, and apparently never got over it. This leads us to the second experiment

BIRCH: Apparently there are two problems in this type of situation, when we begin to deal with the happiness-unhappiness dimension. On superficial examination, it appears that the individuals are happiest when they are (a) either unable to locate a leader in the group situation or (b) where they have a certain degree of probability of assuming that they are determining some kind of leader, whereas in your final situation they have a leader imposed on them, and this leader is imposed on them by the structure of the situation, something which did not arise out of their interrelations but out of your manipulations. Now, given that as the setup, a second point arises: which of these methods would be objectively most efficient in solving a problem? Apparently, from my point of view, I would think that this one central source as a collator would be the most efficient objective method for problem solution, and that people might be happy with such a situation if they were to be permitted to design such a setup for themselves and to select a central individual who would be doing their collating. In that sense they would have a leader who would be their representative and not an individual upon whom they were dependent. I wonder if you began to examine that feature of the relationship? It seems to me that this is somewhat contrived, for that reason.

PITTS: I should think that that could be controlled nicely in one very simple way; namely, one would combine the advantages of symmetry | with the advantages with [12] respect to the position of the participants and those of having freedom of communication by considering the case where everybody can communicate with everybody else. In that case the group can select its own leader if it pleases.

BIRCH: That's right.

BAVELAS: We have run perhaps a half a dozen fully connected groups on this program. The result is complete chaos.

BIRCH: Are they happy?

BAVELAS: They may take as much as a half-hour to do the problem. No, they get frustrated, and there is always a battle going on as to which channels should or shouldn't be used. They want to cut down on the number of channels and they don't have any good way of doing it.

TEUBER: Before you turn to the second experiment –

BAVELAS: Just one comment, first. If you look at the chain, there is no logical reason why all the messages should not go up to the top or why the answer should not come back down. It doesn't take any longer or any more messages than to have them come to the center and back out. As to the degree of imposition of leader, it is not quite as severe as it might appear to be. I know the board has been tilted by the imposition of the network. You are quite correct in that; but it isn't completely determined.

TEUBER: Your speed score indicates how soon the experiment is over, doesn't it?

BAVELAS: The average of trial times.

BIRCH: Does that mean successes or determination?

BAVELAS: Either way; you get the same relative standing.

TEUBER: But that score does not differentiate between the time it takes them to figure out their own group structure and the time it takes them to arrive at a solution of the problem.

BAVELAS: No, except that we know pretty well that the pattern is recognized and known by the second trial in the pattern shown in Figure 5, and by about the sixth or seventh trial in the pattern shown in Figure 4.

BIRCH: So these messages would be correlated so closely that it wouldn't matter?

BAVELAS: No, because here they know the range. The stability in the message-sending doesn't emerge, though.

GERARD: Did you have any data on the abilities of these people, and, if so, did you get any correlation of abilities with what happens?

BAVELAS: Ability?

GERARD: Abilities – plural.

[13] BAVELAS: Of the individual? |

GERARD: Yes.

BAVELAS: We don't think that is a factor because the problem is so low-level. We know it becomes a very important factor if the problem involves not only the collection of data with a very transparent operation to get the answer, but complexities as well.

GERARD: I thought it would have made some difference in the speed and the success with which they discovered the actual network. There was no tendency for a particularly able person to become the acknowledged leader, aside from his position?

BAVELAS: Not for this kind of problem; there certainly was for more complicated problems.

RICHARDS: There is a new preliteracy test that Dr. P. J. Rulon at the Harvard Graduate School of Education recently brought out that is almost the same as your cards. You might find that if you got down to testing on the preliteracy levels, you could find a relevance between the abilities of these participants and their position. It might be worth looking at.

BAVELAS: I am very glad to know that.

ROSENBLUETH: The emphasis seems to be on the way the group is organized, but then you mentioned that you had tried other problems. Would those figures be very much altered? I would expect that they would be altered, depending on the nature of the problem.

BAVELAS: Oh, yes.

ROSENBLUETH: I mean, you are testing two things. One is the way you organize your group, but the other one is the problem you give them.

BAVELAS: Yes. As a matter of fact, it seems to me that the only useful concept here is a kind of task-net concept, and not that of a net which has properties without respect to the task. These things mean something together but not separately.

It has been suggested to me that I should ask you to let me present this companion experiment very quickly, very briefly, so I can then get on with what kind of notions we have as to an explanation of what happened.

You remember that the »circle« group was slow, made a great many errors, but was quite happy. The »star« group shown in Figure 5 was a very fast group, with very few errors, but many of the members were unhappy. Now, two other types of things happened, which suggested the next experiment. If you remember, each subject had in his cubicle a box of switches so that he knew that the right answer must be one of those six. On his card he had five symbols, so he really knew it must be one of those five. Since he was looking for a common symbol, he could just as well send the symbol that was not on his card as send the five that he had. It was just as good a way of [14] getting the answer. But | the task was so stated that it was more »natural« to send what one had rather than what one had not. This insight occurred frequently in each of these patterns. It was always acted upon in the »circle« group and never acted on in the »star« group. It never occurred to the individual in the center of this highly centralized star group that this other method could be used, I think, really, because he was too busy. You see, he was getting all the messages coming in, and he had to get the answer out as quickly as possible. At any rate, he did not get the idea. But the idea occurred as frequently in the other people of this group as it did in the people of the circle group. The messages sent in the star group give some rather amusing interchanges. An individual would say, »We ought to send what we have not.« The answer would come back, »We're doing all right.« The suggestion would come again, and the answer would be, »Forget it,« and so on.

This interested us very much. Another thing that interested us was that in the star group there was a considerable volume of messages suggesting changes or new ways of doing things, new ways of organizing the group. These came from the peripheral members who, you remember, were not very happy with the way things were being done. These suggestions were not accepted by the center man as anything to be acted upon. Also, in one star group where the idea of sending what you have not occurred to two members simultaneously, they began sending this information in spite of the central man's resistance, and consequently they so fouled up his calculations in the center that the group made a great many errors.

It occurred to us, therefore, that perhaps we ought to study groups which had been permitted to »stabilize« and whose operation was then in some way disturbed, and to ask: »Now, what happens? Is there a difference in the ease with which the different groups will adjust to the new circumstance?«

The following experiment was done. All the conditions were like those in the one I described, but instead of showing the subjects a card with five symbols, a little mechanism fed them a box in which there were five marbles. They were colored red, green, white, blue, and so forth. The problem in this case was to find the common marble, since only one color was present in *every* box – every other color occurring in only four out of the five boxes. In this case, however, when the common color is known, the individual picks up the appropriate marble and drops it into a tube which delivers it to a counting mechanism. You will notice that in this case the error is irretrievable.

Each group in this experiment did thirty trials. The first fifteen trials compared roughly with those of the previous experiment. When the group reached the sixteenth trial they had a »shock.« When they | opened the sixteenth box, the five marbles were [15] there, but they were all mottled, a milky, mixed color. If one individual could have seen all five boxes, he could have easily picked the common marble. However, to do this by means of descriptions was another matter. The disturbance might be called »semantic« noise. What happens is interesting.

First of all, the errors per trial for both the circle and the star group increase sharply, and they reach about the same level. The star group goes on making about the same average number of errors throughout the remaining fifteen trials. The circle group comes down very sharply, and by the twentieth trial is doing as well under the new condition as it did under the old conditions.

KLÜVER: The leaderless group comes down first?

BAVELAS: Yes. In the case of the circle group a common language developed; that is, a word is found which fits a marble, and it is used. In the star group no common language seems to develop. What happens in the star group, from what we can tell, is that the messages come in, and the man in the center assumes that the individuals he is working with are unreliable because they are describing the same marble, he feels, with different names. You see, the same marble may be called »ginger ale,« or »amber,« or »light brown,« and so on. He just guesses at what must be the truth. He goes on, very often unaware of the errors that are being committed.

The feeling of the individuals with respect to the problem or the job remain the same as in the previous experiment. The percentage of leader emergence stands the same. What changes mainly is this picture of errors. If you make a study of the speed with which correct answers are turned out after the »shock,« the circle group is, in fact, faster than the star group.

Now, what I would like to do is to tell you a bit of how we are thinking about this problem theoretically, and here I expect to get some help from you. We started originally by looking at these patterns in terms of their topology. We asked the question, »How many ›steps‹ are required to get from any place to any other place in the same network along the shortest Path there is?« For the end man on a chain of five people (Figure 4), the number of steps along the shortest path to all the others is 4 + 3 + 2 + 1. We add it up and say it is ten steps from the end to everyone else. We would do this for each man separately, add it all up, and say: »This chain group has a dispersion of 40, and the circle group has a dispersion of 30.« We felt that that number might bear some relation to the measures we were making of the group's operation.

We went a little bit further. In the chain group (Figure 4) the distance from the [16] next-to-the-end man to all the others was 7, whereas | in the star group the distance from the center man to all others was 4.

What we did was to make a fraction – in the first case 40/7, and in the second case 32/4 – and arrive at a number for the position which we called its relative centrality. We calculated one more number. If you take the most central position in the network and then take the difference between his centrality and the centrality of X (any other position), that difference we called the peripherality of X.

In other words, we asked this: First of all, how spread out is the group? who is closest to everybody? and how far away is anyone from the one who is closest to everybody? We have worked with a limited number of patterns, and it appears that if you take the peripherality index – how far a given individual is from the person who is closest to all – and plot against this index, the morale measurements we were able to make have a very good correspondence.

McCullock[!]: What sort of scatter do you get on your experimental results?

Bavelas: It is quite tight.

Pitts: There are two possible curves you could plot: morale against distance from the most central person; and morale against the man's own centrality.

Bavelas: Yes. Now, the correlation is much better with distance from the most central person than with one's own centrality.

Pitts: You mean it scatters much more if you plot it against his own centrality?

Bavelas: Yes.

Pitts: The curve is the same in general shape, though?

Bavelas: The curve is the same in general shape, yes.

Savage: You say morale goes down if he is close to the central person?

Bavelas: No. The further away he is in distance from the central person, the lower the morale tends to be.

Savage: And yet in the star-shaped configuration, where everybody is proximal to a central person, morale was low in the periphery?

Bavelas: Yes, but the centrality of the central man in the star is 32/4. The centrality of the rest is 32/7. The difference between these two is big. It is the biggest difference we have in these three groups. In other words, the more you centralize your net, the bigger will be the peripherality difference.

MacKay: Have you plotted any index of *morale* against the number of messages sent by each individual?

Bavelas: Yes, and that correlates too. And the number of messages sent, of course, correlates highly with where you are in the group. You see, in the chain the end man sends a message and then he has nothing | else to do, he merely waits for the answer to come back. [17]

MACKAY: I was wondering which was the primary variable, and which might be merely dependent and more or less predictable from it.

BAVELAS: I don't know. We have begun to think of the description of these nets in different terms. What we are doing now is something like this: We plot the topology of a net as a matrix. We put 1's where there is a link, and 0's where there is no direct link. Actually, the important numbers here are the 0's, not the 1's, because the 1's tell you that there *may* be a communication between those people, but the 0's tell you that there will not be. The 0 is a firm figure, but the 1 merely indicates that there is a probability that a message will pass along that link.

We are attempting now to describe the operation of a net in terms of the probability that there will be a message from *A* to *B* in a given time period, and in terms of the probability that an item of information at any point in this net will appear at another point in the net within a certain time. If we accept this way of describing a net, then we can play an interesting game. We play it on paper. This is a calculation, not an experiment.

We make these rules: first, that every person in the net will send a message at every moment of time; second, that he will send all he knows every time; third, that if he has more than one channel on which to send, he chooses a channel at random. The probabilities of choosing any one channel are equal. The last rule is that a person cannot send the same message to the same person twice. If we play out this game on paper, we get almost exactly the operation one gets with an actual group. These rules for operating apparently give the same statistical picture that a group of subjects will.

SAVAGE: Well, there is an ambiguity. If a person is excluded because of this »no repetition« rule, then the probabilities to be assigned to the remainder are the so-called conditional probabilities.

BAVELAS: They are equally partitioned for the remainder.

SAVAGE: Well, suppose his probabilities were already a third, a quarter, or whatever the figure may be.

BAVELAS: No, they are always equally partitioned in this game we are playing.

SAVAGE: Oh, it isn't the real probabilities that occur in the matrix?

Bavelas: No, this is a game, a game being played on a sheet of paper with a pencil and a random number table.

SAVAGE: I, too, refer to such a game, but a more complicated one. Arbitrary probabilities might be assigned in advance, perhaps in the light of empirical work, but you are referring to the special case in which the probabilities are divided equally among the opportunities. | [18]

BAVELAS: That's right.

SAVAGE: I see.

BIGELOW: Why do you rule out the duplication?

BAVELAS: If you don't apply the duplication rule, you get some very nice curves, but they are not the curves which the subjects give you. If you put this »against redundancy« sort of stipulation in, then you get a curve which is very, very close to what the subjects really give you.

BIGELOW: Isn't there an objection to this on philosophical grounds, that you interfere with the randomness and hence put a bias on the results?

BAVELAS: Well, perhaps it is not a good way to do it.

BIGELOW: I am not saying that. I am only saying that if one tries to think of this in terms of a network in which, at certain points, the process is completely random, then perhaps you should allow the duplication.

SAVAGE: It is not that random. Nobody is so foolish as to say the same thing to the same person twice. Only human beings do that.

MEAD: This refers to human beings.

MCCULLOCH: How about children? Might we not get a better fit for very young children if you leave out that requirement?

BAVELAS: I don't know.

PITTS: When you say that the message must contain all the man knows, does that include the fact that he has sent messages to a given person in the past, of a given sort, and received messages from other people?

BAVELAS: No. You see, if he kept every message he received, including the stuff he started out with, in front of him, every message would have all of this on it.

PITTS: I see.

KUBIE: It seems to me there is one assumption here which you may have to make for mathematical purposes, but which is so contrary to the way in which human beings operate that it troubles me. That is the assumption that everything a human being knows is equally available and accessible to him at any particular moment. You are resting on an assumption which is contrary to fact, as the basis of operation of your experiment.

BAVELAS: Well, by »everything he knows,« all I meant was what he knows about these five symbols.

KUBIE: Yes, that is what I mean.

BAVELAS: These are available to him.

KUBIE: Not equally.

BAVELAS: All I meant is that they are on a piece of paper in front of him and that he [19] can copy them. |

PITTS: He might also send a random sample of the bits of information he has in front of him. He might have only one item of information exchanged at every stage.

BAVELAS: Yes. If we take a group in which there are no restrictions placed upon linkage and which sends at each moment the very best message possible to exactly the right place, then the average amount of information across the entire net from moment to moment is given by a curve which is described by $\log_2 n$, »n« being the number of nodes.

Using the rules I first gave you and playing the game, you can, for any given pattern, find who gets the five items first, on the average. And when this person gets those five items, you can discover how much information is present at the other positions. This profile correlates very well with the emergence of the recognized leader and with morale. So this leads us to ask whether, for a group that is working in this laboratory situation, and who have a very specific task to perform, saying that So-and-So is a leader really means that he is a person from whom information which can lead you to successful completion of the task regularly comes. In other words, the man recognized as leader is the man at the position where information accumulates most rapidly.

This line of thinking leads us to make the following general theory. An individual at any time has certain hypotheses that are very important to him, and some that are less important. He is interested in optimizing the probabilities that these hypotheses are correct – for instance that he is a successful person or that he is a loved person or that he is a good person. He does this primarily through information from other people. The sources of such information are very important to him.

Imagine a research group that has been working on a program, and imagine that they are very discouraged and that they feel they are falling. Now we observe the following phenomenon: a man walks in; a smile and a word from him and the picture changes. They don't feel so discouraged. They feel that they may succeed after all. What I would say has happened is that an information process has occurred by which the probabilities that certain hypotheses are correct have changed, and that –

KUBIE: Why information probabilities?

BAVELAS: Because I would like to define a change in probability as being an indication of an information process occurring.

KUBIE: That is a circular assumption.

BAVELAS: Yes, of course.

KUBIE: But because this is a circular conference doesn't mean that all circular assumptions are valid.

BAVELAS: Let me try to explain what I am about in this way: If I set up a situation in which a person must try to pick the one of eight boxes | which has a $10 bill in it, and [20] the distribution of choices tends to be rectilinear, then I would say that he isn't getting any information with respect to this problem. But if I have a person look through a window at him and make faces, and this event changes the distribution, I say that, by definition, information has been received, and that I should be able to calculate how much was received. Now, I am beginning to think that what we call social needs are hypotheses, culture-given hypotheses. The individual who accepts the culture tries to make the probabilities of correctness of these hypotheses as good as he can. Many of the things that happen in a group process depend on how this information is being transmitted and received, whether an individual is below a tolerable level of uncertainty concerning these hypotheses. I can tolerate considerable uncertainty with respect to whether or not my dessert will be chocolate ice cream or vanilla because this hypothesis is not as important to me as some others; but I am not willing to accept that level of uncertainty for other things. The uncertainty concerning hypotheses of social relations between myself and others can be changed mainly – if I am a normal person – by information from others. In other words, an individual tries somehow to make the probability that certain hypotheses are correct as high as possible. Some of these cannot be made higher than a certain amount, but if they are high enough he behaves as though they were really so. I cross the street as though it were certain that I would reach the other side. I know that it is not certain; but when the uncertainty is great enough then I begin to behave »queerly.«

I am suggesting, therefore, that a good deal of the behavior one observes in organizations is not apart from the communication network that obtains in the organization. A person may well be in a situation where he gets no information about important hypotheses, although the leaders of the organization may feel they are giving him a great deal of information. The pamphlets written by the boss or by a personnel expert are distributed to the workers with their pay checks. These pamphlets, found in the gutter outside the plant, contain information – probably from the boss to himself – that is not of much value to the workers.

Now, I think I have talked enough to give you some feeling as to the direction in which we are trying to go theoretically, and I would be interested to hear what you think about all this.

GERARD: I would like to go back to your second experiment, if I may, and ask you two questions about it. Did you try the mottled marbles with the group without the preliminary runs on the clean-cut situation?

BAVELAS: That is being done now. When a group starts with the mottled marbles, and [21] then goes from the mottled marbles to the solid- | colored marbles, these marbles now may appear as very, very different. He is looking for really minute differences, which always exist, such as bubbles in the marbles or slight differences in size. He may, in effect, find he is alerted to search for differences.

GERARD: You are answering a somewhat different question from what I had in mind. Was the performance of the different types of networks with the mottled marbles used from the start the same as that when the mottled ones followed the solid-colored ones?

BAVELAS: The difference is not as great, but they are in the same relative position.

GERARD: In other words, even without a predisposition, the central man does badly?

BAVELAS: Yes.

GERARD: Is the variation between groups greater with the mottled-marble problem as a function of the central man? In other words, I am again asking about this factor of ability.

BAVELAS: Personal ability in the level of problems we have used does not seem to enter as a factor.

GERARD: Did some of your star groups do perfectly with the mottled marbles, for example?

BAVELAS: None did.

GERARD: Never?

BAVELAS: That's right.

SAVAGE: I should like to call attention to a fact brought out by your later discussion; namely, that it is essential to your experiment that the individuals communicate serially. When one writes a note, he can't write it with carbon paper to everybody he is in a position to write to, but must take quite a bit of time writing to one after another. Though some real-life situations may be subject to such a limitation, the situation is often quite different. Consider, for example, the number of people to whom I can simultaneously transmit information simply by talking out loud. I suppose that an experiment parallel to yours, but permitting such simultaneous communication, might be rather dull, for in such a situation it would be sensible for a person simply to keep everyone with whom he could directly communicate up to date on all information which reached him, and the actual experiment might very well simulate such ideal behavior. Though this situation could not be expected to lead to so interesting a set of experiments as those on which you have been reporting, it may be a model for as many, or possibly even more, real-life situations.

While I have the floor, I should like to say that there is a chapter in modern statistics which I think is very intimately associated with communication within groups. I don't [22] want to tell it now, but I do want to | ask that if ten or fifteen minutes can be found at this conference, I be permitted to discuss this topic.

MCCULLOCH: Can we hold that, then, until we come to a place where it might fit?

SAVAGE: Any time you like. I am just asking now, while I think of it, for time to say it.

MCCULLOCH: Right. I should like to hear from Shannon and from MacKay, if they will, as to the extent to which the notion of information, as you have used it, Bavelas, is the same as it is in their scheme of things. Will you take it, Shannon?

SHANNON: Well, I don't see too close a connection between the notion of information as we use it in communication engineering and what you are doing here. I have a feeling that the problem here is not so much finding the best encoding of the symbols, because you do not have a limited amount of paper to write on, which corresponds to the engineering problem of encoding messages to use the smallest amount of channel, but, rather, the determination of the semantic question of what to send and to whom to send it. I don't see quite how you measure any of these things in terms of channel capacity, bits, and so on. I think you are in somewhat higher levels semantically than we who are dealing with the straight communication problem.

BAVELAS: May I just give an example of how one of my students is trying to measure this? What he says he wants to do is this: He says that the group starts out with a probability of one-fifth, that any one of the five symbols each individual has is the right one. Now, at the end of »X« minutes the experiment is halted, and he says to a participant, »Now, of these five symbols, which can you, with certainty, cross off the list?« He hopes in this way to get some indication of how much information has really been received.

SHANNON: There was one remark you made which intrigued me a bit. You relate information to the change in probabilities. I gather by that you mean subjective probabilities, the probabilities a person estimates for such-an-such an event – whether or not there is any valid reason for his estimate. From the communication point of view, the subjective probabilities do not enter at all. I wonder if we shouldn't somehow distinguish between valid information and information that the man thinks he has and acts on, perhaps, but which isn't based on any logical reasoning.

BAVELAS: Of course, I am biased in this direction, because I think the change in the subjective probabilities is the thing that is worth talking about, if you are thinking about people at all. You know the demonstration that is frequently done in classrooms. You come to class and put a deck of cards on the table. You put a $10 bill within the deck of | cards and say, »Now, gentlemen, you are going to line up.« Let's assume we [23] have at least 52 people. »Walk by and take a card off the deck. You take the one that is on top when you get there. Now, will you indicate in a secret ballot which position you want in this line?« You get a distribution. Of course, nobody wants to be first or last. Most people want to be between 15th and 20th in line. Then you spend as much time as is necessary in proving to them that it makes no difference where they stand in line. After they are »convinced,« you ask for another secret ballot, telling them that this ballot wilt determine where each one will be in the line. Well, how much do you think you will have shifted the distribution?

SAVAGE: I don't think that is a fair example. If you had taken an example in which the information imparted had practical consequences, I would have been impressed. But here, you see, it has none. Suppose you argue yourself blue in the face that position in the line doesn't matter, then I say to myself: »Yes, perhaps it doesn't matter, but, after all, the guy might be wrong. If he's right, it follows that there is no harm in my playing my hunch; if he is wrong, I might much better play it.« If you take a situation where the information has a practical consequence – for example, you could arrange one where the first place is actually best, but looks worst – then you will arouse a real conflict in the subject after instruction. He will say, »Gee, the guy says I should go first. It looks as though I should go last. Maybe he's got an argument.« In such a situation you might shift somebody, but when you tell me that my prejudices are simply irrelevant, that they are superstition, and so on, then I might as well follow them, because it doesn't cost a thing to have the best both of your world and of mine.

KLÜVER: I have a question of some practical importance. Some years ago I came across a French book in the field of personnel research. After several hundred pages the author concluded that a man running an army or an industrial empire should never communicate or deal with more than five or six people. Any attempt to have personal contacts with more than six people would turn out to be a disaster either for army or for industry. I wonder whether your work is ultimately going to throw light on such practical problems.

RIOCH: Along the same line, Dr. Bavelas stated at a recent seminar that the size of the groups under study was limited to five, since with six or more the number of possible configurations became unwieldy. It occurred to me later that in team sports we subdivide teams of more than five members into three, four, or five subgroups; also, in other group activities we similarly deal with a small number of subgroups.

GERARD: What happens when you lecture to a class?

[24] RIOCH: You deal with just one, then. You are dealing with a class. |

GERARD: You call that an individual, not a subgroup?

RIOCH: When lecturing I find I usually pick out one, two, or three people in the class and talk to them, treating the rest of the class as a unit.

GERARD: I am not convinced.

MACKAY: I wonder if I could come back to Bavelas's concept of information, because I don't agree with you, Shannon, that it is essentially different from yours. I think what is different is the type of ensemble with respect to which you measure it. Suppose we define »selective« information, which is the concept used by Shannon in communication theory, as *that of which the function is to specify a selective operation:* that is to say, you think of yourself, the receiver, as having prepared a filing cabinet of possibilities, and the receipt of information enables you to make a *selection* from it or, at any rate, to narrow down the probability-distribution over the possible commands that you have to give to your own selective mechanism. Then, I think Bavelas's results are a consistent illustration of this process, particularly when he introduces symbols differing from one another by nonsignificant characteristics, and then goes back to the original symbols. When the nonsignificant characteristics appear, their effect is to enlarge the space of possibilities out of which the receipt of information enables you, the receiver, to select. Originally, you started with a space with just five possibilities in it, or six possibilities; later, you had to widen your space of possibilities to include such things as mottling, and so on. Then, when you come back to the original symbols, the selective operation you are performing at first is still one on the larger space, until you learn that certain dimensions of it are »no longer being used,« so to speak. I agree with Bavelas that it is the subjective probabilities that determine the ensemble from which we estimate amount of selective information here, but, with that proviso, I think it is the same concept as Shannon's. In communication you ask: »In the ensemble of expected messages, how rare is this one which has been selected?« In this problem, we ask: »In the ensemble of expected responses, or expected commands to respond, how rare is the response that I have made?«

SHANNON: I would certainly agree that information is being transmitted here. What I was trying to say was that I don't feel that this group of people will tend necessarily to the most efficient way of doing it. This makes it difficult to apply any of the results of information theory. I would expect that they might settle – in fact, some of your experiments indicate that they do – on a means of communication other than the most efficient. For instance, they do very poorly on the connected graph, which surely should be better from the straight communication point of view, if it weren't for

[25] psychological factors. |

MACKAY: But I thought our question was whether, when Bavelas talks about »information,« he means what you and I do. I think if we use the word »selective« to distinguish it, then we are all three talking about selective information.

BAVELAS: I must admit that the concept I tried to describe is very attractive to me, but it always leads to a very puzzling conclusions. If I may do so, I will cite one.

The following experiment was done by some students. Five individuals were placed in separate cubicles without linkage one to another. They had no communication among themselves in terms of direct linkage. All the subjects could see was an indicator. They were told that the experimenter would cause a number to appear here, and that it would be some number from 0 to 25. All could see the number. Each of them was to write on a slip of paper a number from 0 to 5. The experimenter would collect the contributions and add them up. Then the sum of their contributions would appear, and they would hear him say: »Your target was 17. Your summation was 15. Try again.« They also wrote down why they picked the number.

An interesting result here is that when the subjects are thus given the size and the sign of the error, they don't do as well as a group which is merely told, »You were wrong; try again.« Certainly, it looks as though you are giving them more information when you tell them the size and the direction of the error; but if you look at the »real« array from which the choices are made, the array expands when you give them this additional »information.« Without the size and sign of the error, the subjective array from which they select is smaller. Now, in the first variation, are they really being given more information?

BIGELOW: No matter how much time they have, that is true?

BAVELAS: There is no time limit at all.

BIGELOW: Another point of the same sort: In the previous case, where you were discussing the precision of the process, it isn't clear to me under what circumstances the people were allowed to take as much time as they needed. For example, in the case where you had the ring, were they allowed to take as much time as they needed with mottled marbles? Also, were the people with the central sitaution[!] ever given conditions where they could take as much time as they needed in order to build up the nomenclature?

BAVELAS: No. In those experiments the instructions were: »Get it right. Get the right one as quickly as possible.«

BIGELOW: But has anything been done to explore what happens when you eliminated the necessity to do it quickly?

BAVELAS: Only with the five symbols; in that case, I told you what the facts were. But you see in this last example what this may mean. The | linkage has something to do [26] with how much information the group can use. When the group was told only that they were wrong, they were selecting from a limited array – let us suppose the target was 17. A subject would reason, »If I put in 3 and the others do, too, it will make 15.« Some subjects – most of them – put in 3 or 4. An occasional 2 was contributed, but 0, 1, and 5 were never selected. When they are told the size of the error and the direction of the error, however, the reasoning may go this way: »We're too low. We should go up. But everybody will think of that; therefore I should put in less than I did last time. But everybody will think of that, so maybe I should put in 5 to overcorrect.« So you can see that the range from which choices are made has been increased.

BIGELOW: No matter how long you give them? It is amazing that they don't realize that they should take the desired result »n« and divide it by 5, and come closer to it? There is no tendency to do that? If a person is above the »nth« portion, and the result is still too low, he certainly ought to stay where he is and let the other converge. Did they never learn that point?

BALEVAS[!]: But this is what they do. If the target is 17, they know there are five of them.

TEUBER: Do you finally get a correct result?

BAVELAS: Oh, yes.

TEUBER: After how many trials?

BAVELAS: Oh, the number of trials tends to get shorter; in other words, they do learn something. But it might be three or four trials.

TEUBER: Only three or four?

VON BONIN: Suppose they all put in 5. All right, it will be 25. The experimenter says, »You have 25 instead of 17?«

BAVELAS: Yes.

VON BONIN: Do they undercorrect then? Do they all go down?

BAVELAS: The difficulty is, you never get 25 if the target is 17.

VON BONIN: Just by way of example –

BAVELAS: What happens, very quickly, under either condition, is that some individuals, although they have information as to what the others are doing, begin to structure the group. They say, »My job is to do so-and-so, and somebody else is doing that.« They get very complicated hypotheses about how the group operates. They assume a role and give other people roles, almost as though they couldn't stand the situation if it didn't have some social structure.

BIGELOW: Does anybody take the position that they should stay constant to allow others to adjust the small difference?

BAVELAS: That is what some of them do. A subject may put in 3 ten times in a row, [27] regardless of what the summation is. He won't change. |

Even though the target is 16, and the sum comes out to be 15 five times in a row, he won't change. He will say that his job is to stay constant because somebody else will adjust. Some subjects take the opposite attitude. They keep trying to compensate for what the others are doing.

MACKAY: May I return to the question as to whether or not the error-feedback procedure is giving information? The whole thing is completely consistent with the information picture. The general point is that the specification of more possibilities enlarges the space. That increases the number of bits necessary to perform a given selective operation, and so you would expect that your variance would go up if what you are dealing with is some kind of random process which is being modulated, as it were, in its probability peaks by the information that is coming in. I don't see any paradox. What is happening, if you like, is that the information that is coming in is »structural« in its effect. It is »structural information;« that is to say, its effect is to increase the number of dimensions of the information space. Then, of course, you have a much larger number of possibilities, so you would expect the normal flux of information to have a proportionately smaller selective effect.

SAVAGE: I don't see it at all. You say if I tell people their error, that should naturally make them do worse than before, and you are not a bit surprised?

MACKAY: Yes.

BAVELAS: There is no linkage here, you see.

MACKAY: Let's put it this way: By naming something, you raise in the man's mind a complex array of possibilities corresponding to what you have named. If you name the range of errors, you modify the structure of the filing cabinet from which the almost random mental process is making its selections, and it is not in the least surprising that the result is to increase the spread of the total. It is evidence, if you like, of the often ignored distinction between human *thinking* and human *logic*.

GERARD: Isn't it much like what you said about the difficulty of selecting the clear-colored marbles after they have worked with the mottled ones, that they are looking for more difficulties?

BAVELAS: Yes, the space has been enlarged.

KUBIE: Because we so often fool ourselves that we are basing our choices on purely external criteria when in reality we are choosing in terms of our own unconscious processes, the probability of error increases as we go from random choice to unclear *ad hoc* hypotheses. This was proved in the selection of pilots for naval aviation before the last war. Attempts were made first to select pilots on the basis of a variety of theoretical assumptions, but it was found that choosing | men at random worked at least as well as [28] did these initial attempts to guide the choices in accordance with some pet theories.

Next I want to record my disagreement with Dr. Bavelas's first statement, that this is not an »elegant experiment.« On the contrary, I think it is both beautiful and elegant. I derived from it the same pleasure that I get out of any precise and adroit and clear intellectual or physical activity.

BAVELAS: Well, I can easily demonstrate that it is not precise.

KUBIE: That may be. However, apart from either its beauty or its precision there are elements in it that bother me. In the first place, in addition to the point which I have just raised about the shift from random choice to *ad hoc* hypotheses, we must include among possible sources of error one which is so disturbing that I hate even to consider it. Even if it should turn out to be true, I frankly do not want to believe it. Nevertheless, it must be kept in mind. This is the issue of parapsychology, and whether any extrasensory perceptions exercise an influence from one individual to another in this experimental situation. Actually, it seemed to me that the experiment is so designed that it might even throw light on this problem.

BAVELAS: Would this effect cease if you give them information on the size and direction of the errors?

KUBIE: Yes, because then each individual becomes a restrictive individual when he tries to fit something to his individual hypothesis, whereas if he is working at random choice, if there is such a thing as parapsychological influence – which I hope there isn't, because life is complicated enough as it is – then we give him an opportunity, when all five are working at random, which he wouldn't have when he is obeying his little secret contrived hypothesis. I think an issue comes up in relation to the other one of these experiments.

BAVELAS: In other words, we should measure the departure of the results of this group – without error information – from those derived from the tossing of five six-sided dies.

KUBIE: Correct. My next point, of course, concerns the meaning of »meaningless.« This is most uncertain. It has come up repeatedly in connection with color. We are all aware that color exerts emotional influences and has other subtler significance. But these emblems (for example, the asterisk and the quadrilateral figures) are also not devoid of meaning. In fact, this is demonstrated by the data presented here. I do not know to what extent these unconscious meanings influence the material; but the possibility must be kept in mind, in terms of your conclusions. Certainly, they will have different meanings to different people, to different sexes, and at different ages.

BAVELAS: The thing is whether the effect we are trying to look at is overwhelmingly stronger. | [29]

SAVAGE: This is one of those points of elegance you mentioned earlier. In so far as you have statistical sophistication about the experiment, you will remove the significance of color and of the emblems by suitable randomness and balancing. You have already indicated that some such precautions were taken, particularly with respect to color. With ingenuity and with circumstances at all favorable, you should be able to abolish utterly any of these effects. Strictly speaking, they cannot be really abolished, because they continue to be facts of life, but you can see to it that they cancel out, so that no apparently systematic results of the experiment, depend, for example, on the fact that some particular emblem is phallic, or anything like that.

MEAD: That is fine for the experiment, but I think the important thing here is that we are going to take these experiments and use them in planning for organization in real life. In a sense, once we have been elegant and taken them all out of the experiment, the next elegant thing will be to put them in in such a way that we can learn something about them.

SAVAGE: But you put in the ones in which you are interested. You will never be interested in real life in whatever phallic significance an asterisk may have.

MEAD: I disagree. Have you ever had proofreaders on your hands?

SAVAGE: If I say »never,« I mean »hardly ever.« Particular symbols for emblems which have been chosen for the experiment under discussion are largely accidental so far as any wish to explore the deep psychological significance of shapes is concerned, though that subject is also worthy of experimentation. I therefore reiterate that in *this* experiment you will want to remove the psychological effects of shapes by randomness and balance in design.

KUBIE: May I put in just one point, to make it cohesive?

MCCULLOCH: Go ahead.

KUBIE: Keep in mind the meaningfulness of the whole experimental situation, that is, of the cubicle. To some people the cubicle would be of claustrophobic import; to others it would not be.

KLÜVER: In your experiments, Dr. Bavelas, the items with respect to which communication occurs are such items as circles, triangles, asterisks, marbles, and so forth; that is, items easily named and specified in English. But suppose you deal with such an item as »an irregularly-shaped, wet, smooth object changing rhythmically in brightness.« Perceptually, this is something very simple, since such an object is easily identified and recognized; to designate it unambiguously in English, however, requires telling a rather long story. Items with respect to which communication may occur often differ in »word-nearness« or »word-remoteness,« that is, in linguistic distance. I wonder, therefore, whether you have any data on how the mechanism of communication | [30] is influenced by systematically changing the linguistic distance of the various items entering the communication network.

BAVELAS: We have done nothing on that point.

ROSENBLUETH: But there should be one thing you are trying to measure, to learn or to study with this experiment. I think that it is an extremely elegant experiment, that you correlate certain things, and that you get very nice correlations. But you don't know what you are after. It seems to me that you have given a little too much importance to the notion of information in your judgment of the experiments. Solving the problem is not strictly a matter of conveying information. Here, the nature of the problem is obviously very important for the results that you get, from the data you gave us. You get quite opposite results with the different groups when you give them different problems, because you have introduced different things to solve. I can't quite make out what this is giving you a measure of.

BAVELAS: The question we have been trying to answer is simply this: Does it make a difference if you have groups that must communicate within the constraints of given patterns when all of the patterns under consideration are equally logical with respect to their adequacy for solving a task? Does the pattern itself have any psychological effects which cannot be explained on the basis of anything else but the fact that this pattern was imposed to start with? And does this imposition generate other psychological effects? The answer to these questions is very likely yes.

ROSENBLUETH: Given a specific task, yes, but what about different tasks? How can you generalize, and how can you evaluate what those different tasks have given you when they lead to quite opposite results?

BAVELAS: That, we don't know. We know this, however, that if we change the nature of the task in such a way that certain individuals possess more information regarding that task, then the effects we get here can be explained only on the basis of where the information is at a given time. You see, what we have done here is to guarantee, because of the nature of the experimental task and the topology of the net, that information will accumulate at certain places more rapidly than at others. When you change the task, one thing that changes is the original disposition of relevant information. But again you can say, »What matters here is where is the greatest amount of relevant information and what is the path over which it must travel to get to the others.«

MCCULLOCH: Before we go on, there are now four people whose names I have on the fire for questions. I think Pitts's question should come first.

PITTS: It is purely a numerical one. In the case where information of error is provided, in the second experiment, you said that the groups, | after a time, improved somewhat. How much was that? That is, in the case where you have both the error and the target specified. [31]

BAVELAS: I can't give you exact numbers. There are two ways of looking at it. One is, how many tries –

PITTS: That's what I mean.

BAVELAS: – and the other is, what is the larger error?

PITTS: I mean how many tries.

BAVELAS: It tends to go down. How much, I don't know, but there is a general decrease of the number of tries on the average.

PITTS: Does it ever fall below the number of tries taken by the group with only the target and no additional information about their own errors supplied to them?

BAVELAS: Not consistently.

PITTS: Never?

BAVELAS: No.

SAVAGE: That group also learns. There is learning in both cases; is that right?

BAVELAS: Yes.

TEUBER: I was trying to ask Dr. Rosenblueth's question; let me try it once more: In all these situations you have two things. Your group has to communicate; that is, they have to talk to each other, and they have to solve a problem. Now, to rephrase our question: What difference does it make if their problem is merely to talk to each other and to understand what they are saying, without using this communication process to solve a problem imposed upon them by the experimenter? I suppose you know the experiment that George Miller did with the Harvard articulation series (1). He has a triad, a group of three people, hooked up in different ways by telephone. With one hookup any person can talk to every other person, and they can reply to one another; with another hookup, they have to talk around the triangle, but cannot reverse the flow of communication, and things of that sort. He comes out, I believe, with results that are somewhat different from the situation in which the problem would be not just to understand what the message is, but to solve a problem on the basis of messages.

BAVELAS: Yes, but, you see, it isn't so different, really, is it?

TEUBER: That is what I am asking.

BAVELAS: In an articulation test I think you have a list of words. Something comes over the wire, and that is supposed to help you make a proper or correct choice. In those terms I don't think the problem is very different. This group also has a task which consists of selecting an item out of an array, and the means are signals from others.

[32] TEUBER: But isn't the outcome different? |

BAVELAS: Not very.

TEUBER: The same connectivities, the same effects?

BAVELAS: In general. He finds in that situation that the three or four patterns that he used cannot be distinguished until noise is introduced, and then they are distinguishable in their results.

RIOCH: I was wondering, in this last experiment you described, if that had any relation at all to the class demonstration attributed to Raymond Pearl. He asked his class in statistics to guess the length of a line drawn on the blackboard, and showed that the average of the guesses was always closer to the »measured« length of the line than any one person's guess.

SAVAGE: I don't think that has any real relation at all, because it does not really seem to refer to psychology. If the line had been measured by an instrument such that human error could be considered negligible compared with instrumental error, the results would have been the same. We are dealing here with the phenomenon of statistical stability, the tendency of averages to be close to their expected values.

BIGELOW: I should like to rub in a few points very hard that occasionally seem to have been missed in this group. We are a group of very mixed backgrounds, as everybody knows; and I should like to say that to me this experiment which has been described seems particularly elegant and has in it a drastic lesson for all of us, outside of the particular form in which it appears. The lesson, I think, is that in approaching a very complex field, what one does is to take a very simple model, which is almost always picked purely on an intuitive basis. The intuition may be good or bad, depending upon elements that nobody can judge, so far as I know. Then you set the model up, look at it a while, and sometimes you come to some concrete, although narrow and specific, facts. Now, these may constitute the most useful part of the experiment, rather than what you originally intended the experiment to suggest. That is sometimes useful in a different way, but the most definite and concrete part of the experiment may be the fact that it does or does not show information on one specific point.

Now, one of the questions which was directed to the speaker, I think, concerned the manner in which this experiment is related to the real world. Suppose there are affect values and relationships to the symbols, and so on. I think that it ought to be understood clearly that such questions are really irrelevant; that in making this measurement, the person who is making it is presumably not trying to say, »This is a model of the real world,« when nobody can make a model of the real world. If you try to make a model which contains enough to describe what all of us feel when we see these cards, it is obviously impossible. But lo and behold, this, for example, tells us that human [33] beings under | certain circumstances are worse than the most incredibly bad computing machine you could possibly imagine. They cannot add 2 and 2, so to speak, under certain circumstances of duress. This is a concrete fact and, as far as I am concerned, it is a new one.

I think that the people in the social sciences, the people in the more difficult sciences, ought to realize the tremendously valuable lesson uncovering facts of this sort. It is from these that people in the cleaner and easier sciences have constructed fairly substantial theoretical buildings. To worry whether this is a matter of the real world is fine and has a purpose; but this experiment has shown some facts, and they are the really sound part of it. The facts are that people under certain circumstances apparently cannot do simple computation. The circumstances enormously affect their ability. This is, furthermore, a very good illustration of the difference between what one thinks of as a computing mechanism and what one thinks of as a neurological process that goes on. The neurological processes are very well known to be much more elegant in some respects than a computing machine. On the other hand, they have these shortcomings of breakdown during duress. It is a very concrete and very definite point.

PITTS: I have a suggestion which I think opens up all sorts of interesting possibilities. So far, in these experiments, you have imposed a common purpose on the group in the sense that the members of it are all trying to discover which is the common symbol on their list, and then transmit that information to all the other members. Now, suppose you change it in this way: offer a prize to the person who first discovers the correct answer in rank or order. Then you have all sorts of possibilities of coalitions forming against given people, and so forth. This situation has the great advantage that there exists a mathematical instrument for analyzing it.

SAVAGE: It doesn't analyze it at all.

BAVELAS: May I give an example of something in our own venture close to the game theory. We set up what we thought was, and what I still think is, an example of the classical duel situation. This is set up for two people; in order to get the »bugs« out of the machine, we intend it to be a game between two groups. Two individuals sit on opposite sides of a table. They are separated by a panel on which there is a big green light, which flashes on certain occasions which I shall explain in a moment. In addition, there are sixteen small bulbs across the panel. Under each bulb is a button which can be pushed. An electrode is attached to each individual's arm. The setting is the same for both subjects. At the start of the experiment, the subject is told that one of these sixteen buttons is the correct one, and if he pushes the correct button he will shock the other person and also trip a relay so he him|self can't be shocked. At the [34] beginning of the experiment all the lights are on. Every five seconds, one of the lights goes off; therefore, if he waits long enough, he will know which button is the correct one; but the other fellow's chances are getting better, too. If one subject presses the wrong button, the big green light flashes, and the other subject knows he has missed.

We ran a series of trials under the condition that each man would have only two »shots«; we ran others with four »shots.« We let each pair play long enough under one condition so that they really came very, very close to doing what the mathematical analysis gives as the best time to press the button. This interesting thing happens, however; if they have four shots to fire, the first shot – and we placed particular importance on that one for our measurement – is fired at the seventh or the eighth light, that is, when there are seven or eight lights remaining. If, however, you start this game, the same game, not with sixteen lights but with ten lights, they don't fire at seven or eight; they fire later, although, of course, the probabilities are the same when there are seven or eight lights left!

PITTS: Oh, yes. I would say the game theory in relation to a situation of that kind is roughly analogous to information theory in relation to your other case; namely, where, in both cases, the theory provides maxima or theoretical optima, without, of course, providing any mechanism for analyzing cases where people don't behave in that way; but if you do apply a rank order for the people who arrive at the correct answer first, you have a very interesting combination or conflict of motivation with information, since, of course, information from the other people is necessary for anyone to have the answer. I think that might possibly be important, while still retaining the great advantages of a controlled system where the communication can be analyzed very carefully.

Bavelas: It would be interesting, as a matter of fact, to set up groups in which, if everyone in the group gets the answer, there is a certain payoff to everyone, but with provision for an individual to get a private reward even if the group falls.

Pitts: Yes, make the individual find the correct answer himself first, which he must do, of course, by means of information from the other people.

Rioch: I want to say just one thing about this concept of motivation. I don't think we have any basis for using the term, because there is no differentiation made between »motivation« which is increased alertness and »motivation« which is deterioration of adequacy. We speak about »motivation« when we have a physical feeling, but such [35] somatic feelings may represent any one of several personal operations. There is at | the present time no consistent operational definition of the term »motivation.«

Bigelow: Can you tell us what the increase of adequacy is?

Rioch: If a person is faced with a difficult decision, for example, differentiating shades of color with the expectation of punishment, for example, electric shock, in case of error he may respond with being »alerted« or he may respond with »anxiety« – Kurt Goldstein's catastrophic reaction. The physical sensation of tension may be quite similar. Our concepts of motivation are clouded because they are not adequately differentiated from the conscious sensations or feelings which accompany the act.

Pitts: I expect that when an experimental psychologist uses »motivation,« he is in essence defining it by the experimental arrangement he sets up to produce it in a given way, such as the $10 bill or electric shock. He means the external situation which he has imposed, not the individual's reaction to it.

Rioch: But he doesn't measure it. That is just the point. He doesn't use methods which are operationally directed towards investigating this problem.

Bigelow: He tends to eliminate the abstraction »motivation« by utilizing the end result, by measuring the end result.

Pitts: Certainly, the experimental psychologist in general simply provides what he has reason to suppose is a desirable prospect or the chance of avoiding an undesirable one, and then simply defines motivation as a state produced in his subject by that, without considering that it may vary considerably, depending upon his other conditions.

Rioch: Yes, having defined his desirable and undesirable by what he thinks about it, not by what the rat thinks about it.

Bigelow: Aren't there plenty of examples of learning as a result of hunger?

Birch: Oh, yes.

Bigelow: Aren't these perfectly good continuous variable measurements in the sense that he is –

Birch: No, no.

Bigelow: To some extent?

Rioch: No.

Bigelow: To a considerable extent?

Rioch: No.

Birch: Well, the problem is a rather difficult one. Let us take some of the systematic attempts to study the quantitative variation in motivational states in animal learning or in animal performance – the Columbia studies in the early thirties, for example, that [36] were done under the aegis of Warden, in Warden's laboratory (2) (3). Under | those conditions you tried to compare such things as the relative strengths of maternal drive, sex drive, hunger need, and so forth, in the frequency and in the rate of the animal crossing of a grid, let us say. Warden found that he could get certain curves that would be related to the presumed intensity of drive; actually, however, he was not discussing drive but deprivation, which is something else again. He was discussing what he was doing to the animal rather than the effect of what he was doing upon the animal, so that a number of problems arise.

For example, if you deprive an animal of water, what happens to its sensitivity to shock, independently of how many times it crosses the grid? In other words, the problem of sensory threshold and sensory threshold changes under conditions of deprivation as derived or as an ancillary effect of the condition of deprivation have never been studied; what has been assumed is that once you have deprived the animal of something, you have automatically a direct relation between this deprivation and the amount of motivation which it itself has. It does not follow. Therefore, you get a quantitative relationship between two abstractions that you have developed, rather than a quantitative relationship between two processes that may be occurring for the animal.

RIOCH: As a matter of fact, what we need is a definition of motivation in terms of information theory, because the whole concept of motivation is in terms of energy mechanics.

BIGELOW: You mean, there are no ways to eliminate secondary effects upon the animal itself, of any of the usual incentives, when varied quantitatively?

MCCULLOCH: That is the awkward thing –

BIGELOW: Is that true, sir?

MCCULLOCH: – that motivations to and from turn out not to be added or subtracted but entirely haywire in relation.

BIGELOW: But you might suspect that, for example, one might take the height a rat will jump over a fence one or two or four hours after food to get to a given target –

MCCULLOCH: Well, in three hours, the rat is practically dead. The rat must eat oftener than that.

BIGELOW: But there are no ways to eliminate these secondary effects in experiments?

BIRCH: You get peculiar reversals, too, which apparently indicate that what is happening is that a different qualitative process is becoming involved rather than that a nice homogeneous continuum is being expanded. For example, in one study of my own on the effect of differential food deprivation on the problems of behavior in chimpanzees, you find that you get a greater predicted effect for six-hour deprived | groups than [37] you do for the twelve-hour deprived groups. This was a function not of the time of deprivation but of the time of day that it was possible to introduce such a period of deprivation and still have the animals work. The twelve-hour period could be introduced only overnight, because the animals would not work at certain other times, in accordance with their own rhythms. They had a life history in which they had not been fed during the night, anyhow. Therefore, under those conditions, you had a noncontinuous relationship between the periods of deprivation and the effects of these removals upon the animals.

GERARD: I am sorry, but all this seems to me irrelevant to Bigelow's question, as the whole discussion is irrelevant to the paper. What you gentlemen are saying is that poor experiments have been done, not that the experiments can't be done properly, which is what he means.

MCCULLOCH: I think they can't be done properly, and I am going to say it very simply. Under absolutely standardized conditions, you can have organisms that prefer A to B and B to C and C to A. The consequence is that the strength of motivation with respect to the three items is multidimensional and cannot be measured by any one quantity. I think that some time we had better sit down with the problem of motivation, because it is a tremendously complicated one.

GERARD: I will agree with that, but that is still not the point Bigelow is making – whether you cannot get graded, instead of Yes and No answers. I think these experiments are potentially designed to give graded answers.

BIGELOW: It is not a question of whether they are easily separable but rather whether they are separable –

RIOCH: They are separable on the basis of the frame of reference in which you are looking at them.

BIGELOW: This is my question –

MCCULLOCH: Will you hold it one second. There is a question of Shannon's still on the floor.

SHANNON: Well, the remark was about the theory-of-games possibility. It seems to me that all of the experiments you did could be interpreted as games. The chief trouble is that they are not zero-sum two-person games, which is the simple case. Most of them are nonzero sum, because co-operative effort gains for everybody; also, they require more than two persons. This type of game is much more complicated in the sense that there can be several strategies, several systems of imputations among the various players. I think that perhaps what you are seeing emerge in some of these experiments are those patterns of imputations, particularly in the experiment where people were trying [38] to match a given sum, and where one person would take on a certain | role and another person a different role. The various ways that these roles could be assumed correspond to different good strategies for the game.

I also think that the poorer results you obtained when they were given the amount of error were really an indication of the irrationality of people, because in a game situation the additional information certainly could not hurt you if you were perfectly rational; that is, if you choose the best strategy, additional information can only make it better from the point of view of reducing errors, if you play the game by the most rational means. If you had five Von Neumanns sitting in your cubicles, the answer should be better or at least as good with the additional information.

BAVELAS: You mean if there were five Von Neumanns playing the game, No. 1 would know what No. 2 would do?

BIGELOW: Not by identity.

SHANNON: Each one would choose the best strategy for this game, and if that involved making use of the additional information, that could only help them. If it didn't, they would completely ignore it.

MCCULLOCH: There is one question of Savage's, I believe, still on games. Do you want to bring it in now?

SAVAGE: Yes, I do. When Pitts alluded to the possibility of doing a modified Bavelas experiment, which would be a many-person zero-sum game, as a check on Von Neumann's theory of such games (4), I tried to speak then and there. I wanted to say that so far as I have been able to understand, Von Neumann's theory simply doesn't make testable predictions about many-person games. Though a lot of mathematical machinery is constructed in this connection, Von Neumann, neither in writing nor in conversation, seems to me to make at all clear what empirical consequences this machinery may suggest. The situation is totally different from that for two-person games, about which Von Neumann's writing suggests quite definite consequences. Thus, though such an experimenter as Bavelas can and does test the consequences of the two-person zero-sum theory, I think he would not even know what to look for in the case of many-person games.

While I am on the subject, I should also like to say a word about the experiment Bavelas did do, as a game. Shannon has said, and I think at least partly correctly, that it could be considered as an example of a non-zero-sum game; but as such it would not fall under Von Neumann's definition, because there are artificial constraints on the communication of players. Von Neumann would presumably prefer to consider it as a one-person game, in much the same way as he considers ordinary bridge as a two-person game, neither player of which knows the whole of his own mind – is schizophrenic, so to speak. | [39]

The really interesting aspect of the experiment, though, is its departure from the Von Neumann concept. Thus, the team is not permitted to get together before the whole experiment, armed with foreknowledge about its general nature. In contrast, suppose five of us, having heard Bavelas's lecture, were to go up to the Massachusetts Institute of Technology on the train to participate in this experiment. We would sit in the parlor car and discuss how we should work, what codes and shorthands we should use, and in general make agreements as to how to behave in the face of the various contingencies which might arise. We might then go up there and beat the pants off Bavelas. But the experiment as it is actually done carefully precludes all such prearrangements.

Therefore, interesting and invigorating though Von Neumann's theory is in many contexts, it seems here to have been dragged across our path in two different directions as a red herring.

BIGELOW: If time urgency were eliminated, though, wouldn't this be possible? If they were allowed all the time they wanted to devise a strategy, then this communication with pieces of paper would do that.

SAVAGE: Then you might say that early parts of the game should not be considered as a game; they should be considered in the light of the prearrangements made on the hypothetical trip up to M.I.T.

BIGELOW: There, I agree with you.

SAVAGE: Where there is a long breaking-in period, the participants could begin playing the Von Neumann game, which, as I have said, ought to be considered as a one-person one. But, of course, a one-person Von Neumann game is rather dull.

PITTS: That is exactly the reason why I suggested introducing the conflict of motivations, because, in effect, it does become a real game on the assumption that everybody knows the structure, which you could easily make, but has a motive different from the others.

SAVAGE: But that's a poor game.

PITTS: I know; it's terrible. In the first place, you can say that the theory of games naturally cannot make any empirical predictions as to what any group of people playing a game will actually do. Being mathematical theory, you can't have any empirical consequences directly; but it does provide certain extreme possibilities which you can list; then you can inquire whether the actual behavior of the group shows a tendency to approximate those, and, so to speak, provide you with reference points and a mode of characterizing particular sorts of situations.

GERARD: I want to go back to a comment that Bigelow made. He stated that, to him, the most important consequence of this research was the demonstration that human beings under certain conditions calcu|lated less well than the most primitive calculating machine. It worried me at the time and is still worrying me, and I am going to ask Bavelas if he agrees with that statement. The reason, it seems to me, why it is not a correct interpretation is that when one starts with a rational series or set of operating conditions, as in the paper-solving of these problems, one finds that the subjects do precisely the same thing that happens when the problem is being solved in reality. In other words, I wonder whether it isn't a matter of the people doing the best that is [40] possible with the information available at any time? Would you accept the conclusion that Bigelow drew, that human beings are worse than the calculating machine?

BAVELAS: In these circumstances I would, because in the experiment where you have the individuals in a circle, it can be demonstrated that the problem can be done in just three transmissions; but we've never had a group do it even under specific requests to attempt it.

SAVAGE: But six machines can't do it in three transmission movements; one machine could. It doesn't seem to me the individuals here are guilty of miscalculation. They are guilty of international warfare or something like that; that is, they are guilty of not getting together into the best combines and subteams.

ROSENBLUETH: Given the same instructions that you gave your five people, I suspect five machines would not do it any better.

BIGELOW: I certainly do not. I mean, if you give them more information than they need, the machines will certainly not do worse as a result; but the people did.

ROSENBLUETH: I mean, give them the same information.

BIGELOW: This is a case where excessive information would produce a negative result.

ROSENBLUETH: Excessive information of the wrong sort.

MACKAY: It seems to me that this is a two-level problem, and that is why we are arguing. There is, first, the problem of formulation. You see, these people are locked in, and the problem is formulated only in vague terms, and they've got to be able to develop for themselves a model of their situation in terms of which they may draw their deductions. Second, there is the question of drawing the deduction. If you put a man in the situation in which the problem is so formulated that the question is as simple as, »What is 2 and 2?« – if that is simple – then he will make his deductions as reliably as the computing machine. But the first problem here is essentially one of abstracting a structure from a pattern of experience, isn't it? and any child is better equipped than a normal computer for that job. It is possible that a suitable machine could be made to do better than a human being at both tasks, but it must meet a criterion of performance different from that of [41] | deductive ability. If the mechanism operating here is some kind of random search process in the mind or the mechanism of the individuals concerned, then we should make comparisons not with the ideal logical mechanism to deduce logical sequences, but rather with the ideal random mechanism to induce structure.

BIGELOW: Well, it certainly starts off with a random input in the sense that you never know to whom you are going to talk. But beyond that, the question would be whether these people, on the basis of information that they have at any given time, make choices which are less than optimum. Each individual may or may not make optimum choices with the information he has. Now, I would guess that the evidence here is strongly that they do, and that an individual computing machine, given any information – they may start out with random search and then increase the information in a certain flow pattern which is not exactly the same in each case – would make inferences from these which are far more reliable than those of human beings.

MACKAY: Yes, I think the importance of Shannon's point is that we should ask ourselves with what type of mechanism it is fair to compare a human performer. In attempting to make sense of Bavelas's results, where the subject has to decide whether to fire on the fifth or the seventh light, our analysis will be quite different according to whether we assume the performer to be designed as a logical calculator or as a mechanism adapted to the kinds of situation that confront a human being. If we ask whether a human being is as good as a machine, or whether an action is »unreasonable,« the answer will depend on which data and assumptions we want our ideal mechanism to use. Psychologically, it appears that the subject treats the array of lights as a uniform assembly and picks something near the middle as a typical point in it. He is presumably using a random mechanism whose chief law is, »Thou shalt show no bias.«

BIGELOW: Surely, but that is the point, that here is a striking example where the human being is not behaving like a computing mechanism.

MACKAY: I am sorry, but I think it is a case in which the human being *is* behaving, certainly not like a normal computing mechanism, but like a random-search-computing artefact, if you like.

BIGELOW: Fine, but then it is not a computing mechanism in the ordinary sense.

MACKAY: Not like the conventional type of computer, no; but not necessarily inferior (or superior) to it. It has to be ready for a wide range of quite different tasks.

PITTS: The computing machine can certainly be designed to carry out processes of random search in a way which minimizes the number of expected steps to acquire the desired information, and certainly such a | computing machine would do better than [42] a group of human beings. It could not do any worse.

MACKAY: Yes. The question is, first, whether, in the present problem, there is anything paradoxical, and, second, whether we are justified in comparing the performance of the human operator in this situation with that of an optimum computing machine, which might not have nearly as good an *all-around* performance.

GERARD: We are all talking about simple computing machines, not rather complicated ones.

FREMONT-SMITH: It is evident how intrigued we are with the problem that comes up again and again: whether the computing machine would do better or worse than the human being. I think that the degree to which that intrigues us is one of the most interesting things.

SAVAGE: That used to be our title subject. We used to be a seminar or meeting on the subject of computing machines and we naturally do revert to the theme of the analogy between computing machines and human and social behavior, because at one time, at any rate, it was one of our most important theses, that we might find something fruitful in that analogy – not that the machines would be like people, because they were made by people – after all, we don't expect a neon lamp to be like a person because it is made by a person – rather, we had at that time some hope, or some of us had some hope, and it was widely advertised, that computing machines would be like people because they did so many human things. It seemed to us worth while to examine these analogies, especially to discover, if ever we could, limitations to the analogy.

We have had, throughout the existence of the group, the important problem of seeing if there is anything that people do, that can be precisely stated, that may not be done by a machine. We have always known that that hope was in some sense chimerical because there is the famous theorem of Turing, which we used to hear about at every meeting, to the effect that if you could only tell precisely what you want, you could make a machine that would do it. And yet we have continued to hope in spite of this theorem, which in a sense should end all our hopes that we might discover some kind of behavior which does not deserve to be called mechanical. This has not prevented us from simultaneously hoping the opposite, so we have repeatedly sought mechanical analogues to the various aspects of human behavior. But it is by no subconscious accident that we revert to a theme which, as I say, used to be incorporated in the very title of our group.

RIOCH: It is interesting to note the historical concepts of this. Hanns Sachs wrote a paper, »Why the Delay in Civilization,« making the point that the machine operationally represents a magnified part of the | human being, and that the building of a [43] machine is copying the human being over a narrow range of the human capacities but multiplying the human capacities in that particular narrow range. He described the tremendous resistance to making machines previous to the time of the Renaissance. This anxiety is represented in all sorts of way, both in terms of the danger of copying the human body and in terms of the danger of dissecting or experimenting with the human body. The Koran, for example, forbids the copying of the human body, and a good Mohammedan has no portraits.

In another very interesting scientific revolution, which occurred at the end of the nineteenth century, we have the curious situation of the machine coming between the observer and the phenomenon. The manometer comes between the physiologist and the blood pressure, and the camera comes between the astronomer and the star; along with that, we get a great deal of anxiety about the machine. The exact verbalized form that it takes is different from that which it took in the Renaissance, but the general problem is stilt there. However, I think it would be a horrible comment upon the people who build computing machines if they couldn't build one which, over a narrow range, was better than the human being over the same limited range.

PITTS: That is perfectly true in principle. I think we have made a wrong comparison, really, one which is unfair to the human being. We have compared the human being to a machine which is designed to perform a particular purpose, and naturally the machine will perform that purpose as well as it is possible in principle to do so. We really should make an analogy between a machine designed to perform a particular purpose and one designed for a purpose different from the one we are considering.

TEUBER: You could rephrase that and say that the machine might do better in any individual situation, but an organism of the complexity of man might still answer with relative adequacy to a much wider range of situations than any presently constructed machine. I think Bavelas's concern was not so much that, but the question whether we can use these particular models, at this point, as a measure of rationality, or as a measure of optimum performance against which we can measure the actual performance of the group. Or, if the group tends towards optimum performance, how can we use these mechanisms and mathematical theories in predicting how a group will act under certain simplified conditions? I think that is a somewhat different approach.

GERARD: Along the same lines, it seems to me that, although I may not recognize a good deal of unconscious anxiety as to whether we are or are not better than our machines (which may lead to a preoccupation with it), the conscious basis for preoccupation is with the extent to | which we can get useful clues from studying the machines, as to how our brains or our social groups work; and – [44]

TEUBER: Exactly.

GERARD: – that is not a matter of saying, »It is,« or, »It isn't,« but of saying, »In what respects is it?«

MEAD: May I make one point more? Dr. Bigelow made the point about the experiment that is limited and simple but gives us some clues on the relation between the natural and the social sciences. I think we had a very nice illustration of both uses here. When Bavelas produced the patterns, Lawrence Kubie showed points that might be relevant. Now, it is perfectly true that you can take those out; you can set up an experiment and randomize them so that they are not relevant. Part of what you said, Dr. Savage, was fine, that you should have different experiments to study these deep psychological relevances as a step to doing something else. You also said, »Who cares about an asterisk?« I believe. Weren't you the one who said that?

SAVAGE: I did, and I said I was misinterpreted.

MEAD: That is the point which I think is dangerous. We have to distinguish here between the fact that you can do experiments, you can simplify them, you can make nice tight experiments, and you can use information about the unconscious to be sure that you have randomized the right elements; but if you carry that over into real life, with a statement that a thing like an asterisk is of very little importance, and then make an estimate about the amount of error that you will get in proofreaders in a year in the world today, with the amount of printing that there is, you would make an enormous error. We keep vacillating between the two.

BIGELOW: The danger point is where you make use of such a specific conclusion from an experiment.

MEAD: But that is one of the difficulties that comes between the groups, that when people introduce »real life« they are only introducing »real life« to make the experiment tighter; say, you had better randomize this, or you had better take this into account, but don't generalize from the experiment in such a way as to say, »Who cares about an asterisk?«

SAVAGE: Let me defend myself about the asterisk, if I may. I said what you said I did, and you criticized it rightly. What I should like to have said is this: Who will pretend that the six symbols which happen to enter into this experiment are the best six to study with respect to their psychological import? The six are accidental and, in that sense, no one cares about them, although, of course, one or more of them might turn out to be important in some psychological context.

[45]

COMMUNICATION BETWEEN MEN: THE MEANING OF LANGUAGE

IVOR A. RICHARDS
Graduate School of Education
Harvard University

ALL SORTS OF doubts have waylaid the preparation of these remarks, doubts little and great. As an example of the smallest-scale doubt, there was the queer classification in the advance notice for this meeting. If you remember, we were originally going to study language in bees, vertebrates, and men, which set me to feeling my backbone, but I know enough about this group and its history and its interests not to be too much distressed by that. I knew there would be no special creation of Adam lurking in the background of this picture.

A larger doubt, of course, is the appalling scope of my assignment – »Communication Between Men, and the Meaning of Language.« Obviously, I could only sketch some aspects, and I tried to pick out aspects which might engage with, or fit into, the special interests of the varied groups here present. I should like to stress for a second the variety that must be present. You can fill it in better than I can.

It has been some comfort to me in listening to this morning's discussion to see how many of the topics that I had taken up have actually engaged discussion this morning already. From time to time, I will point them out, if I can remember to do so; but you won't, some of you, have any difficulty at all in seeing preoccupations which have already showed their heads this morning. One of them, of course, is the thing I just referred to – the heterogeneity of intellectual interests in a group like this. The problem is whether this is one group or whether it is a number of groups, and if so, how you would separate them into groups within the group. I can indicate some of them, for I think it will help the strategy of what I am going to say.

Language I take to be pre-eminently the learned activity of man, and learning itself to be the major point of procedure of evolution in a conjectural world process. I am struck, accordingly, by the vigorous efforts which language-theory specialists – and here I had better go into inverted commas – or »linguistic scientists« (as they are proud to call themselves, with a great deal of punch behind the word »scientist« as opposed to any other description) have made to achieve something like autarchy, or independence, of the whole intellectual confraternity, if we may give it such a pleasant name. There has been an attempt at isolationism among linguistic scientists – it keeps recurring in different | [46] modes – and among theorists of the study of behavior. There is a revolt against, and a withdrawal from, other studies that amounts at times to a breaking off of diplomatic relations.

Now, I take philosophy to be the over-all name nowadays of the diplomatic agency which endeavors to keep studies in some touch with one another. I think this attempt to break away, to secede, on the part of new sciences – and linguistics and behavior are two new sciences we might think of in this connection – is dangerous for them and for others. Both dream, I think, of intellectual world conquest. It seems a strange sort of design to see behind a proposal for autarchy or isolationism, doesn't it? And to the observer of these things, the diplomat, the linguistic scientists, and the students of behavior alike show a certain young ruthlessness and disregard.

Now, all this is just to explain what follows and why. I am concerned with the diplomatic-philosophic strategy for the further study of language rather than with any attempt at an over-all presentation or any report on any investigation of my own or anyone else's. As I proceed, I drop the suggestion that among new sciences that may emerge there might be a theory of analogies, systematically developed. There is this great analogy between the state, the political world, and man as the epitome of the political universe – the kind of thing done by Plato, which I suppose has be[e]n the most fertile analogy in the Western tradition. As always with such analogies, it works both ways. There is traffic in both directions. The political concepts developed on the one side shaped beliefs about personality structure on the other, and these in turn are reflected. It is hardly possible to discuss one set apart from the other.

The only other analogy that strikes me as having an equal past, an equal present, and perhaps an equal future is the analogy with which you as a group have been especially concerned and about which we talked this morning – the analogy between an organism and a machine. That, again, is one of these prodigious analogies which should be under special study in relation to other analogies which may be used for similar purpose.

To come back to my doubts: a penetrating doubt, truly a bosom doubt, concerns the sort of language with which one may profitably try to talk about language. I think the most shocking understatement that I have ever met in my reading came to my attention quite recently. It appears in C. L. Stevenson's engaging but somewhat tangled book, *Ethics and Language* (1). There he says, »Language about language must share some of the complexities of all language.« I do think that is a prize winner of an understatement. Of course, it shares; but not so much share them as it must shoulder them, and shoulder them at their | worst, shoulder all of them. It has to represent all [47] the tensions and troubles of language, and these in turn represent all the troubles and tensions of the most complex modes of life.

I have used the word »represent« here, and begged with it, inevitably and typically, typically and inevitably, most of the questions that anybody could raise about what language is and how it works. Does it work through representation? What in the world is representation? Perhaps if we considered those uses closely enough and with an attention of enough resolving power, we should see that they had to be different, in meaning and, even more important, in type of meaning.

My point is simply this: The very instruments we use, if we try to say something which is not trivial about any aspect of language, embody in themselves the problems we hope to use them to explore. The doubt comes up, therefore – and it is a very familiar doubt, if we may linger with it for a second – as to how far we can hope to be understood or even to understand ourselves when we use such words (as we all inevitably do, all the time); and in the lucidity of this doubt, the literature of the subject, I think, can take on a queer appearance. If I may quote one sentence only from the last time I tried to write on this topic, in the *Philosophic Review* of March, 1948 (2) – quite a time ago, I realize – I had this to say: »It is odd indeed how the artificialities of these ancient rituals« – the rituals of philosophic and linguistic discussion – »are maintained, how writer after writer will lay on at his opponent with words like ›know‹ and ›true‹ and ›say‹ and ›be‹ and ›mean‹ and ›believe‹ and ›understand,‹ as though such strokes of tongue or pen could hit anything, and as though finding out how those words may in fact work were not after all and underneath all the forever neglected, though ostentatiously paraded, aim of the procession.«

Well, now, this situation which I have tried to highlight with that picturesque antic is not, of course, peculiar to the study of language, as we all realize very fully. All studies suffer from, and thrive through, this: that the properties of the instruments or apparatus employed enter into, contribute to, belong with, and confine the scope of the investigation.

That, again, is a topic that I am happy to notice arose this morning, fleetingly. I am going to put a great deal of stress upon this. The properties of any apparatus used – and the apparatus will include preeminently the language of discussion and whatever else you like to put behind the interpretation of the language – enter into the investigation, and not only enter into it but belong essentially to it and contribute to it and form and shape it, and, I suspect, confine it.

A way to put a little familiar backing behind this, and to show the general relevance [48] of this strategic problem of how we are to talk about | language, will be to quote to you from Oppenheimer's popular survey of science for fifty years in the *Scientific American,* in its jubilee number of September, 1950. Here is Oppenheimer talking about Bohr's Principle of Complementarity:

»The basic finding was that in the atomic world it is not possible to describe the atomic system under investigation … [*a*] in abstraction from the apparatus used for the investigation; [*b*] by a single, unique objective model« (3).

I have intruded upon Oppenheimer's sentence with *a* and *b*. Oppenheimer does not separate those two things; for his purposes he does not need to. But there is a Principle of Complementarity which applies to the limitations of the model, and another principle that it would be well to have a name for. I haven't thought of the right name. It might be the »Principle of Instrumentality«: you cannot describe any system under investigation in abstraction from the instrument used for the investigation. Oppenheimer goes on to a familiar point:

»Rather, a variety of models, each corresponding to a possible experimental arrangement and all required for a complete description of possible physical experience, stand in a complementary relation to one another, in that the actual realization of any one model excludes the realization of others, yet each is a necessary part of the complete description of experience in the atomic world.« Well, all that is in terms of the atomic world, so Oppenheimer very properly goes on to say: »It is not yet fully clear how characteristically or how frequently we shall meet instances in other fields, above all in the study of biological, psychological, and cultural problems« (3).

What I am going to do essentially is to point out some places in the field of the over-all strategy of linguistics where you will find principles of complementarity and instrumentality intruding. Here, I am very much at your mercy, but it is amusing to speculate for a minute or two on the sequence of fields in which recognition of some such principle as the Principle of Instrumentality would arise.

Where would you most expect to find it recognized? I conjecture – and I speak very humbly here – that mathematics may have been the earliest study forced to ask itself about its own intellectual viewpoint, and the influence of its symbolism on its scope. This may suggest that the more abstract the properties of the instruments, the easier it may be to take account of their presence and not overlook them. If so, we might get the familiar complexity sequence of events and their corresponding subjects: mathematics, physics, chemistry, biology, sociology.

Should psychology or anthropology come next? You may like to change them around. After that, we are sure to have a quarrel. What is more complex and concrete [49] still than anthropology? Poetics, and | then whatever you put above poetics, getting very near the highest of heights, which I am going to list as dialectic. That would be the conventional increase of complexity scale for the history of the universe. Furthermore, it might be held that the higher you go in this scale of complexity, the more of the mind we bring in as apparatus or instrument of the inquiry. Here we have a word which we've got to put in quotes, and you will hear the quotes in my voice, I hope. Later I am going to do some vocal antics to show what kind of quotes I am using, and »mind« is a very suitable word for this sort of experiment. I am going to suggest that we leave the problem of what we might mean by it, and say that the more complex the whole investigation is, and the more complex the subject matter (if you can separate the subject matter from the modes of inquiry), the more necessary it is to bring in more of the mind.

I am suggesting that the mathematician, as a mathematician, uses one branch only. It may be a prodigious branch of human activity, but, in comparison, the anthropologist has to reveal many more sides. As for the student of poetics – I venture to rely on Coleridge here – he is the only student who has to bring in the whole of the mind, with the exception of the dialectician, who manages to bring in a little more.

You will remember that Coleridge, at the end of the fourteenth chapter of *Biographia Literaria,* said, »The poet described in ideal perfection brings the whole soul of man into activity, with the subordination of its faculties to each other, according to their relative worth and dignity.« Well, that is what the great investigation is. What are the relative worths and dignities of all the instruments, the faculties, that together make up that which may and must, perhaps, enter into any investigation and be its instrument?

Now, corresponding to all these studies are characteristic uses of language; and poetics, I suggest, is faced by the most complex of subject matters and is required to undertake the most complex activity itself in the process of the investigation. I have put dialectic above it. Dialectic is above it because it is concerned with the relations of all the subordinate studies in a way in which poetics is not. So dialectic is the supreme study, as it were – with philosophy serving as a sort of Harry Hopkins.

I want to linger for a moment, if I may, with anthropology, because of the close ties and great influence that anthropology has lately had with linguistics. There has been a bit of a »ganging« around there. The chief methodological problem, I take it, of anthropology is very closely paralleled in linguistics, if it is not identical; though perhaps linguistics, by its humbleness of scope, by severe limitations put upon itself as to what it shall and shall not take into the investigation, by confining | itself largely to [50] morphology, and by something that sometimes looks like a kind of panic flight from semantics, has been making the better progress.

This chief methodological problem I think I can state by another quote taken from the same handy source, the *Scientific American,* the same number, in a later article. Kroeber says, »Anthropologists now agree that each culture must be examined in terms of its own structure and values, instead of being rated by the standards of some other civilization exalted as absolute« (4). He puts in a historical comment on that which I think would not stand up very long. He says, »– which in practice, of course, is always our own civilization.« I think study of the curious ways in which modern culture will judge, not according to their own standards in the least, but by Judaic, early Greek standards, and so on, would make trouble for Kroeber. But still he says, »This principle leads, it is true, to a relativistic or pluralistic philosophy, to a belief in many values rather than a simple value system. But why not, if the facts so demand?«

I venture to suggest that the basic principle of the relativistic approach which he is preaching here is plainly a halfway house, not permanently tenable. It is a methodological self-denying ordinance. And better not say a halfway house, for, good heavens! it is an early staging place, not nearly halfway, perhaps not even a millionth part of the way, if one can imagine such a progress being metaphorically measured. It is a negative, defensive step, this relativistic approach, an anti-imperialist move, necessary and desirable, of course, but not at all a sufficient principle for an over-all comparative study. It is a parallel to the linguistic principle that the structure of a language is not safely described in terms of the structure which has been devised for describing some other and different language. English language must not be described in terms of Latin, or Hopi in terms of English grammar.

What the linguistic scientists have been doing is to fashion a growing system of instruments for comparison, an apparatus for overall survey among languages, able, they hope, to put diverse languages into a common frame. They are not, of course, doing what Kroeber says they should do if you transfer it to the linguistic field. They are not attempting to examine and describe Hopi in Hopi, or Kwakiutle in Kwakiutle. They don't know the languages well enough to do it in most cases, I believe. Anyway, it would be very difficult, because in Kwakiutle, we are told, you cannot say, »The farmer killed the duckling,« without saying something in Kwakiutle about the space relations of the farmer, the duckling, its owner, and so forth, to the speaker and the listener. You have to put all those specifications in or you can't say it at all. | [51]

No, the linguistic scientists are working out an over-all apparatus which can, they hope, be used for an examination and description of all language. It may very well be that no one such apparatus will prove to be feasible, in which case you will be back in what will then be a very familiar Principle of Complementarity situation, with a number of different systems of description, each valid within its limits but not compatible with one another, or as yet reducible to an over-all united system.

So far the linguistic scientist's apparatus keeps within very narrow limits as to which of the features and functions of languages it is yet ready to give an account of or even examine. Here is a neat statement of this:

»The linguist focuses his attention upon those selected aspects of language which he believes his methodological equipment gives him the authority to investigate.« Limitation by the equipment. He doesn't attempt to study the totality of language phenomena. Again, »The sharp separation which exists between linguistics and other disciplines is, at best, an arbitrary one based on the development of an isolated methodology and not on any empirical division of subject matter« (5).

This brings me back to what might be called the marked »Ishmaelite complex« that haunts linguistic scientists and students of behavior. Their hand is against all men. I don't think all men's hands are against them, though they seem sometimes to think so, which is enough. But, above all, their hand is against others who like to think about meaning with methods other than their own. That may be explained by the difficult labor they are having in working out an over-all descriptive technique for all the languages which are constantly, of course, showing more and more variety as more and more people know more and more languages. In any case there is – and this is, again, something which popped its head up in our discussion this morning in another context – a hostility and even contempt frequently shown by them for those who talk about meaning, and there are all sorts of rather virulent blasts that they emit against »mentalism.« They share that with some students of behavior very markedly. I think there may have been in the background some academic or, at least, intellectual persecution of both these young studies, but that should surely be over now. The boot, in fact, is rather on the other leg. That is just an aside.

I am coming to some rather deep water in a moment. I am no doubt influenced toward this frivolity by the trivial fact that the book I once wrote with Ogden has been about equally attacked for being mentalistic – and a type specimen would be J. R. Kantor's *Objective Psychology of Grammar* (6) – and for being behavioristic, as recently as two years ago, by Max Black whom I mention because I wish so much that he were here. | [52]

I have come back, you see, to my initial bosom doubt: With what language should we talk about language? And typically, should it be mentalistic or behav[i]oristic? I had better calm your worst fears at once by saying that I do not propose to solve this problem this afternoon. I hope instead to refrain either from dismissing it or from assuming any answer to it, unless some analogue to Bohr's Principle of Complementarity would be an answer or, rather, a staging place on the way to further inquiry.

I have not forgotten that this is a conference on feedbacks and circular mechanisms; but before I show you that I have been trying to prepare for discussion of these things as they appear in language itself and pre-eminently in the learning of language, let me say a word or two about the possibilities that open up of inventing new languages or new language devices, perhaps for use in linguistic theory.

There are the quotes, of course, we use to prevent some routine patterns of behavior from being exhibited by readers. It is a very important device, but we are unregulated in our uses of it. Some time ago I suggested some specialized uses of such quotes, replacing inverted commas by $^{?-?}$, $^{!-!}$, $^{r-r}$, $^{NB-NB}$, $^{SW-SW}$, etcetera (7). In point of fact, this is an excellent teaching device. In a few moments I shall try out some vocal equivalents.

Format linguistic description does, of course, use many more or less routine types of descriptive terms, many names such as phonemes, morphemes, tagmemes, lexemes, etcetera – nearly all nouns, please be it noted. I don't think much experimentation has been done with other categories of technical language, technical verbs. But if we take to heart – and I don't know whether we should take it to heart; I defer to anyone who can check it – what Benjamin Lee Whorf (8), for example, had to say about Hopi handling of numeration, time-space, substance, and matter, and notice his remark that in Hopi »generality of statement is conveyed through the verb or predicator, not the noun,« you will see what an enormous field the construction of speculative instruments other than our ^{r}Standard Average European^{r} patterns might give us.

It is the hardest thing for any study – I think this may be universal – to notice and then abide by the new conditions set up by its own advances. Scientists very often cut their own throats and then go on, proceeding as though their throats had not been cut. It is usually not a success. The problem is how to find out what we have done, take account of it, and proceed in the light of what we have done.

Here is an advance – the recognition that the linguistic structure developed in a group of languages, in the Standard Average European, say, may have structured even the fundamental concepts of their users. | [53]

That recognition is something that we should look into more seriously, for as language study advances, the probability that it will upset fundamentally the procedures of its own inquiry is high, and, if so, the chances of Principle of Complementarity situations arising are good.

Meanwhile, pending such ambitious developments, there would be gain if we merely appended to an important word indices showing, for example, how literally, and with reference to what defining connections, we are using it. »Mind« would be an excellent subject for such work. We tag it, when we can, to show what sort of use we are inviting our hearers and ourselves to make of it. Indices showing the status – empirical, hypothetical, or systematic – we are giving a word in an exposition would be useful, or indices showing the mode of believing we consider appropriate to the utterance in which it occurs. We are not ready to admit that there are a great number of varieties of believings, that the belief attitudes are almost as diverse as the belief objects, or that they can be so treated with advantage, if we are ready to depart from the framework of Standard Average European patterns.

In the Yana language, we are informed, you have to show by the words you use whether a statement is known to be true and vouched for by you, or whether it is made on someone else's authority. Now, if that were made mandatory for all remarks about language and languages, I can tell you something of the silence that would ensue.

To come now to circular and feedback mechanisms: in what I am going to say now, I suspect that if we could allow for the influence of the language I shall use and its vast difference from the languages that almost any of you, I think, would use – different though your languages be from one another and great though the strain we experience in talking at all in the same words, except from the speller's point of view or the phonetician's – in spite of all that, we may have a good degree of agreement. I don't know; recall the embarrassing situation that came up in our discussion this morning. Am I chiefly informing myself? Is it a case of information given by myself to myself? Or am I really talking to people who have to take what I say, put it into very different language through different transformations in many sciences, fundamentally recode it, and then come back, but through another recoding process, to try to put me right?

You see, now we are coming to a nest of things that we talked about this morning. If the properties of the apparatus employed do enter in, and belong with and confine the scope of an investigation, the problems I attempt to explore will be in certain respects essentially different from the problems that you would explore in your intellectually native and acquired languages. How different they must be and will be, and how to [54] investigate that are the methodological problems. |

All this while, of course, I am exemplifying, and I would fain say illustrating, circular and feedback processes in everything I say, as none of us can avoid illustrating them amply and continuously in everything that we say. There is no escaping that. I need not insist here, of course – in other places I would – that what we say next will be in part controlled by feedback from what we have said up to that point, and the feedback from the reverberations of what we have said up to that point. And at this point I must say that throughout I have been preparing myself and you, I hope, to be receptive to, to expect, to attend to a certain peculiar stress on something which does not, I am surprised to note, appear nearly so often as it used to. There has been a waning, commented upon this morning, in some of our attention to the original curiosities that brought members of this group into contact with one another. I am going back, perhaps, nearer to those original interests.

Perhaps this thing on which I want to put the spotlight will be considered to be included in some ingenious way under the word »feedback.« But what I am going to stress stands in an obvious and superficial opposition to »feedback,« and it will, in certain frames of thought, be given nearly, if not quite so much, importance, and sometimes more importance than feedback itself in certain connections. It is certainly as circular. You have no doubt fed forward enough to see that what I am going to talk about from now on is feedforward. I am going to try to suggest its importance in describing how language works and, above all, in determining how languages may best be learned.

Before going into that a bit, I should point out a few instances of feedforward at prelinguistic levels. We are familiar with them at all times. There is nothing novel in the least in what I am going to say, but let's attend to it for a moment.

Let me put a tag on a key word here – »activity« – which is capable of extremely diverse interpretation, as we all know. I put a vocal tag on that and pronounce it »acteevity.« The Scots have a reputation – which may be excessive – for knowing what they mean. Also, I am indebted for my theory of »acteevity« very largely to Stout (9), and I fancy that Stout came from the Shetland Islands.

Take the acteevity of looking for something, hunting, searching. We don't find anything, if you put the right mental-vocal tag on ^NB^find,^NB^ unless in some sense we are looking for it. You may happen on something, but in my technical tagged sense of the word NBfindNB you do not find it unless you are looking for it and – now, mind you; don't be too precise about that – in some sense, at least, know it when you find it. You can't give an account of NBfind,NB with my tag on it which doesn't bring in the problem about NBknowNB with my tag on it, and vice versa. They are correlative terms. Finding is the end phase, in both | main senses of »end« – to the »acteevity« of searching. Something is fed forward by coincidence with which that »acteevity« reaches its terminus and goal. [55]

This word I have been using (NBknowNB), can't be treated too warily. It is a great traitor. I don't suppose one should mean anything more here than I have already said about it in terms of the behavior involved in searching – the feedforward kind of thing. Nothing need be implied as to how it is done, whether it is done through an image or through a substitute for an image, a schema, or anything else you like to think up, anything more generic and more symbolic. I know, generically, in these sorts of terms, now, the grave doubts that many people here must be feeling as to the sort of thing I am saying. My knowing them in some sort of schematic way is my main guide, the feedforward through which I try to find further things to say, which may in some measure meet them. So focus your doubts and bring them, apply them, if you will, to – well, any sort of example.

I am now going to feel in my pocket to see if I have, among the pennies which are there, a dime – a typical acteevity of searching. I am going to know that by touch, possibly aided by hearing – with a bit of difficulty. I have fed forward certain things and – yes, here is the dime. I am too far off and the room's acoustics aren't so good that you can hear what you have fed forward equally with me as the identifying mark – the scrape of my thumbnail on its milled edge – by which I have recognized my dime. That, I think, is an instance of characteristic feedforward behavior. I am trying to draw attention to what is the distinctive mark of what I am calling feedforward.

Very probably some of you will be thinking that there is nothing here but »taping« – if I may use that as a technical term here, and I think I can – a special taping for a special run of the manipulative-perceptual mechanism by which I conduct that particular bit of behavior of finding the dime and not the penny. It is taping plus resort to a memory store, containing what happens when you scratch the edge of a penny, and so on.

You may say that there is nothing distinctive about feedforward, and that this sort of general account in terms of a memory store and an *ad hoc* taping covers all that need worry us about all these very great words – about »purpose,« »intention,« »foresight,« and the rest, and »feedforward« among them. If that is so – and I am not disposed to dispute it – that turns the question simply into one about taping, about *ad hoc* tapings, their sources and their dependence upon more generic tapings. One of the things I want to offer for discussion is the dependence of *ad hoc* taping upon more generic frames of tapings within which it has a place. Tapings, I am suggesting, are hierarchic. All | this is extraordinarily obvious. The only thing that may be puzzling about it, perhaps, is the language I am trying to put it into. Tapings seem to be hierarchic, or to form an enclosure series; the widest, most inclusive, or over-all tapings being least determinate as to what will end them, what will be their goal, their terminus. The higher, more inclusive tapings, I suggest, issue for their own maintenance and execution, to preserve themselves and to achieve their goals, subordinate, narrower, more specific instrumental tapings; and these do so again and again and again and again, in the hierarchical pattern. For example, consider an animal hunting for food. Let him be a sizable animal that we can watch with ease, say, a grizzly. There is a very general, all-inclusive taping – the search for food. As he scents this or that possible source of food, subordinate tapings are issued. Suppose he scents ants. He subtaps himself to follow up [56] the scent. Suppose the ants are under a boulder. Then enter subsubordinate sets of tapings, *ad hoc* tapings, based upon the size and shape of the boulder. When at last he gets to the ants, consummatory behavior occurs, and at that point there is coincidence with the goat-end point of his subordinate tapings, and he stops; or else the smell of some berries sets him off on a new cycle of the same sort of pattern.

All this is a very familiar line of speculation, oh, so familiar! What I am hoping it will say for me is this: with feedforward – and with that I am trying to name the peculiar character of tapings which arise in the service of less definite, more inclusive tapings – the adequacy of any description or evaluation of any »acteevity« depends upon recognition of the sources of its feedforward.

Now, that has all been teleology and old stuff. I have some more old stuff to add to it, and then I shall be making my way toward an end. I am a little alarmed lest I go on too long and too continuously, but I think that in a very few minutes from now I shall have sketched the over-all picture; and, of course, on my own theories, it would only be within the framework, the over-all taping of my general presentation that any of the points in it could properly receive attention. That is an uncomfortable consequence of this sort of doctrine of composition.

This is the moment for me to trot out my two pet professional prejudices. Everybody's pet professional prejudice is ego-idealistic in this connection; as an investigator they govern him. Well, mine are like this: I am by profession a critic, concerned with the value of uses of language, and a pedagogue, principally concerned with teaching the very beginnings of reading, A-B-C stuff, and the very early stages of teaching a second language. Criticism and pedagogy, thus, for me constitute two fairly high-level feedforward systems which tend and guide two extensive worlds of relatively *ad hoc* [57] activity. And in both, in | criticism and in the pedagogy of teaching the beginnings of a subject, a flexible fitting in of means to ends – in more solemn language, design – is all-important.

For criticism, which is radically an evaluative »acteevity,« the difference between better and worse utterances is in design. Poor speech and writing are poor either because they are not attempting anything worth trying or because they are inefficient. You can reduce the whole thing to efficiency, I would maintain. And the design that is there in the background, I should insist, may be, but need not be, witting.

This principle of efficiency is, I think, no more than a recognition of the enclosure series I have been mentioning – the hierarchical service, the *ad hoc* tapings, serving wider aims, which serve wider and wider aims up the hierarchy. And, of course, since language is, I think, inescapably a social activity which only comes into existence and owes its whole character to mutualities between men and within communities, study of language is inevitably dependent upon ethics. The study of language is concerned endlessly with the better and worse of utterance. It is normative through and through, as characteristically normative and inevitably normative as, for example, the study of medicine is, if you will give a sufficiently inclusive, over-all taping to what you put behind medicine, that »acteevity.«

How many other studies must also be normative? I am not certain that they must not all be. How about biology? Isn't it normative, radically normative throughout? I mean, in the sense that about each organism studied the student must ask, »How far is this a typical, normal specimen? How far is it representative? How far is it usable in the purposes of a general investigation?«

This is the situation. Surely, it is very familiar. I am merely going to illustrate it with one glance again to Coleridge. Here is another declaration of the relativistic approach which ties in, I think, as a provisional staging place with the Principle of Complementarity and establishes esthetics and unites it with other evaluations. This is a scrap of

Coleridge's which can be found in his *Miscellaneous Criticism,* his Shakespearian criticism. It is in the first volume, on page 196 (10). He says:

»We call, for we see and we feel, the swan and the dove both beautiful. As absurd as it would be to make a comparison between their separate claims to beauty from any abstract rule common to both, without reference to the life and being of the animals themselves – say rather if, having first seen the dove, we abstracted its outlines, gave them a false generalization, called them principle or ideal of bird-beauty and then proceeded to criticise the swan or the eagle – not less absurd is it to pass judgment on the works of a poet on the mere ground that | they have been called by the same class- [58] name with the works of other poets of other times and circumstances, or on any ground indeed save that of their inappropriateness to their own end and being, their want of significance« – and then he upsets the whole thing with a very strong word – »as symbol and *physiognomy.*«

I think esthetics is there shriveled down to the judgment you commonly pass upon somebody as to whether or not he or she is good-looking. That doesn't mean that he or she has standard patterns of feature, physiognomy. You judge the part played by any element in the whole with reference to the whole, typically, in the case of appreciations of physiognomies. Anyway, here we have the anthropologist's relativistic approach back again in the service of esthetics, and I am venturing to suggest that all studies whatsoever – and this is a very long flight – are normative in this sense, by the very fact that they use definitions and that their statements work only through agreements among users; to use each word in such and such a way and not otherwise. Insofar as anyone does not use them, then the over-all taping, the purpose of the discussion or meeting, is not served. And that gives, unless I am in need of a very great change of view, a sense in which all studies whatsoever are, I believe, both relativistic and normative, insofar as they depend at every step upon *ad hoc* tapings for the definitions of all the terms in the language used in the exposition; and those choices of meanings are controlled by the hierarchy of tapings, *ad hoc* tapings, wider and wider and wider in the hierarchy – all good tapings that we jointly share, I should say, with those which we have as human beings, and under those common tapings we have as employers, let us say, of the Standard European language patterns.

Now I can progress quickly toward an end. You see, in terms of that sort of picture, why I started with a sort of naïve scientism which set up autarchic policies for languages studies and for behavior studies. There are vast areas of purely descriptive linguistics which are a danger at present to all the over-all purposes for which we use language. Here I get on to rather less speculative ground, ground on which I feel myself much more secure, my professional ground. There are, I say, techniques – and standards derived from those techniques – which are a threat to education, to the conduct of language generally. There is the appeal to usage as sanctioning a mode of language. This very frequent appeal I think is vicious. It illustrates all the dangers. Every useful feature of language was not in use once upon a time. It had to come in. Every degradation of language, too, has a starting point and a spread; and behind usage at all times is the question of efficiency. Inefficient features of language are not sanctioned by prevalence. We do not consent to allow them in our own special fields. Anybody who is | teaching or inquiring or searching in an area he thoroughly understands has no [59] patience for a second with the idea that because it is said widely enough, it is right and that is the way you should talk. That is not the point of view of the over-all study of language. That is, I have suggested, inescapably normative. It is concerned (as every speaker and listener is always concerned) with improvement in the use of language. So this »scientific objectivity,« of which so many linguistic scientists are so charmingly vain (like a boy with his first bicycle), is out of place when it interferes, as it does, with education or criticism.

This usage doctrine rules at very humble levels. I am just in the middle of a weird and wonderful course at Harvard, where I have freshmen, two hundred of them. I asked them to write on subjects which you would think they couldn't possibly fail to be interested in and speculative about. But the freshman lacks curiosity. I don't think he knows a curiosity at the linguistic or educational or intellectual levels. The curiosity he most shows is career curiosity. That is very active. But on an intellectual level, if you ask him why he believes twice 2 are 4, two-thirds of them haven't the beginning of an idea of how to speculate about it. All they can say is, »Teacher told me, and I would have got dreadful grades if I had said ›5.‹« They are ruled, you see, by a purely usage doctrine on the multiplication technique. I think that is deplorable; and I notice, I am sorry to say, that this pattern (*ad hoc* examination taping) rules their intellectual behavior very widely. What they want to know is what is thought. They don't want to know any »whys« or what it means.

Now, here is the point, you see, where philosophy, the diplomatic agency of dialectic, must intervene. It has to protect studies from the interferences of other studies; but it has to do more than that. It has to go into studies and protect them and help them out of self-frustrations owing to their neglect and ignorance of what other studies are up to. And it is purposive, the whole activity of thought is as purposive throughout as you can make it.

To come to my final topic: feedforward in teaching – especially in the teaching of the first steps in any subject, for example, beginning reading. I am up against the discussion we had this morning about motivation. I am going to dodge it, possibly, by saying »reinforcement,« where, if there hadn't been the discussion this morning, I might have wanted to say »motivation.« My point is this: all language use, and preeminently all language learning, depends on feedforward confirmed and regenerated, reinforced, by – well, the nearest word I can find for it is »success« – by enhanced power, a very general thing indeed – ability, if you like, increase of ability. Success is the great general motivator, certainty for beginning studies. In the beginning stages of [60] a new activ|ity, it is all-important to avoid drops in what we might call »morale.«

How to teach reading comes down to this: Keep avoidable mistakes to a minimum. There is an elaborate technique by which they can be kept to a minimum, and almost all current practice keeps them absurdly high. There are many sides to that. I have just stressed the motivation side, or, if you like, the reinforcement side – that success is the great generalized reinforcer. Compared with that, such local things as the supposed interest of the child in the narrated doings of some children with their pets don't weigh in the balance in the very least.

But there is another side to this keeping away from mistakes in the early stages of the subject, and it may go up high in the subject. A mistake is a permanent source of weakness. When enough fatigue or enough strain through new problems comes along, the mistake is very likely to occur again; you can see that in any systematic observation of early teaching.

We prevent mistakes by simplification, and here comes in the technique of simplification, say, for beginning reading. Twenty-six letters may not seem a lot, but try seven instead. We are all aware, at least I am acutely aware, of the problem that arises when you meet twenty-six persons for the first time, all together. How long is it going to take me to stand, as it were, in a corner of the room and instantly see who is talking with whom in groups of three, four, five, or six, in various parts of the room? If, on the other hand, I had met seven people and had lived with seven people for several days, and then a stranger had been added to a group, there would have been no generation of uncertain choice points, hazardous speculation, as it were. Mere routine could take care of it. When you throw any letters of the alphabet along with any other letters of the alphabet before a beginning reader, you are putting a strain on him that he is going to suffer from for the remainder of his life, unless he is a person of extraordinary happy conditioning up to the moment, or in an unusual state at the moment. There are some people who can do it, but the great majority of the ordinary population cannot.

The early history of reading for most minds is a history of frustration, of great personality strain, all of which is irrelevant to a task which can be made simple. I speak with some feeling on this because we do teach people to read when nothing else can teach them. When you take boys of fifteen or sixteen, who have been hanging about, becoming delinquency cases because of the frustrations due to their failure – the Ishmaelite complex of being the only persons in the society who hadn't learned to read – put them into another room and, a few weeks later, after they have taken the decisive step and have been taught to read, you find you have a roomful of new people, of people | changed as people, if you can imagine such a thing. And in most cases, we have [61] found so far, unless there is something very wrong, there is no reason why they shouldn't learn. What stops them is the harm that has been done to them by bad practice.

Well, now what is good practice? Suppose we take seven letters – *a, h, i, m, n, s,* and *t* are seven good letters to take for various reasons – all of which illustrate important theoretical considerations. With these seven letters you can offer the beginning reader all the short sentences he needs. All these statements should have to do with a concrete situation; they should be sentences whose meaning he can see, sentences whose meaning he can act. All the materials you need for this very novel activity of learning to read – the optical control, the first discriminations, all this tremendous accustoming himself to a new type of visual attention – can be supplied through seven letters which present minimal opportunities for mistakes. That is the formula. Now, make up some sentences with those letters. You will find there are many that you can make. We add another letter fairly soon, and then another not too long after that; but we keep for a long time to only half the alphabet. It is far better to begin teaching with a film strip than with a book. With a film strip you have a public-meeting atmosphere, a screen and a focal point, and an enormous release of tension.

I must not linger on this, but the first fifty frames of the first film strip use only twelve letters, and an elaborately graded presentation can develop in words and sentences which use only half the alphabet. Which letters should be included? Obviously, you don't want symmetricals. Any child is taped – I don't know how far back it goes – to see that this and this [turning a pipe in four directions] are the same thing. It is still the same thing, but outline it on a board and it is four distinct letters – *p, b, q, d*. It is just a matter of how you turn it. The child couldn't live life unless he saw a knife, say, as a knife, no matter which way up it was. It is bad technique to make a sudden transformation to script, in which it is all – important whether the *u* is upside down – or is it the *n* that is upside down? We penalize the bright child by setting a whole set of bogus traps for him in the script we begin to teach him. They don't belong to the subject. They just betray him through his biological smartness.

The other set to watch, besides the complete oblique symmetricals, are the familiar enclosure things. Here are what look to the child like three attempts to do the same thing [writing *o, c,* and *e* on the blackboard]. Whether or not you finish them, what does it matter? It is easy to spot the probable sources of confusion between letters, and to design a sequence which will avoid all this unnecessary difficulty. Introduce only

[62] one of the mistakable pairs until that is so well established | that the other position comes with a shock, and then you go on.

What we do also – it is all part of the same avoidance of distraction – is to stabilize the syntax. We use only one verb (is, are, and so forth) all the way through learning to read. We cut down syntax variations to a minimum and reduce meaning to something the learner can actually see: the meaning of the sentence, as it were, through the sentence, and so on, right through.

Above all, the interest of the exercise in these early stages is in learning to read and not in any adventures of Jack and Jill and their pets. That is really quite important.

Well, that is a practical outcome of these speculations. I am not sure that they belong together. Again, I am not sure that the air of belonging together which I tried to give them is more than a product of literary composition. It is efficient in that sense, bless it. If it is inefficient, let's pull it to pieces. Thank you very much.

Klüver: You used the example of an animal looking for food to illustrate how sets of subordinate-subordinate tapings are involved in a general and all-inclusive taping, such as the search for food. Perhaps one of the most beautiful illustrations of this kind of taping is to be found in the hunting behavior of the peregrine falcon as described by Tinbergen (11). After leaving its perch, the falcon may start looking for potential prey in a territory as large as ten miles in diameter. The sight of prey does not immediately elicit the consummatory act; there is instead, as Tinbergen points out, a sequence of more and more specific and restricted types of appetitive behavior, down to the final swoop and the catching of the individual prey. Unfortunately, the recognition of the fact that certain tapings are more specific or more general in the hierarchy of tapings is of no great help to the psychologist. He still has to cope with the difficult experimental job of determining the constellation of properties or the particular categories in terms of which particular tapings are made. It is of interest that work on interspecific recognition in birds, especially on the recognition of predators by their victims, has led to the conclusion that certain predators are recognized on the basis of only two visual characters: namely, outline and movement. When dealing with a swiftly moving enemy, it is apparently better to make use of very few and very general schemata than to be dead.

RICHARDS: Well, the *ad hoc* taping can be quite generic in that case.

KLÜVER: It can be generic?

RICHARDS: Yes.

KLÜVER: What you said about the complexities of language which arise when using language about language reminds me of the complexities that bothered such outstanding brain researchers as von Monakow and Mourgue when they tried to talk about the [63] relation of psy|chological phenomena to brain functions (12). Admittedly, science is not possible without language. Language itself, however, is the final product of an unbelievably complex development of the brain. There will be nothing but confusion, these investigators believe, if we use everyday language to talk about the very brain which has created this language. What von Monakow said in effect was that we should not use ordinary English, German, or Russian in talking about brain functions, but should invent a special language.

SAVAGE: But not with the brain.

KLÜVER: He argued that all words dealing with psychological phenomena have gradually acquired too many meanings and connotations, and that real progress in neuropsychology depends, therefore, on the use of neologisms. It is for this reason that he introduced such terms as *klisis, ekklisis, protodiakrisis, kakon,* and so forth.

RICHARDS: But how does he define them? If you introduce enough new technical terms, isn't there a sort of law of economy about the introduction of technical terms? One technical term is excellent business; two technicalities in the same sentence are all my eye can take. There is the problem of anchoring your new technical substitutes to observations, isn't there? You have to go into cases.

KLÜVER: Perhaps von Monakow was of the opinion that desperate situations demand desperate remedies. Historically, he has not been the only one to insist that the use of everyday words in referring to psychological phenomena will forever impede progress in the field of psychology and prevent a sharpening of its conceptual tools.

RICHARDS: I am very much inclined to experiment with not using conventional terms here. I think the introduction of images at this point would be absolutely disastrous, although many of the observations we rely on to some extent are phrased in images. There is a typical anecdote which turns up – I should be able to give you the reference, but it doesn't matter – of the hunter who is inexperienced, going out looking for deer. He never sees a deer. He does everything he can, until one day he goes out with an old hunter who puts his hand on his shoulder and says, »Now, look, don't you see that deer?« The other says, »No.« The old hunter begins to describe it, and as he describes it, it appears. And then the advice is given, very judicious hunting advice: »Don't go out after anything unless you imagine the animal you are looking for. Imagine it as concretely and as fully as possible.«

I begin to feel doubtful about that. In my own case, I can see that a visual image might act as an obstacle if I relied on it; but that illustrates feedforward suggestion.

MEAD: Dr. Richards, in connection with the point you were making about the dangers of autarchy, I think that this danger is very striking | at present for anthropology. [64] By the isolation of linguistics from the study of the rest of culture – and linguistics is more isolated from anthropology than it has been for a very long time – we have lost in a sense our capacity to look at other systematic aspects of human behavior, so that the argument in anthropology at present, which you summed up by picking up Kroeber's statement (4), is this: Yes, of course we know that language has grammar; language is systematic and linguistics can study it. It is something that is in a nice little box. The rest of human behavior, however, is regarded as subject to some completely nonsystematic set of principles which make cross-comparisons impossible, and one comes out with Kroeber's sort of statement as to the uniqueness of each culture; whereas if one says that language as we know it, and other systems of communication between people, other methods, parts of behavior, all of which involve the whole body, are all systematic, because they can be referred to the human organism; then it is possible to make the sort of cross-cultural comparisons that you are asking for and to use the uniqueness of each culture only as a point of reference for particular observations within the culture, so that false equations are not made from one to another.

RICHARDS: That is the danger, and it is very hard to avoid it, because until you get an over-all descriptive system which has been sufficiently criticized, you are almost bound to make false identifications, aren't you?

MEAD: Well, typical points would be that you might be able to get a cross-cultural or cross-language language which is sufficient to think about languages that are quite different and to think about them together; but if you talk about whether you murder your grandmother or not in one setting or another, without reference to the setting, you end up with cultural relativity.

RICHARDS: Would you agree that all subjects are inevitably evaluative?

MEAD: In the sense that you are saying it, yes.

RICHARDS: It does solve a great number of artificial problems – the classic warfare between strict neutral science which is not concerned with value and poetry which is concerned with nothing else, as I see it today.

SAVAGE: I am very much interested, too, in these remarks about normative aspects of science. You said several things about it; some of them I thought were good, and some of them I don't yet understand. But one of them, at least, did seem to me to be a technical mistake; at least I would so interpret it. You say: »Suppose that the biologist worries about whether this bone in the hand is typical of such bones. He must then face the question whether this is a good bone to describe. Well, that is a normative question. It is a biological question. There|fore, biology is normative.« That doesn't seem to me to be at all fair. Wherever we must make a decision, we have, I would say, a normative question. We must decide what it is right to do. And, in particular, if the advance of anatomy is our problem, then what are the right moves to advance anatomy? But that does not in itself imply that the study of anatomy is per se –

[65]

RICHARDS: Well, may I answer that point? So far as the anatomist is looking for a general study and wants to generalize from his observations, I am saying very little more, I think, than that part of his technique in deciding that comes down to: Is it safe? The question is: Is it safe to generalize a description of this bone?

KUBIE: There would seem to be a subtle transition and overlapping of meanings here, beginning with the statement about the »ethical« element that enters into the acquisition of language. This is different from the ethical element of a language or of a science itself. The process of acquiring the language, or the process by which a student acquires a knowledge of chemistry or anatomy or anything else, carries the ethical values and systems with which he has learned to keep himself clean, to brush his teeth, to obey his parents, and so on. There is a right and a wrong, in the sense of a right and a wrong answer, of being a good little boy or a bad little boy. This invades the whole learning process inevitably. Was there not some confusion here?

RICHARDS: I think you are right. There is a hierarchy of the uses of the word »ethical,« and I think probably Savage understood me to use one that was fairly high up, whereas I am really talking about what we usually call good investigatory practice, the sort of precautions that you do observe, that your evidence is not biased by something you have not considered. Take the anatomist and his bone: if somebody else can come along and say, »That is not a typical bone; and you are assuming a very great deal and I will show you a typical bone,« then the problem is, I think, about value, but very low down in the statement.

By the way, I should like to take up a point about the way of teaching a language. Of course, you are being given the culture; you are being given the ethic. But this is the thing that makes me boil when I see it happening on the planetary scale, with billions of people. There is the question: How do you spell this wretched word? [Writing on blackboard, *idear.*] My Harvard freshmen come from 3 per cent, you know, of the most favorably placed people in the country. Many of them write *idear.* They have made a mistake. They have not assimilated one of the curious things about the culture. Of course, they spell it phonetically. It is what they hear – *idear.*

KUBIE: It is how he feels.

[66]

RICHARDS: Actually, it is not a crime; but bad spelling is put into a | worse category of cultural crimes than inefficient use of words. But I am really punishing that boy with disapproval, with social scorn, which is the whip. What I do is to run through their own compositions and throw them on the screen, and the whole two hundred howl. Of course, they are all like wolves, to tear a bad speller to pieces. But if a boy comes along with one of the most important tools of all Western morality, all world morality, and writes »disinterested« where he meant »uninterested,« does anyone wince? When the culture takes one of the most important instruments of discrimination we possess – differentiation between disinterested behavior and uninterested behavior – and throws it into the trash can, that class won't bat an eyelid. Has anything that mattered happened? You can't persuade them. If you could find out how to teach the beginnings of the great subjects, the problems that we are vexed by in the upper hierarchies would be vanishing.

SAVAGE: It seems to me that precisely because words like »normative« and »ethical« can be, and are, used at different levels, there is some sense in referring to some disciplines as normative and others as non-normative. But it is commonly said that medicine is ethical in a sense in which biology is not, and I think that is correct. Such a statement should not be regarded as nonsense and contemptuously thrown away just because we find some normative activity among biologists.

HUTCHINSON: I think one needs at least three distinctions. There is the normativeness of medicine and the normativeness of good practice in any investigation; and then there is a third one that is, it seems to me, escaping us a little bit, and that is the normativeness of whether it is worth while to make statements at all, however well the investigation has been carried out. If that is lacking, then the whole subject is condemned. In mathematics, it is called trivial. I don't know what you call it in biology, but I think you generally say that it is the kind of thing that a man in a museum would do, or that the work is »mere.«

RICHARDS: In learning, »botanizing« is usually the description.

MEAD: »Blind alleys,« too.

HUTCHINSON: Yes. And then there is the very interesting phenomenon that most of the »mere« aspects of biology of twenty years ago now are exceedingly important.

BIGELOW: To some extent that is true in mathematics also.

SAVAGE: What – the trivial parts of mathematics have become important?

BIGELOW: Well, the basis of axiomatics, for example.

RICHARDS: This is uncovering what is behind this word »important.« We throw the word »important« around with great ease because we haven't defined it as ethical. Now, my little talk was a matter of saying | that if you looked at »important« carefully [67] enough, you would find a good ethical tag on it.

MEAD: Well, my understanding, when I answered you that I agreed with your use of the word »normative,« was that it not only applied to good practice in the specific sense, but that it also applied to the fact that you had to consider all of life in the end as a whole and all the sciences in relation to one another, and that these mere, trivial, though worth doing, considerations are part of the total pattern. One gets, therefore, the distinction that the difference between medicine and biology is a difference in level; but they would both be normative in that sense.

SAVAGE: It seems to me tremendously important not to lose sight of something that I think our ancestors gained in formulating the concept of pure science. Though we have heard about it all our lives, it may be difficult to express what is meant by this notion, but the minimum is this: One should not confine one's attention in science to what is superficially useful. The judgment of the scientist, the taste, and even the instinct of the scientist should be allowed a good deal of play and a good deal of time in which to express themselves.

The advocacy of pure science means, for example, that a bacteriologist should be free to study those organisms which seem to him biologically interesting, without special reference to those which sour milk or those which cause mumps.

RICHARDS: I am going to agree with that, of course, but I am also going to try to appeal to this over-all grand analogy here. There is something that corresponds to sovereignty in a separate study. It alone can settle certain things about what it should do. It is its own authority; but, on the other hand, much that it does – and your bacteriology is a good case – is the concern also of other sovereign studies. And here comes in your diplomatic service, not to dictate to people inside bacteriology what studies they should pursue, but to represent public health.

KLÜVER: You referred to the fact that the same object in four different positions is still the same object and that a child, therefore, may have difficulties in learning that *d, b, q,* and *p* represent four distinct letters. Psychological investigations have shown that reversals, rotations, and spatial displacements are characteristic of the perceptual world of certain children or of certain age groups (13) (14). In such children the process of learning to read may involve special difficulties.

RICHARDS: I have looked into it a little bit, and I don't think that has any connection with the symmetricality of certain letters. But, on the other hand, I think this queer reversal, for instance, this tendency to reverse reading and writing, might possibly have some connection with the general stage of intelligibility of the form that is being stud- [68] ied in | the early stages of reading. What one wants to achieve, if one can, is a sequence of steps into reading, or anything else you like, which is maximally intelligible.

KLÜVER: Reversals of letters, if they should occur, are undoubtedly not isolated phenomena but are correlated with other perceptual developments.

RICHARDS: I agree.

VON BONIN: I should like to bring up something which troubles me. You talked about Kroeber and his anthropological ideas about studying cultures, and the idea or the conception of studying each culture from its own point of view as being a halfway house. When you talked about literary criticism afterwards, I think you mentioned that the two things that mattered were the thing the author wanted to express and how efficiently he did it. Isn't that exactly the sort of thing that Kroeber (4) tries to do in culture? I am not quite clear about that.

RICHARDS: Yes, you are right. I think it is quite probable that literary theory is at a very early staging place, and it probably does bring up these characteristic complementarity situations. I don't feel very strongly that there is one there at present. I think you nearly always have two grounds for criticism, two prime questions which you ask in judging anything: What do you want to do? and, Can you do it? I don't think that is quite the same as the Kroeber situation. I want to go back to my remark that that was a defensive utterance of Kroeber's. He wanted to damn the people who were invading a given culture with presuppositions taken from another culture and describing it with their tools.

FRANK: I hope this question is relevant. Can we say that we are becoming increasingly concerned with problems of communication in recent years largely because the traditional usages, the long-accepted meanings of words have begun to break down? As we try to develop new ways of thinking and new ways of communicating what those new ways of thinking are, we are faced with some of these difficulties. Would it give us any better perspective on the very problem you have been putting before us if we thought of ourselves as engaged, here in these meetings, in trying to create a new climate of opinion, a new way of thinking where we are moving from some of the old static, analytic, linear ideas to a way of thinking in terms of context and dynamic processes? We are trying to establish the word »process,« for example, with the assumption that the same process may produce different products, depending upon where and how it works, as an attempt to get away from the static nouns and to put our generalities and abstractions in terms of verbs. It seems to me perhaps that might be one way in which we are trying to move out of an older way, and do what the Hopis have done for a long time; namely, to use only verbs for abstractions. | [69]

RICHARDS: I was struck this morning by a remark about the way in which function and structure can change places as the point, temporarily, of maximum interest and significance. I think there is something very similar going on with process and structure in this language. There are aspects of language which are, for people like that, purpose-serving, rather than perhaps promoting, aspects which have been neglected and should come in. I think there is a very long way to go.

But as to your remark on the general breakdown of communication inside this culture, I agree we've got to take it terribly seriously. I am raw with irritations caused by my Harvard representation of the intellectual »cream« – and I couldn't give you examples that would frighten you as much as I have been frightened. I am terrified. I think it may be that they have been selected by an improper selection process, and, if so, that is very serious.

BIGELOW: Is there any evidence that it is worse now than it has ever been?

RICHARDS: It is all gossip and impressions. I don't know whether it is sound or not. I should like to know. Of course, we have come to a short way of educating, as everybody knows, with so many more students and, on the whole, education in larger classrooms, and we are trying to teach them more about more things. All this background philosophy of mine about the over-all united world view and about getting a headache every morning concerning every grief and trouble there are on the planet – we do that intellectually, too. I don't think it really helps the young initiate in the culture to learn what will grow best in him.

MEAD: But don't you think also that what is happening in a period where we change as rapidly as this is that as teachers, instead of our getting better, in a sense we get worse, because we can't learn by experience from teaching generation after generation of approximately the same pupils because our pupils are changing all the time. Now, I go away and I come back to teaching, sometimes with intervals of four or five years. The thing I am progressively struck by is that students can understand now that which they couldn't understand ten years ago.

RICHARDS: A hopeful thought on which to end.

HUTCHINSON: Well, there was just one parting thought: we are actually, in all this discussion of language, throwing away a great deal of what we know. Lawrence Frank was talking about the substitution of ideas of process for static subjects. Every time we draw a graph with a time axis, that is what we are doing. We have been doing it since the beginning of the eighteenth century. The only thing that we don't do is to recognize in this group that that is what we have been doing half the morning. It is all in front of us, but it is compartmen|talized; and too many people in the literary disciplines regard it as a crude and scientific and inhuman thing to do. But it is perfectly standard and well-integrated practice, and merely has to be spread out a little bit over a few more Harvard and Yale freshmen and other people. [70]

MEAD: And you have to recognize that moving your eye is something that is relevant.

VON BONIN: I think Heraclitus first said, »Everything flows.«

ROSENBLUETH: One of the statements you made, Professor Richards, which impressed me very much and which I want to ask about, is this: You said that the first trouble you found, or the important one, was what language to use to talk about language. That is an extremely important question if we wish to talk about anything. On the other hand, when you told us why you had that trouble, you said, »Would you use mentalistic or behavioristic terms?« Is that really very important, as long as somebody makes the statement, »I am using these terms with this general connotation, with this general meaning?«

RICHARDS: Yes. I think you can get out of it. I think that twenty or thirty years ago that was a barrier to a lot of people's honest thinking. I think now, somehow, they have got accustomed to the situation and they see they can put their inverted commas, their special meanings, on it, and continue. I don't think that would be an important point. There are others for which that would only serve as a model, or as an example. I am not sure I should try to elaborate very much, but in places where you do see a rather crucial choice as to conception, it turns out to be a choice of language that you might employ.

ROSENBLUETH: The language which you employ would, of course, depend on your desires, on what you think is the right way to criticize something, the language you think is the most useful one or the one that fits the purposes which you are going to follow. It will always be nothing but a tool, an important tool in general. The best language, however, is nothing but a poor tool. But once one realizes that, I don't think there is any special trouble that arises. We just have to try to refine it.

RICHARDS: But there is a point where its defection may sometimes be its merit. That is constantly the line of progress in this kind of speculation: the breakdown of the tool. That really is the important part of the observation. We are not trained to note that. We blame the language, and we should, instead, focus on the language.

If I could have a last word, it would be this: if we could only take account of our constant, habitual skill with language and translate that into a general theoretical understanding, we would be very near where we wanted to be. We constantly know [71] things in practice about language which we are blind to, completely, intellectually. |

PITTS: I think that what is important in this is that what can easily be a matter of convenience can become a matter of principle; that is, it is perfectly possible for us in English to express every shade of indecision or truth in a statement by prefacing it with words like, »It is hardly certain that ...« We commonly do that. I think the extreme case of that occurs in mathematics. As Russell and Whitehead have shown, it is possible for every mathematical computation to be expressed in ordinary logic, in ordinary words containing no special mathematical notion such as number, and so forth. But if you do so, the simplest arithmetical computation becomes a very complex computation spread out over not less than thirty pages of text, and resembling legal arguments about the Constitution which nobody can possibly hope to follow. If mathematics had not been developed separately, we can be quite sure that although everybody, in principle, who has a training or intuition for syllogistic logic could make that inference, it would actually never occur to them under any circumstances to do that. If some tribe of American Indians has a language which causes them to express universal appearances by inflecting the verb instead of by qualifications attached to the noun, it is perfectly true that we can say in English everything they could say. We can translate their descriptions into English. But if their natural and brief descriptions tend to become intolerable prolixity in English, the chances are we will not see that aspect of the situation which to them is the simplest because it is the most simply expressed.

RICHARDS: I entirely agree. It is almost a doctrine for the people I am thinking of, the linguistic scientists, that anything can be covered in any language. It may not be so.

PITTS: Perhaps it can, but not in any natural way.

RICHARDS: They may be simplifying their notion of covering.

BIGELOW: Do we believe it can?

RICHARDS: If you are willing to go to enough lengths to say it.

BIGELOW: Perhaps it can be done by the people who translate poetry.

ROSENBLUETH:. I can give you an example in Spanish: *cursi*. I would have to write a book to tell you what that means in English.

SAVAGE: What does it mean?

ROSENBLUETH: It would take a book to tell you.

PITTS: But in any book in which you use that word, at the expense of enough prolixity, you could translate it.

ROSENBLUETH: But not the feeling. You would not know what I am talking about. And if I gave you a good example, it wouldn't even tell you.

MEAD: I think you have a different point, though, the point that Bateson made in his paper on pidgin English. You *can* say anything in | pidgin English (15). I have translated part of *Alice in Wonderland* into it. It can be done. But the point is that you don't want to. There is a way to say it formally, yes. You can translate Sapir's old statement that Kant could be written in Eskimo in such a way that it is formally true, probably, but it leaves out of account the fact that you won't ever want to. Conversely, of course, after we have become acquainted with these concepts, such as the one you are mentioning or those I find in each language I have learned, I have added a few things that I need. We can use our knowledge of other languages to shake us loose. That really is the point: not to take over the Hopi ways of saying things, but to shake our own system loose, so that we can build new ones. Won't you agree to that? [72]

RICHARDS: Yes, I think that is true.

TEUBER: There was a German humanist, for whom I have the greatest respect, Stenzel, who wrote page after page on single Greek nouns, explaining that they could not be translated, and paraphrasing them in a great many ways, trying to show that *logos*, after all, expresses what *logos* expresses, or that *arete,* meant, well, *arete,* and that nothing you could say in German would explain it (16) (17). He went all around the bush in trying to give you a general evocative meaning for *arete* or *logos*, as he felt it should be understood, in the context of the Athenian democracy of such and such a year, even part of a year, when it was used by such and such an author. When you looked at some of the translations that he gave in illustrating his points, the majority of the words were Greek words, interspersed with a few German prepositions, and then he went on to show that *arete* differed from *dikaiosyne,* and there was more Greek to it than German.

MEAD: Before we end this discussion, I should like to go back to one other aspect of what Dr. Richards said. It seems to me that he was also dealing with this question of teaching and learning in the very young, and that what we know about all sorts of things can be translated into pedagogy in such a way that we will be able to educate individuals whose communication potential will be so much higher. It goes back to a question I asked Dr. Bigelow about two years ago: If he could re-educate himself with all he knows now, would he do it differently? As I remember, he said he would settle for twice as large a brain. That was a conversation over coffee, and it may not be fair to quote it. But the description that Von Neumann gave of memory at the time he talked about it, and the possible cross-referencing between filing systems and senescence, for instance, and senescent memory, raises the problem in connection with what Dr. Richards is saying here. (I think you still all hold to the hypothesis, don't you, that we use only about a tenth of our brain in some way or other?) We could build our sys | tem of abstractions in the right order, as we are educating children, and take Dr. Richards's point that we must not let them make a mistake, because the mistake will dog their footsteps forever. The best Illustration I know is still that old study of bilingualism in Welsh and English published in 1923 (18), in which bilingualism showed up as a defect worse at the college level than at the kindergarten and elementary level. We could [73] begin to use all these concepts of selectivity and types of filing and hierarchies of generalizations. What you were giving us, as I understand it, was a model at the educational level.

RICHARDS: Yes. It is work that has been going on very laboriously for a long time, and I am quite ready to say that we have evidence now, in teaching reading and English – I am not sure yet about French and Spanish, but English – that the order of the operations, the order of presentation of sentence-situation units, is decisive.

MEAD: But I should like to add to that, now that we have a little bit of evidence – it is only case-study evidence – that infants who have been reared on what was erroneously called »self-demand« get a generalized capacity to reduce confusion.

RICHARDS: That is quite right.

MEAD: So you get this sort of analogy, for instance, that if you try to teach the child who has learned, who has been fed in relation to its own rhythms, too many transportation systems at once – for instance, three bus routes – he will say: »Stop. If I learn any more bus routes now, I will be confused.«

RICHARDS: They are right.

MEAD: That is at six and seven years of age; so it is possible that we can begin this question of order virtually at birth, and then follow it along.

BROSIN: In your functions as critic, and more so as an educator, do you have systems whereby a person, an interrogator, and his vis-à-vis can establish communication over barriers with more economy? This room can be used as an example. Can I, or other clinicians, when talking to a patient, use these devices which set a theater for the beginning of our exchange and serve as a basis for further exploration? It is useful to remember Sullivan's phrase that the patient is a stranger (19).

RICHARDS: Yes, exactly. I can give you two instances which will fit into what you have asked. We are almost certain now that it is a more efficient, satisfactory procedure to give a beginning class of children, learning to read or learning a second language, every sentence-in-a-situation on a screen in the artificial screen space, not in a book. And it is certainly more efficient to give them the sentence, not as spoken by the teacher, with all attendant complications, but from a recording. They will do things to [74] a recording in the way of parodying it, guying | it, that they would never dream of doing to a teacher's face, and they get a beautiful pronunciation in that way. That is a side product. The actual point is that you have a release of exploratory action in the child or the learner which is constantly not present even with a very good teacher. It is very interesting.

BROSIN: May I follow that? How do you establish communication with, say, an English-speaking person on a train, about subjects upon which neither of you has knowledge of each other's theaters of action? What are the formal steps?

RICHARDS: You mean, how do I do it, if I meet an English-speaking stranger?

BROSIN: Yes. I do not mean in any trivial conventional sense now; but rather, what are the serious operations that you employ to get past these barriers and become acquainted, in the sense of exchanging maximum information in a short time?

RICHARDS: As I understand it, I have no problem in this country, but I have a horrible problem in England. It is perfectly easy to talk to most people here, but it is very hard to get talking to anyone in England. This is a technical problem of communication.

VON BONIN: The same problem, it seems to me, arises in literary criticism, on a somewhat different level. Somehow or other, you have to know or you have to feel or establish what the goal of the author was. Of course, there are his printed words, but one has to read between the lines. The same thing arises for a historian when he wants to evaluate whether Caesar did his job well or not, and things of that sort. The same thing arises in trivial and less trivial situations, but that is the problem.

RICHARDS: I think that is the heart of it. That was easy, relatively, when you had a stable culture in which, when you encountered a new text, you could assume that the people who were meeting it would have had a common literary experience leading them to it, even if it were only the Bible and Homer, and so on. Now, when you can't have any confident expectation that anything you mention will have been met before, criticism is almost lost.

MEAD: Don't you mean that before you can talk to another person, a chance – met and totally unknown person who, however, does speak your language – those are your conditions, aren't they –

BROSIN: Yes.

MEAD: – that you have to work out his probable image and understanding of who you are?

BROSIN: Yes, indeed.

MEAD: And his willingness to talk to that image.

BIGELOW: As you did beforehand? | [75]

BROSIN: You must project this universe of expectations. I would request Dr. Richards to review his statement that criticism is almost lost, and with it the writing at the level of, or with, the certainties of a Goethe or the seventeenth- or eighteenth-century critics, or the Coleridge which you quoted. I can understand your position in comparison with the mid-twentieth-century author, but, actually, each of us is a critic every day in every piece of communication.

RICHARDS: Oh, surely.

BROSIN: I am asking the impossible, of course, but that is the purpose of this meeting. In your experience, what are the formal barriers and how do you operate, either from past experience or by very careful pedagogic logic, to establish a theater, a frame of reference for more meaningful communication in terms not only of the superficial values but of the nuances, the deeper appeals to past experience?

RICHARDS: I can deal with just a little of that in a moment. We had, though, when I said criticism was almost lost, exactly the same situation that arose with Savage and me over »normative« and the bone and »ethics.« You see, you can send criticism to astronomic heights, and it becomes a sort of peculiar privilege of the literary critic to be concerned with criticism. Most people say, »Oh, no, not us!« But you are right. Everybody is being critical, in the most fundamental sense, with every utterance that he makes himself or takes from anyone else. Now, what can break down the barriers? The kind of thing I hope I have illustrated by comparing the ethics situation with this criticism situation.

All the main troubles of language are endlessly recurring. They have certain common patterns. We continue with *ad hoc* analyses and improvisations, to deal with misunderstandings as they come up. But where is there the generalized theory of misunderstanding, or the generalized technique for developing skill in knowing what are the probable meaning variations around a given utterance, taking the utterance just as it comes? The hope of doing it would be through a structuring of the field of misunderstanding. If you could do that, you could begin to teach people to understand one another better. That is my hope.

RIOCH: I would be interested in another aspect of language, taking language as a central nervous system function. It seems to me to have an entirely personal significance, quite apart from any social significance, though the personal significance is almost certainly developed by the social function. But probably there are equivalent activities in forms or species without language. It is pretty clear that a rat confronted by two doors sets up a hypothesis, orienting himself; he goes to one door or the other. The difference between that and the personal function of language is that in some way we can remember something about the situation which calls up a word. That word then organizes the whole; | that is, the function of the word, or the activity of the central nervous system which is this word, then organizes the whole of the central nervous system to be oriented to a situation according to that word. It is an entirely personal thing, and in that sense not –

[76]

Richards: I have a doubt about »entirely personal.« It has been learned in interpersonal relations.

Rioch: Yes, but now you have to consider learning instead of taking the activity as it occurs.

Richards: I don't think they are really separable, are they? That is my point.

Rioch: I think we have to separate them if we are going to think clearly about them, because we have to know what point we have arrived at in learning.

Richards: I am worried about the central nervous system. I want the rest, any nervous systems you can provide, everything. I think it is nearly everything that comes into language.

Rioch: If we make it that broad, then it is very difficult to differentiate anything from it, whereas I think we can select factors out of it, especially when we see the effects in cases in which there are certain deficiencies.

Richards: I want very much to hear how animals do communicate, because that would throw a great deal of light, I think, upon how we do.

Rioch: We can tell a lot more about human beings and how they communicate than about animals. A very curious thing about language is that there is a continuous communication which is going on through time. When a word is used with respect to that communication, then time is stopped and we no longer deal with time, because the word is now either so or not so; and if it is so, it is always so, and if it is not so, it is never so. By the use of words we have introduced a digital system instead of – I don't know whether you can call it an analogizing system – a system which is a continuous communication through time, with continuity of interaction. We put in a word which now destroys time, and that is the only way in which we can deal with past or future time. That, I think, is one of the major functions of language in communication – to destroy time – and it is the thing that at the present time has made language such a dangerous thing, because it has destroyed time in certain directions instead of in other directions, and people no longer trust what other people say.

Brosin: Has it destroyed time? Meaning is independent of time and space. While I am talking, that is a process in time. The meanings that I have reference to have nothing to do with either time or space.

[77]

Rioch: Your tone of voice is something proceeding in time, but the | content of the words you use is timeless.

Brosin: The media for the expression of my meanings is a process occurring both in time and in space.

Mead: To that extent it is analogical, isn't it? The communication that goes on between people that involves their whole persons, the stresses in their shoulders and the lifting of their eyebrows, and all the rest of it, doesn't have this digital character.

Rioch: No. When you talk about communication between people, then you have to divide things up much more than that. Let us consider language and vocalization. If we take a series of categories of either complexities, obscurities, or elaborateness of vocalization, it doesn't matter which category we use, we will at one end of the series have that situation which is probably the simplest, that in which there is a double integration of the organism with the environment, probably mediated by brain-stem mechanisms. In this case the vocalization is a part of the total activity of the organism as a unit.

Now, when one proceeds through a series of categories of complexity of vocalization, one arrives at the other extreme, in which, regardless of the tone of voice, the style of handwriting, any gesture, or anything else that goes with it, the entire communication is contained in the content of the words as defined by Webster. You have everything in between. But whenever a word is used, then you can separate the significance of that word in terms of its content, as something that stops time.

McCulloch: Once for all?

Rioch: Once for all; and that is its main function, which is something about language which one notices when one deals with people who have difficulties with language. I think you notice it with children. You certainly notice it with patients. You can see a patient come from a situation in which the content of the word is essentially of no significance – the tone of voice is of tremendous significance – to a situation in which the content of the word has entirely personal significance in a very limited situation, indicating what he is going to do in terms of some very limited thing; to what we speak of as normal where the content of the word is very important, but in the sense in which the language is socially learned. One of the most interesting things we ran into recently was a lobotomized patient who apparently did not have a concept of what was going to happen in continuous time; that is, there was no feeling that if he hit somebody the other person wouldn't like it. Partly as an experiment, we decided to give him a language formula as a time-binding tool. We tried saying to him, »If you hit somebody, they won't like it.« Within two weeks, the hitting of people stopped. I don't know if the formula really had anything to do with the change | in the patient, [78] or whether other implications or factors were responsible. Be that as it may, the time-binding quality, either getting rid of time or making time permanent – it is the same thing – is very curious in language.

Richards: May I put a note in here? It is just a footnote to this asterisk[!] business. We are hearing one another make a series of what appear to be assertions in the full indicative, yet I think if you asked people you would find they were really talking in another mood altogether. It is customary to talk in the indicative, but might it not be that what you were saying was only the equivalent of an optative? How would it be to conceive that …? There is all this going behind both of your remarks. Dr. Rioch, when you said »destroys time,« I thought I detected around the table a reaction, the sort of reaction that occurs when people use a phrase which is characteristic of poetry in a situation which causes them to expect prose. I know I had to adjust myself for a moment.

The question came up the other day whether there could be a Complementarity Situation, »mentalism« and »behaviorism.« It is much more than that; it is that we have modes of expression which we don't know how to replace but which do work, as it were, taken from the integral action of the mind.

Rioch: But sometimes they don't work.

Richards: This one did work – in time – but it took several people in the room, I think, some time to tolerate this »destroying time.«

Bigelow: I still can't tolerate it!

Rioch: There is another aspect to vocal language which is quite entertaining in terms of the information conveyed. Harry Stack Sullivan illustrated it as follows: »Good night, Mrs. Thomas; it has been a perfectly foul party this evening, and you have been a horrible hostess.« If you put the right intonation on it, the words don't matter and you can actually get away with it.

SAVAGE: But can you?

RIOCH: There is one other aspect to this problem of language I should like to mention; that is, the magic of communication that has nothing to do with the content. Going through the right formula does two things: It keeps the situation structuralized; but it also does another very important thing in that it limits the range over which communication may occur. You bump into somebody and say, »Oh, I'm so sorry.« You're not sorry. It is not something you can be sorry about. You may feel chagrined, you may regret; there are several things you may feel, but you are not sorry. But you say, »Oh, I'm so sorry.« It keeps the situation structuralized and limits the man as to [79] what he can answer you. |

SAVAGE: He may say, »You're not sorry at all.«

RIOCH: It still limits the thing that you can say. Now, this curious function which is carried to, we might say, a pathological extreme in the obsessional neurotic state is, I think, one of the chief feedback defense mechanisms to prevent complete breakdown and anxiety. But it is a function of language in which language is not being used in terms of its content or in terms of its intonation. I think in your teaching experiments you are getting a very important use of this function. The child gets the adequate response from the other person, which is predicted by the child when he reacts to the symbols properly. This may have nothing to do with the child and the symbols, nothing to do with the child's orientation to the content of the symbols. You point that out when you say that Jack and Jill are not important. It is very much concerned with being able to do something that will get the personal response that is predicted.

RICHARDS: Oh, yes, that is all-important.

RIOCH: And that may be in a different direction from the direction of the content.

RICHARDS: Oh, quite. I quite agree with that.

ROSENBLUETH: This states that there are languages other than the spoken or written word, and that language can be used for purposes other than the transmission of information. I would certainly be in complete accord with what you say.

HUTCHINSON: If one takes that point of view too seriously, any repetitive learned movement becomes a language. One learns a repetitive movement of the legs because it is the same form as that used on a previous occasion. If there is going to be a personal element that does not involve social communication, then all repetitive actions throughout the whole universe would have to be regarded as linguistic.

SAVAGE: That is a *tour de force,* isn't it?

MCCULLOCH: I think we had better get on. There are two more people who have asked for the floor. One is MacKay; but just before he speaks I should like to say that he was pointing out to me the other evening the importance of –

BIGELOW: Who is »he«?

MCCULLOCH: Donald MacKay – the importance of maintaining discipline or order, or whatever you want to call it, in the British Parliament; that no matter how foul the remark or how personal the man may want to be in his attack on another, he may only address the chair. The very indirection makes the speech tolerable. Will you come in now, on the main subject?

MACKAY: I really wanted to get back to the question of »language about language,« [80] and the way it ties up with what has just been said, | because I think there is an important distinction between a new language and a *shorthand.* We can devise a shorthand for talking about language, in which we make abbreviated noises that are really equivalent to strings of statements in the ordinary language. On the other hand, a new *language* must, I suggest, have a different set of referents or must include at least some referents different from the other.

Now, if one thinks of the function of language, of communication, as the instruction of the hearer to replicate in his brain or mind – put it as you like – a representation of something in the mind of the communicator, if you think of any linguistic body of information as an instruction, the intended response is essentially a *selective operation* from the junk box of component parts ready to hand in the mind of the hearer. Therefore, if you are dealing with someone from a different background, you may require a complete book in order to represent a word, as in Rosenblueth's example, simply because there does not exist in the ensemble of component parts from which you are instructing him to make a selection any one elementary thing, or a compresence to use the whole spectrum of a Fourier series to describe an impulse if you are confined linguistically to the logical space in which only frequency can be defined. I shall be talking more about this sort of thing tomorrow, so perhaps I ought to say no more now. But the question of formulating a language about language does bring up this distinction, and in fact it seems that the most we can hope to do in this case is to formulate a shorthand.

Now, that doesn't get around the other difficulty that I wanted to ask Dr. Richards about; namely, whether he has considered what happens when a logical sequence itself becomes the subject of the logical sequence. In a deductive machine, for example, that would correspond, or could correspond, to an unstable situation leading to oscillation or blocking or something silly like that. In some logical disciplines it emerges in things like Russell's paradox; and Popper (20) has raised a related question, namely, whether in principle a thinking machine could predict its own state. He shows there, I think in a watertight way, that it is in principle impossible for a calculating machine, given all the classical data, to predict its own future state; and hence he deduces that determinism, of the sort which asserts that a computing machine of sufficient size could completely predict the future, is in principle untenable.

This seems to me to tie up very closely with what Dr. Richards was saying; but this is a case where you have feed*back* in language instead of feed*forward*. I wondered if he had anything to say about it.

RICHARDS: It is the crux, of course. I tried to do a little with that, pointing out that the attempt to use language to represent language, | with the language itself representing something else, caused the paradox. But I should like for a moment to put a picture on the board which I find helps me a little. Suppose you tried to represent the language functions, perhaps, like this (Figure 7). I have a suspicion that you couldn't do with less than six, for the purpose of this afternoon's discussion. For other purposes, you could do with a very much simpler scheme. I think there has been a great confusion in the field through trying to bring in a »reference function of language« and »emotive language« and »promotional,« and so on. But I want to suggest that [1] you have what is essentially a *pointing* function. You pick out, you select, something you are going to talk about. This function is normally not distinguished. [2] You have *characterizing*. You characterize what you pick out in some vague or precise fashion. [3] A function, which can be dropped, if you wish, is *realizing*. Your mode of selection or your mode of characterization plus something else invites you to *realize* what is presented to you by the other person more or less vividly. [4] Here you have *appraising*. [5] Here you have *influencing*. I suggest they *all* normally go together. [6] No. 6, which is left over, is an *organizing* activity. This is the supreme praesidium, as it were; this is the ultimate authority; this is the United Nations. It is still very unsuccessful. But it endeavors to keep these various simultaneous components of the whole language purpose from interfering with one another. Interference comes to a height in a battle

[81]

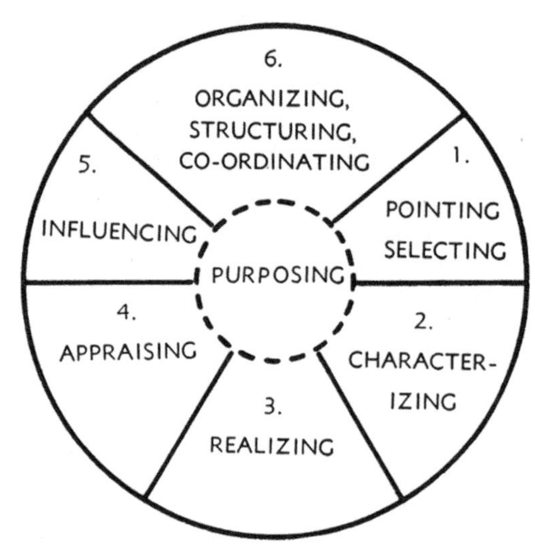

FIGURE 7

between the subjects, but it is actually, I should say, stirring, as a potential self-frustration, inside anyone who says something to himself at any time. | [82]

Your point was: What arises if you put in a proposition about this function of organizing itself? You get opportunities for all the headaches that Russell grappled with. I am not sufficiently interested, really, in struggling with the precise processes that people have used symbolically to get out of that fix. I only know you must be able to get out of it somehow. Somebody ingenious enough will find a way in which you don't get anomalies. But if one could remember the complexity of every utterance – though some of these functions sometimes drop to very low values – it would help.

PITTS: The formal deductive systems which are capable of discussing themselves in the sense of describing their own structure, and the deductions that can be made from them, have, of course, been extensively studied by such people as Gödel. The conclusion they come to is not that there is in any sense an irretrievable paradox in having a logical language which speaks of itself, but that whenever a logical language is rich enough to do so, there is a fundamental incompleteness in that system; that is, the possibility of describing a language in the language itself always means that the language itself is incomplete.

RICHARDS: Would that have anything to do with any hierarchy of tapings? Supreme tapings, you see, would be too much.

PITTS: It is hard to say. It always means that there is a more inclusive one outside.

VON FOERSTER: We can't escape Gödel's Theorem.

PITTS: His theorem depends upon a method which enables you to construct a logical method within that system, using certain things, such as numbers, to describe the symbols in that system; it enables you to formulate the logic of that system in arithmetic relation to numbers. Then you can do this: You can construct a sentence and have it assert that it is not provable, or, rather, that is what it appears to do. What it really does, formally, is to assert that the number of that particular sentence, when you calculate it, is the number of a nonprovable sentence. As you can easily see, that must be a sentence which is not decidable by the means of logic at the disposal of that system, because if it were provable it would not be true; so the assertion is not true. If you could prove the contradictory of it, well, then, it would be true, and then you would have arrived at a proof of it. But, of course, it asserts that it is not provable, so what follows is that that particular sentence which asserts its own unprovability is indeed undecidable either way, by the means at the disposal of the system. You can enlarge a system by simply adding that as an axiom; but then, in the new system, you could construct the sentence which asserted that it was not provable in the uses, even with the additional axiom, and that would become an additional undecidable sentence. You could go on enlarging it, but you would never reach an end. | [83]

MACKAY: A simple way of saying it would be that within the rules of a game you can never find the justification for playing it

PITTS: Whenever you have a language that is rich enough to talk about language, you always have an incomplete language.

SAVAGE: I hope Walter's speech can be kept out of *The New Yorker.*

PITTS: Well, it is always a one-way sequence. In the sense that if language *B* contains all the logical apparatus necessary for describing the structure of a language *A*, then you don't need language *A*. You have all the mechanics at your disposal in language *B,* so the language is necessarily more inclusive. It is a never-ending series.

BIGELOW: I should like to ask some questions essentially on data from Professor Richards. It struck me as very interesting – the process of simplification that you have been going through in teaching language – and the following question occurs to me: Suppose that the students to whom you teach the language develop an ability to read rudimentary sentences in which certain letters have been completely excluded. The question would be whether or not these students had a residual tendency to delete or to eliminate certain characters permanently in their habits of reading and writing –

RICHARDS: No.

BIGELOW: I am not quite through. One of the reasons one might ask this question is that it is very well known that language is highly redundant, especially the English language, and one might very well suspect that a person who is taught systematically this way, and perhaps advanced to a considerable level by such instruction or such techniques, would actually learn to read most of those letters on a page by a process of scanning in which he would never know whether or not the thing was complete.

RICHARDS: Well, he does, and we have to do something about it.

BIGELOW: He does learn to read in that fashion?

RICHARDS: He does learn without any knowledge of what letters the words are composed of.

SAVAGE: No, you are not in mesh.

RICHARDS: Oh, I'm sorry. I missed it.

MEAD: As I understand it, your question is whether learning this smaller set first out of the total alphabet will not establish a predisposition to see only those letters and read, ignoring letters that are acquired later.

BIGELOW: Or a preference of some sort.

RICHARDS: We have not noticed a trace of it, and the answer as to why it shouldn't appear is that no particular attention is paid to the letters in early stages. They are only components, as it were, of phrases.

BIGELOW: But you are selecting them in a particular way.

RICHARDS: You have to work out what letters you are going to use. | You don't worry the children about that. As far as they know, they are just using ordinary sentences. [84]

BIGELOW: But what happens in the case of a language in which individual letters completely murder the meaning of the sentence?

RICHARDS: Well, they don't know that. Beginning readers are only too ready to take it.

BIGELOW: But suppose you are teaching Italian, where the last letter means singular or plural; don't you find some difficulty, if you start out by choosing letters, later on in focusing attention on the individual letters you add, and find there is a hierarchy of preference and that certain ones tend to be excluded?

RICHARDS: Well, except this, that each new letter that comes into the system comes in as a welcome entertainer, as a new problem, a new diversion. All through this, we put the letters, whatever they are, in a sort of cellar box to the film strip frames, so at first there will be *a, hi, nin, st,* with intervals left for the other letters. Then suddenly in comes *d,* and, soon after, *e,* in the box next to it. After a bit the children are fascinated by the great question, »When are we going to get another letter?« Well, we say: »You don't get another letter until you can read what we are giving you. When you can read what we are giving you, we will give you another letter in the picture.« Where does it come? It comes in one of these spaces.

TEUBER: I wonder if this would help Bigelow: If you are not very familiar with Chinese faces, you think they all look alike. Actually, you know they don't, but it is very hard for you to distinguish them unless you have a lot of contact with Chinese. Now, you may have one or two Chinese students, perhaps only one or two over a year, but they have been associated with you over quite a time. From then on, you will no longer have the same amount of trouble – and it is not a matter of specific knowledge of those two faces. You now go out and see a group of Chinese; they all look very different, and it is now much easier for you to recognize them, reidentify them on the subway individually or in groups. It is quite a difficult thing to explain, but I think it is a very common phenomenon.

MEAD: But what you have learned there is a cultural point, isn't it – that the Chinese are identifiable and have »human« faces instead of looking all alike?

TEUBER: I don't know whether that is the main thing. Don Hebb (21) likes to tell about a similar experience he had in Lashley's laboratory. At first, all chimpanzees looked alike to him, but after working with a few chimpanzees for a while, in a very concentrated fashion, he could no longer conceive of anyone confusing any of his chimpanzees.

[85] BIGELOW: My question is not exactly a matter of failing to recognize | one of a species of which you have been taught to examine the members more closely, but the question is whether one might conjecture a redistribution in the process of interpreting a printed page which would result from this ordering of the information. For example, what Professor Richards obviously does is to choose to eliminate right, left, up, down in the order of a given category. He does not choose to eliminate, for example, the ordering of *A* before *B.* because, obviously, the child must learn »no« before »on.« »No« is different from »on.« Then there is a definite preference to the components of language which are being fed in here. This would, presumably, have a certain significance (a) if it were true that this is the optimum component ordering; it would have some significance in determining what are the essential characters of a language that one must learn in order to understand it fastest and best; but (b) one could explore later on and see whether vestigial traces do exist in later performance. If the ordering does affect the preference for any appreciable length of time, then one ought to examine very carefully how it is done, in order not to change the structure and technique of reading.

GERARD: Concretely, you would expect, having started with these original letters and added the *g* much later, that the child would have some difficulty in distinguishing »sin« from »sign.«

BIGELOW: I don't know, but in reading a page rapidly, this might be a vestigial process that would crop out.

MEAD: As I understand what Dr. Richards said, what the child learns is that letters make sense and that you can discriminate between them. The major learning is not the particular letter, but the fact that reading can be done; it works, and letters are discriminable. What you are trying to teach first is the possibility of discrimination and then, within that, adding discriminable objects. If that is the emphasis, then the effect you are talking about would come.

BIGELOW: I am talking about an indirect effect that would come from doing it in that particular, selective way rather than by throwing all the letters at the child and letting him pick at random to form words from them.

RICHARDS: You could get somebody to explore them. That would be the only way, I think. We couldn't do it in our heads.

RIOCH: There is – in a sense it isn't quite comparable – that observation of Nielsen's (22) that children who have been taught to read by words, without learning the alphabet, never develop alitralia when they get aphasia.

RICHARDS: I didn't know that. That is interesting.

SAVAGE: What does that mean in English?

RIOCH: That means they never show the symptom of not being able | to read single letters. Patients who have as children been taught the alphabet and then to read will be unable to read words but will be able to read letters, whereas those who have not been taught the alphabet but taught to read words, according to Nielsen, will either be able to read words or not be able to read anything. [86]

BIGELOW: The elementary bits, then, are words, somehow?

MEAD: Yes.

SAVAGE: An aphasic knows only what he has never learned.

KLÜVER: I have a question which, I am sure, Dr. Richards, with his wide experience in teaching languages, can readily answer. It is generally maintained that a young child put into a different linguistic environment will quickly acquire a new language and soon speak Finnish, Chinese, or any other language with great fluency. Are there any exceptions to this? Or are there at least some languages which are more difficult than others for young children?

RICHARDS: I don't know any, but the anthropologists would be more likely to have such evidence. I am not aware of any, but I may be quite wrong.

MEAD: We have no evidence at present. Now, I have to add how bad our material is on ages of children. Most anthropologists are men who don't pay a great deal of attention to small differences in age of children, and we don't have birth records, so you have to add that. But if you add both those difficulties, we have no evidence that suggests that it is harder for a child, say, to learn Navajo than it is to learn Samoan. Looked at from the outside, the difference in difficulty in those languages is enormous. One of the very curious things is the fact that any human being will learn any human language if he starts as a small child. There is no difficulty about it.

SAVAGE: Didn't you say that in some cultures the older children of ten, twelve, or thirteen spend a noticeably disproportionate amount of time teaching the little children the refinements of the language?

MEAD: But that is not a result of the difficulty of the language; that is a matter of the attitude toward the language in the culture. But, for instance, a great deal of time is spent teaching Manus, which is a very easy language, and there is no such amount of time spent in teaching Tchambuli, which is a very difficult one. That is another aspect of it.

KLÜVER: In view of the relations between thought and language, and in view of the enormous differences in the structure of languages, it seems remarkable that such differences should never affect the process of language acquisition in young children.

MEAD: Well, it is very striking, if you compare how easy or difficult it is for an adult as contrasted with a child. The differential ease for an adult doesn't seem to be a function of the language one has learned [87] | before, either. Some languages are just harder than others.

RICHARDS: Sure, they are.

TEUBER: By the same token, I am inclined to believe that when language breaks down, in dysphasia or aphasia, it does not break apart into such neat categories as some of the clinical reports indicate. I haven't seen too many forms of aphasia, but I have never come across a single case of »pure« dyslexia or alexia in which the recognition of letters was affected, and not the recognition of words, or vice versa. I think these are matters of degree. We need much more empirical evidence before we can be convinced that the early reading habits or the ways in which reading was taught, really show up later in the way in which speech and reading might break down.

BIGELOW: Isn't there considerable psychological evidence that practically everything else you are taught early shows up later?

TEUBER: No.

BIGELOW: There is not?

TEUBER: I wouldn't say so.

BIGELOW: Maybe I am reading too much Freud.

HUTCHINSON: If you could introduce, hypothetically, into a language an element that would make it almost impossible for children to learn the language, children would not learn the language. Presumably another language modification would grow up that they could learn. It seems to me it is exactly what you would expect, within a rather narrow limit – that a sort of natural selection keeps the language learnable. Otherwise, no one is going to talk it. In Japan and Turkey, where literacy is becoming required, there is a great tendency to moderate the most difficult properties of the language.

GERARD: Except that after it has once been learned by the group, it does evolve and become more difficult and depend on more cultural influences.

HUTCHINSON: Yes, but if it were really difficult, then still nobody would learn it.

ROSENBLUETH: Many centuries have gone by, and learning to read Chinese is still terribly difficult.

HUTCHINSON: Yes, but you don't need to be able to read Chinese to be a Chinese. Presumably it is reasonable to suppose you would have to be able to talk to be a member of the social group.

RICHARDS: It is true that in some countries illiteracy is almost a crime.

RIOCH: Deaf mutes are one group that we have with us, open to study, and nobody has studied them. That is really the place where one could study something about language.

[88] HUTCHINSON: They use some sort of sign language. They can communicate. |

TEUBER: There is some evidence on »language« of deaf mutes, mostly in German – on what can be learned from what they use for a language (23) (24).

BIRCH: It seems to me that, despite the complexity of our discussion, we are dealing much too simply with the problem of language. Apparently, language has more than one function. It has more than a specific symbolic communicative function, and if we restrict ourselves merely to a communicative function, then we begin to be able to deal abstractly with language and treat language as a logical system for which another logical system may be developed, for which in turn a further logical system may be developed, and so forth. I would submit, however, that in doing this, we lose the richness of language and we lose the richness of the psychological phenomena that are involved in either communication or in language which is a broader area than the area of communication as such.

Communication really represents but one function of language and is not identical with language. For example, if we take Dr. Rioch's remarks on the question of time and the stopping of time, what I think he means is that words can come to take on representative meaning, in a sense that they can become the equivalent of previous situations; that in using a word, then, one can rearouse the psychological processes attached to previous situations, and that therefore language in a sense frees us from the boundaries of specific time scales. But once we say this, we must mean that in communication we must be dealing with organisms that have a community of representative function in language, and that words are simply not words but are attached to experiences, and that they become the equivalences of these experiences and, as such, become capable of being interchanged. If we then see that, we see that we not only have the problem of representational function of language, but we have also the problem of the conventional function of language; which means that the words that an individual learns, if they are to be used in a communicative sense, must be words that other people have learned in connection with a communal body of experience that has a certain conventional structure and conventional meaning.

Further, we use language in a different way. We may use it symbolically, we may use it as self-stimulus, we may use it in a variety of manners; so that if we begin to discuss the representative function of language, or the symbolic function of language, or the conventional features of language, or the contextual features, in which only specific representations may be produced, or other representations, depending upon contextual circumstances, then we begin to deal not with a specific logic but with an abstraction process. In a sense, we deal with a process whereby generalized experiences have been converted into a form which | permits them *(a)* to be manipulated, *(b)* to be transmitted, *(c)* to be mutually understood, and *(d)* to produce an interaction between individuals, which interaction gives rise to new experiences. These experiences in turn then give rise to new symbolizations, new contexts, and the development of an increasing and expanding language. [89]

It seems to me very difficult to study language as language. We must begin, I think, to differentiate, as Dr. Rioch, I believe, suggested, the process characteristics involved in a consideration of language, and study these in and for themselves for a moment, if we are then going to be better able to see the specific aspects and features of this overall end-result phenomenon – not process – that we referred to as language. In other words, language is not a single unitary event or a homogeneous event, but an agglomeration of events which we have in our own linguistic systems categorized under the framework of the term »language.« In that sense, to discuss language is to engage in a discussion not of process but of a normative procedure that we have developed; it is to discuss our tool of categorization rather than the processes which we have attempted to categorize.

MACKAY: In a way, that was the point I wanted to make earlier. I had been feeling that we ought to ask for a definition of language for this reason: Consider the series: *(a)* a man making a speech; *(b)* a sound-cinefilm showing the whole of the man while he talks; *(c)* a record of his voice; *(d)* a written transcription. At each stage along that sequence we may say we have a representation of his speech. But at what stage do we agree that we have reduced the representation to terms of *language?*

In music, we are not above putting in expression marks as additions to the mere symbolic minimum necessary to begin to represent what happened. We put in rises and falls and expressions of all sorts. If we had equivalent ways of doing that in writing, we would be expressing something which is a genuine part of language, in a sense. But at what point are we going to stop? Because essentially the act of replication that is evoked in the man includes the response to the rise and fall of the voice, and so on. I know more of what a man means than is conveyed by the words he uses, if his voice is rising, because it evokes in me something corresponding to what I would feel if I were to make my voice rise. How do we define language? Where do we draw the dividing line?

RIOCH: Along that line, some work of Jurgen Ruesch (25) in California was very important. When he played a sound-recording of an interview which was just a straight-forward interview to an audience of medical students, it would have no particular effect upon the audience apart from its content. If he played a record of an [90] interview in | which the one interviewed becomes anxious – and frequently also the interviewer – then the audience listening to the record became anxious; but they did not become anxious on reading a typed transcription. Therefore, I think we have to separate various functions of language which are per se functions of language, but yet have to do with different performances.

I should like to speculate along the following line: We can assume a double integration of the organism with the environment as a starting point. With cortical function we get differentiation of different parts of the environment. Now, suppose we differentiated a particular part of the space as separate from other parts. We want to subdifferentiate that particular part of the space; that is, I want to pay more attention to, say, this glass. How do I keep this glass in the space and keep my attention on the glass, keep the glass down where it is, instead of the glass beginning to occupy all of consciousness? That is the problem I am faced with. I am going to pay attention to the glass. How do I keep the glass from expanding into all of consciousness, so that there is nothing but the glass?

Actually, I have seen patients who have reported the latter experience, for whom the glass comes to occupy all of consciousness, and the rest of the room only a little fringe around the periphery of their visual field. Those patients will then use a word, »The glass is on the table,« and immediately everything is back to normal relations.

KLÜVER: It is characteristic of the effects of mescaline that consciousness may be narrowed down to the experience of one sensory or imaginal detail, for instance, to the experience of one scar, one thread, one key, or one plate. Such a detail then expands and finally becomes »everything.« The mescalinized person who has lost all spontaneity feels that it is only this scar or this thread that he is still conscious of. In such a way the boundaries between the world and me, between subject and object, disappear and the individual feels that he is really identical with the object, for instance, the scar (26).

HUTCHINSON: In Buddhic mysticism that is a very well-known technique for demonstrating the unreality of the world. You take an image of a goddess, blow it up, and fill the whole universe; then you bring it down to a point infinitely small and then fill it up again. After you have done this a sufficient number of times, you realize that the goddess is an illusion.

TEUBER: I have a pedestrian concern. We said, somewhere, »communication and language,« implying that we are using two different words for two different referents. I just want to ask – I mean, nobody in particular has done it, but in general, do we mean to say that we all know what communication is, but that we need to define what [91] we mean | by language? Apparently, we mean a special case of communication. If so, which special case, and would it be sufficient at this point, rather than after tomorrow night, to try to say what particular aspect of »language« should be used for the purpose of this discussion? I would try to test that definition, then, by asking the additional question: If that is what you mean by language, would you impute it to animals or only to man?

MEAD: Well, I should like to go back to Dr. Birch's point. It seems to me – probably this is historical, and I am merely making a historical reconstruction for purposes of discussion and not saying that it did happen – but it is conceivable that very early in human history, for some reason, or accidentally, some people thought of categorizing language as the learnable aspect of other people's behavior. That is one of the most significant things about it. We have no people so primitive that they are not able to say, »Those people have a different language.« Sometimes in New Guinea they don't know where the »talk turns« because it modulates sociologically, but they know »that is a different language from ours; I speak it or I don't speak it; I hear it of I don't hear it.« Conceivably, instead of thinking that, one might have said the postural system of another people is the learnable aspect. I don't know what those sounds they make are. I don't know what the sounds we make are. That isn't the point. But I can go into their group and I change my tonus, my shoulders, and so forth. Or take the cookery of another people, which is at least as complicated as many languages. Instead we labeled language. In the whole of human culture, language has been labeled in this peculiar way.

Now, there is a possibility, and it is only a possibility, that the difficulty with language is not this point of using language to talk about language, but that historically we have selected language as a segment, and that has been perpetuated throughout human civilization, when it is not as different from other parts of culture as it appears to be.

[92] # COMMUNICATION BETWEEN SANE AND INSANE: HYPNOSIS

LAWRENCE S. KUBIE
Department of Psychiatry and Mental Hygiene
Yale University School of Medicine

THE SUBJECT THAT I was asked to discuss here broadened with each successive letter that I received about it. First, it was the problem of communication in hypnosis. Then it broadened to include communication in hypnosis and in the psychoses. Therefore, I feel justified in taking an even broader base; and this despite the fact that as a consequence what I say will be a collection of fragmentary, scattered observations tied together by a few questions with few answers. These observations are in the main clinical experiences; and since not many of you have had an opportunity to share such experiences, I will describe and discuss examples of the type of empirical clinical data which seem to me to be relevant to the broad question: *What happens to the processes of symbolic communication in different states of consciousness?* (1) (2)

I had hoped to begin with one living exhibit; but the patient did not show up. He is a key member of our group, and an extraordinarily gifted human being. For me it has been one of the pleasant sideshows of these conferences to watch him do something which I had often heard described but had never seen before. During the course of a heated discussion among his fellow conferees he will be deeply asleep, snoring, in fact; and suddenly he will break off in the middle of a snore to come out fighting, literally leaping to speech in the middle of a paragraph. What is more, that paragraph will be relevant to that which had been under discussion. Evidently you have recognized our good friend Norbert Wiener. This amazing phenomenon is something to give us pause; and it has a relevancy which points in several directions at once. Before going into them, however, let me give a few comparable, but not identical, examples.

I once had as a patient a likable young college student. One Saturday night, I met him out dancing. He was with a crowd of young folk: I was with friends. It was a typical New York dance floor, half the size of a postage stamp. I suppose we rubbed shoulders a hundred times that evening, as we danced through the crush. He looked me in the eye repeatedly, but never batted his; so I assumed that he was embarrassed, and did not greet him, either. I did not want to upset his peace of mind. That was on a Saturday night. On Monday he came [93] | to his session and told me that he had had a most remarkable dream. In one of its scenes he had been out dancing; and he added, »Just imagine this – in the dream I met you out dancing.« He had not the slightest idea that this had actually occurred, or that he had recorded the fact of seeing me and then had reproduced it in the dream, without distortion, and all without conscious participation in the process at any step until the last one.

Another example centers on an office lamp of green imitation bronze, with an inverted bell-like shade which opened to the ceiling. I had owned this for about a month, during which time it stood near the foot of the couch in my office where no patient could fail to see it during the analytic hour. One day a patient told me a dream, which was of critical importance both in the development of her treatment and in her relationship to me, although the manifest content of the dream had nothing to do with either. The dream was of a field full of »strange green daffodils … even the blossoms were green; and instead of drooping over, they pointed straight at the sky.« As she finished describing these daffodils, she said, »I have never in my life seen anything like them,« and with these words pointed at the lamp and exclaimed: »My God, there are my flowers. How long have you had that lamp?« Evidently »communications« can be recorded unconsciously, both in sleep and in the waking state, and can then be reproduced recognizably in subsequent waking thoughts (Norbert Wiener), or in dreams, either with some degree of distortion, as in this case, or without distortion, as in the dance-floor dream.

I can give a similar example of something which happened to me when I was a medical student. On Sunday mornings I used to go on hikes with friends who lived down the street. Our usual signal was to whistle from one window to another. One of the group was musical, and always whistled some erudite theme. One Sunday morning in spring I heard a whistle; and when I put my head out of my window this friend was waving from his. As I walked down the street toward his house, a melody was going through my head, a melody which was familiar, but which I could not name. I said to myself: »I must ask David what that tune is. He will know«; and as I reached him, I asked, »What is this tune?« and whistled it. He looked at me in astonishment, and said, »My God, I was whistling that for half an hour, and you paid no attention to me. Then I finally changed to another tune and you stuck your head out of the window.«

Evidently communication can occur effectively without full conscious participation in the process. This point is basic. Even during states of sleep and of absorbed attention, an automatic unconscious process can occur by which we record sensory experiences, with equally unconscious | subsequent consequences. I stress this point in order [94] to indicate that the barriers between the psychological processes which go on in the sleeping state and in the waking state are relative and not absolute; because parts of us are asleep when we are awake and parts awake when we are asleep. By this I mean that sleeping processes are going on during the waking state, and waking processes during the sleeping state. Sleep and wakefulness are relative and not absolute concepts or states; and the phenomena of hypnosis (which is the challenge that was given to me first) probably lie somewhere between.

Moreover, not only is intake possible, but we can also *do* many things in sleep whose close relationship to what happens in the waking state is not always clearly recognized. There is the random activity of sleep which has never been adequately studied (and never will be until we study infra-red movies of human beings during sleep). Then there are the more organized phenomena of sleepwalking and sleeptalking which deserve study. Finally, there is the form of sleepthinking which we call the dream, which is the only one of these which has been closely studied. It is interesting to contrast the emotional discharge which accompanies these forms of action in sleep and in the waking state. In the waking state the emotional processes are least adequately discharged in thinking, more so in talking, and perhaps most completely in action. In some measure the reverse relation obtains in sleep. The dreamless sleepwalker, the rare pure somnambulist, although he occasionally acts out violent emotions, more often is relatively emotionless. The sleeptalker comes in between; and, as we know it, the dream is the most highly charged.

Let me summarize this by emphasizing that a process of communication can occur in sleep, in states of deeply absorbed attention, and in the hypnotic state, a process of communication which depends on all of the techniques of communication that we use in ordinary speech. Furthermore, although in full sleep such communications are predominantly a process of intake, they can become a two-way process. Consequently, there is such a thing as a *communicative sleep,* which comes close to being what is meant by the hypnotic state. (3)

The next point about communication in hypnosis is the fact that under hypnosis our automatic capacity to *record or recall* minute-by-minute experiences is more photographic and phonographic and more inclusive than seems to be true in the waking state. This is the so-called »hypermnesia« of hypnosis. It refers to a well-known phenomenon; namely, that when anyone tries to tell what he has seen or heard, he can give at best a limited report; whereas under hypnosis that same individual will produce a wealth of additional verifiable memories, without his having even been aware of the [95] fact that he had recorded | them. This experiment has been repeated many times; and we can deduce that we always record more than is readily accessible to us in the fully waking state.

We do not know whether we *record* more fully under hypnosis than in the normal waking state. For instance, in the fully waking state do we screen out a certain amount of incoming percepts? In hypnosis is such selectivity at the intake in abeyance? We know only that we call *recall* more fully under hypnosis than in the waking state.

I emphasize all of this because if we should limit our consideration of the problem of communication to conscious speech and our use of and reactions to it, we would be oversimplifying the problem. The data given above demonstrate that we use and record and respond continuously and at all times to an inarticulate or subverbal form of communication, both on conscious and on unconscious levels. Since I am speaking to mathematicians, I hesitate to use figures; but certainly it would be no exaggeration to say that our conscious and unconscious psychological processes deal with an enormous number of things simultaneously. Therefore, it would seem to me that the first challenge to the mathematicians and to the makers of mechanical models is to arrive at some approximate impression of the numerical limits of this. There must be such limits; but these limits have never been established. Certainly the amount of data that is taken in and recorded is infinitely greater than that scanty sample which we can reproduce consciously. The same is true of our psychomotor and emotional responses, to which, in turn, we react automatically, thus compounding our inner psychic processes by geometric progression. Consciousness then represents only a small fragment of our total psychic state. This fact must be included in any study of communication.

It can be studied in the waking state; but as I have indicated, this presents special difficulties because of the selective processes which the waking state involves. With those individuals who are hypnotizable, it can be studied in the states of hypnosis; but this brings up its own technical difficulty. Nobody has yet been able to determine what it is that determines whether or not an individual is hypnotizable, or why some individuals are hypnotizable and others are not. Consequently, the use of hypnotizable subjects introduces a selective factor, the nature of which we cannot evaluate accurately. Communication can also be studied in sleep; but this, again, introduces a selective factor, since not all individuals are accessible to either one-way or two-way communications during sleep.

Let us carry further our consideration of the study of communication in hypnosis, even though it will be seen to be even more complicated. There have been three inter- [96] esting types of experiment which indicate | that under hypnosis individuals can translate things which they can neither translate nor understand when they are not under hypnosis. This concerns the translation of automatic writing, of automatic drawing (even doodling), and of the symbols of dreams (4) (5) (6) (7). I have a sample of automatic drawing which I will pass around. This is an automatic drawing which in part was translated under hypnosis by the subject who had made it, and in another instance by another individual under hypnosis (4). The same thing was done with the automatic writing of another patient (5) (6). Then there are the interesting experiments of Farber and Fisher (7). In these they described to Subject A under hypnosis a painful »experience,« and then asked him to dream about this implanted »experience.« The subject thereupon reported a dreamlike, disguised reproduction of the »experience,«

transposed into more or less classical symbols, almost as if this naïve individual had read a dictionary of dream symbolism. Then the experimenters showed Subject A's hypnotic dream to Subject B while under hypnosis, and asked Subject B under hypnosis what the dream of Subject A had meant. Thereupon Subject B translated the dream of Subject A back into the essence of the unpleasant story which had originally been told to Subject A.

That is more than a trick. It is a very important experiment which has been repeated often enough to make us sure that one human being can on a dissociated level translate accurately the symbolic representations of the unconscious content of thought and feeling of another human being, thus exposing and uncovering repressed experiences which the other has had in the past. This is a critical and conclusive demonstration of the power and specificity of unconscious psychological processes, and of their importance in the communications which pass between men.

I have had a closely comparable experience in another equally interesting but wholly different form. I must say that it surprised me, because it was wholly unexpected. Some years ago, Bela Mittleman and Ruth Munroe organized a small group which met several times. The group consisted of clinical psychologists who were interested in the meanings of figure drawings, in graphology, the Rorschach, and the TAT, plus a couple of analysts like myself. I sent several of my analytic patients for a Rorschach to Ruth Munroe. Each time Ruth asked the patient for his reactions to the Thematic Apperception Test, asked the patient to make figure drawings (Machover), and then sent a sample of the patients' handwriting and the other material to the other participants in the test, with no information except, »This is a twenty-seven-year-old woman,« or »This is a thirty-three-year-old man.« Each participant was asked to bring an independent interpretation of the | subject's material. When all the reports had [97] been presented and compared, I presented the analytic material which I had gathered over months or years of analytical work.

I was interested to see what different aspects of the personality were delineated by the different tests; but what seemed even more extraordinary was the accuracy with which in certain instances the life history was described, particularly by the graphologist. For instance, in one case, from the study of the handwriting alone of a young woman whom the graphologist had never even seen, the graphologist said: »I think this young woman recently had a baby. I also suspect that there has been a recent suicidal attempt.« Both of those things were true. Of course, a few isolated instances of this kind prove nothing, except to indicate how heavily overweighted are the multiple determining factors of some seemingly simple and stereotyped acts, and consequently how revealing such stereotypes can be.

This leads to the problem of universal symbols in general, which in turn will bring us to the question of communication in the psychoses. Neither problem has been studied or explored systematically. Nor has there been any systematic study of the closely related problem of the influence of culture on universal symbols. Only fragmentary experiences have been noted. These experiences are often amazing, and always interesting and significant, especially when they come through the words of small children who have had no opportunity to be contaminated by the ideas and the literature of their elders. I see Dr. Mead frowning at me already. Perhaps she will agree at least that this parental influence is not in a form which is specific enough to determine their use of classical symbols. Let me give you some examples (8).

A little girl of four, an only child, puts pencils in the drawers of her dolls and calls them penises. Has parental influence produced her conflict over whether she wants to be a little girl or a little boy, or has our culture determined her particular way of symbolizing her problem? The same little girl pins safety pins in the front of her chemise and calls them her penis. She has hardly been instructed to do this either by her parents or by deliberate cultural influences in the society in which she has grown up. Or what of the little boy grown to puberty, whose curiosity about the genitals has been stimulated by the fact that he finds his best friend peeking through the windows of the girls' washroom in school. He has a fight with his best friend about this, and then has nightmares and sleeps badly and comes home from school with an eye-blinking tic and worrying about what will happen to his eye. Then he has a long discussion with his father about the normality of bodily curiosity, after which he dreams about an airplane [98] with | strange bulges, whose appearance is so curious that he cannot tell the front from the back or the top from the bottom. The day after his dream this little boy awakens without his eye-blinking tic. Such commonplace episodes raise issues which are far-reaching in their implications for communications. What role, if any, does culture play, and how does it operate to determine any youngster's use of a flying machine instead of a serpent or a bird for the translation of his instinctual conflicts into symbolic language?

Or what of the little girl who night after night has the same pleasant dream; namely, that a little mouse gets into her bed, and snuggles down beside her? She tells the dream each morning. Both the dreaming and the telling are meaningful. One day, after about two weeks of this, she happens into her father's bathroom just as he has come out of the bath. He is standing there naked; and she comes to him and examines him with close attention. This little girl has seen her brother naked many times, and she knows the names of the male genitals. Nevertheless, as she backs away from her father's nakedness she looks off into the distance and makes up a language of her own, saying, »I have just seen a little mouse.« What determined the translation of this small girl's experience into animal terms, and the substitution of the name of an animal for the known name of a known part of the body?

We could multiply such experiences many times. I do not understand this process. I know only that it happens constantly, and that it forces on us the conclusion that it is impossible to talk about anything in the *outside world* without at the same time having it refer to something in the *inside world*. Thus with every word we utter, with every gesture, with every expression we talk simultaneously about the inner world and the outer world, with the result that all language units have multiple meanings in which the relative roles of internal and external objects and of conscious and unconscious determinants are in a state of continuous and unstable dynamic equilibrium (9).

These are demonstrable facts. Furthermore, we find that the meanings of symbols are not always idiosyncratic, as one might expect, but often are highly stereotyped. Indeed, our ability to make any general translations of psychological tests depends on this stereotype, whether it is graphology, the TAT, the Rorschach, figure drawing, or the Szondi, and so forth. Their meaningfulness is specifically dependent on the stereotyped universality of the conscious and unconscious symbolic implications of their forms. (I must say that of the tests the Szondi puzzles me most. As with the Wassermann, not one word of the underlying theory makes sense; but for some reason the [99] test works none the less. I have no idea why.) |

Finally, these considerations lead us to the problem of psychotic insight (or more specifically, to insight as it occurs among schizophrenics) and its bearing on the problem of communication. Here we are up against the most puzzling manifestation of what I am trying to bring together; namely, the fact that schizophrenics (who psychologically are among the sickest people we deal with) know most about their own unconscious. Indeed, they can often translate their own symptoms, their own behavior, and their own symbols. They can say, »I know that when I wear a knitted tie, it means that I feel *X* about my body, whereas if I wear a silk tie, it means *Y*,« and so forth. The ordinary human being never has this sort of subjective symbolic insight spontaneously. Yet where it arises spontaneously in the psychotic, this type of insight for some reason seems to be devoid of therapeutic leverage.

Let me add another point. It may well be that in steep there is a continuous flow of active psychological processes. We cannot be sure that this is true. The flow may be intermittent, or it may occur only during the processes of falling asleep and of waking up; and it may cease at that level which we figuratively call »deep« sleep; or at this level it may become so inactive as to be unimportant. However that may be, we can be certain that even in sleep a symbolizing process occurs which can have destructive consequences. The fact that destructive psychological processes can occur during sleep is a daily proof of the operation of unconscious psychological forces.

People are naive about sleep. They think of it as always being a heating, resuscitative, and constructive force. Yet actually we know that it can be a destructive experience. People can go to bed well and wake up sick. People can come out of sleep in a paranoid-schizophrenic psychosis. There are patients who waken out of sleep in deep depressions, in acute manic states, in stormy somatic disturbances. These may be the mild transitory depressions of waking, or they may be the emotional states of those more fortunate human beings who waken from sleep in transient, hypomanic states. These may not last long, but at least they get us out of bed. All of this is evidence that much goes on in sleep. Insofar as we can make contact with this, it is represented in the most purely symbolic language that we know; namely, the language of the dream. Further evidence of the activity which goes on even in sleep is the extraordinary time sense that some people have in sleep, but which, for some reason or other, seems to be lacking in others.

I do not want to overburden this. Therefore, at this point I will interrupt this gathering of fragmentary and puzzling empirical data. Subsequently I should like to return to the basic issue of symbolic thinking. It would be better to interrupt at this point, however, so as to open these remarks for discussion and questions. Later, if there is | time [100] and interest, I can go on to the other problems, or else reserve them for another conference.

McCulloch: May I abuse for a moment the privilege of the Chairman and ask two or three questions? I spent a considerable amount of time and effort during my senior year in Yale, back in 1921, attempting to evaluate how much can be recalled under hypnosis. We were by no means convinced that an infinite amount could be recalled, but only that an amount which was vastly in excess of what is available in the waking state could be recalled – I would say perhaps a thousand times as much; still it has an upper limit.

Pitts: Did you try simple material such as a string of nonsense syllables?

McCulloch: We tried this sort of trick: We took master bricklayers who laid face brick and had them recall the seventh brick in the row, or something else like that, in a given year. They were able to recall any one such brick – thirty or forty items at the most. That was a brick that had been through their hands some ten years before. It is still not an infinite amount. A master bricklayer can lay only a certain number of bricks per diem even when his entire attention is riveted on laying bricks. The amount is not infinite.

Savage: What would you call an infinite amount?

McCulloch: A thousand bits per minute at the most, or something like that, that you can get back. It is not infinite. I think that is critical –

Savage: Well, who could believe it could be literally infinite? And if it is not literally infinite, who could believe it would be as much as you say?

McCulloch: There are two things here which are equally important: first, that it is vastly more than one can remember in the waking state; and, second, that it is not infinite.

Pitts: You could take nonsense syllables or present a board filled with letters, and determine how many could be remembered.

McCulloch: It is extremely difficult to get a man to attend to this, but a master bricklayer looks at the face of the brick when he is laying it.

Mead: Did you find differences between two master bricklayers in which thousand items they would notice?

McCulloch: No, not significantly. He mostly noticed three things –

Rosenblueth: How do you go back to check?

McCulloch: These things are verified by checking the bricks. They are master bricklayers. That means they are laying face bricks. That means that even ten years [101] later, you can go to that row and look at the | brick. The only things you can't check are things on the opposite side of the brick or an angle off the side of the wall.

von Bonin: How many things can the normal person remember?

McCulloch: The estimate is that the maximum is 10 frames per second.

von Bonin: No, no, I don't mean that; the normal person.

McCulloch: On the normal person, the best man known on receiving communication in the United States Navy could give you a hundred letters in sequence at the end of having received a hundred in 10 seconds. He had to wait until he had passed through a period during which he could not recall, and then he would give you all the hundred letters.

Bigelow: What kind of communication was it?

McCulloch: Semaphore.

Pitts: If you were to hypnotize him –

McCulloch: That is not more than five bits per letter.

Bigelow: But it is essentially controlled by the semaphore process.

McCulloch: Well, this can be sent as fast as you will. It is sent by machine.

MacKay: What was the redundancy in the information about the bricks? Or, to put it otherwise, how frequently did the features recalled crop up normally in bricks?

McCulloch: In bricks, it is rather rare. The kind of things men remember are that in the lower lefthand corner, about an inch up and two inches over, is a purple stone, which doesn't occur in any other brick that they laid in that whole wall, or things of that sort. The pebble may be about a millimeter in diameter.

Bigelow: How could he possibly remember thirty of those features on this one brick?

McCulloch: Oh, they do. It is amazing, when you get a man to recall a given brick, the amount of detail he can remember about that one brick. That is the thing that is amazing. I note, as a result, that there is an enormous amount that goes into us that never comes through.

Kubie: This is comparable to the experiments under hypnosis, in which the subject is induced to return to earlier age periods in his life (10).

McCulloch: I have never done any of those.

Kubie: In these you say to the subject, »Tell me about your seventh birthday,« or his ninth birthday, or something of that kind; and he gives you verifiable data about the party, who was there, and so on.

McCulloch: I have seen my mother's uncle, who was an incredible person in the way he could recall things – he was law librarian in Washington for a few years – testify that he had looked at such and such a | document some twenty years before, and shut [102] his eyes and read the document and the signatures under it.

SAVAGE: I don't have absolute recall, but I can remember hearing that story more than once in sessions of this group.

McCULLOCH: They said, »But the document does not read that way; it reads so and so from there on,« and he replied, »Then the document has been altered and you had better check the ink.« And when they checked it, it was a forgery.

SAVAGE: It's the same story, all right.

GERARD: What about your statement, Dr. Kubie, that a person going into the room under hypnosis can remember still more?

KUBIE: I am not sure about that. The evidence is not conclusive. It has been hard to quantify these things.

McCULLOCH: May I ask one question about the hypnotizable? And I ask it in all humility. There are two groups which, I would admit right away, are extremely hypnotizable, in the sense that any of us with sufficient effort can hypnotize them. They are very distinct. The other day a group of students who had been working on it at the University of Chicago came to us to discuss this. They had two hypnotizable groups – one of which, afterward, says, »I did this, I did that, I did the other«; these subjects perceive the whole experience as something in which they played a part. They are always ready to come back to be hypnotized again. The other group experiences the world as »this happened, that happened, and the other happened,« completely depersonalized. It is very difficult to get them back for a second trial in hypnosis. Bagby and I ran into the same thing at Yale, long years ago, with our techniques. Then, we had only about 20 per cent hypnotizable, whereas, according to the group in Chicago, about 30 per cent of college students are hypnotizable. Do you know what happens with a group of people who are not hypnotizable? Do you know how such people feel about the attempts?

KUBIE: No.

GERARD: Incidentally, Willey and von Borstel, whose work you have mentioned, have found an index now that seems to set off those groups.

McCUL[L]OCH: Which groups?

GERARD: The active participators from the nonhypnotizable, and the passive ones.

McCULLOCH: The »I's« from the »It's.«

GERARD: The number of breaks in the galvanic skin resistance curve.

McCULLOCH: Yes, that is what the criterion was.

BIGELOW: During the process, or what?

GERARD: No, quite aside from it; that is, under ordinary conditions, or while the operator is trying to hypnotize them. These experimenters | couldn't separate the [103] groups in terms of the height of response or the length of response or of any other one thing other than the total breaks in the curve; the general damped character, if you will, of the autonomic response. This is the way they are tending to interpret it.

McCULLOCH: It is extremely damped in the »I« and the recurrence is extremely high in the »It.«

GERARD: That's right; those are different from the normals.

McCULLOCH: No, the normals are not distinguishable from the »It's« in there. Do you remember?

MACKAY: There was a difference of 2.5 in 24, or something like that.

McCULLOCH: Statistically, it is not distinguishable between the hypnotizables and the nonhypnotizables. The hypnotizables split very sharply into the two groups – the »I's«

and the »It's.« The »I's« have the very low number, about 12. The »It's« have about 26 – that is, psychogalvanic wobbles.

PITTS: Do you mean fewer of them?

MCCULLOCH: There are very few of them in the »I's« and very, very many in the »It,« but the »It« is not statistically different from the nonhypnotizables. Isn't that right?

MACKAY: Yes, that was as they described it.

BIGELOW: Will you read the reference to this into the notes of this meeting?

MCCULLOCH: It has not been published yet, so far as I know.

GERARD: These fellows are just working on it.

MCCULLOCH: Yes; they are trying to get it out now. Now, the next thing I want to say about this is communication during sleep. Communication to somebody who is asleep and communication from somebody who is asleep may be quite different problems.

During World War I, while I was in the Navy, part of my job was to see to it that every man got out of his hammock when a bugle blew. Every old sailor, when the bugle blows, throws a leg out of his hammock. That is known as the »shore leg.« When he has shown his leg, it is not permissible either to paddle his bottom with a wooden slab provided for the purpose or to cut the lashings of his hammock, which drops him on the floor. Those men were as sound asleep by every test after they had shown a leg as they were before. There is no question but that they were able to communicate that they were awake in spite of the fact that they were otherwise sound asleep. That is Item No. 1.

Item No. 2: During World War I, when the Navy went into – yes, I think it was only the Navy – a long study of the training of men in dot-dash Morse code while they were asleep, they found that men could be taught the Morse code while they were asleep to such an extent that although they were never awake, so far as we knew, [104] during the inter|val, when their name was called and they were told to report to station they understood it in Morse code; so that there is no question but that communication to people goes on while they are officially asleep. But I think one has to distinguish between communication to and communication from them.

ROSENBLUETH: In line with what we are saying, I don't see that there is any puzzle here because of the fact that these things occur in certain people under certain conditions. The fact that somebody can, at a given moment, recall many things which he did not recall under other circumstances is not particularly puzzling. The puzzling thing is the problem of memory. I don't think Dr. Kubie's examples differ from any other type of memory, so far as I can see. The problem that we have to discuss, if we are going to discuss it, is the problem of memory.

There is one assumption – it wasn't all data that Dr. Kubie gave us – that we should consider. I may be wrong, but I thought he implied that there must have been impulses registering unconsciously. I think this is unacceptable, because it is a contradiction in terms. Either something occurred or it didn't occur. Do we mean by »mental« anything a person can report at a given moment? Then the experiences were mental. If he cannot, then they were not mental. It becomes mental only when he can report it. If that is the definition of the word »unconscious,« its use in any other connection, except to say that something has not been mentalized so far, means nothing, in my opinion.

RIOCH: I don't think it is possible to use the word »unconscious,« but it is possible to use the word »conscious.« I think one can give an operational definition of »conscious,« but not of »unconscious.«

MEAD: Can't you think in terms of the steps it is necessary to go through to make something conscious?

ROSENBLUETH: Again, on the problem of sleep, I wish to make more explicit what Rioch said, because I wouldn't entirely agree with him. What is puzzling is that somebody in sleep registers certain things. The problem is the problem of sleep itself, about which we know nothing. The phenomenon is puzzling only because we don't understand the problem of sleep. I don't think it is any more puzzling to grasp the fact that I can say something to somebody who is awake and he can repeat it, than the fact that I can say something to somebody who is asleep and he can repeat it. I don't think there is any important difference there.

GERARD: The problem there is why, under hypnosis, it is possible to bring these things into awareness.

RIOCH: I think it merely makes the situation more complicated.

BIGELOW: I agree with that.

RIOCH: You can have lots of fun with it but you don't know what you are talking about. | [105]

BIGELOW: If you don't know any of its specific properties, how can you be surprised when such properties are exhibited?

PITTS: If I recall, the EEG of a man under hypnosis is not in the least like that of a man asleep. Instead, it is like the case of a man who is concentrating intensely on something, either a thought of his own or something you tell him to look at fixedly.

BIGELOW: The same difference in EEG as might exist, for example, between a man who is conscious of doing some trivial arithmetic and one not conscious of any thought? It might not relate to his memory properties at all.

KUBIE: May I put in a few words before you go too far with this? The whole discussion reminds me unhappily of the ancient discussions about behaviorism. The behaviorists also hated the concept of conscious and unconscious psychological processes. They did not even like to use the words »thinking« or »not thinking.« But to get away from such terms, they had to devise a new pair of words: »verbalized« and »unverbalized.« These were John Watson's. They also used »laryngeal posture tonus« and, I suppose, »laryngeal unpostures.« All of this is plain silly. There are many terms in science, and in human life in general, that are difficult to define, yet this need not mean that we do not know what we are talking about when we use them. We know what we are talking about when we talk about steep, yet no adequate definition of sleep can be given. The same is true of awake, or of the process of hypnosis. We know that these are two states; and we can think of them either as two absolutely opposed and absolutely different states, or we can look upon them as relative terms, for example, points along a continuous spectrum, as I have suggested. It seems to me that there are many reasons why the differences between sleeping and waking are relative and not absolute, and that hypnotic phenomena fall somewhere in between, with some of the characteristics of one and some of the characteristics of the other. This is advanced, however, not as a fact but as a working hypothesis.

MCCULLOCH: Can we reserve that question, Dr. Kubie, for when we come to an attempt to sharpen up the question of hypnosis itself? I think you should be in on that later.

KUBIE: Certainly; but I want to repeat that it is nothing short of silly to waste time in a verbal quagmire over whether we have the right to think of »conscious« but not of »unconscious,« or of »unconscious« but not of »conscious.« This is like saying, »I can think of white but not of black.« We know that there are processes with full attention and awareness which have psychological consequences, and that there are others which have equally important psychological consequences and which operate in the same way, but which are below the level of con|scious attention and awareness. Call [106] them anything you please. Call them »X« and »non-X« if you prefer; but the phenomena remain; and the simplest operational and descriptive characterization of them is as Conscious and Unconscious.

MEAD: But, Dr. Kubie, I think you get a complication here by including under one heading instances of attentiveness, relative attentiveness, under sleep, under hypnosis, and in ordinary states.

KUBIE: This is done on purpose, because the phenomena of attention are closely related to the problem of hypnosis (3).

MEAD: Wait a minute. You didn't let me finish my sentence. Those three are paired, or triplicated together, or tied together, or whatever. We get the question of symbolism where, for instance, one subject is able to interpret the artistic product of another subject. I think that it is an artefact of our own society and our own methods of exploration, at present, that we recognize a particular kind of thinking, if you want to call it that, most prominently in the dream. We also recognize that it occurs in poetry and in the arts. We oppose it to rational, conscious, logical thinking in our culture, and therefore there is a tendency for this sort of thinking that you have talked about to occur more in states that are locked off, or in sleep.

If you look at some other cultures, you don't necessarily find that same contrast. We can usefully make a separation in our thinking about this whole problem of consciousness or attentiveness or degree of receptivity to imprint and ability to recall – and I agree with Dr. Rosenblueth that it is a question of memory and attention in one form or another – putting on the one side the question of types of thinking that are modeled on the external world and, on the other, the types of thinking that are modeled on the body. We can use this shorthand for considering these two sorts of thinking, one in which you have a whole series of proprioceptive and bodily images which one person can interpret from another very easily and of course they do it better under hypnosis. I have had a subject put under hypnosis and then showed her films of Balinese trance, and she would pick accurately the place where the Balinese went into trance. She can do it perfectly well, and I don't see why she shouldn't, because under that state she is using a common system of thinking, which is what the psychotic uses when he seems telepathic. Now, I don't believe that the association of these two sorts of thinking, although you find it all the time in patients, is necessary in this society. If we thought about the two things separately, I think we would get further.

MCCULLOCH: Right. That is one of the real questions to my mind, Dr. Mead. The symbolism required for translation, retranslation, and the likeness to the original may very largely, in any population that we [107] | deal with here, be culturally determined. There are things which make me think otherwise about some aspects of it, but I will come to those in a minute. I think it is rather crucial that we keep in mind that there may be symbols. For example, a large amount of my dreaming is in verbal symbols.

BIGELOW: How do you know that?

MCCULLOCH: Because I wake up laughing. Having managed not to say something during the day, I then put the people back into their situations and am saying it when I wake up in a gale of laughter.

BIGELOW: Is this sufficient proof?

MCCULLOCH: Yes. I have very clearly the meaning of all the remarks that were passing at the time.

PITTS: Do you mean verbal in the sense of actual speech?

MCCULLOCH: Yes, verbal in the sense of actual speech, but not audible, or only rarely are they audible. Now, this type of dreaming clearly depends upon puns, and so on, which are only possible in the English language in many cases. They could not be done outside the English language. In German they would be impossible; in Latin they would be impossible. I think, therefore, that one has to take care of the extent to which the cultural component is determinate.

MEAD: I should like to make clear, though, that that is not what I was saying. I would regard those things as vocabulary. Surely, if you are English, you dream in English. If you have never seen a certain thing, the chance of your using it as a symbol is low. But the point I was making, on differences between cultures, was the difference between which sorts of thinking were the correct kinds of thinking for people to employ. Now, in the Trobriands it is incorrect to recognize what we regard as logical relations, so that when you are being a rational, conscious person, you know that copulation has nothing to do with procreation, that food has nothing to do with growth, and that the planting of seed has nothing to do with the plant growing. But you continue to act on the assumption that all those things happen. The only possible assumption, therefore, that I can see in the light of our present knowledge, is that the Trobriands make »conscious« something that we do not, normally, and carry in this other set of imagery the sorts of knowledge that it is necessary to plant plants and produce children and all the other things they do, in a perfectly accurate and efficient way.

McCULLOCH: That is what I mean by the »forbidden« meaning of words. It doesn't seem to me that this is fundamentally different.

MEAD: Except that I am assuming that the mechanisms of symbol use are probably universal. The question isn't: Which symbol? I thought it was misleading if you just used the content there.

KLÜVER: Dr. Kubie, you repeatedly referred to the fact that items | from past experience may be reproduced, for instance in a dream, without the person, when telling his dream, being consciously aware that the item in question represents a revival of past experience. It seems to me that it is the study of eidetic imagery which has furnished the most striking illustrations of such mechanisms. I spent several years studying eidetic individuals and found again and again that the subject, when observing and describing the eidetic image in front of him, was greatly surprised in seeing objects or details he could not recall as having seen in the picture previously shown to him (11) (12). However, it was obvious that the objects or the details seen by the subject were often merely eidetic reproductions of details or stimuli presented by the experimenter a few minutes, weeks, or months ago. This leads to a second question – [108]

KUBIE: May I ask a question? To your mind, does this provide adequate experimental proof that we can register and record perceptions without conscious participation in that process, and then reproduce those perceptions without conscious participation in the process?

KLÜVER: The relation of eidetic imagery to ordinary visual imagery has been the subject of numerous experimental investigations (13) (14). Most investigators in this field agree that visual memory and eidetic imagery are, on the whole, independent. The fact that certain objects, forms, colors, and so forth, have been consciously perceived and are subsequently remembered in great detail does not necessarily imply that the subject can reproduce them eidetically. Furthermore, his intention to revive them eidetically may be entirely ineffective. On the other hand, elements or items, for instance, in a picture shown to the subject may appear in the eidetic image of this picture, although the subject cannot recall such items as parts of the original stimulus. Or these items may later appear »spontaneously« while the subject is observing an afterimage or describing eidetic images of entirely different stimulus objects. In eidetic reproduction certain selective factors are operative, just as they are in other forms of reproduction. Past experience is often eidetically revived by way of very specific spatiotemporal fragments. This leads to another problem.

In analyzing the behavior of eidetic images we often find a translocation of objects or a transfer of certain properties of these objects, fusions and composite formations, substitutions, the appearance of parts instead of wholes, the nonappearance or the belated appearance of objects, reversals of right and left, above and below, and of other directions. Differently expressed, we find condensations, displacements, reversals, and other mechanisms supposedly typical of dream formation (15) (16). Because Schilder himself was an eidetic, it is not surprising that he alone among the psychoanalysts was [109] able to appreci|ate the structural similarity of psychoanalytical mechanisms and mechanisms of eidetic imagery (17) (18). In this connection a truly horrible thought has at times occurred to me; namely, that Freud himself, like Schilder, was an eidetic. It is true that Zeman published an experimental study in which he insisted that eidetic imagery is commonly found in all Viennese children between the ages of eleven and sixteen, and that 61.5 per cent of all high school students in Vienna – or even 88 per cent when including »latent« cases – are eidetic (19). I have never seen any data on the percentage of eidetic individuals in the adult Viennese population. It is also true, of course, that Freud was not born in Vienna. Whatever the merits may be of this terrifying idea of mine, at least as an amusing psychological exercise I recommend an examination of the edifice of psychoanalysis in the light of the »as if« *Fiktion* that the builder of this edifice was an eidetic.

I should like to return for a moment to the problem of »symbols.« Dr. McCulloch mentioned facts of translation, retranslation, and transformation in connection with symbol formation. This reminds me that Storfer once wrote an article on »psychoanalytical animal psychology,« and that Imre Hermann described »models of Oedipus and castration complexes in monkeys« (20) (21). Hermann, for example, referred to the simian equivalent of the strict father preventing sexual intercourse of the son with the mother, or to such observations as the son having intercourse with the mother immediately after death of the father. To be sure, he made the most of such simian demonstrations or »models.«

I am wondering, Dr. Kubie, whether these psychoanalytical writers did not miss a bet. It occurs to me that by an equally strenuous exercise of their imagination they could have found far better »models« of psychoanalytical mechanisms in the field of bird behavior. I am referring here particularly to the so-called »displacement activities« of birds. In recent years, Armstrong in England has been especially concerned with studying such transposed behavior patterns and the way activities become displaced and ritualized (22). Apparently there is no doubt that these displacement activities have reached their highest development in birds and that they play a far less important role in mammalian behavior. Armstrong has even considered the possibility that the high development of the striatum in birds may be related to these ritualized movement patterns and displacement activities (23). In pursuit of this delightful speculation, I should like to ask Dr. Kubie or a clinical neurologist whether symbol formation or the character of a neurosis has special features if the neurotic happens to have at the same time one of the basal ganglia diseases. Has this ever been studied?

[110] RIOCH: Not that I know of. The function of repetition of behavior | is, I think, a property of all nervous systems; that is, there is a strong tendency to repetition of preceding patterns of function in situations of conflicting stimulation. I do not know how far back it goes in the phyletic series, but certainly to forms simpler than birds.

KLÜVER: It has been often pointed out that displacement movements and other displacement activities become more easily ritualized or formalized in birds than in any other group of animals. In fact, it has been suggested that there are relations between speciation and displacement proneness.

RIOCH: I can't say.

VON BONIN: May I put in a word of warning? I am quite sure that the striatum in birds, whatever it is, is hypertrophied; it is a huge affair that fills up practically everything, but it undoubtedly has an entirely different functional role than the striatum in mammals. In mammals the striatum is coupled with the cortex. Birds haven't got a cortex, so it can't be coupled with it. The striata may be much more independent and may be a much higher mechanism in birds and may actually replace to some extent the cortex in mammals.

RIOCH: I will argue that.

VON BONIN: Well, you can't argue that the birds don't have a cortex.

RIOCH: No, but I will argue that the striatum has this high form of function.

VON BONIN: Oh, I will retract that any moment.

RIOCH: I think the large striatum in the bird is probably secondary to its large superior colliculus.

VON BONIN: I would take exception to that.

KLÜVER: I could rephrase the question.

RIOCH: I should like to toss something in here that may be unrelated. We have seen cats and dogs go through running motions in sleep, and we say they are »dreaming.« Well, with complete transection of the brain at the level of the superior colliculus, cats and dogs make the identical movements in the identical way; so if those are dreaming movements, then dreaming is somewhere around the level of the cerebellum.

MCCULLOCH: Or below.

HUTCHINSON: There are cases, I think, in insects that are very similar to Armstrong's displacement activity where you have presumably a whole central nervous system which is of the order of magnitude of a very small part of the cord. The only difficulty there is that they appear to be phylogenetically developed and not things that are alternative for actual behavior within the species. One example I can think of occurs among the Empid flies *(Empis, hilara),* where the male normally catches another insect and wraps it up in silk and hands it to the female. | The female apparently does not eat [111] this offering, and in some species the male just wraps up nothing in silk and hands it over.

BIGELOW: That is a jilting trick.

VON BONIN: Do they get away with it?

HUTCHINSON: The other point I want to make, which may have a sort of higher relevance, is that it does not seem adequately to be appreciated that there are certain connections between birds and men which are not apparent between birds and the dog, or man and the dog. They include the pre-eminence of vision, particularly color vision, and the probable importance of vision in sexual behavior, particularly in ourselves and in birds, which we don't share with so many mammals that depend more on olfactory stimuli. It may be that we would expect the kind of psychoanalytic situations that occur in man to be more closely modeled in birds than in lower mammalia, for some reason connected with that. I don't think that that has been very adequately considered, except that possibly Armstrong may have talked along those lines.

BIRCH: I should like to make a remark or two about Armstrong's observations and general considerations. Essentially, he has great freedom in developing any interpretation, mainly because we have no systematic knowledge of the phenomenon under consideration. This is true either in terms of the discussions of display or discussions of displacement or discussions of intention movement or what have you. All of these concepts developed essentially out of the interpretation of certain kinds of naturalistic phenomena and out of naturalistic observation. Whenever we begin to have some beginnings or glimmerings of an attempt to examine experimentally the meaning of these behaviors in birds, I think that we begin to find that more primitive and less humanistic or anthropomorphic or psychoanalytic analogies or interpretations are in order. I am thinking, for example, of a striking study that appeared just a few days ago, or came to my attention a few days ago, a study by Ramsey, in the course of which he raised some of his birds in an incubator in which there was a box. Usually, in naturalistic situations, when the chicks are startled in certain ways, they run to the mother. Under these conditions, when they were startled, they ran to the box and huddled in the neighborhood of the box – or a shoe, or in the neighborhood of a number of other objects that had been systematically introduced into the general home-territory area. What they are responding to is not necessarily what we would tend to respond to, and may be – and not only may be, but I would submit *is* – on a very much simpler basis, and is only analogically and not homologically or homologously related to the phenomena under consideration.

[112] Second, I should like to re-emphasize and perhaps extend the point | that Dr. Hutchinson made. In looking at these behaviors, whether they be some of the behavior mimicry in birds, in the Australian bower birds or in any of a number of other bird groupings, we find that these behaviors represent phylogenetically or evolutionarily determined mechanisms of responding, which have certain positive significances and are phylogenetically determined rather than ontogenetically determined. I think that when we begin to develop these analogies, we begin to fall into the classical error of an ontogenetic-phylogenetic-recapitulation type of theory, and I don't think the mechanisms involved in the ontogenetic theory of symbolism at the level of the higher mammal are necessarily duplicated in the process of the phylogenetic development of certain of these displacement or intention movements or any of a number of other phenomena, such as display phenomena, at the level of birds.

HUTCHINSON: The whole question about the box and the incubator seems to me to be entirely irrelevant. Lorenz, for instance, has a case in which a parakeet copulated with a ball poised in space some centimeters from a board. That was undoubtedly some sort of sexual act. It was not a purely arbitrary affair that happened to be connected with a ball, although it might be denied that it had anything to do with reproduction. I think it would be extremely silly to insist that it had nothing whatever to do with sexual behavior and just happened to be something that occurred in that particular experiment. It seems to me that the only problem is: How does the chick recognize the kind of thing that normally is a function fulfilled by the mother? And if it is a box that is big enough and just about the right shape, then we are justified in concluding that from the point of view of the chick the box is the mother – just as a chick will react to two balls, provided one, if I remember rightly, is not more than one-third nor less than one-half the diameter of the other. It is merely a question of defining what the sign stimuli are (to use Tinbergen's term), and it merely obscures the issues to argue whether it is anthropomorphic to call it the mother bird or a substitute for the mother bird, or a box. They are all identically equal in the language.

BIRCH: I would disagree with that entirely. I think that what confuses the issue is Tinbergen's and Lorenz's tendency to ignore that there is a difference between the head and two balls, and that although certain features of stimulation in the environment may release certain behaviors in the bird, or in other organisms, I think that our understanding of the level of psychological function of these organisms is determined by the kinds of things which are equivalent to the head or to the so-called »mother organism« for the organism under study. If these are merely signs, stimuli, or releasers, [113] or anything you wish to call them, | then the bird or the organism under consideration is not responding to another bird as a bird; but it is so organized and has a nervous system so organized that certain activities are produced when it is stimulated by any of a number of congeries of external energy organization or of stimulation, which is not true for higher organisms or organisms higher in the scale. Thus, two balls are not identical with a head for me, nor for you, nor for many other organisms. I think that the study of what types of equivalences exist gives us an understanding of the meaning, in a sense, of the action of the bird.

KUBIE: I am going to argue about that. We have moved so far from where we were that we are no longer talking about communication at all. There is a reason for this. We do not mind talking about behavior, so long as we stay away from communication for the simple reason that we cannot talk about communication without facing up to our most difficult problem; namely, the problem of symbolic function. Communication depends upon the symbolic function, and in turn the symbolic function is the essential hallmark of man. It is what differentiates man, sick and well, from all other animals. Therefore, we cannot continue to evade its challenge forever. I purposely have not discussed the data from lower animals, although I am quite willing to admit that some of that data are more empirical and clearer than some human data. Nevertheless, I keep to material from human beings. Why? Because man is the only animal with whom we can communicate symbolically, and who can communicate with us on that level. Therefore, it is only in man that we can even attempt to estimate in some measure what the symbolic process actually does and how it can become distorted.

Not all symbolic functions are identical. They range themselves in a spectrum. At one extreme are those symbolic processes in which what is represented is known in its entirety to the individual who is communicating. At the other extreme are the symbolic functions in which what is represented is wholly unknown to the individual who is communicating. This will vary and fluctuate with every individual, and from moment to moment. But if we conveniently delude ourselves by talking about communication only among lower animals, we turn our backs on that aspect of the problem which is essentially human. We cannot make progress by any such artifice (24).

RIOCH: I disagree with that, Dr. Kubie, for the reason that I think it is only when we get down to simple situations where we can begin to get some relatively complete observations that we can begin to work out this problem.

KUBIE: But the lower animal is not a simple situation. Actually, it is in certain ways far more complex to analyze, and far less accessible | to observation, precisely because the [114] lower animal's lack of symbolic speech deprives him of the ability to communicate to us his inner experiences, his thoughts, purposes, anticipations, memories, and feelings. Therefore, whatever of such experiences he may have on a conscious symbolic level is inaccessible to us.

BIGELOW: May I make a point which I think comes back to the first round of the speaker's discussion? I think you were speaking, Dr. Kubie, of communication, of whether or not it exists during sleep. There are some very primitive forms of this that one can point to, which you didn't do, but I am sure everybody knows about them. I should like to reiterate one of them. I think it is fairly easy to set up an experiment in which you show that a person can do operations which are fairly subtle, of the following sort: You put a person in a room with the Third Avenue Elevated outside his room, and after a week or so, he gets so that he sleeps perfectly well when a train goes by. You can put a noise-meter in the room and measure the DB level of noise. You can actually change the schedule of the trains so that they are relatively aperiodic, so that the person doesn't really learn to block out his hearing sense during the intervals, and the person will apparently behave like a normal human being and get his sleep and go to work and live perfectly all right.

You can also introduce a noise which is a soft footstep in the room, at an enormously lower level, and yet, very often, that person will wake up instantly, or almost instantly, upon this cue. That certainly calls for some sort of interpretation process which is certainly, or probably, not of a high intellectual center of the brain; but it certainly involves something like acoustical triangulation and the recognition of some coded message which is not a part of the environment.

I think you can point to many examples like this, in which a person who is in that state which we commonly call sleep, actually is capable of receiving messages in a very distinct and numerical fashion, in essentially the same way that he can when he is awake.

RIOCH: I can show the same phenomenon in the decorticated cat.

BIGELOW: In a decorticated cat?

RIOCH: Yes.

BIGELOW: Well, I am not trying to raise the question as to exactly where in the nervous system it exists. Is it the opinion of the group that the fact that you can exhibit this in a decorticated cat removes all interest in this behavior of a human being?

MCCULLOCH: Certainly not.

BIGELOW: That seemed to me to be the implication, and I don't see why this particular point has a more significant aspect.

[115] RIOCH: I don't think it bears upon the problem of sleep. |

BIGELOW: I think it bears upon the question of whether communication can exist between a human being who is asleep and his outside surroundings, does it not?

TEUBER: Is it the problem of communication or the problem of sleep?

VON BONIN: Exactly what is that experiment?

BIGELOW: I am merely exhibiting what I think is a valid example of communication which exists in the average person, let us say, in that state which we call sleep.

TEUBER: So sleep doesn't have graduations?

MEAD: But if you have a low-voiced call instead of a footstep, we avoid the problem of whether it is communication or not. A mother will wake to her own child's voice, though it is much lower and a long way off. It can fit in just as well.

VON BONIN: May I ask exactly how that cat reacts? If there is constant noise, does it wake up then; that is, to the slight rustling of a mouse or something of the sort?

RIOCH: No.

VON BONIN: Does it react merely to low-pitched noises, not very loud noises?

RIOCH: It will give no reaction at all to the people cleaning its cage, but if you go in while they are cleaning the cage, very quietly, the cat will immediately prick up its ears and turn its head toward you.

BIGELOW: Is this a question of frequency response of the animal?

RIOCH: You have a frequency spectrum in the room that is far beyond that. You don't get this response to low frequencies.

BIGELOW: Yes, but it is well known that cats have a peak in the frequency response which is in the high region; I mean, that can be easily demonstrated.

RIOCH: You can also do it with a whistle of a thousand per second.

BIGELOW: But this may constitute a lot more energy input due to selectivity of the frequency response.

RIOCH: Except that there are plenty of those frequencies in the room at the time.

BIGELOW: But the question is whether or not this sound suddenly produces a cod[e]able message in the animal, independent of its attitude. That is the question I was trying to raise. The Third Avenue Elevated does not wake the man, but a sound that requires a certain amount of decoding to know that it is abnormal does wake him up.

RIOCH: I will add this to it: You can have a whistle going continuously, and if you add a new whistle to it, the new whistle will evoke a response that the other whistle did not.

MCCULLOCH: May I ask one question before we go back, just as a matter of fact? Does it make any sense to talk of a midbrain cat, | everything above the midbrain [116] gone, sleeping and waking?

RIOCH: I think so. I think you've got to make a section at the lower border of the midbrain that is behind the inferior colliculus and down to the rostral margin of the pons to get rid of all the rhythmical sleep-wakefulness change in threshold. Now, I am not even sure of that. It may have to be a lower section. But I am sure you have to cut the midbrain out to get rid of it. On the other hand, when the transection is made from the anterior margin of the superior colliculus to the posterior margin of the mamillary bodies, the result is a decrease of the behavior ordinarily interpreted as sleep. The onset is sudden, and the threshold for bringing the animal back to what appears to be complete alertness is very low. With the hypothalamus and the central nuclei of the thalamus intact, then, to arouse the animal, you sometimes have to shake it; but with the midbrain animal, just a shift in the position of the hind leg is enough to bring it into a very changed state.

MCCULLOCH: Let's go back to a discussion of schizophrenics because the ostensive definitions of schizophrenic terms are patent to anyone who works long enough with a given schizophrenic.

RIOCH: I should like to tell you about a patient I saw briefly in consultation because it illustrates the kind of thing I run into. This patient was a boy who had cut off the end of his middle finger and sent it to a friend. He told one observer he was doing an experiment in pain, that he was proving one could disjoint a finger without conscious sensation of pain, which was true. He told another observer that this was an act of getting rid of his mother, getting her out of him.

KUBIE: What was that?

RIOCH: An act of getting rid of his mother out of him, one of the classical schizophrenic forms of communication. I did not, however, find out the other probable significant life experiences which this act of self-mutilation also symbolized, nor did I find out the implications of the formulations which he offered. The practical problem presented is that of determining the relevant interpersonal experiences and problems symbolized in such a condensed way by schizophrenic patients.

In contrast, there is the form of communication with neurotic patients and in neurotic dreams. The place where you hear this form of communication probably in purest culture is in certain moderately low-grade bars when the boys are getting a little bit tight. At such times everything round in shape is either a breast or a buttock. Everything pointed is a penis, and everything that has any depression in it represents the female genitalia. The game is to see how picturesquely one can deal with such references and be understood.

Now, there is one thing I would like to ask you, Dr. Kubie, about dreams. I warn all the patients I see against giving me any dreams, | because I don't know what to do [117] with them, but once in a while they come up anyhow, and the impression I have is this: the severe schizophrenic is very much like the normal in respect to dreams; namely, in both, the dream life and the waking life are very close replicas. In between there is a wide range, including neurotics in whom there is this very picturesque type of dreaming which is like conversations in bars and is like the way kids will play when they are not observed. In this dreaming there is what may be called conceptual-clang associations (not clang associations referring to sound).

PITTS: Excuse me, but what is a conceptual-clang association?

RIOCH: I call conceptual-clang associations phenomena such as using a long or pointed object to refer to a penis in communication. This conceptual-clang communication is, I think, universal. Hindu religious writings, particularly in describing the goddess Kali as the productive mother earth and as the force of destruction, present practically every psychoanalytical symbol in much the same way as these symbols are interpreted in our culture. This symbolic language is very much limited to the personal objective view of the child, that is, to anatomical structures of the child and his immediate associates. The communication is in a verbal form which presupposes a certain response from the environment in terms of understanding and an answer in a similar form. This is not what one gets in optimal normal performance nor in schizophrenic performance.

HUTCHINSON: May I inquire why one doesn't get it in optimal performance?

RIOCH: Because you see through it and you are interested in something else.

KUBIE: I can't make out at the moment whether you agree or disagree with me or want to ask a question.

RIOCH: The question I wanted to ask you was whether this impression I have is correct – whether you have any definite observations on schizophrenic dream states and (optimal) normal dream states as contrasted with the neurotic form of dream content.

KUBIE: I cannot answer that. A lot of people talk glibly about schizophrenic dreams. I have many reservations about them. I have never been convinced that I could recognize a dream as being, in some special way, schizophrenic.

RIOCH: Those are what I would class with the neurotic dreams. Some people claim they can recognize schizophrenic features in a neurotic dream. I never could make out what they were referring to. I am talking about the kind of dream a severe schizophrenic reports.

KUBIE: Well, the other implications of what you say really carry us far afield from my topic. Nevertheless, it may be worth while to [118] | say that in a severe schizophrenic it may be difficult to know whether he is reporting a dreamed experience or a waking fantasy, because his waking fantasies are almost as completely overlaid with obscure symbolic implications, connotation, and language as is his dream life; and in such patients the borderline between waking and sleeping, between experience and fantasy is obscure. Therefore, here again I cannot draw any hard and fast lines.

BROSIN: I wonder if it would help you at this time, in defining communication in many different states, such as the sleeping state, hypnosis, and the different clinical conditions, to use as a springboard the concept of interpretations of many types. These interpretations may vary from the more rigidly logical to those which we call intuitive, that is, dependent on nonverbalized learning or past experience. Can interpretation as a method be used to examine these latter phenomena in a manner different from a method now being more sharply defined, that is, the examination of the interpersonal relationship between the particular organism in question and its social matrix?

That may not be very clear, but it may be the difference between the interpretive methods utilizing models, such as interpreting dreams by various patterns or by looking for the universals in dreams, in hypnotic states, in myths, or in symbolisms of various cultures as opposed to – perhaps they are overlapping – the participation in the direct experiences of a person communicating with his fellows, whether these communications be internalized or actually outside.

KUBIE: I don't get you, Dr. Brosin.

BROSIN: Well, in its crudest form, I suppose, this is the problem of intellectual interpretations versus emotional relationships. Let's take the patient-doctor relationship which has so many ramifications involving problems of communication. I was hoping that we could discuss the basis for the types of communication, verbal and nonverbal, intellectual and emotional – to use overly simplified terms – to describe these complex acts. As the problem of what language is came up from time to time today, we seemed to have multiple definitions available to meet our needs. Various members referred to these multiple levels of activity, and we know without dictionary definitions that there are many levels of communication. One is to intellectualize the symbols used, whether they are letters or pictures, and to handle them at various levels, and the other is to examine much more closely the direct relationship of the organism in question to the persons with which it is most concerned, and the emotional relations between them.

MEAD: I can give you a model for that, I think. It seems to me that the distinction is, again, a cultural one, because we have learned how to intellectualize these things. You take three girls in different | stages of adolescence, making little model gardens, taking [119] flowers and shells and building lovely little structures with a nice room with a red carpet. Then one girl takes a leaf and puts it in the doorway so you can't see the red carpet. Then there is a discussion whether she is going to have the driveway come up to the house or whether motor cars are to be allowed in, or whether she should just have stepping-stones up to the house and the garage down there, and so forth. You can look at that in formal intellectual terms if you are used to reading that kind of material, for instance, in terms of ordinary child analysis, and see the stages of adolescence of the different girls and what is said by that picture. People can do it. Eric[k]son, for instance, could take photographs of such play and say: That girl is probably a year before menarche, this girl is just at menarche, and this one is a year after it. Or, after one of these little girls has carefully made a house, and considered whether she would let a road come up to the house or keep the cars down in the corner, you can watch a boy who is a couple of years younger come over and dig a well in the back yard of her house, very carefully. Then she says, »It isn't a well at all; it's an incinerator.« At that point, you are having communication between the two. The boy is certainly saying something very definite to the girl: they may not be aware of what they are saying, but you can measure it and check it against many things. They are using a language which one can handle intellectually, also, so that I don't see that these statements as to usage of a conceptual model which is mythology or religion or symbolism or poetry, in contrast to the relationships between the therapist and the patient, are exclusive at all. If we hadn't overdone the rational approach to life, we wouldn't have to make that distinction.

KUBIE: I will go back a bit. I fear that we are losing our ability to communicate with one another here, and are bogging down. I suspect that this is owing in part to the nature of the data, and partly to the fact that we are trying to bridge a wide gap in experience. Those of us who have had clinical experience have no reluctance in considering the highly symbolic data which are to be found at one end of this bridge, whereas those others who lack clinical familiarity do not like to deal with data of this kind, because they cannot be corralled in a reproducible experimental situation, as indeed all of us would like to do. There are methods by which this can be done ultimately, and in the not too distant future, for that matter. But it has not been done as yet; and it never will be done if we continue resolutely to deny the existence of the data themselves, merely because they make us uncomfortable by being elusive and hard to pin down.

Therefore I should like to return to an elementary and simple restatement of my own fundamental working hypothesis (25) (26). The sym|bolic function characterizes [120] the human being as homo sapiens, setting him apart from all other animals. By this I do not mean that no other animal is capable of any symbolic functions, but that between the symbolic potential of man and that of lower animals there is an enormous qualitative and quantitative difference. Indeed, if it were not for our highly developed symbolic function, none of the things which we are doing here today, and none of the other processes which characterize human cultural development, would be possible. The human symbolic function involves the capacity to make abstractions from discrete individual experiences, to condense these in various ways, and then to represent these condensations of abstractions with gestures, with facial expressions, with the spoken word, and with the written word, that is, through the whole paraphernalia of symbolic language by which we communicate to one another our thoughts, our feelings, our memories of past experience, our hopes and our plans.

Let me backtrack and say that without this, the highest level of psychological experience to which the human being could attain would be a pallid, sensory reverberation, a sensory afterimage of previous experiences, a simple wish-fulfilling dream (such as a little child might dream), or something like a phantom limb. I realize fully that these words themselves are symbols and therefore subject to distortion, but I still think that all of us know what I am talking about and what I mean when I say that because of our capacity for abstracting from experience, plus the ability to represent those abstractions symbolically, everything that characterizes human culture becomes possible.

Early in life, however, something of great importance happens to the development of the symbolic representation of abstraction. Later in life the same thing can occur in the process of falling asleep, in the process of waking, in the dream itself, and in the hypnagogic revery to which we are subject from early childhood. In these states symbols become primarily visual rather than verbal, probably because they had originally been predominantly sensory and visual before they were verbal. These symbols represent more than one thing, however. They are multivalent. Furthermore, under certain circumstances the symbolic representative becomes dissociated from that which it represents; and psychopathology starts with this dichotomy, that is, with the dissociation between the symbol and what it represents. It is my working hypothesis that the neurotic potential out of which everything that is neurotic in human life and human behavior develops arises precisely here. This is universal, because we have not yet discovered how to avoid or limit this fundamental and basic dichotomy in early life.

What does that mean for communication? I did not expect to have to reaffirm my [121] fundamental thesis about the neurotic process here, be | cause we have been over it several times; but the discussion makes it clear that it is necessary to clarify this before discussing what it means for the problems of communication. Actually, it means that every symbol we use, whether it is a word or a gesture or an expression or a tool, is multivalent; and that some of the determinants of the relationship of the determinant to the symbol are known and some unknown. Consequently, if we had an instrument for measuring the relative roles of conscious and unconscious determinants it would show that when I use such a word as »chair,« on some occasion, 90 per cent of the determinants of my use of the word would be matters of which I was aware, and that 10 per cent would be something I was unaware of. Whereas, on another occasion when I used the word »chair,« 90 per cent of the determinants might be unknown to me and 10 per cent of the determinants might be known. Therefore, my symbol sometimes represents the known to me, and at other times the unknown; but it is always a mixture. If this is true, then in language, in speech, in thought, in poetry, in anything that we are doing, we are dealing with a symbolic process in which the relative roles of »conscious« and »unconscious« determinants, that is, of known and unknown determinants, are constantly shifting. That, I think, is fundamental and basic.

A few other relevant points concern the relationships which can be established between human beings in a certain state. It is hard to characterize this state. Operationally, however, it is a state which results from a certain type of procedure. We can call it *hypnosis,* or an induced, controlled, dissociated state, a communicative state of semisleep, a sleepoid state, a communicative hypnoidal state – anything you want. What is important is not the name but the fact that there is such a state, and that in this state it is possible for one of the subjects to be more aware of the unconscious determinants of the actions of another subject (or of himself) than is pos[s]ible in the fully alert, fully awake state. *Furthermore, something of the same kind may be possible between and among psychotics.* These interesting points have never been studied fully: I mean the ability of one psychotic to understand another, or of one hypnotic subject to understand another. We know that there are some psychotics who understand themselves; but we have not investigated beyond this the important question of whether one psychotic can translate another's psychotic products.

That is as much of a summary as I can pull together in a hurry, without trying to tie up every loose string of my own perplexity about this problem.

May I give examples of two other important points? The capacity for communication is linked closely to the developmental processes of the ego, that is, to the ontogenetic process. This can be proved and has | been proved in experiments in which [122] regressions to earlier age periods are induced. For instance, an Austrian who learned English when he was twelve or thirteen was regressed to the age period before he knew English, and could no longer speak English. At the same time, he would make figure drawings like those of a child of eight or nine, write like a child of eight or nine, spell like a child of eight or nine, and react to the Rorschach test, to the Szondi, to the Bender Gestalt, as though all of his symbolic processes had returned together to the age period to which he had been regressed (10).

Another interesting experiment, done during the war, illustrates similar things. (It was restricted, and therefore has never been published.) There was an air crash in which there were no survivors. For various reasons, it was desirable to find out what had happened. As the plane was falling, there had been a brief interchange between the pilot and the radio operators at several fields. The pilot's message was an explosive, excited rush of unfamiliar words, not the usual stereotyped communication as to direction of wind, altitude, and so on. This was before automatic recordings of such communications were made. Afterward, when questioned, every radio operator in every field said, »I don't remember a word he said.« We were asked whether we could disentangle out of these quickly obliterated memories of barely perceived, unorganized, unanalyzed auditory impressions something that would make it possible to reconstruct what had happened?

I am not sure whether or not we succeeded. The engineers of the particular outfit that asked me to make the study and who worked with me were sure it had been reconstructed. All the auditory symbols were so charged with emotion that even under hypnosis and with drugs, the radio operators could not talk about it. Therefore I asked them to write about it. The words came along fairly clearly. Then came the word »deicing,« and the curve on which they wrote was the falling curve of the plane [demonstrating on blackboard]. It seemed as though they were translating into an arm movement their unconscious knowledge of what they believed had happened; namely, that somebody had tampered with the mechanism to warm the carburetor.

McCulloch: I have two fag ends I wish you would clean up for me. First, you several times referred to the universal symbols, or the universality of the symbols that occur in dreams, and I am not sure what the terms »universal« or »universality« mean in that connection.

The second thing is that you have twice spoken of insight and therapy as though they were either necessarily correlative or as though one was necessarily consequent to the other. I did not quite get the connection. Would you clear up those two points for me?

[123] KUBIE: Gladly, because I would not like to leave any misunderstand | ing about my own feelings on either of these issues. In the first place, I do not know whether there are universal symbols. I have always had a bias against them. But as the years have gone on, I have run into enough disturbing examples of identical translations of elements in dreams in subjects of extremely varied ages and cultural backgrounds to make me hesitate to dismiss the idea. Certainly there are some areas of symbolic implication which seem to be, if not universal, at least broadly distributed in our own cultures and in many different cultures. I do not think we have any right to speak with confidence of universal symbols, any more than I think we have any right to speak with confidence of inherited symbols; but I must say again that these are two areas that have not been fully investigated (8).

MCCULLOCH: You mean, common to all men, regardless of their culture or background, something of that sort?

KUBIE: That is right. I find myself with a bias against such a notion, yet unable comfortably to dismiss it as impossible. There may be some types of human experience which are so fundamental that they are universally represented. For instance, one of the most fundamental of all trauma is the business of being a small child in a world of grownups. This may be represented not only in such fables and myths as Jack, the Giant Killer, but also in many other ways.

VON FOERSTER: If there were such universal symbols, could we ever detect them? Because even if we come up to that age where we are able to express these universal symbols, we have already merged with the cultural pattern in which we have to think. Therefore, I don't know – and I would only raise that question – of any way in which we could even detect them, if there are such things.

KUBIE: Fair enough.

MCCULLOCH: May I put the question another way? Is it that the same symbol is used by everyone, or is it that the same state is symbolized by everyone? It is a very simple question, first, in my mind.

KUBIE: Dr. McCulloch, I should say –

MCCULLOCH: Or is it that there is a lock of the two?

KUBIE: It is at least conceivable that both may occur; that is, in different individuals a state may be symbolized by one and sometimes by differing symbols. Your second question arises out of a misunderstanding. I said specifically that I do *not* know the relationship of insight to therapy. I wish I did. One of the most puzzling problems in psychotherapy is why insight sometimes seems to be powerless and at other times has extraordinary therapeutic leverage. By »insight,« I mean that a person who has not previously understood the meaning of his own behavior, thoughts, or feelings acquires a full knowledge of them, their meaning, their roots, their ontogenetic history in his [124] own life. It | is characteristic of many schizophrenics (although not all, nor is this confined to the schizophrenic) that the acquisition of insight seems to have little therapeutic effect.

PITTS: What do you mean by the »feeling of understanding« of the subject?

KUBIE: I mean that the patient has not merely an intellectual realization that something may be true, but an awareness which involves some actual memory of the experiences which underly his behavior, plus enough emotional participation in that awareness so that in some measure it is as though the experiences themselves were being relived. Such insight is a combination of intellectual understanding, plus memory, infused with appropriate feelings.

RIOCH: I wish to make a point with regard to the problem of these universal symbols. I think to a large extent similar symbols are used in many cultures. That doesn't mean that the thing that is being said by them is similar at all.

MCCULLOCH: Exactly.

RIOCH: Especially in that type of communication in which these symbols are used, the actual communication as to what the situation is – in terms of what the interpersonal relations have been and what they are going to be – has to be determined not in terms of the content of the symbols, but by the sequence of the real behavior occurring at the time. Infrequently the content of the symbols due to their particular cultural significance gives an indication of the predictive value. The actual communication can be determined only by the sequence of events, not by the general or personal significance of the symbolic content. I think that is where, in so far as we differ on this question, the main point lies on which we differ.

The problem of therapy is so complex that I don't think it can be used as data. I had one patient, for example, who was cured by Christian Science and then broke down again. He described in great detail the »cure,« the break, and the results, and the »cure« had all the characteristics of insight that I have had in patients whom I have »cured.« In one sense, at least, that which is called insight is a matter of having come to a situation of communicating with another person in a reasonably reliable, predictive manner. This gives so much relief that those defenses against anxiety that were used before are no longer necessary. »Insight« in this case refers to developing a common content for communication. Whether we ever get anything more than that, I don't know. I think we may get different degrees of it.

VON FOERSTER: Have you ever observed symbols of self-communication? I have the strong suspicion that some people create certain symbols which only they understand, for instance, in the sense that they | make themselves understandable or establish for [125] themselves a place in the very complicated pattern of the world in which otherwise they would not know how to fit.

BIGELOW: Isn't self-communication a contradiction in terms?

VON FOERSTER: No, I don't believe so.

RIOCH: The »self« being a symbolic social structure and not something inherent in the organism, it is not a contradiction in terms. What is actually going on is that the person, using some kind of idea of the »self,« experiments with that until he finds somebody who can understand what he is trying to get across. As soon as he finds somebody who can understand what he is trying to get across and who answers appropriately, the »self« mechanism changes and he proceeds to orient to people in a much more adequate and elaborate form.

BIGELOW: Does the mechanism display any external manifestations when this is taking place?

RIOCH: The one that came to mind did.

BIGELOW: Then is it really external communication, even at that level?

RIOCH: Oh, yes.

BIGELOW: Then how is it self-communication?

RIOCH: These particular formulations are done very privately, and you have to demonstrate with the person that there is an acceptance of the situation before he will ever let you in on it. It is like the schizophrenic who goes fifty years before he tells anybody about his hallucinations.

Kubie: Mr. Chairman, because we are hearing so much nihilism, I am going to do what I ordinarily do not do; that is, give an actual clinical report. It is important to realize the sort of therapy that can occur. As I have listened here for the last hour or so, this has kept coming to mind; and I think I will »spill« it.

I have in mind the critical moment, the turning-point in the cure (and I say »cure« advisedly) of an active homosexual who is homosexual no longer. He had been in intensive analytic treatment for only a few months. A couple of years earlier I had sent him to Stockbridge for several weeks. Before his intensive analytical therapy began, he had been able to establish himself in a job and had begun to make a few friends of both sexes. He would oscillate between homosexual pickups (with all the degradation and risks which that involved) and trying to establish sound human relationships with young men and women. A moment came in our work when I felt that it was safe to impose an absolute taboo on any further homosexual contacts which did not grow out of pre-existing sounder human relationships. This blocked his blind wandering around [126] Central Park, the making of blind pickups, and | the danger of his landing himself in extremely unsavory circumstances. He accepted this willingly, but he came in with a dream.

This dream was so interesting that I took it down verbatim, and have used it in teaching on several occasions since. The dream follows: »I was in a red something or other. I can't remember what it was. Then I was in a room with some young women and, somehow, I was accepted by the young women more than I was accepted by the young men; because there were also some young men there. It is funny. There's a lot more to the dream, but I can't remember anything else about the dream. I am puzzled about it, and bothered about it.« Then he talked about a lot of other things, going back to some of the material of the previous day, and to the experiences of the previous evening. As he did so, it became clear that the dream was a transparent and literal extension of that evening's events. Then he came back to the dream and repeated it, puzzled by the gaps; but there was no change in his recital. Nothing was omitted, nothing added, nothing changed.

I said to him: »You realize, I am sure, that it is quite usual among young girls« – I started with girls purposely – at some point in their lives to fantasy that they might be young boys, so that they would be free to be with their father or their big brother or a boy cousin in the boy's locker room when the boys are dressing and undressing or taking a bath or going to the toilet, just to satisfy their natural curiosity. And it is equally natural, of course, for young boys to imagine that they are young girls for the same reason.« I had gone only that far, when he broke in: »My God, I remember the rest of the dream! My mother was there; and she was naked except for some blue lace panties. She came up to me and I looked at her and she had a penis; and then suddenly she walked around behind me, from right to left, and when she came around in front of me she was just the way she really is.« Then came a series of memories about his mother, about physical contacts with her in his very early years, about his feeling about her body, and especially about a portrait that had been painted of her, a small copy of which he always carried in his wallet. In this portrait she wears a blue lace brassiere, just the color of the panties in the dream; and in the portrait the bra is cut so low that one breast was almost completely exposed. This started the unraveling of his intricate relationship to his mother. Such experiences are our daily diet, but not yours. Nevertheless, you cannot leave this out of account in your approach to this problem.

Bigelow: Dr. Kubie, is there a certain element of feeling that it is all right to have such dreams but not all right to have such conscious thoughts, and therefore one actually gets some statements which are in the form of dreams but may actually not be [127] such? |

KUBIE: Oh, yes, constantly. My patient had been able to describe every last detail of his sexual experiences with men; but until this dream opened the door, not even a fantasy about women had been able to enter into his conscious thoughts and feelings.

BIGELOW: However, perhaps his mother was something more sacred than all of these.

SAVAGE: You would have to be rather elaborate, Dr. Bigelow, not to be able to remember this dream, if the dream was explicit. If the dream was a conscious device to communicate with Kubie about something which he didn't want to do, certainly, it was really very elaborate for him not to remember the dream for twenty minutes.

VON BONIN: Isn't it the Freudian sense that we are talking about now?

SAVAGE: Dr. Bigelow is talking about a very naïve sense in which I call up a doctor and say, »Doctor, a friend of mine has athlete's foot; how do you cure it?« and it is my athlete's foot, of course. It is the matter of trying, perfectly directly and explicitly, to trick the doctor, and I do think that is really ruled out fairly well by the evidence here.

MEAD: Yes. Isn't it worth picking up the last remark Dr. Kubie made when he said that this was an everyday experience to him but not to some of us? It seems to me that one of the difficulties we often encounter in this group is between the natural scientists and the psychologically trained. To go back to your word »chair,« which in a given context is 90 per cent known or conscious and 10 per cent unconscious. The natural scientists have been trained in this group to pay attention to the part that is known and to leave out the part that is unknown, to keep it out and rule it out and make a set of rules that make people more or less sit tight and use the same words and play a game very rigorously.

BIGELOW: I think we are perfectly willing to conjecture about this part, but we like to say it is conjecture.

MEAD: But you are still acting in the conscious realm when you say that. When you say it is conjecture, you still want to act according to one set of rules. Now, the psychiatrists listen to the other set. They are trained to listen to the other set. Nobody seems to suggest that you might do both at once, that there isn't any reason in the world, except the habits of our present discipline in the scientific world, why you can't use visual images, kinesthetic images, olfactory images simultaneously. You can think a great deal faster if you do, as well as using words – which is what probably everybody here does when he is being creative. However, the rules of the game have been, in the natural sciences, one set, and in psychiatry a different set, not really different, because they split it off too – | [128]

RIOCH: You are correct, but the rules are not necessarily different. Why should we take psychiatry and confine it to the operational interpretation of symbols? Why not deal with events according to the rules of the natural sciences? Dr. Kubie gives a beautiful example of the sequence of events in getting into a relationship with this boy. (Intuitively Dr. Kubie knew the time to say, »I, as a person, am going to give you support by giving you a rule of thumb.«) He does that and what happens? Now comes a flood of memory material which appears in a relationship which hasn't existed between this boy and anyone else before. You can study the sequence. When you are treating the schizophrenic, if you don't pay attention to the sequence you are lost. The neurotic patient is perfectly willing to go along with you. He is so tickled at his capacity to see implications that when you find implications that agree, he will go along with you and give up symptoms. But I think that the same principles hold, that we want to treat psychiatry as a natural science. We have to deal with all the communication in as accurate a description of the real setting (which is the sequence of events), and in as extensive a sequence of events, as we can get. A great part of the time we don't know just what the relevant sequence was.

KUBIE: I agree with that completely, of course. That is why for the last umpteen years I have been arguing that we should reproduce therapeutic sessions in their entirety, so that groups of experienced therapists could sit down together to study them. Only in this way can we rule out the artefacts that are introduced by any one individual's isolated observations. In the past, the lack of such records has set us apart from any other science that I know about. Every moment is ephemeral. Once it passes, it can never be recaptured by the written word or even by the tape recorder alone. Too many things are going on at once. It Is only in the last few years that we have had a mechanical means for reproducing the entire phenomena for subsequent study and analysis (25, 26).

Naturally any such method of reproduction introduces its own artefacts, which must themselves be studied for their distortions. We will never meet your justified requirements for an objective scientific process until we can reproduce analytical interviews so that we can sit down and study them over and over again, just as a dozen people can look at a specimen under the microscope, or listen to a heart.

ROSENBLUETH: To come back to Dr. Mead's comment. I don't think there is any real distinction to be made between the language of psychiatry and that of the natural sciences except in the rigor with which we want to do our thinking. I don't think the natural sciences are playing a special game. I don't think we are neglecting any data. [129] There are problems in physiology which are not formulated precisely, and yet | one works on them. But what one tries to do is to get to the point where one is able to formulate precise questions which are susceptible of precise answers. Until we can achieve that, we don't feel that we have gotten very far.

With other topics it may be more difficult to focus on them in such a way that one is able to formulate precise questions and then to carry out observations that can give precise answers. The only difference I can see is that, perhaps because the material with which we deal is simpler, we manage to select those problems that are amenable to accurate analysis and then work on them. We would not be content to give an answer to the problem on which we are working when such answer turns out to be far too vague, according to our own standards.

RIOCH: There is another angle to it, and that is the capacity of a patient or any organism to arrive at a certain state, taking any one of many possible paths. If we are dealing with an experimental situation, we don't mind repeating and repeating and testing one part and then another. When you are dealing with a therapeutic situation, you are limited in terms of the patient getting what he »wants,« because then he stops. You can't check it again. A large part of the time, the thing we did that we think worked was only some vague reflection of what really happened.

ROSENBLUETH: No, let me make this clear by returning to something that Dr. Kubie misunderstood in what I said originally, and which I was not planning to go back to; but it becomes pertinent in view of Dr. Mead's comment. Within physiology, or within other types of science, there are many things we cannot define. We point to them; we recognize them. Sleep is not only a psychoanalytical or a psychological condition; it is also a physiological condition. When I said we did not know anything about sleep, I was really talking as a physiologist, but that does not change the comment I made on the distinction of certain phenomena when they occur during sleep and when we see the same phenomena while a person is awake. Granting we all know what we are talking about and that it is not necessary to define those terms, Dr. Kubie's distinction was artificial because he did not give one single fact that pointed to qualitative differences between the two phenomena, except for one word, which is the word »unconscious.« You may say, »I was totally unconscious when these phenomena, these things occurred,« or else the patient or the subject being examined may report,

»I was not aware.« You may use these statements to exemplify the unconscious. But the term has an unfortunate connotation, which is the only reason I objected to it. If we consider the interpretative connotation that the word has, then it becomes objectionable. Something which is not in the mind cannot be conscious, and I am | using [130] »mind« with the quotation marks that Professor Richards used, and also marks around »conscious,« again with the understanding that we cannot define this term, but I think we know what I am talking about.

BIGELOW: I should like to add one more remark. I don't think anybody in the natural sciences, as Dr. Mead calls them, would attempt to make a rational basis or even come anywhere near such an idea in connection with something like psychoanalysis. I don't think anybody here could object to a description of the processes and the exploration of their vague and ephemeral character. It is only that there is a certain crossing of purpose when it comes to using terms which appear to have specific meaning but which are vague to the group, and I think that is what Dr. Rosenblueth was also saying. At this point I think it is perfectly reasonable to get down to brass tacks and say, »I mean to state this pragmatically, my definition of ›unconscious,‹ and work with it, so please let me continue.« That is perfectly all right. But unless there is something specific to hang on to, or unless it is just a useful notion, loosely tied, which we choose to put up for lack of something better, one doesn't know where one stands, so one doesn't know whether to draw inferences from this or to see a picture. That is the state I find.

KUBIE: By now we should have gone far beyond this hackneyed issue. It has been batted around for fifty years. It has been argued so often and in so many places and in so many contexts, even here in our own conferences, that by now we should be able to take this working hypothesis for granted, as we strive to understand one another. All psychological functions (such as communication) are stratified or layered; and there are variations both in the degree of awareness of these layers of psychological processes and in their accessibility to our own conscious self-examination. We must also include in our approach the fact that there are two kinds of *unaware* psychological functions, which have been called by various names. William James spoke of the »fringe of consciousness.« Freud calls this the »preconscious« or the descriptive »subconscious,« as opposed to the more dynamic »unconscious« processes which are walled off from our conscious introspection by forces which are actively resistant to our understanding of them. Such a term as »walled off« is obviously an allegorical or figurative term to describe the clinical fact that in order to penetrate to these processes, and in order to reintegrate the various »levels« of psychological function, one must overcome certain opposing forces. On the other hand there are no such barriers to conscious perceptions of our »preconscious« or »fringe« material.

Anyone who has any familiarity with pathological psychological phenomena, who has ever encountered the processes of repression, who | has ever attempted to pene- [131] trate selective memories and thus to reintegrate the dissociated components of human experience (as has been done in the laboratory and under controlled conditions) cannot think of human psychological processes as though only that small fragment is important of which we can be conscious. Indeed, that is the smallest part of the continuous psychological flux which goes on within us at all times. It is a regression to the Dark Ages to hear mature scientists still haggling over whether to acknowledge that there are levels of psychological functioning which vary both in their degree of conscious awareness and in their accessibility to simple direct introspection, and to which one can penetrate only with special techniques of which the pioneer technique, and still the most important, is psychoanalysis. Heaven knows that it is a clumsy instrument; but it is still the best one we have. Therefore, if we are to communicate with one another, we must say at least that this is a fundamental working hypothesis in any study of psychological function.

MEAD: Who said it wasn't? Did anybody say it wasn't?

KUBIE: Yes, Rioch and Rosenblueth said it wasn't. They could not even talk about unconscious processes. They said if it is not conscious, it is not in the mind.

ROSENBLUETH: Excuse me, but I objected to the term. You probably are all aware of Poincaré's theory of creative imagination in mathematics. It is one instance in which I think that one of the most intelligent people who ever existed in the world went thoroughly haywire. It just doesn't make sense. It is not a theory; it doesn't explain anything; it doesn't lead to any further experimentation, merely because he got bogged down by the word »unconscious.« He used it thinking he was saying something. He referred to processes going on without the person being aware of them; they were unconscious. All the possible solutions were threshed out unconsciously and were offered to the conscious only when they satisfied certain esthetic criteria. The suggestion has no explanatory and no predictive value. Therefore I consider that it constitutes a very poor scientific hypothesis, and I think that damage was done by the word »unconscious,« which is interpretative. I don't mean to object to the phenomena. They should be studied. And certainly I don't want to quarrel with any one of Dr. Kubie's observations. If I say he uses a word merely to describe them, the use is clear, but –

KUBIE: I would agree with the way Dr. Rosenblueth uses the word also.

BIGELOW: I think you just did the defining a moment ago that we would have liked earlier in your talk.

[132] KUBIE: I am sorry. I thought that because of our many previous | discussions of this problem another systematic review of it would be superfluous by now.

RIOCH: I find that definition quite inadequate. There are a large variety of phenomena one has to differentiate.

KUBIE: May I say one further word? I would agree with you. I said that simply to characterize one of the outstanding features. I am wholly uninterested in definitions.

RIOCH: My objection to »unconscious« is that we talk about layers and we have no idea what sort of layer it is. Is it a longitudinal layer? Is it a horizontal layer? Is it an interlacing layer? I don't think that the methods that have been used to investigate the so-called »unconscious« have been adequate. There are phenomena in the schizophrenic which are nonverbalizable, which are totally different from phenomena in the neurotic. Again, if you go into animal experimentation, it is possible to formulate a set of categories of behavior which one can also apply to the human. Then one can define central nervous system activities which get down to simpler levels than those we have been considering in this, and also define more complex forms. I think what we need is some kind of an attempt at an operational definition, so that the operation can be described so that somebody else can do it –

MCCULLOCH: Hear, hear!

KUBIE: This is precisely what I have been attempting to give you: an operational definition in terms of a procedure to meet certain concrete clinical phenomena. In doing so, I use a terminology which describes in recognizable terms both the clinical human challenge and the procedure accurately enough to make it possible for others to duplicate the experience. This provides a constant frame of reference for further investigations of the subject by other observers, and this is precisely what I have offered you here. Now, for heaven's sake, let us get on with the job and end this childish hugger-mugger over words.

PITTS: I would like to say that I maintain the scientific method is something more than a cultural prejudice, along with Margaret Mead –

Mead: I didn't say that.

Pitts: – that is, if the methods which the psychoanalyst uses in dealing with this material are not scientific, it is up to him to make them so, not for us to admit that his methods or modes of dealing with them are just as good as ours, if we are scientists. There is a terrific difference between the –

Mead: That isn't what I said at all. I said that one group here is used to dealing with a certain kind of rigor. They use the word »game« with approval. I object to that out of my background, and Dr. Rosenblueth objects to it out of his; but it is an evaluative term, not a devaluative term, when I use it. There is another group of people here | [133] who are used to dealing with another section, and we are not accustomed to using them both at once. Now, that is not saying that we are prejudiced. The scientific method is certainly a way of behaving and of being very rigorous and listening to everything.

Rioch: The scientific method is an attitude, not a way of thinking.

McCulloch: Gentlemen, we won't settle it here tonight. We are adjourned.

[134]

COMMUNICATION BETWEEN ANIMALS

HERBERT G. BIRCH
Department of Psychology
City College of New York

When I was asked to speak to this meeting on the problem of communication in animals, I had, at the very beginning, a negative reaction to the topic, and that negative reaction, instead of abating in the course of preparing my talk, has grown ever more intense, until at this point I would say that I am unwilling to discuss the problem of communication in animals for a variety of reasons, but I am willing to discuss the various ways in which animals interrelate one with another. The reason for that kind of resistance and for that differentiation stems from a general difficulty that applies not only to the problem of communication but applies also to any general category in animal behavior that can be used or tends to be used in a consideration of the problems of the way in which animals adapt to the world or to one another.

I would lump the problem of communication with such problems as feeding, mating behavior, migration, or what have you. These behaviors are really not behaviors at all, and we are not discussing behaviors when we are discussing communication. We are, rather, discussing certain phenomenal similarities that may appear in highly different forms of behavior. We are concentrating not upon the behavior process as such but, rather, upon the end result of any of a number of different kinds of behavior processes. When one concentrates upon similarity in end results, it is especially important to concentrate next upon the differences in underlying processes, both physiological and psychological, that may be producing the end result. Otherwise, we fall into the pervasive trap in science; namely, the trap of analogy rather than the method of understanding that is known as the method of homology or the examination of processes which have a systematic relation, one to the other.

Perhaps I can illustrate this point best by taking a few examples from the problem of feeding behavior and of so-called »migratory« behavior in animals. There is no doubt that an amoeba feeds. There is also no doubt that a human being feeds. There is no doubt that a salmon migrates. There is also no doubt that the human nomad migrates. But I would submit that there is a fundamental difference between the amoeba, extending its pseudopod in the direction of a small food particle, and Oliver Twist, extending his porridge bowl to the cruel dispenser at the orphanage. In the case of the [135] amoeba, and in | the case of Oliver Twist under propitious circumstances, both organisms obtain food and are sustained by this event, but the amoeba is responding now not to food as such but to a weak chemical stimulation, and it will so respond to almost any of a wide variety of weak chemical stimuli. Thus, if the minute food particle is replaced by a minute particle of potassium permanganate, what happens, as in the case of Hyman's and Child's experiments (1) (2), is that the amoeba ingests the potassium permanganate in the same manner and with the same alacrity that it ingests the ordinary food particle in its environment. However, the consequence for the amoeba is not sustenance but a vital staining process which produces a dead amoeba, beautifully stained.

I would further say, in the case of a human organism reaching for food, it is food that is being reached for. The organism has some notion of what it is that it is responding to. In short, the equivalences in the environment for the human organism are not at all the same as are the equivalences in the environment for the unicellular organism.

If we take the problem of migration, we will find that the salmon migrates, it is true, but basically the migration of salmon rests upon certain endocrinological changes which occur in it, which endocrinological changes produce certain pigmental changes, which pigmental changes are related to a sensitivity to light, which sensitivity to light produces certain behaviors at a later point; then further endocrinological changes create a change in the metabolic rate of this organism, increase its need for oxygen, and drive it in the direction of fresh-water streams once again; so that the behavior of this organism is based in large part upon biochemical temperature and endocrinological changes in its own internal structure.

The behavior of the human nomad, for example, as our anthropological colleagues have sometimes told us, is not based upon such changes primarily but, rather, is based upon the ways in which the food substances or the means of sustenance of the community are produced, and the behavior of the nomadic individual is based upon these technological problems of his existence rather than upon the specific nature of biochemical changes.

Now, then, if we start from a broad, general, end-result category, we must be especially careful to seek out and to enunciate the dissimilarities and the discontinuities of process which may underlie the described behaviors. With this caution in mind, we can perhaps formulate the problem of communication between animals. Let us start with a broad definition of »communication« as the effect of the behavior of one organism upon the behavior of another organism. It is clear that this category is so broad as to become almost meaningless and almost useless as a means for differentiating communication function from so-|called »other« functions of organisms. But I think we [136] can use it because it leads us to a consideration of the ways and the levels at which animals interrelate, one with another.

I should like to start with the story of the scallop and its relation to the starfish. The scallop, as you know, is one of the food sources for the starfish. If a starfish is placed in the environment of a scallop, it rather quickly elicits a flight reaction on the part of the scallop. Is the starfish communicating with the scallop? Well, most certainly the scallop is fleeing because of stimuli that have impinged upon it and which have emerged from the starfish as another organism. The behavior of the scallop has been affected by the behavior of the starfish. The starfish has now come within a given range of the scallop, and within the framework of this broad definition we would be forced to call such behavioral interrelations communication.

However, if we do what several investigators have done and take a starfish and boil it and make a soup, a nice rich starfish soup, and then extract this soup just a little bit longer and have a very strong broth of starfish and place this starfish broth in the environment of the scallop, the scallop at once exhibits the flight reaction. Under these conditions, are we justified in saying that the broth of the starfish now is communicating with the scallop, and how does soup communicate with the individual? I would submit that under certain social conditions, an individual may communicate with soup, but that is, again, another problem.

I think that what we find here, then, is that at the level of the lower invertebrate organisms there have evolved in the course of the history of animals certain sensitivities, which sensitivities permit of the survival of organisms within their customary environments.

This should not surprise us. It should not surprise us at all to find that the scallop has certain adaptive characteristics. It should not surprise us, because with these characteristics not present there would today be no scallops. What we find here, then, is that such a variety of animal as the scallop has survived because it possesses such sensitivities that throw it into flight reactions when chemical characteristics, such as the chemical characteristics emitted by the starfish, come into contact with its sensory apparatus.

There we have, in an elementary sense, a variety of interrelation between animals, but I think we will see that there are a number of things absent from this that we customarily assume as existing in such phenomena as communication, particularly at the human level. I should like to give another example.

In several recent studies on the mating behavior of mosquitoes (3), Roth, among [137] others, has found that the mating response of the male | mosquito may be elicited; in fact, it tends in nature to be elicited by certain tonal vibrations or vibration frequencies that are transmitted by the female mosquito of its species or of certain related species. Now, the mating behavior of the male mosquito is affected, that is to say, it is brought into play and into action by the noises which are transmitted by the female mosquito, but would we say that the female mosquito is communicating with the male? Perhaps some of us would. I, for one, would not, because in Roth's experiments he showed that if you present frequencies of vibration by means of tuning forks or other arrangements, you elicit the mating behavior of the male mosquito now to the tuning fork or now to certain other objects located in space, none of which has any specific relation to the female mosquito, except in that the tonal frequencies transmitted are those characteristic of the wing-beat frequency produced by this female mosquito.

Most recently, we have been amazed and entertained and enlightened by a series of researches on the communicative interrelations in insects that have been reported as the result of the work of Professor von Frisch on the behavior of bees (4). Dr.
at Cornell, in his introduction to von Frisch's volume on the bees, claims that no sensible scientist or no competent scientist ought to believe what von Frisch has reported, upon first reading. He goes further and implies in his foreword that these phenomena as reported by von Frisch reveal a kind of complexity of relationship between insects that should force the students of animal behavior to a complete revision of their conceptions. This idea is echoed in a more explicit form by Dr. Thorpe in England (5). He, upon viewing von Frisch's work, claimed to have been amazed, astounded, entranced, and a number of other things, and even went so far as to visit von Frisch's field station to have reproduced for him certain of the behaviors that von Frisch reported for the bee. Upon his return to England, he wrote an extremely enthusiastic article, in which article he claimed that the behavior of the bee was so complex that it should revolutionize our conception of the evolution of behavior as such. He felt that this complex behavior was so complicated that it could only have been subserved by something that we might call intelligence; and, again, I shall use Professor Richard's inverted commas, except that I use a different direction. Thorpe always puts this in commas, as does Griffin. But putting it in commas is not the important thing. The important thing is, what are these people getting at?

The assumption that underlies their thinking is expressed in other articles of Thorpe's in which he gives the animals at the level of the insects the capacity for insights; that is to say, complete reorganizations of their experience and an understanding of the world in which they live – really a kind of conscious understanding. | [138]

Now, I should like to point out that whenever a complex interrelationship among insects is described, and before it is fully explored, there are always individuals who will leap upon these phenomena and use them to advance a notion of high levels of intelligence and high levels of psychological function in insects. This is not unique to Griffin and Thorpe. The same situation existed in regard to the interpretation of the extremely complex behaviors of the army ant, before Dr. Schneirla (6) (7) performed a whole series of investigations that revealed that these complexities of behavior are not based upon any high-level intellective function but, rather, are based upon very simple insectlike features of psychological function – more about which a little later.

Briefly, I wish to tell you about von Frisch's experiments. I am sure that many of you are familiar with them, but perhaps a brief review would be in order. What von Frisch found – and this is what is most important for communication – is that the activity of the finder bee, that is, the bee that first finds food, when this finder bee returns to the hive, influences the amount of activity of secondary bees who go to the food in a variety of ways. First of all, it influences the distance which these bees will tend to fly in their search for food; second, it influences the kind of food materials that they will tend to be responsive to; third, and most astonishingly, the behavior of the initial finder bee influences the direction in flight – and this is what is apparently most astounding – of the secondary bees.

The first problem is relatively simple to deal with and to describe. The finder bee, when it is eating food, may accumulate certain of the chemical characteristics of this food, either on its abdomen or in its »social« stomach. Insects are peculiar in having a »social« stomach. I am using Forel's designation there. By »social« stomach we mean a stomach that regurgitates easily. It is a stomach that regurgitates to antennal contact upon the part of another insect. Thus, when the finder bee comes back with its »social« stomach gorged with nectar, it enters the hive and begins to have antennal contacts (which have a history of their own) with other bees, other worker bees in the hive. These young ladies in turn stimulate the antennal receptors of the finder bee. The stimulation of these antennae now produces regurgitation almost reflexly in the finder bee, and a droplet of nectar is transmitted from the finder bee to the secondary bee in the hive. This pattern of behavior is characteristic not only of the bees but also of a variety of insects, such as the ants and others.

The nectar itself, I should point out, has chemical characteristics. Thus, for example, if the nectar has been obtained from a given kind of flower, it contains not only the sugars that are in solution at the | base of that flower, but also some of the aromatic [139] substances in solution that characterize this flower; in other words, the nectar itself is scented, and the secondary bee now is not merely being stimulated to a higher level of activity by this food exchange, but also by a second kind of activity that the finder bee engages in, a dance which is a kind of figure-8 type of dance, when the food is near the hive. When such secondary bees leave the hive, they will tend to fly around, and if there is a variety of different flowers present, for example, if you have phlox in one place and another kind of flower in another place and the finder bee has been drinking at the phlox, then the secondary bees will move toward the phlox patch and will there ingest the nectar, after which they will return; so that two things now occur: first of all, the finder bee excites other secondary bees to go out; second, the finder bee presents certain chemical stimulation to these secondary bees, which chemical stimulation leads to selectivity in the subsequent behavior of the bee.

It has been shown by von Frisch that you can perform this sort of thing experimentally, so that if you have two dishes of sugar water, one of which is scented with lavender, for example, and the other of which is not scented with oil of lavender, and you get the bees from one hive to feed upon the lavender-scented materials and from another hive upon the unscented materials, the finder bees that drank from the lavender-scented material will transmit to the secondary bees in that neighborhood the scent or the odor of lavender, and there will be preeminent feeding of such bees at the lavender dish rather than at the unscented dish.

SAVAGE: Do you mean literally unscented? Can bees discover a literally unscented dish?

BIRCH: Well, you can do a number of things. You can shellac over the sensory organs involved in olfaction, both at the tarsi and at the antennae tips, and under those conditions you can guarantee they cannot be responding to the olfactory materials. There is a variety of experiments here, and I don't think there is any real question on that point at all. I had thought initially that Professor von Frisch was going to be here to discuss this, but that was impossible.

While these things are interesting enough in and of themselves, certain further things happen to the behavior of the secondary bees and to the behavior of the finder bees in connection with the foraging act. It has been reported by von Frisch, and both Griffin and Thorpe have said, that they have repeated the following kinds of observations: If the food dish is within a hundred meters of the hive, then the behavior of the finder bee is characterized almost entirely by this figure-8 type of ground dance. [140] However, if the food source is more than a hundred | meters away from the hive, then a peculiar change appears in the activity of the finder bee as it returns to the hive. It engages now not only in this round dance but also, between the two segments of the round dance, it engages in what von Frisch has called the »waggle« dance. It is a relatively straight-line run which has a given directional orientation, and then the other half of the round dance is engaged in.

A second point appears: the rate at which this combination round and wiggle-waggle dance is engaged in, that is, the number of times it is engaged in per second, is roughly inversely proportional to the distance of the food source from the hive; that is to say, that if the food source from the hive is at two hundred meters – or, rather, if the bee is two hundred meters from the hive, it makes some twelve or fourteen of these per unit of time. If it is at a distance of a kilometer or five hundred meters, half a kilometer, from the hive, the number of turns in its activity goes down; so that per unit of time it may now make only four, five, or six of these activities in the time period.

BIGELOW: What is the method of observation? Photography?

BIRCH: No, the method of observation is direct observation on a glassed-in hive, laid out so that all of its combs can be seen through glass.

VON FOERSTER: And also motion pictures.

KLÜVER: Perhaps you should point out whether these observations refer to dances on vertical honeycombs.

BIRCH: Well, some of the observations, yes; that is why I didn't say that, because under certain conditions that we shall come to in a minute, these hives are placed vertically, but under other conditions, the hives are placed horizontally.

MCCULLOCH: May I describe the setup for a moment? I saw it not long ago. He uses a glass-sided box which can be rotated from the vertical to the horizontal position, so that he is able to observe and to photograph the behavior of the bees in the hive.

BIGELOW: Does he actually take a long string of motion pictures or something?

MCCULLOCH: Yes.

BIGELOW: And take a given bee and put an X on it and follow its path and gather statistics on it?

MCCULLOCH: Oh, yes.

BIRCH: Yes, he labels each of the bees in the feeding situation by putting a small drop of quickly drying lacquer on them, or pigment and shellac, and under those conditions it is possible, through a numbering system that he has developed on the basis of dots on different portions of the insect, to identify the insects that are involved.

[141] BIGELOW: In other words, there is no reasonable doubt about his data? |

MCCULLOCH: None.

VON FOERSTER: May I refer at this point to a formula I tried out once and which correlates the wiggle-frequency f in wiggles per second with the distance S in meters of the feeding-place from the hive. The formula has a hyperbolic form and reads:

$$(S - S_0) \cdot f = c$$

It fits astonishingly well for different experimental results as far as I could find those in some publications. The first constant S_0 seems to be the minimum distance for which the bees decide to use the wiggle-dance communication-technique. The other constant c has the dimension of a velocity. If you evaluate from the experimental results the numerical value of this constant, you come to a very surprising figure: c becomes about 330 meters per second which is very close to the velocity of sound in air. It seems to me that the velocity of sound may be for the bees the same universal constant as is the velocity of light in man's electromagnetic philosophy.

SAVAGE: The experiment should be carried out in different atmospheres or something, to see if there is any truth in that.

VON BONIN: How do you choose the S_0?

VON FOERSTER: In the formula above, S is the distance from the hive to the feeding-place. But actually the bees begin to interpret distances of the feeding-place with the wiggle-dance when the feeding-place is a minimum distance away. This minimum-distance I call S_0 and has the magnitude of about 100 meters, as it was pointed out before.

BIRCH: While von Frisch, Griffin, and Thorpe have all said that this waggle dance is simply appearing full-blown at a hundred meters, which gives you your S_0, C. G. Butler (8), for example, of the Department of Bee Behavior in England, has made thousands of drawings of the bees' dances after having fed them at different distances from the hive, and he finds that this waggle dance does not appear full-blown out of nothing but, rather, that as the distance increases from 0 to 100 meters, there is a greater and greater separation between the circles and they become more and more distinct, so that gradually what emerges is two separate circles, which more and more are joined by a straightaway, so that the S_0 does not represent a true starting point, but simply represents that nodal point at which the wiggle-waggle phase of the dance is most unmistakably present. So there is no question about the validity of using S_0 at that point, except for purposes of computation at this moment. Therefore, we have the formula $(S - S_0) \cdot f = 330$ meters per second.

That is just one phase of the problem that requires much fuller investigation, of course.

While these things are amazing enough, I think that a consideration | of the wiggle-waggle dance yields certain other interesting results. It is found usually that the hive of the bee is in the vertical plane, so that when a bee is standing on the hive it is standing vertically. In flight the bee is moving in the horizontal plane, in the main. The interesting point is that these bees indicate direction or provide a basis for directionality in horizontal flight to the secondary bees on the basis of the direction of the waggle dance itself. This is done in terms of a conversion of horizontal directionality which is related to the position of the sun into a system of vertical directionality which is related to the gravitational field, where the gravitational pull apparently becomes the equivalent in direction indication to the sun itself. [142]

You can begin to examine the bees' relation to the sun, as von Frisch has done, by tilting the hive from the vertical to the horizontal, under which conditions there is no gravitational component to which it may orient. Under these conditions, if no light is permitted to enter the hive and the bees are observed through red glass, that is, red light is permitted to go in, we can see it, but it is outside the visual spectrum of the bee, then you find that these dances simply become randomized and have no specific direction. However, if sunlight is permitted to enter the hive, then these dances at once are oriented in direction and begin to be pointed toward the food place as it is related to the direction of the sun itself.

But what would happen on a cloudy day or on a day when the conditions are such that no direct vision of the sun's position in the heavens itself would be observed? Well, von Frisch finds that the bees are still capable of maintaining this kind of directional orientation as long as a small patch of blue sky is visible to them, and by »blue« he simply means clear. I don't know whether it is blue for the bee.

This led him to extend certain notions that had been initially developed in connection with the study of ant behavior; namely, a notion that the bee may not simply be responding to the directionality of the sun's rays as such, but may now be responding to the polarization of light as such, and there is a relationship between the directional position of the sun and the polarization matrix which may exist; so that apparently the bee now is responding not to the sun itself, but may be responding to the polarization characteristics of light. This can easily be tested.

If one were to take a piece of polarized glass, for example, as von Frisch did, and place it over this opening, and rotate it in this opening, then, if the bee is responding to the polarization characteristics of the light, the direction of pointing in the waggle phase of the dance should be changed; and this is exactly what does happen. The bee [143] then is responding and maintaining its orientation toward the food place or | toward the place at which it has been fed, in relation to the polarized characteristics of the light, and it is then transforming this into activity in relation to a gravitational field.

When the finder bee engages in its wiggle-waggle dance, it excites other bees in the colony, and these bees, too, now begin to go into activity. They begin to engage in their so-called »dancing,« and they tend to orient toward the abdomen of the finder bee and to direct a straight-line movement of their own in relation to the wagging abdomen of the finder bee itself. Thereafter, when they reach a certain stage of excitation, they leave the hive; and now, rather than moving around in a random distribution, in all directions around the hive, which is the case when merely the round dance is engaged in, the bees now take off in the direction pointed to by the wagging bee. Well, certainly, here we have, in a complex animal social organization at the bee level in the society, interrelations between bees, and we have the behavior of the secondary bees consistently affected by the behavior of the finder or primary bee.

But does this necessitate a revision of all our concepts of animal behavior? Does this mean that we now have to consider bees as geniuses? Does it mean that we now have to impute certain high levels of psychological function like abstraction or what have you to the bee itself? I would submit that we do not.

KLÜVER: We should have more information on the mechanisms involved in imparting information in the dark hive, that is, in the absence of polarized light and under vertical conditions.

BIRCH: Yes, under vertical conditions.

KLÜVER: Perhaps you have some ideas as to what really happens.

BIRCH: That is what I meant when I said that the polarization field can be converted and is converted into the gravitational field.

RIOCH: I wonder if there are data on several other points. Does Dr. Birch have data upon the number of bees it takes to find the food after the finder bee comes back? That is, can one other bee respond to the finder bee? Then, another question: Does the finder bee do the dance if there are no other bees in the environment, that is, if the hive has been emptied? Is it the hive or the presence of other bees that evokes the dance, or is it something like the distance back from the food that evokes the dance? And also, are there any data that give an idea as to how long the finder bee can remember the direction? That is, if one were to put a tarpaulin over the hive so that the bee had to fly without any polarized light to give him direction for any length of time, how long could he remember that? You see, you have this fantastic business in insects of types of memory. Your bee knows its own hive on the basis of the one experience of coming out of it just after he has | been hatched, or at least so I understand. [144]

McCulloch: That's right?

Birch: That is not so, but go ahead.

Rioch: Well, then some time I should like to hear some of the data on it.

Birch: I will agree that it is fantastic.

Rioch: But if this bee has to fly under trees or something for a while, what distance can he fly and still remember which way the polarized light was? Those are the questions I had.

Bigelow: My questions are very much in the same category. Mine concern exactly what response the other bees in the hive give when you perturb the directional axis of the dance by polarized light or whatever means you wish, whether the bee himself shows any disturbance of his normal pattern in finding his way back; and whether anybody has ever carried out an experiment of keeping a bee in the air by some subterfuge for a certain period of time corresponding to a different distance of flight, to see whether or not this affects his dance régime into Mode 1 or Mode 2?

Brosin: I have an extension of Dr. Rioch's question. Is there more information about the social organization or the »needs« of the bees that alter the relationship between the finder bee and the subsequent behavior of the secondary group?

Pitts: I think there is one factual point I might make about Dr. Rioch's question. If you look at any square centimeter of sky and measure the direction and the degree of polarization in it, you can tell where the sun is. In very rare cases there may be two positions for the sun that would be compatible with the amount of polarization in any given place; but those are rare cases. In general, you can tell where the sun is by looking at a piece of sky, however small, in any direction.

Birch: If you are sensitive to polarization, yes. Certain humans are, in small part.

Pitts: No, I mean the data are determinable in the apparent field.

von Bonin: There is one more question. You talk about the finder bee coming back. Suppose the next bee goes out; does it also go through the dance when it comes back? Does it become a finder bee by virtue of –

Birch: Oh, yes.

von Bonin: And some more come up?

Birch: Oh, yes. The finder bee is not a special individual. It is simply the bee that has been to the food.

von Bonin: It doesn't matter that it is the first one, but anyone that comes back is so proud of it that he has to tell about it?

Birch: Well, I wouldn't put it that way. Anyone that comes back | engages in certain [145] behaviors that excite certain subsequent behaviors upon the part of the other bees. Now, I would be proud of that, but I don't know about the bee.

Savage: Well, does it not come to pass that so many are coming back that the pattern is broken up, that there are so many dances going on at once that nobody has an audience?

Birch: A hive is a pretty big place, and there are many dances that can go on at any point in time. There may be some fifty thousand individuals present, or more, and certain groups of them will orient with regard to one of these finder bees, and others to another one.

Now, may I answer these questions as they have been given? In the first place, about the memory of insects and orientation toward the nesting place itself: you find that this orientation is something which the bee or the ant or any of the other social insects has to learn after it emerges from its incubation state. When it appears as a nymph, it makes small migrations and it begins gradually to increase this distance of activity from the hive; it begins to orient toward specific features of the environment about the hive or about the nesting place. For example, there are many studies at the level of the ant, where you can take a tree, for instance, that previously bore a given relation to the nest and move it, dig it up and move it a certain distance, or establish an artificial potted tree, so that you can move it more easily. Under those conditions, you find that when the insects begin to come into the region of the hive itself, or the nest itself, they begin to orient in relation to those landmarks and those features rather than to the specific position of the nest itself; so that they do learn to respond to specific features of the environment. Once they get within the range of those environmental fields, they will tend to return.

RIOCH: Do you have any data on that, as to the number of experiences that are involved?

BIRCH: It has not been studied in a sufficiently systematic way to do that. I can extrapolate, of course, from some of Schneirla's researches on learning in the ant itself, in a maze situation in which the animal must traverse a given kind of maze before it arrives at a feeding place and then traverse a different kind of maze in the return from the feeding place to the nest itself. Under these conditions you find that the learning of the ant is an extremely slow process; that is to say, if you control the laying down of chemical pathways by removing the linings of the maze itself, then it takes the ant many, many trials – forty, fifty, or sixty trials – to begin efficiently and relatively errorlessly to traverse a relatively simple maze, and this learning, further, goes through at least three stages: first, a provisional orientation stage, in which the ant learns not to walk on the glass cover of the maze but, rather, on | sections of the floor and things of that sort; then, a phase in which blinds are omitted in a very piecemeal way. Blind alleys here are not eliminated in the manner that you find in many animals.

[146]

RIOCH: I didn't want to get off into learning of insects, except in this –

BIRCH: That was the question you asked.

RIOCH: No, no. I asked a question about a particular learning. You see, we have been dealing with the problem of things that are learned with one experience, and things that are learned that will require multiple experiences, and things that require particular situations, internal situations in addition to the experience. This is the question I asked: Is there in the insect a certain type of experience which, with one experience, produces a durable change? We know that in the bird there are such situations.

BIRCH: Yes. I would say – well, a durable change in terms of how long? For example, I think that the chemical stimulation of the insect, or of the bee in this situation, produces a change which will persist for a finite period, long enough in a number of cases to lead the organism to a given kind of feeding place. But whether or not this is something which it has learned, or something which provides a residuum of continued stimulation on the basis of the nectar which it has ingested, on the basis of residual chemical effects after its eating, is another question. We don't know whether it has learned that in one trial or whether it is now simply responding to a continuing situation of a stimulating kind.

MCCULLOCH: May I in part answer what I believe is your question? I think the best data on learning from a single experience by any of the hymenoptera is Baerends's work on wasps, the solitary wasp. I refer to the work of G. P. Baerends of Gröningen[!]. There is no question that a single sight of a landmark is enough to enable them to find it again.

RIOCH: Can they forget it?

MCCULLOCH: If that landmark is moved, they don't find it again. This is a matter of something which persists over days or even weeks.

RIOCH: But can they forget it? That is, can they learn another landmark which means forgetting the first one? I mean an organization of the central nervous system, whatever the –

MCCULLOCH: The answer to your question is »Yes, I think so,« but I would have to look up the data.

BIRCH: But with difficulty, if the other landmark is still present.

RIOCH: Because the forgetting is much more important a problem in the central-nervous-system organization.

ROSENBLUETH: The important thing is to find another landmark the next day. | [147]

MCCULLOCH: Will you go ahead to the next point, Dr. Birch?

BIRCH: Yes. Are other bees necessary? Although there has been no systematic examination of that problem, there is some incidental evidence, I think, that other bees are not necessary for the dance to occur. Thus, for example, a finder bee may alight on the entrance portion of the hive and there be stimulated in various ways, and begin to go into a dance on the platform in front of the hive rather than in the midst of the other bees.

PITTS: Be stimulated in what way?

BIRCH: For example, you can tap its antennae.

PITTS: That is, be stimulated in a way that is normally done only by other bees?

BIRCH: Certain of the stimuli may be given, or are usually given, by other bees, but under certain circumstances, too, there can be this spontaneous activity. It would be an interesting question to examine a bee going into a hive that has previously been emptied. It has not been studied in any detail. Like many of these processes, at this moment, what we have is a description of the complex behavior and not any thoroughly detailed analysis of the specific actions and of the specific interrelations between the individual organism and the environment; in other words, we are at a preliminary stage in the investigation of this problem.

ROSENBLUETH: The behavior of the bee is not very different from that of a warrior who gets home and begins to dance by himself.

BIRCH: I would say the bee is very different.

ROSENBLUETH: The bee behaves similarly but looks different.

BIRCH: The warrior behaves differently and looks similar. I would say there is a phenomenal similarity but a difference in –

ROSENBLUETH: They are both excited and stimulated and dance by themselves.

MACKAY: I think a closer analogy is that of a man singing in his bath.

BIRCH: It is again an analogy and, as such, I will accept it – as long as it stays an analogy.

MEAD: But the present data do suggest that you need something of the order of the stimulation of the antenna to set off the dance?

BIRCH: No, the present data do not indicate what you need to set off the dance. It sometimes apparently occurs spontaneously. Under other conditions it is a further consequence of subsequent excitation and things of that sort.

MEAD: Although the usual pattern is the stimulation?

BIRCH: Well, the usual pattern is the stimulation because a bee coming into a hive [148] with fifty thousand other bees present is inevitably in | contact with those other bees. I will grant that. It's awfully hard; it's like not touching your neighbor in the subway rush hour.

KUBIE: The danger of superficial analogizing is very great. Let us suppose that a college lad goes into church and starts to dance –

VON BONIN: Church?

KUBIE: Yes. Perhaps he is doing this to pay an election bet, or as a hazing stunt; or perhaps he is doing it because of a delusion that he has had a message from Rome which directs him to do so. Perhaps he is doing it because of the pressure of an unconscious obsessional-compulsive drive which must be obeyed, even though he knows it is foolish. Evidently the psychological significance of the behavior derives neither from its superficial nature nor from its effects, but from that subtle balance of conscious and unconscious processes which have produced the act.

BIGELOW: Don't we all grant that the analogies now being made are superficial?

KUBIE: Just a second. We may or may not. The point is that –

RIOCH: No, these are absolutely fundamental.

KUBIE: Viewed superficially, the behavior may be sensible or foolish, useless or useful, creative or destructive, good or bad; but in human beings, at least, we know that it always serves several concomitant purposes, some of which may be manifestations of deeply unconscious forces. When, on the other hand, we turn to such an animal as the bee, we can know nothing about the relationship even of his hypothetical conscious purposes to his activities, much less of his unconscious purposes. Therefore when we make any simple analogies between human behavior and that of the bee, we are in danger of getting in hot water.

BIGELOW: It is all right to do it, though, isn't it?

KUBIE: I doubt it. Let us consider so simple a human function as eating. We know that we eat to serve innumerable conscious purposes – out of fear, anger, sorrow, loneliness, or to celebrate. On top of that, we also know that we may eat for a wide variety of unconscious purposes, in the service of which eating itself becomes a purely symbolic, symptomatic act, as much a symptom as a hand-washing compulsion. If we could not talk, we would know nothing of this, all of which can be represented in the simple hieroglyphic or shorthand behavior of compulsive eating, phobic eating, delusional eating, and so forth. What access do you have to any such data for any one of the subverbal animal forms?

BIGELOW: Yes, but it is all right to do it if you know you are doing something risky.

[149] GERARD: All these questions have been directed to different aspects of | complex human behavior, and I think Dr. Birch is anxious to tell us how they have been broken down experimentally. I am sure many of these casual remarks were made partly in fun, to keep the thing alive, but I should like to hear what the facts are.

RIOCH: They were not made in fun.

GERARD: Well, let's hear the rest of the answers, and then we can argue about them.

RIOCH: There is one other question to answer; that is, if there are any data upon the duration for which the bee will remember.

BIRCH: I shall try to answer it. The problem next is: What happens if you restrict the activity of the bee over a period of time? Does it still maintain its orientation? There only the sketchiest experimentation has been done at the bee level, but extremely interesting experimentation has been done over a long period, a long time ago, at the level of the ant – another insect. Santschi, for example, in some of his early experiments, and some investigators working with the driver ants in Africa, as well as

Goetsch (9) in many of his investigations of the ant have shown that if an ant is following a trail – if this ant is a visually oriented ant rather than a chemical trail follower – if you put a box over the insect and keep it in this box, away from the sun, that is to say, away from light stimulation, for a period of time, and then remove the box, it takes off at an angle which is related to its previous orientation to the sun as such. Apparently a continuing process persists, and the ant now is orienting toward the sun and maintaining this kind of orientation over a period of several hours.

PITTS: He doesn't allow for the motion of the sun?

BIRCH: He doesn't know about it. He is now simply responding in terms of his previous orientation to the sun, so that I think this is more important for Dr. Rioch's purposes, I believe, than the notion that the ant may or may not have heard about the movement of the sun in relation to the earth; so that you do have the persistence of such effects over a relatively long period in insects. I am glad you raised that question, because it is especially important, I believe, in regard to the beginning of an investigation of the directional orientation of these bees.

But before I get to that, let me try to answer a couple of other specific questions that were raised. There was something about time of flight. Who asked that?

BIGELOW: I did.

BIRCH: What was the question?

BIGELOW: The question concerns the following conjecture: Possibly the bees changed from one mode of dancing to another on the basis of some physical fatigue, not distance but fatigue.

BIRCH: Oh, yes, I remember the question. Well, there is some evi|dence on that. For [150] example, if we were to assume that the bee is merely responding to distance, then, independently of the time required to traverse that distance, it should give the same temporal sequence of dances. But von Frisch himself has shown that when there are winds present, for instance, the bee behaves as though the distance were greater – in quotes now. When there are tail winds, under those conditions, the bee behaves as though the distance were less; in other words, the bee is apparently responding to its own effort and to the metabolic expenditure that is involved in the movement from the feeding place to the hive.

BIGELOW: You would grant, then, that his energy output is in fact his distance indicator?

BIRCH: His energy output tends to be his distance cue.

PITTS: But which trip – from the flower to the hive or from the hive to the flower?

BIRCH: From the flower to the hive.

PITTS: That is what he measures in time?

BIRCH: Yes, because in arriving at the flower – for instance, if you were to follow the flight of any bee, you would see that it will meander in many, many different directions and eventually wind up at a feeding place. Now, let's say you take a place that is denuded of flowers, a big football field or something like that, and at the end of it establish a feeding place for bees, and at the other end of it (as I myself once did as a sort of check experiment) have a hive. Under those conditions you will find that the bee traverses a huge amount of distance from the hive to the feeding place.

PITTS: It certainly is more intelligent for him to do it the other way.

BIRCH: And his response now tends to be oriented toward the place where the bee filled its belly rather than toward the hive itself. There is a relation –

VON BONIN: And then does it make the proverbial beeline home?

BIRCH: Yes, depending upon the availability of orientation cues.

VON BONIN: The football field?

BIRCH: Yes. He proceeds relatively straightly.

SAVAGE: From what you say, I gather that in the presence of a wind the finder bee systematically deludes all the other bees?

BIRCH: That the finder bee then gives – no, I'm not going to fall into that. That the finder bee excites the secondary bees in such a manner that there is a lesser accuracy in their activity than is the case under more neutral conditions.

BIGELOW: I'm sorry, but isn't it relatively accurate? Because he assumes they are replicas, and he is using a yardstick whose value changes with energy output, but when [151] applied to the same flight-energy process, | in the next bee, it gives the same result?

BIRCH: But, after all, the finder bee now has a tail wind to the secondary bee, so you begin to have compulsory action.

ROSENBLUETH: I want, for my information, to know whether you think »excites« belongs to a lower category than »informs.«

BIRCH: Oh, yes.

RIOCH: The bee must be able to make allowance for head wind and tail wind. In Maine an old method of finding a beehive in the woods is to use a trap with two compartments in one of which is a bowl of sugar. After getting a bee in the trap, the hunter walks in the direction from which the bee came, and then releases it. When it returns it is again trapped and the process repeated. In this way the bee leads the hunter to the hive. The bee must make some judgment as to distance on each trip.

BIGELOW: This method would result in finding the beehive if the bee just knew which of the two 180° sectors of a circle is correct.

BIRCH: The other factual question was: What are the needs of the bees? The problem of what different bees would begin to do under controlled differences in any motivational state or in food-deprivation state is a question yet to be investigated. It has not been fully investigated. We find some small amount of evidence. We know this, though, that the giving of nectar by one bee to another functions to excite it, and to prod it into activity.

KLÜVER: This is perhaps the time to say a few words about the behavior of bees in New York City. In the experiments reported by von Frisch, it seems remarkable that the upward movement during the straight component of the wagging dance corresponds to the direction of flight toward the sun, and that the bees, aroused by the dance on the vertical honeycomb in the dark hive recognize the angle of the dance relative to gravity. For instance, they recognize, if the straight portion of the dance is pointed 60° to the left of the vertical, that this means that the feeding place is located 60° to the left of the sun's position in the sky. This implies a truly remarkable interrelation of gravitational and optical factors. More than twenty years ago, Selig Hecht, in the Laboratory of Biophysics at Columbia University, wanted to measure the visual acuity of the honeybee, but decided not to make use of such conditioning or training methods as had previously been used with such great success by von Frisch and others. He looked for a method involving no training of animals. It is known, of course, that many animals respond to a sudden movement in their visual field. If the visual field consists of a pattern of dark and illuminated bars of equal size, an animal will respond to a displacement of the whole field only if it is able to distinguish the components of [152] the pattern, that is, if it can | resolve the black and white bars. If it cannot do so, that is, if the whole field appears uniformly illuminated, it will not respond to a movement of the field. In Hecht's setup the experimental bee was confined in a glass compartment. Light reached this creeping department from the bottom after first passing through two pieces of glass held in a frame; namely, an opal glass plate and a transparent plate with opaque and transparent bars. In this setup it was possible to compute the visual angle from the dimensions of the bars and the distance between the bars and the center of the bee's eye. Hecht hoped, of course, that a sudden movement of the black and white stripes of the visual field below the glass compartment in which the bee was crawling would induce a sudden change in the direction of the crawling. Unfortunately, this did not happen at all, and I well remember that Ernst Wolf, the collaborator of Hecht, became almost desperate when he tried in vain, month after month, to elicit a response of the bee by moving the visual field. However, all troubles were over one morning when he hit on a slight modification of the experimental setup. This modification consisted in simply tilting the bee's creeping compartment and the movable black and white stripes of the visual field at an angle of about 30°. The moment the whole system was tilted at an angle of 30°, the bee in the glass compartment, because of its negative geotropism, tended to crawl upward in a straight line, and deviations from its linear progression were easily observable each time the visual field below the glass compartment was moved. In other words, Wolf finally found an elegant way of measuring the visual acuity of the bee by taking into account both gravitational and optical factors (10).

BIRCH: This shows the importance of the gravitational factors, and other directional indicators, and one that in some way is reciprocal and equivalent to the visual system. Just what that relation is will have to be determined.

KLÜVER: I am sure it will be necessary to study not only certain aspects of the behavior of the dancing or sender bees in greater detail but also – and this is probably even more important – to analyze the behavior of the receivers or of what you have called the secondary bees.

ROSENBLUETH: What kind of labyrinth, or the equivalent, do bees have?

MCCULLOCH: May I answer that? The bee has a material which is apparently piezoelectric, and which is capped with a smallish weight. Work on similar piezoelectric devices, particularly in fish, is now under way in Gröningen[!], in de Vries's laboratory.

ROSENBLUETH: With electric signals? I mean, is it converted into electric signals?

MCCULLOCH: It is probably not very dissimilar from all of our devices | for picking [153] up mechanical motion or accelerations.

KLÜVER: Dr. McCulloch, since you have just been to Groningen, I am wondering whether you can report on any progress made by de Vries, one of the physicists in Groningen. In 1844 Haidinger discovered that the human eye can detect the direction of polarization of linearly polarized light. In October, 1950, de Vries told me that he was engaged in some experimental work on the properties of the human eye with respect to the so-called »Haidinger brushes.« I wonder whether he has any new data bearing on the interpretation of these polarization brushes

MCCULLOCH: No, and I don't think they were far enough along with it then to be ready to say anything. That was just a few months ago.

BIRCH: I would like to deal analytically with some of these fascinating phenomena that have been revealed by von Frisch. In the first place, I should like to deal with the problem of how the insect maintains a kind of orientation with the polarization characteristics of light rather than to the specific light intensity directions themselves.

This question von Frisch answered (4) by analyzing the nature of the eye of the insect, and by pointing out that this compound eye represents an ideal kind of analyzer for a polarized system; so that if you were to take polaroid glass and arrange a model which would, in a simplified way, represent the directions to which these various eye units of the bee are pointing, you could then begin to analyze the way in which a given bodily orientation toward the direction of the sun could produce different effects that are the result of the polarization characteristics of the light. If you take such a model, as von Frisch did, made out of about six units of polaroid glass and then rotate these in the direction of light, much as an artificial eye may be rotated, then you find that what is projected by such a system is a pattern of brightnesses characteristic of the direction in relation to the polarization phenomenon; so that if the bee now turns in a different direction to the sun, a very different brightness pattern is established in its visual field. Its maintenance of visual orientation, then, is based in large part, or may be based in large part, upon the existence of such brightness patterns, and differential brightness patterns, depending upon the orientation toward the polarized features of light.

The big and key question is: How in the world does the bee translate this kind of orientation, based upon the polarization features of light, into the gravitational field itself? When it is in the hive, the hive itself is quite dark; there is no light entering the hive, and you can observe only if you create artificial conditions in which you look at the bees through red glass. Under these conditions, it is found that these animals translate the direction maintained toward the sun into a set of | directions related to the gravitational field. [154]

Frankly, I have no specific answer to this question, nor do I feel that von Frisch does or Griffin or Thorpe. However, that we have no specific answer to the question is not, to me, a sufficient basis for assuming that such a translation could only be able to occur on the basis of some higher-level intellective process wherein the bee understands the relationship of gravity as a directional feature to polarization as another directional feature, and so on. I think that such a position can be taken only if we ignore all the other kinds of finding about the essential simplicity of the behavioral process in insects.

I think that a productive line of inquiry here would be the examination of what the intersensory equivalences are in an organism such as the bee. To what degree does an orientation process which is established by chemical means or by light means or by gravitational means become one which the organism can now engage in independently of the specific kind of receptor system which is being stimulated?

In that sense, then, I would begin to ask the question: How equivalent for the bee is a visual stimulation with a chemical stimulation with a gravitational pull of various kinds? What is the way in which such different sensory inputs have for the bee the same effect upon it? That would require a detailed and minute investigation of the sensory physiology of the bee, which investigation has not yet been engaged in in this form. There are certain items of investigation, but our information certainly is incomplete.

The reason I say that I would move toward such an examination of the problem rather than toward an interpretation based upon the notion that the bee now knows what the light direction is and then knows what gravitation is like and then tells the other insects what is going on, is that when any complex insect phenomenon has been investigated – and I refer here particularly to Schneirla's research on the raid of the army ant, which is an extremely complex pattern involving the development of an expanding feeding front and the development of ancillary flanking movement, so called, all of which have an extremely useful function in the survival of the army ant – when these are analyzed, they are found to be based upon the characteristics of sensory organization that cause this ant to do the most »stupid« kinds of things under other conditions of environment. Thus, for example, if, in the laboratory, you take a bell jar and turn it upside down and let a group of army ants begin to walk around it, and then lift up the bell jar, what happens is that you have a circular mill established, which continues until either these insects die or until dessication factors force a breakup. Therefore, in the army ant, the enslavement of the ant, in a sense, to the chemical features of stimulation is such that this same enslave|ment under given eco- [155]

logical conditions produces the complex kind of behavior that we find in a swarm-raid formation and in bivouac formation and other things, and, under different ecological conditions, produces just this same kind of self-destructive behavior. We find this happening not only in the laboratory but also in our own kitchens.

I had this happen with a group of ants that was closely related to the Ecitons. I was living at that time in a town in Florida. To the discomfort of my wife, a circular mill was established on the ceiling of our kitchen. She wanted to break it up, but I spent a whole day trying to observe this mill, so we had a rather difficult condition for a little while. But the features that produce this milling are the same features that produce the stereotyped kind of performance that really characterizes the insect as an organism.

With this, I should like to move away from the insects for a little bit and begin to deal with some of the problems at the vertebrate level. Of course, we could stay with arthropods all day if we wanted to, and begin to discuss such things as the so-called »display phenomena« in spiders and things of that sort, but I prefer to make this presentation synoptic rather than inclusive.

At the vertebrate level we have a number of interesting phenomena in the lower vertebrates, that, under a general heading, could be classified as communication. Of course, almost any of the mating behaviors of the vertebrates represent a way in which the characteristic activity of Animal A affects the characteristic activity of Animal B, but let us try not so much to find examples of instances in which the activity of Animal A influences the activity of Animal B, but, rather, to find the kinds of examples which permit us to understand what the basis for the activation of B by A happens to be. I should like to deal with a couple of phenomena in fish, for example.

We have the phenomenon of schooling in fish, and the phenomenon of milling in fish. Various fish will begin to establish schools, and still other fish will begin to establish mills, like the herring or the mackerel. In the schooling behavior of fish, the investigations of Breder and Nigrelli (11) (12) (13) (14) primarily have indicated that this behavior is related to the visual system of the fish, as such; that it is related to the kind of visual angle which the fish has, and to its directly determined responses to certain kinds of visual stimuli. There is a certain optimal position of visual fixation on objects between Fish A and Fish B and Fish C, such that a change in the distance between them produces a distortion of image, and the fish then tend to maintain relative positions, which are positions of relative optimal fixation.

If you have these fish in an aquarium and have a school going and turn out the lights, the school breaks up. If you turn the light on again, | you see the fish scattered [156] here, there, and everywhere, and then you begin to see them responding to the visual stimulation of the movements of other fish, and they begin to establish a spatial relation of a neat, orderly kind, a repetitive kind, in a sense, which we refer to as a school. Therefore, the schooling in fish in itself is a relatively simple phenomenon in which fish whose behavior is almost completely directly determined in a stimulation sense – a given stimulus is presented and the fish will tend to respond, if it has given intensity characteristics – now form these schools on the basis of their sensory situation rather than upon their liking for other fish or anything of the sort. You can establish schooling formation or positional relations, therefore, by putting other kinds of objects in the aquarium, and by moving those along so that equivalent stimulation may be given to the fish.

With regard to the phenomenon of milling, the establishing of circular mills in fish, Parr (15) developed a hypothesis which I think is still the best for consideration of circular mill formation in fish. He refers it to the problem of fixation, again the problem of distance, and to the problem of the panoramic visual field of most fish, so that in the schooling fish you have one eye that looks to the right and another eye that looks to the left, and the twain do not encompass the same portions or very much of the same portions of the visual field. This fish [indicating on board] is responding to a visual object here, and this fish to a visual object here, and the mill is established in the sense of maintaining a continuous relationship in space and in fixation, again, between Fish A and Fish B; so that the phenomenon of schooling and the phenomenon of milling simply represent two different forms which can occur as a result of the same characteristic features of the sensory organization.

PITTS: When you talk about the position of maximum fixation, I am not quite clear about that, as to exactly what pattern of stimuli on its visual field –

BIRCH: I didn't say »maximum fixation.« I spoke of optimal fixation and clarity.

PITTS: Well, what does it do to try to maintain itself in its visual field?

BIRCH: Apparently it maintains, or functions to maintain, a sharpness of vision.

PITTS: Of a single fish, or of several?

BIRCH: Of single ones, usually.

PITTS: It tries to get the next fish in a given position in its visual field and in focus?

BIRCH: That's right. A very interesting experiment suggests itself. It would be a direct test of Parr's hypothesis. What would happen if the fish were blind in the right eye and [157] the left eyes of all fish were | used? Well, according to Parr's hypothesis, what you would expect to find would be mills going in one direction from the left-eyed fish and mills going in the other direction from right-eyed fish, and if you were simply to use shellac or other things to blind them, and then reverse the blinding, if Parr's hypothesis is again correct, then you would expect to find a reversal in the mill formation. It would be a neat experiment and easy to do. It just hasn't been done. It suggests itself quite naturally from the phenomena and from the proper explanation of the phenomena.

A second kind of behavior in fish is one that customarily is referred to when we speak of the way in which behavior of Fish A affects the behavior of Fish B. This is what happens in some of the fish that »establish« nests or regions where they deposit eggs, and so on. Some of these fish, as Tinbergen has pointed out, and as others have pointed out, the male fish, by its orientation toward a given place, now causes the female to respond and to thrust her head into the given region. This occurs even when females and males have been entirely separately reared, and it also occurs when such organisms have never had any chance to learn specifically to respond to a nesting place as such.

You find that this can be done with artefacts, too. If you take a small piece of wood and point it in the direction of the nesting place, the stimulus orients the female fish and she again begins to thrust her head in there. It need not be a nesting place, I should like to point out to you; you could take the stick and point it at a side wall of an aquarium, and the fish would begin to bang its nose against the wall of the aquarium. Or you can point it down or you can point it up, and the orientation that is established through this visual stimulus is persisted in within limits by the fish, independently of what the specific environment is.

Of course, under natural conditions this has positive survival value. It has survival value because it does tend to orient the animal toward a given kind of nesting place where certain spawning will take place. But I think it is clear from these little experiments that the male fish is not doing this because it wants the female to look there. Rather, it is doing this and the female looks at it. What we have here is a very interesting example of the way in which, in the course of evolution, survival has been promoted by the appearance of certain characteristic behaviors rather than an example of the nice intentional behavior of a lower vertebrate with regard to its mate.

I should now like to jump away from fish for a minute. Because of the lateness of the hour, I will not deal with bird display – song and things of that sort, and I will skip the reptiles and the amphibia for the moment, although I should like to deal in detail with Armstrong's analysis, for example, of bird display (with which I disagree heartily), and with Tinbergen's analysis of bird display (with which I disagree | even more vigorously, if that were possible). I don't think I will take the time to do that here, because it would be expressing more a pet peeve of my own than dealing with the problem of communication in the vertebrate series. [158]

If we skip all the way from the lower vertebrates to the mammal, and particularly to the primate mammal, we begin to find certain very real changes in the basis for interbehavior in animals. In the first place, mammals in their relation one to the other generally have a characteristic which is not present in many of the lower vertebrates, although it begins to appear in birds a little bit; that is, that the behavior of the sender animal or the transmitter animal is aimed or directed. Second, it is aimed at Animal B under certain rather specific conditions. Third, the activity of Animal A in regard to Animal B is something which does not simply exist in the structure of the organism; that is, it is not something that is simply built into the organism or wired into it or whatever term you want to use, but it is something which the organism has to acquire in the course of its relations with other organisms or with its environment. In other words, we begin to move from what I would call a phylogenetic kind of determination or an evolutionary kind of determination, in the historical sense now, of mechanisms of interrelationship, to what I would call an ontogenetic development of a communicative relationship; that is to say, when we begin to deal with the higher mammal, and particularly with the primate, we begin to deal with learned patterns of activity that have been acquired by the given animals in the course of their existence. We begin to move toward a condition which is more homologous with the kind of thing we mean by communication between individuals, such as ourselves.

What we mean, essentially, is this: the animal's behavior tends to become directed; second, in this activity, certain content is transmitted; third, this content predictably elicits for the organism sending it certain kinds of activity in other organisms, that is to imply, the organism that is now sending has an expectation. It anticipates a behavior upon the part of the other organism. Further, certain of these activities become conventionalized, and they begin to take on meaning in the specific life histories of the specific organisms that are being dealt with. Thus, for example, if we examine some phases of the behavior of primates, I think the studies that were performed by Crawford and by Nissen on co-operative problem solving in the chimpanzee are extremely interesting in this connection (16) (17).

First, the two chimpanzees learned to pull rather light boxes and obtain food from them. The next task that they were given, after a varied training series – I am skipping a lot of steps here – was one in which they were placed together and given a box that was too heavy for a | single chimpanzee to pull. Under those conditions, the first animal pulled at the box and didn't get it in. The second animal pulled at the box and didn't get it in. The first animal returned and pulled at it. Then, over a long and complex course of interbehaviors, the animal learned to develop a solicitational relation with the second animal, and the second animal developed a solicitational relation to the first animal, in the course of which, when such a circumstance arose, Animal A would wave to Animal B, or go and grab Animal B by the hand and bring him to the situation. In some cases, he took his hand, put it on the rope, and then the two of them would pull. Under conditions in which A would be pulling and B wouldn't pull, A would reach back and wallop B, under which conditions B would begin to pull, too. Therefore, in a kind of co-operative-work situation, a set of gestures began to take [159] on not only sign characteristics, that is, directional-producing activity features such as we have in the lower vertebrate, but they also began to take on significant characteristics. They began to have meaning in terms of the interrelation between the two animals.

We find that in field observation of the primates, too. In a study both of the howling monkeys and of the red spider monkeys of Panama, Carpenter (18) has shown how, in the course of the life of many of these primates, you have the development – the ontogenetic development now, the learned development – in the specific life history of the given organism, or a relationship between experiences and gestures, between features of the environment, let us say, and activity, which begin to take on conventionalized significance. By »conventionalized significance« I mean that these gesture-environment relations begin to be the property, not of any specific individual organism, but become the joint property and the joint »learned« features of life in more than one organism.

Given this as a general sort of situation, we can also see this kind of behavior in the feeding situation of young primates in captivity, who establish a play group. We had one chimpanzee at Orange Park who, when feeding time came around, would always stay in the far corner of the enclosure and would not come up for his milk. This went on for a period, until a very good social interrelation had been established among the animals, including food interchanges as part of the social relationship. Under these conditions I observed that the chimpanzee who tended to stay in the corner had developed a very close relation with another highly active chimpanzee. When this had occurred, when feeding time came, this partner chimpanzee would now dash off to the far corner, grab his pal by the hand, and drag him or carry him, practically, across the enclosure to the feeding place, and push him up against the wire mesh at the time [160] of feeding; so that you had not merely the | way in which an animal's activity, occurring for its own reasons and independently of another organism, affects the behavior of another organism by modifying the general environment, but the interrelations between two individuals as well, where there is an interaction now between the organism sending the signal, the organism receiving the signal, and then, reciprocally, between the receiver and the sender.

I would submit that in a legitimate sense the kind of communication that we usually refer to as communication at the human level represents the extension and the elaboration of such ontogenetically developed components rather than the extension of behavior of the phylogenetic, stereotyped kind of components that we find either in the invertebrate series or in the lower vertebrate series of organisms, and that actually true communication, in my sense at any rate – not necessarily in the sense of the communications engineer or in the sense of the physicist, but in the sense of a student of animal behavior – would be represented by this kind of level of interdependent communication that has direction, that involves the process of anticipation, and that involves the process of conventionalization of sound.

This means, then, that the study of the evolution of the communication process at the mammalian level requires not the examination of communication itself, but the study of the learning process, of the perceptual process, and of the process of social interrelationship. It means, then, that we have not one problem but a series of problems which must be investigated.

What kind of learning is necessary for an organism to generalize, to abstract from concrete experience, certain gestural relations which are relevant to these experiences? So we have, then, the problem of the study of the evolution of intelligence in organisms; that is to say, the study of the development of modifiability. Further, we should like to know what is the perception of the world by these different organisms. How do they see it?

I was engaged, while working as a Behavior Research Fellow for the New York Zoological Society, in a series of studies involving an agglomeration of primates at the Bronx Zoo. One thing in which I was very much interested was this: Is the perceptual world of the gibbon, for example, like the perceptual world of the chimpanzee? Is the perceptual world of the young orangutan like the perceptual world of young chimpanzee? What does it see? What does it respond to? What is the organization of the world? Can we analogize from our own perceptions to the world of these animals? I thought we could not, so I started to test them on patterned string problems of various kinds, and I got the most astounding differences. Unfortunately, the zoo is not a good place to do systematic research, but in a preliminary way I ob|tained for myself, at any rate, [161] the very deep-lying prejudice that the initial world for these organisms, the perception and the interrelation of forms and space, of directions and space, are not the same. We have a whole series of problems that we must investigate if we are going to look at communication. We want to find out what it is that an animal is perceiving. We cannot assume that because Hayes's chimpanzee down at Orange Park can say »Papa« that that now means »Papa« or anything of the sort. Personally, I think the whole direction of that study involves a failure to understand chimpanzees, but that is neither here nor there. If we study the development of communication in chimpanzees, it would be far more profitable to study the development of gesture symbolization, for example, and a few other things.

Pitts: Could you characterize, even in a vague way, the differences between these perceptual worlds?

Birch: I will be very vague, because my data are entirely preliminary. This is work conducted from five o'clock in the morning until seven-thirty in the morning, through the summer. The problem is that visitors come into the zoo and begin to interfere with your legitimate activities, or you with theirs, after a certain hour of the day. But at least you can get some general notion.

For example, in patterned string problems, in the subanthropoid organism, there appears to be a greater tendency to follow the direction of the relation of the animal's body motion to the food than to follow the visual line which existed between the food and the animal. You see, you had food out in cylinders, and string patterns. Food would be attached to one of the strings, no food to the other, and then there would be cross-strings.

In the lower orders of primates, there appeared to be a tendency to begin to reach rather directly for the grape or the cherry or whatever food it was, and to establish a linear relation between itself and the food, rather than to begin to respond to the lines that I had introduced, the lines of connection that I had introduced into the environment. That kind of process, whether the animal is responding in terms of its own bodily orientation and spatial distance from the food or whether it is going to respond to the organization of visual field as you present it, became one of the problems that I wanted to investigate, and one that emerged out of these preliminary activities. Now, such a study, or such studies, would have to be done, and I think they would be highly profitable, if they were carried through. At the moment I don't know of any place where they are being carried on.

Pitts: How about the differences between the chimpanzee and the orangutan and possibly the young gorilla?

Birch: I don't have enough data. I just have suspicions. | [162]

Pitts: What sort of suspicions?

Birch: What is important is whether it is visual stimulation as contrasted with olfactory stimulation, whether it is taste as a basis for social relationship or touch, or contact. Do you see what I mean? What is the basic structure, the nature of this organism, out of which these complex ontogenetically developed patterns are developed? Those are problems that we still have to investigate.

Kubie: Have you any data on facial expression?

Birch: There are some data on facial expression but, again, not systematized.

Kubie: That seems so primitive in the human infant.

Birch: Yes. Well, it appeared to be a fundamental problem for Darwin (19), but it has not been picked up and carried through, in his whole initial study or rather, discussion of the relation of expression to emotional states and feeling states which you certainly begin to see in the chimpanzee. You have species-type gestures, for example, species-type expressions and general bodily-type orientations which have signaling value to other organisms.

I should like to end with just one further precautionary note in dealing with some of the materials on primates. It is very easy to find certain gestures that are attached to specific situations in the case of primates and then to make the assumption that these gestures then refer specifically to these situations rather than to understand that a specific gesture for the primate is interpreted by a second primate and has meaning to the first primate itself only in terms of the surrounding circumstances and the situation, in a sense, the environment in which the gesture is being made. I should like to highlight this with one example from my own experience.

Students of the social behavior of primates have customarily referred to the »female presentation gesture,« the presentation of the genitalia on the part of either the male or the female, to another dominant organism as a sign of submission, and that once the animal does this, then it is clearly a submissive animal, and the animal that does the mounting is clearly the dominant animal. It seems to me that that is an extrapolation into biological theory of a mid-Victorian or perhaps earlier conception of human sexuality, in which the assumption is that females engaging in sexual behavior, particularly if they are primates, are really suffering some kind of abnegation. There is the assumption that some degradatory phenomenon is involved.

Well, in studying hormonal effects on social relationships of the chimpanzee, I paired a large male chimpanzee with a very actively mating female. The male had never had any mating experience. The female was put into this cage, and the first thing [163] she did was to present | to the male. The male just sat in the corner and crossed his legs. The female moved a little bit closer to the male, presented again, assuming this so-called »subjugated« position. The male ate a piece of food that was lying on the floor and ignored her. The female presented a third time, and the male ignored her further, at which point she turned around and beat the living daylights out of the male. She just tore into him, bit him, scratched him, and kicked, so we had to turn the hoses on them to separate them.

Savage: »Hell hath no fury.«

Birch: Exactly. At any rate, this gesture, as such, may have meaning, ontogenetic meaning, as a submissive gesture, under certain social circumstances for the chimpanzee, but it has significance now as a demand gesture at a different point in the relationship.

Klüver: In connection with problems of »comparative« psychology and, more particularly, the problem of comparing the perceptual worlds of different animals, it is pertinent to recall that most investigations in the field of sensory psychology have been concerned with threshold determinations, for instance, with determining visual, auditory, gustatory, and so forth, acuity. Such studies have shown, for example, that there are no great differences in the sensory equipment of man, apes, and monkeys. There is no reason to believe that even the sense of smell is more highly developed in monkeys than in man. And yet, a careful observer, in watching the reactions of different primates to sensory stimuli, will often gain the impression that different species or genera differ markedly in their reactions to the stimuli of the external environment. Qualitatively, there seem to be some striking differences in the reactions of a marmoset, squirrel monkey, and chimpanzee, for example, to stimuli of the visual environment, and yet, quantitatively, these differences are often not caught by existing methods. Why?

If we examine the methods of sensory psychology, we discover that most of them are concerned with measuring the animal's ability to cope with the discriminable aspects of the external environment. In effect, the intent of such methods is to answer the question of how small we may make certain stimulus differences and yet obtain a differential response. And in trying to answer this question, the classical procedure of keeping all variables except one constant is generally being followed. However, the net result of employing such methods seems to have been that no significant differences between the perceptual worlds of different primates, let us say, between the visual world of a squirrel monkey and that of a chimpanzee, have been discovered. To detect such differences we must choose, I believe, an entirely different point of departure. Instead of testing the ability of the organism to cope with *discriminable* aspects in otherwise homogeneous stimulus situations, we should con | cern ourselves with *identifiable* [164] aspects in heterogeneous situations, that is, we should analyze the ability of the organism to identify or isolate a certain factor in stimulus situations exhibiting widely different properties. Instead of testing the ability to discriminate small differences in brightness, color, size, shape, and so forth, we should inquire whether large, numerous, and complex stimulus differences do or do not interfere with the animal's ability to identify or isolate a given brightness or color or a particular brightness or color relation. Differently expressed, to discover differences between the perceptual worlds of different animals, the most promising approach, as far as I am able to judge on the basis of work done so far, is to study, not the animal's ability to attend to one aspect in an otherwise nondifferentiated stimulus situation, but to study the ability to attend to one or even several aspects in multi differentiated stimulus situations. It is for the latter purpose that I developed »the method of equivalent and nonequivalent stimuli« and, when employing this method, had no great difficulties in finding differences, for example, between durukulis, marmosets, squirrel monkeys, spider monkeys, and rhesus monkeys in their reactions to stimuli of the visual environment (20).

Gerard: I have been waiting for the floor for a long time. May I have it? I want to ask three questions of the speaker that are relevant to the things he has developed. The first one is immediately in line with what Dr. Klüver was just pointing out.

While I would confess freely that the detailed physiological mechanisms of the translation of visual orientation to gravitational orientation, or vice versa, are not known, it somehow doesn't bother me particularly because I think it is equivalent to other situations with which we are familiar, and therefore don't worry about, or, in another sense, worry about all the time. It would, in principle, be no different from the reaction of posture to vestibular impulses; it is built in, in some way. But what does worry me is one of the experimental aspects of this, unless I misunderstand you; namely, that when the hive is horizontal, and appropriate polarized light is allowed to enter, the bee then makes his dance, not by translating his previous visual cues into gravitational ones, but now in terms of a visual cue?

Birch: That's right.

Gerard: And that is not normal. This is, therefore, something that seems to me to present a very real problem.

Birch: It is normal because the bee, when landing at the entrance to the hive, for example, will not infrequently go into a dance and indicate direction that is related to the polarization characteristics of light. It is now on the horizontal. It is not vertical. That is natural, in a sense, therefore. All that von Frisch has done by putting it on the [165] hori|zontal is to give himself an opportunity under controlled conditions to observe the way in which changes in polarization may affect the orientation.

Gerard: Well, I agree, if it occurs naturally, it is not the same sort of problem. It still seems to me rather a neat problem that the bee can do it in either of the modalities. That takes a bit of doing.

Another, unrelated, question is with respect to the fish-school story. I once spent a moderate amount of my time lying on the pier at Woods Hole, watching the fish schools go round, and one of the problems that has always intrigued me is the cue and the timing involved in a sudden shift in direction. As far as I could tell from observing them visually, there was no wave of change from a leader or from any other one fish. They all moved simultaneously.

I was once able to check that in the case of birds. A flight of birds was going along parallel to my car, so I could time them. I happened to be watching them as they all veered away, and I would certainly have seen one bird go forward or drop back relative to the others if its timing was off. As I remember, I calculated there was less than five milliseconds possible time for cueing from one to another. I would like any evidence on that question that you have.

And let me ask my third question now, so it doesn't get lost. This turns to primate behavior and its relation to human beings. I suppose everybody is worrying about it because it is the basic problem. Is there any valid evidence in animal groups, not of learning, that certainly occurs, nor of teaching, as a man can teach a chimpanzee or as one chimpanzee can teach another, but of socially propagated teaching?

Birch: Socially, from generation to generation?

Gerard: Primarily that.

Birch: You mean a culture?

Gerard: That's right. In other words, social inheritance of habit patterns. Is there any positive evidence of that?

Birch: I will leave the first question, which we discussed a moment ago, about the naturalness. Let's deal with the problem of schooling and cueing in schooling. I left out one factor that has been referred to, at least by Breder and some of his associates, in the development of schooling itself. He once took a group of young jewel fish, which were schooling, from one pond environment and brought them into a museum or aquarium environment. Under those conditions, the school was re-established. He inferred from that that he had a very interesting problem – which I must point out to you has not been systematically investigated. All we are doing today, probably, is opening up areas for investigation more than we are answering questions. But apparently [166] there are environmental features, as well as interindividual-stimulated | features, which produce these characteristic school or flock relationships in these organisms as well.

Our attention has been directed, because we have been interested, in part, in the problem of communication, at the way in which activity of animals within the school influences the behavior of other animals within the school. An equally legitimate question would be the way in which identical behaviors, which would then maintain a relationship within a school, could be produced and are produced by environmental changes having identical effects on the different organisms, whether it be your car or any other feature.

GERARD: Well, say one millisecond.

BIRCH: If one of them could respond within one millisecond, there is no reason why another one could not respond within one millisecond. What I am talking about now is that each may be responding individually; that is to say, to the same feature of the environment rather than to one another.

GERARD: Yes. All I am saying is that the simultaneity must be there. I don't know the time of response.

BIRCH: Simultaneity can be the function, and usually is the function, of a given environmental change which is simultaneously affecting the different animals, under which conditions the simultaneity of response or the lack of simultaneity of response is a function of the reaction time of each of the individuals therein.

GERARD: I would question – and I could be completely wrong; indeed, I don't know what other answer there is beside the kind you are giving – but I would question whether, when you have trained a group of like organisms to respond to a particular signal, you would find the response of the whole group so beautifully synchronized as that.

MACKAY: I am not quite sure why you feel it must be a millisecond, or as short as that, because, suppose you have a statistically scattered response, you have at the same time continuous feedback from all the positions of all the neighbors, and any slight lag in one creature would very soon be fed back.

BIGELOW: You still need terms like »millisecond,« though, not to deviate too far. There must be a lag-correcting operation there which is very close to this sort of magnitude.

MACKAY: In a nonlinear system of this sort the rate of change is the important thing. It is really a question of how many bits of information you need, and how fast.

BIGELOW: But to get back to the point, isn't it true that if you fired off a gun in this room, everybody would jump within a millisecond?

GERARD: We wouldn't within a millisecond. I doubt if we should all jump. | [167]

BIGELOW: There would be an appropriate electrical pickup on every person.

SAVAGE: Perhaps within a millisecond of each other, not within a millisecond of the gunshot.

GERARD: Yes, within a millisecond of each other. That is just what I doubt.

SAVAGE: How long does it take a man to respond to a shot like that? How long from the gunshot to his response?

RIOCH: Two-fifths of a second.

SAVAGE: Well, a couple of hundred milliseconds, so it implies a synchronization of a half per cent or so.

GERARD: That is so, and that is an extremely abrupt and vigorous stimulus, which apparently was not happening with the birds.

KLÜVER: If I remember correctly, the observations on birds you mentioned a minute ago were described by Gerard in *Science* several years ago. Is that correct?

GERARD: That is correct.

KLÜVER: Some ornithologists have studied the way birds flying in flocks take to wing synchronously. Apparently the taking to wing is preceded by a great deal of mutual stimulation, that is, by certain initiating or »intention movements« which are »understood« by all fellow members of the species.

GERARD: That is right; so you are suggesting that there must have been some further communication?

KLÜVER: As Lorenz undoubtedly would express it, the finely graded intention movements and movement patterns preceding the flight are »releasers« switching on the flying reaction in social birds.

GERARD: Which is what I would suspect.

BIRCH: I would raise the question of your perceptions. I think part of what Dr. Klüver is saying is this: if you take photographs or moving pictures of bird groups starting off, and then do frame by frame analysis, you find that what you viewed perceptually as a whole group going off simultaneously shows you birds in many, many stages of activity, all of which would indicate that there is a wider range of reaction times in those organisms.

May I answer the last question on culture? This question has intrigued many people, and Yerkes as you may know, for years made a systematic effort to find a culture or a prototype of culture in the higher primate. He ended up with one instance that I can think of; that is, that in the wild, the chimpanzee tends from generation to generation to sleep in a sort of bower-type hut, in which branches are bent over and sat down upon, but males raised in captivity do not have this, and it does not apparently develop [168] spontaneously when they are left in wild | situations. However, nobody has ever studied the captive-raised chimpanzee in a woods situation under conditions in which it could spontaneously develop this kind of activity over a long-enough period to make it meaningful. Further, the captive chimpanzee is not at home in the woods.

RIOCH: How do you regard the training of the young primiparous chimpanzee by the multiparous chimpanzee in the acceptance of the first baby, and the training of the young mate by the old female to copulate?

BIRCH: First of all, I would deny the training of the primiparous by multiparous. I know of no good evidence whatsoever that that takes place. I have, in my own observations of chimpanzees, observed that when an old multipara or any other chimpanzee comes near a mother chimpanzee with its infant, the response is not one of training but is a fighting kind of response.

RIOCH: This was at the time of birth that you observed them?

BIRCH: Oh, yes, I have made observations at birth, where animals were in adjacent cages. Some were primiparas and some were multiparas, and there is not that tutelage relationship that has been somewhat romantically described. In the learning to mate, all you have to do is watch a young male begin to mate. What the female does is to maintain a position in response to the male stimulation and keep orienting toward the male. What the male learns, he learns on the basis of his own errors and his own activities rather than on the basis of the specific tutelage. I don't think you can read tutelage into it.

ROSENBLUETH: I find myself in disagreement with many of the statements that Dr. Birch made because I think they are irrelevant to the problem he was considering. There is a false distinction, I think, in his introduction, in the criteria which he adopted. The distinction between the feeding behavior of the amoeba and that of the human being I find quite impossible to maintain. He decided that they were different types or modes of seeking food, because the amoeba could be fooled into seizing a particle of potassium permanganate. Of course human beings are also fooled into seizing material which is not food. The notion that feeding is something specific, that it is not just a movement in relation to something in the surroundings of an organism, I don't think can be upheld. A child can eat things that are not food, and even adults have been poisoned because of eating material that was not proper food. Both the amoeba and the human being respond to certain stimuli which may have not been identical in the two cases. That is not surprising. We are dealing with different organisms. All they were doing was responding to certain stimuli and then reacting with the acts that corresponded to their own organization. But the distinction in terms of seeking food – and I take it that means there is a conscious process that goes | on which [169] leads the human to that goal, and that this mechanism is very different from the mechanism used by the amoeba – I find quite impossible to sustain. Now, if you can give me some criterion by which I can apply that distinction, I should like to hear it, because I cannot find it myself. I made my first point, which was that it seems to me that some distinctions are rather artificial. I don't see how one can give them any sense unless one introduces into one type of behavior notions that cannot be used or measured. That brings me to my second comment. Dr. Birch spent a lot of time trying to prove to us that animals do not have »intelligence« – in quotation marks – in the sense in which he was using this term. He even used the word »genius,« which needs more quotation marks than the word »intelligence.« Nobody can prove or disprove that an animal or a man does or does not have intelligence. It seems to me absolutely irrelevant to the problem being studied – the problem of communication. The question of intelligence is something that is going on in the mind. I don't see what bearing it can have on the problem of communication. The only way we can get together with other people and observe other people and make ourselves come in contact with them and receive contacts from them is in terms of their behavior. That is the only thing we can see, that we can judge; and that is going to make an impact on us.

Dr. Klüver mentioned the danger of gross analogies. Well, I don't know; I personally feel that the problem of other people's minds belongs in that same realm. It is one of the grossest analogies. It is an indispensable one. I think we all assume these minds; that is why we are here. On the other hand, it doesn't belong with the problem that we are considering. We can dismiss it entirely.

As to the question of whether communication among animals is in[!] intelligent communication, first, it cannot be discussed because nobody can either prove or disprove intelligence. I don't know of any operation by which we can judge whether or not such a thing exists. Second, it is quite irrelevant to the problem. When we describe the behavior of lower organisms or of machines (as has been done very often in this group), we use terms which can be qualified as mentalistic. I don't think there is any special objection to this if it is understood that the terms are used merely for convenience. When Pavlov tried very seriously to dismiss any terms with a psychological implication in his description of the studies he was carrying out on animals, there was one point at which he broke down, at which he violated the law that he had established, and which led him to become separated from several of the collaborators that had joined him in his work, because they were not able to avoid the use of terms of that sort, and they became very upset because they were not used in the study. He broke down when he used | the term »experimental neuroses.« But, of course, it is [170] quite clear from Pavlov's writing that all he meant was a verbal shortcut.

KUBIE: He even used the word, »unconscious« (21).

ROSENBLUETH: Then he broke down twice. He may have broken down many times, but he could have done it quite consciously and there wouldn't have been any objection.

If we say that a machine has a memory, what we mean is that in so far as we can describe in objective terms what we mean by memory, that can be put into a machine. When we say the machine learns, what we mean is, again, that in so far as we can state in objective, accurate terms, independent of our own personal and private experience, what we mean by learning, that can be put into a machine. There is no objection to using those terms so long as it is understood that they are going to have a particular realm of application and a particular set of meanings, so that actually –

BIGELOW: May I add to that when you are through?

ROSENBLUETH: Go ahead.

BIGELOW: There is one further point I should like to air a little bit, on the story about putting a soup made of a starfish near a scallop. This story is very amusing, and it seems to reduce the question of communication to an absurdity. On the other hand, I don't think that one can argue that the fact that communication is reduced to a mechanically ridiculous process means that it is not communication. The definition lies elsewhere than in that fact.

BIGELOW: A phonograph record is certainly a mechanical device which can elicit communication from a human being.

ROSENBLUETH: I think also that the definition adopted by Dr. Birch is much too restricted and too narrow. It would be inapplicable even to human beings. If he is going to restrict it to that, he is going to eliminate certain things which we all want to include in the group of interrelations.

If we should adopt his criterion, the fact would remain that there are other types of messages sent by organisms of the same or of different species which are usually included under the general term of »communication,« and which are worth studying; we should not eliminate them. We know the physiology of some reactions, and if we are going to postulate that this knowledge implies that the reactions are no longer communications, but belong to some other category, I don't think we gain anything. It seems to me that the proper way to approach the problem would be to take a very broad definition of communication. If one wishes, one might adopt one of the expressions that Dr. Birch used, such as »interrelations between organisms.« That is one possibility. Or we might define it as the influence of the behavior of an organism *A* | [171] on that of another, *B*. By that, I take it we can mean any organisms we wish. With either definition, it may be that we will find there are different types of communication, and that some are inborn and some are learned. I am sure that in the human being, among human communications, there are many which are not learned, which are inborn. Certainly, some of the sexual behavior reactions, some of those exhibited at moments of emergency, and many other reactions belong in that category. That is why I said that the restriction to learned behavior is quite arbitrary.

It may be interesting from one standpoint to know what may be the responses to inborn messages as opposed to learned reactions, and it is an important distinction from many standpoints. But it is not a particularly important division from the standpoint of the problem of communication. The group of messages studied should be quite inclusive. And then we may perhaps classify it, but if we do that, it is not desirable to adopt a very sharp and fine distinction made largely on the basis of something like »anticipation« or »direction« or »content.« Those are terms which I don't think Dr. Birch or anybody else can define.

GERARD: I should like to comment on all three points. I simply want to say, on the first point, I am inclined to agree with Dr. Rosenblueth that there is a good deal of unnecessary verbalism in the distinction. On the second point, I should like to defend Dr. Birch. I don't think he did any of the things you accused him of, Dr. Rosenblueth, and when he spoke of a genius in the chimpanzees, he was talking figuratively, just as when he said the male went in the corner and crossed his legs. I don't believe that actually happened, either.

BIRCH: It did.

GERARD: Well, then as one of the other things the chimp did, not in the sense that a man would.

I think »intelligence« was used behavioristically and that it is a perfectly good word to use to describe such behavior. There is no necessary imputation of what is going on inside.

ROSENBLUETH: I think that is a hopeless task. You can neither prove it nor disprove it.

GERARD: You said in your last sentence or two that which I think invalidates much of what you said in the rest of your comment. It is purely a semantic matter – whether or not one wants to call something communication. What is important is that we recognize various categories, and that is what Dr. Birch was trying to do. Whether one calls them all different phases of communication or calls one communication and one not communication is, I think, a trivial matter.

BIRCH: I think that the last remark of Dr. Gerard is the most perti|nent one. What I [172] was saying did not mean that there were no unlearned behaviors in the human being. It simply meant that at different levels of the evolution of animals, we have pre-eminently present methods of interrelation with other animals which are not the same, and that there is a difference between the ontogenetic acquisition of communicative devices and the phylogenetic emergences of those. Now, what I was trying to do was to give a picture of the way in which interrelations between animals could take place in a variety of ways, and not to deny that such interrelation was taking place. If the human is fooled, he is fooled in a different way, and because of different mechanisms from the way in which an amoeba is fooled. If Dr. Rosenblueth cannot distinguish between these ways, that is unfortunate; nevertheless, these ways are different.

PITTS: Not always.

RIOCH: But that is not the problem.

MCCULLOCH: Well, let's get Bowman's data before us now.

BOWMAN: There is a black beetle which is found in fairly large – I won't say social groups, but in large colonies – under rotten stumps. You will find at least ten and sometimes up to several hundred of them, if you find one. Along with the adults, there are always various immature stages. The beetle itself probably has very few enemies. It is very heavily armored and has extremely strong mandibles, but the larvae and pupae are practically defenseless. If you open a stump that contains these beetles, you will hear a rather high-pitched hiss, a sort of whistle, at which time all of the soft-bodied forms will head toward the center of the stump, and all of the adult beetles will face outward. The adults themselves can make that sound. If you just wait and watch, they will soon resume normal activity, but then you can, without further disturbing them, give that same note and they will perform that same protective defensive act.

PITTS: Which family of beetles is this?

BOWMAN: A species that falls in a family all to itself – passalus. They are big black things.

BIRCH: What is the frequency?

BOWMAN: It is around 6,000 cycles, I should guess. You can make it with your mouth. It is quite sharp. If you whistle a slide over a range, you get response at that one pitch.

[173] # PRESENTATION OF A MAZE-SOLVING MACHINE

CLAUDE SHANNON
Bell Laboratories, Murray Hill, N. J.

THIS IS A maze-solving machine that is capable of solving a maze by trial-and-error means, of remembering the solution, and also of forgetting it in case the situation changes and the solution is no longer applicable. I think this machine may be of interest in view of its connection with the problems of trial-and-error learning, forgetting and feedback systems.

As you can see (Figure 8), there is a maze on the top panel of the machine which has a range of 5 × 5 squares. The maze can be changed in any desired manner by rearranging the partitions between the twenty-five squares. In the maze there is a sensing finger, which can feel the partitions of the maze as it comes against them. This finger is moved by two motors, an east-west motor and a north-south motor. The problem facing the machine is to move the finger through the maze to the goal. The goal is mounted on a pin which can be slipped into a jack in any of the twenty-five squares. Thus you can change the problem any way you choose, within the limits of the 5 × 5 maze. I will turn it on so you can see it, in the first place, trying to solve the maze. When the machine was turned off, the relays essentially forgot everything they knew, so that they are now starting afresh, with no knowledge of the maze.

SAVAGE: Does than mean they are in a neutral position, neither to the right nor the left?

SHANNON: They are in a kind of nominal position. It isn't really a neutral position but a meaningless one.

You see the finger now exploring the maze, hunting for the goal. When it reaches the center of a square, the machine makes a new decision as to the next direction to try. If the finger hits a partition, the motors reverse, taking the finger back to the center of the square, where a new direction is chosen. The choices are based on previous knowledge and according to a certain strategy, which is a bit complicated.

PITTS: It is a fixed strategy? It is not a randomization?

SHANNON: There is no random element present. I first considered using a probability element, but decided it was easier to do it with a fixed strategy. The sensing finger in [174] its exploration has now reached | [Figure 8] | the goal, and this stops the motors, lights [175] a lamp on the finger, and rings a bell. The machine has solved the maze. I will now run the finger, manually, back to the starting point, and you will see that the machine remembers the solution it has found. When I turn it on, it goes directly to the goal without striking the partitions or making side excursions into blind alleys. It is able to go directly to the goal from any part of the maze that it has visited in its exploration. If I now move the finger to a part of the maze that it has not explored, it will fumble around until it reaches a known region. From there it goes directly to the goal.

Now I should like to show you one further feature of the machine. I will change the maze so that the solution the machine found no longer works. By moving the partitions in a suitable way, I can obtain a rather interesting effect. In the previous maze the proper solution starting from Square A led to Square B, then to C, and on to the goal. By changing the partitions I have forced the machine at Square C to go to a new square, Square D, and from there back to the original square, A. When it arrives at A, it remembers that the old solution said to go to B, and so it goes around the circle A, B, C, D, A, B, C, D, ... It has established a vicious circle, or a singing condition.

GERARD: A neurosis.

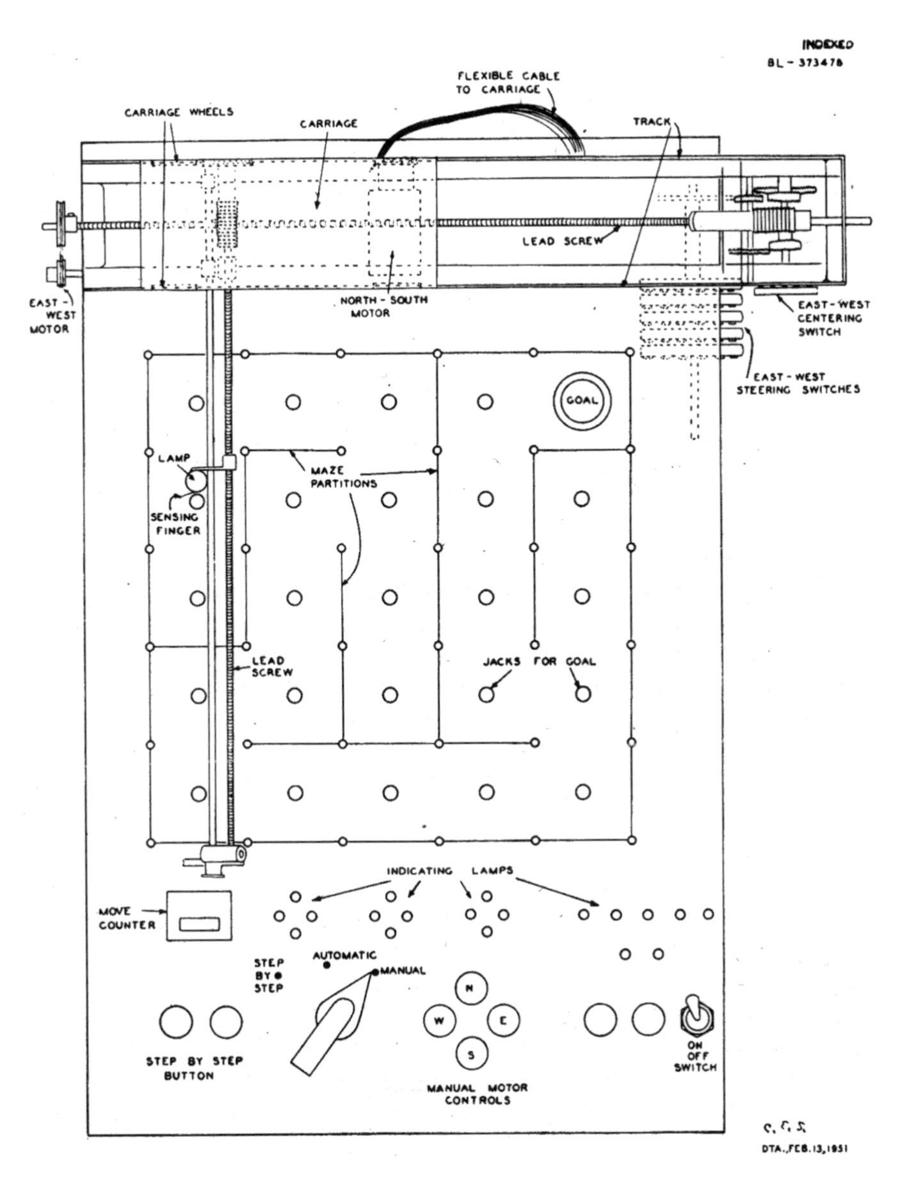

FIGURE 8

SHANNON: Yes.

SAVAGE: It can't do that when its mind is blank, but it can do it after it has been conditioned?

SHANNON: Yes, only after it has been conditioned. However, the machine has an antineurotic circuit built in to prevent just this sort of situation.

MEAD: After it has done it a number of times?

SHANNON: After it has gone around the circle about six times, it will break out. The relay circuit includes a counter which stops this behavior at the twenty-fourth count.

FRANK: How many relays are there in it?

SHANNON: All told, there are about seventy-five relays.

SAVAGE: It doesn't have any way to recognize that it is »psycho;« it just recognizes that it has been going too long?

SHANNON: Yes. As you see, it has now gone back to the exploring strategy.

TEUBER: Now, does it have to relearn the entire maze, or can it still utilize some form of it?

SHANNON: No. As it stands, it can't utilize any information it had before.

[176] SAVAGE: But it is trying to utilize it, I suppose. It is moving as it would move. |

SHANNON: As a matter of fact, the old information is doing it harm.

BIGELOW: I think it's getting to it.

SHANNON: Yes, it is gradually working over toward the goal. I should like to spend the rest of my time explaining some of the things which are involved in the operation of the machine.

The strategy by which the machine operates can be described as follows: There are two modes of operation, which I call the »exploration strategy« and the »goal strategy.« They are both quite simple. The exploration strategy is used when it is first trying to find the goal. For each square in the maze, there is associated a memory, consisting of two relays. These are capable of remembering one of four possible directions: north, east, south, or west. The direction that is remembered for a square is the direction by which the sensing finger left the square the last time it visited that square. Those are the only data the machine remembers about the course of the finger through the maze. There are some other memory functions in the computing part of the circuit, but these remembered directions are the data which allow it to reproduce its path at a later time.

Now, let's call the remembered direction for a particular square, D, considered as a vector. In exploration strategy, the machine takes the vector D and rotates it 90° as the first choice when it comes into a square. For example, suppose it left a square in the easterly direction at the last visit. If it comes to that square again, it will try the northern direction as the first choice. If it hits a barrier and comes back, it again rotates 90°, because it has just put this northern direction into the memory, and, advancing 90°, it tries the westerly direction, and so on. The choices progress around counterclockwise, starting with the direction by which it left the square last time – with one exception: it also remembers the direction by which it came into the square at the current visit, and on the first rotation of the vector D, it skips that direction of entrance. This is to prevent the path repeating too much. Before that feature was installed, there was a tendency to explore up to a new square, go back through the entire maze, and then go one square further, and so on; and it took a very long time to solve the maze. It required about three times as long as it does now, with this skipping feature added.

When it hits the goal, a relay operates and locks in, and the machine then acts according to the goal strategy, which is also based on this vector D.

In the goal strategy, the machine takes as its first choice direction D, which is the direction by which it left the square on its last visit. This is very simple to do, and it has many convenient features for maze solving, because it cancels out all blind alleys and [177] circular paths. Since | a blind alley must be left by way of the same square through which it was entered, the direction D retained for that square will necessarily lead to the goal directly rather than by way of the side excursion into the blind alley. In a similar way, if the machine follows a circular or re-entrant path in exploring its way to the goal, the direction retained for the last fork in this path must be that going to the goal rather than around the side loop. As a consequence, the machine follows a fairly direct path to the goal after it has first found its way there.

The final feature of forgetting is obtained as follows: After reaching the goal, suppose we move the sensing finger to a different point in the maze and start it operating. The machine then starts counting the number of moves it takes, and if it does not reach the goal within a certain specified number of moves, which happens to be twenty-four in this case, the machine decides that the maze has been changed or that it is in a circular loop, or something of that sort, and that the previous solution is no longer relevant. The circuit then reverts to the exploration-type strategy which is mathematically guaranteed to solve any finite solvable maze.

There are a few other points about the machine which may be of some interest. The memory is quite undifferentiated in the sense that I can take the group of wires leading from the rest of the circuit into the memory, shift them over either in the north-south or east-west directions, and the machine will still operate correctly, with no significant change, although the data corresponding to a square are then stored in a different part of the memory.

Another point is that there are, of course, a large number of feedback loops in this system. The most prominent is the feedback loop from the sensing finger through the circuit to the driving motors and back to the sensing finger, by mechanical motion of the motors. Normally, if you have a feedback loop and change the sign of the feedback, it completely ruins the operation of the system. There is ordinarily a great difference between positive and negative feedbacks. This maze-solving machine, however, happens to be such that you can change either or both of the signs in the feedback connections, and the machine still operates equally well. What it amounts to within the circuit is that the significance of right and left is interchanged; in other words, the effect on the strategy if one of the feedback loops is changed is that the advance of 90° counterclockwise becomes an advance of 90° clockwise. If both of them are changed, the strategy is not altered.

VON FOERSTER: If there are two different ways to reach the target, certainly the machine is only able to find one. Does the possibility point to its making a choice of the better way?

SHANNON: No, it does not necessarily choose the best way, although | the probabilities are in favor of its choosing the shorter of two paths. Incidentally, the exploration strategy of this machine will solve any maze whether it be simply or multiply connected. Some of the classic solutions of the maze problem are satisfactory only in the case of simply connected mazes. An example is the method of keeping your hand always on the right-hand wall. While this will solve any simply connected maze, it often fails if there are closed loops. [178]

SAVAGE: This cyclical feature that you illustrated occurred because the machine was not then in really searching condition?

SHANNON: No, it was in the goal strategy rather than in the exploratory.

SAVAGE: A goal strategy is to go the way you last went, but what are you to do if the attempt to do that is frustrated?

SHANNON: Then it returns to the center of the square and advances 90° and tries that direction. But it still remains in goal strategy.

SAVAGE: I see. When it gets into the next square, it tries to go ahead in the accustomed direction?

SHANNON: That's right. The purpose of this is that it may have learned most of a maze in its first exploration, but not quite all of it. If we put it into a square it has not visited, it explores around by trial and error until it reaches a familiar square, and from there goes directly to the goal. The previously unknown squares have by this process been added to its previous solution.

BIGELOW: You can then put new loops on any known path; it will learn those new loops immediately and not get into trouble. Is that right?

SHANNON: That's right.

BIGELOW: Because when you come back to the main stream, the search goes in the right direction, if it recognizes that square.

SHANNON: I am not sure I understand what you mean.

BIGELOW: It forms a single-directional path. Now, then, if you introduce a new path which brings it out of the known path into strange territory, back into the known path again –

SHANNON: Such a side path is completely canceled when it has gone into the goal strategy.

BIGELOW: But once you start it around that circuit, then the procedure is correct after the starting point.

SHANNON: If it is in goal strategy, yes, but not in exploratory.

BIGELOW: What would you have to do to minimize running time – in order to make it learn on repeated trials eventually to take the shortest possible path in a more complex maze?

SHANNON: I think that would require a considerable amount of memory in the form [179] of relays, because of the need to store up a number | of different solutions of the maze as well as additional computing relays to compare and evaluate them. It surely could be done, but it would be more difficult; it would mean a much more complicated machine than this.

SAVAGE: And it would have to decide when to invest the effort to seek a new path. That is really a very important problem in any kind of real human learning. If you can already peel a potato, why should you take the trouble to find a better way to peel it? Perhaps you are already peeling it correctly. How do you know?

VON FOERSTER: What happens if there is no goal?

SHANNON: If there is no goal, the machine establishes a periodic path, searching for the goal; that is, it gradually works out a path which goes through every square and tries every barrier, and if it doesn't find the goal, the path is repeated again and again. The machine just continues looking for the goal throughout every square, making sure that it looks at every square.

FRANK: It is all too human.

BROSIN: George Orwell, the late author of 1984, should have seen this.[1]

VON FOERSTER: And after that? For instance, if you put a goal into the path after the machine has established such a periodic motion, what happens then?

SHANNON: When it hits the goal, the machine stops and changes into the goal strategy, and from there on it goes to the goal as placed there. Incidentally, it is interesting to think of this – if I can speak mathematically for a moment – in the following way. For each of the twenty-five squares, the memory of the machine retains a vector direction, north, east, south, or west. Thus, as a whole, the memory contains a vector field defined over the 5×5 maze. As the sensing finger moves through the maze, it continually revises this remembered vector field in such a way that the vectors point along possible paths of the maze leading to the point currently occupied by the finger.

TEUBER: If you rotate the field through 180°, would it continue to function?

MCCULLOCH: Suppose you reverse the connections and leave the motor, so that you reverse your direction of rotation; can it still find its way?

1 Orwell, G.: 1984. New York, Harcourt, Brace & Co., 1949 and Signet Books, 1950. No. 798.

SHANNON: Only if I reverse some switches within the machine which tell it what square it is currently occupying. If I reverse the motors, I must change these switches to compensate. Otherwise, it would think it was moving one way and put that in the memory and actually be moving in a different direction. | [180]

GERARD: That would be like cross-suturing the motor nerves of animals and getting flexion when you want extension.

BIGELOW: Have you considered how difficult it would be to have a circuit which, instead of forgetting everything, goes back to the origin and remembers what it did at the first square but tries something else, say, the opposite search sequence? When that produces no new solution, go back where it was, in the second square, but try the opposite, therefore asking for the possibility of replacing each square in its memory as it goes systematically through. In other words, this would require a very small addition of memory because it need only remember the entire past pattern once, but then, having reached the state where goat behavior is no longer a solution (which it knows by exceeding »N« trials), then, instead of erasing its entire thinking, you have a switching technique where it goes back to the origin, and then tests each hypothesis in turn, and finds the particular one to replace.

SHANNON: I haven't considered that, but I think it would be rather slow, because there is a great deal of backtracking in that procedure, back to the origin, as it tries out different hypotheses.

BIGELOW: If it knows how to get from the origin to the target, does it not always know how to get from the target back to the origin, by a very simple reversal of the switches?

SHANNON: No. You see, this vector field, if you like, is unique in going in the direction of the vectors, but going backward, there are branch points, so it does not know where it came from.

SAVAGE: Does this vector field flow into the target from every point?

SHANNON: Yes, if you follow the vectors you will get to the goal, but, going in reverse, you may come to branch points from which you may go in any of various directions. You can't say where the sensing finger came from by studying the memory.

SAVAGE: It is not organized around any particular initial point; and that is one of the features of it, that once it has learned the maze, if you start it anywhere where it has been on its way to the maze, it continues; if you start it where it hasn't been, it finds one of those places where it has been, and then continues.

MCCULLOCH: Like a man who knows the town, so he can go from any place to any other place, but doesn't always remember how he went.

[181]

IN SEARCH OF BASIC SYMBOLS

DONALD M. MACKAY
King's College, University of London

I DON'T KNOW whether you people feel agreeable, but if there are any questions which you believe are not absolutely vital, such as matters of error, I should be very grateful if you would make a note of them and bring them all up at one time rather than interrupt the flow, because I think it may be difficult to get across enough to make sense of what I want to say at the end.

What I want to do first is to present a way of looking at the problem tackled by general information theory which finds a place for, and shows the relationship between, different concepts which have been labeled »information« by different people, which finds a place for concepts such as *meaning,* and which I think links on to the domain of symbolism and language.

In common speech we say we have received information, when we know something now that we did not know before; when the total of »what we know« has increased.

If then we were able to measure »what we know,« we could talk meaningfully about the »amount of information« we have received, in terms of the change it has caused.

General information theory is concerned with this problem of measuring changes in knowledge. Its key is the fact that we can *represent* what we know by means of pictures, logical statements, symbolic models, or what you will. When we receive information, it causes a change in the symbolic picture, or *representation,* which we would use to depict what we know.

We shall want to keep in mind this notion of a *representation,* which is a crucial one. Indeed, the subject matter of general information theory could be said to be the making of representations – the different ways in which representations can be produced, and the numerics both of the production processes and of the representations themselves.

By throwing our spotlight on this representational activity, we find ourselves able to formulate definitions of the central notions of information theory which are *operational,* with more resultant advantages than that of current respectability. In any question or debate about »amount of information,« we have simply to ask: »What representational activity are we talking about, and what numerical parameter is in question?« and we eliminate most of the ground for altercations –| [182] or we should do so, if we are careful enough!

We can cover, I think, all technical senses of the term »information« by defining it operationally as *that which logically enables the receiver to make or add to a representation of that which is the case, or is believed or alleged to be the case.*

When, on the other hand, we come to measure »amount of information,« we may expect to find ambiguities. We shall expect two people to differ as to whether A or B has given them more information, unless both have the same representational activity in mind, and are estimating the same parameter. Our expectation is not disappointed: few topics can arouse stronger debate. The problem is simply one of a deficiency of vocabulary, and we have the same reasons for confusion as we should have if we lacked the linguistic means to distinguish between volume, area, and length as measures of »size.« I am afraid, therefore, that our first concern must be to make enough distinctions and provide ourselves with an adequate vocabulary to avoid major trouble. To help supplement what now must be a very condensed presentation, I shall, if I may, append to our published proceedings an integrating survey of the nomenclature of information theory prepared for last year's *London Symposium on Information Theory.*[1]

Representations commonly can originate in two distinct ways. The difference between these is the essence of one of the most important distinctions in information theory, between the *theory of communication* on the one hand, and what, for want of a better term, we may call the *theory of scientific information* on the other. Both a communication process and a scientific observation process result in the appearance of a representation in the »representation space« of the receiver or observer. But what distinguishes communication, I suggest, is the fact that the representation produced is (or purports to be) a *replica* of a representation already present to (with, in the mind of) the sender. Communication is the activity of *replicating representations.*

This is to be contrasted with the typical activity of physical scientific observation of which the goal is the making of a *new* representation, representing some additional knowledge of that-which-is-physically-the-case concerning some unique space-time tract not heretofore represented anywhere.

We might put it crudely as the distinction between the replication and the formulation of knowledge. The problems raised in the two cases are in some respects quite different, and give rise to different »measures of information.«

An example will illustrate this point. Two people, A and B, are | listening for a signal [183] which each knows will be either a dot or a dash. A dash arrives. A has made various measurements, represents what happened by a graph, and remarks that there was »a good deal of information« in the signal. B says: »I knew it would be either a dot or a dash. All I had to do was to make a single choice between one of two *prefabricated* representations. I gained little information.«

A and B, of course, are not in disagreement. For lack of a vocabulary, they are using the phrase »amount of information« to refer to different measurable parameters of the different representational activities in which they engaged.

A (to whose activities we shall return) was concerned with representing what had actually happened, as a new, never-before-described spatio-temporal pattern of relations. B was concerned with replicating the sender's representational symbol – a dash. For him, what happened was merely a determinant of a choice between preconceived possibilities.

Preconceived possibilities: that is the key phrase in communication theory. The communication engineer assumes that the receiver possesses a filing cabinet of *prefabricated* representations, so that for him a signal is an instruction to *select* one from the assembly or »ensemble« of possibilities already foreseen and provided for. His representational activity is not a constructional but a *selective* operation.

You are all familiar, I expect, with the way in which »amount of information« is defined for a selective operation. We imagine ourselves playing a game of »twenty questions,« in which every question may receive only the answer Yes or No; and we define *amount-of-selective-information* (the adjective, I think, is essential) as the minimum number of such questions logically necessary to determine the selection. To identify one out of N possibilities, for example, we require at least $\log_2 N$ independent yes-or-no answers.

When some possibilities are more likely than others, we keep proportionately more replicas of them in our filing cabinet, so that (with the optimum selection mechanism) it takes us fewer questions to hit on those which are more often required.

If one of these, say, the ith, occupies $(1/N_i)$ of the filing cabinet, it will require roughly $\log_2 N_i$ questions and answers to locate it.

1 See Appendix I

We say that its selection has required (or provided) $\log_2 N_i$ »bits« of selective information. The average number of bits per selection will evidently be the weighted mean of $\log_2 N$ (as defined above) over all possible selections. But we have supposed that each possibility occupies space in the cabinet proportional to its own frequency of occurrence. The weighted mean is thus simply $\Sigma(1/N_i)\log_2 N_i$, or in terms of probability (pace the rigorists): $\Sigma p_i \log_2 p_i$.

[184] So much for selective information in communication theory. Claude | Shannon and others, of course, treated the whole matter in greater detail some years ago (1) (2), and I have given this outline only to help us to see where it fits into the general picture. Amount of selective information is evidently a measure of the *statistical rarity* of a representation and has no direct logical connection with its form or content, except in cases where these affect its statistical status. One word which was unexpected could yield more selective information to a receiver than a whole paragraph which he knew he would receive.

Now it is evident that in *any* situation in which what is observed is thought of as specifying one out of an ensemble of preconceived possibilities, the amount of selective information so specified can in principle be computed. The concept has, therefore, a much wider domain of usefulness than that of communication theory. The point is that it is always a relevant parameter of a communication process, because successful communication depends on symbols having significance for the receiver, and hence on their being already in some sense prefabricated for him. The practical difficulty, of course, is to estimate the proportions of the appropriate ensemble, when these are determined by subjectively – and even unconsciously – assessed probabilities.

But now let us turn to this other problem, which faces, say, the physicist; namely, making a representation of that-which-is-physically the case concerning some tract of space-time. This I have discussed at length elsewhere (3) (4), and I want now only to indicate the different and complementary senses of the term »information-content« to which it gives rise, and to outline the kind of formalism which is useful to represent the processes concerned.

Here we are not usually in a position to select from a filing cabinet of preformed representations; we have to produce our representations *ab initio.* Our scientific representation is in general compounded of elements asserting certain relations between the magnitude of a voltage and a particular point on a time axis, or between the intensity of transmitted light and a particular co-ordinate intersection in the field of view of a microscope, for example. We say »the voltage was 10 volts at time t_1, 10.5 at t_2« and so forth.

Our ability to name operationally a certain number of distinct coordinate values such as t_1 and t_2, enables us to prepare in advance the same number of distinguishable, independent »blank statements« of the form: »The magnitude had the value such and such at co-ordinate point q_n (or q_{n-1}, q_{n+1}, or what have you), or rather, »The magnitude had the average value such and such over the co-ordinate interval Δq around q_n (q_{n-1}, q_{n+1}), and so on.« The blanks in these statements or »propositional functions« we [185] then fill in as a result of our observations. |

We are thus clearly faced with a twofold problem: First, we must be able to *define* distinguishably in operational terms the blank statements which we want to prepare. In other words, something in the design of the experimental apparatus or procedure must enable us to identify and distinguish between observations if we want to call these observations »independent« or even »distinct.«

Then, second, we must collect *evidence* for our statements by observation of events. We »plug in« observed data, so to speak, into the blank spaces which we have for them in our previously prepared propositional structure. If we boll a typical statement down to the oversimple form, »Value X relates to interval Y,« then our two problems are the *operational definition of Y and the collection of evidence for X.*

I think there is a fair analogy of the first problem, by the way, if you imagine a fly crawling on a perfectly blank, infinite white sheet. If you want to make a description of the movements of the fly, you are wordless unless you have some means, by projecting the co-ordinate system of your eye, or some way or another, of identifying co-ordinate points on the blank sheet. In order to utter your description, in order to make any scientific statement at all about the movements of the fly, you must somehow or other have means of labeling the fly's position. In a sense, therefore, it would be defensible to say that that which enables us to name, to formulate our propositions, is *»information.«* Or at any rate we can define a measure of *information content* in a certain sense as the number of independent propositional functions which we are enabled by a particular experimental method to formulate. To distinguish this from other senses of the term, we shall call it the *structural* information content of our representation. This could be described as the number of logically distinguishable degrees of freedom of the representation. Each of the blank statements we were talking about a moment ago represents ideally one independent respect in which the representation could be different.

I don't think that in this gathering it would be appropriate to go too far into technicalities, but I do want to mention that Gabor (5), in the field of communication, defined what he called the »amount of information« in a signal in such a way that he was essentially talking about the number of independent propositions necessary to define its amplitude over a given period of time. Let's look at an example:

Suppose we want to represent the voltage of a signal coming through a channel of a certain band width, as a function of time. At certain intervals, we want to take »new« readings to provide »new« ordinates for our graph. Obviously, however, if we take two too close together they are practically *the same reading,* since the inertia of the system prevents very rapid changes. Gabor showed by an elegant method for | the ideal case [186] that there is a minimal separation in time between readings, below which (according to a certain criterion of independence) they cease to be »practically independent.« This minimal separation – let us call it Δt – is related to the band width Δf by a very simple relation of the form $\Delta f \cdot \Delta t \geq K$, where K is a constant depending on convention, but of the order ½. There is incidentally a rather intriguing way of looking at this (3) that I can't go into now, which brings out the fact that the size of the interval Δt *is* limited really by our inability to name a smaller interval in the language whose terms are operationally defined by the apparatus we are using. So the »uncertainty principle« here is in essence a logical truism.

But the point now is that in time t, apparatus with a band width f enables you to formulate just about $2f \times t$ independent propositions about the signal amplitude, no matter how you chop up your frequency-time area, so to speak. You could either have a lot of channels of narrow band width, in which case each signal would take a long time to be succeeded by its next practically independent signal, or you could have a wide band width; then you would have many independent readings close together in time; so that his definition of »amount of information« was, again, a measure of the number of labels or blank statements with which his experimental method provided him, a priori, before the performing of the experiment. It is the structural information content of his ultimate description of the signal. A given band width, which is to be available for a given time, provides him before the experiment begins with knowledge that $2ft$ practically independent propositions (as he defined independence) could be formulated about amplitude.

Let's go on now to the complementary problem; namely, the collection of evidence by performing the experiment, or, if you like, the acquisition of a measure of *confidence* in the propositions which we are going to make, in this case about our signal voltage over a given period of time. Here we make contact with the thinking of R. A. Fisher, who, back in the early thirties or before, but in particular in his book, *The Design of Experiments,* in 1935 (6), defined what he called »amount of information« in such a way that, in the simplest case, it is measured by the reciprocal of the variance of a statistical sample. In other words, if we take the case of communication, Fisher made »amount of information« depend on the amount of noise present in a signal – to be precise, the »noise power,« or at any rate the variance of the amplitude – taking the reciprocal of it as his measure.

Well, this quantity is not dimensionless. Again, I don't want to bother going into detail too much, but it is true for a certain class of measurements that if you take the [187] ratio of the magnitude itself to the | noise amplitude and square it (the variance being the square of the noise amplitude), you get something which we can call the *amount of metrical information,* which certainly increases as the reliability of your measurement increases. If you have a voltage of 10 volts with noise of 1 volt, then that gives a more reliable reading than the measurement of 10 volts with a noise of 2 volts; and it is also, intuitively at any rate, more worth while than a measurement of 5 volts with a noise voltage of 1 volt. The signal-noise ratio, of course, is familiar to electrical engineers as related, at any rate, to the notion of reliability; and I think if we pass from this particular illustration, we can agree that, in general, this is a legitimate and distinct use of the phrase »amount of information« to represent the *amount of evidence* we have for the statement we are making about a reading. Passing further from the description of readings to the general notion of the assertion of propositions, we can say that in a representation which we have been enabled to make by certain observations, by devising an experimental situation in which a given number of structural propositional functions are provided, then we can define a measure of our total evidence for the propositions which we eventually formulate, as the total amount of metrical information provided by the experiment.

We can symbolize this in quite a simple way. Fisher's measure, or my modified form of it, which in this case is the square of the signal-noise ratio, is additive in the sense that the metrical information content of a combination of readings (such as their mean) can equal but never exceed the sum of their individual information contents. Assuming for the sake of argument that we are measuring a steady voltage, then if we take two readings and combine them and calculate the amount of metrical information in the mean, we shall find that we have just twice what we had in either of the two individually. Each structural proposition *adds* its contribution to the metrical information content of the resultant summary statement.

Now a set of independent propositions can be represented or symbolized by a set of perpendicular axes in a multidimensional hyperspace. So we can represent this additive process by a convenient geometrical vector model in which for each new independent proposition we add one dimension to our hyperspace-our »information space.« We then can take *distance* in each of those dimensions to represent some function of the amount of metrical information associated with each corresponding proposition. If each structural proposition is represented by a vector whose length is the square root of its metrical information content, then the total information content, struc[t]ural and metrical, is represented by the vector sum of the individual components.

[188] For example, if we had just two propositions, we could define their | total information content by drawing a single vector whose two perpendicular components are the square roots of the amount of metrical information in each. In the particular case of

voltage measurement, the two propositions concern two successive independent readings, and these vector components are actually proportional to the signal-noise voltage ratio. In that case, of course, the square of the length of the resultant is the sum of the squares of the lengths of its individual components and is proportional to the total energy, and so we get a representation in which additivity is preserved.

How does all this relate to our initial notion, the one which is familiar to all of us, the definition of information which Claude has given in communication theory, in terms, roughly speaking, of the statistical rarity of a representation? Well, I would suggest that if we had to communicate to somebody else a representation such as the one we have developed, we could think of our activity as instructing him to select out of a certain number of possible positions for the information vector, one representing the result that we have obtained. The tip of the vector can be represented as occupying one of a number of cells into which the space is divided or quantized. In that case (on the assumption that each position is equally probable for the sake of argument), we can take the logarithm (base 2) of the number of possible positions out of which our result has selected one, as a measure, first, of the number of binary decisions to which this selection is equivalent and, hence, as a measure of the amount of information in Shannon's sense, which you remember we distinguished by calling it the amount-of-*selective* information. From this standpoint, when we are talking about information content, we are now thinking of the problem of *replicating* by a certain procedure, a procedure in which we have a filing cabinet of all possible representations of this sort; and we have to pick one out of it, and we assume for the moment that these are all equally likely and all equally represented in our filing cabinet. In that case, the logarithm of the number of possibilities represents, you remember, the minimum number of successive questions in a game of twenty questions by which we should arrive at the point which has been specified. I won't bother to go into the case where the probabilities are not equal, beyond saying that if you like to picture the cells as deformable, then the logarithm of the number of cells will still give you the selective-information content if you warp your space so that all cells remain equally likely to be occupied.

Recapitulating, we have seen that *information* can be defined generally as that which enables us to make or add to a representation. We then distinguished between the problem of *communication* (which is the production here of a representation already in existence somewhere, | in prefabricated components at least) and the problem of *scientific description,* where your own procedure must provide you by ostensive definition with the symbols that appear in your representation. [189]

In communication between human beings and possibly between animals, the problem is ultimately the production in one reasoning mechanism of a representation – a pattern – already present in another reasoning mechanism. In the human case communication theory is interested especially in the most economical way in which we could conduct the selective operation that evokes the appropriate pattern; and since we can do this by coding, we are always prepared in principle to take the logarithm of the total number of possibilities as our measure of the amount of information given, irrespective of the properties of the pattern signified.

In the other process, the process of scientific description in which you are confronted by a situation about which you are initially wordless, your experimental method, your mode of approaching the situation, provides you with (*a*) the conceptual possibility of formulating – giving distinguishable significance to – a certain number of propositions, and (*b*) as a result of observing events, the ability to adduce evidence for these propositions.

Now, what about the concept of meaning? Suppose we forget for the moment about signals, which are very often symbols for something else, and just take the case of two propositions. In ordinary mathematical logic, one could say that if you asserted two independent propositions A and B, you have said something which is equivalent to the logical combination of these two, which could be symbolized, therefore, as a point in a diagram with four possible positions (Figure 9). In position 1, you have said both. In position 2, you say »A« but not »B,« and so on. And one can say in a rough way that the statement you make could be defined by a vector – the vector linking these points to the origin-which has four possible quantal positions.

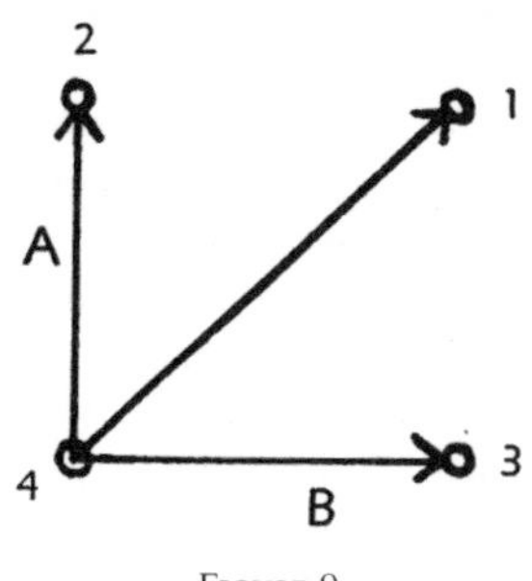

FIGURE 9

[190] | It is common experience that we do not define many of our concepts in terms of a set of unique propositions. Someone mentioned, I think yesterday, that the concept of a chair is not definable simply by enumerating a certain number of characteristics, because we all know that if a chair doesn't have a leg, we may still judge that it is a chair which has lost its leg, or something like that. I have heard of some work going on in Cambridge, England, on this point with, I believe, the conclusion that the most you can do may be to enumerate a set of possible characteristics of chairs, of which any adequate subselection constitutes a chair where found. What we would like, it seems, is some way of symbolizing the *partial* participation of one or more characteristics in a definition. The black and white of atomic yes-or-no components is too coarse for our everyday terms.

Well, now, can we sharpen up this notion of »partial participation?« What could we mean by saying that a certain term means »a little of A and a lot of B,« instead of accepting the four-way choice offered to us by conventional logic in Figure 9? Suppose we try to say that the meaning of a term represents, not a discrete selection from a set of yes-or-no characteristics, but a selection of each *in a certain proportion*. What reasonable meaning could we give to that?

What we could mean, of course, is that our definitions are ostensive – that their meaning is defined in terms of experience. Then, in the past over which a given term has acquired certain associations – to which we point, implicitly or explicitly, in order to give our ostensive definition – we have found that A was associated only 10 times, say, and B, 100 times, so that the term *means* »a little of A and a lot of B.« If you like to picture the elementary characteristics as spread out along a scale, the meaning of a term becomes a kind of spectrum, a spectral distribution, over the scale, the relative frequencies of occurrence of the different elements being symbolized by the height of the spectrum over the corresponding points on the scale.

I wonder if I am making that clear? The idea is that you think of chairs as things sometimes having four legs, backs, and what have you; then, again some chairs are backless, some chairs do not have four legs, and so on. So, if we accept these (for the sake of argument) as simple yes-or-no characters, then over a long experience of the word »chair,« we should build up a concept of »chairfulness« which could be defined by the proportions of different characters in the ensemble of all chairs experienced.

The interesting thing is that introducing this possibility would correspond in our vector model (Figure 9) to attributing significance to all orientations of the vector. Instead of having merely the possibility that the vector is vertical, horizontal, or at 45°, we now have the possibility | of conceptually infinitesimal gradations of orientation of [191] the vector. Quite precisely, what we mean by the *meaning* of a given term which is definable in terms of this space of basic vectors (elementary component characters) is the *orientation* of its representative vector – the direction which defines the proportion in which those elementary components enter into *our* experience of the ostensive definition of the term.

PITTS: You are assuming that it is an ellipsoid?

MACKAY: Let me say it again. Given two propositions, A and B, I am suggesting that a proposition which is neither A nor B nor A and B equally, or, if you like, a word which is defined neither by the character A nor the character B alone nor by both equally, but by the two in certain proportions statistically, can be defined by a direction in this space (Figure 10).

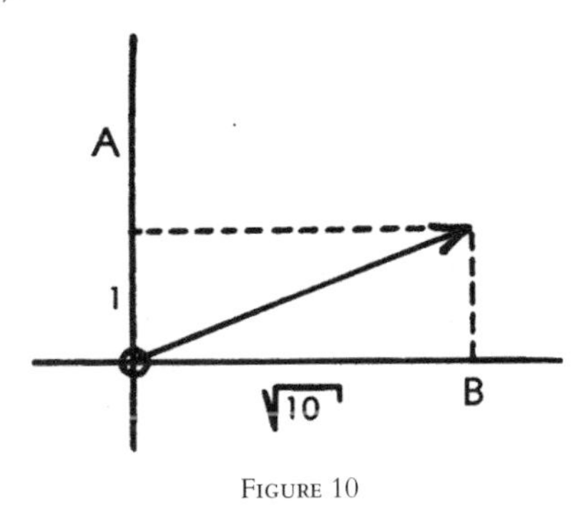

FIGURE 10

PITTS: They combine additively? That is what I wondered.

MACKAY: Yes.

PITTS: That is the point.

MACKAY: All right. Now, you can see that the concept –

MCCULLOCH: Might I make this clear for myself? »A« here is a line which represents one item of structure on a logon principle, essentially, and »B« is another proposition. The distances out on each of these are the metrical strengths?

MACKAY: Yes, the square root of metron content, to be precise. It is possible, of course, to use different metrics here. You could arrange it so that as the *probability* of inclusion goes from 0 to 1, so length only goes from 0 to 1, as in quantum theory; but if you are dealing in terms of metrical information you make no restriction, and the distance could go to infinity (which would mean in theory an infinite number of past experiences).

The choice of metric for our purpose is not particularly important. The point is that with a given metric you can give a precise significance to the *meaning of a statement for a given individual*, remembering | that each individual receiver will have his basic vec- [192] tors defined by his own receiving apparatus and a few other things – which we shall come to in Part II.

The notion of *amount* of meaning, therefore, is scarcely meaningful, and certainly not precise. So I suggest that we ought not to talk, as Wiener does, about »amount of meaning,« but that we ought to keep the concepts of information and meaning quite distinct. I would suggest that we can, in terms of this diagram, for any proposition or any body of data or any representation, define quite unambiguously what we mean by (*a*) the amount of structural information, dimensionality, number of degrees of freedom, if you like; (*b*) the amount of metrical information, the degree of confidence in, or the amount of evidence for, the assertion represented by the naming of the propositions or the making of the statement, and (*c*) the amount of selective information, measuring the statistical rarity of the selective operation specified by the naming of the total proposition. From the last point of view, when this (Figure 10) comes up, I think of it as an instruction to me to make a selection from the repertoire of my past experience to represent it. To take this very simple case, if there are only two possible choices, and if one of them has in the past been elicited ten times more frequently than the other, then one could imagine that over the course of time I would evolve a means of selecting »B,« which involved fewer binary choices than the means of selecting »A,« so that in a sense I should say that receiving the message that elicited »B« gave me less information than receiving the message that elicited »A.« Then I should be talking about *selective* information; I should essentially be referring the incoming representation to my ensemble or assembly of past-experienced situations in which the same representation arose.

McCulloch: Will you tie up which of these is now related to the concept of entropy?

MacKay: It is rather intriguing, really, because it shows how bewilderment arises if one does not define one's ensemble. I made it clear, I think, that selective information is defined only relative to an ensemble – to a filing cabinet. It is a measure of the amount of trouble it would give you to pick out the thing you are talking about in the filing cabinet if the filing cabinet is designed according to optimum principles.

Now, suppose that in the case of voltage measurement we take the definition of metrical information content as signal-noise power ratio, and suppose we inquire how much entropy change is involved in measuring a single independent reading, using the definition that one independent reading requires an »uncertainty product« $\Delta f \cdot \Delta t$ of [193] the order of 1/2. We can suppose that we have loaded our source with a matched | resistance at temperature *T*, so that the »noise power« is given by $kT\Delta f$, where *k* is Boltzmann's constant and Δf, as before, the band width. Per unit of metrical information, therefore we shall require signal-power dissipation at least equal to $kT\Delta f$, and therefore energy dissipation of $kT\Delta f\Delta t$, where Δt is the minimal duration appropriate to the band width Δf. If we now ask what the entropy increase corresponding to this transfer of energy at temperature *T* is to be, we find its lower bound, at least, by dividing the energy transferred by the temperature at which the transfer occurs – if this is in the range for which classical Boltzmann statistics hold, of course. Since $\Delta f \cdot \Delta t$ is of the order 1/2, we see that a physical entropy increase is required which cannot be less than about 1/2 *k* units. The amount of entropy change in general is proportional to the amount of *metrical* information, so –

Pitts: This is the minimum amount of entropy?

MacKay: Yes.

Pitts: But the actual – this does not establish a connection between the actual entropy and metrical information.

MacKay: This establishes the actual entropy change and the actual metrical information content, if you extract all the metrical information.

Pitts: No, the entropy change can be greater than that.

MacKay: True, but it can be equal to it under optimum conditions.

This, you see, equates the minimal entropy change to the square of the length of our information vector. On the other hand, the amount of selective information, as we saw earlier, is definable as the logarithm of the number of distinguishable positions of the information vector, which is not the same thing.

This might appear to be a paradox because, of course, when you define amount of selective information in terms of probabilities, you arrive at something which has the same form as the definition of entropy in statistical mechanics. The point is that we have begun by assuming each of those positions of the vector to be equally likely, in computing our amount of selective information. We have done so because we are prepared to operate in such a way that at our receiving end we can regard each of those signals as equally likely. Consequently, we are referring our question as to the amount of selective information to an ensemble appropriate to the assumption that all of those states are equally probable – in which, if you like, all possible states are equally represented.

When we calculate the amount of physical entropy, on the other hand, we are referring to the ensemble appropriate to a physical system in equilibrium at temperature *T*, for which not all possible states are equally probable. And I think that all the debates and paradoxes | which keep cropping up as to the relation between Shannon's amount [194] of selective information and the concept of physical entropy disappear if one asks precisely what assembly is being used for the computation of the amount of selective information. (I should be interested to know whether you agree with that, Dr. Shannon). You get the physical measure if you use an assembly defined for thermodynamic equilibrium; and you get quite a different measure, of course, if you use the artificial assembly (the filing cabinet of the receiver) that regards all states as equally likely. In that case it is the metrical information content and not the selective information content that correlates with physical entropy increase. Does that answer your question?

McCulloch: That answers my question. Are there any other questions? If there is nothing from the floor, then I think we had better develop the question of symbolism and then throw it open to discussion. I believe there is one question from Von Foerster.

Von Foerster: Can the two vectors A and B, which are also to be considered as unitary vectors, also be considered as results of two other vectors, let us say, alpha and beta, and so on?

MacKay: Yes, indeed. That does bring up a point I should have made. One naturally asks with respect to what are those different measures invariant? Well, the first measure, the structural measure, we saw was invariant with respect to subdivision of a given band width and time tract. No matter how you chop up your frequency band or time tract, you can never get more than 2 *ft* independent signal ordinates. (You can certainly get more distinguishable *values* of each ordinate, but that is a different matter.) The second measure, the metrical information content, we saw was invariant with respect to coalescence of independent readings, say by averaging – which the third is not. The third measure represents the invariant upper limit to what you can do by ingenuity in coding. Is that fair, Dr. Shannon?

Shannon: Well –

MacKay: Perhaps you could put in a sentence what the third is invariant with respect to.

Shannon: There are two things I might argue a bit here. In the first place, the structural measure is really a measure of dimensionality and, as such, is not invariant under anything you do. For example, you can map a two-dimensional region in a one-to-one manner into a one- or three-dimensional region. You cannot do it with a topological mapping, so I would prefer to characterize the structural measure as being invariant under all topological mappings. These do preserve dimension as a basic theorem of dimension theory states.

It seems to me that another way of characterizing selective information is that it is [195] invariant under all one-one probability-preserving map|pings of a space into itself. Concentrating on the distribution function of the ensemble, which is the primary concern of the selective information, it is invariant under any such measure-preserving mapping, because it is defined just in terms of the probability measure.

MacKay: Yes. You would agree, though, that from the point of view of the experimenter, who does not think so much in terms of mappings as of design, an experimenter who knows that he can play with chopping up band width – multiplexing band width and multiplexing time – the first measure is invariant with respect to his ingenuity in design. It represents the upper limit to the number of independent physical readings you can get by utilization of a given frequency band width in a given time.

Shannon: It seems to me it depends on how much chopping you are allowing; that is, there are these one-to-one mappings of a three-dimensional space into a two-dimensional space, which correspond to ways of compressing band width or enlarging band width, either way, which preserve everything – I mean, if your experimenter is rugged enough with his equipment, he can actually do that, so it is not invariant under anything he can do but only under relatively continuous things that he can do.

MacKay: But it represents an upper limit, surely; basically it means that you require $2ft$ independent ordinates to specify a signal (within a precisely rectangular band width f) which persists for time t; and there is no question but that you cannot get more ordinates out of that particular specification, because the signal is defined when you set that number of ordinates.

Shannon: Yes, that's right.

Bigelow: That isn't clear to me. In general, that is true, but is it always absolutely true?

Shannon: I think it is, but aren't we talking about a technicality here which has no bearing on the main point?

McCulloch: The point is that in one way, the first way, it preserves – what?

MacKay: The first way preserves the number of independent propositions you can make – the number of independent readings of amplitude you can take – against your ingenuity, within wide limits any way, in chopping up the band width or time. The second preserves the measure of total statistical reliability, roughly speaking – the total amount of evidence – against coalescing of two adjacent readings to become one. That is where your question comes in, Dr. Von Foerster, because we can think of any information vector whose length is $\sqrt{n}$ and whose amount of metrical information is therefore n, as the resultant of, or as equivalent to, n unitary components. In other [196] words, if you have a | space, an n-dimensional space, and a unit vector along each axis, then the square of the length of the resultant vector is n. Then you can rotate your basis so that this vector lies along one of your new axes, where you can think of it as representing just one unit of structural information carrying n »metrons« (as one calls them) – the individual units of metrical information which have gone to make up the total evidence for the structural proposition. These n metrons can be thought of, if you like, as what might have been, alas, the »stuff« of n independent propositions, but they have all been run into one: the point being that you are no longer able to distinguish *order* in this group of n units combined to form one resultant vector. You can't distinguish one from the other, and that is why they represent only log n units of selective information. They are not *bits*, They only represent log n bits because you have »folded up the umbrella,« so to speak, by combining the contents of n dimensions into one resultant vector.

Well, this business of symbolism is a long story, and time is short. Perhaps I can best introduce it by presenting to you the bare bones of a possible reasoning mechanism (7), which we must think of as dealing not only in black-and-white logic, but as capable of handling or transducing or reacting appropriately to the kind of stuff we have been talking about – information in the general sense. In terms of this concept, you remember, we find precise interpretations for terms such as *meaning* – even *shade of meaning* (both related to the orientation of our vector) and *relevance*. The last, I perhaps ought to mention, is quite precisely the square of the cosine of the *angle* between two directions in information space, in this sense: if you ask how much metrical information (evidence) is afforded to a dependent statement by a given body of information, the answer is found by squaring the projection of the information vector on the ray which defines the statement.

There isn't time to expand on this, but you can see that it does correspond fairly well to our notion of the *relevance* of data to a statement.

PITTS: What space is this, incidentally?

MACKAY: Co-ordinate Cartesian.

PITTS: I know, but when I say »space,« I mean, what do the axes represent?

MACKAY: Independent propositions.

PITTS: But here you have just a number of them. You were talking about two dependent events.

MACKAY: No, I'm sorry. These are not events. I am assuming that you have made a certain observation or statement which has a certain metrical reliability indicated by the length of this vector, and a certain meaning indicated by its orientation relative to the vector basis, the basic component propositions. And then you want to make a deduction | which amounts to, or is defined by, a different proportionality of those. [197]

PITTS: Well, have you defined how you determine the meaning which projects it along the core of axes?

MACKAY: Yes, it is just another way of representing the spectrum of meaning that we were talking about. The height, or some function of height of the spectral lines at each point representing an independent elementary component proposition, defines the orientation of the vector.

VON FOERSTER: Are the components negations? Because they are certainly independent from each other; for instance, red and nonred or something like that.

MACKAY: Those two? [indicating two perpendicular components] Oh, no.

VON FOERSTER: They do not have to be negations, because with a negation you could certainly divide the whole world into two different parts.

MACKAY: They are independent characters which are not necessarily present in unit quantity, because in the statistical ostensive history of the definition they may not have occurred with equal frequency.

SAVAGE: What does »independent« mean? Just that one can occur without the other, or that they are statistically independent?

MACKAY: The assertion of one does not affect the probability of the other.

SAVAGE: How can you tell?

MACKAY: Oh, let's say they're defined to be so. Right? Let's get on, then, to consider the kind of thing which could handle what we have called information; that is to say, essentially a probabilistic mechanism in which trains of thought (if you like the term) correspond to successive transformations of information vectors, or changes of state in which each state represents, or can be thought of as representing, a particular total proposition symbolized by certain information vectors –

VON FOERSTER: Excuse me, but if some concept is 20 per cent area and 30 per cent length and 50 per cent volume, and so forth –

MACKAY: Oh, not volume!

VON FOERSTER: I am simply picking random ones, any one you like. If another one could be tied in in the same way with different percentages, then in what sense can you talk about an angle between measuring the conditional probabilities of one on the other? You see, I don't know whether you are dealing with concepts, statements, or what, exactly.

MACKAY: A thing can't be 20 per cent volume. You may make an assertion about its [198] volume which has 20 per cent reliability, or something like that. |

VON FOERSTER: Oh, but this is an unspecified object, of which we know the probability, that it has these characters, and that is what we are plotting.

MACKAY: That's right.

VON FOERSTER: If we are talking about two different objects, how can the angle between them say anything about the probability that those statements are true?

MACKAY: If you are talking about two *different* objects, then the statements you are making about them are independent, and then their vectors are at right angles, I don't think there is any obvious fallacy.

In any case, the essential things we are going to need are, first, some means of symbolizing probability, of introducing probability into our mechanism; and, second, some means of identifying the basic vectors, the elementary components, the basic symbols, out of which such a mechanism could build its representation of a universe of discourse.

I think, perhaps, since my chief concern today is with the second problem, we might leave the probabilistic question to the last, and consider first the problem we are confronted with in asking how an organism reacting with its environment derives its vocabulary of elementary symbols. As I see it, we have a choice between (*a*) the use of the *incoming stimuli* or filtrates thereof as elementary symbols, and (*b*) the use of the elementary («atomic«) *acts of response* to received stimuli as the symbolic components for the description of what is perceived. So this is our question: Does one's internal representational mechanism describe a concept by simply asserting that it means so much of this received stimulus and so much of that received stimulus? Or are the elementary symbols going to be something else?

Again for the sake of brevity, I will jump straight on to suggest a mechanism on the second principle, without discussing now the reasons that have led me to favor it in preference to the other.

Suppose we have a device that reacts to incoming stimuli by an act of *symbolic replication*. At the very simplest level, let's consider one which has an irritable surface and is designed so that if it is excited here [indicating palm] at a single point, its reaction is initially a random hunting motion which eventually elicits a success signal. Call this, if you like, a scratch-response mechanism. It is easy to see that if this stimulus recurs often enough, you can devise a self-molding statistical mechanism (I can go into details later on, if you like) which will increase the probability that on the next arrival of such a stimulus the scratch mechanism will tend to hunt in its neighborhood. You can see that if any particular irritation is a sufficiently consistent feature of the flux of incoming stimuli, it will quite soon elicit the successful scratch response by a series of ele- [199] mentary operations between which the transi|tion probabilities are higher than they would have been originally. We think, in other words, of some form of mechanism which, beginning with random attempts to scratch and having a success signal, or more generally an evaluating mechanism, which controls the probability of a given sequence being tried the next time, arrives at what is essentially a symbolization of the incoming stimulus; namely, the sequence by which it is successfully replicated.

Suppose now that we have several of those stimuli – say, three forming a triangular pattern – and suppose that we have a mechanism whose constant activity is to »doodle« in speculative movement from one to the other; in other words, it is built to respond with initially random attempts at movement from one to the other, We will equip it with means of determining that it has successfully finished all its scratching. It goes from here to there, from there to there, and gets the sensation that it has no more scratching to do. And we will suppose again that triangular patterns recur with sufficient frequency in its incoming stimuli. Then just the same process of natural selection, which I think you can easily envisage, is all we need invoke in order to make a device that automatically discovers for itself, and names, an abstraction – a device which, when stimulated in this case with a triangular pattern, would raise the probabilistic status of the sequence, »make a line, turn, make a line, turn, and so forth,« so that we could think of the sequence as becoming one of the elementary symbols of the »experience« of this device – its name for »triangularity.« Out of the welter of all possible symbolizations, this one has acquired the dignity of a universal through recurring with sufficient frequency among the successful responses to the flux of incoming stimuli.

One can go on from this to think of *any* incoming pattern (which is, again, sufficiently recurrent or which persists for long enough at a given time to evoke a satisfactory sequence of transitions from one elementary component act of response to another). We can consider any invariant in this way as gaining the status of a »universal« in the world of discourse of this device. You see, I am going *behind* the problem of devising a machine to reason deductively, which we can take for granted well enough, to the problem of transforming the information carried by incoming »sensory« stimuli into a symbolic linguistic form suitable for such a machine.

Forgetting the probabilistic aspect of it for the moment, and just supposing it is a question of symbolizing propositions, our typical problem is this: How are you going to insure that a deductive mechanism presented with, say, a triangle in the field of its optical receptors will always make the deductions appropriate to the presence of triangularity, irrespective of size, shape, or orientation? What I have been describing | is [200] one possible way of solving the problem which in detail, I suspect, may not be very realistic; but you can see what will happen if you design the machine properly. Its definition of a particular complex universal – its *name for it* – is a compresence and/or a sequence of the elementary acts involved in responding to it.

Now comes the probabilistic aspect. These components will not always recur in association in equal proportions. Therefore, a complex universal becomes defined not by simply *enumerating* the elementary acts which go into its symbolization, but by enumerating those *plus* the *relative frequencies* with which they have had to go into its replication. And here, of course, our old information space and the cloven hoof of probability appear again; because manifestly what we have done, in effect, is to define our universal by a vector in a space which is not quantal (at this macroscopic level at any rate), but has the possibility of representing practically continuous gradations or shades of meaning. The meaning of this universal is defined by the *orientation* of the information vector, by the statistical spectrum, if you like, over the elementary acts of response which exemplars of this universal have evoked in the organism in its past.

Well, the story goes on and on. Obviously, this merely takes us past the first step. We have discussed an artificial organism which can recognize pattern in the flux of experienced stimuli – which can abstract from among its received data those relationships whose recurrence or invariance gives them the status of universals.

The next step is to consider the possibility of recognizing pattern in the flux of *perceived universals* – of abstracting from among the evoked *acts of symbolic response* those relationships whose recurrence or invariance give them the status of universals.

This, I think, is the essence of the making of *hypotheses:* to predicate the pattern of a group of abstractions. Can an artificial organism spontaneously formulate hypotheses? I think it can, certainly in the sense that we can devise a second-order probabilistic mechanism analogous to our first, but one in which each individual basic symbol now represents one complete abstraction; and the orientation of the representative vector in the corresponding space of basic vectors indicates the relative frequency of occurrence of total response sequences (in past experience if this hypothesis is to be based on experience). I believe this would be at least the simplest way to do it. It's evident that we have here the possibility of a *hierarchy of abstraction,* which could itself become the subject of discourse; but perhaps I've said enough to show you the possibilities.

Here, our device, designed probabilistically, makes its abstractions by a process of [201] natural selection. It chooses a means of (internal) re|sponse which is invariant with respect to the transformations that leave the abstraction invariant, out of the random trial of such responses, by a self-guiding process in which the statistical configuration for each next attempt is dependent upon the success of the last. Experience elevates the statistical status of certain response sequences, and these can then appear as definitive elements in the internal logical vocabulary. And then second-order abstractions (which we call hypotheses), or even *n*th order abstractions, are in principle just a repetition of this process at the next level or at a higher level.

This kind of approach to the concept of reasoning might, I think, find a place for most of the concepts that we arrive at from the other direction. Consciousness, for example – if I dare stick my neck out – might be introduced in this way: We might say that the point or area »of conscious attention« in the field of view – in a field of data – is the point or area under active symbolic replication, or evocative of (internal) response. When a man speaks to another man, the »meaning« of what the man says is defined by a spectrum over the elementary acts of response which can be evoked in the hearer. If the hearer is such that when the man raises his eyebrows something inside the hearer happens which cor[r]esponds to imitation – the internal initiation of part of the sequence that would normally lead to the raising of eyebrows, and so on – then, clearly, the *meaning* of what the man says can only be fully symbolized in terms of the full vector basis (basic-symbol complex) defined by all elementary responses evoked, which must, for example, include here the initiation of the »Internal« command to raise eyebrows, and may include visceral responses and hormonal secretions, and what have you. So, along these lines, I think one would say that an organism probably includes in its elementary vocabulary, its catalogue of »atomic propositions,« all the »atomic« acts of response to the environment which have acquired a sufficiently high probabilistic status, and not merely those for which verbal projections have been found.

Now, I shall have to stop somewhere, only briefly throwing out the suggestion that such concepts as emotional bias, and the like, would have obvious analogues in such a mechanism, in the shape of the alteration of the thresholds, or, if you like, the distortion of the probability amplitudes, appropriate to the basic vector components affected by the »bias.« The recognition of things like this as equivalent to things like that [indicating letters *d* and *p* on board], mentioned yesterday by Dr. Richards, would be expected in such a device if, before it reached the level at which these were meaningful as letters, the reasoning mechanism had already developed the habit of doing that [202] [drawing a semicircle adjacent to a line as in *d* and *p*] by some self-guiding process | leading to satisfactory replication, and regarding the corresponding sequence of internal commands as definitive of a universal. Because, you see, if you think of a servo mechanism set down on this white line (on letter *d*), and if you stand outside and listen to the commands that the servo mechanism gives itself in following the line, the command sequence will be, »Go straight ahead; turn (right) with a certain curvature until you're back on the line.« That sequence of commands *defines* the universal for that particular simple mechanism; and so you would expect that a mechanism that used this method of naming (recognizing) patterns would not distinguish between this, that, and the other variants (*d, b, p, q*) unless some experience indicating their nonequivalence caused a search so that discrimination of direction came in and added a new dimension, if you like, to the space in which the meaning was defined.

The same sort of principle would apply to the recognition of Chinese faces which we were discussing earlier. If the act of recognition involved active response by symbolic internal replication, then a sufficient experience of the nonequivalence of samples of Chinese faces would evoke elementary operations for – would raise to the status of symbolic elementary operations – the drawing in of those identifying features which are peculiar to Chinese faces. Once that had happened, then either never again, or certainly not for some time afterward, would you cease to be able to recognize them; because although they would not be internally »named« by you in faces which did not have the features, your »descriptive space« for faces would include the necessary dimensions, and you could be conscious of their absence.

Along these lines, I think one could go a very long way toward simulating what appears to be the ordinary conscious behavior of human beings. On the other hand, of course, if one were to ask whether such a mechanism could ever be built, I would take refuge for the moment in the blessed phrase »in principle,« and say that *in principle* I see no reason why it shouldn't; but I would not myself be surprised were one to attempt to devise a probabilistic mechanism with the same mobility, and so on, as Homo sapiens, if one would have to go in for mechanisms in protoplasm instead of mechanisms in copper. That seems to me to be one implication of some of the very neat tricks that we find in the central nervous system.

BOWMAN: I should like to return for a little bit to the entropy and the filing cabinets of the first half of Dr. MacKay's talk. If we start with a box that has an imaginary partition in it, and we know that somewhere in that box there is a particle, the entropy of that system is the same if the particle is on the left or on the right. It makes no difference. The information, however, is different if we have a filing-|cabinet-coded key. If [203] the fact of the particle being on the left means something to an intellect by previous arrangement, by education, then we have a system here that distinguishes quite sharply between entropy and information.

MACKAY: What kind of information?

BOWMAN: Let me use the term in quotes first, then gradually build up a definition of the term »information,« as I am using it now. Perhaps I am using it in an anticipatory sense.

If we have two things in the box, they may be on the left side or right side, or one on each; and, again, a change in their position will not affect the entropy but will affect information, if you have a proper code. With a third particle, I believe you get something different for the first time, in that there appears a sort of objective information. There is a fundamental difference in the probability of an arrangement like this, and one where you have one of the particles over on the other side.

PITTS: Excuse me, but the two halves of the box are not distinguishable. We can't tell if they are turned around?

BOWMAN: You can't tell.

PITTS: Oh, that is the point?

BOWMAN: Yes.

MACKAY: Could you define the person making the choice? That is very important. You get an answer depending on which you take.

BOWMAN: May I go on for a bit?

MACKAY: I'm sorry. I think that was Pitt's question.

BOWMAN: I should like to see if I could introduce the idea now of useful information, perhaps as a brand-new kind, as distinguished from what we have discussed before as a number pair. The usefulness of the information is a measure of coincidences between two measures. A book in your native language can convey information to you that is potentially useful. A book in a language you do not know does not, and yet may be a translation and contain inherently the same information as the book you can read. For the usefulness of information, then, we must look to coincidences between number pairs – the one, the stored, the filing cabinet, the education, if you like, and the other, what we ordinarily think of as a code which stands for something.

Now, you can easily set up some extreme examples of that. If I gave you a book in very fine print consisting all of zeros, it wouldn't mean anything, and yet, under conceivable codes, that might be a very remarkably large amount of information. On the other hand, if I had made an elaborate tentative plan with any one of you and said, »Well, now, I don't know at this time whether or not this will go through, but here is the plan,« and then on the following day, I said one word, »Yes« or »No,« that one bit [204] of information, in fitting into a coin|cidence with a whole subsequence previously stored and matched, conveys a great usefulness.

I should like to see if we can't recognize a double algebra: the filing cabinet (as Dr. MacKay termed it), or, as I would term it, the code or the education, on the one hand, and the communication or replication on the other hand. The coincidences between those two, and only those, are of value. I do not believe that we can speak in a purely objective way of information as such. Information must have meaning. It must be understood to be useful. Information without a code, or at least with some code that is so deeply innate in us as to be unrecognizable, is perhaps what we do have in the entropy in the sense of Clausius – the simple »*dq* over *t*.« There, you need no code. That is as objective as any physical quantity is. A simple statement of what it is in other objective terms would enable one skilled in the art to measure the entropy of something. I don't think that that objectivity can be applied to information. There is a subjective half, half of the number pair, that we have in our head.

SAVAGE: I should like to applaud what you have just said, and, now that I have the floor, I should like to speak about the last portion of Dr. MacKay's talk; but first I must know clearly whether he did or did not mean to say a certain thing. Did you, Dr. MacKay, intend that the automata you discussed should contain in themselves random gadgets, or did you mean only to say that they are in some sense capable of doing inductive inference? Do they operate on random strategies in the sense that Shannon's mechanical rat does not?

MACKAY: Oh, yes, that was one of the bits I had to miss. They are essentially statistical in their operation; that is to say, they contain devices of which it is meaningless (unless you go down to the molecular level) to predicate predictability. You simply have a set of statistically defined transition probabilities as, for example, in a network of thyratron tubes (gas tubes) whose bias is very near the threshold, or – actually a much more likely example – a nervous network.

SAVAGE: That, then, is what I should like to make a few remarks about. Modern mathematical statistics are largely concerned with the critique of inductive inference. Statisticians, as such, do not ask how an animal behaves, but how he should behave.

More specifically, they ask how a human being should find things out. Some statisticians, apparently beginning with Fisher, have advocated the use of random strategies in finding things out. Indeed, this is the position of almost all statisticians today, and I myself am inclined to share it. Nonetheless, it is my opinion that no one has yet clearly shown how the employment of randomized strategy can help a person isolated from others in his search for truth. | [205]

New interest in random strategy has risen from the theory of games. When one is faced with an opponent, a relatively clear case may be made for the use of them, but our dealings with nature are a different matter.

I would therefore suggest that in your automata, random strategies play a superficial role except, perhaps, as a little gloss or decoration tending toward realism. Thus, you might object to Dr. Shannon's machine on the grounds that it never makes a mistake, but it doesn't seem to me that any really deep function is necessarily served by incorporating random strategies into the automata. We should, at any rate, give fundamental thought to what role random strategies would, or do have, in behavior before assuming glibly that the automata which best mimic human behavior do necessarily rattle around inside.

MacKay: There is nothing glib about this. If you consider that the organism is designed to interact with a statistically fluctuating environment, in which it is precisely because two events are never similar that the organism must only have a probabilistic response, then I think you can see that the thing can be more efficient as a business proposition if you design it to have a spectrum of probability of operation which matches the spectrum of frequency of demand for that.

Savage: No, I cannot see that at all. I, and others, have carefully considered the behavior of an ideal statistician embedded in just such a statistical environment as you allude to, and, according to our considerations, he has not the slightest incentive to rattle around. The point is a technical one and would have to be sketched out technically, but, as I say, the conclusion is not altog[e]ther a casual one.

MacKay: Yes. Well, there are two points, of course. The first question is whether you are trying to produce something that resembles human behavior, and that certainly is one of the reasons why I think we ought to make our mechanism a statistical one. But the second question is whether a statistical mechanism could pursue an optimum strategy. Now, if your mechanism had to be prepared for one of two possible situations, of which one occurs just slightly more frequently than the other, then it seems to me that to arrange that the device always adopt one of those, instead of now trying one and now the other in appropriate proportions, is going to make it more difficult for it to operate in situations where it can only learn by sometimes making the other decision.

Savage: I can only reiterate that I believe the situation to have been explored in some technical detail, and that it has been established that the most efficient results can be achieved by following deliberate nonrandomized strategies. Randomized ones can be at best as good as the most efficient nonrandomized ones, but not better. | [206]

MacKay: That is right; they can be as good.

Savage: But not better.

MacKay: They can be devised much more simply, much more economically. They certainly can be as good. You have shown that, in effect, Dr. Shannon.

Savage: That they can be as good is obvious, because nonrandomized strategies are by definition special instances of randomized ones.

MacKay: Therefore, if you are trying to produce something which has any parallel with the mechanism of human thinking, it would be rather foolish to start with a complicated deterministic one.

GERARD: I should like to point out that, if I have followed this elegant and useful formulation, the point I raised last year about the synapse itself being analogical and not digital, and therefore operating near the threshold of response, is relevant. This is exactly what is needed to get this kind of behavior out of the nervous system, so perhaps the nondigital performance of the synapse is what makes imagination possible.

PITTS: I should like to ask Savage a rather obvious question because, well, you can see why it is obvious by what I am going to say –

SAVAGE: Is the answer also obvious? Otherwise, I would rather not try it.

PITTS: No; that is, just a superficial inspection of why the statistician uses random behavior in planning an experiment is simply, of course, this: he wants a mode of procedure which he can be fairly sure is uncorrelated with some given variable, and he feels more sure of being able to secure that requirement if he uses the random arrangement than if he uses any systematic one. At least, isn't that the usual reason given?

SAVAGE: The usual reason given by statisticians is that they, unlike MacKay, don't believe in an entirely probabilistic world. They believe in a kind of unknowability and an unknowingness which is not probabilistic. They believe in, and often allude to, a sort of absolute ignorance which defies description in probabilistic terms. In such an unprobabilistic context, the strategy proposed by Bayes, which would otherwise be the ideal one, is simply not meaningful, so here statisticians look around for something else. In particular, they sometimes find an advantage in introducing mixed strategies. This enables them to inject probability into situations in which, from their point of view, it would not otherwise occur.

BIGELOW: Does your conclusion disallow any possibility that the individual who is operating against the outside world produces a change in his environment by his action?

SAVAGE: No, his action cannot influence the facts of nature; but his fate depends, typ- [207] ically, not only on the facts of nature but also on his action. |

BIGELOW: That is to say, that the outside world actually plays a game against him?

SAVAGE: No, it isn't to say that the outside world plays a game against him. It would be absurd and ruinous for him to assume that the outside world is aiming at his downfall and destruction, for him to regard it as a competitor.

BIGELOW: But the point I make is that, essentially, if you use random strategies in the game theory, you somehow make use of a type of decision which is unknowable to your opponent each time you use it, and yet you yourself know the event of the particular decision on each occasion that you use it.

SAVAGE: Well, as I said before, in the theory of games you sometimes have a special reason to adopt a randomized strategy, for you are struggling against an opponent who, in some sense, might know what you are going to do unless you incorporate randomness in your behavior.

BIGELOW: Unless you get something of which he has zero information, essentially.

SHANNON: There are a couple of remarks I wanted to make, if I can remember what they were. In connection with this last discussion, it seems to me that a random element in the machine, in a theoretical sense, may be necessary because of such results as Church's theorem in symbolic logic, that there are some mathematical theorems that you cannot prove according to any given determinate strategy of proof you set into the machine, although a human mathematician might go directly to the result. If the random element were in the machine, it leaves the possibility open of arriving at a proof in a similar manner.

Actually this is a rather theoretical question, but on the practical end I don't think there is too much difference between a very complicated determinate machine and a truly random one. It is really a matter of the complexity of the determinate part of it, compared to the length of life of the machine. If we are constructing random numbers, for example, as Dr. Bigelow has suggested, he might, in the Maniac computer, by multiplying a pair of ten-digit numbers together, and by taking away the first and last five digits of the product, leaving the middle ten, and then repeating the process, get something which looks more or less like a random sequence, although it is perfectly determined, and you could calculate the entire sequence, knowing the first element. It will look random over a period comparable to 10^{10}, if you are working with ten-digit numbers. After that, it begins to repeat. If the numbers are being used only for that length of calculation, it doesn't matter whether they will repeat at some later point or not.

There is one other point I want to make while I have the floor. In connection with Dr. Bowman's remarks, it seems to me that we can all | define »information« as we [208] choose; and, depending on what field we are working in, we will choose different definitions. My own model of information theory, based mainly on entropy, was framed precisely to work with the problem of communication. Several people have suggested using the concept in other fields where in many cases a completely different formulation would be more appropriate. In the communication problem, entropy is the precise concept you need, because the particular problem you are interested in is, »How much channel do I need to transmit this information?« and entropy is a quantity which measures or determines the amount of channel required. So long as you ask that question, that is the answer. If you are asking what does information mean to the user of it and how is it going to affect him, then perhaps such a two-number system might be appropriate.

BOWMAN: I believe that the use of something like entropy as a measure of information is a perfectly valid basis for communication problems where the recipient of the information is assumed to have had infinite education; that is, if you are talking to a person on the telephone and the person talking to you uses words that you know, without exception, then I would say that the use of the entropy function, or something like it, as an information measure is perfectly good. If, on the other hand, you are working with a digital calculating machine that feeds information back into itself and has a very limited built-in system of codes, a small filing cabinet of education, then I think you have to regard the usefulness of the information not as a number but as a number pair.

PITTS: Perhaps you could simply alter the definition of »noise« slightly and call everything which is not in the receiver's filing cabinet »noise.«

BOWMAN: It is, to that particular receiver, in a subjective sense, but it might not be noise to a different receiver or a different recipient.

PITTS: Well, when you have the information, you are always counting it as transmission of information; that is, you are always counting it between one definite place and another place, and to say you transfer it to another receiver or a different channel so that it might not be the same amount of information is perfectly possible. It is a practical question, of course.

BOWMAN: You could take that point. I would prefer to look upon the information as transmitted as something measurable. The information from a radio program as broadcast is a measurable number of bits. The meaning of that to the receiver is something that he takes two numbers to specify: first, the number of bits that were broadcast and, second, some measure of the comprehension capacity of the listener.

PITTS: Well, you see, I was speaking exactly from the listener's point of view. If one considers nonlinear loss of information or combinations | with noise or perhaps simi- [209] lar distortions – distortions, say, of a word beyond a certain amount result in some loss of intelligibility, and result in its effective conversion from a signal into noise –

BOWMAN: I would make a big distinction, though, between a word all hashed up with noise and a word unknown to the recipient. There is a big difference there.

MCCULLOCH: Hold it a minute. Is this point fairly clear, that one of us was thinking about a sender and the other about a receiver? If you take both into account, then you are going to deal with a number pair.

PITTS: Certainly both must be considered, but the question as to how you want to calculate the information numerically depends upon the exigencies of the particular situation, of course. I am sure we are all agreed on the necessity for considering both factors. With respect to Church's theorem which Shannon mentioned earlier, just for the record I should like to say one thing about that. It is not impossible to make a machine that will prove provable theorems; but what Church's theorem asserts is that it is impossible, given the theorem, to set any upper boundary to the time it may take. It is very easy to show that you can make a machine to print all theorems because you can write out the axioms in a finite list as they are generally constructed, and you can reduce the rules of procedure to be applied to those to a small finite number. Then you can simply classify all the theorems as those which result from one application of the rule, those which result from two applications of the rule, and so forth; that is, the machine can print all theorems in order, starting from the axiom. The only point is, if you are given a theorem, you don't know how long it will be before that particular theorem shows up. A random process doesn't help because there, again, although you may be able to be sure with Probability 1 that every theorem may occur sooner or later, still you can place no upper limit to the bounds which it may take for a given theorem, so it doesn't help you there. But that wasn't one of your important points.

SHANNON: Perhaps I misunderstood the theorem, but I didn't have that impression of it.

PITTS: Well, you see, in the case which I mentioned – in the sense that the common systems to which Church's theorems apply can be listed that way – since the theorem is defined as the end result, and since the single steps are each of a mechanical character, of course, all can be obtained.

MCCULLOCH: Dr. Klüver, do you want to speak of Shannon's point? Because Shannon has two more points to bring up.

KLÜVER: I want to make sure that I got a certain point straight. You talked about the important problem of how concepts, abstractions, and hypotheses are arrived at. Did [210] you wish to imply that such mechanisms | of concept formation as you have described are really involved and occur in human thinking?

MACKAY: I am suggesting that this possibility is present in the components which we have to play with in speculative models.

KLÜVER: For instance, the formation of the concept »chair« – at least, the way I understood it – comes down to a sort of statistical consideration of its components?

MACKAY: Yes, in the shape of the modification of threshold amplitudes.

KLÜVER: I should like to make one general remark. In listening to physicists and engineers, I am generally impressed by their optimism; but in hearing Dr. Bowman today, I believe that I detected a somewhat pessimistic note.

BOWMAN: It was intended.

KLÜVER: It looks as if the human organism is often viewed here as merely a marvelous device for registering incoming stimuli, for receiving and coding of information, and for doing a large number of equally remarkable things. For the psychologist, the picture is unfortunately more complex; unfortunately he cannot see such simple outlines. To be sure, in this picture we have the influence of past experience, we have the storing of items in filing cabinets. But experience can enter the picture only because we are able to catch it by means of certain schemata. What are these schemata? And must we assume that the very schemata by means of which we catch experience can in turn be influenced by experience? At best, for the psychologist, the picture resolves itself into the formulation of a large number of unsolved or only partially solved problems. I am glad that Dr. Bowman introduced a somewhat pessimistic note when considering the properties of the human receiver.

BOWMAN: It ties in, perhaps, with some of the things we discussed yesterday, in the sense that a phenomenon observed depends on the way in which it is observed. The instrument used to observe it is an extension of the observer and is limited by his education, so that the observation made and the conclusions reached depend upon the instrument and the education of the observer in a sort of, well, as we said yesterday, complementary way.

MCCULLOCH: I think MacKay is shaking his head. When Bowman is through, you can go ahead.

MACKAY: Well, I wonder if I shouldn't make one point clear about the old business of engrams. I think the only *stable* threshold modification one can envisage would be a variation in the diameter or length of fibers or something like that. I am not suggesting chemical traces or anything similar. I am not sure whether that question was in your | [211] mind; and by talking about variable threshold, I do not mean just variation at a synapse, but anything such as facilitation and inhibition due to proximate volleys, anything which affects the probability of transmission at branch points, and various mechanisms of that sort, which affect probabilities of excitation.

I am sorry I had to condense this so very much. I probably expressed it very badly. The idea is, if you follow this through, that you can develop an evolutionary theory of perception, evolutionary in the sense that it relies on self-molding statistical processes to abstract complex percepta from this activity of elementary replication. There will be two complementary possibilities: first, quite conceivably, that what you might call the natural choices (in the way of activities of replication) are those which, because of the *hereditary* structure, have a high probability of excitation as responses; and then, second, that those will be molded and modified by the actual success evaluations and threshold feedback evoked by successful replication. That is really why I gave the example of the triangle arrived at by »doodling.« A mechanism which begins by doing nothing but doodling could evolutionarily evolve a symbol for triangularity by this mechanism. I don't see that there should be any great difficulty in going from that to higher things, particularly since one can go to the next level of abstractions about abstractions, and on up.

SHANNON: I should like to add a word to Pitts which just occurred to me with regard to the random element in a machine for proving mathematical theorems. What I really had in mind was a finite state machine, that is, a machine with a finite number of possible internal states. In such a case the machine can only go through a periodic sequence from one state to another because there are only a finite number of states that it can be in, and each subsequent state will be definitely determined by the previous one, if it is a determinate machine. On the other hand, if you have a true random element inside, it will not in general be a periodic sequence and could presumably arrive at any possible theorem.

SAVAGE: Well, conceivably, you would have technical work to do, to show that, considering that some theorems, for example, have very long statements, let alone proofs, and the machine would presumably have to bear the whole or large parts of a proof in mind in order to complete it.

SHANNON: I think it probably couldn't go through everything –

SAVAGE: But the scope could be enlarged.

SHANNON: It could come out with any sequence of output symbols, presumably. It may not have proved that it was a true theorem.

[212] PITTS: But that is the point. First, if the theorem is too long, the | memory of the machine, clearly, could not contain it. Second, if any of the necessary steps on the way to the machine is too long for the machine, it could never contain it.

BOWMAN: Isn't it quite possible for a finite determinate machine to compute a transcendental number to any desired number of significant digits? That is certainly not a periodic output.

PITTS: No, that is not possible, unless it refers to the numbers which it had previously calculated.

BOWMAN: Oh, you exclude that?

PITTS: Yes. I am trying to exclude this infinite or potentially infinite memory which it can use for calculation or for reference.

BOWMAN: Oh, yes, I see.

PITTS: It can only refer to its internal memory, so it must certainly be a periodic output.

BIGELOW: The question of the invariance of this vector system he had on the board –

MCCULLOCH: Were you interested in the problem of the invariance, as to what was invariant under what transformation?

BIGELOW: He had a propositional vector system, and then said he had put all these vectors into it, or put in something which was a mapping from one direction to another.

SHANNON: The point I was raising there was, perhaps, that old result of Kantor's that you can map a square, for example, into a line in a one-to-one way, so that what MacKay calls the structural information, Gabor's notion, is not invariant under anything you can do.

MACKAY: You mean, not invariant under *all* things you can do?

SHANNON: Yes, under arbitrary remappings. You can take two vectors, or two numbers given to an arbitrarily large number of places, and condense them into one number which contains all the information of the two which you are able to construct from this one, and vice versa.

BIGELOW: Yes, that contains all but the selective information. Your measure of selective information isn't at variance at all if you consider also as information the details of the condensations.

PITTS: You have lost the structural information, in a sense, from the composite number. Suppose I have two ten-digit numbers and I make the contention that I can make a twenty-digit number out of them, the even ones being taken in order from the first and the odds from the second. If I am given the composite number, I know not only what the two ten-digit numbers were, but also which way they were combined. I can retranslate them without loss in both directions.

BIGELOW: If you notice another additional factor, of course.

MACKAY: I have said it twice, but I have said it badly each time. This structural information-content is invariant with respect to a priori | manipulation of the design of a [213] physical method. If you can devise an experimental method by which a given frequency-time domain can give you more than $2ft$ data on amplitude, you are contradicting the theory which Shannon and Gabor and others have worked out. I think we are constantly confusing this *physical* question with what you can do to a *logical* structure by way of mapping without loss of selective information. There are any number of ways in which you can code something to get back from one representation with few dimensions to another with more, but that is a problem of communication, of specifying a selective operation, not of constructing a representation *ab initio.*

The problem you run into concerning structural information is our problem of the fly and the need for a co-ordinate system. That's typical of all. You have your crawling fly. Somehow or other, you have to be able to utter words about its position. It is the old business of Wittgenstein and, »Vovon[!] man nicht sprechen kann, darüber muss man schweigen« (If you can't name something, you can't talk about it). Therefore, you must somehow be able to call out co-ordinates that label or identify a statement uniquely. You are just using the same »blank statement« twice if you are not doing anything to distinguish two statements except in the trivial sense that you make them at two different times. For example, if I say, »The height is 6,« »The height is 5,« I am simply contradicting myself by using the same propositional function twice. But if I am able to say »The height at x = 1 cm. is 6,« and, »The height at x = 2 cm. is 5,« all is well. Do you see what I mean? Now, in a given band width and a given time, it can be shown that you have a finite number of independent propositions which you can make and must make to describe the amplitude of the particular signal you observe during that period.

PITTS: Not if the noise level is zero. If the noise level is zero, then I don't think that is true.

MACKAY: But surely what I'm saying is an old story? What you are saying is that in that case the number of *cells* in the information space is not finite.

PITTS: Well, let us ask Shannon. Does his theorem hold when the noise level is zero?

MACKAY: We are not talking about Shannon's theorem. This is a question of independent measurements of amplitude, not a question of what you can code. I think Shannon's theorem surely holds with respect to that because he takes »$2ft$« as the power of his brackets in deducing capacity. Let's put it this way. All I am saying is that the dimensionality of the space is finite and determined by f and t. The number of distinguishable points in the space, on the other hand, is a function of the noise. The noise goes down, and though the dimen | sionality doesn't change, you get more distinguishable points for a given amount of power or energy; therefore, of course, you can *symbolize* (in communication) a larger selective operation; and if you know your code you can always map from the received representation to another which shows no sign of the dimensionality of the signal. That, I take it, is what you are saying. But, you remember, I began by distinguishing between the problem of communicating representations, in which you already have the representation in file at the receiver and you want to make some sort of code wiggles on a communication line which will instruct him to draw from the filing cabinet – [214]

PITTS: When you are talking about dimensionality of the space being finite, you are using »dimensionality of space« in a rhetorical sense. What space is this?

MACKAY: It is the space of which the axes are defined by the independent readings, or the necessary definitive amplitudes, of a function limited within a frequency band f, and limited to a time period t.

PITTS: If the band width is finite, then no two readings are strictly finite, in the rhetorical sense.

MACKAY: Let's put it this way: What Shannon and Gabor and others have established is that a given band-limited function, persisting for a time t which has passed through a band width f, requires $2ft$ amplitudes to determine it and no more. Now, it is quite true that there may be instances in which the amplitudes you measure are not independent, in the sense that there is »a bit of one contributing to the reading of the other«; but between them, they do define a $2ft$-dimensional space.

BIGELOW: The accuracy of the observation here must enter somehow. The accuracy with which you are able to determine each of these discrete components –

MACKAY: No, I am sorry, but the $2ft$ theorem is a truism, as I understand it. It arises from the definition of what you mean by »frequency.« It is the same thing which gives you the quantum error.

PITTS: You say $2ft$ amplitudes will define what?

MACKAY: The $2ft$ equally spaced ordinates wilt define a function for you, which has passed through a band width f, over time t.

PITTS: Well, suppose I increase t by a number; suppose I increase the band width by some small part of the epsilon, what happens to it? Do I get one more amplitude or not? Suppose $2ft$ is not an integer?

MACKAY: Then it is a limit. The point is that you can't have more than that. But the essential thing, you see, is to distinguish between the problem of devising physical means of labeling a proposition; that is to say, confronted with initially blank experience, acquiring the ability to label uniquely something in experience, and the problem [215] of making | some conventionalized code of representation of something which is already namable, in such a way that the man from the other end can reproduce it. There are two distinct problems which have given rise to different concepts of information, and both, I think, are relevant. They can all be covered by the definition of information as »that which enables one to make a representation.« But when you talk about *amount* of information, you get the same kind of variation (though it is not a good analogy) as you get between area, length, and volume, as measures of »size.« It is a very poor analogy, but one would expect or could expect that kind of diversity in the concept. As for perception, remember I said, »unless you project the optic co-ordinate system in some way onto the screen.« What you say is exactly to my earlier point, because the perception of the movements of a fly (for which we all know perfectly well you don't need a grid on the screen), the naming of it to ourselves, would on this theory consist in the internal act of imitative or symbolic response to the movement of the fly. You see? The »basic words« of the organism are formed automatically by the acts of response.

PITTS: Let me ask this again. It seems reasonable to me at the moment, although I don't recall it, that what comes out is something with a rectangular, well, with a strictly rectangular impedance function, that is, as a periodic function, as a superposition of approximately $2ft$, or distinct periodic components. However, the first remark is that physically it is impossible to realize anything with an impedance function of that kind.

MACKAY: I agree.

PITTS: It must tail off at the ends, and, as a matter of fact, it must tail off not too fast at the ends, in a certain well-defined sense. Now, as soon as it tails off slowly enough at the ends to be physically realizable, then no longer would any finite number of amplitudes be specifying what will come out.

MACKAY: Well, what you are saying, in effect, is that you have to consider noise when you want to determine *effective* band width if you don't have the ideal case I specified. I was deliberately simplifying this so that people who weren't particularly interested in the technicalities would get a concept of structural information without worrying about which approximate number we take to measure it in nonideal cases.

PITTS: It is not relevant.

MACKAY: Noise is completely relevant if you ask how many of those amplitudes are far enough above the noise to specify one quantum of metrical information.

PITTS: No, that is not the point. It has to do with structural information. You would define »structural information« strictly for the case | which is not physically realizable [216] in principle; namely, where you have a sharp cutoff. What the engineer commonly does, of course, when he concerns himself with realizable cases and inquires what the band width is, is simply to select a certain low level of the impedance function on both sides, cut it off there, and say that that is the effective band width. But then you have an infinite number of amplitudes.

MACKAY: I did not define »structural information« in terms of frequency or time at all. I instanced that particular theoretical case to illustrate what I meant by »structural information.« I'm not at all interested at the moment in defining frequency band width for the nonrectangular case, which is a technical red herring.

GERARD: What are f and t in terms of ordinary propositions?

MACKAY: I don't quite know what you mean, but if you mean, what are the definitions of f and t, f *is* the frequency band width over which an idealized receiver is flatly responsive and cuts off at the edges, and t is the time during which you make the observation. But, you see, any –

BIGELOW: Frequencies from 100 to 10,000, say, are absolutely completely passed through, with no distortion, and with unit-multiplication factor, and everything below that and above that, say, is absolutely deleted? I am putting this in a terribly crude fashion. The band width is how wide? Is it from 1 to 100,000?

GERARD: I did understand that, but I thought MacKay generalized this specific illustration to propositions in general; the number of items that you add up with experience to make the word »chair.« What happens to these?

MACKAY: You don't have to think of frequency and time. Perhaps I should never have brought them in.

GERARD: But what are their equivalents in your structural proposition?

MACKAY: Well, consider an organism which has only two possible modes of response, by way of replication, doing this and doing that. Let's call them »pip« and »pop.« Well, then, the evocation of pip corresponds to the perception of pip in the stimulus, and the evocation of pop corresponds to the perception of pop; and something which has in the past evoked each at one time or another, 60 per cent pip and 40 per cent pop, you see, is represented by something which is not set by merely saying »pip-pop,« but saying it with modulation (Figure 11). Now, we can't do that verbally, and that is the difficulty, I think, in a great deal of discussion – I am not thinking of our discussion

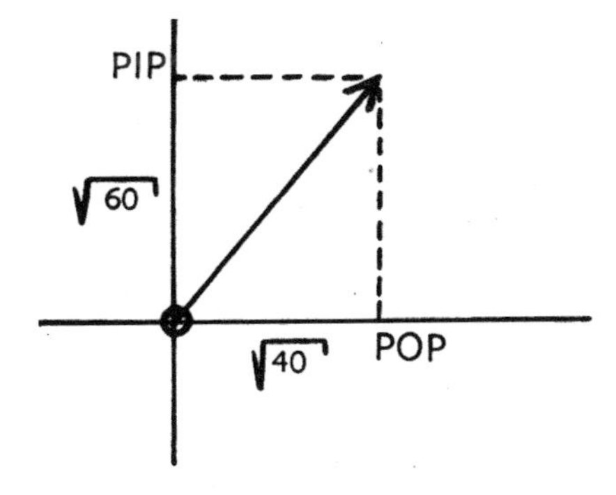

FIGURE 11

now but, in general, of human discussion – the inability to modulate the metron content (the metrical information content) of those propositions.

McCulloch: You mean you should say something like »*pip*-pop« or pip-*pop*« to do [217] it? |

MacKay: Yes. Disagreements between two people as to shades of meaning, I suspect, could often be described and clarified in those terms.

Pitts: The probability points to »pip.«

MacKay: We want a language with two dimensions, really.

Savage: I wanted very much to say a word or two about Bowman's second concept of amount of information. This theme has come around again and again. Dr. Shannon has said, »Well, I have a kind of information, but I invented it for a special purpose and it would be foolhardy to suppose that it will really work for every purpose.« Bowman has hinted that there must be a kind that somehow tells what the information is worth to the person who gets it. He says, »Well, it depends on what the person already knows.« I think he might also have said, »It depends on what the person wants to do.«

If I tell you something, however intelligible, but something you don't want to know, as when I tell you exactly how some industry is carried out in some far corner of the world, supposing you are not curious about such things, you may get all I say but you find it worthless. I think, in the last analysis – and I do not mean this facetiously, though I am saying it in a compact form – the *value* of information is its cash value, understanding cash in a rather metaphorical sense. If you want to talk seriously about what information is worth to the receiver, you must have in mind what it is the receiver wants to do, and you must have in mind some quantitative description of his well-being. Then you must ask yourself, »How does this information contribute to his well-being?« There are two questions to be distinguished here. I may be about to get information; for example, I may be about to do an experiment. Then I can say, »Well, what is it worth to do this experiment? What would I pay to do the experiment?« Again, the experiment being done, I now occupy a different situation, am poorer or [218] richer. If, for | example, the experiment consists of opening a letter from the Fuller Brush Company, to whom I have recently applied for a job, if, on opening it, I find I've got the job, well, then I am ten times as rich as I was.

Therefore, there are these two: There is the cash value of an opportunity to get information, and there is the shift in your cash assets, or in your personal evaluation of your cash assets, on receiving the particular information. There is no shortcut avoiding such analysis to that, I believe. Statisticians have sought one for over fifty years now, and in the last eight or ten years they seem to have been settling down to the view which I am expressing; namely, that only a detailed economic analysis of the situation can properly appraise the value of information.

McCulloch: I am afraid »value« in your sense would turn out to be a multidimensional affair, with very little chance of its being simplified to a measure.

Savage: No, it is simply one-dimensional. The values to which I have referred as »cash« may be measured in actual cash, or lives saved, or whatever it is in the given situation which properly measures the well-being of the individual concerned. This value is, in fact, to refer you to a theme familiar to us, the Von Neumann-Morgenstern utility (8).

McCulloch: Very familiar and very illusory.

Pitts: The El Dorado.

Savage: In the two hundred years since the concept of utility has been introduced, it has been known by several names, but surely, especially since Von Neumann is one of our members, we will continue to call it by the name preferred by him.

There is one misunderstanding. I said »cash« because that is a short, vulgar word, but the Von Neumann-Morgenstern utility theory is sophisticated enough to capture other things, including even the joy of discovery, and, in principle, there need not be a multidimensionality to the situation; that is, in principle, in so far as the Von Neumann-Morgenstern theory really works, it captures and compares in one measure, and in a way successful for this context, cash, admiration, a satisfaction of discovery, the saving of life, and all the sort of things that we do value. It is an elaborate, rather farfetched construct, but there is some satisfaction in it, too.

BIGELOW: I should like to raise a point to see if MacKay and Shannon would comment on it. It has always seemed to me that the concept of information that is used in these theoretical analyses of communication does not really ever speak about information *in abstracto* nor define it *in abstracto,* but is concerned with something like the capacity of a channel or the capacity of a system or the capacity of a structure; situations which are conceptually somewhat *concrete.* But if you once | discuss a communications channel and determine that it has a certain information capacity, then, in various ways you can produce paradoxes resulting from attempts to think about information *in abstracto* as actually flowing through it; for example, if the channel is in any static state, one is at a loss to know whether, as a matter of fact, information is continuing or not. Every state of the channel constitutes information transmission and every state of the channel may constitute full use of the information capacity. [219]

Actually, what is done here is to define a capacity rather than to define the concept of information in any concrete or abstract sense. Would you comment on this point?

MCCULLOCH: Will you hold your answers to that and let's take Pitts's question; then we will answer all of them.

PITTS: I was just going to make one very short remark. I think the missing character comes not between the transition from capacity to past information and the amount of information, but in supposing that we can divide the compound phrase »amount-of-information,« into »amount« and into »information.« What Shannon has defined is »amount of information.« I think that should be regarded as a specific phrase, and possibly there should be some Greek word invented to convey it by itself.

SHANNON: I agree with what Pitts has just said, and I should like to add that we should also remember that this kind of information is an ensemble concept. It is not a statement about a proposition, if you like, or a fact, but a statement about a probability measure of a large ensemble of statements or propositions or facts. It is a measure of a kind of dispersion of that probability distribution.

I think perhaps the word »Information« is causing more trouble in this connection than it is worth, except that it is difficult to find another word that is anywhere near right. It should be kept solidly in mind that it is only a measure of the difficulty in transmitting the sequences that are produced by some information source.

Now, there is one other point I want to bring up in connection with the ideal filter. It is not possible to obtain a sharp cutoff over a limited band width with zero phase, that is, with no delay. Bode's loss-phase relations show that this is impossible. But it is possible to obtain this sharp cutoff if you are willing to allow delay. As you allow larger and larger delays, you can come closer and closer to a square gain function over a limited frequency band. This means that we cannot determine these co-ordinates that we are talking about, or perform these ideal operations instantaneously, but to do so we must wait for some time until these tails for that delay have been reduced to a tolerable amount.

PITTS: Except that keeping the same rate of decline at the edges, you | can increase its absolute rate by increasing the delay, but with any finite delay, of course, the approach [220] is fixed independent of the delay. For example, you can't realize the function of a square with any amount of phase.

SHANNON: No, but we can realize a nearly square function which has a total area outside its square function less than an arbitrarily small epsilon, which means that only epsilon power is getting through from outside the frequency range in question.

PITTS: But you can't make the rate of decline at the edges much faster than –

MCCULLOCH: All right; we get the point. Will you close, Dr. MacKay?

MACKAY: Well, when I called my talk, »In Search of Basic Symbols,« I meant »search,« and I suppose it is partly my own fault that we did not pursue our search very far. I should like, however, to suggest that if those who are more competent can give some thought to the problem of identifying the basic symbols in the human thought mechanism, or, rather, the basic physiological response acts in the human being which might conceivably form and define the basic vectors in his symbolic information space, it would be a matter of some interest, as representing or suggesting possible tests.

As to the discussion about information, let me say first that I was not presenting in any way an alternative theory to Shannon's, but a wider framework into which I think his ideas fit perfectly. In fact, as I tried to make clear, what I was doing was to show how Shannon's ideas were validly applicable, I think, universally, though such an analysis is not necessarily *appropriate* to all problems. I think he is too modest in his disclaimer, because as long as you define your ensemble, then his definition of the entropy or selective-information content of the ensemble, it seems to me, is always applicable without any contradictions. The trouble people get into, I think, is always due to the fact that they do not define the proper ensemble; and that is particularly difficult when the ensemble is defined by the internal statistical structure of the man receiving information; that is to say, where only subjectively apprehended probabilities define your ensemble. But if you do define your ensemble, I can only say that I haven't come across any cases where that concept cannot be validly applied, and usefully applied if the problem is one of communication. Naïve equation of selective information content with *physical* entropy, however, is generally unwarranted, as we have seen.

When dealing with digital computing machines, and so on, the problem of defining the ensemble is particularly difficult, and it is possible that the best interpretation of amount of information is obtained by simply totting up the number of atomic propositions in a situation, | [221] because the machine presumably is working into a whole host of different ensembles corresponding to the expectations of the different people who are reading it. There would, I think, be no more reliable way of getting a measure than by totting up the propositions. My own theory, incidentally, in so far as it has been different from what has been done by other people, is characterized by attending to this notion of *number of atomic propositions* as suitable measures of each of those three senses of »amount-of-information.« I do agree very much that the expression should be hyphenated. I suggested (3) (4) »logon content« and »metron content« as names for »amount-of-information« in the structural and metrical sense to avoid using the mischievous word. One could talk about »number of bits« for Shannon's measure, but there doesn't seem to be any obvious word, as he says. I would make a plea that if we are going to talk about information, we should use adjectives – perhaps mine are not good – but some adjectives such as »selective,« »structural,« and »metrical,« to distinguish what we are talking about. It is my experience that as soon as one becomes accustomed to thinking in those terms, the majority of arguments one hears about »information« dissolve; and it is sad if clarity is lost in any way through not using distinguishing symbols for different concepts.

The only other thing I should like to say is that thinking in those terms, I believe, provides one with a sharp definition of »communication« in terms of the replication of representations, which I feel ought to be used as a razor in situations where people attempt to talk about the physical world »communicating with you« in an experiment. I think there is an inevitably anthropomorphic connotation in the word »communication« which bedevils discussion, and that such a use of the term would certainly be invalid in this case, and should apply only in cases where a representation is already present at the sending end of the process. That, of course, is only a personal plea, but I feel that we are in danger of getting this subject into disrepute among those sketchily acquainted with it if we ourselves do not make as many new symbols, in the shape of technical terms, as the diversity of our concepts requires. I don't think we should be frightened off from doing so by the fact that it may take some time to be understood.

APPENDIX I THE NOMENCLATURE OF INFORMATION THEORY[1]

DONALD M. MACKAY
King's College, University of London

THE EXPLANATORY survey of nomenclature which follows is not a glossary in the sense of an agreed list of standard terms in Information Theory. No such agreement yet exists in this new subject. Its purpose is rather to collect and collate as many as possible of the terms which are in current use, and to define tentatively the ways in which they are related and the senses in which they may be interpreted without conflict. Only time will show whether these interpretations will be adequate or acceptable; but if it succeeds only in demonstrating the complementary and noncompetitive relationship of different current approaches to the problem, this somewhat Augean task will have been worth attempting.

The glossary is not of course meant as an »introduction for beginners,« so much as a logical framework in which terms may be seen in perspective as they arise. The proportions of space devoted to different aspects of the subject are therefore no indication of their relative importance, being dictated merely by the exigencies of exposition. Although the approach is bound to be a personal one, a considerable effort has been made to base it on logical consideration of what has actually been meant by various authors, in consultation as far as possible with other speakers. The particular terms »metrical,« »structural,« and »selective« information content may possibly be thought unsuitable; but the logical distinctions for which they stand appear to be essential. Confounding of these is believed to be the source of much unedifying debate.

In everyday speech we say we have received information when we know something that we did not know before: when »what we know« has changed. If then we were able to *measure* »what we know,« we could talk meaningfully about the *amount* of information we have received, in terms of the measurable change it has caused. This would be invaluable in assessing and comparing the *efficiency* of methods of gaining or communicating information.

Information Theory is concerned with this problem of measuring changes in knowledge. Its key is the fact that we can *represent* what we know by means of pictures, sentences, models, or the like. When we receive information, it causes a change in the symbolic picture or representation which we would use to depict what we know. It is found that changes in representations can be measured, so »amount of information,« actually in more than one sense, can be given numerical meaning. It is as if we had discovered how to talk quantitatively about size through discovering its effects on measuring apparatus. We should at once find that it had the quite different but complementary senses of *volume*, *area*, and *length* – if not others. The analogy is potentially misleading, but may show us what to expect. |

Right at the start the term »information« takes on two different kinds of meaning in answer really to different kinds of question. An example will illustrate the point. Two people, A and B, are listening for a signal which they know will be either a dot or a dash. A dash arrives. A makes various measurements, represents what has happened by a graph, and asserts that there was »a good deal of information« in the signal. B says: »I

1 Prepared for the Symposium on Information Theory held in London in September, 1950. Revised in April, 1951.

knew it would be either a dot or a dash. All I had to do was to choose one or other of these prefabricated representations. I gained little information.«

A and B are not, of course, in disagreement. For lack of a vocabulary, they are using the term »amount of information« as a measure of different things. A is using the term in the sense of what we call »Scientific Information Content.« This in itself has two aspects, relating roughly to the number of independently variable features (structural information content) and the precision or reliability (metrical information content) of the representation he has made. The knowledge which he says has increased is knowledge of what has actually happened and been observed.

What of B? He was not waiting to observe everything that happened. He already knew that *for his purposes* only two kinds of representation would be needed, and he had prefabricated one of each. The knowledge he acquired was knowledge of *which representation to select.* B was therefore using »information« in the sense of »that which determines choice,« which we may call *selective information.*

One word which was unexpected would yield B more selective information than a whole message which he was already sure he would receive.

A's approach is typical of the physicist, who wants to make a representation of physical events which he must not prejudge. B's is typical of the communication engineer, whose task is to make a representation, at the end of a communication channel, of something he already knows to be one member of a set of standard representations which he possesses. His concern is therefore not with the size or form of a representation, but with its relative rarity, since this will govern the complexity of the »filing system« he should use to identify it. Each, however, may on occasion find both approaches relevant to different aspects of his work.

To sum up, if we ask how much information there is in a given representation, we may mean: »How many distinct features has it? How many elementary events does it describe?« in which case we require answers in terms of *scientific* information content; or we may be ignoring questions of the size and complexity of the representation, and thinking instead of the complexity of the selection process by which it was identified, meaning: »How unexpected was it? How small a proportion of all representations is of this form? In how many steps were you able to identify it in your ›filing cabinet‹ of possibilities?« In this case our question refers to *selective* information content. Rarity here is the touchstone, as against logical structure in the first case. It will be realized – and this may be an important help to the understanding of the subject – that the term »information« means something quite distinct from »meaning.« If the reader begins by divorcing the two completely, he may find it easier to trace the connections in any subsequent reunion.

EXPLANATORY GLOSSARY

1. The Scope of Information Theory

Representations

1.1 Information Theory is concerned with the making of *representations* – i.e., symbolism in its most general sense.

1.1.1 By a representation is meant any structure (pattern, picture, model), whether abstract or concrete, of which the features purport to symbolize or correspond in some sense with those of some other structure.

1.1.2 The physical processes concerned in the formation or transformation of a representation are thus distinguished from other physical processes by the element of *significance* which they possess when conceived as representing something else.

1.1.3 For any given structure there may be several equivalent representations, defined as such through possessing certain abstract features in common.

1.2 It is these abstract features of representations which are of interest in Information Theory. Its aims are (*a*) to isolate from their particular contexts those abstract features of representations which can remain invariant under reformulation,[1] (*b*) to treat quantitatively the abstract features of processes by which representations are made, and (*c*) to give quantitative meanings to the several senses in which the notion of *amount* of information can be used.

1.3 The scope of Information Theory thus includes in principle at least three classes of activity:

Scientific Information Theory

1.3.1 Making a representation of some *physical* aspect of experience. This is the problem treated in *Scientific Information Theory.*

1.3.2 Making a representation of some nonphysical (mental or ideational) aspect of experience. This is at the moment outside our concern, being the problems of the Arts.

Communication Theory
Communication Channel

1.3.3 Making a representation in one space *B,* of a *representation* already present in another space *A.* This is the problem of *Communication Theory, B* being termed the *receiving end* and *A* the *transmitting end* of a *Communication Channel.*

Space

1.3.3.1 By a *space* is meant any physical or abstract mathematical co-ordinate framework or manifold, of any number of dimensions, in which the elements of a representation can be ordered.

1.4 These classes are not, of course, exclusive. The problem of communication in particular is seldom separable from one or both of the first two. In its present state of development, Information Theory is concerned mainly with (1.3.1) the problem of representing the physical world, and (1.3.3) the problem of communicating representations (of any kind). It is communication theory (1.3.3.), however, with its immense practical importance, which has received the greatest attention; and it is only the log-

1 This aspect of the subject is already an established branch of mathematics under the name of Representation Theory or Abstract Group Theory.

[225] ical priority of the other two which prevents it from coming first on the list. |

2. Information

Information

The foregoing definition of the scope of Information Theory provides the necessary background for the definition of *Information.*

2.1 In all its senses, the term can be covered by a general »operational« definition (i.e., a definition in terms of what it *does:* as [e.g.] *force* is classically defined in terms of the acceleration which it causes or could cause). The effect of information is a *change* in a representational construct.

2.1.1 Information may be defined in the most general sense as *that which adds to a representation.*

2.2 This leaves open the possibility that information may be true or false.

2.2.1 When a representation alters, we define the new information as *true* if the change *increases* the extent of correspondence between the representation and the original.

2.2.2 The information is said to be *false* if the change *diminishes* the extent of this correspondence.

2.2.3 Strictly, the truth or falsehood is an attribute of the resultant representation, but it is customary to attribute it to that which has given rise to the change in the representation.

3. Measurement of Information Content

Two quite different but complementary approaches are possible to the measurement of Information Content, and have given rise to uses of the term in quite different senses.

3.1 From a quantitative analysis of what a representation *portrays,* we can isolate fundamental numerical features common to all its equivalent representations, and can say that they constitute the »corpus of information« which it contains or represents.

3.1.1 »Information Content« in this context is a numerical index (in one sense or another) of the »size« of the representation itself.

3.2 But if instead of asking »How many elements, and so forth, are there in this representation?« we ask, »In how many stages, and in what way, has it been *built up?*« we arrive at a different kind of measure. This becomes clear if we consider that the same representation could be constructed in a number of different ways according to the amount of prefabrication used.

3.2.1 »Information Content« in this context is a numerical index of the complexity of the *construction process.*

3.3 In the following paragraphs 4 and 5, these two approaches will be discussed. The first has given rise to two complementary definitions of what has been called »amount of information,« and the second to a third. These, however, are not rivals, but are autonomously valid measures, appropriate in answer to different questions. |

4. Analysis of Representations

[226]

4.0.1 Representations communicable in a two-valued (yes-or-no) form are necessarily *quantal* in structure, since an »imperceptible change« in a two-valued logical form is by definition meaningless. All the *changes* are discrete, therefore the elementary concepts of logical representations are discrete and enumerable.

4.0.2 In fact, a large class of such representations can be reduced to a form made up only from identical elements, so simple that their only attribute is *existence* (1). This fact provides a basis for the quantitative analysis of such representations (2). (Representations not amenable to precise logical description have not so far been considered in the theory, though a large class of these might be handled in terms of approximate quantal equivalents representing »upper and lower bounds« to their logical content.)

4.0.3 In general a pattern reduced to such fundamental terms will contain a certain number of distinguishable groups or clusters of elements, the elements in each group being indistinguishable among themselves. There are thus two numerical features of interest: (*1*) the number of distinguishable *groups* or clusters of indistinguishable elements in a representation, and (*2*) the number of *elements* in a given group or cluster.

4.0.3.1 The number of groups, and the numbers of their elements, may be thought of as respectively analogous to the number of columns and the number of entries per column in a histogram.

4.1 *Structural Information*

Structural Information Content

4.1.1 The number of *distinguishable groups or clusters* in a representation – the number of definably independent respects in which it could vary – its *dimensionality* or *number of degrees of freedom* or *basal multiplicity* – is termed its *Structural Information Content*.

Logon

4.1.2 The unit of structural information, one *logon* (3) (2), is that which enables one such new distinguishable group to be defined for a representation. Thus structural information is not concerned with the *number* of elements in a pattern, but with the possibility of *distinguishing* between them.

4.1.2.1 For example, if we are counting identical sheep jumping a gate and have no sense of time, our result can only be represented by a certain total number; but if we have a clock, we can now define what we mean by »the number in the first minute« and »in the second,« and so forth, and represent our result by a set of *distinguishable* subtotals. The clock has provided *Structural* information. In a similar way the ability to distinguish (e.g.) spatial position along the gate would provide distinguishable subtotals and hence increase the structural information content of our representation.

Logon Content

4.1.3 *Logon content* is a convenient term for the structural information content or number of logons (number of independently variable features) in a representation (e.g., the number of independent coefficients required to specify a given wave form over a given period of time). |

[227] **Logon Capacity**
synonym:
Logon Density

4.1.4 The number of logons provided by apparatus per unit of coordinate space (centimeter, square centimeter, second, etc.) is termed its *logon capacity.*

4.1.4.1 For example, a channel whose band width permits of X independent amplitude readings per second has a logon capacity of X; in a microscope, logon capacity is a measure of *resolving Power;* and so forth.

Structural Scale Unit

4.1.5 The reciprocal of logon capacity is termed the *structural scale unit* for the apparatus.

4.2 *Metrical Information*

Metrical Information Content

4.2.1 The number of (indistinguishable) logical *elements* in a given group or in the total pattern is termed the *Metrical Information Content* of the group or pattern.

Metron

4.2.2 The unit of metrical information, one *metron,* is defined as that which supplies one element for a pattern. Each element may be considered to represent one unit of evidence. Thus the amount of metrical information in a pattern measures the *weight of evidence* to which it is equivalent. Metrical information gives a pattern its weight or density – the »stuff« out of which the »structure« is formed.

4.2.3 In a scientific representation each metrical unit may be thought of as associated with one elementary event of the sequence of physical events which the pattern represents.

Metron Content
synonym:
Information Content

4.2.4 Thus the amount of metrical information in a single logon, or its *metron content,* can be thought of as the number of elementary events which have been subsumed under one hand or »condensed« to form it. For example, in the case of a numerical parameter, this is a measure of the *precision* with which it has been determined.

4.2.4.1 Notice that these elements are indistinguishable, so that their number is *not* the number of binary digits (5.1.3) to which the logon is equivalent.

4.2.5 When we come to represent the results of physical observations, we are often interested in magnitudes which are not directly proportional to the metron content. The representations we use do not then show the metron content explicitly. It must be clearly realized that metron content as defined is a measure of the number of elements appearing when what is believed to have happened is represented *in its most fundamental physical terms.*

Conceptual Scale

4.2.6 Thus, if an estimate is made of a parameter from a statistical sample, the elementary events concerned are the arrivals of »unit contributions« to the sample. These could be represented by the number of intervals occupied on a *conceptual scale* proportional to metron content. | [228]

4.2.7 On the other hand, the usual representation shows the magnitude of the parameter concerned, and not generally the metron content, on a linear scale, graduated in elementary intervals which in the useful limit are just large enough to give the representation of the magnitude (scale reading) a probability of ½. Such a scale is termed a *proper scale.* Now the probable error, in a normal population, is inversely proportional to the

Proper Scale

square root of the size of the sample. Hence the magnitude in such a case would be shown as occupying on the proper scale a number of elementary intervals proportional only to the square root of the number of elementary events. The number of metrons is (in this special but common case) the square of the number of occupied intervals shown in this less fundamental physical representation.

Numerical Energy

4.2.7.1 In connection with radar information, the term *numerical energy* has been used to represent what is essentially the metron content of a signal. It is the ratio (Total Energy)/(Noise power per unit band width).

4.2.8 In general, then, a clear distinction exists between (*a*) the fundamental representation on a conceptual scale showing the invariant number of logical elements, and (*b*) the representation of the magnitude which is of practical interest. In fact, the connection between the two is little closer than that between the precision with which a given variable can be measured, and its magnitude. Precision increases monotonically with metron content, but few quantities are *linearly* related to metron content. *Power* and *energy* in the classical case are among the few exceptions; this accords with their apparently fundamental status among physical concepts.

Scale Unit

4.2.9 The *scale unit* of a magnitude is the minimum interval in terms of which the scale can usefully or definably be graduated.

4.2.9.1 For a magnitude imprecisely known, it is defined as above (4.2.7) to be equal to the probable range of error. A magnitude supported by a single metron occupies just one interval on such a scale. In practice larger units are often used – e.g., range of standard error.

Coincidence Relation

4.2.9.2 But it should be remembered that in *theoretical* representations the size of the scale unit is generally limited by our inability to *define* a smaller unit in terms of the *coincidence relations* out of which physical representations are constructed.

Metron Capacity
synonym:
Metron Density

4.2.10 The number of metrons per unit of co-ordinate space is termed the *metron capacity* or *metron density* of a physical observation system (*cf.* 4.1.4).

Conceptual Unit
undesirable synonym:
Metrical Scale Unit

4.2.11 The co-ordinate interval over which one metron is acquired is termed a *conceptual unit* or (undesirably) a *metrical scale unit* of co-ordinate.

4.2.12 It will be noted that metron content is necessarily positive.

4.2.13 Returning to the example of the jumping sheep, let us now suppose that we are trying to determine a figure for the average value of some parameter of a sheep in each group which we are able to distinguish. Assuming for the sake of illustration that the parameter is normally distributed, the metron content of each esti|mate would be proportional to the number of sheep per group, and the probable error in each estimate would be inversely proportional to the square root of this number. Hence the number of proper scale intervals occupied by the estimated parameter would be proportional to the square root of the metron content of each group. [229]

4.2.14 The term »amount of information« in a sense analogous to this was first used by R. A. Fisher (4), who defines it as follows: Suppose we have a probability distribution function $f(x, x_0)$ showing how in a given population a variable x is distributed about a parameter x_0; e.g.,

$$f(x, x_0) = Ae^{-(x - x_0)^2/2\sigma^2}$$

The »amount of information« in n samples from the population is defined as *n times the weighted* mean of

$$\left(\frac{\partial \log f}{\partial x_0}\right)^2$$

over the range of ∞; i.e.,

$$n\int_{-\infty}^{\infty}\left(\frac{\partial \log f}{\partial x_0}\right)^2 f dx$$

Equivalent forms are

$$n\int_{-\infty}^{\infty}\frac{1}{f}\left(\frac{\partial f}{\partial x_0}\right)^2 dx$$

and

$$-n\int_{-\infty}^{\infty}\left(\frac{\partial^2 \log f}{\partial x_0^2}\right) f dx$$

In the case of a normal distribution, this reduces to the reciprocal of the variance, provided that the range is independent of x_0. It is thus a direct measure of precision, though it is not dimensionless unless suitably normalized.

4.3 *Representation of Information*

The Information Content of a given representation is specified by setting down the metron content of each logon. This may be represented in various ways.

Information Vector
Information Space

4.3.1 One convenient method is to use a multidimensional *information vector* in an *information space* of which each axis represents one logon. The squares of the components of this vector are the metron contents of the respective logons. Thus the square of the length of the vector itself is the total metron content, the sum of the individual metron contents. |

[230]

4.3.1.1 In this representation the angle between two directions has a direct interpretation as a measure of *relevance.* A dependent statement is defined by a ray in the space, and the metron content afforded to it by the information is found by squaring the projection of the information vector on the ray.

4.3.1.2 A new *complete representation* may be set up by supplementing this dependent statement (4.3.1.1) by a set of others represented by orthogonal rays. This process amounts to a *rotation of axes* which leaves us with a new total metron content equal to the old.

4.3.2 The same processes can be represented in terms of matrix algebra, if the metron contents of logons are set out ini-

Information Matrix

Trace
synonym:
Spur, Characteristic, Diagonal Sum

tially as the elements of a diagonal *Information Matrix*. Dependent statements now define vector functions, and their metron content is found by forming scalar products of the form $\Phi' \cdot I\Phi$, where I is the information matrix, Φ a vector function, and Φ' its transpose. Under all complete (»unitary«) transformations, the *trace* or sum of the diagonal elements of I remains invariant, being the total metron content.

4.3.3 An alternative geometrical representation suitable for some particular cases employs a three-dimensional histogram having a co-ordinate and its Fourier transform (e.g., time and frequency) as its two basal axes. Since the number of logons provided by given apparatus is proportional to the product of band width *(q.v.)* and conjugate co-ordinate, the base is divisible into equal cells each representing one logon. On each cell is erected a column having a height proportional to the logarithm of metron content. This gives the total *volume* of the histogram the same qualitative significance as the logarithm of the volume spanned by the information vector of 4.3.1.

5. Communication: Replication of Representations

5.1 The problem of communication usually concerns representations of which all parts *exist already* in the past experience of the receiver. In other words the receiver already possesses prefabricated components of the representation.

Ensemble

5.1.1 In fact it is generally assumed to be known that the complete representation to be replicated is one member of a finite *ensemble (q.v.)* of possible originals, some of which have in the past been communicated more often than others. We may then say that these more common messages »give less information« than the others, using the notion of »information content« now in the important sense of Para. 3.2.1.

5.1.1.1 Our message is here thought of as telling us to select one prefabricated representation.

5.1.1.2 We are not now asking »How big is it?« or, »How much detail has it?« but rather, »How unusual or unexpected is it?« »How much trouble will it take to find it in my ensemble?«

5.1.2 A convenient measure of information content in this sense is the *negative logarithm (base 2) of the prior probability* of the representation concerned (5) (6). | [231]

5.1.2.1 The base 2 is chosen because a selection among a set of n possibilities can be carried out most economically by dividing the total successively into halves, quarters, eighths, etc., until the desired member is identified. The number of stages in this process is then the integer nearest to, and not less, than $\log_2 n$.

5.1.2.2 The information measure so defined equals the *number of independent choices between equiprobable alternatives* which would have to be determined before the required representation could be identified in the ensemble of which it is a member. (The prior probability measures the fraction of the members of the ensemble which are of the required kind.)

5.1.2.3 This, like the first two measures (Paras. 4.1.1 and 4.2.1), represents a number of logically »atomic« propositions;

but this time they define, not the representation, but the selection process leading to it.

Selective Information

5.1.3 Information in the above sense of *that which determines choice* may be termed *selective information.*

Binary Digit
synonym:
Bit

5.1.4 The *unit* of selective information, one *binary digit* or *bit,* is that which determines a single choice between equiprobable alternatives.

Entropy

5.1.5 In a long sequence of different representations of which the *i*'th kind has a prior probability p_i (and hence an average frequency of occurrence p_i), the average amount of selective information per representation is evidently the weighted mean of -log p over all kinds of representation, or $H = -\Sigma p_i \log p_i$, which, apart from some ambiguity of sign in the literature, is also the standard definition of the *entropy* of a selection.

5.1.5.1 Where representations take the form of continuous functions, H takes the form $-\int f(x) \log f(x)\, dx$, where $f(x)$ is the probability distribution of the representative variable x. It is thus the weighted mean of [$-\log f$] over the range of x.

5.1.6 In practice the receipt of a communication signal disturbed by noise merely alters the form of $f(x)$ (generally narrowing it), and does not specify x uniquely. The amount of selective information received is then defined as the *difference* between the values of H computed before and after receipt of the signal.

Conditional Entropy

5.1.7 When two representative variables, x and y, say (discrete or continuous), are in question, knowledge of the value of one may affect the prior probability of the other. In the above notation, $f(y)$ depends on x, so that the entropy H of y will also vary with x. If we picture an ensemble in which values of x occur in their expected proportions, we define the *conditional entropy* of y, $H_x(y)$ as the average value of the entropy of y (calculated for each value of x) over all members of this ensemble.

5.1.7.1 This may also be described as the »weighted mean entropy« of y, weighted by the probability of getting the different values of x. It therefore measures our uncertainty about y when we know x.

[232]

5.1.7.2 An analogous conditional entropy $H_y(x)$ can be defined for x when y is known. |

Equivocation

5.1.8 Where x and y represent respectively the input and output of a noisy communication channel, the conditional entropy $H_y(x)$ is termed the *equivocation*. It is a measure of *ambiguity.*

Capacity

5.1.9 The number of bits per second which a channel can transmit is termed its *capacity.* For the case (5.1.8) it is defined as the maximum of $[H(x) - H_y(x)]$.

Relative Entropy

5.1.10 The ratio of the entropy of a source to the maximum value, which it could have while using the same symbols is called its *relative entropy.*

Redundancy

5.1.10.1 *One minus the relative entropy* is termed the *redundancy.*

5.2 These considerations suggest a more economical method of communicating a representation.

5.2.1 Instead of transmitting a physical representation of the representation itself, we may transmit a representation of the *selection process* by which it may be identified in the ensemble of possible representations which is assumed to exist at the receiving end.

Code System

5.2.2 A system whereby a representation is defined by a selection process is termed a *code system*.

Code Signal

5.2.3 The corresponding representation of the selection process transmitted is known as a *code signal*.

5.2.3.1 As a physical sequence the code signal will itself have metrical and structural features as discussed in Para. 4, and will be definable by a vector in an information space. But its structure need not have anything in common with that of the *representation* which it identifies.

5.2.3.2 On the other hand, the ordinary case of making physical representations *could* be thought of formally as a special case of coding, one-for-one.

Selective Information Content

5.3 It follows that the result of an experiment, as well as a communication signal, could be analyzed in terms of its *selective information content*.

5.3.1 This is a relative measure, depending on the number of distinct results which were regarded as equally probable by the observer. The result observed is thought of as specifying one of a number of possibilities already contemplated by the observer as forming an ensemble in defined proportions.

5.3.2 The amount of selective information derived from the experiment can then be computed in the same way as for a message, (5.1).

5.4 The entropy of selective information content of a selection should not be facilely identified with the physical entropy of thermodynamics. The two are equivalent only in the particular case where the ensemble from which the selection is made is a physical one defined for a state of thermodynamic equilibrium. | [233]

5.4.1 For example, if all n distinguishable voltage-levels of a transmitted signal are regarded as equiprobable, the selective information content per logon is proportional to *log n*. On the other hand the physical entropy increase is proportional to, or must exceed n^2 (2) (7).

5.4.2 Here the correlation is between *metrical* information content (Para. 4.2.7) and physical entropy increase. In fact, metron content can be thought of here as the number of unit increases of physical entropy – i.e., of elementary events – which have been subsumed under one head, thereby losing their distinguishability and potentiality of serving as »bits.«

[234]

ALPHABETICAL INDEX OF TERMS USED IN INFORMATION THEORY AND RELATED COMMUNICATION THEORY

References are to Paragraphs in the Glossary

Band width: In general terms, the region of Fourier space *(q.v.)* to which the output of an instrument is confined. In particular, the effective frequency range (conjugate to a given co-ordinate) to which it responds.

Binary digit:
Bit: (5.1.3.) Unit of selective information content.

Capacity: (5.1.9) Number of bits transmissible per second.

Code system: (5.2.2)

Code signal: (5.2.3)

Conceptual unit: (4.2.11) The co-ordinate interval in which one metron is acquired. Reciprocal of metron density.

Ensemble: A set of possibilities each of which has a defined probability.

Entropy: (5.1.5) *(a)* In statistical mechanics, the weighted mean of the (negative) logarithm of the probabilities of members of an ensemble. *(b)* In thermodynamics that function of state of a body or system which increases by

$$\int_1^2 \Delta Q/T$$

in a reversible process between two states 1 and 2, where ΔQ is the heat taken up by the body or system at temperature T. [Definitions *(a)* and *(b)* are equivalent.]

Conditional: (5.1.6)

Relative: (5.1.10)

Equivocation: (5.1.8)

Fourier space: The space whose dimensions represent variables which are Fourier transforms of co-ordinates (e.g., the frequency of conjugate to the time co-ordinate).

Information: That which alters representations (2.1).

Metrical: (4.2.) Specifying the number of elements of a representational pattern.

Selective: (5.1.3) Specifying the unforeseeableness of a pattern.

Structural: (4.1.) Specifying the number of independently variable features or degrees of freedom of a pattern. | [235]

Information matrix: A matrix in which the metrical and structural information content of an experiment are specified.

Information space: The space in which independent logons are represented by orthogonal rays, and their metron contents by the squares of distances along these rays. (4.3.1)

Information vector: The vector whose components in information space are the distances just mentioned.

Logon: Unit of structural information content *(q.v.)*.

Logon capacity:
possible synonym:
Logon density: (4.1.4) Number of logons per unit of co-ordinate space.

Logon content:
synonym: Structural
information content: (4.1.3) Number of independently variable features.

Metron: (4.2.2) Unit of metrical information content *(q.v.)*.

Metron capacity:
synonym:
Metron density: (4.2.10) *Cf.* logon capacity.

Metron content:
synonym: Metrical
information content: (4.2.4) Measures the amount of evidence to which a representation is equivalent.

Numerical Energy: (4.2.7.1) Ratio of (Energy)/(Noise Power per unit band width). Analogous to metron content.

Proper scale: (4.2.6) A representational scale on which equal intervals are equiprobable.

Redundancy: (5.1.10.1) One minus relative entropy.

Representation: (1.1) A symbolic picture, model, statement, etc.

Scale unit: (4.2.9) The minimum interval in terms of which a scale can definably or usefully be graduated.

Metrical: (4.2.11) Undesirable equivalent of *conceptual unit*. Reciprocal of metron density.

Structural: (4.1.5) Reciprocal of logon capacity.

APPENDIX II
REFERENCES

[236]

COMMUNICATION PATTERNS IN PROBLEM-SOLVING GROUPS

1. Miller, G. A.: Paper read at Psychological Round Table, December 3, 1950, Elizabeth, N. J. (Not published)
2. Birch, H. G.: The relation of previous experience to insightful problem-solving. 1. *Comp. Psychol.* **38**, *367* (1945).
3. Warden, C. J.: *Animal Motivation.* New York, Columbia Univ. Press, 1931.
4. Von Neumann, J., and Morgenstern, O.: *Theory of Games and Economic Behavior,* (2nd ed.). Princeton, Princeton Univ. Press, 1947.

COMMUNICATION BETWEEN MEN

1. Stevenson, C. L.: *Ethics and Language.* New Haven and London, Yale Univ. Press and Oxford Univ. Press, 1945 (p. 80).
2. Richards, I. A.: Emotive meaning again. *Philosoph. Rev.* **57**, *145* (1948).
3. Oppenheimer, J. R.: The age of science 1900-1950. *Scient. Am.* **183**, *20* (1950).
4. Kroeber, A. L.: Anthropology. *Scient. Am.* **183**, *87* (1950).
5. Newman, S. S.: Behavior patterns in linguistic structure: a case study. *Language, Culture and Personality.* Spier, L., Hallowell, A. I., and Newman, S. S., Editors. Menasha, Wisc., Sapir Memorial Publication Fund, 1941 (p. 94).
6. Kantor, J. R.: *An Objective Psychology of Grammar.* Bloomington, Indiana, Indiana Univ. Book Store, 1936.
7. Richards, I. A.: *How to Read a Page.* New York, W. W. Norton & Co., Inc., 1942 (p. 66).
8. Whorf, B. L.: The relation of habitual thought and behavior to language. *Language, Culture and Personality.* Spier, L., Hallowell, A. I., and Newman, S. S., Editors. Menasha, Wisc., Sapir Memorial Publication Fund, 1941 (p. 75).
9. Stout, G. F.: *Analytic Psychology.* New York and London, The Macmillan Co., Swan Sonnenschein & Co., Ltd., 1902 (Vol. II).
10. Raysor, T. M., editor: *Coleridge's Shakesperean Criticism.* Cambridge, Harvard Univ. Press, 1930 (Vol I).
11. Tinbergen, N.: The hierarchical organization of nervous mechanisms underlying instinctive behavior. *Symp. Soc. Exp. Biol.* **4**, *305* (1950).
12. von Monakow, C., and Mourgue, R.: *Biologische Einführung in das Studium der Neurologie und Psychopathologie.* Stuttgart and Leipzig, Hippokrates-Verlag, 1930.
13. Stern, W.: Über verlagerte Raumformen. Ein Beitrag zur Psychologie der Kindlichen Raumdarstellung und Auffassung. *Ztschr. angew. Psychol.* **2**, *498* (1909). | [237]
14. Jaensch, E. R.: Über Raumverlagerung und die Beziehung von Raumwahrnehmung und Handeln. *Ztschr. f. Psychol. u. Physiol. d. Sinnesorg.* **89**, *116* (1922).
15. Bateson, G.: Pidgin English and cross-cultural communication. *Tr. New York Acad. Sc. Ser.* II, **6**, *137* (1944).
16. Stenzel, J.: Über den Einfluss der griechischen Sprache auf die philosophische Begriffsbildung. *Neue Jahrb. f. d. klass. Altert.* **47**, *152* (1921).
17. —: Metaphysik des Altertums. *Handbuch der Philosophie,* Bäumler, A., and Schröter, M., Editors. Munich, Oldenbourg, 1934 (Sect. I, D. 1-196).
18. Saer, D. J.: The effect of bilingualism on intelligence. *Brit. Psychol.* **14**, *25* (1923).

19. SULLIVAN, H. S.: *Conceptions of Modern Psychiatry.* Washington, D. C., W. A. White Psychiatric Foundation, 1947.
20. POPPER, K.: Indeterminism in quantum physics and in classical physics. *Brit. J. Phil. Sc.* **1**, *173* (1950).
21. HEBB, D. O.: *Organization of Behavior. A Neuropsychological Theory.* New York, Wiley, 1949.
22. NIELSEN, J. M.: *Agnosia, Apraxia, Aphasia. Their Value in Cerebral Localization.* (2nd ed.) New York, Hoeber, 1946 (p. 292)
23. ISSERLIN, M.: Die pathologische Physiologie der Sprache. *Erg. Physiol.* **33**, *1* (1931).
24. REUSCHERT, E.: *Die Gebärdensprache der Taubstummen,* Leipzig, H. Dude[n], 1909.
25. RUESCH, J., and PRESTWOOD, A.: Anxiety; its initiation, communication and interpersonal management. *Arch. Neurol. & Psychiat.* **62**, *527* (1949).
26. KLÜVER, H.: *Mescal.* London, Kegan Paul, 1928.

COMMUNICATION BETWEEN SANE AND INSANE

1. KUBIE, L. S., and MARGOLIN, S.: Therapeutic role of drugs in process of repression, dissociation, and synthesis. *Psychosom. Med.* **7**, *147* (1945).
2. KUBIE, L. S.: Value of induced dissociated states in therapeutic process. *Proc. Roy. Soc. Med.* **38**, *681* (1945).
3. KUBIE, L. S., and MARGOLIN, S.: Process of hypnotism and nature of hypnotic state. *Am. J. Psychiat,* **100**, *611* (1944).
4. ERICKSON, M. H., and KUBIE, L. S.: Use of automatic drawing in interpretation and relief of state of acute obsessional depression. *Psychoanalyt. Quart.* **7**, *443* (1938).
5. —: Permanent relief of obsessional phobia by means of communications with unsuspected dual personality. *ibid.* **8**, *471* (1939).
6. —: The translation of the cryptic automatic writing of one hypnotic subject by another in a trance-like dissociated state. *ibid.* **9**, *51* (1940).
7. FARBER, L. H., and FISHER, C.: Experimental approach to dream psychology through use of hypnosis. *Psychoanalyt. Quart.* **12**, *202* (1943). | [238]
8. KUBIE, L. S.: Body symbolization and development of language. *Psychoanalyt. Quart.* **3**, *430* (1934).
9. —: The neurotic potential, the neurotic process, and the neurotic state. *U. S. Armed Forces M. J.* **2**, *1* (1951).
10. SPIEGEL, H., SHOR, J., and FISHMAN, S.: Hypnotic ablation technique for study of personality development; preliminary Med. report. *Psychosom. Med.,* **7**, *273* (1945).
11. KLÜVER, H.: The eidetic child. *A Handbook of Child Psychology.* Murchison, C., Editor. Worcester, Clark Univ. Press, 1931 (p. 643).
12. —: The eidetic type. *Research Nerv. & Ment. Dis. Proc.* **14**, *150* (1934).
13. —: Studies on the eidetic type and on eidetic imagery. *Psychol. Bull.* **25**, *69* (1928).
14. —: Eidetic phenomena. *ibid.* **29**, *181* (1932).
15. —: Fragmentary eidetic imagery. *Psychol. Rev.* **37**, *441* (1930).
16. —: Mechanisms of hallucinations. *Studies in Personality.* Turman, L., Editor. New York, McGraw Hill, 1942 (p. 175).
17. SCHILDER, P.: Psychoanalyse und Eidetik. *Ztschr. f. Sexualwissen[s]ch.* **13**, *56* (1926).
18. —: *Mind: Perception and Thought in Their Constructive Aspects.* New York, Columbia Univ. Press, 1942.

19. Zeman, H.: Verbreitung und Grad der eidetischen Anlage. *Ztschr. f. Psychol. u. Physiol. d. Sinnesorg.* **96**, *208* (1924).
20. Storfer, A. J.: Über psychoanalytische Tierpsychologie. *Psychoanalytische Bewegung.* **3**, *40* (1931).
21. Hermann, I.: Modelle zu den Odipus[!]- und Kastrationskomplexen bei Affen. *Imago* **12**, *59* (1926).
22. Armstrong, E. A.: *Bird Display and Behaviour.* London, Drummond, 1947.
23. —: The nature and function of displacement activities. *Symp. Soc. Exp. Biol.* **4**, *361* (1950).
24. Kubie, L. S.: The concept of normality and the neurotic process. *Practical and Theoretical Aspects of Psychoanalysis.* New York, International Univ. Press, 1950 (p. 12).
25. —, *et al.*: Problems in clinical research, Round Table, 1946. *Am. J. Orthopsychiat.* **17**, *196* (1947).
26. —: The objective evaluation of psychotherapy, Round table, 1948. *Ibid.* **19**, *463* (1949).

COMMUNICATION BETWEEN ANIMALS

1. Hyman, L.: Metabolic gradients in amoeba and their relation to the mechanism of amoeboid movement. *J. Exper. Zool.* **24**, *55* (1917).
2. Child, C. M.: *Physiological Foundations of Behavior.* New York, Holt, 1924.
3. Roth, L. M.: A study of mosquito behavior. *Am. Midland Naturalist* **40**, *265* (1948).
4. von Frisch, K.: *Bees.* Ithaca, Cornell Univ., 1950. | [239]
5. Thorpe, W. H.: Orientation and methods of communciation[!] of honey bee and its sensitivity to polarization of light. *Nature* **164**, *1 (1949).*
6. Schneirla, T. C.: Social organization in insects, as related to individual function. *Psychol. Rev.* **48**, *465* (1941).
7. Maier, N. R. F., and Schneirla, T. C.: *Principles of Animal Psychology.* New York, McGraw-Hill, 1935.
8. Butler, C. G.: Bee behaviour. *Proc. Roy. Inst. Gr. Brit.* **34**, *244* (1948).
9. Goetsch, W.: *Die Staaten der Ameisen.* Berlin, Springer, 1937.
10. Hecht, S., and Wolf, E.: The visual acuity of the honey bee. *J. Gen. Physiol.* **12**, *727* (1929).
11. Breder, C. M., Jr.: Certain effects in habits of schooling fishes, as based on observation of Jenkinsia. *Am. Mus. Novit.* **382**, *1* (1929).
12. —, and Halpern, F.: Innate and acquired behavior affecting aggregation of fishes. *Physiol. Zool.* **19**, *154* (1946).
13. —, and Nigrelli, R. F.: The influence of temperature and other factors on water aggregation of sunfish, *Leponis Auritus. Ecology* **16**, *33* (1935).
14. —, and Roemhild, J.: Comparative behavior of various fishes under different conditions of aggregation. *Copeia* **1**, *29* (1947).
15. Parr, A. E.: A contribution to the theoretical analysis of schooling behavior of fishes. *Occ. Pap. Bingham Oceanogr. Coll.* **1**, *1* (1927).
16. Crawford, M. P.: The cooperative solving of problems by young chimpanzees. *Comp. Psychol. Monogr.* **14**, *No. 2* (1937).
17. Yerkes, R. M., and Nissen, H. W.: Pre-linguistic sign behavior in chimpanzee. *Science* **89**, *585* (1939).
18. Carpenter, C. R.: A field study of behavior and social relations of howling monkeys *(Alouatta palliata). Comp. Psychol. Monogr.* **10**, *No. 2* (1934).

19. DARWIN, C.: *The Expression of Emotions in Man and Animals.* New York, Appleton, 1872.
20. KLÜVER, H.: Functional significance of the geniculo-striate system. *Biol. Symposia* 7, *253* (1942).
21. KUBIE, L. S.: Pavlov and his school (review). *Psychoanalyt. Quart.* **10**, *329* (1941).

IN SEARCH OF BASIC SYMBOLS

1. SHANNON, C. E., and WEAVER, W.: *Mathematical Theory of Communication,* Urbana, Univ. of Illinois Press, 1949.
2. WIENER, N.: *Cybernetics.* New York, Wiley, 1948.
3. MACKAY, D. M.: Quantal aspects of scientific information. *Phil. Mag. (Ser. 7)* **41**, *289* (1950).
4. —: (See Appendix **I**, this volume)
5. GABOR, D.: Theory of communication. *J. Inst. Elec. Engrs.* **93**, *429* (1946).
6. [240] FISHER, R. A.: *The Design of Experiments.* Edinburgh, Oliver & Boyd, 1935. |
7. MACKAY, D. M.: Mind-like behavior in artefacts. *Brit. J. Phil. Sci.* (Aug. 1951).
8. FRIEDMAN, M., and SAVAGE, L. J.: The utility analysis of choices involving risk. *J. Polit. Econ.* **56**, *279* (1948).

THE NOMENCLATURE OF INFORMATION THEORY

1. WITTGENSTEIN, L.: *Tractatus Logico-Philosophicus,* London, Kegan Paul, and New York, Harcourt Brace, 1922.
2. MACKAY, D. M.: Quantal aspects of scientific information. *Phil. Mag. (Ser. 7)* **41**, *289* (1950).
3. GABOR, D.: Theory of communication. *J. Inst. Elec, Engrs.* **93**, *429* (1946).
4. FISHER, R. A.: *The Design of Experiments.* Edinburgh, Oliver & Boyd, 1935 (p. 188).
5. SHANNON, C. E.: Mathematical theory of communication. *Bell Syst. Tech. J.* **27**, *379* (1948).
6. WIENER, N.: *Cybernetics,* New York, Wiley, 1948.
7. MACKAY, D. M.: *Entropy, Time and Information,* Information-Theory Symposium, London, September, 1950.

CYBERNETICS

CIRCULAR CAUSAL AND FEEDBACK MECHANISMS IN BIOLOGICAL AND SOCIAL SYSTEMS

Transactions of the Ninth Conference
March 20-21, 1952, New York, N. Y.

Edited by
HEINZ VON FOERSTER
DEPARTMENT OF ELECTRICAL ENGINEERING
UNIVERSITY OF ILLINOIS
CHAMPAIGN, ILLINOIS

Assistant Editors
MARGARET MEAD
AMERICAN MUSEUM OF NATURAL HISTORY
NEW YORK, N. Y.
HANS LUKAS TEUBER
DEPARTMENT OF PSYCHIATRY AND NEUROLOGY
NEW YORK UNIVERSITY COLLEGE OF MEDICINE
NEW YORK, N. Y.

Sponsored by the
JOSIAH MACY, JR. FOUNDATION
NEW YORK, N. Y.

PARTICIPANTS

Ninth Conference on Cybernetics

MEMBERS

WARREN S. McCULLOCH, Chairman
Research Laboratory of Electronics, Massachusetts Institute of Technology; Cambridge, Mass.

HEINZ VON FOERSTER, Secretary
Department of Electrical Engineering, University of Illinois; Champaign, Ill.

GREGORY BATESON
Veterans Administration Hospital; Palo Alto, Calif.

ALEX BAVELAS[1]
Department of Economics and Social Science, Massachusetts Institute of Technology; Cambridge, Mass.

JULIAN H. BIGELOW
Department of Mathematics, Institute for Advanced Study; Princeton, N. J.

HENRY W. BROSIN[1]
Department of Psychiatry, University of Pittsburgh School of Medicine; Pittsburgh, Pa.

LAWRENCE K. FRANK
72 Perry St., New York, N.Y.

RALPH W. GERARD
Departments of Psychiatry and Physiology, University of Illinois College of Medicine; Chicago, Ill.

GEORGE EVELYN HUTCHINSON
Department of Zoology, Yale University; New Haven, Conn.

HEINRICH KLÜVER
Division of the Biological Sciences, University of Chicago; Chicago, Ill.

LAWRENCE S. KUBIE
Department of Psychiatry and Mental Hygiene, Yale University School of Medicine; New Haven, Conn.

RAFAEL LORENTE DE NÓ
Rockefeller Institute for Medical Research; New York, N.Y.

DONALD G. MARQUIS
Department of Psychology, University of Michigan; Ann Arbor, Mich.

MARGARET MEAD
American Museum of Natural History; New York, N.Y.

F. S. C. NORTHROP
Department of Philosophy, Yale University; New Haven, Conn.

WALTER PITTS
Research Laboratory of Electronics, Massachusetts Institute of Technology; Cambridge, Mass.

ARTURO S. ROSENBLUETH
Department of Physiology, Instituto Nacional de Cardiologia; Mexico City, D.F., Mexico

LEONARD J. SAVAGE[1]
Committee on Statistics, University of Chicago; Chicago, Ill.

THEODORE C. SCHNEIRLA[1]
American Museum of Natural History; New York, N.Y.

HANS LUKAS TEUBER
Department of psychiatry and Neurology, New York University College of Medicine; New York, N.Y.

GERHARDT von BONIN
Department of Anatomy, University of Illinois College of Medicine; Chicago, Ill.

GUESTS

W. ROSS ASHBY
Department of Research, Barnwood House; Gloucester, England

JOHN R. BOWMAN
Department of Physical Chemistry
Mellon Institute of Industrial Research, University of Pittsburgh; Pittsburgh, Pa.

DUNCAN LUCE
Research Laboratory of Electronics, Massachusetts Institute of Technology; Cambridge, Mass.

MARCEL MONNIER
Department of Medicine, Laboratory of Applied Neurophysiology, University of Geneva; Geneva, Switzerland

HENRY QUASTLER
Control Systems Laboratory, University of Illinois; Urbana, Ill.

ANTOINE REMOND
Neuropsychiatric Institute, University of Illinois College of Medicine; Chicago, Ill.

MOTTRAM TORRE
Personnel Division, Mutual Security Agency; Washington, D.C.

JEROME B. WIESNER
Research Laboratory of Electronics, Massachusetts Institute of Technology; Cambridge, Mass.

J. Z. YOUNG
Department of Anatomy, University College; London, England

JOSIAH MACY, JR. FOUNDATION

FRANK FREMONT-SMITH, Medical Director

JANET FREED, Assistant for the Conference Program

1 Absent

JOSIAH MACY, JR. FOUNDATION CONFERENCE PROGRAM

As an introduction to these Transactions of the Ninth Conference on Cybernetics, I should like to outline what it is that the Foundation hopes to accomplish by its Conference Program. We are interested, first of all, in furthering knowledge about cybernetics, and to this end the participants were brought together to exchange ideas, experiences, data, and methods. In addition to this particular goal, however, there is a further, and perhaps more fundamental, aim which is shared by all our conference groups. This is the promotion of meaningful communication between scientific disciplines.

The problem of communication between disciplines we feel to be a very real and urgent one, the most effective advancement of the whole of science being to a large extent dependent upon it. Because of the accelerating rate at which new knowledge is accumulating, and because discoveries in one field so often result from information gained in quite another, channels must be established for the most effective dissemination and exchange of this knowledge.

The increasing realization that nature itself recognizes no boundaries makes it evident that the continued isolation of the several branches of science is a serious obstacle to scientific progress. Particularly is it true in medicine that the limited view through the lens of one discipline is no longer enough. For example, today medicine must be well versed in nuclear physics because of the tracer techniques and the injury which can result from radiation. At the other extreme, medicine is certainly a social science and, through mental health, must be concerned with economic and social questions. The answer, then, is not further fragmentation into increasingly isolated specialities, disciplines, and departments, but the integration of science and scientific knowledge for the enrichment of all branches. This integration, we feel, can be encouraged by providing opportunities for a multiprofessional approach to given topics.

Although the fertility of the multidiscipline approach is recognized, adequate provision is not made for it by our universities, scientific societies, and journals. And perhaps the presence of other hindering factors must be admitted. Partly semantic in nature, they may also to some degree be psychological. Admittedly, it is oftentimes difficult to | accept data derived from methods with which one is unfamiliar. By making free and informal discussion the central core of our meetings, we hope to achieve an atmosphere which minimizes as much as possible these semantic and emotional barriers.

Thus, our conferences are in contrast to the usual scientific gatherings. Presentations are not designed to present neat solutions to tidy problems, but rather to elicit provocative discussion of the difficulties which are being encountered in research and practice. We ask that the presentations be relatively brief and emphasis is placed upon discussion as the heart of the meeting. Our hope is that the participants will come prepared not to defend a single point of view but, with open minds, to take full advantage of the meeting as an opportunity to speak with representatives of other disciplines in much the same way as they talk with their own colleagues in their own laboratories.

During 1953 under the Conference Program conferences will be held on the following topics: administrative medicine, adrenal cortex, aging, cold injury, connective tissues, consciousness, cybernetics, infancy and childhood, liver injury, metabolic interrelations, nerve impulse, renal function, and shock and circulatory homeostasis.

When a new conference group is organized, the Chairman, in consultation with the Foundation, selects fifteen scientists to be the nucleus of the group which will hold annual meetings for a period of five years. Every effort is made to include representatives from all pertinent disciplines. From time to time, however, new members are added by the group to fill gaps in viewpoint or techniques. A small number of guests is invited to attend each meeting, but, for the purposes of promoting full participation by all members and guests, attendance at any meeting is limited to twenty-five. During a conference's prescribed lifetime we cannot possibly include more than a small fraction of the key investigators in the field, and one of the difficulties in forming a group, such as this one on cybernetics, is that it is necessary to exclude so many investigators we should like to include.

The transactions of these meetings are recorded and published. This is done because the Foundation wishes to make current thinking in a field available to all those working in it. Logic is a vital aspect of science, but equally essential is the intuitive or creative aspect. Research is as creative as the painting of a portrait or the composing of a symphony. Although logic is, of course, necessary in order to rearrange, to test, and to validate, research thrives on creativity which has its source in unconscious, nonrational [ix] processes. Unfortunately, however, | in the research reports which are presented to the world in scientific journals, this integral part of scientific endeavor is shriveled by the cold, white light of logic. By preserving the informality of our conferences in the published transactions, we hope to portray, more accurately what goes on in the minds of scientists and to give a truer picture of the rote which creativity plays in scientific research.

FRANK FREMONT-SMITH, M.D.
Medical Director

A NOTE BY THE EDITORS*

To the reader of this somewhat unusual document, a few words of explanation, and caution. This is not a book in the usual sense, nor the well-rounded transcript of a symposium. These pages should rather be received as the partial account of conversations within a group, whose interchange extends beyond the confines of a two-day meeting. The account attempts to capture some of this group interchange in all its evanescence because it represents to us one of the few concerted efforts at interdisciplinary communication.

The members of this group share the belief that one can, and must, attempt communication across the boundaries which separate the various sciences. The participants have come from many fields: they are physicists, mathematicians, electrical engineers, physiologists, neurologists, experimental psychologists, psychiatrists, sociologists, and cultural anthropologists. That such a gathering failed to produce the Babylonian confusion that might have been expected is probably the most remarkable result of this meeting and of those which preceded it.

This ability to remain in touch with each other, to sustain the dialogue across departmental boundaries and, in particular, across the gulf between natural and social sciences is due to the unifying effect of certain key problems with which all members are concerned: the problems of communication and of self-integrating mechanisms. Revolving around these concepts, the discussion was communication about communication, necessarily obscure in places and for more than one reason. Yet the actual outcome was certainly more intelligible than one might think, so that the editors felt enjoined to reproduce the transcript as faithfully as possible.

The social process, of which these transactions are an incomplete residue, was not a sequence of formal »papers« read by individual participants and followed by prepared discussion. With few exceptions people spoke freely and without notes. Unavoidably some speakers produced inaccurate memories of their own facts, or of those of others, and trends of thought were often left incomplete. The printed record preserves the essential nature of this interchange in which partial associations were permitted on the assumption that closure would take place at some other time, producing new ideas or reenforcing those that were thought of in passing. If the reader wishes a formal statement of the work and point of view of individual participants, he will have to consult other sources. This can be done with case, since most of the contributors have provided references to previously published material. | [xii]

The reader should be warned that the presentations and discussion tend to be responsive to previous meetings of the group. Some statements were designed to answer questions asked months or years before, or designed to evoke some long-anticipated answer from a fellow member. Radical changes in the manuscript would have been necessary had we attempted to rid the group of its history. Such changes would have been distortions, and would prevent the reader from noticing the unfinished state of the group's affairs.

Our editorial procedure, nevertheless, involved some revision of the transcript prior to publication. Each participant was supplied with a mimeographed copy of his transcribed remarks, and was given the chance to revise his material for the sake of clarity and coherence. Not all of the participants availed themselves of this opportunity, so that our working copy represented a mixture of revised and unrevised contributions. In the unrevised passages the editors corrected only those statements which seemed to them obvious errors in recording. For the rest, they confined their censorial activities

* Reprinted, with minor changes, from the Transactions of the Eighth Conference, 1951.

to occasional deletions of overlapping or repetitive passages, and of a few all-too-cryptic digressions. Most of the asides, such as jokes or acidities, were preserved as long as they seemed intelligible to people outside the group.

It is noteworthy that our invited guests cannot be identified by any obvious differences between their vocabulary and that of the regular members. One of the most surprising features of the group is the almost complete absence of an idiosyncratic vocabulary. In spite of their seven years of association, these twenty-five people have not developed any rigid, in-group language of their own. Their idioms are limited to a handful of terms borrowed from each other: analogical and digital devices, feedback and servomechanisms, and circular causal processes. Even these terms are used only with diffidence by most of the members, and a philologist, given to word-frequency counts, might discover that the originators of »cybernetics« use less of its lingo than do their more recent followers. The scarcity of jargon may perhaps be a sign of genuine effort to learn the language of other disciplines, or it may be that the common point of view provided sufficient basis for group coherence.

This common ground covered more than the mere belief in the worth-whileness of interdisciplinary discussion. All of the members have an interest in certain conceptual models which they consider potentially applicable to problems in many sciences. By agreeing on the usefulness of these models, we get glimpses of a new *lingua franca* of science – fragments of a common tongue likely to counteract some of the confusion and complexity of our language.

[xiii] Chief among these conceptual models are those supplied by the | theory of information.[1,2,3] This theory has arisen under the pressure of engineering needs: the efficient design of electronic communication devices (telephone, radio, radar, and television) depended on achieving favorable »signal-to-noise ratios.« Application of mathematical tools to these problems had to wait for an adequate formulation of »information« as contrasted to »noise.«

If noise is defined as random activity, then information can be considered as order wrenched from disorder; as improbable structure in contrast to the greater probability of randomness. With the concept of entropy, classical thermodynamics expressed the universal trend toward more probable states: any physicochemical change tending to produce a more nearly random distribution of particles. Information can thus be formulated as negative entropy, and a precise measure of certain classes of information can be found by referring to degrees of improbability of a state.

The improbable distribution of slots in a slotted card, or the improbable arrangement of nucleic acids in the highly specific pattern of a gene – both can be considered as »coded« information – the one decoded in the course of a technical (cultural) process, the other in the course of embryogeny. In both instances, that of the slotted card and that of the gene, we are faced, not only with carriers of information, but with powerful mechanisms of control: the slotted card can control a long series of processes in a plant, without itself furnishing any of the requisite energy; the gene, as an organic template, somehow provides for its own reproduction and governs the building of a multicellular organism from a single cell. In the latter case, mere rearrangement of submicroscopic particles can apparently lead to mutations, improving or corrupting the organism's plans as the case may be. Such rearrangement may indeed be similar to the

1 Shannon, C. E.: A mathematical theory of communication. *Bell System. Tech. J.* **27**, *379-423* and *623-656 (1948)*.

2 Shannon, C. E., and Weaver, W.: *The Mathematical Theory of Communication*. Urbana: University of Illinois Press, 1949 (p. 116).

3 Wiener, N.: *Cybernetics or Control and Communication in the Animal and the Machine*. New York; Wiley, 1948 (p. 194).

difference brought about by the transposing of digits in numbers, 724 to 472, or by transposing letters in words such as art and rat.[4]

Extension of information theory to problems of language structure has been furthered by psychologists and statisticians.[5,6,7,8,9] There are | unexploited opportunities [xiv] for additional applications of the theory in comparative linguistics, and more particularly in studies of the pathology of language. Yet available work is sufficient to show how communication considered from this standpoint can be investigated in mechanical systems, in organisms, and in social groups;[10] and the logical and mathematical problems that go into the construction of modern automata, in particular the large electronic computers,[11] have at least partial application to our theorizing about nervous systems and social interactions.

A second concept, now closely allied to information theory, is the notion of circular causal processes. A state reproducing itself, like an organism, or a social system in equilibrium, or a physicochemical aggregate in a steady-state, defied analysis until the simple notion of one-dimensional cause-and-effect chains was replaced by the two-dimensional notion of a circular process. The need for such reasoning was clear to L. J. Henderson, the physiologist, when he applied the logic of Gibbsian physiocochemical systems[12] to the steady-states of human blood,[13] and to integration in social groups, down to miniature social systems.[14] Quite independently, social scientists had been tending in the same direction, as witnessed by the work of the functional anthropologists Radcliffe-Brown,[15] and Bateson.[16,17] In ecology, the concept of circular causal systems has been employed by Hutchinson,[18] and further applications in statistical biology and genetics can be expected.

The remarkable constancy in the concentration of certain substances in the fluid matrix of the body, led Claude Bernard originally to posit the fixity of the »milieu interieur,« as one of the elementary conditions of life.[19] Cannon[20] designated as

4 Gerard, R. W.: *Unresting Cells.* New York: Harper, 1940 (Re-issued 1949).

5 Frick, F. C., and Miller, G. A.: Statistical behavioristics and sequences of responses. *Psychol. Rev.* **56**, *311 (1949).*

6 Miller, G. A.: *Language and Communication.* New York: McGraw-Hill, 1951.

7 Miller, G. A., and Selfridge, J. A.: Verbal context and the recall of meaningful material. *Am. J. Psychol.* **63**, *176 (1950).*

8 Newman, E. B.: Computational methods useful in analyzing series of binary data. *Am. J. Psychol.* **64**, *252 (1951).*

9 —: The pattern of vowels and consonants in various languages. *Ibid.*, 369.

10 Bavelas, A.: A mathematical model for group structures. *Appl. Anthropol.* **7**, (part 3), *16 (1948).*

11 Von Neumann, J.: The general and logical theory of automata. *Cerebral Mechanisms in Behavior (The Hixon Symposium).* Jeffress, L. A., Editor. New York: Wiley, 1951 (pp. 1-41).

12 Gibbs, J. W.: On the equilibrium of heterogeneous substances. *Transactions of the Connecticut Academy* III, Oct. 1875 – May 1876, 108-248; May 1877 – July 1878. Reprinted in: Gibbs, J. W.: *The Collected Works, Vol. 1. Thermodynamics.* New York: Longmans, Green and Co., 1928 (pp. 55-371).

13 Henderson, L. J.: *Blood: A Study in General Physiology.* New Haven: Yale University Press, 1928 (p. 390).

14 —: Physician and patient as a social system. *New England J. Med.* **212**, *819 (1935).*

15 Radcliffe-Brown, A. R.: *The Andaman Islanders. A Study in Social Anthropology.* Cambridge, England: Cambridge University Press, 1922 (pp. XIV and 504).

16 Bateson, G.: *Naven: A Survey of the Problems Suggested by a Composite Picture of the Culture of a New Guinea Tribe, Drawn from Three Points of View.* Cambridge: Cambridge University Press, 1936 (p. 286).

17 —: Bali, the value system of a steady state. In: *Social Structure. Studies Presented to A. R. Radcliffe-Brown.* Oxford University Press, 1949 (p. 35).

18 Hutchinson, G. E.: Circular causal systems in ecology. *Ann. New York Acad. Sc.* **50**, *221 (1948).*

19 Bernard, C.: *Leçons sur les phénomènes de la vie communes aux animaux et aux végétaux.* Two volumes. Paris: J. B. Baillière, 1878-79.

20 Cannon, W. B.: *The Wisdom of the Body.* New York: Norton, 1932 (pp. XV and 312).

»homeostasis« those functions that restore a disturbed equilibrium in the internal environment: the complex self-regulatory processes which guarantee a relative constancy [xv] of | blood sugar level, of osmotic pressure, of hydronium ion concentration, or of body temperature. Many of these processes are at least partially understood, but, as Klüver[21] has pointed out, we know next to nothing of the physiological functions which underlie our perceptual »constancies.«

Normal perception is reaction to relations, to »universals« such as size, shape, and color. Perceived objects tend to remain invariant in their size, while distance from the observer varies; perceived shapes and colors retain subjective identity in varying positions, and under varying illumination. This crucial problem for the physiological psychology of perception is rarely faced,[22] and the neural correlates for our reaction to universals are still *sub judice.*

Recent attempts at identifying a possible neural basis for our reactions to universals[23] have adduced hypothetical sustained activity in neuronal circuits as one of the prerequisites for the central processes which guarantee perceptual constancies. Persistent circular activity in nervous nets had been postulated on theoretical grounds by Kubie,[24] over twenty years ago, thereby anticipating the subsequent empirical demonstration of such reverberating circuits by Lorente de Nó.[25] The importance of Lorente de Nó's disclosures for neurological theory lies in the fact that, earlier in the century, many investigators considered the central nervous system as a mere reflex-organ; the mode of action of this organ, despite all the evidence to the contrary, was viewed as limited to the relating of input to output, stimulus and response corresponding to cause and effect. The possibility of self-sustained central activity in the nervous system was overlooked. Thence the denial of memory-images in early behaviorism, the emphasis on chain reflexes in attempts at explaining coordinated action.

To this day, many psychologists tend to see the prototype of all learning in elementary conditioned reflexes, a tendency which cannot be understood unless one assumes, with Lashley,[26] that these psychologists are still handicapped by »peripheralistic« notions – the unwarranted idea that central nervous activity cannot endure in the [xvi] absence | of continuing specific input from the periphery. Undoubtedly, the action of reverberating circuits can be overgeneralized, and has been abused. Long-range memory may need more permanent neural changes, but the notion of such circuits has suggested models of neural activity which are potentially testable and therefore of value.[27,28]

Activity in closed central loops thus has to be carefully distinguished from the older and simpler notions of neural circuits, circuits which join periphery and centers, as in

21 Klüver, H.: *Behavior Mechanisms in Monkeys.* Chicago: University of Chicago Press, 1933 (pp. XVIII and 387).

22 —: Functional significance of the geniculo-striate system. *Biol. Symposia* 7, *253*(1942).

23 Pitts, W., and McCulloch, W. S.: How we know universals; the perception of auditory and visual forms. *Bull. Math. Biophys.* 9, *127 (1947).*

24 Kubie, L. S.: A theoretical application to some neurological problems of the properties of excitation waves which move in closed circuits. *Brain* 53, *166 (1930).*

25 Lorente de Nó, R.: Analysis of the activity of the chains of internuncial neurons. *J. Neurophysiol.* 1, *207 (1938).*

26 Lashley, K. S.: Discussion. In: *Cerebral Mechanisms in Behavior (The Hixon Symposium).* Jeffress, L. A., Editor. New York: Wiley, 1951 (p. 82).

27 McCulloch, W. S.: A heterarchy of values determined by the topology of nerve nets. *Bull. Math. Biophys.* 7, *89 (1945).*

28 Hebb, D. O.: *Organization of Behavior: A Neuropsychological Theory.* New York: Wiley, 1949 (pp. XIX and 335).

the classical conception of postural reflexes. Sir Charles Bell[29] spoke of a »nervous circle which connects the voluntary muscles with the brain.« Through such circuits, muscles in a limb maintain a given tension, as long as motor impulses flow into the muscles according to the sensory signals which issue from these same muscles. The idea antedated the discovery of the sense organs (muscle spindles) which monitor a state of tension in the muscle and, by increasing their rate of centripetal firing, set off centrifugal volleys which shorten the muscle in response to imposed stretch.

Numerous analogues for such recalibrating mechanisms can be found in those modern electronic devices in which input is regulated by constant comparison with output. The automatic volume control circuit of a radio receiver prevents »blasting« by decreasing the volume as the signal is increased, and counteracts »fading« by increasing the volume. A speed control unit slows a motor down when its revolutions exceed a desired value and speeds it up when revolutions fall below this value. Such »feedback« or »servomechanisms«[30] are man-made models of homeostatic processes. They are not exclusively found among electronic devices. In the days of the thermal engine, Maxwell[31] developed the theory of the mechanical »governor« of steam engines. Small versions of this governor are still found today in old-fashioned phonograph turntables. Two massive metal spheres are suspended by movable links from a vertical shaft which spins with the main shaft of the machine. As speed increases, centrifugal force drives the metal spheres apart and increases the drag on the shaft; the machine slows down. Again, with the spheres sinking low, the drag on the shaft is decreased and the machine speeds up. Such a governor insures approximate constancy of speed in the engine, by the simplest mechanical means, and, in contrast to many more complicated devices, the mechanical governor shows little likelihood of going into uncontrollable oscillations. | [xvii]

Recent complex electronic devices are not only »error-controlled« (like a mechanical governor), but can be so built as to »seek« a certain state, like »goal-seeking[«] missiles which predict the future position of a moving target (at time of impact) by extrapolation from its earlier positions during pursuit. Such devices embody electronic computing circuits, and the appearance of »purpose« in their behavior (a feedback over the target!) has intrigued the theorists[32,33] and has prompted the construction of such likeable robots as Shannon's »electronic rat,« described in the previous volume.

The fascination of watching Shannon's innocent rat negotiate its maze does not derive from any obvious similarity between the machine and a real rat: they are, in fact, rather dissimilar. The mechanism, however, is strikingly similar to the *notions* held by certain learning theorists about rats and about organisms in general. Shannon's construction serves to bring these notions into bold relief.

Recent emphasis on giant electronic computers as analogues of the human brain should perhaps be considered in the same light. The logical and mathematical theories demanded by the construction of these computers raise problems similar to those faced on considering certain aspects of the nervous system or of social structures.[11] It is no accident that John von Neumann, a mathematician who is currently concerned with the theory of computers, should be more generally known for his analysis of human

29 Bell, C.: On the nervous circle which connects the voluntary muscles with the brain. *Proc. Roy. Soc.* 2, *266 (1826).*

30 MacColl, L. R.: *Fundamental Theory of Servo-Mechanisms.* New York: Van Nostrand, 1945.

31 Maxwell, C.: On governors. *Proc. Roy. Soc.* 16, *270 (1868).*

32 Rosenblueth, A., Wiener, N., and Bigelow, J.: Behavior, purpose and teleology. *Philos. of Sc.* 10, *18 (1943).*

33 Northrop, F. S. C.: The neurological and behavioristic psychological basis of the ordering of society by means of ideas. *Science* 107, *411 (1948).*

interaction in games and economic behavior,[34] and that Norbert Wiener, after working on computers and guided missiles, turned to the consideration of the social significance of these mechanisms.[35] Brief consideration of computers may therefore be in order.

Computers are constructed on either of two principles: they may be digital or analogical. In an analogical device, numbers are represented by a continuous variation of some physical quantity – a voltage, say, or a distance on a disc. A digital device, however, represents numbers as discrete units which may or may not be present, e.g., a circuit that may be open or closed, and the basic alphabet of the machine may be a simple yes or no, zero or one.

Peripheral neurons act on an all-or-none principle, and synapses in the central nervous system are frequently considered to act similarly. Theories of central nervous activity have consequently often paralleled those required for digital rather than analogue machines.[23] The ap|plicability of digital notions to the actions of the central [xviii] nervous system has been questioned,[36] but the calculus worked out for handling them is certainly applicable to electronic digital computers,[37] and the very fact that testable theories of nerve action have been proposed is due to the availability of the electronic models.

We all know that we ought to study the organism, and not the computers, if we wish to understand the organism. Differences in levels of organization may be more than quantitative.[38] But the computing robot provides us with analogues that are helpful as far as they seem to hold, and no less helpful whenever they break down. To find out in what ways a nervous system (or a social group) differs from our man-made analogues requires experiment. These experiments would not have been considered if the analogue had not been proposed, and new observations on biological and social systems result from an empirical demonstration of the shortcomings of our models. It is characteristic that we tend to think of the intricacies of living systems in terms of nonliving models which are obviously less intricate. Still, the reader will admit that, in some respects, these models are rather convincing facsimiles of organismic or social processes – not of the organism or social group as a whole, but of significant parts.

How this way of thinking emerged in the group is difficult to reconstruct. From the outset, John von Neumann and Norbert Wiener furnished the mathematical and logical tools. Warren McCulloch, as the group's »chronic chairman,« infused it with enthusiasm and insisted on not respecting any of the boundaries between disciplines. The Josiah Macy, Jr. Foundation, through Dr. Frank Fremont-Smith, provided the physical setting but actually much more than that: the social sanction for so unorthodox an undertaking. The confidence of the Foundation, and of Dr. Frank Fremont-Smith, made it possible to obtain a type of cross-fertilization which has proved rewarding over a period of seven years.

The gradual growth of the principles can be recognized from dates and titles of successive Conferences. The nucleus of the current Cybernetics Conference was formed

34 Von Neumann, J., and Morgenstern, O.: *Theory of Games and Economic Behavior.* Princeton: Princeton University Press, 1944.

35 Wiener, N.: *The Human Use of Human Beings.* Boston: Houghton Mifflin, 1950.

36 Gerard, R. W.: Some of the problems concerning digital notions in the central nervous system. *Cybernetics.* Von Foerster, H., Mead, M., Teuber, H. L., Editors. Trans. Seventh Conf. New York: Josiah Macy, Jr. Foundation, 1950 (p. 11).

37 McCulloch, W. S., and Pitts, W.: A logical calculus of the ideas immanent in nervous activity. *Bull. Math. Biophys,* **5**, *115 (1943).*

38 Schneirla, T. C.: Problems in the biopsychology of social organization. *J. Abn. Soc. Psychol.* **41**, *385 (1946).*

in May 1942 at the Macy Foundation Conference on »Cerebral Inhibition.« Among the participants were G. Bateson, L.K. Frank, F. Fremont-Smith, L. Kubie, W. McCulloch, M. Mead, and A. Rosenblueth. All of these later became members of the continuing group devoted to the discussion of »cybernetics.« The | publication of the [xix] article on »Behavior, Purpose and Teleology« by Rosenblueth, Wiener, and Bigelow in 1943,[32] focused attention on several of the problems which led to the organization of the first Macy Foundation Conference in March 1946, devoted to »Feedback Mechanisms and Circular Causal Systems in Biological and Social Systems.«

The fall of 1946 found the group very active. Two meetings sponsored by the Macy Foundation followed rapidly upon each other: first, in September, a special meeting on »Teleological Mechanisms in Society;« then, in October, the second Conference on Teleological Mechanism[s] and Circular Causal Systems. Next, the group formed the nucleus of a formal symposium on »Teleological Mechanisms« held under the auspices of the New York Academy of Sciences.[39]

In the following year, 1947, the third Macy Foundation Conference was held, retaining the title of the second meeting; the fourth and fifth conferences, in 1948, were entitled Conference on Circular Causal and Feedback Mechanisms in Biological and Social Systems. The fifth conference concerned itself particularly with considerations of the structure of language.

With the publication of Norbert Wiener's book *Cybernetics* a term appeared which was unanimously chosen as title for the sixth conference in the spring of 1949. The title *Cybernetics* was maintained for the seventh, eighth, and the present ninth conference 1950, 1951, and 1952, with the subtitle *Circular Causal and Feedback Mechanisms in Biological and Social Systems.*

Through the fifth conference, no transactions were published. With the sixth conference (1949) our program of publication began, so that the three preceding conferences, the sixth,[40] seventh,[41] and eighth,[42] are available in print. The reader might suspect that this urge to fix the group process in printed form is the beginning of fossilization. He may be right, but we prefer to think of it in terms more favorable to the group.

For well over two thousand years, the »symposium« has been a setting for the matching and sharpening of ideas. Evolved from the Attic stage, the literary form created by Plato has persisted through the Middle Ages and the Renaissance. Until recently, it was unrestrained by stenotypists and tape recorders. Few of the classical symposia were anything but prose poems of one man's making. They brought a simple message, stated in contrapuntal fashion. Now we have our modern | devices for the [xx] recording and storing of information. Communication transmitted has been infinitely multiplied in volume, but the thinking of simplifying ideas has not kept pace.

Whether the meetings here recorded contain such simplifying ideas, the editors would not presume to say. Some of us believe we can see such ideas here and there. For this reason, we preserved the record, and exhibit it to others for their judgment.

Pressure of time made it impossible in the earlier years to sum up the historical background and to formulate an editorial policy. The published records of the sixth and seventh conferences, in 1949 and 1950, therefore appeared without any intro-

39 Frank, L. K., *et al.*: Teleological mechanisms. *Ann. New York Acad. Sc.* 50, *189 (1948).*

40 *Cybernetics.* Von Foerster, H., Editor. Trans. Sixth Conf. New York: Josiah Macy, Jr. Foundation, 1949.

41 *Cybernetics.* Von Foerster, H., Mead, M., Teuber, H. L., Editors. Trans. Seventh Conf. New York: Josiah Macy, Jr. Foundation, 1950.

42 *Cybernetics.* Von Foerster, H., Mead, M., Teuber, H. L., Editors. Trans. Eighth Conf. New York: Josiah Macy, Jr. Foundation, 1951.

duction. We hope that this note will serve as a preface to the earlier reports, as well as to the present publication.

HEINZ VON FOERSTER
MARGARET MEAD
HANS LUKAS TEUBER

New York, N.Y.
January 29, 1952

THE POSITION OF HUMOR IN HUMAN COMMUNICATION [1]

GREGORY BATESON

To DISCUSS the position of humor in the equilibration of human relationship, I shall build up from things that have previously been talked about in this room.

Consider a message of a very simple kind, such as, »The cat is on the mat.« That message contains, as has been emphasized here, many other things besides the piece of information which may be defined as the »Yes« or »No« answer to the question which would be created by inverting the same words and adding an interrogation mark. It contains a series of things of which one set would be answers to other informational questions. Not only does it give the answer to: »Is the cat on the mat?«, but also to »Where is the cat?«, which is a much wider question. The message also contains, as McCulloch has stressed, something in addition to a report about the cat, namely, a mandatory aspect; it urges the recipient of the message to pick the cat up, to kick the cat, feed it, ignore it, put it out, according to taste, purpose, and so forth. The message is a command or stimulus as well as being a report.

There is a further range of implicit communication in this message, two additional categories of implicit content. One category includes the implicit communication betwe[e]n A and B that the word »cat« shall stand for a particular furry, four-footed thing or for a category of furry, four-footed things. People are not necessarily in clear agreement about what their messages mean. The senders have their rules or habits in constructing messages; the recipients have their rules and habits in interpreting them; and there is not always agreement between the rules of the sender and the rules of the recipient. One of the most important uses of messages, and especially of their interchange – the single message doesn't mean much or do much in this respect – is to bring the two persons or the many persons together into an implicit agreement as to what the words are to mean. That is one of the most important social functions of talking. It is not that we want to know where the cat is, but that we terribly want it to be true that both persons are talking the same »language« in the widest sense of the word. If we discover that we are not communicating in the same way, we become anxious, unhappy, angry; we find ourselves at cross-purposes.

Ongoing interchanges are very useful in building up among a group of persons the conventions of communication. These conventions range from vocabulary and the rules of grammar and syntax to much more abstract conventions of category formation, such as the conventions for | structuring the universe and the conventions of [2] epistemology. The conventions of communication include the material of linguistics at the simplest level, but also under this heading comes material which is the field of study of psychiatry and of cultural anthropology. When I, as an anthropologist, say there is something different about those English or those Balinese, I don't only mean that they eat their vegetables in a rather uncooked form or that they go to boarding schools. I do not refer to a set of simple descriptive statements of action or a set of descriptive statements at the vocabulary level. I mean also that their actual conventions of communication are different from those of some other culture (1).

I classify together the simplest conventions of communication and the most abstract cultural and psychiatric premises, and insist that a vast range of premises of this sort are implicit in every message. For example, I believe that the world is »agin« me and I am in communication with some other person, the premise about the world being »agin«

me is going to be built into the way in which I structure my messages and interpret his. In a sense, a philosophy of life is describable as a set of rules for constructing messages, and the individual's culture or *Weltanschauung,* call it what you will, is built into his conventions of communication.

There is another set of implicit contents in such a message as: »The cat is on the mat,« namely, implicit statements about relationship. We are trying to tell each other that we love each other, that we hate each other, that we are in communication, that we are not in communication, and so on. The implicit statements about the conventions of communication are messages about the »how« of communication, but these (about relationship) are messages about the fact of communication. »We are communicating« is a statement by two persons.

You meet somebody in the street and he turns and looks into a shop window. You noticed that he saw you coming; you observed that he turned and looked into the shop window. He may be transmitting the very peculiar message: »We are not communicating.« Whether he is or is not communicating is a question which brings us to Epimenides' paradoxes.

One of the rather curious things about *homo sapiens is* laughter, one of the three common convulsive behaviors of people in daily life, the others being grief and orgasm. I don't want to say that they do not occur at animal levels, partly because I am not competent to say such a thing, partly because I suspect that there are prefigurations in certain mammals but all three phenomena certainly are not developed among mammals to the extent that they are among *homo sapiens.* Because they are involuntary, or partially so, one tends to think of these [3] | phenomena as lower functions, animalish functions, but since the full development of these phenomena is characteristically human, it seems that laughter, sobbing, and orgasm are perhaps not lower functions in a simple neurophysiologic sense but have evolved because of the hypertrophy of the upper levels and the resulting peculiar relationship between the cortical-intellectual processes and those which go on below.

These three phenomena, and also the convulsions of epilepsy and shock therapy, have the characteristic that there is a build-up, a so-called tonic phase, in which something called »tension« – which it certainly is not – builds up for a period; then something happens, and the organism begins quaking, heaving, oscillating, especially about the diaphragm. I leave it to the physiologists to discuss what happens.

These three convulsive phenomena are subject to impairment in mental illness. The inability to weep, the impairment of orgasm, and the impairment of laughter are among the indices of illness that the psychiatrist looks for. If those three things are functioning nicely, the individual probably is not doing so badly. If one of them is hypertrophied, or two or three impaired or absent, then the psychiatrist knows that something is not functioning right.

Of the three types of convulsion, laughter is the one for which there is the clearest ideational content. It is relatively easy to discuss what is a joke, what are the characteristics that make a joke, what is the point of a joke. The sort of analysis that I want to propose assumes that the messages in the first phase of telling the joke are such that while the informational content is, so to speak, on the surface, the other content types in various forms are implicit in the background. When the point of a joke is reached, suddenly this background material is brought into attention and a paradox, or something like it is touched off. A circuit of contradictory notions is completed.

There is a very simple and not very good joke going around – for some reason, those who discuss humor from the scientific point of view always use rather dull jokes: A man working in an atomic plant knew the guard at the gate slightly, and one day he comes out with a wheelbarrow full of excelsior. When the guard says, »Say, Bill, you can't take that out,« he says, »It's only excelsior, they throw the stuff away, anyway.« The guard says, »What do you want it for?« Well, he said he wanted to dig it into his garden because the soil was a bit heavy, and the guard let him go. The next day he comes out again with a wheelbarrow full of excelsior. This goes on day after day, and the gateman is increasingly worried. Finally, he says, »Bill, look, I'm going to have to put you on the suspect list. If you tell me what it is you're stealing from this place, maybe we can keep it quiet between us, but I'm perfectly sure you're stealing something.« Bill says, »No, it's only | excelsior. You've looked through it every day and dug to the bottom of it. There's nothing there.« But the guard says, »Bill, I'm not satisfied. I'm going to have to protect myself by putting you on the list if you won't tell me what this is all about.« Finally, Bill says, »Well maybe we can get together on this. I've got a dozen wheelbarrows at home now.« [4]

We have talked a good deal at these Conferences about figure-ground relations. If we name something as a person, a face, or a table, or whatever, by the fact of naming it, we have defined the existence of a universe of not-this, a ground. We have also discussed, although not, I think, as much as we should have, the Russellian paradoxes, especially the class of classes which are not members of themselves. These paradoxes arise when a message about the message is contained in the message. The man who says, »I am lying,« is also implicitly saying, »The statement which I now make is untrue.« Those two statements, the message and the message about the message criss-cross each other to complete an oscillating system of notions: if he is lying, then he is telling the truth; but if he is telling the truth, then he is not lying; and so on.

The paradox of the class of classes which are not members of themselves arises similarly from examining the implicit message. The first step toward building the paradox is to say that the man who speaks of elephants is thereby defining the class of non-elephants. The possibility of the class being a member of itself is then introduced via the class of non-elephants, which class is evidently not an elephant and therefore is a member of itself. The circuit of ideas which is the paradox is closed or completed by treating seriously the background: the non-table, the non-elephant. The ground is a part of the implicit information. It just is. You can't ever really get away from it.

The hypothesis that I am presenting is that the paradoxes are the prototypic paradigm for humor, and that laughter occurs at the moment when a circuit of that kind is completed. This hypothesis could be followed up with an analysis of jokes, but rather than do that, I should like to present to you the notion that these paradoxes are the stuff of human communication. As scientists, we try very hard to keep our levels of abstraction straight; for instance, in these conferences we have gotten into very great trouble when the levels of abstraction became tangled and the theory of types showed itself. In ordinary life, as distinct from scientific talk, we continually accept the implicit paradoxes. If the psychiatric patient says, »I dreamed,« and then narrates his dream, he is making a set of statements within a framework not unrelated to that of Epimenides. If an artist paints a picture and says, either implicitly or explicitly, »This picture is a truth, this picture is an attempt to convince you,« he is, if I may say so, probably not an artist but a | scientist or a propagandist. If he says, »This picture is in an Epimenides frame,« he is a »real« artist. Or consider the old difference between Ruskin's true and false grotesque (2). The true grotesque is, I suggest, created by the man who says frankly, »I am lying,« and who goes on to create a thing whose truth is that it is created. The man who says, »This is a horrible dragon,« and tries to make his work of art into a factual statement is the one who produces the false grotesque. He is the propagandist. [5]

The setting of the psychotherapeutic interview has a peculiar relationship to reality. Is it real or is it not? The fantastic exchanges that go on within it are paradoxical. The patient who says, »I walked around the grounds this morning and I said, ›I will be honest. I am going to get something straight,‹« fairly certainly will not achieve much that day. The likelihood of his making an advance depends much more on his ability to say to himself, »Let me freely imagine what I want to imagine and see what comes up.« Indeed, the whole free association technique is an attempt to give that freedom.

But the therapy situation is not unique. It is, perhaps, a specialized version of what, after all, goes on between us all the time. The therapy situation is a place where the freedom to admit paradox has been cultivated as a technique, but on the whole this flexibility exists between two people whenever, God willing, they succeed in giving each other a freedom of discussion. That freedom, the freedom to talk nonsense, the freedom to entertain illogical alternatives, the freedom to ignore the theory of types, is probably essential to comfortable human relations.

In sum, I am arguing that there is an important ingredient common to comfortable human relations, humor, and psychotherapeutic change, and that this ingredient is the implicit presence and acceptance of the paradoxes. It appears that the patient (especially the Freudian analysand) makes progress via the mental flux, confusion, or entropy stirred up by paradox, that, passing through this state of inner disorder, he is partly free to achieve a new affective organization of experience or new premises for the codification of his thoughts.

The alternative to the freedoms introduced by paradox is the rigidity of logic. Logic is a very peculiar human invention, more or less timeless. We say, »If A, then B,« but in logic, the word, »then« does not mean »at a later time.« It means that statement B is synchronously implicit in statement A. But when we speak of causes and say, »If I drop the glass, then it will fall,« the words »if … then« refer to a sequence in time and are quite different from the »if … then« of logic. When logic encounters, the theory of types and paradox is generated, its whole exposition breaks down – «Pouf!« It is perhaps some terror that mental process may go »pouf« which compels many patients and [6] | persons at large to cling to logic. But casual systems do not go »pouf« in this way. As in an electric buzzer, there is sequential contradiction, and the system merely oscillates.

One of the hypotheses in this group is that mental processes can appropriately be described in terms of causal hypothesis with all due qualification of the word »cause.« I would suggest that these processes absolutely cannot be described in terms of timeless logic. The study of mind through the causal approach, however, will lead us into accepting the paradoxes of thinking, which are related to humor, which are related to a freedom to change the system of thought related to humor, and in general are related to mental health and human amenity.

I think that opens enough subjects for discussion, but there is just one other thing I should like to speak of. I want to refer back to some talk which we have had in the past over the words, »unconscious« and »the unconscious.« Conventional theories about humor usually refer to repression, release of repression, *Schadenfreude* – the pleasure which we feel in somebody else's pain – and so on. I want to say that the various types of implicit content of messages constitute what I personally would understand by the content of the unconscious. Those are the items which, when we think only of the cat and its location, we are likely not to notice as messages which we have received. It seems to me that the *Schadenfreude* theory, which, after all, is classic for this subject, arises because the implicit enjoyment of another's pain is among those things which we prefer not to notice. It is a premise which we leave implicit among those messages which we receive without noticing that we received them. All or most of the cultures of the world have some degree of restriction and taboo upon hostile expressions and hostile actions, and, therefore, in all cultures of the world that type of material is likely to be sidetracked into the implicit and to be unnoticed until a joke is completed. And that is as near as I can get to an explanation of why people make *Schadenfreude* theories about humor.

FRANK: Gregory Bateson referred very briefly to the figure-ground concept. We could further our thinking by emphasizing the selective awareness and patterned perception of each person, and some of the problems which seem to be involved. For example, we were talking in this room earlier this week[1] about the primary discrimination of self and nonself in the child, discussing the fact that primary discrimination is not to an outside objective reality but is always to an idiomatically highly-patterned nonself. Later on, the child may have to learn to modify that objective nonself and accept the social-cultural definitions of the environing world. Some children do not wholly accept these cultural definitions, as we know, and perhaps that is how psychiatric pa | tients develop, from those who have not made the transition from the purely idiomatic to the public world. [7]

The figure-ground concept is further illuminated if the joke is thought of as involving a shift between the figure and ground, where the figure is altered or the ground is reconstituted or a reversal of the figure-ground situation takes place.

Another aspect that may be worth examining is to think of the figure-ground in these terms: that the figure is a cognitive pattern perception, selectively chosen because of learning, constitutional susceptibility, and so on, while the ground is that to which an affective re[s]ponse is made. In all experience, we selectively perceive, define, and impute meanings to the different figures that are largely personal, idiomatic versions of socially and culturally patterned ideas and beliefs. Concurrently, in every situation we respond affectively without being aware of it. If we can use the concept of people growing up with highly conflicting responses, one, a cognitive, meaningful one to the figure, the other an affective response to the ground situation, which is in conflict to the first, we might get a chance to make some kind of an interpretation of what we call »emotional conflicts« and the »unconscious« bias in perception.

BATESON: I think I am responsible for a possible misunderstanding at this point. There is a danger which one has to be aware of all the time in the psychological sciences, namely, the danger of taking a dichotomy, such as figure-ground, and equating it with every other dichotomy, such as affect-cognition or consciousness-unconsciousness. I set the stage by, using the yes-or-no answer to the question, »Is the cat on the mat?« as in some sense a primarily conscious, figure-ish item, and I defined the other things as background items. But it is important to insist that that was a purely arbitrary selection on my part.

In talking about the character structure of a certain individual or about the thought habits or the communication habits distinguishing a certain culture, it may be important to say which categories of content appear in the forefront of consciousness. There are, certainly, many people who are enormously more conscious of some of the items which I labelled as »implicit« than they are of the concrete information. After the conversation, they don't know whether the cat was on the mat but they do know whether somebody loves or hates them, and so on. I don't think it can be said that affect is necessarily the more unconscious component.

FRANK: I didn't want to separate affect and cognition. I merely wanted to point out, in discussing and conceptualizing the picture, that the affective reaction might be looked upon as analogous to the way we adjust to the temperature and barometric pressure in this room without being aware of it, that is, they are part of the ground in which this meeting is taking place. | [8]

1 Conference on Problems of Infancy and Childhood, sponsored by the Josiah Macy, Jr. Foundation.

May I make just one other point? I think you would agree, wouldn't you, Gregory, that the individual is not only communicating to somebody else but at the same time he is trying to reaffirm and re-establish his own idiomatic version of the word?

BATESON: Surely.

FRANK: There is, then, the problem of whether the individual is consciously aware of trying to communicate or of his attempt to reassure himself as I suggested at one of our earlier meetings, we should discuss internal speech because that is a highly significant aspect of this problem.

VON BONIN: In the joke that was told, all of a sudden the figure-ground relationship switched over into another constellation. The wheelbarrow was background and was not noticed, but I don't think it had any affective tone. I can't see that the background was anything to which we reacted emotionally.

BATESON: I cut down the affective tension of that joke, if I may use the word tension, knowing that I don't mean it, by saying that the man with the wheelbarrow and the gate guard were friends. By making it obvious that they were going to get in cahoots, there was no serious danger in the situation. There would have been more laughter after that joke had I not said that.

VON BONIN: I don't think it matters much whether you say that or not. I heard the joke before in a slightly different version, and it evoked the same laughter because one simply does not think of the wheelbarrow and it makes a completely different structure of the whole situation.

As you told that, I thought of another. It is not a good one. We were in the north woods and a man drove into the camp with a huge; sixteen-cylinder Cadillac. The Indian guide said, »Big car.« The man said, »Yes, very big car; sixteen-cylinders.« The guide said, »Can go fast?« and the man said, »Yes.« The guide spit on the ground and said to me, »Every time a cylinder misses he saves a dollar.«

Again, the point can be made that what one first has in view is a battery of cylinders as a complete whole, doing certain things. Then, all of a sudden, attention is directed to an individual cylinder. You've never thought of sixteen cylinders as sixteen individuals, so the situation becomes completely restructured. The man on the banana peel is the same sort of thing, although I think Bergson makes the point that the essence of a joke is when the laws of gravity or the laws of the inert universe suddenly apply to something that lives and topple it over.

YOUNG: Couldn't laughter be defined as the sign of sudden agreement? A smile is the sign of agreement. Laughter appears when there is sudden agreement, for a variety of reasons. It may be recognition of a nonmember of the group, for example. It may be reversal of figure and ground, as mentioned. But it is a communication sign; it is the [9] sign of a sudden achievement of communication. |

BATESON: I would agree, but I would narrow it to say that laughter is the sign of agreement that X is both equal to Y and not equal to Y. It is agreement in a field in which paradox has been presented.

QUASTLER: Isn't it true that you have introduced, surprisingly, a new dichotomy between Z and non-Z, with no reference to the Y and non-Y dichotomy? It turns out that X is equal to Z, but it still is equal to Y; the man still has the excelsior.

BATESON: Yes, he's still got the excelsior. The previous figure is not denied; only its relevance is. We know that the figure is the excelsior. Suddenly, we are told, no, it is the wheelbarrow. But it is still the excelsior, too. The original figure survives, and it is that doubling, I think, which promotes laughter.

PITTS: One of the essences of humor consists in the restructuring or reversal of the figure-ground relationship, but, of course, there is a great difficulty in explaining why not all of these cases are jokes. It is one of the most frequent components of our experience that what we did not attend to, we now attend to, and what was not important becomes important. But, certainly, the vast majority of these transitions are not regarded as humorous by us; thus, there must be something else which is a common characteristic of humor beyond the reconstructing of the figure-ground relationship or the distribution of tension.

BATESON: There is a rather poor joke going round the West Coast about two men playing golf. A couple of women are on the course ahead of them, playing very slowly. The men want to pass, and one fellow says to the other, »You go forward and talk to those gals and ask permission to pass them.« He goes forward, returns and says, »Gee, I can't talk to them. One of them is my wife and the other is my mistress. You do it.« So the other guy goes forward and he comes back and says, »It's a small world.« Now, it is practically impossible to tell that joke without somebody guessing that that particular reversal is going to occur, and it is less of a joke because it has that leak in it.

MCCULLOCH: There is no surprise.

BATESON: The surprise of the point is lost. I have now heard it told twice and I have told it twice, and none of those four tellings has taken place without leakage.

GERARD: There is a joke which exemplifies all the points made so far, except for Walter's question of why the shift is not always humorous, which I think is a critical one. A fellow says to his friend, »Do you know these ice cubes with the hole in them?«; and the reply, »Know them? Hell, I'm married to one.« That has the sudden inversion, the carrying of the inanimate to the human, the problem of tensions and expression and suppression.

I told this joke deliberately to raise the question of the difference | between ordinary jokes and so-called dirty ones. There is a very real difference in the kinds of things that elicit laughter and the kind of laughter that is elicited depending upon the setting, the group, and so on. The reaction of this group is illustrative. I have told that story twice to small groups this morning and they laughed uproariously, right here in this room. I have now told it publicly, in the presence of a woman, and the guilt feelings almost suppressed any laughter. [10]

KLÜVER: What about the relationship between humor and irony?

BATESON: Do you mean irony in the classical sense, such as occurs in Greek tragedy when the final disaster is implicitly or explicitly predicted in the beginning by a speaker who doesn't know what he is predicting? Or do you mean irony in the sense of saying the opposite of what is meant?

TEUBER: One would be the irony of the situation of Oedipus who does not know what everybody else knows, and the other would be the Socratic irony. Socrates insists he doesn't know what everybody else presumes to know …

PITTS: No, he doesn't want to say he does, but the other person doesn't, either.

TEUBER: He knows one thing that the other fellow doesn't: he knows that he doesn't know.

PITTS: And the other man supposes he does, and the irony is directly implicit in the fact that the other man doesn't, either.

VON BONIN: May we know how the Greeks defined irony? They talked a lot about it.

PITTS: In relation to the tragedy.

VON BONIN: Yes.

MEAD: Just a moment. Why are we getting so literary?

PITTS: Well, who started it?

MEAD: I am just raising it as a question. Why this outcrop of literary-historical erudition here?

GERARD: Maybe we haven't anything constructive to say.

MONNIER: Why does laughter not exist in animals? Laughter implies a comparison of the code of one individual with the code adopted by the group. Laughter arises, for instance, when the individual observed does not behave according to the code of the observers. A man walking on a curb is expected to see the edge and to step to the street properly. If he behaves like an automaton, does not see the edge of the curb, and falls, the observer laughs. Bergson pointed out the biological function of laughter, that it tends to protect society against egocentric mechanical behavior of individuals at variance with outer reality.

MEAD: I would be willing to accept that laughter can occur when there is a contrast [11] between the code of the collectivity and the individual | event or remark, but not that it necessarily requires that something has gone wrong; there is also the laughter when something goes right. Laughter is one of the easiest human responses to evoke by someone saying what everybody is feeling but nobody has expressed it or is quite willing to say it in that way. It isn't that the remark is wrong to make, but that there is a discrepancy between what is correct to express and what everybody feels. The discrepancy is the thing that produces the laughter. People laugh when the cork is pulled from the bottle.

YOUNG: Children's laughter.

WIESNER: People often laugh when they are upset or nervous. The situation in itself is not humorous, but when the relationship between the external and internal world is not quite right, laughter is one way of bridging the gap.

MEAD: So there is again a discrepancy.

WIESNER: The discrepancy seems to be a common thing.

YOUNG: Humor is only one of the situations that evoke laughter. That is what we want to say.

BATESON: Yes, and the situations should be subject to formal analysis., We should be able to say how we would construct a cybernetic machine of some kind which would show this characteristic which would be thrown into some sort of oscillating condition by certain types of contradiction.

WIESNER: It would laugh whenever the input and the coding did not match properly.

BOWMAN: There can be a very simple network of two tubes in such form that if one conducts, it cuts off the other. A circuit of that type may have two stable states. If it is put in any state, it will asymptotically approach one of the two stable states and stay there. On the other hand, with the same components in slightly different values of the circuit constants, it can oscillate.

BATESON: I am always prepared to say that an electric buzzer is laughing.

BOWMAN: It has no stable state.

BIGELOW: I don't understand what we are trying to do here. Are we trying to construct a definition which will be adequate for all types of humor?

MEAD: No; we are not studying humor.

GERARD: We seem to be trying to equate humor and laughter.

McCULLOCH: We are trying to study the role of humor in communication.

VON BONIN: I am guilty of this digression, for I wanted to speak about figure-ground and used a joke as an example because it seemed to me to illustrate the point more [12] clearly than the cat on the mat. | Throwing in another joke got the discussion off on a tangent, May I bring it back by bringing up another point. In language, there are not only the actual words which are announced but there are also the overtones in the language.

In studies being done in Chicago, the experimenters are putting forward that there is a difference between laryngeal and oral speech. It has been shown that you can frequently understand the emotional state of a person even when you don't understand a word he says. We have had, for instance, a man talking in Hungarian, which none of us understands, but we have gotten a faint idea of what he said.

BIGELOW: What could you tell?

VON BONIN: Whether he related a story, whether he was trying to express his displeasure, whether he approved heartily – that sort of thing.

MEAD: That won't stand up cross-culturally.

VON BONIN: I don't think it will at all. For instance, you can't ask a question in Chinese by raising your voice at the end of the sentence because the last syllable would mean something entirely different.

MEAD: What does stand up cross-culturally is that in every society that has been analyzed so far there seems to be a tendency to symbolize certain states by certain sounds. The sounds are not constant but they have enough physiological congruence so they may recur.

WIESNER: May not there be physiological changes in the mechanisms of speech which can be universally recognized and deciphered? For example, when an individual is angry, his muscles tighten so the format structure is very different, thus changing his tone.

VON BONIN: That is the problem the Chicago group studied, whether the voice can be meaningful without an understanding of the words.

PITTS: That is, do all people in all cultures raise their voices when they are angry.

BIGELOW: Does the aspect of information content involving emotion remain across cultures?

MEAD: No.

BIGELOW: Can you enumerate cultures in which these overtones do not contain, essentially, emotion but some other information?

WIESNER: In other words, do people always talk faster when they get excited?

MEAD: As far as is known at present, there are no universals of that order. The universal is that every culture, if the language is properly analyzed, includes what Trager and Smith are coding as superscripts; that is, every language has a recognizable intonational pattern. Similarly, every culture has a code of emotional expression but the code differs from one society to another.

BIGELOW: But is it emotion in every case? | [13]

MEAD: The best example I can give are the shouting signals of the Arapesh, in which they use words. The words may be, »Somebody is coming,« but nobody hears the words. Some words are shouted that nobody can understand, that communicate only a degree of affect by their loudness and their frequency. The people hearing the shouts sit down and figure out what is meant entirely in terms of their knowledge of the probabilities of the situation, which are quite reasonable. They translate a message which has the form of information but which never gets across. They sit there and say, »Now, that came from there. Who do you think would be there now? Who would be shouting that loud and that often? And if it were he who is shouting, what does it mean? Does it mean that his mother-in-law who has been quite sick has died?« They work up a whole series of probabilities and then they set out to the funeral.

BIGELOW: In Such a case as that, then certainly the overtones contain something else besides the usual emotion; they contain a lot of information separate from emotion.

BATESON: The tone languages and the use of drum signals should be mentioned. There are languages in which words have significance on a flat tone or a rising tone or a falling tone. In Chinese and in many of the African languages, this occurs. The pitch or pitch structure of a word discriminates that word from others which would otherwise be homonyms.

This problem of homonymy arises in reverse in African drum signals (3). The Bantu spoken languages have significant pitch, but in sending messages by drum, only the pitch can be transmitted. This would lead to serious homonymy except that it is avoided by transmitting whole phrases instead of single words. Thus the word »girl« is conventionally replaced in drum messages by the phrase: »The girl will never go to the *linginda* fishing net.« (The use of this type of net is a traditionally masculine occupation.) The long tonal sequence provided by the whole phrase precludes homonymy.

For the purposes of this discussion, the important thing is to treat the word »language« as including all of this. We should drop the idea that language is made up of words and that words are toneless sequences of letters on paper, although even on paper there are possibilities for poetic overtones. We are dealing here with language in a very general sense, which would include posture, gesture, and intonation.

KLÜVER: First, I should like to remind you that Yerkes once pointed out that the chimpanzee resents being laughed at by man or other animals. Second, I wonder whether what has been said here should not be related to more general considerations. The factor of discontinuity which has been emphasized in this discussion is, of course, [14] | characteristic of many psychological phenomena. For example, all our dealings with inanimate and animate objects, with humans and animals, involve processes of »typification.« One may doubt whether personality »types« exist, but one cannot doubt that processes of »typification« constantly occur in our response to environmental objects and events. The great psychological and sociological significance of such »typifications« was recognized long ago by philosophers, such as Simmel (4).

It seems to be the fate of many »typifications« to suffer sudden breaks or reversals. You encounter a man on a beach and after talking to him for a while you learn that he is, let us say, a priest or a colonel. As a result, the whole field may suddenly become restructured and reorganized. Or let us consider our reactions to objects of the visual environment. We are in optical contact with an object, and we may go to the trouble of performing numerous and diverse motor reactions to stay in optical contact. However, it happens again and again that the contact is broken since the appropriate movements either cannot be performed or cannot be performed quickly enough. Thus, optically induced behavior constantly involves discontinuities and breaks resulting in loss of contact or coherence between ourselves and the object.

More generally speaking, life seems to be a sequence of jokes, the humor of which we often fail to recognize.

ASHBY: Perhaps this repeats what has just been said, but the language is sufficiently different to suggest that there may be some more general principle behind both. I want to consider the question of an observer getting information from some physical system, either an inanimate system or another human being. Every physical system lives in a physical universe. The system is surrounded or supported by a great number of variables that are in some effective contact with it. The observer can profitably study only systems in which these surrounding variables are constant. If the surrounding variables are held constant, the constancy is sufficient to isolate the system, and the observer can get useful information out of it. But because the surrounding variables are constant does not in any way prejudge what values they are constant at. Thus, when the observer is studying the system, this is one of the first things he must find out. In ordinary language, he must find out what the person takes for granted.

The number of surrounding variables is usually uncountable. If one started to write down what we are taking for granted this morning, for instance, that we are talking in

1952 A.D. and not in 1952 B.C., the list would get sillier and sillier but it would have no end. Consequently, all the information that is coming out of here this morning is related to these values, even though they can't be given explicitly. | What may happen [15] is that the observer, taking for granted that a surrounding variable has the value of, say, zero, may go on collecting information about the system until suddenly some astonishing event shows him that the variable must really have been at one all the time. He suddenly has to re-interpret all his past information on a new basis. That is the critical moment, when he realizes that the variable which he had assumed to have one value evidently must have some other value.

WIESNER: This is the situation you have when somebody talks at you in a foreign language and you don't realize it for a moment; then you suddenly switch. If you go to England and expect an accent that you have to adjust to, and a man talks French to you or German, it sometimes takes many words before you realize it and make the translation and get information.

BATESON: The social scientist is not only in the sort of position that Ashby has suggested for his observer but, worse, he is investigating a dynamic system more or less in the dark with a flexible stick, his own personality, the characteristics of whose flexibility he does not fully know. There is, therefore, a set of unknowns in the observer, which are also subject to investigation. Every statement we make about the observed derives from premises about the self. I say this glass of water is there because I can touch it with my hands and feel it there with my eyes shut. In order to make this statement, »It is there,« I have to know where my arm is, and, on the premise that my arm is out in that direction, I conclude that the glass is there. But the premise about myself is built into my conclusion. The whole gamut of projection phenomena follows.

There are premises about one's self, in terms of which one understands something else. But the events in interaction between oneself and the something else may lead to a revision of premises about one's self. Then, suddenly, one sees the other thing in a new light. It is this sort of thing that leads to the paradoxes and to a good deal of humor, I would suspect.

ASHBY: A paradox might start in this way. You begin by thinking that parameter alpha is at zero, but, after you have gone on for a time, you suddenly realize it must be at one, and you start to re-explore on the assumption it is one. If the system has something rather peculiar in it, it might force you back to the deduction that alpha is zero. Obviously, if you go on without any further change, you are caught because you will go on changing in opinion backwards and forwards. What it means is that, simply from the physical point of view, the two, observer and system, have gotten into a cycle. There is nothing strange in the physical aspect, although it may be disturbing to the observer. | [16]

BATESON: And if those are two human beings, when that point is reached, laughter is likely to occur.

ASHBY: Very likely.

TEUBER: Wasn't it Gregory's point that it is quite desirable for the benefit of the process of communication – to let jokes, or riddles of a certain sort, point up the schematism that is shot through all of our communicative processes and without which we could not communicate ?

BATESON: A schematism which we cannot communicate by itself.

TEUBER: Yes. There have to be schemata; we cannot talk or communicate, even in nonverbal forms, without some schematism. At the same time, I want to point out, and this, I think, was also Klüver's point, that the schemata are quite limited. We have constantly to pick and choose, shift or be pushed from one to another. Whether the sudden transitions are frightening or exhilarating probably depends on very many things that have not been enumerated. But I think it is no accident that jokes and riddles tend to appear together in child development. When the child begins to make jokes, he usually will ask riddles for the first time in his life. Similarly, the so-called primitive riddle seems to lie somewhere between the pun and the prototype of a lyrical metaphor, These riddles exist in all sorts of languages and cultures, although I would not know whether they are really universal.

MEAD: No, these riddles are not universal, Some people do not have them.

TEUBER: Still, those that do exist are surprisingly similar in structure. For example, »bird without feathers flies to a tree without leaves.« The answer: »fire consumes a log.« Such a primitive riddle seems to play at making a definition.

PITTS: Is not the definition of a good riddle that its answer is a good joke?

TEUBER: Certainly, or a poem. All these forms of expression have this in common: they point simultaneously at the value and at the limitations of all schemata. They force us to realize that the communication process is what it is – it cannot do without the schemata. They make communication, for a moment, about communication.

YOUNG: Laughter is the recognition of the achievement of that communication.

MEAD: But Walter made the point that all such occasions do not provoke laughter, for instance, Dr. Ashby's picture of the scientist who has worked for years and then he discovers he has made a mistake in attributing a certain value to a variable. The response there might well be convulsive sobbing instead of laughter. I think if we keep laughter in the context originally suggested, of a tension release that is related [17] | to other tension releases, we shall do much better. In such a context, laughter has the function of a safety valve.

REMOND: That brings up the point of the emotional status of the individual at the times when humor has a possibility of occurring. For instance, A can say a particular phrase to B, and in a certain emotional state, it will not be humorous; at another time, because of what has been said before or what he has lived through before, B will laugh uproariously. There is, therefore, a very important difference between the reaction of a human being and a machine. Man adapts to the moment and a machine should be, at all times identical to itself, not changed by emotions built up for a variety of reasons not absolutely relevant to the joke being made.

Some people laugh very easily. They see something to laugh at immediately in everything. Some people, who are extremely cold or who are sad for some reason, will not laugh at anything. But sometimes laughter depends on things other than the emotional state. For instance, the meaning of some phrase can be well understood but the phrase does not carry the humorous message it should. I am thinking about the fractured French jokes on napkins. Since I am French, I was interested in them. My emotional state at the time I saw them was quite adequate. I was at parties; I had been laughing already; I had been drinking, and I was set to laugh easily. But the fact that those jokes were not made for French people and that I had to make an effort to understand them put me in an intellectual attitude rather than a humorous one. I had to be led to understand that in America such and such a phrase was pronounced with such and such an inflection or such and such an accent so that it could refer to such and such a situation. But I wasn't happy with it; it wasn't funny.

WIESNER: Well, I, as an American, don't find them very funny, either.

REMOND: Sometimes I can see that some are funny, but I have to analyze their positive meaning to understand them and I don't feel them really, which is quite different.

BATESON: The diaphragm is not really involved.

GERARD: And that factor vitiates a great deal of the discussion that has gone on this morning. There is something quite unique and explosive when the diaphragm gets out of control, but most of the discussion has not dealt with that semiphysiological aspect of it. Laughter may become as uncontrollable as the other two elements you mentioned, or as a fourth one that I think is probably related, the yawn.

McCULLOCH: Domarus worked up a set of Jokes ranging from those which will make a man laugh under almost any circumstances to those which are so dull and boring that you just don't see how anybody | could laugh at them. He told these deliberately [18] and systematically to people in various degrees of fatigue, and found that the ease of provoking laughter was dependent in large measure on fatigue. Dusser de Barenne and I were among his guinea pigs. He would never forewarn us, of course. He would simply be around while we worked. We were really horrified that, at the end of seventy-odd hours of work without more than a few minutes snatched in sleep, he could tell us that one and one made two and we would burst into laughter. We became furious with ourselves at the ease with which laughter was evoked when we were tired. The physiological state of the organism is crucial, but just how, I don't know.

MEAD: The most laughter I have ever gotten was when I gave the last lecture to a group of social workers who had had a week's conference. They laughed at anything. It didn't make the slightest difference. They laughed virtually before I opened my mouth. But there was something in what I said that gave them permission to laugh, just as when a joke was told to you. All the cue you needed at that state of fatigue was, »It's all right to laugh.« A comparable situation is when one has been repressing yawns with a terrific effort. The minute the chairman says, »Let's have some coffee,« the yawn will burst out in that same way.

FREMONT-SMITH: There is another element in Warren's situation, that he had been trying for seventy-odd hours to focus his attention on a problem. He really wanted relief from that. The »one and one makes two« provided a situation for a withdrawal of attention and a moment's relief and relaxation.

One point that seems to me important is suddenness of shift; I don't know whether there is such a thing as a slow development of a sense of humor. I suspect that what happens is that a series of sudden steps must be involved rather than a gradation.

Another thing I should like to bring up is, shall we put a little more attention on the humorless person and on the person who is at a given moment humorless? It has seemed to me that the humorless person is the person who lacks perspective or lacks the capacity to see something in several different perspectives. Isn't that the figure-ground situation again? The humorless persons sees things only in a very narrow frame of reference, and therefore he cannot shift.

TEUBER: For that reason, if we are working on a difficult experiment, we ordinarily don't appreciate any sudden increase in difficulty as humorous.

McCULLOCH: If a man already has investigated those possibilities and you bring up one of them, he isn't likely to laugh.

PITTS: I should like to say several things, of which a number are | meant as a sum- [19] mary. First, I should say that we are probably agreed that, in some sense of the term, a restructuring of the situation is necessary to a joke, and we should probably also agree that a certain suddenness is required if it is to produce an effect. The restructuring will explain Dr. Fremont-Smith's case of the man who is humorless because of his incapacity for restructuring his point of view, and the suddenness will presumably explain Tony's case of the joke whose point cannot be perceived without a considerable intellectual application, that is to say, not except by a relatively slow process.

In addition, I still maintain, in agreement with Dr. Klüver, that some additional quantum is required to make something into a joke. I would like to deviate from that, however, for one further point, namely this, that one must, although this is not the kind of thing I customarily say, not suppose that a joke, every time it is said or every time it is heard by a given person, is necessarily the same joke. The joke must be considered in the context of the person who hears it, and his past. The fifth time you hear a joke, you rarely laugh. Naturally, the reconstructuring of the situation in your case is in that case absent because, well, you can predict the future course of the joke, and so, when you begin hearing it, you have the whole situation in mind and that simply persists without any restructuring, all the way to the end.

With respect to the additional quantum, there is only one suggestion as far as I can see, namely, Gregory Bateson's, that there is a kind of self-reference of the type seen in the logical or pre-Socratic paradoxes which is superimposed on the restructuring of the situation to produce the humorous element. However, that is something I can't easily understand and, consequently, I should like to ask him how he would apply this additional element in the case of the joke he gave. I don't think there is any process of self-reference in the story about the man with the wheelbarrow and the excelsior.

BATESON: When the story is told, the hearer is invited to identify himself either with the gate guard or with the man with the excelsior. »If you were in that situation« is the premise which is introduced. That is one part of the problem of self-reference. The other part is related, I think, to a peculiarity of human communication, which I think was implicit in what you said, Dr. Monnier, that when two human beings are talking or communicating in any form, there is a mutual awareness of the fact that they are communicating. It is not clear that similar mutual awareness is always present among animals. In the courtship of sticklebacks, for example, there is an exchange of signals in quite a complex sequence. The male has to do A and in reply (as we say) to A, the female does X; and X sets free the next step in the male's behavior which is B; which [20] sets off the next step in the female's | behavior which is Y; and so on: A-X, B-Y, C-Z; ending with a completion of the driving of the female into a nest which the male has built, where she lays her eggs and he looks after them. A, B, X, and Y are various sorts of perceptible behavior, exhibitionism, as we might say: raising the spines, exposing the colored belly, etc. But it is fairly doubtful in such sequences how much each communicates or is adjusting his communication to the circumstance of whether it is or is not perceived by the recipient. The male will, I think, start doing his belly dance in parts of the aquarium where the female can't see him.

When human beings try to communicate with each other, we raise our voices, for example, according to the distance that the recipient is from us. We modify our speech in all sorts of ways and include in our speech all sorts of messages about how the speech is to be interpreted. At the end of the message, we say, »Over,« in some form or other. We punctuate. We stop and ask at a given moment, »Have you got me so far?« We watch the faces of the people we are talking to, to see whether the message is getting through, and what they do with their faces is a very important contribution to the communication because it tells us about the success of the communication. The faces give us a message about communication at this higher level of abstraction. In human communication, the essence of it, almost, is the fact of a mutual awareness of the other person's perception. Often, it gets distorted; often, we don't behave rationally in terms of this awareness. We may repeat and repeat when we know very well that the other person got the message. But that mutual awareness seems to me to be very important in human communication.

YOUNG: Why do you say »awareness« rather than »repeated exchange of signs«?

BATESON: Because I want to stress again the implicit content. Many of the implicit messages are about that awareness.

PITTS: But what about all this as peculiar to a joke?

BATESON: The involvement of self in a joke is the thing I was getting to. You can't stand to hear a joke more than three or four times. By the fifth time, you don't laugh. However, a very large number of people will laugh at a joke the twentieth time they tell it.

GERARD: Well, that was a shift. Tell it or hear it?

PITTS: There was a shift.

TORRE: That is the point now.

BATESON: The teller of the joke is able to be self-involved in the joke because he can hear it as if it were new. Granted he hears through his Eustachian tubes and not as a simple recipient, but he can identify with the hearer of the joke as a creature who has never heard it before and therefore he can laugh. | [21]

FREMONT-SMITH: Two elements come in there. One is the business of contagion; very often, somebody who has not heard the joke or has not understood it at all will laugh if the group laughs. But the man hearing a joke for the fifth time does not laugh because the element of surprise or suddenness is absent.

MEAD: But the significant thing still is the conditions under which laughter will or will not be evoked as they relate to the question of identification that Walter brought up. Humor is a playful change of identification, which is safe. One of the things you communicate to an audience, when you keep them laughing, is, »It is safe to think like this, it is safe to think like me, it is safe for a minute to say it like that. Nobody will keep you there. You can get back. You can move around. It is play. It is free.«

FREMONT-SMITH: And something you wanted to do before.

MEAD: As to grief, if one takes Erich Lindemann's studies of grief (5), there is, again, identification involved. His studies, which are the best that I know of, are cases where the total identification with the person who was lost was such that it was unbearable. Tension was built up to an unbearable point and was released in a different type of diaphragmic breakdown. Identification is required before there is grief or laughter, but in one case it may be something that is terribly dangerous.

Once, I was presiding at a conference of dreadfully solemn people on family life. It was just before Mother's day, and everyone was tired. Our P.T.A. delegate had announced she was going home to take up her duties as a mother, and I wanted to give the audience a sense of not being worried if people went out early on this last morning so I said, as chairman, »Our principal mother has already gone because she wanted to be home on Mother's Day, and we will all understand that this is the day before Mother's Day and anybody who leaves is going home to be a mother.« And then I thought, well, I have to deal with the men, and I said, »Or going home to help their wives be mothers.« The audience roared with delight. If I had said it knowingly, they would not have laughed because they would have been frightened. You can't have chairmen, you know, at a conference on family life who make dirty jokes.

FREMONT-SMITH: The audience laughed at you. There probably was in that situation a recognition that you had slipped without meaning to, and they were enjoying your discomfiture.

MEAD: No, the essence of it is, surely, a safe recognition of the communication of sex, which is one of the funniest things. I think the element of relaxation when it is safe is the pertinent thing. The release of tension when unsafety has built up, ties in with what hap|pens in grief and, in a sense, in orgasm, because orgasm is a problem of safety, too, of trust. [22]

PITTS: I will accept that as an explanation rather than identification. Many of the most amusing things people say are not said with the intention of being funny.

BIGELOW: Isn't there some element of personal discovery?

YOUNG: Or group discovery.

MEAD: If it isn't too painful.

GERARD: To follow up a point that Frank made about the contagion of laughter, you probably all have heard these »laughing« records. If I hear one by myself, I am quite able not to laugh; but in a group, when laughing starts, I cannot avoid an uncontrollable laughing response. This is a case, then, of laughter itself provoking laughter, without any symbolic or conscious or logical or other meaning.

MEAD: Yawning, too, provokes yawning.

FREMONT-SMITH: Laughter has memory meaning, and therefore symbolic meaning, I would think.

GERARD: I don't know that it has to have any memory meaning.

FREMONT-SMITH: Would someone who has never laughed go off that way? I think it almost inevitable that hearing laughter and seeing other people laugh would evoke memories of laughing situations, unconscious memories.

GERARD: It would be interesting to try it out on somebody who has never laughed, if such a person could be found.

FREMONT-SMITH: There is contagious coughing at a concert or in a whooping cough ward; if one person starts to whoop, they will all whoop; and if somebody has tears come to his eyes and you watch him, tears are very likely to spring to your eyes.

YOUNG: Have we sufficiently recognized the place of laughter in communication signs? The difference between man and the stickleback is that we have specific signs to indicate communication in general; a series of those, which are very complicated, start on the face. I wonder if there is any significance in the proximity of the face area, the mouth area, and the laryngeal areas of speech in the cortex. Is it an accident that the smallest communication signs appear in the face and are part, almost, of the speech mechanism itself? From the face, a whole series of communication signs for use in expressing more emphatic and sudden achievements of communication spread down. The diaphragm has been mentioned, but convulsions of the entire organism may be used to indicate sudden and important intercommunication, as, for example, in dancing.

KLÜVER: In connection with Dr. Young's remarks, it is a very interesting point that [23] many animals communicate with the face of man | instead of some other part of the human anatomy. It may be worth while to study this form of communication and also to get some information on animals which do not communicate with the face. As far as our own reactions to the human face, it is somewhat surprising that we speak so often of sweet, sour, and bitter faces. There seems to be a strong tendency in man to communicate in terms of gustatory qualities.

VON BONIN: I think most emotions are contagious, whether they appear in the face or not. If somebody cries, many will start crying. You may not and I may not, but very many people will.

GERARD: At least you won't go around giggling, chuckling, or laughing.

VON BONIN: The question as to how we participate in and how we perceive the emotional state of another being is a large problem which I don't believe anybody has tackled very clearly.

BATESON: When I was talking of mutual awareness of perception, I was leading up to empathy.

VON BONIN: Mutual awareness of perception ?

BATESON: Yes, in human communication.

PITTS: It does not generate laughter.

WIESNER: One does not laugh hard where there is not the possibility of feedback. If you are listening to the radio by yourself or reading a book, you will chuckle, whereas the same stimulus, in a group, may evoke enormous laughter.

MEAD: A complete sequence can be proposed from the smile to the socialized dance or to copulation, but then grief cannot be handled in it. Grief, in a sense, would have to be regarded as a failure in social interaction. The sobbing that goes with grief is not dependent on the presence of another Person, and yet it has the same convulsive aspects.

In the conference on »Problems of Consciousness« held last week, one of the problems raised was the protective function of breaks in tension.

PITTS: Does anyone know what the word »tension« is a metaphor for? I think that is the most promising avenue of approach, but this is the difficulty that strikes me first.

MEAD: It is an idea that has arisen in the course of studies on epilepsy. If all convulsive states could be regarded as having protective functions in breaking rising tension, then they could be differentiated in terms of how much need of protection one has. Laughter protects in a real communication system with other people. Grief protects against a moving out of communication, against such an identification with the dead that one is no longer in communication at all. They both are protective and they both are comments on communication, but one of them occurs in a real intercommunication system and one occurs outside it. | [24]

MONNIER: I have the impression that the physiological basis of these two expressions, laughter and grief, is different. Both these expressions have different physiological inductors. There are cases in which paroxysmal laughter leads to loss of tone, patients who, when laughing at a joke, lose their tone and fall prone. This is called catalepsy and may be the result of a generalized emotion or tension. In grief, as we know from primitive societies, a generalized emotion may end in rhythmic vocal expression and not in a collapse of tone. In both cases, relaxation of tension is obtained.

MEAD: But either control or loss of control is possible. Grief can be controlled; Mourning can be patterned so that it is highly stylized and has a rhythmic quality which is reassuring, or it can be of the type that moves more and more towards loss of control . One can be helpless with sobbing or helpless with laughter. There are two possibilities in the same system, really, either to achieve oscillatory steadiness or to move toward the point where people throw themselves on the ground and no longer have any control at all.

FREMONT-SMITH: The small child so frequently goes back and forth between laughter to crying.

MCCULLOCH: Well, isn't it true that with most people, if they get to laughing very hard, are apt to end up weeping, too? I don't think the two mechanisms are completely independent.

FREMONT-SMITH: I think it is interesting, after what Dr. Monnier said, to touch on narcolepsy. There are patients who have a lesion in the hypothalamic area and are constantly dropping off to sleep. They are relieved of this sleep tendency by the benzedrine group of drugs but they cannot go to the movies frequently because the comics throw them into unconsciousness.

PITTS: Do you know anything of the effect of grief on such patients?

FREMONT-SMITH: No.

MCCULLOCH: I had to go over the literature about four years ago. At that time, there was no recorded case in which grief precipitated sleep, at least none I could find. On the other hand, I myself have seen cases, and there are several instances in the literature, in which anger precipitated it.

FREMONT-SMITH: And conflict. I have seen emotional conflicts in the narcoleptic precipitate the sleep state in exactly the same way as any other psychosomatic phenomenon was precipitated.

QUASTLER: What happens to the narcoleptic if you make him laugh just by tickling him, without any humor being involved at all?

FREMONT-SMITH: I think they lose their tone and may go right into sleep.

[25] QUASTLER: It is the laughing that causes it? |

FREMONT-SMITH: Yes.

VON BONIN: *Lachschlag,* in German.

BATESON: Tickling for some reason hasn't been mentioned, or the relation between laughing and the scratch reflex. I wish somebody who knows about such things would speak about them.

BATESON: We use tickling metaphorically; we laugh when »tickled.«

KLÜVER: So does the chimpanzee.

MCCULLOCH: And the orangutan.

MONNIER: The common feature of the two conditions which produce the tickling sensation and laughter is the repetitive action of very slight, or even subliminal, stimuli. This gives rise to a spreading process, which activates consciousness. We spoke, in the meeting on consciousness (6), of the ascending activating reticular system, which has been identified by Moruzzi and Magoun and which induces the arousal reaction. The mechanisms which increase consciousness, pain or laughter produced by a tickling sensation, have something in common. They are put in action by repetitive stimuli and they induce a generalized excitatory state. If the increase in tension becomes too great, it may suddenly be cut by a protective mechanism which produces, in one case, loss of consciousness or tone and, in other cases, rhythmic vocal expression. But these various forms of expression are always the result of repetitive stimuli, ending in a widespread (irradiated) paroxysmal excitation.

MCCULLOCH: There are two varieties of tickling. We use one word for two entirely different things, I am sure. There is tickling in the sense in which a fly tickles you or a straw up your nose tickles you, and there is the tickling produced by a rather strong stimulus of a fluctuating kind, which results in laughter. That kind of tickling can rarely be done to oneself. The kind with the straw up one's nose certainly can. They differ in the self-reference component in them. The one that produces laughter loses its effect in many postencephalitic patients, while the other does not. Postencephalitic patients do not laugh, and almost all of them show also a remarkable reduction in sexual activity. Those who have lost laughter have lost sex, for the most part. It is the common mechanism involved.

VON BONIN: Does the straw ever evoke laughter in anyone?

PITTS: It is rather more like itching than tickling.

MEAD: You have a problem here, Warren, if you equate repetitive tickling with various varieties of sexual foreplay that act as sexual stimulant, for that can be something self-administered or other-person administered.

MCCULLOCH: That's right, it can be; there is only the question of whether it must be [26] brushed off or whether it switches over to sexual | excitement. But the kind of tickling that evokes laughter is lost in the postencephalitic whose sexuality is also down.

BOWMAN: The straw can cause a sneeze. Is that an allied effect?

MEAD: Quite.

BATESON: Do you think one could discriminate between these two sorts of tickling in a dog?

McCulloch: Yes, very decidedly, and in the cat it is even easier.

von Bonin: You can tickle the ear of a cat.

McCulloch: Yes, and the ear starts to snap, to get rid of the tickle, and then the paw comes up.

Bateson: That is one type. How about the other?

McCulloch: The other type is produced usually by stimulation in the small of the back of a rhythmical kind. The cat will start arching and its tail goes up. The dog is ticklish in the same region, and it is in this region that man is also most ticklish.

Pitts: It seems to be wholly pleasant, though, in the case of the cat or dog, whereas we don't usually enjoy being tickled.

Teuber: Oh, it can end by the cat biting. The transition can be sudden.

Mead: There are cases where tickling is a definite form of foreplay and other contexts where tickling is regarded as unpleasant. Take the tickling that occurs among adolescents, for instance, where it is very common. This is an age that goes in for a great deal of tickling. If it cannot be allowed to go to a sexual conclusion and it is unpleasant, it becomes a rejected activity, but in an approved situation of very rough forms of courtship of certain sorts, tickling goes right into a developing sexual sequence.

Klüver: We have discussed a number of situations in which a sudden break, reversal, or discontinuity leads to a restructuring or reorganizing of the whole field. Such situations occur on all levels of behavior, ranging from the perceptual to the emotional. It seems impossible to discuss all these situations profitably in a general way without recourse to a scientific analysis of particular situations. Only such an analysis can specify the properties of a given structure as well as the conditions in the external and internal environment related to this structure and governing the transition from one structure to another. Let us suppose such an analysis of a concrete situation, for example, of a certain phenomenon in the field of laughter, has been successful in specifying the numerous psychological, physiological, and other factors involved and let us suppose the results of such a scientific analysis are handed to Dr. Bateson. The question I wish to raise is whether at this point there are any problems left unsolved? And if so, what are these problems? | [27]

Bateson: Yes. I opened the discussion with the focus on laughter and humor, but the thing that I would be interested in from such a study would be to use the occurrence of laughter as an indicator, a sort of litmus paper. This would be helpful in studying the implicit content of communication. It is an extraordinarily hard thing to study, actually, because we do not know what is in the mind of the communicator or what is aroused in the mind of the recipient. It seems to me very, very important for sociocultural investigation and for psychologic and physiologic investigation to begin from some fairly sharp criterion for what is in the message. Dr. Mead told a story about herself as a president. Von Bonin said that it was a *Schadenfreude joke*. He heard an overtone which Dr. Mead, so far as consciousness is concerned, is prepared to deny, perhaps correctly. She, after all, was present at the meeting and von Bonin wasn't. But it is awfully hard to test any statement of that kind. One uses one's sensitivity and one's imperfect knowledge of his own communicative habits. One predicts. The question is, if one had a satisfactory working hypothesis, or some idea of the types of paradigm which lead to something like laughter – could the occurrence of laughter be used as an indicator for what there was implicit in the communication? That is the question in which I would be interested, not so much in the significance of the laughter as in using its occurrence as an indicator.

McCulloch: May I say that we have two questions still before us. It is fairly clear that one item of value in jests leading to laughter is that the joke sets up some kind of a relation in which it is safe to play. The second thing that is fairly clear is that there is always some reshuffling or restating of the problem, which in itself may be valuable in the transfer of information. But it is by no means clear that these are the only functions that humor may have in communication. There is the double role of the jest, one, the reorganization within the person, and the other, the reorganization between people, and I don't believe this has been sufficiently disclosed. Can we have Bateson say once more what he thinks is communicated besides what formally appears in the jest? Is it the relation of people to one another? Is it the relation of people to themselves in the situation?

BATESON: In human exchange, in general, we deal with material which cannot be overtly communicated: the premises of how we understand life, how we construct our understandings, and so forth. These are very, very difficult matters for people to talk about with precision, but if these premises are out of kilter between two people, the individuals grow anxious or unhappy. Humor seems to me to be important in that it [28] gives the persons an indirect clue to what sort of view of life they share or might share. |

As to the way in which humor does that: Consider some swallows that are migrating, we will say, from London to New York and suppose that we are scientists who face the problem of finding out how the swallows know the route. We invite the swallows to communicate to us how their conceptual world is made up: what sense data they use and how these data are fitted together to enable them to find their way. If we watch the swallows and we find, for example, that they travel on a great circle without error, it is true we know something about the swallows now that we did not know before, but we are left pretty much in the dark on the question of *how* they do it. The only way in which we can have the swallows communicate to us how they know is either by their making errors and correcting them or by our performing experiments which will put them in error and then observing which errors they can correct and which they cannot.

It seems to me that a very important element is added in human communication when B is able to observe what corrections A makes in his (A's) course. One of the questions which the young psychiatrist asks is, »Is it a bad thing to say such-and-such to a patient in such-and-such a situation?« to which the only answer is, if it be a bad thing and the patient react unfavorably to that »bad« thing, and if it be later possible to communicate to the patient that that was the thing to which he reacted unfavorably, then all may be well. In fact, if the therapist is able to correct his course and thereby communicate to the patient some hint of how matters appear to the therapist, the original error may become a very important and useful thing in the communication. A great deal of communication occurs not directly but by the commission of error and its later correction.

It seems to me that the nature of a jest is somehow related to this point, that when the joke breaks open and the implicit levels have been touched, have met each other, and oscillation has occurred, the laughter verifies an agreement that this is »unimportant,« it is »play,« and yet, within the very situation which is defined by the laughter as play, there is a juxtaposition of contrasting polarities, which contrast may be compared to the commission and correction of an error. The laughter lets those who laugh know that there is a common subsumption of how they see the universe. Do I answer the question that you asked ?

MCCULLOCH: Exactly.

FREMONT-SMITH: I wonder if we don't have to go back to the earliest development of laughter or smiling in the infant to get some idea of all the meaning of the shared experience? One of the early ways of communication between the mother and the baby is the mother's smile to the baby, which a little later is responded to by a smile on the part | of the baby. The mother's smile is one of the basic means of reassurance to [29] the small child. It seems to me that when two people are talking and one of them smiles at the other, the smile contains the element of reassurance. The person is saying, »I like you, I like what you are saying, I understand you,« so that it is a sign of the effectiveness of their communication; it is a reassurance. A smile is associated with physiological changes, such as dilatation of the skin vessels, which are opposed to those found in an anxiety reaction. Anxiety is almost always associated with the absence of a smile and with a fall in skin temperature.

BATESON: I think we are clear on the reassurant aspects of laughter, the in-group statements, the affirmation of group membership which is implied when both individuals laugh or smile; and we are clear enough that laughter, especially thoracic rather than belly laughter, is a conventional sign which people use to each other, quite apart from whether it is the »real thing.« Such laughter becomes almost a part of the vocabulary and is almost as voluntary as the use of words, not quite but nearly so. The problem, which I want to push toward, is that of involuntary laughter and its antecedents, rather than the problem of the function of laughter between two persons in melting the ice.

MCCULLOCH: The latter says, »I got you,« and »I got you at the level of premises.«

BATESON: At the level of premises, and it is indicated that the premises are right because there is a crisscross of them. We define a point not by drawing a line but by making two lines cross.

KUBIE: Laughter is in itself a language, and, like all languages, it can say many things. In the rectangle of Figure I are represented two poles of meaningfulness. At one is the unchecked or uninhibited belly laugh, and at the other the inhibited laughter. The major difference between the two poles is that at the one extreme there is a general

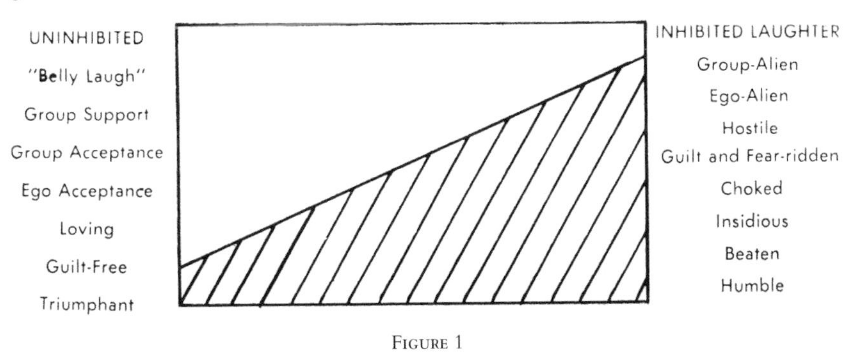

FIGURE 1

sense of group-support and group-acceptance; whereas at the other end, | the laughter [30] is group-alien. Group-supported as opposed to »group-alien« refers to the relationship that is communicated between the person who starts the laughter and the group to which he is talking, or the group that is represented, or that he represents. It may be a group that is present in the flesh or a group that is there only in his thinking and in his own words or actions.

In the unchecked belly laugh, there may be a loving element. Therefore, it is guiltless and is not held in check by guilt feelings; whereas, in the inhibited laughter, as we all know, it is difficult sometimes to tell whether a person is laughing or is grimacing with hostility. It carries an implication of masked hatred, with an enormous guilt factor which stifles the laughter even as one laughs. Finally, there is an element of triumph in the unchecked belly laugh. The person is unafraid, free of all apprehension of defeat and of the fear that inhibits the ordinary tense laughter, with which, I am afraid, we are far more familiar.

One other point: I drew this line slanting in this way purposely to indicate diagrammatically that these never occur in pure culture but that there are always varying admixtures of the two components. It must also be added that these differences can exist on conscious and/or unconscious levels of psychological function. We can be triumphant and loving on a conscious level, yet full of hate and guilt on an unconscious level, or vice versa. This makes the phenomena of laughter as complex as are all other mental acts. Finally, in any consideration of the problems of laughter, as in all emotional problems, we must include a consideration of the role of trigger mechanisms. Laughter is par excellence released by such mechanisms. In this respect, it is closer to a phobic mechanism than is usually realized. This trigger element, which I plan to discuss in connection with the role of feedback mechanisms in emotional processes, is one of the basic elements in laughter which has been overlooked.

BATESON: With shared guilt as a very important element, down in the lower right side of the diagram.

KUBIE: When guilt is shared, you receive some degree of group support.

BATESON: Yes. What I was getting at is that these components of yours keep crisscrossing on each other.

KUBIE: They are all mixed together. To represent all possible permutations and combinations diagrammatically, we would need a series of planes in a three dimensional nomogram.

MCCULLOCH: Larry, how about attempting to state what we are talking about when we speak of the release of tension that comes with laughter? What are we talking [31] about? |

KUBIE: That is Chapter IV of the manuscript I have brought with me[1].

MCCULLOCH: How do we go at it?

KUBIE: I hesitate to leap into the middle of an exposition which requires step-by-step logical elaboration, but in essence, my thesis is that the peculiar attribute of emotions in psychological affairs is that they impose an automatic value-judgment on experience, which does one of two things: this creates an impulse, conscious or unconscious, either to repeat that experience in the future or else to avoid it in the future. Emotions give experience either a plus or a negative sign. I believe that one can group all emotional states in these two categories. Sometimes their influence is relative and the same experience evokes both plus and minus reaction, for special reasons; but, basically, the emotion always falls on one side or the other. By and large, anger and elation are the emotional qualities which tend to be repeated, whereas fear and depression are the emotional qualities that we avoid if we can.

The relationship of an affect to a drive of any kind can, therefore, best be understood in these terms. To put this succinctly, my thesis about tension is that the word is a figure of speech by which we characterize that state which arises within us whenever there is some compelling inner necessity towards some action against which at the same time there are countervailing forces. These countervailing forces can be external or internal or both. They can be conscious or unconscious or both. But where there are no countervailing forces, the mere existence of an impulse towards something does not give rise to that inner experience which we characterize with the particular word »tension.« The countervailing force may be nothing more important than the unavoidable delay which is inherent in the transport of chemicals in any multicellular organism. Tension, like all psychic phenomena, is inconceivable without delay. In human life, one sees this in its simplest form in the infant, where a delay of only a few seconds

1 Dr. Kubie refers to Chapter IV of a manuscript on which he draws more extensively in the next section.

is enough to evoke random discharge which is the infantile precursor of the controlled tensions of adult life. Thus, tension, as we know it in adult life, implies an aggregate of forces moving in one direction opposed by an aggregate of internal and external forces moving in another.

BATESON: When we say that a man is tense, we mean, I suppose, that while his hand is lying on the table, or wherever it is, there is more muscular activity going on in it than need be; that not only is there the necessary tension in the flexor to support the hand in the position in which it is, but also some antagonistic contribution in the extensor. The metaphor of tension is a psychological metaphor but often it is | worked out or [32] exemplified by extensor-flexor opposition in the body.

MCCULLOCH: In other words, it is a rise in tension in the muscle that we are talking about when we say a man is getting tense?

BATESON: Or it is from that rise of tension that we derive the psychological metaphor. I don't want to suggest that language precedes the physiology or vice versa. I don't know about that.

YOUNG: Is it physiological? Is the physiology correct? I think not. The balance, if you are speaking of a balance between antagonists, will not be at different levels, as far as I know.

GERARD: I think what Gregory says is certainly valid in many cases. I don't think it is universal.

YOUNG: I don't think it is the basis of what we mean by tension. It is a false clue, if I may say so.

BATESON: That is the question I was asking. Would it be false or true?

YOUNG: I would suspect it.

BATESON: There are people whose psychological tension is expressed with a general limpness.

MCCULLOCH: People complain of a headache when they report tension, and tension of the scalp muscles can be recorded. It appears in many EEG tracings.

YOUNG: But that is different from his thesis altogether. Certainly, there would be other somatic manifestations accompanying so-called tension.

GERARD: He is equating tension and tonus. What is it specifically that you are objecting to, John? I don't quite understand.

YOUNG: I think the danger is that we should use this relatively low-level metaphor for more than a metaphor. The tension you are speaking of is surely at an altogether different level.

BATESON: We don't know how much the levels echo each other.

GERARD: What I understand Gregory is asking is whether there is a sufficient correlation between this internal state or emotional state that is called tension and a manifest physiological state in terms of muscle tension so that there could be an etiological relationship between them. Now, you feel that is entirely wrong?

YOUNG: I should doubt it.

GERARD: Why do you react so strongly? I would have doubts about it but, on the whole, I would be inclined to be hospitable to thinking along those lines.

KUBIE: I wonder whether one of the reasons why this concept seems so difficult (and I have heard it batted around a hundred times) is because of the implicit assumption that some kind of undifferentiated emotional state can form in us which cannot itself properly be called an | emotional state. It might be called a pre-emotional state, or a [33] larval emotional state, or a precursor state out of which emotional feelings and actions and expressions can be precipitated in various directions. Although in itself it is undifferentiated, out of it can come tears, laughter, anger, elation, depression, fear, and even sleep or an obsessional-compulsive furor. It is this diffuse, undifferentiated state for which we seek a name. It is not the same as alertness, yet it is quite different from a state of sleep or apathy. We must use some figurative word to characterize it. The particular example which Gregory Bateson used is, in some ways, the simplest, because there the tension is expressed in muscular terms and is related to a close balance between aggression and its withholding. Yet, one can also find its expression in speech, or in specific somatic language, such as that of the gastrointestinal tract, among people who, on the somato-muscular side, are quite relaxed. The particular somatic vehicle which is used varies from individual to individual. Nobody has ever found a satisfactory definition of it, but nobody can think in this field without accepting the existence of this phenomenon because subjectively we are aware that there is something which we have to characterize by some such word as tension. Call it »X« if you prefer, as long as we all know that we are thinking and talking about a state which arises in human beings and which can, under appropriate internal and external circumstances, be channeled into any of various directions. Tension is not a bad word with which to characterize it figuratively, and its use crops us again and again precisely because it gives us a sense of knowing what we are communicating about with one another.

HUTCHINSON: I want to add two points: first, it seems to me that the very fact that some people, as Gregory said, show a sort of limpness suggests that this psychological tension can be modified or reversed by a learning process. If so, this leaves the whole thing wide open, so that objections are probably irrelevant until they are further analyzed.

My second point is that, etymologically or semantically, there are probably two things involved: the obvious observation made in many cultures that there is increased tension of the fingers, and, something which continually crops up even in the most respectable writing on comparative behavior, a consideration of the discharge as a release of something like potential energy, so that one particular kind of potential energy, and its release, occurring, in our example, in the musculature of the fingers, occupies a considerable semantic area in discussions of this kind.

BATESON: Would we get on better if, instead of saying we must conceptualize this state that Kubie has just offered us, we said that the important thing might be to build [34] a classification of the resolutions of | such states? Later, we could ask about the states themselves.

YOUNG: That is rather my objection. From a physiological point of view, I would say it is dangerous to simplify, as Kubie suggests, by postulating a central reservoir of tension. I would say that was a dangerous approach for the cerebral physiologist and that, however hard it may be, we must dissect these individual manifestations that we classify as tension and identify their cerebral components.

KUBIE: I am not assuming the existence of a single central mechanism. I am saying only that clinically an extraordinary transmutability among various kinds of tension states is observed. This suggests that there is something which precedes any of the various differentiated forms of emotional experience, acting almost as a common root out of which all can evolve. This inescapable clinical fact has to be included among the phenomena that we are trying to understand and explain.

MCCULLOCH: May I put it somewhat differently? Suppose a man is tense; in that man is there any place where one could look and find a particular change?

KUBIE: I shall counter the question with a question. Let us picture three youngsters. One has an intense eating compulsion. The second has a handwashing compulsion. The third has a counting compulsion. As we have said, if the subject does not fight against his own inner drive and if there is no external person or force which acts against it, the drive will be expressed freely and insatiably. The one will eat voraciously, even until he vomits and after. The other will wash his hands until soap and towels and water are exhausted or until the skin peals from his hands, leaving open sores. The third child will count until there is nothing around to count. As long as one of the individuals is carried on the flood tide of his drive, neither another's observations of him nor his own self-observation will lead to a state of »tension,« whatever that may be. On the other hand, if anyone tries to stop him, or if he tries to stop himself, a state arises in him at once for which the observer, whether he be uneducated or the most highly trained and sophisticated psychologist, will automatically turn to the word »tension.« For this state, we have no other name at present. In this state of »tension,« many different things can happen. The person can have an attack of what the layman calls »hysterics,« and laugh and cry. He can become overwhelmingly depressed and morose. He can go into a state of panic or rage or elation. He can get bowel upsets. He can vomit. Or he may even, paradoxically enough, go to sleep. I am not trying to explain tension. I am trying, rather, to characterize it in all of its complexity, to save ourselves from the seductive tendency to oversimplify nature in the interests of our theories.

YOUNG: How do you identify the state before it has reached the extremes you mention? | [35]

VON BONIN: Being a biologist, I can't talk in abstractions, so take the example of a man who hears a shoe thrown down by somebody who is undressing above. He expects, of course, the next shoe to be dropped too, but the sound never comes. What happens, as I see it, is that he forecasts in his mind the noise of the second shoe falling down. I would look in the cerebral cortex for some configuration which makes that forecast effective, and I would expect that the noise that actually follows destroys the configuration that is forecasted and lets the nerve cells resume their normal rhythm.

YOUNG: I would accept that.

VON BONIN: Whether the thing forecast is a happening in the outside world or something the individual programs for himself, as the boy who washes his hands or wants to count, when something that the brain has made up its mind should happen, either within or without, does not come about, then that release of the neuronal pattern which would come about if the program were carried out is inhibited.

GERARD: Gerhardt, I like that. But why do you call it a biological or physiological explanation?

VON BONIN: The two shoes will fit.

FREMONT-SMITH: From the biological aspect, it is very concrete. You said »when the brain had made up its mind.«

FRANK: May I remind you that Howard Liddell has said that he can distinguish in his experimental animals between an acute alarm reaction and what he has called the state of watchfulness? He has various criteria, both physiological and motor for doing so. In the experimental animal, the expectancy that something is going to happen produces a sort of subacute emotional state, if it can be called that.

MCCULLOCH: The expectancy is definitely revealed by motor manifestations.

FRANK: But some physiological variables were also recorded.

MONNIER: It is hardly necessary to recall what happens to the electrical activity of the brain when a subject suddenly awakens and become alert or excited. There is a real spectrum of changes paralleling the transition from deep sleep to alertness or an excitatory state, or from deep narcosis to wakefulness. The chief changes are accompanied by electrical activities of increased frequency and lower voltage, the so-called desynchronization of electrical patterns. At the same time, the cortex becomes more reactive to afferent stimuli. All organs, including the cortex, which can be considered as a terminal organ, simultaneously show a change in reactivity and functional readiness. This shift may be due to a greater generalization of afferent stimuli, or to a greater reactivity of the sensory, cortical, and motor organs to the same stimuli. On the contrary, in [36] deep sleep there is a decrease in reactivity | on all levels: cortex, muscles, sensory organs. This is particularly obvious in states of increased tension, manic excitement or anxiety. In these cases, electrical fast activities of low voltage are found in increased proportion in the precentral and postcentral region of the cortex, as a symptom of greater reactivity of the cortex to afferent stimuli.

VON BONIN: The precentral gyrus receives radiations which come from the cerebellum, as I understand. Is that correct?

MONNIER: Yes, the whole area, precentral and central, becomes a projection place for afferent stimulation, not only from the primary sensory afferents but also from other parts of the brain.

VON BONIN: Oh, yes, surely.

MCCULLOCH: Even photic stimulation comes through to precentral areas, is that right?

MONNIER: That's right.

VON BONIN: The afferent activity can be picked up precentrally even after the cerebellum has been destroyed.

BIGELOW: I understood the question as to what trace can be found of the existence of tension to be one raised in objection to the use of the concept »tension,« at least to its use as if it were something centralized or local. It seems to me that this is a very weak objection because there are certainly changes which occur in the neurological system, which we know must occur because of exterior evidence, of which we cannot find any direct trace by anatomical means. For example, if a man is multiplying a sum in his head, I challenge anyone to find out from internal changes whether he is multiplying, and yet it can be determined that he is multiplying by the answers he gives to questions. There should be no objection to Kubie's using the word tension as he pleases. His observation that tension, in his sense, is something that is probably widely spread over a number of different locations, of concepts or type situations, is not negated simply by the fact that Kubie can't put his finger on exactly what physiological or neurological change occurs when tension exists.

MCCULLOCH: I am not sure Kubie can't, sooner or later.

BIGELOW: I am not sure, either, but I say this is a very weak way of objecting to the use of the word tension.

VON BONIN: Does anybody object? I thought we had made neurologists out of the physiologists. I thought we took it for granted there is such a thing as a brain.

TEUBER: It was not Kubie but Bateson who started the argument about physiology. Gregory was the one who suggested that »tension,« in Kubie's sense, might be correlated with some measurable tonus, either postural or central. Such correlations have been looked for in many places but, as Dr. Young said, just about every correlation [37] that has been | claimed to exist has turned out to be unreliable. We certainly can't expect any simple one-to-one correlations, no matter whether we use the electromyogram, the galvanic skin response, or even the EEG. The EEG, though, may be a special case. If worked with on the head end of the animal, one seems to get fairly good correlations not with tension but, at least, with relaxation.

It has always bothered me that the most reliable thing an EEG can show is that the brain is not doing anything significant at the time of recording. At such times, the EEG shows characteristic regular activity; but as soon as the brain is doing something (usually it is very difficult to say what), this regular activity disappears. For that reason, I have never been too sure that searching for correlations between mental states and EEG signs would lead very far. But you were challenging people to show some electrophysiologic correlates of multiplication, perhaps with tongue in cheek, and I want to pick that up.

A young lady, Lila Ghent, has investigated the effects of various types of tasks on the slow-wave activity shown by the EEG of patients who have had electroshock. During the rather long periods after the electroshock when the EEG showed slow waves, these patients were asked to perform various tasks, for instance, tapping with a stylus on a drum. Such rhythmic tapping abolished the slow-wave activity for a short time; if they went on tapping, the slow waves reappeared. The picture was somewhat different with patients who were asked to perform more complex tasks. If they were told to go through a reaction time experiment, the slow-wave activity was abolished for quite a long time. Another effective way of abolishing the abnormal slow-wave activity was to ask them to count back from one hundred by sevens. This serial substraction very markedly reduced their slow-wave activity.

BIGELOW: Can you distinguish by that method, say, subtraction from multiplication?

TEUBER: I should suppose not. However, it wasn't tried. There is no reason to believe that division or multiplication would have effects different from addition or subtraction. There was something rather odd though: the most effective way of getting rid of the slow waves was to have through the patients make errors. When a patient made a mistake in counting, his slow waves disappeared for a particularly long time.

BIGELOW: Mental processes may be determined by terminal performance only, perhaps.

TEUBER: Surely.

GERARD: What happens to the Cheshire cat's smile when the cat disappears, in other words.

YOUNG: The danger is, surely, if terminal effects which are similar | are referred to one postulated central source, which then turns out not to be one. [38]

BIGELOW: It depends upon whether the oneness is critical. Is it in this case? It hasn't been demonstrated yet, so far as I can see. I grant it is a possibility, but it hasn't been shown.

YOUNG: That is what we are asking Dr. Kubie.

GERARD: The only person who has made even a presumptive attempt so far to give this any kind of an organic mechanism has been Monnier, who tried to tie these things up to changes in the measurable behavior of neurons, at least through their distant signals of the EEG changes. Nobody else tried to do it so nobody else should be criticized. That is why I did not think your objection, Dr. Young, to what Gregory said was valid. You were reacting to the kind of dangerous verbal analyzing that Evelyn Hutchinson was warning against, the idea of the building up of potential energy. It is hard to avoid this idea. Adrian told me he could not do so. The reason why one gets a little bit apprehensive about it is that we are perfectly sure that the kind of thing a neurophysiologist means when speaking of inhibition and so on, – well, we are sure he does not mean »inhibition« in the psychological sense but perhaps we are not even sure of that! But Gregory's original question, it seemed to me, did not imply a positive answer, merely being an attempt to get at the origin of the use of the figure. I think it was entirely legitimate from that point of view.

MCCULLOCH: Well, may I put the question in a slightly different way? Is the word »tension« simply one name for a host of different affairs or have they some common factor in the sense that in all of them there is some part of the nervous system or of the body which is in a given state or exhibiting a general pattern of activity? I think, for example, we use the word »memory« altogether too loosely. We use it often for processes which are inherently or essentially dissimilar, and I am not sure we may not be doing the same with the word »tension.«

BIGELOW: Isn't it essential, if a word is to be useful, that it cover a class of phenomena which may, in some sense, be different but have some common property? Isn't the answer to the question this: that if »tension« is to be a useful word, it must cover some properties which are in some way different but have a common aspect?

GERARD: I was going to say another word on the physiological side. It seems to me that if we substitute the physiological. term »irradiation,« which is not too well-defined in terms of its mechanisms but is objectively quite measurable, and then think of irradiation as increasing in quantity as an excitation state builds up in neuron pools, it will help. Then when we want to ask, »What do we mean by excitation state?«, we [39] shall have to go back to the concentration of energy-rich | phosphates in the membrane or the number of potassium ions that have crossed it or something like that, in other words, to perfectly real things whether or not we know just which they are.

We are not too far away from this general concept of tension, and that is why I feel there is a good deal of validity in the kind of tie-up Gregory is trying to make. We recognize an increase in tension, subjectively in ourselves and objectively in others, in terms of increasing neuronal irradiation, whether it is increased contraction of antagonist flexors and extensors or whether it is tapping the table with the fingers or whether it is shifting around restlessly in a chair or whether it is performing a ritualistic act or whether it is merely counting mentally a series of numbers. There is greater activity of some sort, greater neurological discharge, spreading over a wider and wider group of neurons, it seems to me. Do any of you physiologists take exception to that in biological terms, and do any of the psychological people feel that that is too far away from what we really do mean by »tension?«

KLÜVER: From a psychological point of view, it is worth mentioning that tension, whatever it is, and the perception of tension are two different things. The fact that one is able to perceive tension in the face of a person does not necessarily imply that the observed person is in a state of tension. Nor does it imply that the observer is tense. Either the observer or the observed person or both of them may or may not be in a state of tension. Under pathological conditions, there may be an inability to perceive tension, sadness, cheerfulness, etc.; that is, there may be an agnosia for physiognomic characteristics. A patient may be able to recognize his wife and see that her eyes are blue and that her mouth is red, but he may no longer be able to recognize tension or sadness in her face. The visibility of emotions is undoubtedly as important a problem as the visibility of colors.

PITTS: I should doubt whether a satisfactory correlation can be made between the psychological concept of tension and the mere number of excited neurons. Consider the case of a boy with a handwashing compulsion. We sit him in a chair and we don't allow him to wash his hands. Presumably, his inner tension increases as he sits there. Then we set him free and he promptly goes and washes his hands. As soon as he washes his hands, allegedly his inner tension declines very sharply, but a large number of neurons, namely, those involved in washing his hands, now accelerate, so he may have a greater number of neurons discharging per se than he had when the state of tension was at its height.

GERARD: Excuse me, but you imply total number, which I did not. Irradiation is not [40] just a volume-conductor type of thing. It is usually along a defined path. |

PITTS: It is usually along a definite line of activity in which the person engages and is accompanied by a reduction of tension.

GERARD: That can no longer be called irradiation.

PITTS: Then irradiation excludes channelization.

YOUNG: Would it be fair to say that your attempt to use the concept of irradiation and to give it a quantitative meaning is the best one can do with physiological terms, but that you would not regard it as a completely satisfactory statement of the cerebral process involved?

GERARD: Of course not.

YOUNG: You are putting up a preliminary model.

PITTS: Then you must mean by irradiation something more than the mere engagement of a large number of neurons in the process.

KUBIE: Something akin to the old Pavlovian concept of a diffuse overflowing irradiation of some kind of activating or inhibiting process.

YOUNG: To my mind, there is a danger there.

GERARD: No, I don't like that either, Larry.

MONNIER: The word irradiation is misleading because it has been used in many different senses. The process responsible for such changes has something to do with an increased propagation of impulses; the Germans call that *Ausbreitung.*

GERARD: Yes, a spread.

MONNIER: But it is probably in this meaning that you use the word irradiation loosely.

GERARD: I was avoiding bringing this down to the individual neuron because I think that does impinge on the next level. This is not simply total number of neurons, but number and pattern. If that is your point, Walter, I agree with you.

PITTS: The spread is perhaps all-important.

KUBIE: I have two complications in mind. One concerns the basic feedback function of emotional processes. I am thinking of a patient who is an exceptionally effective, competent, and able person, who thinks problems through extremely well, reaches decisions, and then acts on them. At present, he is juggling ten different balls in the air at once and doing it well. But the moment of reaching and implementing a decision precipitates in this patient an obsessional furor of doubt. Consequently, after a decision is made and after appropriate action is taken, when he reaches the very point at which he should be able to relax, heave a sigh, take a drink, and be comfortably free from tension, a storm erupts. This storm is a reaction to the fact of having made a decision and acted upon it, which arouses fear and guilt and an obsessional furor of extraordinary severity. Doubts go round and round in his mind like squirrels in a cage, with an enormous piling up of something that can be described only with this same figurative word. | I describe this clinical phenomenon as another example of the complexity of [41] the manifestations of the feedback systems in the emotional sphere.

The second complication centers around the fact that there are such things as chronic emotional states. Up to the present, our discussion has dealt only with acute emotions, as though emotions were always sharp processes. What about those individuals who seem to have a fixed center of emotional gravity to which they always return, no matter what forces swing them temporarily away from it? They function as though some persistent emotional set or emotional potential formed the center of gravity of their emotional lives. Sometimes, this is a pleasant and comfortable center which they do not want to disturb. The chronic hypomanic is an example. (Unfortunately, however, in the end this usually catches up with them; but that is another story.) Sometimes, the emotional center is a chronic rage state, a disguised temper tantrum. I have known patients who lived out their entire lives in disguised temper tantrums, masking these in a thousand different ways. Sometimes, it is chronic depression, which may arise in very early years and last throughout life. I know two eighty-year-old patients who face today the problems with which they were dealing when they were four years old. Indeed, they have lived with their reactions to these problems as their fundamental emotional base or potential throughout their lives. Clinically, this is an inescapable, basic, and puzzling fact.

How can we put this in terms which are descriptively accurate? The first requirement for such a term is that it shall be an adequate representation of observable phenomena in nature. The second is that the term should at the same time lead one's mind to explore possible explanations while avoiding figures of speech which beg all essential questions. For me, a term such as »chronic emotional potential« or »chronic emotional set« meets these requirements perhaps a little better than »tension.« Yet it does not help us to escape the word tension, because, although in these particular cases there is a chronic emotional set with a specific quality, there are also other clinical states in which the emotional set is undifferentiated, with no qualitatively differentiated feeling tone, but out of which the more highly differentiated emotional states can precipitate. Thus, there would seem to be two contrasting clinical manifestations: the differentiated and the undifferentiated chronic emotional tensions.

YOUNG: But these particular words are very valuable, aren't they, because they give us a picture? One could imagine that they would equally well describe chronic states of activity of parts of the nervous system. You could really cover everything you said [42] without using the word »emotion.« |

KUBIE: Only by paraphrasing it with some neologism; and in the end that is no gain.

YOUNG: One could visualize a condition of parts of the brain being responsible for these states throughout life, by virtue of the particular activity of one or another aspect of cerebral physiology.

KUBIE: Isn't there a danger that that may also beg the question, although it is possible, of course, that an undifferentiated tension or potential existed first, subsequently and for special reasons acquiring specific coloring.

YOUNG: We do know that local lesions may produce, in both man and animal, syndromes of that sort. A lesion in the midbrain of the cat produces the syndrome of obstinate progression, as it has been called, in which the cat just walks and walks and walks. That could be described in terms of an emotional state.

GERARD: This is going back a little bit but I think it may be useful in pointing up to our friends who deal with the more difficult levels of the brain that we too sometimes run up against difficult and seemingly insoluble problems of analysis at a level where we would not expect it. I could not help but think, as we discussed the building up of tension, of a strict physiological analogy, one which points up the irradiation problem.

Nerve paths descend on each side of the brain stem from the respiratory centers in the medulla to the upper spinal cord, from which come the two phrenic nerves that innervate the diaphragm. If a cut is made halfway across the neuraxis on, say, the left side, the corresponding half of the diaphragm stops. The right side goes on working perfectly well. If the right phrenic nerve is then cut, so that the right half of the diaphragm cannot respond, the left half starts again. This is perfectly simple to understand. Because the animal has lost its aeration, it becomes progressively asphyxiated, there is a change in the carbon dioxide and oxygen situation in the brain nourishment, the cells become more irritable, and messages coming down the brain stem, not quite able to break across at the ordinary level of excitability, now do break across from right to left, across the midline, and set off the left phrenic. The only trouble with this simple explanation is that it is not true. As shown by Arturo Rosenblueth, if the right phrenic is blocked (by a current, which stops nerve messages as fully as a cut but can be turned off again and the experiment repeated), the very next respiration comes through on the left. Thus, the switchover is not due to an accumulation of carbon dioxide, or to any other slowly built-up change.

Here, then, is a case of a building of tension until it escapes, if I may use that word, and a case of sudden irradiation. It would be very nice and very simple to interpret this in a perfectly mechanistic way, in | terms of a change of threshold of neurons and of [43] the gradual accumulation of summated impulses until they can escape, neuron by neuron. It just happens not to work. If anybody has yet come up with an explanation of this that is physiologically acceptable, I have not heard of it. It is a mystifying, very real phenomenon that any student can repeat at will.

BIGELOW: Are there no local cross fibers there of any sort?

GERARD: No. There are many of these intriguing neurophysiological paradoxes. For example, after denervating the lower cord, changes in the reflexes of the fore limbs are still produced by cutting away some of the denervated lower cord.

MCCULLOCH: The interesting thing about it is that this happens in certain animals but not in all. The dog and the rabbit work one way and the cat the other, or vice versa, which means that there must be either an anatomical or physiological substrate which is different in the two kinds of animals.

BIGELOW: Is there anything else that characterizes the two animals?

VON BONIN: The cat has much larger cells than the dog or rabbit.

MCCULLOCH: There is a possibility, of course, that we are dealing with some »pup« coming back up the nerve when the main impulse goes down, that there is a backfiring, for when we have actual collaterals, it is quite a different story. »Pup« is laboratory slang for back impulses over the motor nerve. If there are axonal collaterals, then, in the case in which there is a return volley of this kind from the muscle, far more impulses in the axonal collaterals would be expected than otherwise.

GERARD: Oh, obviously, it is explicable sooner or later. It isn't gremlins.

MCCULLOCH: That's right, but there must be a new way to attack it.

GERARD: There must be another way of patterning it besides the simple interaction of neurons and axons.

MCCULLOCH: I don't think so.

FREMONT-SMITH: Would we gain anything by going back to a state of »un-tension,« examining it, and then moving on to consider the state of so-called tension? I should like to start off by saying I don't believe there is any state of absolute »un-tension« other than death; in other words, the organism is constantly reacting to its internal and external environment. The closest it comes to an absence of tension, presumably, is in deep narcosis. From that level, there is a progression through varying states of activity.

MCCULLOCH: May I bring us back for a moment? The crucial thing that we are talking about here is tension in the sense in which it is somehow a trouble in communication between people, directing our | attention to our own carcass or our own brain, [44] making us heed our own effort instead of heeding what the other man is saying.

KUBIE: Because it is relevant, I want to remind you of the work of Barach (7, 8). It bears directly on this matter of tension, even though his observations were made during studies of a quite different problem. He was evaluating a method of producing complete respiratory rest by placing patients entirely within a chamber in which alternate increases and decreases of the pressure of the air cause sufficient diffusion of O_2 and CO_2 betwe[e]n pulmonary alveoli and the blood stream to maintain respiratory exchange without any actual motion in the diaphragm or chest wall. For some reason, not all patients can stop breathing in this chamber, an interesting fact which has not yet been explained. What is more important, a large number of those who can stop breathing soon enter into a curious state, as close an approximation to a completely relaxed hypnoidal state as has ever been achieved without hypnosis or drugs. It is even more complete, I think, than are those hypnagogic reveries which Margolin and I used to induce by having patients listen to their own respiratory sounds brought back to their ears through throat microphones and an amplifier (9, 10).

Those of Barach's patients who achieve this nearly complete respiratory rest and who go into the hypnoidal state also have certain chemical changes (8). In this state, patients lie motionless for long hours, without any sense of the passage of time, without restlessness or movement. Afterwards, they report that little, if anything, was going on in their thinking Processes, although they were not asleep

KLÜVER: Do these patients, instead of reporting that little or nothing has happened, ever say that a given period of time appeared infinitely long, like an eternity?

KUBIE: I do not know. They have not been fully explored psychologically as yet. This phenomenon calls our attention to the relationship of the central respiratory nuclei to the level of activity in the nervous system as a whole, and also to the influence of the ascending reticular substance, which has been studied by Magoun (11). These investigations give us clues as to certain processes in the central nervous system which may influence levels of tension or of activation.

MCCULLOCH: Do you happen to know what the electroencephalograms of patients in this state look like, and do you know whether they are more or less responsive to information at the time?

KUBIE: There have been technical difficulties about getting good electroencephalograms under these circumstances. It has not been done as yet.

MCCULLOCH: Using earphones or signal boxes to communicate with these patients is [45] their reception better at such times, with the tension | down, than it is at a time when they are attending to something?

KUBIE: They can communicate with you, but I don't know the exact answer to that.

FREMONT-SMITH: Larry, doesn't it take some time for people to go into this hypnoidal state?

KUBIE: Some go very promptly, some very slowly.

MCCULLOCH: If they have a familiarity with the situation, do they go in much more rapidly?

KUBIE: Yes, usually.

FREMONT-SMITH: I was in it once, and it is a surprising thing to discover that one doesn't have to breathe; but nobody told me that I went into a hypnoidal state and I wasn't aware of it if I did.

REMOND: There may be a state of tension in an individual even when unconscious, deeply unconscious, in coma. If, while taking the electroencephalogram of a comatose person, some sort of sensory stimulus is produced, a noise, for example, a K complex can be recorded, just as in sleep, a change in the encephalogram which is quite recognizable. If the stimulus is repeated after a certain time, the response will be less marked. If repeated a third time, it will be barely apparent. But if, at the time when the reaction has become unnoticeable, the stimulus is altered, if, instead of making a noise, there is a sudden, important change in the lighting of the room, then once again there is a strong response in the electroencephalogram, which will vanish with repetition of the stimulus. When stimuli have been given with less and less response, and if new kinds of stimuli are no longer efficient even at their first introduction, the name of the patient pronounced very softly may »awaken« him. But that patient is absolutely »unconscious,« and he will not remember at all what happened. Nevertheless, he has some sort of attention, he is able to be attentive unconsciously, and he loses that state of attention when getting accustomed to the stimulus.

WIESNER: If a particular stimulus is repeated at a later time, will there be a response?

REMOND: Yes, if there is a wait of a long enough time, say, half an hour, to let the patient lose his adaptation to stimulation.

PITTS: I wonder if anyone would be interested in a somewhat frivolous, dynamic analogy to the concept of a state of tension? It seems to me that the proper correspondence to make is not between tension and potential energy but between tension and the second derivative of the mean rate of change of potential energy.

When tension reaches a critical degree, apparently the state of the organism begins changing in a rather violent way; the actions of the individual change rapidly, but in what way is not determinate from the value of the tension. Suppose we consider the simple case of a | marble in a cup, a perfect analogy with the most general dynamic [46] instances. Naturally, if we consider small deviations from the position of equilibrium at the bottom, the rapidity with which the marble will return to its equilibrium position depends, in essence, upon the curvature of the cup; the more curved the cup is, the smaller the deviations produced by any given disturbing force will be, and the more rapidly the marble will return to equilibrium. But what very often happens with dynamic systems is that their character depends upon some sort of external parameter. We might suppose there was an external force, for example, which went through a series of fixed values, and this external parameter, as it varied, would change the curvature of the cup, so that, say, when the external force, A, was equal to zero, the cup might possibly be extremely highly curved. As A vanishes, it varies between zero and one; the curvature of the cup decreases gradually until finally, when it reaches one, it is flat. And, say, when A is greater than one, it even inverts.

As soon as it reaches this point, of course, the situation is quite different from any deviation from equilibrium. As soon as A reaches the value of one, or possibly slightly beyond it, then a slight push, of course, is going to send the state of the dynamic system off to a different position of equilibrium, or, in any case, to some completely different form of behavior. Exactly what will happen is not determinate simply by knowing the value of A when it approaches one. There are several possibilities. But if you know the initial position and you know that the disturbing forces are not too great, as long as A has values between zero and one, there will be an equilibrium position which can be fixed in advance. It can be said that if the particle is not there, it will at least be there very soon, or it will oscillate a small degree about this position, and so forth. But assume, roughly, that, as soon as the curvature of the pocket in which it is becomes zero, it inverts, then, of course, this system behaves in quite a different way.

I suggest that the kind of dynamic variable which tension, in this sense we are using it, is really analogous to is not the value of a potential energy but of something like this curvature. This is a perfectly general sort of situation. Consider the case of rotating liquid masses, for example, rotating stars, and assume the velocity of rotation and the mean angular momentum would constantly increase. Up to a certain point, there is a gradually increasing deviation from the spherical shape. But as soon as it reaches a certain point, the rotating liquid mass becomes unstable, and, thereafter, small deviations in its shape cause it to break up or to have a furrow which increases in size, and one can no longer say, from merely knowing its angular velocity, what its subsequent history will be. As long as the velocity of rotation is smaller | than the critical amount, [47] then, if one knew nothing else about that sphere of liquid except that it was rotating with that angular velocity, it could be said it would have a certain shape and would stay very nearly about that shape.

The critical parameter there would be what corresponds to the curvature of the cup in the example of the marble, namely, those coefficients of the second derivatives of potential energy that determine the stability in characteristic grooves. I should say the tension in this case is really something like the reciprocal of the absolute magnitude of the real part of the largest characteristic groove; that is, it is a number which measures the tendency of the system to return to equilibrium after a small disturbance, and when the tension becomes too large, it corresponds essentially to an inversion, to the case where there is instability because the curvature turns out negative. I would say that tension is essentially a measure of the rate of return to an equilibrium after small disturbances rather than potential energy itself. If this analogy is exact, potential energy is a bead sliding on wire. The potential energy, of course, is proportional to the height of the wire from the ground. But what matters in the case of tension, so to speak, is the curvature of the wire rather than its absolute height from the ground.

BIGELOW: Walter, you don't really mean that the rapidity with which the system returns to equilibrium is a function of the curve, do you? It is not a function of the coefficient of the second derivative, but a function of the decrement, of the dissipation factor.

PITTS: In part, naturally; if it is moved to a small degree and the system is conservative, of course it will keep oscillating indefinitely.

MCCULLOCH: May we hear from Larry Kubie and then we will stop.

KUBIE: I want to explain why I brought up the example of the extremely efficient person who becomes upset precisely at the point at which, if he was strictly analogous to any simple physical system, be ought to achieve equilibrium. At this very point, the unconscious symbolic values of his decisive behavior throw into action a new set of forces which disturb the equilibrium all over again. That is the kind of event which makes life difficult for the psychologist.

THE PLACE OF EMOTIONS IN THE FEEDBACK CONCEPT

[48]

LAWRENCE S. KUBIE

IN THESE conferences, my role is one to which the psychiatrist must often reconcile himself. He is always a troublemaker because he must insist on the complexity of the phenomena of the mind, a complexity over which not only the laity but even fellow scientists would prefer to gloss. The layman wants these phenomena to be simple so that his pet beliefs and biases and prejudices will not be disturbed. The experimentalist wants them to be simple because otherwise the experiments which he can devise in the laboratory, his laboratory models, and his mathematical formulae will be an inadequate facsimile of that which they aim to reproduce.

In multiprofessional gatherings, the psychologist, the psychiatrist, and especially the psychoanalyst functions as a naturalist, reporting on the facts of human nature as observed by him, facts which are dismayingly complex. The experimentalist and mathematician then offer their explanations, whereupon, the naturalist presents additional observations which confront the experimentalist and the mathematician with an even more complex version of natural phenomena. With each recurrence of this cycle, these new complexities are accepted with increasing reluctance and skepticism. The skepticism is justified because each step brings its own new error, but this skepticism, when friendly and open-minded, keeps the psychologist on his toes. Throughout all scientific exchange, the role of the psychologist remains essentially the same: to make sure that we do not become so enchanted by mathematical models that we reject nature in favor of theoretical constructs.

In any broad survey of the effects of experimental procedures on emotional processes, the various areas of pyschic[!] life are so interdependent that no one of them can be assayed alone and apart from the others. Consequently, any experimental procedure has to be studied with respect to its effects on:

1. All fundamental functions:
   - *a)* Perceptual processes, conscious and unconscious.
   - *b)* Conceptual superstructure, conscious and unconscious.
   - *c)* Symbolic representation of conscious processes (language) and of unconscious processes (symptom structure).
   - *d)* The affective accompaniments of all of these functions.

2. On etiological mechanisms:
   - *a)* On conscious and unconscious drives and their related affects.
   - *b)* On defenses against these drives, and the associated affects. | [49]
   - *c)* On the resulting conscious and unconscious conflicts and their related affects: (1) when the defenses are successful; and (2) when the drives break through them.

3. On the dichotomy between conscious and unconscious processes.

4. On ego boundaries and related problems, such as hypnotizability, etc.

5. On something called loosely »vulnerability« to external and internal stresses.

In an effort to achieve clearer definition of such problems as these, I shall ask many questions but answer none. Because of the special focus of these conferences, I shall start with the relation of affective processes to feedback circuits, following this central theme out into its various ramifications.

There are at least four classes of feedback relations which influence behavior:

1. Circuits which tend to maintain energy relationships or balance by means of self-signalling. These serve processes of memory abstraction and symbolic representation.
2. Circuits which are corrective as to the precision with which conscious or unconscious goals are pursued, i.e., through reflex adjustments.
3. Circuits which are corrective as to the balance of pleasure-pain and determine, therefore, the tendency either to repeat or to avoid past experiences. This type of feedback is directly relevant to the phenomenon of emotions
4. Circuits which are corrective as to the psychosocial consequences of thought and action, utilizing rewards and punishments, virtue and guilt. This again plays into the pleasure-pain emotional balance.

THE ROLE OF EMOTIONS IN THE FEEDBACK MECHANISMS

Let us begin by considering the basic significance of emotions in human life. Although emotional states are themselves products of complex psychological processes, they are also causal in that they exercise a vitally important feedback influence on psychic processes. In this circular or feedback function, they are like the governor on a machine; indeed, this is the major key to an understanding of the role of emotions in psychic life. Under normal circumstances, one group of emotions (i.e., anger and elation) lends a quality to any psychological experience which makes one want to experience it again. In general, elation tends to have this effect consciously and anger relatively unconsciously. By contrast, and still within normal limits, depression and fear give a quality to any psychological experience which makes one want to avoid its repetition, [50] depression exercising this influence consciously and fear | tending to exercise its damping, feedback influence relatively unconsciously.

All emotional states can be grouped under these four major categories (Figure 2), each of which varies qualitatively within itself without losing its fundamental quality, and each of which can be combined in various more complex emotional states with one or more of the others. Certain combinations occur more frequently than others, but there is no combination, even of seemingly opposite pairs, which is unknown both clinically and in daily life. In Figure 2, anger and elation are above an arbitrary dividing line, fear and depression below the line, with cross-relations indicated by dotted lines. Various maneuvers, such as psychosurgery, may release or facilitate the expression and/or the intrapsychic influence of those emotional states which lie above the line, i.e., of those emotions whose circular influence on behavior is reinforcing, and they tend to decrease the influence and/or the expression of fear and depression, i.e., of those emotional processes whose circular influence on behavior is inhibitory. This does not always occur, however; nor can any generalizations be made as to why

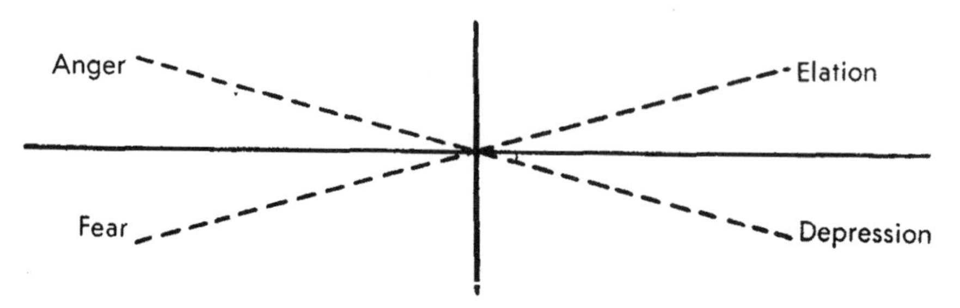

Figure 2. Reprinted, by permission, from Kubie, L. S., and Glasser, G.: Affectivity and psychosurgery. *Proc. Second Research Conf. on Psychosurgery. Public Health Service Publications* No. 156, *67 (1952).*

this sometimes happens and sometimes fails. It must be borne in mind that the effects of any experimental maneuver on the expression of an emotion and on its circular or feedback influence need not be identical from person to person or from one time to another in the same person.

From these facts, it can be concluded that the most fundamental index of the influence of any procedure on the role of emotions in psychic life would be to compare, before and after, the effectiveness of anger and elation in causing repetitions of experience, and, conversely, the effectiveness of depression and fear in causing avoidance of such repetitions. For example, we would expect that after psychosurgery the subject, even in the face of fear or depression, might be | relatively complacent about repeating experiences which because of this fear and/or depression he would previously have avoided. This is the basic change to look for, not merely the influence of the operation on the overt display of emotional fireworks. [51]

CHRONIC AFFECTIVE TONUS

The general affective status of an individual, whether normal, neurotic, or psychotic, is determined by the influence of a complex balance of forces. Chronic emotional tensions arise in some form whenever there is an unconscious, and therefore compelling, inner necessity to do something, the performance of which is at the same time blocked by the influence of unconscious or conscious depression and/or fear. Note that, for this to be true, some important component of the drive must be unconscious; whereas, the opposing forces may be conscious, unconscious, or both. Where some inner rearrangement occurs, as a result of which the internal blockade is lifted, it would be expected that the tone of the chronic emotional tension state would change from the unpleasant quality of depressive tension and fear to the far pleasanter state of aggression and elation, and, actually, this is precisely what does occur (as has been pointed out recently by Lewin in his book *The Psychoanalysis of Elation*).

In chronic affective states, the influence of any procedure would have to be measured in two directions: (*a*) whether the procedure alters only the role of depression and fear in inhibiting the activity toward which there is a compelling inner drive; or (*b*) whether it alters, as well, the underlying unconscious, compelling needs themselves. In other words, does psychosurgery, for example, affect unconscious drives directly, the unconscious conflicts which arise over those drives, or only those affective by-products which in turn exercise a feedback influence on the expression of the conflicts?

Let me illustrate the problem by putting together a composite picture of a hypothetical patient. He is a young man who suffers from a vigorous but unconscious homosexual compulsion, which in turn is the storm center of violent, conflicting feelings of guilt and fear. Because of this conflict, he is in a constant state of chronic emotional tension, colored by undercurrents of anxiety and depression, which are more or less continuous although fluctuating in violence and intensity. Whenever he happens by chance to see a photograph of a young man who attracts him, or to read about some young man who stirs him, or to pass one in the street, his drives are momentarily intensified. This occurs especially when he remains unaware of the stimulus, because at such times his guard is down. In response to chance stimulus, his conflict | is correspondingly activated, and when this happens, his symptomatic affective tensions are aggravated and become explosive in several directions. In such a patient, what might be the hypothetical noxious or therapeutic influence of any maneuver such as psychosurgery? Where could it affect this constellation? Conceivably, it might lessen the compulsive homosexual drive and the resultant play of conscious and unconscious homo- [52]

sexual fantasy, and in this way it might raise the threshold of vulnerability to chance homosexual contacts. All of this is possible, if not wholly probable; it should be subjected to experimental investigation. Alternatively, the maneuver might diminish all restraining impulses as well as the resulting conflicts with their attendant affects. Or, anatomically, it might alter the connections between those areas in the nervous system which subserve the conflict and those which subserve emotional states, thereby diminishing or even eliminating the affective by-products of the conflict, making it possible for the patient to live more comfortably in spite of an unaltered underlying conflict. That we still tack precise instruments with which to define the point at which such a chain is interrupted is a challenge to clinical psychology.

Or let us assume that a patient deals with his homosexual problem in a different way as, for instance, by developing a cat phobia. Through a series of relatively accidental, secondary circumstances, the image of a cat has become a symbol of everything which to this homosexual male patient is unacceptable about the female genitalia. He has the same fundamental dynamic problems as the first patient but they are incorporated into a different psychological structure in such a way that he is relatively if not absolutely free from tension or anxiety except when a cat enters his presence. In short, he has buried (repressed) the homosexual conflict more successfully and completely than the first patient, and the symbolic cat has drained off and/or circumscribed some of his conflict-driven tensions. Under these circumstances, where might the influence of any maneuver hpyothetically[!] be sought? Clearly, as in the other case, it might diminish the initiating phobic reaction to the female genitalia, or the associated compulsive homosexual drive, or the conflict over the drive. Anatomically, it might do this by altering the pathways linking the areas which subserve such a conflict to the cortical neurons which presumably must link the image *cat* with his unconscious fantasies of female genitalia. Alternatively, the maneuver might sever the link between cortical fields which subserve the unconscious symbolic connections of *cat* to the deeper centers which subserve emotional states. These alternatives are not mutually exclusive.

Finally, let us consider a third alternative. The patient has developed an active delusional system, the outstanding component of which is a delusion about cats, which [53] represent to him female witches disguised | as animals. Under these circumstances, where and how might a therapeutically effective procedure hypothetically influence the patient's condition? Again, it might lessen the power of the underlying compulsive homosexual drive, or of its counterpart the phobic reaction to the woman's genitalia. Or it might lessen the conflicts arising over the homosexual compulsion, or the activity of the unconscious fantasies about female genitalia, which in turn are represented by the delusional misinterpretation of the significance of a cat. Anatomically, it might isolate the cortical fields subserving the systems of conscious and unconscious fantasies by severing the links between different cortical areas, or between the cortex and basal structures which subserve the emotional reactions to fantasy and delusion.

The symptom picture of any patient may alter after various disparate therapeutic attempts, depending upon which links in these complex chains have been altered by the procedure. For instance, if the link between the delusional structure and the emotion has been severed, we would expect the patient to retain his delusions but to have a »So what?« attitude toward them, something that everyone has seen in clinical practice. On the other hand, if the procedure alters the underlying storm center of conflict, then the whole symptom structure could disappear.

CERTAIN THEORETICAL ASPECTS OF AFFECTIVITY IN GENERAL

Certain general aspects of the phenomenology of affectivity are not covered by the above considerations or by the hypothetical examples. The phenomenology of affectivity can be subdivided into a number of logical components, in all of which quantitative and qualitative variables, both physiological and psychological, are conceivable. Yet, the mere fact that the phenomenology of affectivity can be subdivided into logical components does not mean that each one of these components actually functions independently of the others. They may, rather, be mere facets of a single process. The significance of this will become clearer as we proceed.

a. The Stimulus

The arousal of any emotional state requires an affective stimulus. Such a stimulus can be either external or internal in origin, conscious or unconscious in operation. Usually, it is mixed with respect to both polarities. It can be an adequate reality stimulus in the sense that it represents an intrinsically important frustration or threat or gratification. Or it can be adequate only in the sense in which a highly charged symbol (such as a melody or a flag) is adequate. Furthermore, a realis | tically adequate stimulus can at [54]

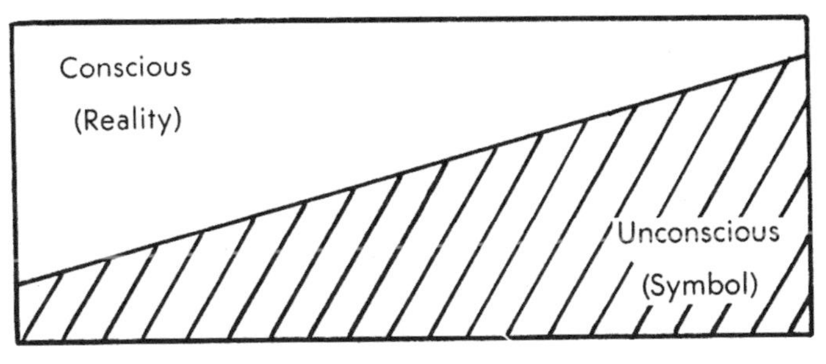

FIGURE 3. Reprinted, by permission, from HILGARD, E. R., KUBIE, L. S., and PUMPIAN-MINDLIN, E.: Problems and techniques of psychoanalytic validation and progress. *Psychoanalysis as Science. Hixon Lecture.* Stanford Univ. Press, 1952 (p. 51).

the same time be symbolically supercharged, as, for instance, when an odor rouses nostaglic[!] overtones. (For me, the odor of boxhedge rouses overwhelming feelings that go back to my earliest years.) Finally, of course, the symbolic overtone or supercharge of an affective stimulus may be either conscious or unconscious. Usually it is both, in varying mixtures.

b. Trigger Stimuli

When the stimulus is predominantly symbolic rather than realistic, when its symbolic significance is predominantly unconscious, and when its impact has a short latency, we speak of trigger or phobic mechanisms. In this connection, it is important to realize something which is generally overlooked, namely, that there are trigger mechanisms for any and all emotions. The pathologic nature of unconsciously triggered affective states is generally recognized only for the phobias. Yet, anger is even more frequently set off by unconscious triggers than is fear, and the same is true of laughter. Because we usually rationalize the anger and enjoy the laughter, even analysts have failed to realize fully how pathologic either or both can be. In contrast, because they are painful, the pathologic nature of depressive states is, if anything, exaggerated. Actually, many depressive states which are viewed as pathologic are the inevitable and logical outcome of painful situations and events. Pathologic depressions almost always have a relation to unconscious trigger mechanisms, which has been overlooked even in technical psychiatric literature. Under analysis, these trigger mechanisms become more apparent, and it sometimes seems as though a man lives each day through a succession of subliminal impacts from trigger stimuli, which batter him incessantly, inducing [55] abrupt emotional swings, until his | emotional state is like a particle in a state of perpetual Brownian movement.[1]

c. Time Relations of Emotional Triggers

Trigger mechanisms for emotional states which are instantaneous in their effects are easily recognized, whereas those triggers whose effects are delayed for minutes or hours, even for one or more days, tend to be overlooked. There are some indications that nighttime and sleep may circumscribe sharply the duration of the influence of individual trigger stimuli. This is a matter which requires experimental clarification. The extent to which sleep limits the duration of the influence of trigger stimuli on emotional states is an issue of prime importance for our understanding both of emotional processes and of sleep itself.

What is the effect of experimental or therapeutic procedure on all trigger mechanisms: on specific and individual trigger mechanisms, on the level of reactivity and responsiveness to them, on their duration, on the formation of new trigger mechanisms, and on their fate during sleep? None of these questions has ever been asked, much less explored, either clinically or experimentally.

I shall not attempt here to discuss how such triggers are established, whether they take root only early in life or subtly throughout life. But if the latter is the case, then we must ask not only what is the effect of any procedure on trigger mechanisms which are already established but also whether new trigger mechanisms, i.e., »conditionings« of a special type, can be formed after such procedures.[2]

THE AFFECTIVE PROCESS

[56] Affective phenomena in human life are made up of several com|ponents or phases. These components can be logically separated into three aspects: (*a*) an affective tonus or latent affective potential, (*b*) an affective process, and (*c*) the affective state or dis-

1 »The ancient confusion between description and explanation in psychological theory takes several forms. One of these is the failure to differentiate between a trigger and a cause. The thing that pulls a trigger releases the bullet but it does not impart to the bullet its energy. There are many trigger mechanisms in psychological affairs. The phobia is the most familiar example, merely because it is easy to recognize. Laughter, however, is another triggered response, although its species similarity to a phobia has been overlooked. Tears, sudden waves of depression, acute elations, obsessional and compulsive furors, sudden regressions, sudden rages, equally sudden states of apathy, lethargy, or even sleep, acute dissociated states, and sometimes states of partial or complete multiple personality, any of these may be set off by subtle and inconspicuous moments which, partly because of their symbolic meanings, acquire a trigger action. In all such episodes, the trigger is not the cause. Nor is it in any strict sense analagous[!] to a catalyst. Triggers are effective stimuli primarily because of their unconscious symbolic implications. Clearly, therefore, there is a vital difference between a cause and a trigger, but their clear definition and their experimental isolation has not been achieved.« (1)

2 In personal communications, Dr. Howard Liddell and Horsley Gantt have written me that there are as yet no adequate experimental data on the effects of psychosurgery on them, whether in humans or lower animals.

charge. These terms imply, however, several unsolved questions, all of which must be considered separately.

Is there a diffuse emotional tonus or potential which is basic in the structure of every personality, determined by some inherited or acquired physiological idiosyncracy? Or is there an emotional tonus determined by early, overwhelming psychological experiences which flood the individual with a dominant emotion from which he spends the rest of his life in a vain effort to escape? Certainly, one encounters individuals whose entire lives seem to be colored by a chronic depressive bent begun in the pain of a separation in the first few years of life, others with lifelong states of masked or overt terror against which they struggle. There are equally chronic states of masked rage, and also chronic hypomanic states, but these tend usually to be accepted with greater complacency because they cause less pain to the patient than to the environment. One asks, therefore, whether every life has a dominant emotional color which acts as a center of equilibrium to which the personality returns from any excursion into other emotional experiences. If this is true, the influence of any experimental procedure must be evaluated in terms of its effects on this hypothetical underlying emotional tonus, and on the strength and automaticity of the tendency always to return to it.

It is probable that another way in which chronic affective tensions arise is through the presence of an unconscious, compelling, inner necessity to do something, the performance of which is blocked by some countervailing inner force. This situation gives rise to an affective tension which can determine both the quality and level of the general affective state and the quality of its discharge. Consequently, in any experimental study of emotional processes it would be necessary to determine whether the unconscious inner necessity or its counterforce or the emotional reaction to the conflict, or all of these, were altered by the experimental procedure. The devising of experiments to this end is far from simple.

These considerations lead directly to a third unsettled question concerning the phenomena of affectivity. Is there such a thing as an undifferentiated affective or emotional potential out of which, under suitable circumstances, any one of the more differentiated affective states may precipitate, to wit, anxiety, sadness, elation, or anger? If there is such an undifferentiated potential, to what extent is each differentiated emotional state the expression of some specific type of chronic, unresolved, internal conflict, or to what extent is it a product of external situational variants? Alternatively, would it be more accurate to de|scribe the differentiated affective states as transient expressions of [57] some dominant »master mood,« such as anxiety, of which all others are mere derivatives? For instance, is elation always an escape from anxiety, depression a reaction to being swamped by anxiety, rage a response to being confronted by anxiety, etc.? Here the question is whether from both the physiological and psychological points of view there is one primary master affect, with numerous secondary derivative affective manifestations; or whether there are a number of primary, affective processes, each of which may respond differently to experimental procedures: drugs, psychotherapy, psychosurgery, etc. These ancient questions are of basic importance; yet, they have been argued in academic psychology more than in clinical psychology and psychiatry, and still have not been subjected to precise experimental investigation.

In discussing these problems, how helpful is it and how essential for clarity and accuracy to differentiate an affective process from all affective state, and an affective state from the affective discharge, and, finally, the phenomena of episodic discharge (catharsis?) from the true resolution of affects? In short, is there a process, whether physiological or psychological, which, once initiated, runs a predictable course through a series of phases, each of which must be considered both separately and in conjunction with others? This question must be kept in mind as we search for the point at which any manipulative procedures, whether psychosurgical or otherwise, influence emotions.

Such terms as mastery, discharge, and resolution, unless used with care and precision, are mere question-begging metaphors which require far more thought than they customarily receive. Indeed, in our current concepts or terms there is little clarity about: (*a*) the circular influence of affective processes on behavior, (*b*) the expression of affect, (*c*) the episodic discharge of affect, (*d*) the resolution of affect, and (*e*) the relationship of any or all of these to what is called catharsis. Because of this lack of clarity, we have never even asked with precision whether, after successful therapeutic interference of any kind, a previously disturbed patient has become more or less prone to generate new affective tensions, more or less capable of experiencing affects subjectively once they are generated, more or less capable of expressing them, nor what happens to his capacity to resolve his affects spontaneously. Each of these general questions must be brought down to concrete clinical and experimental issues. Probably we would be well-advised to start by developing techniques for determining in the normal human being how man generates affective states; under what circumstances he then experiences the affect consciously or unconsciously; what determines the freedom with which he expresses his affect either episodically or continuously; the relationship of all [58] of this to the »discharge« or »resolu | tion« of the affective state; and, finally, the circular or feedback influence of emotional process on behavior.

SOMATIC INVOLVEMENT

Certain problems concerning the expression of affective states and processes have been alluded to, but not their psychosomatic implications. Some individuals seem consistently to use only one internal organ as the outlet for every affective state as though all other somatic, neuromuscular expressions of emotion toward the outside world were inhibited, except insofar as these merely serve the intrinsic vegetative components of emotional processes. For instance, when angry some individuals neither fight nor flee nor argue. One merely walks to the toilet to defecate; another vomits; another urinates; another eats and bites; another breathes heavily. But another individual, when angry, may hold his breath, retain feces, withhold urine, or have gastrointestinal spasms. These physiological processes rarely become conscious components of the pattern of emotional expression except in certain psychotic states. Except in children and psychotics, they can be recognized only when a patient is under analysis.

The vegetative components of emotional states are the dominant emotional outlets in infancy and early childhood when they constitute the preverbal body-language of infantile emotions. In the study of adult life, the technical problem is how to determine what happens in later years to this preverbal, emotionalized body-language of infancy, and especially how to recognize and measure its masked influence after it has been repressed to the point of becoming wholly unconscious. It should be found out how someone who expresses hate through defecating or biting expresses love or fear or elation. Do other emotions find expression through the same organ as hate does, or are some of us (or all of us) Johnny-one-notes in the somatic expression of our feelings? Is there in some of us just one affective body instrument for all emotions, while other individuals express different emotions through different organ systems? If human beings vary as to the ways in which their internal organs are involved in their emotional processes, corresponding variations in the impact of all experimental and therapeutic procedures on their affective processes, and especially on the somatic concomitants, would be expected.

Perhaps someday it will be possible to record continuously and simultaneously the ebb and flow of secretion and of tonic innervations in the several organ systems and to correlate these with the interplay of conscious and unconscious components in the affective processes. Until recently, however, most of the work in the field of psychosomatic medi|cine has been confined to the correlation of transitory physiological [59] changes with transitory changes in conscious emotions. Margolin and his co-workers (2) have recently made a brilliant start toward a more basic attack on this problem, correlating chronic affects and their unconscious components with simultaneous, subtle physiological changes.

QUANTITATIVE CONSIDERATIONS

The next problem concerns the ultimate challenge of quantification. What if anything is meant by »strong« and »weak« emotions? What do variations in apparent »strength« of an emotion signify? What are the variables of feeling or behavior which can be used as indices of the strength with respect to any single component in the total emotional constellation, or with respect to the conscious emotion itself? Is it the violence of the way the individual feels subjectively and consciously? or the violence of how he behaves? Is it the repetitive persistence of an emotional state under varied circumstances? or its apparent ability to overcome both conscious and unconscious controls? Is it, rather, the repression of the emotion, i.e., its disappearance from consciousness and/or from direct expression, which measures its potential strength? (One often reads of hate so »strong« that it has to be repressed, a phrase which implies several unproven assumptions). Or is it the tendency of a repressed emotion to erupt again into consciousness, either before or after repression? Or is it the spread of the emotion or of its controls into seemingly unrelated fields (presumably equivalent to what Pavlov speaks of as »the diffusion of cortical excitatory and inhibitory processes«)? Which of these measures quantitative differences in emotional phenomena, and how are these quantitative indices influenced by various procedures?

If the Pavlovian concept corresponds to reality in any degree, it must be asked whether an experimental procedure (e.g., surgery) blocks such a spread by increasing cortical inhibitory processes or by the direct interruption of intra- or inter-cortical pathways. In this connection, how the anatomical effects of the various procedures may differ would also have to be considered. Do some interrupt the short association pathways within and between cortical and subcortical fields, which would presumably interrupt those reverberating circuits which may subserve both unconscious fantasies and also the unconscious inhibitory processes? Alternatively, do others interrupt the longer feedback pathways, which link the cortical and subcortical circuits to deeper centers of emotional experience and emotional discharge? Such possibilities are speculative at present, but they are not inaccessible to experimental investigation.

To summarize, in any study of the effects of therapeutic or experi|mental proce- [60] dures, quantitative estimates of changes are important, and ultimately essential. The preceding sections of this discussion have dealt theoretically with various efforts to define units which can be quantified, as follows:

a) One might attempt to quantify the effective stimulus, both its conscious and its unconscious, its realistic and its symbolic components (as indicated in Figure 3).

b) One might quantify the patient's tendency to express emotions directly in action as a partial index of the balance between sensitivity and control.

c) One might quantify the patient's tendency consciously to characterize in speech the affects he can feel, while withholding any direct expression in action.

d) One might quantify the patient's tendency to develop some form of affect-like process of which he himself remains unconscious, but which he expresses indirectly in derivative form through qualitative or temporal displacements. This would lead to a quantification of the various possible transformations of affects: (i.e., changes in the affective quality itself, changes in objects towards whom affects are directed, changes in content, changes in somatic components, and, finally, displacements through delay).

Conceivably, each of these variables can be influenced quantitatively as well as qualitatively by surgical or other procedures. It remains to be seen whether it will prove to be practically feasible to explore separately the qualitative and quantitative variations of each.

OBJECTS OF EMOTIONS

These many possibilities are surely complicated enough; yet, the problem is even more complex since the question remains, toward whom are affects directed. In other words, there is not only the organ and the aim to consider, but also the object. In some individuals, the pattern of affective vulnerability and its somatic expression through characteristic idiosyncratic organs can be triggered off by anything or anybody. In others, the effective trigger figure or trigger situation may be quite limited and quite specific, such as an older man, a yellow-haired girl, a sick child, and the like. Any experimental procedure which omits from its controls this problem of the direct linkage of emotions to specific objects will be inherently faulty.

Finally, we must ask how to characterize the distinction between normal and abnormal affective states. This issue is related in turn to the question of whether the conventionally accepted psychiatric concept of an affective illness is itself justifiable. Is there [61] an illness which is | primary with regard to some aspect of the affective process, or is the generally accepted concept of an affective illness based on tacit but unproved assumptions which arose because of the dramatic quality of certain illnesses in which affective symptoms are prominent? Before considering this question, we must offer a pragmatic characterization of the distinction between normal and abnormal affects.

a) Is it adequate to make this distinction in terms of the relative roles of conscious and unconscious components of the affective state, components having to do with its precipitating stimulus, its content, or its expression? (v.s. Figure 3)

b) Can the difference be characterized in terms of the significance of the conscious content of the affective state? What is the actual role of some problem that an individual thinks he is depressed about, afraid of, angry at, or elated over? Specifically, does this manifest content play any role in causing the affective state, or is the manifest content rather a product of the affective state, just as the lion of a child's nightmare is the product rather than the cause of the child's terror? Similarly, is the apparent content of a mood little more than a pseudorational explanation of the affect? And if this is so, can we say that an affect is normal when the manifest content is the true major cause and that it is abnormal when the manifest content is a minor contributor or a secondary by-product of the affect? If so, how can we determine with precision whether the manifest content of a mood is its cause or its result?

c) Perhaps emotions are neurotic when the free flow of the emotional process is obstructed, because while they may be positive on a conscious level, they may be negative on the unconscious level, or vice versa.

d) Again, does the distinction between a normal and an abnormal affect depend upon the extent to which the affect remains circumscribed (both in content and in its secondary effects) as opposed to the extent to which it spreads (Pavlov)? Is the tendency to remain circumscribed or to spread a quantitative variable which is due to variations in something which might be called the strength of the affective process, or to its specificity? Or is this a function either of the strength or the specificity of the circumscribing defense mechanism? If the latter, then an affect may be quite as abnormal when it is circumscribed as when it spreads.

RETROSPECT

I began this discussion by stating that I would ask many questions and answer none. This is precisely what I have done and I offer no apologies, especially since many of these questions have never been | asked before. Furthermore, each of these questions [62] and their ultimate answers bear directly, and I believe importantly, on all experimental and therapeutic considerations of affectivity. Consequently, if these ruminations lead ultimately to clearer clinical and experimental studies, I will be content.

McCulloch: May I say, before I throw it open for general discussion, that there are three terms which you have used again and again which have not sufficiently sharp definitions, namely, »feeling,« »affect,« and »emotion.« These three have chased each others' tails round and round without my being able to get hold of them. You apparently mean different things by them but I am not at all sure what the differences are.

Kubie: A definition is the answer that comes at the end of the book, never at the beginning. It is like insight in psychotherapy. You cannot start with it. You have to achieve it at the end of a long, hard journey. I am entirely happy, therefore, to say »affect,« or »emotions,« or »feeling,« and say they are all approximate ways of identifying an experience. If I try to define these more precisely, I would be going through a pseudoscientific dance that would be meaningless.

McCulloch: But I am not sure whether you mean different things by them. I think you do.

Kubie: Of course, I do. These are multidimensional experiences, no dimension of which should be excluded at this time by overprecise definitions.

Fremont-Smith: Then you are saying, Larry, that for the purposes of this discussion, you don't mean very different things?

Kubie: The meanings are overlapping. I shift from one term to the other only so as not to be boring by repeating the same term over and over.

Fremont-Smith: So it is not very different.

McCulloch: Elegant variation.

Kubie: Yes, that's right.

Bigelow: Dr. Kubie mentioned the possible utility of being able to quantify these things. In other fields of the natural sciences, the business of introducing a metric is always a very important step forward, and I admit there is at least this analogical reasoning to support the suggestion here. On the other hand, this field is diffuse and contains many factors whose interactions we do not understand. It frequently requires considerable analysis just to be able to name attributes. And it is not clear to me exactly what you could do if you could demonstrate that man A is angrier than man B. Exactly how would you use the measurement if you had it?

Kubie: There is no human biogenetic need that operates in a psycho | logical vacuum. [63] We do not breathe only because we need oxygen. True, if we do not breathe, we will die, but that is not why we draw any specific breath. I do not take this inspiration because at this moment I suffer from an oxygen debt and need to take in just so much oxygen. Rather, there is a mixture of complex psychological processes which play around my respiratory pattern in an array of subtle compulsions and phobias.

Biogenetic appetites range in a hierarchy from the simplest to the most complex, the most complex being, naturally, that old troublemaker, sex. A psychological supercharge of compulsions and phobias operates on all, but increases steadily in importance from one end of this spectrum to the other. But we have no methods as yet for measuring the relative roles of biogenetic need and psychological superchargers in any single act. If I drink a glass of water, it may be easy to recognize qualitatively the concurrent activity of a biochemical and of a psychological superstructure, but there are no methods by which we can estimate their relative roles with quantitative precision.

BIGELOW: Can't you measure it by external effects to the extent that you want to measure it?

KUBIE: No, I do not think you can.

FREMONT-SMITH: Measure it on other people, for example, measuring anger in terms of physiological changes in blood pressure?

BIGELOW: No, I mean, if a man is angry, it is important only in the sense, I suggest, that a psychiatrist worries about it, namely, if his anger interferes with his behavior. Now, the extent to which he is angry is represented by the extent to which it interferes with his behavior. Isn't this all the measurement that is needed? Is this not a satisfactory working technique? What is wrong with that as far as your field is concerned?

KUBIE: A great many things are wrong with that, but that takes us into a very different and complex issue. Men do an extraordinary amount of valuable and useful, work for all kinds of complicated reasons of which they are unaware. A man's anger may be completely neurotic and yet productive. Take a man who was a patient of mine. He probably did as creative work in his field as any human being in this country, yet the man has lived in a temper tantrum ever since he was a child of about four or five. Outwardly, he has often masked this successfully for months at a time, but people in general do not know some of the things he has done in secret, only his unhappy victims. Furthermore, his buried rage finally caught up with him and tumbled him into a depression of suicidal violence. He has been angry, without adequate current or initiating reasons, all his life and this self-destroying anger has actually energized his entire [64] career and made him enormously pro|ductive. Yet it has nearly killed him and has made him in many ways a very sick man.

BIGELOW: It seems quite likely to me that measuring anger quantitatively, which I understood you to suggest would be desirable or useful, would in fact be an impossible thing. It may be an irremovable obstacle to your technique, if you feel that it is an essential step in order to go forward, because there seems to me very little likelihood that there would be any invariant measure of anger from one man to the next, ally more than there is any invariant measure of practically any other sensory process. There is no reason to think that I get the same sensation when I feel a certain temperature as another man does. The sensation is a subjective experience which is a function of my background and of my system. All one can possibly hope to do is to judge this in terms of its effect upon my behavior. As a closed system, I have a certain routine of activity, a certain life, and whether or not I am angry can be judged by whether or not my behavior has been disturbed from its equilibrium. I think it is impossible to hope to judge whether I am as angered as my neighbor by a certain event. This is a fundamental obstacle which can possibly never be removed, so one may have to orient one's thinking in terms of measuring entirely in the terminal sense.

HUTCHINSON: Isn't a great deal of the trouble that is always coming up when we talk about this due to using the word quantification in a rather vague sense, using it to mean either knowing intuitively that something is greater or less than (and conceivably, very rarely, equal to) something else on the one hand, and the process of constructing an actual calibrated scale on the other? If we could get some relationships by pretending that there is a nice calibration scale showing that one thing should be greater than or less than another, then we could go back to observation and, having made our prediction, perhaps check it, without ever having had a real quantification but having played at a sort of pseudoquantifying?

BIGELOW: I would see no hope of doing that in this case because I don't think there is an invariance among individuals.

HUTCHINSON: No. It could be done, though, with a single individual. It could probably be said that a particular person at a given time is more angry than he is at another time.

BIGELOW: But only in terms of a measurement involving a deviation of his behavior from his normal.

HUTCHINSON: Yes, but if you can't do that, then you can say almost nothing at all. With no standards of reference whatever, it can't be said whether the people that we have just been hearing about were really sick at all or whether they weren't a good deal more normal than everybody we have ever met. | [65]

BIGELOW: You can say that a man's behavior is pathologic compared with the average person, but you can't say anything more than that.

HUTCHINSON: You don't think you can say that a person's behavior at a given time is more pathologic compared to the average person than at some other time when you could hardly call his behavior pathologic?

BIGELOW: I doubt it. I think there is a strong technical difficulty here.

HUTCHINSON: I would like to know whether that question can be answered affirmatively or negatively, Dr. Kubie.

KUBIE: The question indicates how hard it is to get away from certain habits of speech and thinking. In a sense, my whole purpose was to avoid formulating the question in an unanswerable form by breaking the problem of emotional processes into its fragments, some of which can be subjected to various kinds of precise scrutiny, including at some points useful quantifiable data. As I said before, when we talk about one person being angrier than another of being angrier at certain moments than at other moments, we describe a confused, complicated process which has so many variables in it that I do not believe it can be quantified as a whole. On the other hand, if we can break it into components which are more precisely defined, then it can be dealt with in various quantifiable forms, such as thresholds of vulnerability, number of stimuli, the variability of stimuli which evoke certain comparable emotional states, the time during which an individual remains in one emotional state before it becomes transformed into something else.

BATESON: Could the quantification problem be approached from a quite different angle altogether? One of the things which is of first-rate importance to the human organism is that the premises upon which it acts should be correct, its premises about itself and its premises about the environment. In an individual's premises about himself, information is the means he uses in order to achieve a correspondence between what is in his head and what is outside himself. Goal-seeking is a change which he tries to make in the environment so as to have it correspond to something which is inside his head. It seems to me that a lot of the neurotic and psychotic phenomena that Dr. Kubie has been talking about are instances where an individual has taken a wrong course, in a sense, in his attempt to make his information correct. As I said earlier, I know that glass is there because I touch it, and I know that it is there because I know my arm is. The premise about my arm is a part of my knowledge about the position of the glass. If the premises about ourselves upon which we base our conclusions are somehow false, and we are unwilling or unable to change them, we then have to further falsify our view of the outside world in order to get back to making any sort of sense again. We have to force an unnatural solution upon the world to account for that [66] in which we are wrong about ourselves.

A thing like chronic anger, it seems to me, could be considered as an extrapolation from an early traumatic anger situation which becomes generalized into an expectation that the world is such as to make the person angry. I am angry, and therefore the world must be so constituted as to make me angry, and I shall be anxious if it isn't. Or the example of the man Larry Kubie spoke about this morning, the man who falls into anxiety whenever he is successful in his enterprises. If he has a major premise that he cannot be successful in his enterprises and he is successful, then there is something discrepant about the world and he must falsify himself in some way; he must force himself into depression, for instance, which will validate his major premise about himself and the world in which he lives.

The question of quantifying phenomena of that kind I would think to be a question not of quantifying information in bits, a process such as we are familiar with from other fields, but, rather, a question of quantifying distortions of systems of bits. How can one measure how angry a chronically angry person is, or the chronic disturbance related to the anger in this person? Is there conceivably a quantification which could deal with the distortion of his information system? Or perhaps we should ask about rate of distortion, because the system is circular; the more wrong he is, the more wrong he has to be, and so on.

BIGELOW: I am attempting simply to make a conjecture of the following sort: that in a field as extremely complicated, because of the variety of data entering it constantly, as is Dr. Kubie's, the things described as anger may not have sufficiently determinate coordinates with respect to which to make a meaningful measurement; that anger varies from one person to the next so vastly that there is nothing which is an invariant thing called anger in man A and man B; and that, in fact, an attempt to establish this is simply wasted time until more is known about the specificity of these components.

BATESON: Yes, but could you conceive of having a more abstract variant appearing as anger, appearing as anxiety, appearing as fear?

BIGELOW: In a field like natural science, such things as this have been useless. Measurements which are useful can only be taken when the thing is so narrowed down that it can be said precisely what the coordinate axis is: exactly where is the evidence of anger in this man or that man; exactly what is it that is to be measured? It must be describable with a precise set of coordinates, then it can be measured. But this is done only in very simple systems. It is never done in anything as complicated as I understand psychotherapy to be.

HUTCHINSON: But how can you understand anything about psychiatry if the word anger in the way Kubie uses it doesn't mean approximately the same thing every time [67] he uses it? |

BIGELOW: It can mean the same thing every time he uses it, but we still may not know enough about it to measure. For example, you might merely say it is present or is not present, the way you say an animal is a dog or is not a dog even if you don't know how big he is.

EDITOR'S NOTE: Dr. Bigelow has the following comments to add to his remarks made at the conference:

> To me, this point in the discussion illustrates a crucial and longstanding impasse in communication between workers in mathematical or physical fields and those who work in biological, social, or medical fields. Any thoughtful physical scientist is completely amazed

when confronted by the fact that an experimental technique like psychotherapy demonstrably works, amazed because of the subtlety, complexity, and heterogeneity of the domain in which the experiments are carried out and the intangible nature of the methods of exploration and of producing modification. On the other hand, the researcher in the non-physical fields, aware of the overwhelming complexity encountered in his field, turns hopefully to mathematics for help since mathematical methods are known to aim at handling complex situations by reducing them to terse statements through analysis.

Because it is true that the methods of the mathematical-physical sciences do successfully reduce situations of apparently great complexity to simple formalizations, it is difficult to persuade non-mathematicians that these successes are due to essentially simple artifacts and apply to special situations not generally found in the real world of experiment. Sufficiently general mathematical-physical techniques capable of handling complex, heterogeneous, and interrelated data and of reducing them to concise and information-preserving formalisms do not exist today and may still be a long way off. Those methods that do exist are quite special, and are informative only when the particular artificial assumptions and processes involved are understood and are constantly held in view against the background situation concerning which inferences are to be drawn.

Misconceptions as to the nature of this limitation are frequent. It is proverbial that »measurement is the basis of science,« and non-physicists frequently believe that measurement is more successful in the physical sciences because »true« quantities can be measured directly, for example, force, mass, and acceleration in Newton's law. This is today not believed to be the essential reason for the success of measurement (and mathematical formulation) in the physical science; and, in fact, it is thought to be a fundamental misconception that significant quantities can ever be measured directly. In general, they cannot be. It is always done through some more or less indirect manifestation, i.e., the force exerted upon a material object when accelerated or held stationary in the earth's gravitational field as the measure of the mass of the object.

Measurability (and hence mathematical reducibility to simpler formalism) is today believed to be more directly related to experimental repeatability than to any other simply stated property. If anger could always be measured by measuring, say, blood pressure and if | these two were constantly and reproducibly related, then there would seem to be no reason why this gauge of anger would not be as reliable as any other, the measurement as valid as any in the physical sciences and as suitable a basis for mathematical processes. It seems a fact, however, that such effects are not simply (not to mention, constantly) related and that this is, in fact, the reason one is unwilling to take blood pressure as a gauge of anger, rather than because it is an indirect gauge. [68]

To repeat, the suitability of an externally measurable quantity as a gauge is not determined by whether it is direct or indirect – all are indirect – but, rather, by whether it is simply and reproducibly related to the phenomena in question. If the relationship is complex, not expressible by a smooth curve, for example, but by one with jumps or loops in it, or if other uncontrolled factors such as constitution, heredity, or metabolism enter, then the gauge is apt to be unsuitable for measuring the phenomena.

It would seem to me that the difficulty in measuring quantities of interest to a psychotherapist lies in the complexity, and inconstancy, of their interrelationships and that the first problem is to trace out these connections before attempting to measure. In some miraculous way, psychotherapy seems to be a successful technique for exploring connections in a very complex and inconstant field; therefore, perhaps it is essentially correct in method. Measurement (and metric method) is not the only avenue, even in the physical sciences; frequently, identification of objects, their enumeration and the exploration of their interconnection is a very successful observational procedure and may profit from the application of special, non-metric, combinatorial or topological branches of mathematics. It is, of course, true that objects (e.g., »fixation« or »tension«) cannot be identified without implying a decision as to their being present or not to a »significant« extent; nevertheless, this decision implies much less than quantitative measurement. It seems to me, in the present state of ignorance concerning the psychophysical processes, futile sophistry to imply that any particular branch of mathematics, combinatorial, metric, or other is the particularly appropriate tool.

KLÜVER: As regards emotions such as anger and fear, the hormic psychology of von Monakow, with its concepts of *klisis* and *ekklisis,* considers the impulses and tendencies associated with these emotions to be the most fundamental tendencies operative in behavior, namely, the tendency to approach, touch, attack, or fight, on the one hand, and the tendency to avoid, retreat, or flee, on the other. Hediger, a Swiss zoologist, obtained quantitative data on flight or escape reactions, (3) although he did not study emotions in animals such as man or the domesticated Norway rat, animals which in the opinion of Curt Richter are alike in having an inadequate equipment for meeting stress (4). Hediger, studying wild animals under natural conditions, found that, when encountering an enemy, a wild animal shows an escape or flight reaction as soon as the [69] enemy approaches within a certain distance. The | flight distance varies with the species; it may be, for instance, two, twenty-five, or five hundred yards. Hediger believes that the impulse to escape from its enemies, the ever-present flight tendency, is far more powerful in the wild animal than impulses associated with sex or hunger. In fact, a sudden reduction of the flight distance to zero, that is, captivity, may lead to death.

I should like to ask Dr. Kubie a question. You mentioned psychosurgical procedures upon the brain in connection with emotions. I am wondering whether you want to make specific statements about the anatomical substratum of emotions.

In 1937, I showed a film before the American Physiological Society illustrating striking changes in emotional behavior produced by a brain operation in a rhesus monkey. The changes following the operation included marked differences in sexual and oral behavior. These behavior alterations were brought about by removing major portions of the hippocampus, uncus, amygdala, and temporal neocortex. A few months later Papez (5) published a paper which is now generally considered one of the most brilliant theoretical papers ever written on anatomical mechanisms in emotion. Starting from an analysis of neuroanatomical, comparative anatomical, clinical, neurophysiological, and neuropathological data, he advanced the hypothesis that certain rhinencephalic structures and their connections constitute the anatomical substratum of emotions, structures which were previously believed to mediate olfactory functions. Further experimental work has lent support to this view that rhinencephalic structures play a very important role in emotional behavior (6). I am curious to know whether, in psychoanalytical speculations before 1937, there was ever any hint or recognition of the fundamental importance of the rhinencephalon for emotions. To most investigators in the field, it was somewhat of a surprise that this part of the brain should play such a decisive role in connection with emotions and personality.

KUBIE: I did not want to discuss psychosurgery; I was simply indicating some of the questions we ask ourselves about the effects of any procedure on emotional processes, the points at which they are vulnerable and alterable, etc. What it really comes down to is the fact that the subjectively-perceived emotional state is only a fragment of complex forces, a conscious fragment which represents the resultant, a kind of algebraic summation.

QUASTLER: Just a small point of semantics. Kubie tried to deal with anger. Both Bigelow and Bateson considered, I believe, the quantization of the effect of anger in the angry man, and this presumably can be done in some way. On the other hand, a number of very different features of anger could also be quantized, in rough approxima- [70] tion. But it won't work to do two quantizations together. |

BIGELOW: There has been quite a tendency in these meetings to assume that the mathematical methods of the physical sciences are necessarily those appropriate for most of the other sciences and other fields, but there are cases where this may not be true. Specifically, cases where the picture is so complex and so variable that the Gestalt figures that are used effectively by the experts in the field may be too variable from case to case for quantitative measurement, and yet configurations may have use.

MONNIER: We cannot understand the quantification process without considering from an anatomic-physiologic point of view the regulators of emotions. Some regulators of emotional patterns in the hypothalamus are known. One area, for instance, induces, when stimulated, anger and defensive reactions. Another area is related to the oral sphere, with its alimentary and digestive drives; stimulation of this area in the cat produces bulimia. Another substrate is related to flight, and another, not too well located, is concerned with the sexual functions. All these patterns, elicited by electrical stimulation in animals, may also be stimulated during epileptic attacks in humans. There are patients in whom a flight pattern is induced, the so-called poriomania, and other epileptic patients in whom aggressive patterns are elicited. The clear knowledge of the anatomic-physiologic organization of the emotional patterns is a necessary condition for every attempt at quantification.

In relation to the same topic, I should like to emphasize the part played by »tension,« a notion which was difficult to bring into the discussion before. A generally increased tension and awareness is the primary condition for establishing communication between individuals. It is, however, by no means a sufficient condition for emotional adaptation. An adequate adaptation requires a selective increase of tension in one or another diencephalic area controlling instinctive behavior, in the areas for aggressive reaction, flight, oral, or sexual drives. In terms of cybernetics, one should go further and correlate with the neurophysiologic patterns the psychologic patterns of emotion. We certainly have in our minds definite codes, which correspond to the emotional patterns of aggressiveness, flight, and sexual behavior. The problem should be discussed, therefore, on both levels.

TEUBER: This remark doesn't fit quite as well now because you have answered part of it, Dr. Monnier, but I agree it may not be particularly fruitful to try quantifying degrees of anger, degrees of fear, and the like, in any direct fashion. It is difficult enough to answer the question whether they are present or absent. This becomes a particularly serious problem when we are trying to describe the effects of certain cerebral lesions in man.

In a series of patients with lesions in the frontal lobes, we have found | quite a few [71] who seem to be just like our control subjects in the performance of so-called intellectual tasks. This group includes men with massive bilateral lesions, and yet it is extremely difficult to find any specific intellectual deficit even though it has very often been claimed to exist. Then people tell us, »But you are looking at the wrong thing. What you should be trying to demonstrate is a change in affect.« This suggestion implies a very interesting possibility: a dissociation of intellectual and emotional symptoms. The suggestion implies that there can be an affective change in the absence of any demonstrable intellectual change. Is that possible? And, if so, how do we go about finding it in our patients?

In some ways, the problem might be simpler in working with animals which have selective lesions in the rhinencephalic structures or electrodes placed in the diencephalon. In these animals, certain behavior sequences can possibly be labeled in a fairly unambiguous way. When you deliver a stimulus through indwelling electrodes in the diencephalon of a cat, the cat might hiss, snarl, bare his teeth, and start tearing around. Most observers would agree on how to describe this behavior. Or, on looking at a cat with electrical stimulation in a slightly different diencephalic region, people might agree that the cat acts as if it wanted to devour the wooden table on which it is standing at that moment. But we are off the deep end, if we call that behavior analogous to a human eating compulsion.

Generally, everything becomes more complicated when we attempt to evaluate affective changes in humans. In our series of frontal lobe cases, some are apparently much more irritable than before the injury. Others seem to be rather apathetic, and very, very many seem not to have changed at all. Are they really unchanged?

KUBIE: I think the primary problem is what is meant by the affective state itself. Probably all of us can agree that a certain man is angry or sad or gay and have a usable idea of what we mean. The question of quantification comes up as soon as we ask has anything made him more angry or less angry, more or less gay. Are we going to use as an index the man's self-description of how his mood feels to him; or are we going to use the overt manifestations of his mood? Here we come up against the fact that some people show their feelings, whereas others hide them. Then we ask what about the mood itself. Is this mood some kind of a self-sustaining continuum, or is it due to the fact that man is subjected to a continuous battering, which I have likened to Brownian movement, by scarcely noted stimuli which constantly throw him into emotional states? And can his vulnerability to such a battering, which may be occurring all the time, be altered?

[72] In other words, we can observe an emotion as a state, describe it and | agree on it. Then we can agree that some kind of a change has occurred in it. Probably all would be in full agreement up to this point, until we take the next step and say, in what does this change consist? How has the mood been altered? As soon as we get beyond the descriptive levels, the problem becomes extraordinarily complicated. Yet, we have to devise ways of thinking about this clearly, or at least of describing more precisely the areas in which changes occur. Until we can do this, we are unable to measure any of them.

MONNIER: In other words, the problem is to find the psychologic correlation of physiologic patterns, i.e., to solve the code.

FRANK: What Dr. Kubie said is very reassuring because his statement, »Let's talk about an emotional process,« is a great advance over the usual practice of picking out and reifying different states of intensities. We know we are dealing with a continuum, but before we can attempt quantification, we may have to try to state some of the dimensions, whether or not they can be measured. Are we dealing with acute or chronic situations? Are we dealing with a process that is provoked by the immediate, present situation or by something that happened a long time ago where some feedback process seems to be operating? Are we dealing with a process that is irradiating or is it localized? Are we dealing with an expression that comes out in some sort of overt motor activity, symbolic language, or in somatic visceral disturbances? Is the target of expression another person, a symbol, activities, or the self? It seems to me that we can build up a series of possible dimensions of the emotional process and then begin to see where we can tackle them. I take it that is what you had in mind, isn't it, Larry?

KUBIE: Yes.

HOMEOSTASIS

W. ROSS ASHBY

THE PROBLEM I want to discuss is how the organism manages to establish homeostasis in the larger sense, how the learning organism manages to reorganize its cerebral neuronic equipment so that, however unusual a new environment may be, it can learn to take appropriate action. This approach has the advantage of providing a clear-cut and essentially objective type of problem. We can consider the living mouse as being essentially similar to the clockwork mouse and we can use the same physical principles and the same objective method in the study of both.

I shall consider the organism, then, as a mechanism which faces a hostile and difficult world and has as its fundamental task keeping itself alive. I ask, what sort of a mechanism can it be that can do this in an almost limitless variety of environments?

First, I must say unambiguously what I mean by »alive.« I assume that if the organism is to stay alive, a comparatively small number of essential variables must be kept within physiologic limits. Each of these variables can be represented by a pointer on a dial. They include such things as the animal's temperature, the amount of sugar in its blood, the amount of water in its tissues, and, say, the pressure on its nose, for an animal that runs fast must try to run in such a way as not to get intense pressures suddenly occurring on its nose. The limits are given as soon as the species is given: a cat must keep itself out of water, a fish must keep itself in water. It is not the cat's business to think about wetness but to keep itself dry. In the same way, I shall assume that our organism, whatever it is, is simply given certain limits, and it will be judged according to whether it is successful in keeping the essential variables within these limits.

This, of course, is in no way peculiar to living organisms. An engineer, sitting at the control panel in a ship, has exactly the same task to do. He has a row of dials, some of which represent essential variables in the ship, and it is his business to run things so that the needles always stay within their proper limits. The problem, then, is uniform between the inanimate and the animate.

I assume that the organism has a mass of cerebral equipment to work with; at the moment, I shall not go into what that may be. All I can assume is that information about the state of the essential variables can be sent into the neuronic network. To make the situation real, we must not talk vaguely about an organism with an input and an output and leave the environment unspecified. Our question is how the organism is going to struggle with its environment, and if that question is to be | treated adequately, we must assume some specific environment, which can be as complicated as you please, provided it is definite. Let us assume, then, that a definite environment has been provided and that the organism now faces it.

The relationship between organism and environment has next to be made clear. It is apparent that they act on each other in various ways. The arrangement that I shall consider is the one which seems to me to be the most interesting because it is the most serious; it is the case where the environment has a direct action on the essential variables, where the amount of water in the body, say, is affected by the steady drying effect of the climate. If the organism does not behave suitably, it will pay the penalty of its inefficiency by being killed.

Next, we must allow the organism some way of acting on the environment, so some channel from brain to environment is needed. These may be arms and legs; or, if it is a baby, it has a voice to cry with; or a dog has its teeth to bite with; and so on. The organism, then, with its effectors and receptors, forms, with the environment, *a system with feedback*.

The problem now becomes capable of being stated with precision. Consider the environment as a transducer, as an operator that converts whatever action comes from the organism into some effect that goes back to the organism. Let the environment be represented by the operator, E. The organism's problem is to convert its brain into an operator, which might be represented by E^{-1}. It must be the inverse operator, in a sense, to E, because if a disturbance, starting at some point, throws the essential variables off their proper values, by the time the disturbance has been around the circuit the effect must be negative so as to get the inverse change coming back to the essential variables.

GERARD: May I interrupt a moment? It seems to me that you have changed the meaning of the variables. I thought you spoke of them first as the readings of some of the physiologic constants, such as the blood sugar level, on which the environment need not act directly. You now have the environment acting directly on these, as if they were receptors of the organism.

ASHBY: If you like, we can add further channels of sensory type, which represent a simple informative action from the environment to the brain. That is, as a matter of fact, merely helping the organism generally.

GERARD: O.K. I just wanted to be sure I wasn't getting mixed on it.

ASHBY: I want to have these remain direct effects from environment to essential variables because they make the organism's problem real and urgent.

[75] This formulation with E and E^{-1} is merely meant to state the problem | in a form that is clear physically and mechanistically. One of the things that I would like to ask the members of this group is whether it would be possible, assuming these dynamic networks to be of an extremely general type, with discontinuities and delays whether E and E^{-1} can be specified with sufficient generality to enable the same symbolism to be used rigorously? If that could be done, it would be most helpful, for it would enable the somewhat loosely defined biological problem to be replaced by a rigorously defined mathematical and physical one.

WIESNER: I would be worried about the inverse operator. This is the most disturbing thing.

PITTS: Yes.

ASHBY: I don't mind in the least if the inverse is not unique. That merely means that the organism can solve its problem in more than one way, which is very common.

WIESNER: But this is the difficulty. This is obviously not a neat solution.

PITTS: Certainly, the inverse need not exist.

ASHBY: True enough. Animals discover that sometimes.

WIESNER: But it presents a mathematical difficulty.

ASHBY: The chief thing that I am interested in is the question, if this system is grossly nonlinear and if there are delays – so that one effect goes through quickly and another through slowly – can this operator E be given a meaningful and rigorous form?

BIGELOW: In general, no.

ASHBY: No?

PITTS: Well, presumably, it is a physically existing operator. What the environment does is certainty realizable, because it is realized. The question is what the organism can realize.

BIGELOW: It obviously has to know about the operator.

PITTS: No, his operator, E, is certainly specified. It is whatever happens to be found in the environment. But the question as to whether you can invert it or find an approximate inverse is the important question.

ASHBY: That is it. I want to know whether E can be specified precisely. E^{-1} is fairly straightforward; if E^{-1} does not exist, the organism cannot find a solution and it dies. If it does exist, then the animal may perhaps find several solutions. But what I want to know is whether E can be given a fully rigorous definition.

WIESNER: That is an operation we would agree could be done.

PITTS: It is certain that it could be.

ASHBY: You think it could be?

WIESNER: By definition, yes.

YOUNG: Is this a question put to the biologist or the mathematician?

ASHBY: The mathematician. | [76]

PITTS: Suppose the output of the physical system depends upon the input and always, certainly, upon the past. There are ways of specifying the dependence to any arbitrary degree of possible approximation.

BIGELOW: I think you started out by saying that the environment consisted of everything under the sun.

ASHBY: Assume a closed environment, which we suppose to have only a certain number of variable parts.

PITTS: He is concerned with a compartmentalized environment, set apart from the rest of the environment.

BIGELOW: Well, let's see what this is supposed to do before we go ahead.

ASHBY: It looks as though it might be possible to give this concept a rigorous definition; in which case, we can restate the organism's problem: given E, the organism's problem is to find E^{-1}. If E^{-1} is multiple, it must find one of them. If it solves this problem, it lives; if not, it dies.

Suppose we approach the question by trying to make a machine that will do just that. On what principle must it work? The one thing that we must not do is to know beforehand that we are going to present it with E_1, E_2, and E_3, and then build into it E_1^{-1}, E_2^{-1} and E_3^{-1}, with a switch to throw it from one to another. If we do, the model is still meaningful, but it means that we have made a perfectly good insect that has three reflexes for three situations. That is not what I am discussing. What I want to consider is the problem that faces learning organisms, the mammals particularly, the ones that have the power of developing an adaptive reaction to any one of an almost unlimited number of environments.

The fundamental problem is one of organization, of finding the appropriate switching-pattern. Clearly, the instructions for what is appropriate must come, ultimately, from the environment, for what is fight for one environment may be wrong for another. The problem is how the information from the environment can be used to adjust the switching pattern. What the organism needs is a system or method which, if followed blindly, will almost always result in the switching pattern changing from »inappropriate« to »appropriate.« I have given the reasons for thinking that there is only one way in which this can be done (1) (2). The switching must be arranged, at first, at random, and then there must be a corrective feedback from the essential variables into the main network, such that if any essential variable goes outside its proper limits, a random, disruptive effect is to be thrown into the network. I believe that this method is practical with biological material and is also effective, in the sense that it will always tend automatically to find an inverse operator, an E^{-1}.

[77] WIESNER: Can't you go one step further? Maybe you won't want to | accept this, but instead of having one operator, E, and having one inverse, E^{-1}, you might have available from the environment a whole series of different operators. The inverse operator may turn out to be a randomly found sum of the inverse operators, which somehow is searched to minimize the error. In other words, what I am saying is that, instead of searching randomly over all possible combinations of information that the neurons might generate, we have a smaller number of reflexes. No one of them may be a sufficient reflex or reaction to get one out of difficulty, but the learning process consists of randomly trying these combinations. If one is successful, one survives.

ASHBY: Oh, yes.

BIGELOW: You must postulate a combination, anyway.

MCCULLOCH: What sort of a randomly connected net is inside the box?

ASHBY: I wanted to stop anyway at this point so as to discuss whether the simple method that I have proposed is basically sound. The homeostat I have built suggests that it is (3) (4). But even so, I must, of course, at once admit that there still remain great difficulties. The chief one is simply that of time. The method proposed works quite well when the variables are few, but when they become many, the time that the method is likely to take may well exceed all possible allowance, even all astronomical duration (1). The problem then becomes: if the environment and organism are both large and complex, is there any way in which this basic method of adaptation can be modified so as to produce adaptation in a reasonably short time? I think there is (1), and I shall try to sketch the main points, though I'm afraid I shall not be able to treat the question adequately in the time at my disposal.

A fundamental property of any random network is the extent to which the variables are likely to be constant, unvarying. If we take any dynamic system, a solar system or a telephone exchange or a brain, select a variable at random, record its behavior over a time, and then examine the record, we will find that the behavior falls, apart from mixed forms, into one of four classes. It may be a full-function, varying all the time, like the scalp potential recorded by the EEG; it may be a part-function, varying part of the time but constant over finite intervals, like the potential of a telephone line that is being used intermittently; it may be a step-function, like a relay in a telephone exchange; and, to complete the set, it may be a null-function, unchanging throughout the whole period of observation. In the ordinary way, these constancies tend to be lost sight of, but when one comes to consider a neuronic or similar network where the numbers are statistically large, they become important. Their importance lies in the fact that if any variable, for any reason, goes constant for a time, then during that time [78] it cannot transmit information. Variables that are temporarily constant | act temporarily as barriers to the flow of information and control. As a result, if the network contains many part-functions, it will tend to be cut, functionally, into subsystems of temporary individuality. And if such a system is disturbed, the disturbance affects only some of the subsystems.

In discussing complexity, wondering how long it would take a system with, say, 10^{10} parts to get adapted, we tend to think of the 10^{10} parts as being all full-functions. Such a system would, it is true, be extremely active and complex, and it would also be very difficult to get properly adapted, but I suggest that we are proposing more difficulties than actually exist. In fact, neither the ordinary world that provides our environment nor the brain that reacts consists wholly of full-functions; both contain a large proportion of part-functions, and this fact makes the achievement of adaptation much easier. Two factors in particular tend to make many of the variables behave as part-functions. The first is that where »threshold« properties are common, so will part-functions be common, for if some continuously changing variable drives another variable through a threshold, the second variable will be constant during all those intervals in which the first variable is below the threshold. The second factor is that any variable that is »amplitude-limited,« that has a range it cannot exceed, will be constant while it is acted on by forces that have driven it to its extreme and are tending to drive it further. We may expect, therefore, that part-functions will be common in both organism and environment, and this is confirmed by observation.

If the environment contains many part-functions, simplicities will occur, because, instead of the environment being wholly joined up, like a jelly, so that every variable affects every other variable, the variables will be joined into loosely coupled sub sets, each sub set being smaller and simpler than the set of the whole environment. If that happens, the animal is lucky. It is in an environment that is not too difficult, and it can get control of it by forming one simple operator to control one part and another simple operator to control another part, and so on.

To give an example showing the two types of environment, and which is illustrative of how the one is difficult to control and the other easy, in the early days of the cathode-ray oscilloscope, if you tried to make the spot on the screen brighter, it moved over to the right, and if you tried to make the sweep slower, it went to the top of the screen, and so on. Everything interacted with everything else, and the person controlling it found it a highly complicated and difficult system, precisely because the properties of the spot were all closely and actively related to one another. Since then, the designers have worked on it and now the cathode-ray oscilloscope is reduced functionally to about a dozen inde|pendent variables. A knob is turned for brightness and [79] nothing but brightness alters; a knob is turned for moving it to right or left and the spot moves to right or left without altering in brightness. If you don't look inside the box, if you treat it simply as a black box with certain functional characteristics, you find it to be equivalent to a dozen or so functionally independent parts, and it is precisely this independence that makes the modern cathode-ray oscilloscope easy to control.

WIESNER: I don't see how functional independence is guaranteed by the fact that the function exists for only part of the time. It has the value, 0, the rest of the time.

MCCULLOCH: That isn't necessarily so. It would mean, let's say, that throughout a certain range of change in brightness, the deflection is not being interfered with.

WIESNER: I understand that model, but I don't see that it is representative of the different types of functional environment.

ASHBY: I am using the cathode-ray oscilloscope as an illustration, simply to create a practical picture of what is involved in trying to control the one type of environment or the other. Suppose your life depended on your tracking a moving point while you were at the controls of the cathode-ray oscilloscope, so that you had to make the point follow some given trajectory, with the old type, your life would be in great danger.

WIESNER: Yes, I realize that.

ASHBY: With the new, your life would be safe.

WIESNER: In talking about the random net, I thought you took advantage of the property that these things were conducting actively only a very small part of the time to postulate some functional independence. This may be all right within the network, but I refuse to accept the environment as behaving in this way.

PITTS: I am a little mixed up about the variables which are active only part of the time. This is just a matter of terminology. You talk about the three possible ways in which the variables behave, but there are really several sets of variables in the model. Firstly, there are the variables that the environment actually puts into the dials; that is, what the dials receive from the environment. That, presumably, depends on some operator, which, in turn, depends upon two sorts of things: upon the inputs that it has received from the organism itself, and also upon additional variables from within the environment which are not at all subject to the organism's control in any way. Of course, it is an important question to what extent the output of the environment into the dials depends upon the input from the organism itself and to what extent it depends upon extraneous environmental factors. Now, what I am not clear about is [80] whether this classification of the modes of varia | tion of the variables applies to, well, first of all, to the actual output of the environment to the dials, that is, what the environment provides, or, secondly, to the extraneous factors in the environment that are independent of any control by the organism.

ASHBY: I shall deal with the extraneous factors first. In any given experimental situation, they must be kept under control. Otherwise, one is conducting an experiment with some unknown, arbitrary interference going on; it would be rather like trying to demonstrate an experiment while interfering bystanders are playing with the apparatus.

PITTS: But, you see, there is no problem. If there are no extraneous factors which the organism does not control, then he has nothing to adjust to.

ASHBY: Oh, yes, he has. If he gets his nerve-network wrongly patterned, he will make the thing unstable, in which case the essential variables will all run away, in spite of everything he does.

PITTS: But it is the values of the external inputs that set the problem and change the required behavior, so one should probably introduce those explicitly, don't you think?

ASHBY: Shall I put it this way: A switch can be put in the environment and held constant so that the environment has some characteristic, and you can then say, »Let's see what the organism does about it«. The switch is in a fixed position during the whole process of the adaptation. The organism works out the opposite pattern in its brain, and thus gets its essential variables within their proper range. If the system is then slightly disturbed anywhere, being stable, it will oscillate and come back to within the proper range. That is stage one. Stage two consists of pushing the switch over; that is, making a change of parameter that is equivalent to taking the first environment out and putting in another.

PITTS: Well, of course, the properties of the organism depend upon how rapidly the properties of the environment are changed around, and I wasn't sure whether your classification referred to the rate at which the environment was changed or, possibly, to variables within the organism.

WIESNER: What you are really asking is, can the organism find a stability condition for any given environment. You are ignoring the dynamics of the environment, which I think is fair, but then I don't see why you ascribe to your environment any time parameters, which I understood you were doing.

BIGELOW: He never discussed explicitly the question of dynamics.

ASHBY: What I am suggesting is that many of the variables in the environment are such that if their behavior is recorded over a period of time, it will be found that dur- [81] ing some of that time each has been constant, has been unvarying. |

WIESNER: If you put on a recorder, you are implicitly bringing time, but you previously ruled out time automatically.

PITTS: If these are measurements taken from the environment and the environment is kept fixed, the only thing that is varying is the organism's input to the environment. What these variables are must be a function of what the output of the organism is, in effect.

ASHBY: I am assuming that watching the organism and its environment is an experimenter who, when the whole system seems to be settled, can prod it gently and make it display something of its behavior. The experimenter provokes it into showing whether or not it has learned its trick.

With regard to the part-functions, I am not trying to demand that the environment shall consist of them only. What I am saying is that if the environment consists of variables, whatever they are, they are continuously fluctuating.

BIGELOW: Are there independently varying potentials existing in your environment?

MCCULLOCH: Independently in what sense?

WIESNER: He has frozen the environment. All that can vary is the R in each element of the environment.

ASHBY: I am assuming that while the experiment is going on, while the rat is running round the maze and trying to find its food, it must be left alone. The whole system must be isolated, closed. Then environment and brain can, as it were, fight it out. Each will affect the other. The effects will go round and round, What one looks for, if the animal has learned successfully, is that, whereas at first the output from the brain is of such type that it acts in the wrong way, later it becomes of such type as will bring the essential variables within the proper limits and keep them stable. If the experimenter then gives it a slight disturbance and follows the behavior around, he will find that wherever the system is disturbed, the changes as they are fed round will act so as to keep the essential variables within their limits. If it does that, the organism has learned to keep itself alive.

What I am suggesting is that a system with this corrective feedback will so rearrange itself that it will get itself into such a switching pattern. Then, if in environment and organism there are many variables that, for any reason, lock under certain conditions so that constancies are common, the process becomes much easier; such an environment is much easier for the organism to get a grip on, and the brain can much more easily develop an appropriate switching pattern.

PITTS: You are really considering two sorts of cases, more or less interchangeably: the experimental case, where you plant the environment and see what the organism does, and in that case, of course, you have | secured the relative constancy of those variables; [82] and then, also, more or less alternatively, you are considering the case of the organism in its natural environment, in which there can be variables independent of the organism, variables which may shift just as rapidly as in your first case.

ASHBY: If these other variables do shift rapidly, you may well get an environment that no organism can adapt to. Such would happen if, while a rat was learning a maze, you were to keep moving the food from place to place; no rat could possibly learn to find it under those conditions. If changes are to be put into the system, it must be done with a slowness that is of a different order of slowness from the order of speed of the learning animal.

WIESNER: That is one problem, that these variables do not change the speed of learning. But the second thing which you are actually implying is independence. An environment in which a solution could be found if a second effect didn't occur is conceivable. For example, if you have too much water, you can undertake to evaporate it very rapidly and in the process burn up. That has nothing to do with the time rate of change of the environment. I think you are saying a second thing, that the elements of the environment do not interact, or the effects of your feedback on the environment do not interact with each other.

MCCULLOCH: I think what he is saying is that there must be some ranges in the values of some of the variables within which they do not significantly affect other variables, but that those variables are intermittently constant throughout, one no less a function than the other.

PITTS: That is not necessary generally. It is necessary only for his theory.

ASHBY: The case I have drawn here is one where the environment is peculiarly simple. I was not intending to suggest that such a brain can deal only with environments that are cut into parts, because that is one of the crucial problems: how to deal with an environment that is not cut into parts. What I am suggesting is that there is a form intermediate between the environment in which everything upsets everything else and the environment which is cut into parts. This intermediate type of environment is common and is of real significance here; it is an environment that consists of parts that are temporarily separable, and yet by no means permanently separable. For instance, at the moment, the properties of this piece of chalk I am holding go on, as far as its writing power is concerned, more or less independently of what that eraser is doing; they are to some extent independent. If I move the chalk, the eraser does not immediately burst into flame. If it did and I had to deal with an environment of that crazy type, the whole business of getting control of what was going on around me would be much more difficult. Fortunately, the environment is not always as complicated as that, and I [83] | think it is that simplicity which enables the living organism to take on peculiarities and get control of them.

If you went over Niagara into the whirlpool below and you tried to swim, everything would be on the move; everything that you did would upset something else, and you would find it extremely difficult to get any sort of control.

Another example that will show the difficulty is the game that consists of a box with a glass lid and some objects inside that have each to be got, by manipulating the whole box, into its appointed place. Suppose that the three objects inside the box were actually three living mice so that every action of yours affected all three and each affected the other two. This would be what one may call a class I environment, an environment where every variable continuously affects every other variable, and you would find it very difficult to control.

The ordinary puzzle is class II and is rather simpler. Suppose it has three rings that have to be put on three pegs. If a ring is at one end of the box, you can rotate the box about that ring as axis so as to make the other two move while the first one keeps still. If you watch a person solving the problem and record the positions of the rings, you will find that the rings move intermittently; quite apart from any reason why, it is a fact which can be observed. The possibility that each ring can be sometimes moved and sometimes kept still is something which the solver can take advantage of.

Class III environment would occur if you cut the puzzle into three pieces with one ring and one peg in each sub-box. It can be solved, or controlled, very easily, for each ring can be treated independently. The problem has now become so simple that a simple organism could control it. The problem, as environment, can be said, mathematically, to be »reducible,« for it can be treated as three environments, each one a third the size of the whole. By being cut into three, the complexity of the whole is much reduced; for, though it is difficult to be accurate, the complexity tends to depend on the number of parts as its factorial. If any breaking into pieces is possible, an enormous simplification can occur.

What I am suggesting is that the sort of environment in which we live on this earth, the sort of environment that the living organism has to deal with, is largely of the intermediate type in which the parts are connected only intermittently. Consider the ordinary automobile, for instance; a movement of the brake pedal doesn't make any difference to the steering. The very fact that there is that simplicity, that possibility of temporarily regarding the braking and steering as functionally independent, introduces a simplicity that the beginner can take full advantage of in order to learn how to drive the car. | [84]

BIGELOW: Our environment doesn't consist of that sort of phenomena in very many ways. For example, in learning to ride a bicycle, if you fall to the left, you must learn to steer to the left. The statement you are making amounts to something which mathematically sounds like independence but does not exist in the real world, and one of the phenomena that biological organisms must do is to learn to find their way around in a world in which variables are not independent.

WIESNER: But on a given time scale, there is a certain amount of independence.

YOUNG: When the carnivore catches a piece of prey, it doesn't immediately alter the supply of prey. It ultimately may do so, but there is a certain constancy. Is that the sort of thing you mean?

ASHBY: Yes. When a carnivore catches some prey, for the moment that does not depend on the question of where its drinking pool is and where it will get a drink later.

BIGELOW: Oh, it certainly affects the relative position of the drinking pool. If he has to chase his prey, he has to go away from the drinking pool. His proximity to the prey and to the drinking pool are connected in a related fashion.

YOUNG: There are connections, but they are not significant for the experiment.

ASHBY: Certainly.

WIESNER: Not for this animal, anyway.

ASHBY: Such an example does not imply that the environment is always connected; there are lots of ways in which it demonstrably is not. For instance, while you are speaking, you don't have to bother about the glass of water in front of you. If you want a drink later, that comes in as an entirely independent question. Sometimes the variables are independent; sometimes they are not. I don't beg the question or prejudge the issue in any way. What I do say is that if the variables are closely linked, which is not a common physical situation, then the environment is complex and difficult. I don't say for a moment that the system with corrective feedback will always solve a problem. Such a system, against such an environment, will readily fait; and so do living things, when confronted with the same situation. But this type of environment is rare. More common is the environment in which the variables are partly linked, partly independent, in which, very often, the linkage and the independence are temporary. The organism has some chance of picking out combinations that are almost independent, and can thus simplify its problems. Consider, for instance, putting on one's shoes and getting the knot properly tied, and then going to work at a bench. What you do at the bench and what you do with your shoes are so separate that they can practically be treated as two independent | activities. The organism can seize on that almost complete separation to avoid having to work with one hand at the bench and another hand on its shoes, which would be the sort of thing that would be necessary if every variable interacted with every other variable. [85]

BATESON: Would it be fair to say, for the sake of illustration, that an animal in a natural environment deals with variables of the class 2 type mainly; whereas, we in this room, who have words and communication, especially our chairman and yourself, are dealing with an environment with many variables of the class 1 type?

ASHBY: Yes, I think so. Thus, the very fact that a number in this room are silent means that they are, in effect, constant, which means that the two who are talking have some chance of arriving at a stabilized arrangement. But if everybody tried to talk at once, the whole system would be fluctuating.

McCulloch: Some of the members have thresholds that things have to get over before they join the discussion.

Hutchinson: Actually, that is the way the environment presents itself, as a class 1 environment, unless the organism has some sort of screening device which prevents everything except a really big oscillation from having any effect. When you speak of going over Niagara, nearly all animals live in comparable turbulence.

Gerard: Let me ask something else that is bothering me. I don't know what your further derivation is going to be, but you obviously are going to provide your organism-box with a set of switches so that it can make a proper response under another type of environment. What is going to happen to your organism-box when you do keep moving the cheese around in the maze so that the mouse never gets it? In reality, of course, the organism will not remain in the same random state in which it started.

Ashby: As it stands at the moment, the specification is so simple that it is only beginning to touch the real biological complexities. As a formulation, this concept of corrective feedback is nothing more than a bare skeleton.

Pitts: Maybe we should see what the organism can do before discussing further what it cannot do.

Young: And you are limiting yourself to fairly short periods of time, aren't you?

Pitts: Yes, he is.

Young: It immediately raises the question of longer periods of time, which are adjusted by other methods, genetics, population, and so forth.

Ashby: I am thinking of the sort of thing that occurs in an ordinary learning experiment that lasts long enough to give the organism a chance of dealing with the situation, without assuming that the condi|tions hold forever. [86] I want to show that, given some definite environment, such a brain with corrective feedback can, and will, adapt to it. This adaptation will be useful and life-saving. If later the environment is changed in some semipermanent way, the same powers of adaptation will lead to it adapting anew.

Klüver: Your scheme should be of a particular interest to psychologists because, unlike other models we have encountered here, it stresses environmental factors and the interrelations of animal and environment. I did not quite follow what was implied in your statement that constancies may function as barriers.

Ashby: If a system is composed of variables that often get locked constant, it tends to cut the system into functionally independent subsystems which can join and separate. Instead of being a totally interlaced system with everything acting on everything else, it allows subsystems to have temporary independencies.

McCulloch: It might look a good deal like the game of croquet in *Through the Looking Glass,* in which the wickets got up and wandered around. But while the wickets move, the animal has a chance to shoot at them. Is that it?

Ashby: Yes.

McCulloch: So there are periods in which subsystems make sense temporarily.

von Bonin: Is it quite as important at the moment to talk about the characterization of the environment? Isn't full-function, part-function, step-function, and null-function becoming more important in the construction of the »brain«? The »brain« picks out certain parts of the environment, say, the shoe or the bench, in sequence and deals with that. Couldn't the environment be left more or less unspecified and the concentration be on how the »brain« works? Isn't that the crucial thing?

Ashby: In my opinion, in this sort of study, there can't be a proper theory of the brain until there is a proper theory of the environment as well. The two work together. What must be discussed is not what the »brain« will do but what »the system« will do,

»the system« being the brain and the environment acting mutually on each other. If the environment is left out and only the brain discussed, you find that something is missing. You can't tell whether the brain is doing the right thing or the wrong thing. Without an environment, »right« and »wrong« have no meaning. It is not until you say, »Let's give it E_{47}; let's see what it will do against that,« it is not until you join the brain on to a distinctly formulated environment that you begin to get a clear statement of what will happen.

My opinion is that the subject has been hampered by our not paying | sufficiently serious attention to the environmental half of the process, because the system is a whole and only the whole clear-cut properties. If the variables tend to go constant, for instance, the whole tends to break temporarily into more or less functionally separate parts. But that is a statement about the whole, and it is only afterwards that the effects can be allotted more specifically to the nervous system or the environment. [87]

VON BONIN: To pick up a remark which Klüver made a moment ago, the animal appears to break down the environment into certain patterns which seem to develop in its mind. Something goes on in its brain and then that structures the environment. It perceives the environment in a certain pattern, set by its brain, so that it can deal with it.

ASHBY: I assume that the brain works entirely for its own end.

VON BONIN: Oh, surely.

ASHBY: It doesn't work on an input that we give it, and it doesn't give an output back to us. It simply works for its own ends. It must be given something to work with; otherwise, it has an input with nothing supplied. This means that we must treat the organism's environment as important and worth considering. I think that the »psychology« of the environment will have to be given almost as much thought as the psychology of the nerve network itself.

KLÜVER: A psychologist might insist that the science of animal psychology is concerned with nothing but the determination of the psychological structure of the environment or, differently expressed, with the specification of the behaviorally effective properties existing in the environment of a given animal. A behavior analysis may show that an animal handles the complexities of the environment by way of developing certain constancies or, to put it another way, by becoming incapable of responding to certain variations and changes. Or it might be said that the environment, from the very start, has a certain framework and is structured in terms of constancy phenomena. I wish you would again enlarge on your idea that a given constancy may serve as a barrier.

ASHBY: Suppose there are two armies joined by a telegraph wire; you want to stop communication between them but you have no way of cutting the wire. You can stop all communications absolutely by just fixing a point on that wire at a constant potential. If it is held at that potential, no communication can go through. Constancy is sufficient to stop effective functional linkage between the two systems.

KLÜVER: Behavior implies constancies which, in turn, imply the ability to appreciate certain kinds of relations, fluctuations, and inconstancies.

ASHBY: Here is an example from physics. Suppose you wanted to study heat conduction and you wanted the system to be thermally isolated from its surroundings. You don't try to make a »nothingness« of | heat around it: you put a wall of constant temperature around it. That is a functional isolation. Even though there may be solid matter all the way through the structure from the inside to the outside, a wall of constant temperature creates a functional isolation. The principle applies very widely; it has obviously something to do with information. [88]

BIGELOW: You stated the environment as E_{47}. You say if you put an environment, E_{47}, in it, then you can draw certain specific conclusions. Will you please continue and draw some conclusions for us?

QUASTLER: May I ask one question before you do so. Does your E^{-1} include the previous E? What the animal is going to do depends on its state and its state depends on what environment there was before, doesn't it?

ASHBY: All I assume about the environment is that it is determinate.

WIESNER: Let us assume we have a brand-new animal, so he has no history in the box.

ASHBY: In the brain?

WIESNER: Yes.

ASHBY: It doesn't really matter.

MCCULLOCH: You could feed him a good deal of random information for a while.

WIESNER: Give him shock therapy or something.

MCCULLOCH: Yes.

ASHBY: The system with corrective feedback is intended to work on nothing but success and failure, and failure means that the pattern of switching will be disrupted, so that if the organism has the wrong ideas, the wrong ideas will be knocked out.

PITTS: How is that going to be done? We are in the position of discussing the proper pay scale before we have the employee produce.

MCCULLOCH: All he has to do is describe the gadget that has the properties.

PITTS: It is not a question of a gadget. Anyone can make a gadget. I thought there was some gadget that was particularly efficient in dealing with the environment. Wasn't there a scheme involving random networks that did it particularly efficiently?

ASHBY: Only in a relative sense of the word. The question of developing the most efficient cerebral mechanism is complicated and difficult. It must be, otherwise the brain would have been developed and perfected much more quickly in evolution. As it is, it has taken the zoological world an enormous length of time to build up its nervous system and that suggests that all sorts of constants and parameters have had to be tested and adjusted so that they had values that were optimal for the average environment of the species.

[89] YOUNG: Can you give us some conception, though? |

BIGELOW: Yes; what do you learn from this?

WIESNER: You implied at the beginning that this would be better than taking all possible combinations of the total number of elements in the brain box. That is what we want to see.

ASHBY: One way of getting variables that are sometimes fluctuating, sometimes constant is to use the thermionic valve, for if the grid goes too positive or too negative, it clamps. Suppose then that the brain, or our model of it, used valves. Even so, we would still have to find the optimal conditions. Suppose that the working conditions were such that the valves were very seldom constant; then the machine would be very active but very difficult to stabilize. If, on the contrary, the valves were nearly always clamped, the system would be easy to stabilize but would be mostly inactive and simple in its behavior patterns. Which is better cannot be decided until the environment has been decided.

GERARD: I should like to prod you this much further. Given a sufficiently stable, reasonable environment that the organism can react to, it will develop adaptive responses and solve it. What is the picture you will give us for a situation where the environment is erratic, or unreasonably changed from the point of view of the organism, so that instead of developing adaptive behavior, or remaining random, the organism develops a very definite disadaptive behavior; in other words, it gets a neurosis?

McCulloch: By »neurosis,« you mean an undue constancy?

Ashby: The concept of a neurosis introduces ideas that are rather more advanced than anything that has been put forward so far by me.

Gerard: You don't want to go that far?

Ashby: What I have proposed so far does not contain sufficient complexity to allow the system to develop a neurosis.

Bateson: If it should evolve a recipe about how to try, it would then have a neurosis, wouldn't it?

Ashby: It could, yes.

Wiesner: How *is* your »brain« specified?

Ashby: The brain, natural or mechanical, is specified only in that it shall be filled indiscriminately with switch gear, that it shall have essential variables provided, and that a disruptive feedback shall come from the essential variables to the network, and act when, and only when, the essential variables exceed the given limits. That is a comparatively small specification.

Wiesner: What feature changes the searching that this system would do from complete randomness, which is what I understood you to say would happen?

Ashby: The point you are raising is a difficult one. The question is: how does the organism manage to cut short its searching from eternity | to something less? What I [90] am suggesting is that if the variables in the brain are of the type where there is a lot of constancy, the system tends to be much more easily stabilized. It is chiefly a question of stability. A question I would very much like to have answered is: what is the probability that any randomly assembled system of n variables should be stable?

Pitts: Probably very small.

Ashby: The question, as far as I know, is unsolved. A useful linear equivalent, merely an approximation, would be: given a matrix with elements taken at random from a known distribution, what is the probability that all the latent roots have negative real parts? The evidence suggests that, as n tends to infinity, the probability goes rapidly to zero. The important case occurs when the distributions are symmetrical about zero; then the probability is somewhere about $(1/2)n$ (5), so if n exceeds quite a small number, the probability of ever getting a stable system tends rapidly to zero.

Bigelow: So big brains are unstable?

Ashby: They would be.

Wiesner: The big brains start in a completely random state?

Ashby: They would be impossible to stabilize if their variables were all continuously variable. But because so many variables in the brain are intermittently constant, it breaks up; you no longer have to deal with 10^{10} variables, all in full effective connection with one another.

Young: Can you now compute the probability on a given set of assumptions? That is rather what we want to know.

Pitts: You can do it on some; if the variables are evenly distributed, you can break down the distribution.

Ashby: These considerations show that the actual number of working parts in the brain that take part in any reaction must be quite small at any given moment.

McCulloch: Because they are grouped in temporarily isolated units?

Ashby: Yes.

Bigelow: You mean only relative to the entire number there, which is 10^{10}? Do you mean that the number is absolutely small?

Ashby: No, much smaller than 10^{10}.

Bigelow: Well, like what?

Ashby: Oh, ten to a hundred, shall we say.

Bigelow: That is a big number of combinations though.

Pitts: The conception of stability in the case of a network which is made up of all the elements which are automatically amplitude-limited is somewhat difficult because the definition is rather arbitrary. I can cite a few facts about the cortex, for example; [91] namely, if it is unanesthe | tized and essentially cut loose from anything else, and it is given a big bump in one place, it will spread a wave of near maximum activity almost to the other end. In that sense, it can be termed unstable, in the sense that if a large enough initial stimulus is put in, the result will simply be maximal activity. However, what happens is that this is followed by a period of profound depression. The maximal activity can't be kept up indefinitely. The phenomenon of fatigue begins to come in. Therefore, if the proportion of the cells that are active during a given short time in response to a given stimulus are considered, then the cortex might be called unstable; but if you consider long-term averages of cortical activity, it would have to be said to be stable simply because, of course, the elements in the network will not fire off as often as is theoretically possible.

Bigelow: The stability which you now speak of is primarily a function of refractoriness, which has not so far entered our discussion. Otherwise, I completely agree with what you say.

Pitts: I would say the cortex is unstable in the sense that, in a short time, a small disturbance will grow indefinitely.

McCulloch: In the unanesthetized animal, yes.

Pitts: But, certainly, considering the rate of activity on the average, over any given length of time as a function of an input which has a given mean value over that period of time, it is almost certainly stable.

von Bonin: I am not sure we have to think of units which are actually single nerve cells. Couldn't it be clumps of nerve cells? Couldn't it even be four or five centers?

Pitts: I was taking a certain statistical variable, namely, the proportion of cells in a small region which are firing at any given moment, or in some small period of time.

Ashby: I agree with Dr. von Bonin, for I do not prejudge in any way the question of what the units are in this functional system. I am prepared to find that in some actions the functional parts may be as small, perhaps, as a protein molecule, and in other actions perhaps as large as a complete nerve center acting for the moment in unison. The question of what the learning mechanism corresponds to in terms of practical physiologic details is one that has still to be solved. I only want to leave the question as wide open as possible at the moment.

Klüver: What seems to me so attractive about this approach is the fact that we do not have to worry for the time being about the question of how certain neuroanatomical and neurophysiological details fit into the picture since this approach is chiefly concerned with the relation between animal and environment. We do not have to worry about the problem of what 10^{10} neurons or any number of neurons are doing, or even [92] about the question of whether any neurons are involved at all. |

Some of you have undoubtedly seen the beautiful motion picture films made by Dr. Pomerat of the University of Texas which illustrate contractile elements in brain tissue cultures. There seems to be little doubt that some of these cells with rhythmical pulsatile activity are oligodendroglia cells (6). A famous physician, Schleich, once wrote a book *On the Switchboard of Thought* (7) in which he tried to explain all psychological phenomena on the basis of neuroglial activity. He insisted that thinking, forgetting, remembering, imagination, action, etc., can be understood only by reference to the contractions of »neuroglia muscles.« In view of the facts recently reported by Pomerat, the question must be raised whether Schleich's glia switchboard is merely a wild idea or deserves more serious consideration. As far I am able to see, Dr. Ashby, you are not prejudging the question whether neurons, neuroglia, or some other elements will ultimately turn out to be the most important units for psychology.

ASHBY: No, I am not.

BIGELOW: Are you able to draw any more specific conclusion from this model than that you believe the number of nerve cells acting at any time is small?

PITTS: Or number of glia!

MCCULLOCH: Number of components, actually.

BATESON: If the changes in the environment are achieved by chance, by the box being made to throw the dice, e.g., by a noise phenomenon introduced by the dials, and if these changes include an increase of the number of parts and complexity of the box, is it possible to predict that the brain so evolved will have a great deal of vicarious function? For example, if a piece of cortex in it is cut off, is the rest likely to be able to solve problems?

ASHBY: Yes. It is a peculiar advantage of the ultrastable system that it has some ability to develop vicarious function; in fact, one can work out quite easily what it can stand and what it cannot. Suppose, for instance, its effects, its outputs to the environment, were the flexors and extensors of an arm. If they are crossed over, the organism has to do just the opposite of what it did before. This type of system can and will readapt to changes like that (1). The change-over will reverse the action and will probably send a reversed effect to the essential variables. Whereas before the change, the organism was acting like a thermostat, pulling its temperature always back to the optimum, after the change it will become like a thermostat with its parts reversed, so that it develops a runaway. But the very fact that it develops the runaway means that automatically the corrective feedback will throw the switches about in a random way. Such changes can stop when, and only when, the temperature gets back to the center again; | in other [93] words, it can stop only when the brain develops a pattern which holds the temperature stable.

The self-corrective power of this system can be shown most clearly by a comparison with the automatic pilot. The automatic pilot keeps the aircraft stable by acting on the ailerons, so that when the craft rolls a little to the right, it introduces a change which forces the aircraft to roll to the left. The automatic pilot has to be joined to the ailerons with some care because if it is joined to them the wrong way round, any small disturbance is self-aggravating, and the automatic pilot under such conditions will overturn the aircraft. A system with second-order feedback, however, will not do that. If joined on the wrong way, the circuit will be unstable and it will at first behave exactly as the wrongly connected automatic pilot does: it will start to increase the disturbance. But the very fact that it goes outside the normal limits will force changes in its network, random changes that can stop when, and only when, the roll is back to zero again. If a mechanism of this type were to be made into an automatic pilot and if the mechanic asked, »Which way shall I join it on to the ailerons?,« the answer would be, »It doesn't matter; join it which way you like; the mechanism will sort itself out.«

FREMONT-SMITH: Don't you have to make assumptions about how much time lag there is in the environment?

ASHBY: Yes, very much so.

FREMONT-SMITH: You are assuming at this point in the argument that there isn't sufficient time lag in the environment to disturb the situation unduly; that there will be enough response to the organism's input into the environment, enough response from the environment back to the indicator, in time for the organism to recover itself?

ASHBY: Not really. If the environment acts so quickly that the organism is killed before it can react, then it doesn't matter what the brain does.

FREMONT-SMITH: I am worrying about the slow-acting environment.

ASHBY: The slow-acting environments are no problem. If the brain can react much more quickly than the environment, the brain can get a grip of it very easily.

FREMONT-SMITH: But suppose the input from the environment back to the organism is long delayed?

ASHBY: Then you have the problem that faces that organism when, say, it is rewarded or punished five minutes after its action is over; it is notorious that that makes it very difficult for the living organism to adapt itself to the conditions.

[94] FREMONT-SMITH: So, you are assuming that doesn't happen now, for these purposes? |

ASHBY: I am assuming at the moment that that sort of delay doesn't happen, that action by the organism is followed almost at once by the signal back from the environment.

FREMONT-SMITH: Thank you; that was the point I wanted to get clear.

BIGELOW: Dr. Ashby, is this an existing model?

ASHBY: Yes; I'm describing the homeostat (3).

BIGELOW: Well, which object is the homeostat?

ASHBY: The homeostat is the whole thing, organism *and* environment.

BIGELOW: So the whole thing exists; is that correct ?

ASHBY: Yes.

BIGELOW: I would like to ask some specific questions. How do you establish a random network?

ASHBY: By taking resistors with values given by Fisher and Yates' table of random numbers (8). These resistors were put on a uniselector (stepping switch) and were joined so that as the wipers go round, the rest of the system sees a value that is simply what was given by the table of random numbers.

BIGELOW: But what steps the switch?

ASHBY: The essential variable; in the homeostat, a relay. If a variable goes outside its proper limits, this closes the relay, energizes the coils of the uniselector, which pulls it round to a new random number, and it is this random number that determines the polarity and the size of what goes through the mechanical brain.

BIGELOW: Sorry, but I don't see how this gets to be the random network. I now see how you get a random group of resistor values selected by a mechanism, but how do you make a network out of this which is random?

ASHBY: It is random in the sense that it contains values defined only by what is given at some place in Fisher and Yates' table of random numbers.

BIGELOW: How about its connectivity? You have to connect something to something else. How do you make a random network of connections? How is that done?

PITTS: Through a circuit diagram.

ASHBY: The four units of the homeostat are actually joined in all possible ways, so there are twenty possible circuits around. What unit 2 gets from unit 1 depends on the resistor and on the position of a commutator.

BIGELOW: Am I correct in describing this as a network in which there are random values, but put in a fixed geometric fashion?

McCulloch: Yes, they are put in a fixed geometric fashion, but the selection of which of those is going to be hooked in is random. | [95]

Bigelow: This is not a random system. This is a selection of fixed constants in a circuit of relatively simple connectivity.

Wiesner: The network may not always have the same connectivity, but there are a relatively few number of connections.

Bigelow: The combinatorial possibilities are relatively small. You can enumerate them in five minutes, I am sure.

Ashby: There are three hundred thousand.

Wiesner: If you include the resistor values.

Ashby: They are of functional importance.

Bigelow: But resistors are a different variety of animal from a randomly connected net?

Ashby: The units are always connected; how they vary is whether the connection is strong or weak, positive or negative. The values are evenly distributed between plus 9 and minus 9.

Bigelow: Well, it is a fair question, is it not, as to whether this particular model has any relation to the nervous system?

McCulloch: In what sense?

Bigelow: Well, you see, you don't have a random network. You have various kinds of networks of certain random values produced at certain points.

Wiesner: The network is looking for a variable which matches the outside environment among a group of random values which are possible in a rather definitely connected network. This is very different from having a large number of possible combinations.

Bigelow: It may be a beautiful replica of something, but heaven only knows what.

Bowman: Do some of the resistors have extremely large values that would amount practically to an open circuit?

Ashby: In fact, the resistors shunt the current, so a low resistance means that one unit has little direct effect on the other. Zeros occur, and when this happens the units are, in effect, cut off from one another.

Bigelow: One more question. How are the feedback loops introduced in the circuit?

Ashby: Automatically, by the fact, for example, that 1 affects 2, and 2 affects 1; or, another loop, 1 affects 4, 4 affects 2, and 2 affects 1.

Bigelow: I see. So then the feedback loops have that degree of randomness associated with the connectivity of the system?

Ashby: That's right. The actual geometrical pattern in the homeostat is always the same; what varies is the polarity and strength of the linkage between the units.

Bateson: In the homeostat, are the variables all of type 3? Do they change in steps ?

Ashby: No. It is essential in this class of machine that the variables | be of two types, [96] for the machine is really a machine within a machine. There is the continuously fluctuating type of variable (and by that I mean that if you give to any one of them an alteration, all four units will undergo some form of fluctuation) and there is the step-function, represented by the relay and the uniselector.

The point is that the learning and adapting organism has to produce two types of change, and they must not be allowed to become confused, either in the organism or in our discussion of them. On the one hand, there are the small corrective movements that are made by the adapted organism, or by the correctly made machine when it corrects small deviations from its normal state, the small movements incessantly made by the automatic pilot, for instance, or the trip made by a rat in a cage when, being thirsty, it goes to the water bottle and has a drink. These small corrective actions are normally made in endless succession as each little disturbance is corrected. But this includes no learning; it shows behavior but no change from one form of behavior to another. During learning, the organism must change, but this change is not to be confused with the change that it undergoes during its small corrective movements. The same distinction in the automatic pilot would distinguish between the changes of, say, potential that occur while it is functioning and the change of a resistance that might be made if we were changing its design. The homeostat undertakes *both* these changes, and it is thus apt to confuse at first sight. What happens is that the resistances on the uniselectors are fixed and constant, temporarily. On this basis, the feedbacks can show, by the movement of the needles, whether the whole is stable or unstable. The changes at this stage are continuous and correspond to the continuous fluctuations of the automatic pilot. Then, comes, perhaps, the other change; if the resistances make the feedbacks wrong, making the whole unstable, the uniselector moves to a new position and stops there. (This would correspond to making a change in the design of the pilot.) Then the continuous changes occur again, testing whether the new pattern of feedbacks is satisfactory. It is clearly essential, in principle, that the resistances that determine the feedbacks should change as step-functions; they must change sharply, and then they must stay constant while the small fluctuations test whether the feedbacks they provide are satisfactory. All design of machinery must go in stages: make a model, test it, change the design, test again, make a further change, test again, and so on. The homeostat does just that.

WIESNER: Well, if this is randomly operated, what is the definition of an unstable runaway condition?

ASHBY: Just a minute, I don't like the term »randomly operated.« The machine's behavior depends, determinately, on whether its essential variables are inside or outside their proper limits. | [97]

WIESNER: But if it is outside its limits, you don't prescribe how it searches. You just prescribe that it searches. It continues to search until it finds something that puts it within its limits, and on the assumption that you have defined the system so that there is one such combination, what is the definition of an unstable system?

ASHBY: The homeostat has two fundamentally independent sorts of stability in it. »Stability« must be given rigorous definition when one comes to deal with systems like this.

WIESNER: Can you sketch a little more of the functional system of the homeostat? You have shown the internal mechanism: can you show the way it is connected with its environment and give an indication of how it operates?

MCCULLOCH: Any portion of it can be made the environment.

ASHBY: There are four units that act on each other, and one might think of the machine as being all »brain;« but, by merely locking the uniselectors in some of the units, those units can be regarded as »environment« with which the remainder, the »brain,« is struggling. You can arrange it, if you please, so that one unit tries to control three units, a small brain trying to control a large environment, or so that three units try to control one, a large brain controlling a small environment. As I said earlier, an adapting mechanical brain must be given an actual environment it can adapt to. In the homeostat, any unit can be regarded as »environment« to which the rest must adapt.

WIESNER: Given the three functional environments and one brain, how does the brain get a measure of its particular state? I am just trying to understand the homeostat, which I have not heard described before.

YOUNG: Where are the dials?

WIESNER: No, what tells it, It is not in the proper state and to continue searching?

ASHBY: On top of each box is a magnet which can turn on a pivot. The central position counts arbitrarily as the »optimum;« at 45° on either side is the »lethal« state the brain must avoid, and it is at that position that the relay closes to make the uniselector change to a new position. If it is stable, the needles come to the center; if they are displaced, they will all fluctuate but they will all come back to the center. That behavior corresponds to the behavior of the adapted organism. Suppose, however, that it had been unstable; the usual runaway will develop, the magnets will actively diverge from the center and will thus reach the position at 45°, either just going right over or perhaps developing big oscillations.

WIESNER: Does it recover as it continues the search?

ASHBY: When it gets back to the center, that stops the uniselectors | from trying new combinations. After that, it will always act, when slightly disturbed, so as to get back to the center. If you push it too far, of course, you will knock it out and it will start hunting again. [98]

BIGELOW: Is the definition of a small push the amount of push which carries it to the next resistor value adjacent to where it is?

ASHBY: Well, a small push is a push that doesn't send a magnet to the 45° position. If you add some cold water to a thermostat, the thermostat will alter a little, will turn up the heat, and will come back to normal. That is what I am calling a »push« and a »stable recovery.«

BIGELOW: But this is a discrete system. There are taps, are there not? This thing hunts among a set of discrete values which come out to connectors. It is a switch.

ASHBY: The homeostat has two activities. One is the activity by which it shows whether it is like a properly connected thermostat, whether it will react to a disturbance so as to restore itself to its central, optimal position. The other is the change it makes when it changes from one set of feedbacks, that have been found unstable, to another set. This second activity is of a different order from the first; by the second, it converts itself from an unstable system to a stable system.

BATESON: What happens if a needle is tied or is fixed outside those limits?

ASHBY: That, as a matter of fact, would set the organism a problem that it could not solve. It would just go on hunting.

BATESON: It goes on hunting forever?

ASHBY: Until the machine wears out.

HUTCHINSON: Do you interpret the ordinary sort of random searching movements of one of the lower animals as something that is comparable to the uniselector?

ASHBY: Yes.

HUTCHINSON: It doesn't seem to me that the setup gives any particularly clear suggestion that this is comparable to the way invertebrates actually do behave.

BIGELOW: Agreed.

HUTCHINSON: I would like your thoughts on that.

BIGELOW: The great difficulty with such models is to see how to draw an informative conclusion from them.

HUTCHINSON: It is a matter of time, largely, isn't it? If all this happened exceedingly fast over an area of small size, it would perhaps go totally unnoticed.

BIGELOW: I don't understand how time enters in that sense at all.

YOUNG: It is hunting.

HUTCHINSON: If the searching was so quick and involved a small displacement, it [99] would show up only with very high-speed films of the thing. |

BIGELOW: The answer is that these things do search. It isn't the fact that searching is not perceptible. If you put an ant on a table with some grains of sugar on it, it will certainly search, so it isn't that aspect that I object to.

HUTCHINSON: I am not altogether convinced that that searching is in any way comparable to what we have seen here.

ASHBY: But don't you agree that if you disturb a living organism with some vital threat and the organism takes steps to bring itself back to normal physiological conditions, that is perfectly typical of vital activity?

HUTCHINSON: Yes, but I think when it does that, there is not the fluttering oscillation that you first described. Isn't that inherent in the system?

ASHBY: Well, some sort of random trial is necessary. Surely, random trying in difficult situations has been described over and over again from Paramecium upwards.

YOUNG: How many trials are you thinking of, what order of number? The octopus for example, may learn with one, and very often with two or three.

GERARD: I think these gentlemen are now worried about the critical damping of your system. Can you not arrange it so that the displacement will go slowly enough so it comes right back?

BIGELOW: I'm sorry, but these are two different things. Hunting and critical damping in this particular system are absolutely different. This system must hunt to find a solution, and you can damp it all you please and it will still try to find a solution. Put damping on a system and it will still trace the same pattern with the same zigzags in it.

HUTCHINSON: That is why I asked the question about the small spatial dimensions. Can you arrange to damp to get that?

BIGELOW: I don't think you can do that. This is not a system whose oscillation is determined by the linear laws of oscillation, like a pendulum. This is a system in which a solution must be sought among a number of possibilities.

HUTCHINSON: Yes, but it is a bit arbitrary what the displacement is.

BIGELOW: With this, I absolutely agree. I simply say you can't cure it by talking about damping or time scale.

QUASTLER: I believe Dr. Ashby would not claim that random searching was the only mechanism by which an organism finds a normal or favorable condition.

ASHBY: I don't know of any other way that the mechanism can do it.

HUTCHINSON: I am worried by your remarks about the insect, because it would be in the insects that one would expect to get this kind of situation best developed because [100] there is both a relatively small cen|tral nervous system and, undoubtedly, learning capacity.

ASHBY: I referred in passing to the insect as a type that doesn't learn. I believe it is a fact that they do learn. I would be glad to get that straight.

PITTS: Well, unless the problem were extremely easy to solve, I should certainly think that, as the number of random elements went up, the average time it would take to reach a solution would go up at a fantastic rate. If there were only one possible combination that would solve the problem, the amount of time it would take to find that would go up with incredible rapidity.

HUTCHINSON: Certainly long enough to be observed, anyway.

ASHBY: I quite agree with you. I think with this type of mechanism, the time would go up fantastically, but I don't think it is a fatal objection.

PITTS: One has to have a better search scheme. That is the answer, I suppose. In any case, if the machine has just made a trial and it hasn't worked, that means the next one has to be somewhat more likely to be closer to the goal instead of running through every possible attempt.

ASHBY: The trouble is, how are you going to get that search scheme into a mechanistic system? It would be nice if it could be provided, but I don't think it is necessary.

PITTS: Well, it is necessary if the objection of requiring such an immense time is going to be overcome.

ASHBY: No, I don't think so. I don't think directive searching is necessary. I think it can be overcome provided the environment is not too difficult.

YOUNG: There are a number of things it can do, but the number is limited. If it tries only the limited number of solutions and meets the environment in that way, that alters the whole situation, doesn't it?

ASHBY: For the organism, the greater the number of solutions it has available, the greater its possible powers of adaptation.

YOUNG: Certainly; but in practice there are a limited number of rather stereotyped solutions in the case of simple organisms.

ASHBY: Yes. That, of course, makes our problem easier, but in the Procrustean way of considering only what happens when the organism or mechanical brain is given very simple problems. I would like to get on to the more difficult case of the clever animal that has a lot of nervous system and is, nevertheless, trying to get itself stable.

YOUNG: Actually that is experimentally rather dangerous. You are all talking about the cortex and you have it very much in mind. Simpler systems have only a limited number of possibilities.

ASHBY: I think it can be done in this way: The enormous time that we have mentioned can be cut down provided the environment allows | the adaptation to be broken up into stages. This does often happen. The child, for instance, learns to crawl first, then to walk, then to learn his letters, then to learn words, then to do simple arithmetic, then to do complicated arithmetic, and so on, every reaction being established before the next one is tackled. Thus, the fantastically long time collapses down to a very much lower order of time. If the adaptation can be broken up into stages, so that instead of trying to get all 10^{10} rightly connected in one vast adaptation, with a fantastically low probability of such a thing ever happening, you can actually get, say, 1000 and 1000 and 1000 and 1000, then the time taken becomes far less. The same principle is seen in crystallization. If a crystal could form only by all molecules arriving in position simultaneously, then no macroscopic crystal could ever form: it would be too improbable. But, in fact, a few molecules can get into position, and as soon as they are in position, the next can add on; then the next and the next and so on. In that way, a large crystal can be assembled in a short time, white the probability of the whole crystal occurring simultaneously is practically zero. [101]

BIGELOW: Well, this sort of phenomenon is absolutely different from the other phenomena that you have been describing. The crystal is a case where the successive events are not occurring entirely independently.

WIESNER: If you were trying to arrive at one very specific crystal configuration, the events would be the same, would they not? The laws that govern crystal growth are quite simple compared to the subject we are discussing.

BIGELOW: In this device, the criticism may be aimed, I think, at the fact that it does nothing new in solving a problem that it didn't do the previous time. To this extent, it really does nothing except search.

ASHBY: After a child has been taught simple addition, is it expected at once to do multiplication?

BIGELOW: If the child is taught how to multiply a certain small class of numbers by each other, the child can immediately multiply with higher efficiency an enormously wider class of numbers, and by efficiency I mean loss in time or loss in false motions. This machine cannot. If you give it a certain problem to solve, it will solve that problem once. If you give it a second problem to solve, it will solve it as if the first problem had not taught it anything.

PITTS: Wouldn't the machine work just as well if the resistors had not been taken out of random tables but simply arranged in order from the least to the greatest? Would it not work as well?

ASHBY: It would work, but whether it would work as well, frankly, I just don't know.

[102] BIGELOW: More predictably. |

ASHBY: I have a feeling that a random arrangement in the long run is probably better than a systematic arrangement, but I just can't say. It involves some very subtle problems.

PITTS: I would doubt it.

WIESNER: I think you would have to search for a problem that comes out the same.

ASHBY: What you gain on one run, you are liable to lose on another because now you have systematically arranged them the wrong way.

PITTS: Whenever you choose a particular arrangement, it is no longer random. In a particular one chosen from those tables, there would be advantages to one run and disadvantages to its inverse run, so exactly the same thing can be said. As a matter of fact, it would be very strange to me if it made any difference at all which one you started with. After all, any particular sequence of numbers is on the same plane as any other, and the fact that it was got out of a table of random numbers instead of being other sequence of values I think, makes no difference.

ASHBY: Yes. It is quite possible that the regular arrangements might be better, but I have dealt with random numbers almost deliberately, to show that it can be done the random way. The systematic arrangement was just what I wanted to keep out of the machine.

PITTS: That particular system which you have is a systematic arrangement since it is only one machine. If you built a whole collection of these machines, say, fifty of them, and then selected them out of a table of random numbers, it would make some sense to speak of randomness. With only one, there is no randomness anyway.

MCCULLOCH: When you have four units, and each is selected at random?

ASHBY: The machine is only one of an ensemble that might have been made.

WIESNER: All possible combinations are equally likely.

PITTS: You might just as well have taken from 1 to 50 out of a table of random numbers.

GERARD: What more would you have to do to this to answer the question of »learning to learn?«

ASHBY: I would have to go a long way ahead of where this machine is.

YOUNG: The system would have to change itself as a result of the functioning. That is why it doesn't seem to learn, because no engram is left. That is an important point.

ASHBY: The homeostat would need further development, using, in part, randomness to provide variety and, in part, special mechanisms to help the phenomena show more readily. For instance, the machine could become more powerful if, instead of just having uniselectors with [103] | its present distribution of random numbers, it had several sets of uniselectors, one set with the numbers distributed evenly between minus 9 and plus 9, another set with the values distributed in a different form, and a third set with yet another form. If, then, it had a built-in feedback mechanism so that after many failures with one set it changed to another set, it would show behavior of the type that Harlow reported (9), that is, a learning to learn, a second-order trial and error. The trials themselves go in batches, with a searching for the best batch to supply the trials.

BIGELOW: Sir, in what way do you think of the random discovery of an equilibrium by this machine as comparable to a learning process?

ASHBY: I don't think it matters. Your opinion is as good as mine.

BIGELOW: But I am asking, if you place an animal in a maze, he does something which we call learning. Now, in what way does this machine do something like that?

ASHBY: Well, as the inventor, I am not going to stick up for it and say I think it is homologous.

WIESNER: I would think that Shannon and Merceau came close to having a learning machine.

BIGELOW: That is the point. This machine finds a solution, I grant you – and please call it anything you want; I am not trying to criticize your language. I merely wonder why finding a solution necessarily implies that it learns anything, in your opinion.

ASHBY: I think the word learning, as we understand it in the objective sense, without considering anything obtained introspectively, is based on observations of this sort of thing happening.

BIGELOW: In ourselves?

ASHBY: The dog jumps on the chair; you beat it three times in succession and then it doesn't jump on the chair any more.

BIGELOW: It has learned something, although it still shows random behavior in other respects. This machine does not do that.

WIESNER: If you disturb this machine enough so that it has to go through its selections, it will jump on the chair again as one of the possible combinations.

ASHBY: If you beat the dog every time, whether it jumps in the chair or not, you won't get it trained.

YOUNG: It doesn't show change of state. The essence of learning is that the system that has been through a procedure has different properties than those it had before.

ASHBY: Would you agree that after an animal has learned something, it behaves differently?

BIGELOW: Yes.

ASHBY: Well, the homeostat behaves differently. | [104]

BIGELOW: In what way? Can you describe it?

ASHBY: Suppose you start it off and it is unstable; this means that it is doing the wrong thing, like the dog that jumps into the chair and gets beaten. The machine gets its »punishment« when the needles go out by more than 45°; then it will hunt until it finds a way of behaving that doesn't lead to such punishment. The machine can be said to have learned not to do »this« but to do »that.«

QUASTLER: But if you take it out into another state and then bring it back into the old one, it will learn just as slowly?

ASHBY: Yes, I quite agree. That is a serious fault of this machine: if you disconnect the environment and give it a second environment, and then bring the first environment back again, its memory of the first environment is totally lost. It was to deal with that difficulty that I got into the earlier, rather complicated, discussion of variables that sometimes go constant, because I think that is the way in which the machine can be extricated from these difficulties. But I admit, with the homeostat, as it exists at present, if you teach it one thing, teach it something else and then try it on the first, its first memory is gone totally.

GERARD: Last year, we saw Shannon's tin mouse that did learn in just this sense. It was put in a maze, it explored it, and, after one complete exploration, it ran the maze perfectly. The difference is that it had a stereotyped way of exploration. One could always predict exactly the way it would go about it. What you are trying to do is to introduce a variable procedure, and, in doing that, you have lost the gain of having gone through the procedure. What we are all wanting from you is to combine those two elements, which I think is the sort of thing that Donald McKay was talking about in his presentation last year, and have probability elements in a mechanism that will do the other thing.

YOUNG: Suppose the result of going through a particular sequence is to alter the resistance in a systematic manner; then the thing could be said to have worked.

BIGELOW: Only if the alteration is such as in some sense to reduce the complication of the problem.

YOUNG: Granted, but that is not inconceivable.

WIESNER: It is learning only if on the next trial the searching is not completely random.

BIGELOW: Yes. You see, the thing I find difficult here is to associate this way of finding a solution with the word learning. These two, perhaps, ought to be associated, but I can't see it because I can think of many models in which a device finds a solution and does not learn, but only a few cases in which I would think the device has learned. For example, if you have a box with a hole in the bottom and a ball bearing inside and keep banging at it, the bearing will fall through. That is, in your sense, a way of learning. | [105]

PITTS: The homeostat is a bit better than that.

BIGELOW: Not if you disturb it from rest. That is what I was asking earlier. If you put the resistor one step out of the way, what happens?

ASHBY: The word »disturb« has two different meanings in the homeostat; there is the slight disturbance that makes it wobble a little and shows it is still self-correcting, and there is the major correction which makes the uniselectors move another step and shatters the feedback. If you talk about disturbance, you must specify which of the two disturbances you mean.

BIGELOW: I would call only the second a disturbance. The first one would seem not to be a disturbance in the sense of changing the problem at all.

PITTS: But then you could complicate this machine by putting in a censor which says which box the ball bearing is in, so to speak; which box it is in determines the path it takes to get out of the box.

BIGELOW: You can certainly make a learning apparatus out of it.

PITTS: Only by inferring the input. After all, no cat that learns to get out of two boxes is going to know which procedure to try unless it has some way of discovering which box it is in.

BIGELOW: Do I understand you to say that this is in your opinion a learning process?

MCCULLOCH: Yes, this is a learning process.

PITTS: Yes; as a matter of fact, the classic experiments of Thorndike were concerned precisely with putting a cat in a box.

BIGELOW: I am not referring to that. Do you consider this a learning device as it stands ?

PITTS: Yes, I should say so.

MCCULLOCH: Would you consider Shannon's mechanical mouse a learning device?

BIGELOW: Surely.

MCCULLOCH: Then this is a learning device in just that sense Shannon's mouse learns to run one maze. If you put it in a second maze, it learns the second maze and forgets the first. You now rearrange the maze in the first form and Shannon's mouse has lost it.

PITTS: That is not the problem. The point is that the mistakes help toward the solution. At least, the mouse doesn't repeat the mistakes. But, of course, cats do, when they are put into boxes.

FREMONT-SMITH: Isn't that true in this system?

BIGELOW: There is a significant difference in the behavior of a cat after each step toward learning how to get out of a box.

PITTS: But while they learn, they will make the same mistake any number of times.

BIGELOW: But the number of times is always less as the learning goes on. Isn't this how one defines learning? | [106]

FREMONT-SMITH: But isn't that true in this system ?

BIGELOW: I don't think so.

FREMONT-SMITH: Would it keep on making the same mistake?

BIGELOW: Not the same mistake, but the same category of mistake.

BATESON: I should like to put a question to our ecologist: if an environment consists largely of organisms, or importantly of organisms, is not the learning characteristic of Ashby's machine approximately the same sort of learning as that which is shown by the ecological system?

HUTCHINSON: Yes, definitely it is.

BATESON: One of the questions that came up earlier was the question of the survival of environments. These wibbly-wobbly environments, containing many type 1 variables, are obviously unlikely to go on in the wibbly-wobbly state. They are likely to settle down to various degrees of greater stability.

HUTCHINSON: Yes. All sorts of situations, certainly, essentially produce constant stretches, such as territorial organization, for instance, which prevents the organisms from meeting each other as frequently as they would if they were moving about at random.

TEUBER: Isn't it much easier to build a machine that will act the way Thorndike thought the cat in the puzzle box acted than to build one that acts like a cat? A definite decrease in random behavior, an eventual hitting upon a solution, and afterwards a straight-line progression to that solution is Thorndike's idea of what cats in problem boxes do. However quite a number of careful observations on cats in problem boxes have shown us a complicated situation which is quite dissimilar, but this would not have been noted if Thorndike had not at first created this ideology about the purely random search. In that sense, Thorndike's ideas have been very helpful.

BIGELOW: If you describe the behavior, you can certainly produce a machine which is in some sense analogous to that behavior. It is very difficult, for one thing, to establish whether something is random if there are only a small number of sample trials. By small, I mean even a few hundred. »Random« is one of the nastiest things to ask a statistician to evaluate.

TEUBER: What happens, may I ask, if the solution is not found, or if it takes a very long time? One of the first things that happens in mice, in cats, and in men is, of course, the adoption of certain highly stereotyped forms of behavior, which run for a long, long time. The animal might assume a position habit: say it always goes to the right, irrespective of success or failure. Or it climbs on top of a goal box and lets itself fall down instead of running straight. Shannon's rat does not do that, and neither does Ashby's homeostat. In that sense, these mechanisms are much more appropriate to certain simple learning theories than they are to real rats in mazes. | [107]

BIGELOW: If you could characterize these stereotypes by some description, sufficiently accurate, of course, you could use a machine with this sort of bias.

GERARD: You want this machine to turn its switches according to some hypothesis, instead of at random.

BIGELOW: Yes.

TEUBER: Not just one, in fact, quite a number.

BIGELOW: It may be a very complex hypothesis.

KLÜVER: The psychiatrist Störring once wrote a monograph (10) on »a man without memory of time,« a man who could not remember anything for longer than one second. Although there was complete loss of memory in that every sight, sound, taste, smell, touch, and pain was completely forgotten after one second, there were no disturbances in other psychological functions, but there was even no memory for emotional experiences beyond the time period of one second. It looks, therefore, as though Dr. Ashby's machine behaves somewhat like Störring's gas-poisoned patient.

The interrelations of animal and environment involve numerous and diverse mechanisms that are of possible interest to the psychologist. And it may even be argued that one of the least interesting is learning. No doubt, Störring's unique case is of greater interest than a few thousand conventional learning experiments. If Dr. Ashby's machine is of some help in understanding pathology of behavior, it may be of greater value than merely another model for a learning process.

PITTS: As a matter of fact, I discussed random nets at one of these meetings some years ago, and certainly I think we agreed, if I recall correctly, that the introduction of a random component in a machine that is in some sense to alter its response to relatively short trains of inputs gives flexibility that is not present in a predetermined pattern of search. The pattern of search which is predetermined may exclude systematically certain things that will ultimately appear if operation of the random component is permitted through a sufficiently long period of time. So much we probably can take for granted.

The question, of course, is how to construct the random machine. I am quite sure that a random machine can be constructed that will duplicate practically any learning theory. I think, as in the case of all other cerebral mechanisms, what we have to do is to cease being quite so naive. At the very beginning of these meetings, the question was frequently under discussion of whether a machine could be built which could do a particular thing, and, of course, the answer, which everybody has realized by now, is that as long as you definitely specify what you want the machine to do, you can, in principle, build a machine to do it. I think the same thing is true about random machines. As soon as the | [108] statistical behavior of the system is specified, a network can be designed, composed of some fixed connections and others with parameters which are subject to some degree of randomness, which will exhibit whatever degree of statistical behavior is desired.

Therefore, I think the only control we have is by using whatever small knowledge we possess about the actual anatomy and physiology of the system in question to try to confine the infinite forest of possible machines within enough limits so that we can at least set up a hypothetical network that will have some general implications, a network that can be compared with the way the animal actually behaves, supposing that a complete specification of the animal's behavior was not among the premises. Possibly in the case of some of the simpler organisms, we might find or indicate junctions or types of junctions which are actually present in the nervous system of simple invertebrates and ascribe a degree of randomness to their effective connectivity or their thresholds that could be checked. But, certainly, constructing a machine that behaves in principle in any given statistical way is something which one can always take as a possibility, and it is possible in an infinite variety of ways.

BIGELOW: What variety of ways?

PITTS: Infinite if the number of components are not restricted within previously fixed bounds.

[109] # DISCRIMINATION AND LEARNING IN OCTOPUS

J. Z. YOUNG

FIRST OF ALL let me show you schematically the topography of some pathways in octopus (Figure 4).

The apparatus with which octopus learns to make visual discriminations consists of a series of lobes. The rods of the retina project directly on to the surface of the optic lobe where there is a deep retina, in which the topography of the outer retina is preserved. There is an elaborate system with amacrine cells, resembling the human retina quite strongly. From these, a system of bipolar and ganglion cells projects inward to the center of the optic lobe, where there is an elaborate anastomotic arrangement, which seems to ensure that the effect of what happens at one point on the retina can be mixed with what happens at other points. The optic lobe also receives afferents from the skin of the arms. From the optic lobe, a tract runs to the superior frontal lobe, which also receives fibers from the skin via the inferior frontal lobe. Fibers from the superior frontal run to the vertical lobe, and from this there are pathways leading back to the optic lobe. The motor output runs from the optic lobe and may be considered to produce one of two responses, a positive action of attack or a negative one of retreat. The experiment that Mr. Boycott and I have done (1) (2), consists in the removal of the vertical lobe. After that operation, the animal has it[s] power of retention greatly reduced. You will be able to follow in the film how the octopus learns not to attack when it is shown a crab and a white square from which it gets a shock, but to continue to attack when it is shown a crab alone. That discrimination is learned very quickly, in very few trials. The motor output, M plus, which is the initial one, is reversed to an output, M minus, that produces retreat.

The memory will last for upwards of forty-eight hours in the normal animal. If the vertical lobe is removed, however, the memory is not able to last for more than about ten minutes. But the operated animal can still make the discrimination in its optic lobe. If the situation is represented, say, five minutes after exposure to it, the animal will remember it, but ten or fifteen minutes later it will have forgotten and will attack and receive another shock. Associations can, therefore, be formed in the optic lobe but they rapidly fade.

I suggest I show the film now, and then it can be discussed.

(Film is run)

MCCULLOCH: Do you have any more detailed histologic work on the organs you [110] believe are responsible for the memory? What do they | [Figure 4] | look like? Is it [111] generally a reticulate mess of some sort or very well structured?

YOUNG: Very well structured, indeed. It would take a long time to deal with that in detail. We don't have all the information we would like about the whole arrangement, particularly on the quantitative aspects. We don't know how many of these cells there are; we don't know how many connections there are. There must be something of the order of millions of cells in each optic lobe, perhaps many millions. Most of the things one wants to know, one doesn't know; for instance, about the convergence pattern. There are crisscrossing arrangements of bundles and the bundles interweave. I suspect that the crisscrossing is an important feature of the system. Anyhow, it exists. The arrangement of the bundles is characteristic for each lobe, but we haven't found a way of describing it exactly. The analysis is far from complete.

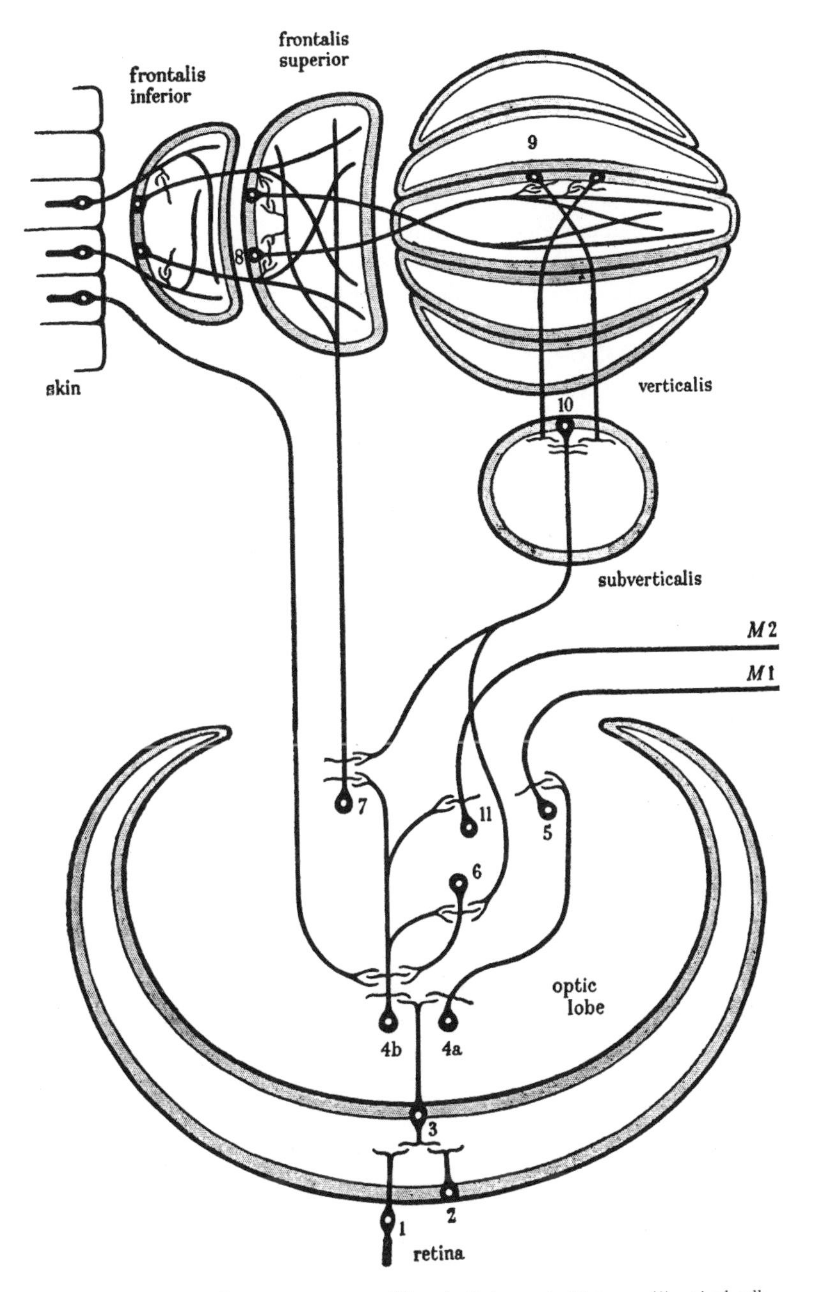

FIGURE 4. Diagram of the Arrangement of Certain Pathways in Octopus. (1) retinal cell; (2) amacrine cell; (3) bipolar cell; (4a)-(5) pathway representing the positive responsive of attack (M_1); (4b)-(11) pathway of the negative response of avoidance (M_2); (4b)-(6)-(4b) represents possible cycles in the optic lobe responsible for short-term learning: (4b)-(7)-(8)-(9)-(10)-(6)-(4b) represents the cycles involved in learning that is longer maintained. Reprinted, by permission, from Young, J. Z.: Growth and plasticity in the nervous system. *Proc. Roy. Soc. B.* **139**, *18 (1951).*

McCulloch: Who would like to ask what kind of question about that structure? I see we do have a structure where memory is occurring. Go ahead, ask him!

Bowman: What is the scale of this and the size of the animal?

Young: The whole brain weighs very little, less than a gram. The vertical lobe weighs only a tenth of a gram, but it contains a vast number of cells, some of the smallest nerve cells I have ever seen anywhere.

Teuber: What other behavioral data are available? Is there anything of the more traditional sort, such as detour behavior, before and after the operation? Are you sure that this is a memory change rather than a change in avoidance behavior?

Young: The general behavior is unaltered after removal of the vertical lobe. We can't recognize an operated animal from a normal one by any criteria except its memory.

Teuber: Do the animals tend to collect more crabs after the vertical lobe removal?

Young: I couldn't say that, no. They are very voracious anyhow. They go on eating very readily after the removal.

Ashby: Would you say, as a matter of impression, that there is always a path from any input to any output?

Young: I should suppose so, but I wouldn't like to be dogmatic about it. It is a highly anastomotic system. But can you imagine now, in that short period, something corresponding to your random selection going on in a system like that?

Ashby: I see nothing incompatible. What impresses me is the enormous advantage the octopus has over the homeostat in being able to hold a memory temporarily until it decides whether or not the memory is worth holding permanently; it can then either fix the memory or scrap it. That puts the octopus a complete functional level ahead of [112] my simple mechanism. |

Pitts: Is it quite certain that memory does not occur in other structures that feed out to the verticalis? Could it merely be that their output has been removed?

Young: I didn't get that.

Pitts: Are there certain lobes superior to the verticalis?

Young: No, it has only a limited input (Figure 4).

Pitts: Could one be certain that the memory did not occur in the superior frontal lobe?

Young: No; removal of this will equally interrupt the memory.

Pitts: Even though the verticalis has an independent input from the optic lobe?

Young: No, no, it has no independent input from the optic lobe! It has only that one input. It is astonishing, with such a large number of cells, that the input is quite homogeneous. It is just the crisscrossing tract passing from the frontalis superior.

Pitts: But the structure of the frontalis superior is much simpler.

Young: No, just different.

Pitts: It could well be either one?

Young: Oh yes, either or both. You can't, of course, investigate them separately.

Ashby: Because if you interrupt the circuit at any point, you have interrupted the output.

Pitts: The frontalis superior has no other output than the verticalis either ?

Young: It has, as a matter of fact. That is incompletely shown in the Figure.

Kubie: Has it been possible to interrupt the tactile input, apart from the visual?

Young: Yes. Removal of the inferior frontal lobe does not interfere with learning, but it has other effects on behavior. The handling power is affected. The octopus can eat the crab perfectly well, but it handles it rather clumsily.

KUBIE: It is interesting that it does not interfere with a learning process which depends upon a tactile afferent.

YOUNG: Well, the actual seizure does, but the preliminary attack does not. That is purely optic.

KUBIE: In other words, there is certainly something equivalent to pain; a painful or unpleasant thing must have access, apart from this tactile input.

YOUNG: It has. The so-called pain, if you like, has a direct track into the optic lobe. The inferior frontal lobe is some kind of form discriminator, which helps in the handling of the crab. It is a very interesting system in itself but not directly connected with the formation of this association. | [113]

PITTS: It is perfectly natural that the pain should go to the place where the learning Occurs, supposedly into where it is stored. The optic lobe itself can do the learning.

YOUNG: Yes. After the vertical lobe is removed, there is still learning but it lasts for a very short time. This can be tested by showing the crab and white plate at various intervals after the octopus has received a shock. If the interval is less than about ten minutes, the octopus will not attack (Table I). The memory lasts, therefore, only for this short time. This suggests that there is a short-term memory system in the optic lobe, which in the normal animal is reinforced, preserved, or »printed« by the vertical lobe system.

VON BONIN: The optical has vision and touch?

YOUNG: There is also a chemoreceptor system but it is not very well developed. It doesn't come into this.

KLÜVER: What is the approximate number of ganglion cells in the retina ?

YOUNG: It is easy to guess at these things, but there is very little precise information available. The density is very high.

KLÜVER: How about movement perception.?

YOUNG: Movement is the prime clue. The octopus detects very small movements.

BATESON: When you say the memory is »retained,« that means the octopus does not get a shock on that trial ?

MCCULLOCH: He didn't try it. He wouldn't go after the crab at all.

BATESON: He did not get a shock, so that the retention is extended from the first trial through to the fourth trial but presumably reinforced by not getting a shock in the interval?

TABLE I

Series of Responses of an Octopus without Lobus Frontalis Superior to Presentation of Crab, White Plate and Shock

Intervals from Previous Shock *Minutes*	Response
3	no attack
5	no attack
8	no attack
10	attack
13	no attack
15	attack
19	attack
25	attack

[114] | YOUNG: No. Each of those trials is preceded by a shock. Those are simply intervals. We can find the half life of the short time memory to a minute in any given animal.

BATESON: Does going through the experience of: »I had better not take that crab because it's got that white thing on,« extend the length of the memory?

YOUNG: Yes, it does. It is an interesting thing, indeed. If you present the crab and plate to the operated octopus at three minute intervals, you can prolong the memory for half an hour or more.

PITTS: In the operated animal?

YOUNG: Yes, in the operated animal.

BATESON: Without its ever getting a shock?

YOUNG: Without its getting a shock at all. And if you keep the white plate in all the time, you can prolong the memory for as long as your patience holds out. You keep moving the plate there in front of the animal and it does not attack. As long as the retinal image is there, the association with the previous shock is retained. As soon as the retinal image is removed, then this association, or whatever it is, begins to fade, because it is not being reinforced by the cyclical system. That is the easy way of looking at it.

ASHBY: There is evidence that one should find, in the random system, that training with a mixture of reinforcement and non-reinforcement should actually proceed faster than with only reinforcement. If the responses are developed as an average, the non-reinforcement selectively removes the less active half of the distribution and makes the response greater. I have not followed the evidence far, but I think it may be found to throw some light on the nature of the conditioned reflex.

YOUNG: I can't see, I must confess, how statistical analysis can treat these cases where the learning is extremely fast, for instance, in this particular case where the animal is learning in a single trial.

ASHBY: The octopus doesn't really have to learn more than one binary decision: whether to attack or not to attack.

YOUNG: That is what I was trying to bring out in relation to your presentation. If you think of an animal as divided up into a series of binary decisions of this sort, perhaps a rather small series, you may be able to combine that with the statistical machinery, but I can't quite see how.

ASHBY: Suppose the animal had to learn that the crab and white card was safe when a red light was shining, but that that combination was reversed in effect if the water had previously been made more salty; then I think the learning would be much slower.

[115] PITTS: We don't know that the animal has learned; that is, we don't | know, roughly, how much something has to deviate from being a white plate in order to be responded to as if it were a neutral object. The smaller the difference allowable in a white plate, the more the animal has in fact learned.

YOUNG: There are experiments on form discrimination. I won't bother you with them now, but the octopus discriminates sizes quite sharply.

PITTS: It may have learned the particular plate, and possibly it may also have included as an element that a rather slightly larger one would not produce the same effect; in which case, it would have learned more than if it had merely learned the presence of some white plate. We don't know precisely what it has learned.

ASHBY: I am not sure that your statement of how much it has learned is the right approach.

PITTS: I was simply counting bits.

ASHBY: You can't break it up into so much and so much and add them together.

PITTS: Well, if you want to call it bits, in the sense of information theory, you roughly have to determine what proportion of possible cases in the universe fall into the two classes of »responded to« and »avoided,« and the fewer the cases that fall into the class of »responded to,« the more bits it has learned.

YOUNG: The operated animals can learn to make a discrimination (Table II). The animal is shown the white plate and crab and receives a shock and then, five minutes later, is shown the crab alone. Ten minutes after the previous shock the white plate and crab are shown again, and the octopus gets another shock. That can go on at five minute intervals regularly, and at first there seems to be no learning. But at the end of a series the octopus begins not to attack on the negative situation. After each period of ten minutes, which is the time limit almost, he has only a little bit left, but, if I interpret it right, it is building up until there is enough memory to persist over half an hour.

TEUBER: If a crab is presented on one side and a plate on the other, simultaneously, in the ordinary discrimination-choice situation –

YOUNG: We haven't done that much.

TEUBER: Would that be feasible?

YOUNG: I don't know, because of the eight arms.

TEUBER: Oh, he would take both?

YOUNG: It is difficult experimentally. It could be done, but I have no data on that.

TEUBER: How about a glass septum between them?

YOUNG: Well, we haven't done it.

KLÜVER: In higher animals, the situations in which I have encountered | learning in [116]

TABLE II

Octopus with Vertical Lobe Removed, Learning to Discriminate Between Crab Alone (Feeding) and Crab and White Plate (Shock)

Time of Trial[a]	Time Taken for Attack	
	Crab Plate Shock	Crab alone
Minutes	*Seconds*	*Seconds*
0	4	
5		4
10	5	
15		21
20	20	
25		10
30	6	
35		10
40	15	
45		58
50	no attack	
55		27
60	no attack	
65		11
70	no attack	
75		6
80	25	
85		30

a. The trials are at intervals of five minutes and the numbers show the time taken for the attack in seconds.

one or two trials seem always to have been situations associated with strong emotions. Your experiments presumably involve the production of pain.

BATESON: Is there any indication at all that after operation, the octopus, which evidently forgets quicker, does in some degree learn more quickly? Is it possible that one shock is sufficient to remind it and give it the answer, whereas, in the initial learning, it might have needed two or three trials? Does it really, after the operation, go back to the initial condition of ignorance of an octopus that has never seen the plate?

YOUNG: I am afraid I couldn't answer that honestly because the octopus learns so quickly. The difference between learning in one and two trials would need a very long [117] series of tests to establish significantly. |

TEUBER: Incidentally, I suppose the octopus is one of the few animals in which you can determine the emotional sequence with the photometer?

YOUNG: Theoretically, yes.

PITTS: Quantitatively?

YOUNG: Theoretically. I wouldn't like to do it in practice.

PITTS: With suitable filters?

YOUNG: I would like to persist just a bit more and ask if anybody has any ideas about what is happening in the vertical lobe with its numerous cells to enable it to reinforce the learning?

PITTS: I understood from your presentation that at one place the synapses are dendritic largely, and at other places they are on the cell body; is that so ?

YOUNG: No, approximately all are dendritic in this animal.

VON BONIN: We have talked a good deal about memory and reverberating circuits. As I see it, there is no anatomical evidence of any reverberating circuit in these lobes.

YOUNG: Oh, yes, there could very well be a reverberating circuit between the optic lobe and the verticalis system.

VON BONIN: I don't see it. Does it show on the diagram?

YOUNG: Oh, yes. I would rather like to think of the memory as largely a function of a reverberating circuit. One circular path can be traced from the optic lobe to the frontalis superior and over verticalis and subverticalis back again to the optic lobe. (Figure 4). But what I want to know is what, in this one presentation, happens. Something happens which sets a particular configuration producing a negative sequence, and then something else happens, perhaps starting up a circuit, which reinforces the process that connects the set into the negative rather than the positive channel. Why should it be that the verticalis contains so many cells? Is that a significant clue?

GERARD: You could test whether this reverberating circuit actually means anything, I should think, by doing almost the experiment you have done, getting the memory in the verticalis, then superimposing upon that a secondary conditioning, if you will, in which the white plate is the signal so that the touch stimulus becomes negative. Then cut the afferent from the retina, and see if the animal can still remember the touch stimulus.

YOUNG: Yes, it might do it.

WIESNER: If you could use a narcotic to deaden temporarily the connection between the verticalis and the M plus-M minus circuit, – approach and avoidance –, it would be interesting to see if there was a reproducible effect.

[118] YOUNG: I can tell you the answer to that; anesthesia does not destroy the memory. |

WIESNER: Even after you have interrupted the connection to the verticalis, anesthesia does not destroy the memory?

YOUNG: You can give the animal an anesthetic for the maximum time it will stand it, which is about a half hour without breaking up what it has learned.

WIESNER: But if you give it a general anesthesia, you can't tell whether it is temporarily in the second condition which results after operation. It would be nice to see the animal go from the »short-span memory« back to its original condition.

YOUNG: It would, indeed.

PITTS: You can block activity with a single shock, I suppose?

YOUNG: Yes.

PITTS: After operation, when the trace persists for only a few minutes, one could see if the trace were destroyed by a large shock which blocked whatever patterns might be reverberating. This second case, where memory persists for only a short time, would be more likely to be pure reverberation than the first, of course.

YOUNG: I do find that the anesthesia results make it rather difficult to allow the reverberating circuit concept. The animal comes out of the anesthetic apparently unchanged.

WIESNER: It retains its memory?

YOUNG: Yes.

VON BONIN: Is there any way of actually checking the electric record of the brains or is it so small, that it is impossible to do it?

YOUNG: We have, of course, tried that, but we have gotten nothing out of it so far.

GERARD: We don't want to get into reverberating memory circuits, but do you remember the experiment in which we took hamsters and refrigerated them, or hibernated them, down to about 4°? We could find no evidence of any electrical activity in the brain anywhere, and yet they remembered a maze that they had been taught before. Similarly, after electroshock treatment these animals also retained the learned maze. This is the reverse kind of stopping of dynamic circuits as that just described on giving a maximal shock.

ASHBY: The analogy with the homeostat could be worked out. Part of it is fairly clear. The feedback from the center that registers pain must have an effect in the frontalis superior, tending to break up, or to reshuffle, the patterns there. The main part of the reaction obviously goes through the frontalis superior because – am I right? – when it is removed, the animal can still show its normal catching behavior. The uniselector part of the homeostat corresponds to the region, wherever it is, where pain impulses, or their consequences, meet the path of the »attack« reaction. I am a little puzzled to know how the verticalis | comes in. Obviously, the octopus has evolved a complete [119] stage ahead of the homeostat, and has some arrangement by which it can use the homeostat's principle, with the uniselector part in the frontalis superior providing random patterns with pain as the corrective feedback; but obviously there is some way by which the verticalis can carry some memory. I can't give the analysis offhand because it needs a good deal of thought, but at least the rudiments are clear.

BATESON: Could I ask one question of Dr. Young before we move on? It seems to me that the question he raised about the rapidity of the learning in this case, arguing that such rapidity of learning does not fit with the random picture, is only a question of monkeying with the probabilities in the uniselector. Imagine a box which, apart from outside noise, gives an attacking response with a high probability that any noise will switch it to some other response. If now, among these alternatives, a majority of the positions of the switches lead to retreat, the system will meet your requirements, won't it?

YOUNG: That is what I would like to think. Then you would have all stages from that to the learning of more difficult situations, such as Ashby has been talking about.

WIESNER: There is a lot of past history which also may have gone into this situation; for example, the animal may have already learned that there are unpleasant situations in the world, maybe it was one of his ancestors who learned it. There is no reason to start with a completely randomized network .

ASHBY: I think you could have made the essential randomness of the octopus' reactions more obvious if you had been less kind to the animal. Suppose, for instance, you had given it shocks if it had *not* attacked the crab; then its sudden retreat, which one instinctively accepts as being the normal response to the shock, would have evoked further shocks, and would thus have been shown to be merely a random way which turned out wrong.

PITTS: Can you say anything anatomically about how close to the efferent side of the optic lobe is the input from the verticalis?

YOUNG: No. I wish I could. That ought to be worked out.

PITTS: That, of course, is a fundamental question.

YOUNG: It is very important, indeed. The whole of the neural arrangement should be worked out in detail.

MCCULLOCH: There is one point which is still reverberating in my mind. This animal learns comparatively rapidly, the probability of finding correct paths comes up, and the duration required to learn something when a large number of randomly connected paths must be gone through. I know Bowman has a point on it and I wish he would make it now, if he will.

REDUCTION OF THE NUMBER OF POSSIBLE BOOLEAN FUNCTIONS

[120]

JOHN R. BOWMAN

I WILL make this quite short. This is to be said from a purely mathematical standpoint. Previously, we were disturbed about the time that would apparently be necessary for random exchange. If we have a black box with a number of inputs, let's say n inputs, and a single output, it can fairly easily be shown that the number of distinct black boxes of that type is:

$$2^{2^n}.$$

The number of distinct Boolean functions of n binary variables is:

$$2^{2^n}.$$

That can be shown by a very simple analysis. There are two 2^n possible types of inputs and two just possible outputs. If, however, we admit the possibility that was explicitly pointed out by Dr. Ashby last night, that we may permute these inputs, we decrease that number a very great deal. If, also, we consider the alternative of taking the complements and reading 1 for 0, or read 0 for 1, we have a still further decrease.

Just as a very brief illustration of how large that number can be, suppose we take the case, $n = 8$. That gives us 2^{256}, approximately the total number of electrons and proton in the cosmos as estimated by Eddington. It also is, though, for $n = 8$, the number of ways that you can build a coded decimal adder.

If you had eight inputs, you would have, in effect, two decimal digits. It takes four binary ones to designate a single decimal digit, and if you combine two decimal digits to get a single binary output, then you have eight in and one out.

BIGELOW: I don't understand how this number of combinations applies to an adder; I don't understand what your assumption about two decimal digits producing a single binary output is. Two decimal digits have a possibility of a hundred states. The binary output has a possibility of two states. What sort of adder do you mean?

BOWMAN: Let me elaborate on this a little bit. The general case would be one has m input and n output channels. The possible combination one gets are:

$$2^{m \cdot 2^n}.$$

If we wish to have a black box that will perform an arithmetical function on two coded decimal digits, we will have 8 in and 8 out. | Each output, in the most general form, must be looked on as a complete reduplication of the whole. [121]

Inserting for m and n, in this particular case 8, one gets an extraordinary large number as follows:

$$2^{2 \cdot 8^2} = 2^{2048}.$$

Just as an illustration, can somebody give me in round numbers, the number of parallel input and output channels for the brain?

YOUNG: There are a million fibers in the optic nerve.

MCCULLOCH: All right; 10^7.

BOWMAN: That gives us 10^7 for m. How many output channels do you want?

MCCULLOCH: Say 10^6.

BOWMAN: This gives us 10^6 for n, and we arrive at:

$$2^{10^7 \cdot 2^{10^6}}.$$

I imagine this would be the largest integer that has ever been seriously considered scientifically.

McCulloch: Yes, I think so.

Bowman: That is the number of kinds of brains you could have. But I mentioned the fact that we can freely permute inputs and outputs. We can freely complement inputs and ou[t]puts. Third, as was also mentioned previously, we can factor, we can seek groups of inputs that are exclusively responsible for the character of groups of outputs, and break the problem down that way.

I don't want to jump the gun here on some very fine research recently completed at the Harvard computation laboratory, but I can tell you this much: a group of men at that laboratory have counted the irreducible Boolean functions invariant under input transformations, a problem that has been an outstanding one in pure mathematics for many years. It is a pretty long-haired kind of a problem, and its treatment involves something somewhere between group theory and number theory.

If the number of independent n-in, one-out Boolean functions are counted and they are reduced to a minimum set, an extraordinarily small number of functions remain. To give an obvious illustration, if n equals 2, say p and q, there are 16 possible Boolean functions, but there is only one fundamental one, namely »not both, p and q,« the Sheffer Stroke Function, from which, by a permutation or a complementing of the inputs, you can get all the remaining 15 so that I believe the random searching process, to find the right Boolean function for survival, is not too complicated.

[122] Pitts: I don't think that you should count them that way because the | inputs are wired into particular places, so the arrangement doesn't have the possibility of providing an arbitrary permutation in the input.

Bowman: If there is a self-correcting loop, which, if it fails, tries something else, I think there is complete freedom of permutation of the inputs and the outputs.

Pitts: Actually, a particular fiber carries information from a particular source, and it is not interchangeable with any other. If you take the brain and wire it up with the inputs connected differently, that is, in effect, a different brain. It would act quite differently on the environment and receive quite different things from one not so permuted. It is perfectly possible that you can divide the random search process into two kinds, the first of which finds the proper reduced operator and the second of which has to find which permutation is necessary to adapt inputs and outputs so as to produce the correct result. But it still has to search two things: firstly, the reduced operator and, secondly, the permutations.

Bowman: I believe that is true.

Pitts: But the permutations are enormous in number, of course.

Bowman: Anytime you can break down a problem into parts, the sum of the parts is a far simpler problem than the whole taken as a whole.

Pitts: Well, this is the product of the parts, really, in the sense of the operations that have to be performed apparently, finding the correct reduced operator would take a negligible time in comparison to the time it would take to find the proper permutations of the input so as to connect it up effectively. The latter number is still enormous.

Bowman: If there is any possibility that the fundamental Boolian functions can be, in a sense, stored and always available, then the main search, or scanning problem is a permutation of the inputs and outputs.

Just one more point, although I don't think it has any particular bearing on Dr. Ashby's comments. If we permit a requirement at an intermediate point in the circuit, it provides this Boolian function switching. The problem is again vastly simplified. To make a coded decimal arithmetic unit, not more than three levels of rectifiers are needed. Rectifiers can be set to provide either the »and« or the »or« function. The outputs of those, then, can be combined as »ands« or »ors.« A third time is all that is needed to provide all the arithmetic operations. Now, if there is any clue whatsoever as to what functions are needed on the intermediate levels in this logical switching circuit, then the problem is vastly simplified and the whole thing can be built up with »ands« or »ors,« two variables in and one variable out. This would be the case as long as we are speaking of single output Boolean functions. | [123]

Consider now the the[!] multiple output case; let's take it as three in and three out; in effect, we need three black boxes. If we build up a circuit with »and« or »or« elements, these can be combined by pairs to give a first-level set of internal circuit points, which in turn, can be combined by pairs to give a second level of internal circuit points. A third is all that is needed to provide any function if there are eight inputs. All that I have tried to bring out is that if the animal has any clue as to the required functional behavior of an intermediate circuit point, the problem is vastly simplified.

BIGELOW: You can restate this in terms of there being three levels of function, but is there not still the high combinatorial problem of picking out the right path?

BOWMAN: It is still extremely high, but if you can say that there is a point which is the »and« of A and B, the mere existence of such a point in the circuit leads to a very great simplification in the discovery of a suitable circuit by random searching.

BIGELOW: There is no question that if some configuration which contains a large part of the problem can be picked out beforehand, then an arrangement can be put together of some of the grosser elements and the problem is simplified. The real question is whether or not this is significant in the present problem. Is there any reason to believe that configurations of this sort can be assumed and that they do take part in the process.

BOWMAN: I don't know. I refer that question to Dr. Ashby.

BIGELOW: The combinatorial aspects of the problem seem to be left unchanged by this statement.

MCCULLOCH: But the process of search would be simpler.

BIGELOW: Only because the items are renamed into larger groups.

ASHBY: Not only because of a renaming; often the environment itself allows it. Get a nonmathematician to make up what he considers to be a very complicated problem, and then examine it. You will almost certainly find that it really consists of several parts, each of which can be solved separately. That makes the problem very much more simple. The fact that it can be broken up enables the solution to be found more easily.

BIGELOW: There is implied in the assumption of randomness a difficulty which is a combinatorial one. That difficulty may or may not be a true one. It may very well be that an organism of this sort, operating as does the nervous system, does not, in fact, have random pathways. I have some considerable doubt as to whether this assumption is proper in that sense. But let's suppose it is. If there is a genuine randomness, then there is a genuine combinatorial problem of search, and that general combinatorial problem of finding connections is not | helped by considering local regroupings and rechristening them and saying you search among them, unless those regroupings contain something which is not random. [124]

If you believe that the randomness is an unfair assumption and you wish to discard it, there are many ways in which one can discard it. This regrouping is certainly one way. One can say, »I don't believe it is random; on the whole, I believe it consists of small clusters which have predetermined functions and which are not random, which represent the possible number of things among which the system ought to search.« One may simplify it that way. It is another way of saying that all the possibilities are not independent, that some of the possibilities exclude others and that they go forward in groups, which, if you are working with one group, means you are doing a certain sort of operation. There are different ways to state this mathematically; there are different ways to realize this in the physical system. It seems to me that we do not genuinely believe in the random network. We believe there is some semisystematic aspect of it and that that semisystematic aspect perhaps facilitates things. More than that, what can we say?

PITTS: This is true physiologically. Most of the communication of the brain is done by various nuclei connected by tracts. These tracts are very small in number of fibers compared to the possible combinatorial configurations of the interacting elements within the nuclei which they connect. This, of course, reduces the number of possible configurations very greatly over the case where, say, two nuclei are connected by a tract, of which the maximum case is that every cell gives rise to an axon leading to the second one. But there is a much more important simplification, or one as important, in reducing the number of possibilities, namely, that the output is never wanted in a combinatorial way. The output is always in the form of a certain degree of muscular contraction, or the rate at which it takes place or its acceleration; that is to say, the neurons simply add up in parallel channels so that the total number which are excited during some interval of time, within the nerve, is all that really matters. There may be some variation in the strength of the muscle fiber to which they are attached, but it is still an enormous reduction of the amount of information that goes over it.

BIGELOW: But don't you suppose there is independence when we remember what a picture looks like?

PITTS: I am talking about the output, not the input.

BIGELOW: But one does remember what a picture looks like. True, you are doing an [125] operation that has somewhere an input and an output, | but it seems to me that there the same integrating effects may not be present.

PITTS: I am talking about the output. What I actually do and say is the result of a collection of motor nuclei, of which all that matters at any moment is the distribution of the number of those which are firing. There is an enormous reduction of information. Of course, the same sort of thing occurs to some degree on the input side. It occurs with respect to most somatic sensations. It does not occur very much with respect to the visual system, which is by far the major source of information to the brain, but nevertheless it does reduce the large number, since complexity depends upon the output. And in addition, the fact that the commutation is actually performed in nuclei which are connected by pathways, which are restricted in the amount of information they can carry, simplifies it further, in the sense of the number of possibilities that one could assume actually are hunted through.

BIGELOW: I agree completely with what you say. I merely tried to emphasize in unmistakable fashion that what we are trying to do now, as far as I can see, is to abandon the hypothesis contained in our own word »randomness,« though which way it should be abandoned I don't know. If you are talking about a real computing machine-like device, then combinatorial numbers of this sort can arise, and there is no way to get around them. If it is a genuine search problem and if the variables which you speak of are geniune[!] random independent selections, you must look among them all. There is no way of faking that.

PITTS: Randomness does not imply independence. It is perfectly possible that metabolic variables affect the thresholds of cells or the probability of a connection being actually effective. As a result an impulse might be able to travel along a path in a way which is independent of the particular information that happens to be in the system. This is a randomness which is not the same as a completely *a priori* randomly interconnected net, in the sense of conceiving of the ensemble of all possibilities, but it has a random element in the sense of how the output depends upon the input.

BIGELOW: I agree to that statement as very likely true and an interesting summary, but I don't think that it is uppermost in the minds of the people here while they have been discussing randomness.

PITTS: That is how I understand it. I would agree with you that that is the only reasonable way.

WIESNER: The theory which continues to reverberate in this room is absurd on the basis of what we know anatomically.

PITTS: Certainly.

WIESNER: It is inconceivable that every neuron in the brain could appear in every possible place. | [126]

FREMONT-SMITH: Are we now saying that randomness for our purposes must have a relativity?

PITTS: Yes. There has to be a stochastic element somewhere in the system, placed under heavy restrictions as to where it may occur, and I think we hope that the insertion of such a stochastic element would be useful.

FREMONT-SMITH: You are discussing what kind of relativity is most useful to think of in a system which contains randomness?

PITTS: Yes, where to put in the randomness and how much and under what conditions it should operate.

BATESON: I find myself wanting to help Ashby's animal by not putting all of those noise-producing dials on the surface. He deliberately made it difficult for his animal by doing that. If some of those dials were controlled by internal variables of the system and were able to fire out noise when certain unfavorable internal states of the animal were approached, the system would have a very considerable degree of choice among randomnesses.

ASHBY: Yes, but you might allow the animal to solve its problem in a merely nominal way by just cutting itself off from its environment, which of course is physically unrealistic.

BATESON: Some people try it.

ASHBY: It would correspond to the octopus' developing a hysterical anesthesia. I think my original formulation is better, where the environment acts directly on the essential variables so that the organism must solve the problem or die. This arrangement sets the organism the really critical problem. If it can solve that, it can solve the others easily. It is a great mistake, I think, to make it too easy for the mechanism.

PITTS: There is a general principle here that we should not put the randomness too close to the input side because then, to repeat, it reduces the information which the organism has available because it could never trust its input. If you put random noise generators in the optic nerve, it is not going to improve matters.

MCCULLOCH: Before we generate too much randomness in the group and cut ourselves off from the outside world, let's pause for a moment and then turn to Ralph Gerard.

[127]

CENTRAL EXCITATION AND INHIBITION

RALPH W. GERARD

WE ENDED an earlier session on a note of requesting some actual kinds of brains, and we have several times come back to the problem as to what actually might be going on among the collection of neurons. I had originally intended to say just a few words about some work in our own laboratory that hits on central inhibition, but finding myself formally scheduled, I will first try to sketch the overall framework, at least as I see it.

Let's start with behavior. This necessarily involves the action of certain neurons, certain motoneurons, at certain times. Let me hastily restate that. It may not involve any one particular neuron. It may not even involve any given motor nucleus. But, sooner or later, individual neurons have to fire or not, in particular order. On this matter of not needing particular neurons or even particular nuclei, I must say a word about the discussion just past.

Animals can achieve the same behavioral objective by using a variety of different muscles or even limbs in many motor combinations, and if something is done to interfere with a given one, substitutions readily take place. But there remains a very large residue of nonsubstitutability, nonlearnability, and in respect to one of the points explicitly made previously, the nervous system cannot do nearly as well or badly, as the case may be, as a homeostat. If the output is crossed in the homeostat, as pointed out, it would readjust to it. If the output from the nervous system is crossed, or if the input is, it will not readjust to it. When a frog's eye is cut out, rotated 180°, and allowed to heal back, the frog, for the rest of its existence, will jump at a fly in the opposite direction from which it should. If the nerves to the flexors and extensors are crossed, a rat from then on flexes when it wishes to extend and vice versa, and gets along by a sort of hopping on its elbows and knees. So there are certain patterns, as has been argued this morning, and one does not start with a tabula rasa.

FREMONT-SMITH: There are exceptions to that, aren't there, Ralph?

GERARD: No exceptions until after I have finished, please.

YOUNG: It is true, by and large.

GERARD: I want to make a rather tight argument, and it will go better for me, I think, if I can continue with it.

We have spent a great deal of time, and profitably, talking about what is happening inside the nervous system before something comes out, or in the black box before something comes out. I must confess [128] | that much of what happens in the black box, I can't follow. I found myself wondering this morning whether there would be one of these great industrial crises of Gluyas Williams if one dealt with a white box.

BOWMAN: They are often gray.

GERARD: But whatever the processes going on in the black box or the brain are, whether there is a three-dimensional nerve net which is not further divided up, or whether neurons break up into assembly loops in three-dimensional patterns, or whether there is a single reverberating circuit, or, to take the other approach, whether there is some general potential field which has its own topography or its own profile, whatever the mechanism may be, sooner or later the internal process, the change in time and in space, presumably has to engage particular motoneurons. It is that problem, essentially, that I have been asked to face in discussing central excitation and inhibition. I may put it this way: A melody can be transposed in pitch and time and on to a different piano and in other ways; nevertheless, individual notes have to be struck at individual times to actually produce it. I ask, then: How is the state of a motoneuron changed, whatever the antecedent processes in the brain?

I shall omit any discussion of a change of state of the motoneuron induced by chemical adventures, such as those due to alterations in the blood stream. Important though they are, let's leave them out. I shall not be concerned, since it is not a critical matter, with the initial state of the motoneuron: whether it is at some uniform level which can be called zero or plus or minus or whether it is in an oscillating state, whether, that is, a rhythmic activity of that particular neuron, pre-exists or not. The alteration in state can obviously be plus or minus. That can be made a little more precise because, I think, the only things that can go plus or minus that will matter at the behavior level of which we are speaking are the actual discharge of impulses, which can be more or fewer, and the actability or irritability (or inverse threshold), which can be raised or lowered. Plus or minus changes can also involve the potentials, metabolism, physical organization, etc., of a motoneuron, but those, I think, are not going to matter in this connection.

What mechanisms can bring about such alterations in a given neuron. First, they can be nonsynaptic: mass field effects, or whatever word is wished. Changes in the surrounding population of neurons can give a change in the total current flow through that given neuron and thereby alter its state. Or a chemical change, over short distances and in contradistinction to those general changes in the blood which I dismissed, could lead to a nonsynaptic effect. Potassium, or what you will, could be released from nearby neurons and, by quick limited diffusion, act upon a given neuron. These mechanisms also will be left | out of the discussion, although the electrical ones, at least, [129] have been proven important; for example, passing currents through neurons, or through masses of neurons, in polarization experiments can start or stop rhythms and increase or decrease irritability. Specifically, as we shall see later in another connection, if one polarizes a ventral root, with current going into the axon and having to come out centrally (either where the axon enters the volume conductor of the cord or by passing through parts of the neuron soma itself), considerable changes are produced in the threshold to synaptic stimuli of the polarized neurons. With the root positive, the neuron normally becomes more irritable and may be pushed into violent rhythmic discharges.[1] So much for the nonsynaptic effects on a motoneuron; we can turn to the synaptic ones, which are much closer to the interest of this group.

Discrete impulses arriving at synapses have very limited parameters with regard to which they vary. Theoretically, different kinds of endings of the presynaptic unit could reach the motoneuron. If, for example, excitation and inhibition are, respectively, associated with two different kinds of chemicals released, two different neurohumors, these processes would necessitate two kinds of endings. There is not any conclusive evidence that I know of regarding qualitative differences of synapses in the central nervous system. Certainly, in the case of nerve itself, the evidence is against two kinds of impulses; impulses which excite and impulses which inhibit are, by all the criteria available identical.

In fact, logically, one would have some trouble in creating a propagation of inhibition. I thought of that when Mr. Bateson spoke of the man who turns away when a friend or acquaintance comes down the street and so communicates to him, »We are not communicating.« It seemed to me that would be the transmission of inhibition, and I preferred to think he was communicating, »Do not communicate with me; I will not communicate back to you.« Or, an aside on another point that came up: Excita-

1 Experiments of K. Frank and S. Ochs.

tion tends to spread, inhibition tends not to spread; laughter is contagious, whereas sorrow is self-contained.

MEAD: No.

GERARD: Maybe that is one of those easy but unsound comparisons.

MEAD: It is not true, no.

GERARD: Within the nervous system itself, there is the so-called spreading inhibition wave across the cortex, but I don't think it is really a spreading inhibition. It strikes me as being a bit like the spread of silence over a group when the chairman begins to tap [130] and people become aware, in progressive sequence, of the cessation of room noises. |

Individuals react to the perception of quiet. The electrical silencing of neurons may spread as a positive stimulus of a different kind. But these items of inhibition are not critical for most of the discussion. There is no reason at present, except for one bit of evidence which has been sharply questioned, to believe that excitation and inhibition involve two different kinds of endings on the motoneuron.

Another obvious parameter is the number of endings. That has been discussed here frequently; it is reasonably obvious, and I shall not say more about it explicitly, except that a variation of intensity at an ending could be subsumed under a variation in number. A third one is the pattern of the endings in space, the position of these on the motor unit, and this I shall speak about later. And the last is, of course, the pattern in time, rather than in space, the frequencies at which impulses arrive and their temporal interaction. This needs more study than it has had, although it has not been entirely neglected. The physiologists here know a number of cases where it has been possible to reverse a reflex response, from positive to negative, merely by altering the frequency of stimulation to a given set of input fibers or by altering the pattern of two sequences relative to each other, so that the impulses shift in intervals of their pairing and separation.

Two important questions need to be considered. First, the response or responsiveness of a motoneuron can be increased or decreased in two separable ways: by increasing the inflow of exciting impulses and driving more strongly an essentially uncharged unit, or by making the unit more reactive to a given inflow. This, therefore, is one way of formulating the problem: Will we find excitation and inhibition, or either, to be primarily presynaptic, involving a change in input, or postsynaptic, involving a change in the responding unit itself? The second question is concerned only with decreases in the response or responsiveness of units: Is inhibition a zero; does lowered activity result only from the decrease of whatever has been causing excitation, whether pre- or postsynaptic; or is it a negative number, adding algebraically the opposite of excitation as an inverse excitation?

Let's look at excitation, quickly, first. (We are not going to be worried about whether excitation is by eddy currents or by the release of neurohumors; this is a different order of problem.) Excitation can lead to the discharge of the motor unit and, once a discharge has occurred, time enters the picture, for the unit goes through a so-called recovery cycle. It is, briefly, completely inexcitable, then excitable with difficulty, then more than normally excitable; and periods of sub- and supernormal excitability may oscillate for some time before the neuron settles down at its normal level. Such post-activation cycles may be important to the rhythmic outputs of the nervous [131] system we | touched on previously, but, of course, they do not come about unless units have actually discharged.

There is normally not only an immediate discharge when a volley hits the neuron, but also an after-discharge. In the great majority of cases, a single input to the nervous system leads not to a single output but to rather regularly repetitive discharges; and in some cases, this lasts for a long time. The continuing discharges could, of course, result from a temporarily maintained increase in the presynaptic barrage reaching that neuron, by reverberating circuits and the like; or it could result from a change in the properties of the motoneuron itself, other than the simple activation and recovery cycle. The rhythmic effects that can be obtained by polarization of the neuron would fit with the second alternative.

Changes in irritability can occur aside from discharge or after-discharge. Facilitation is an increase in irritability of this sort. The cell, as a result of one input, is more easily excited by other inputs. This, again, could depend on a change in the presynaptic barrage or in the motoneuron itself, the former implying a build up of more and more excitation, the latter, a progressive lowering of the amount of excitation needed, that is, of the threshold. These two, I think, can be separated; but this gets us into a long story, which we touched on before in connection with the building up of tension and with rhythmic or other discharges, and I shall not digress into it.

Let us turn to inhibition specifically and explore, first, the question: Is inhibition an inverse excitation or is it an absence of excitation Inhibition as inverse excitation, as a negative number, would, of course, fit Sherrington's formulation of a central excitatory state and a central inhibitory state beautifully; in fact, he developed that picture essentially to fit the observations showing that, in the reflex animal, by and large and in considerable detail, inhibition has the same physiological properties as excitation. It exhibits summation, after-discharge, recruitment or build-up; the times involved are similar for positive and negative effects; and so on. Physiologically, inhibition was the mirror image of excitation. We now know that this is not rigorously true. Lloyd's curves, for example, show excitation starting at a maximum and falling off logarithmically, while inhibition rises quickly to a maximum and then falls off along the mirror image of the excitation curve. Yet, even for the case of excitation of the nerve itself, an opposed process, a sort of inhibition (accommodation), exists, and the equations for E and I are symmetrical.

The second kind of evidence favoring the mirror image interpretation comes from changes in metabolism. So far as I know, there is still only one sharp observation on this, made by my students many | years ago at Woods Hole. If one stimulates the [132] inhibitory preganglionic nerve to the ganglion of the limulus heart, the oxygen consumption and carbon dioxide output of the ganglion decrease. A decrease in output could, of course, be a true lowering of metabolism of the ganglion neurons. As a matter of fact, we tried for years to show that, for nerve fibers, the oxygen consumption was increased by cathodal polarization and decreased by anodal polarization without quite making it come off; but it has been shown finally by Bronk's group, so such changes in metabolism are possible. The decreased respiration of the ganglion could also, of course, be due to the dropping out of some kind of spontaneous activity of inhibited units and thus be a fall from activity to rest, rather than from rest to active inhibition. But, clearly, inhibition could not be due to an increased activity of some units which thereby come into conflict with others, as has been demanded by some of the theories that I shall mention in a moment.

A third line of evidence has been electrical. Ever since Gaskell's finding in the eighties that, whereas an active region of the heart went negative to an indifferent electrode, an inhibited region went positive, there has been an attempt to relate excitation to decreased potentials and inhibition to increased potentials. In the case of nerve, responsiveness is depressed when positive action potentials are large. Eccles has just brought the report from Australia (not entirely unchallenged) that with a microneedle in a single motoneuron in the spinal cord, the membrane potential measured between this and an outer one shows a decrease during excitation and an increase during inhibition. If this holds up, it seems pretty conclusive that, at least in this case, there are changes in the motoneuron unit itself and inhibition cannot all be presynaptic.

It is possible to change the activity of neurons by polarizing them, and if a current in one direction increases reflex or rhythmic discharges, the reversed current stops them. We have often done such experiments. Even in the spreading inhibitory wave, so-called, or the Leão waves across the cortex, there are marked d. c. potential changes. As the wave goes by a surface electrode, the potential goes negative and then sharply positive, for minutes, and spikes are suppressed during the negative shift. In excitation of a cell unit, whether it be a neuron, nerve fiber, or muscle cell, there is a pre-existing membrane potential. In the case of muscle (which we have worked with mainly), this is around 90 or 100 millivolts, with the microelectrode inside and an indifferent electrode outside. We can change that membrane potential by an applied current pulse, and we find that whenever the membrane potential has been dropped to 57 millivolts, the muscle contracts, a full propagated response occurs. Moreover, if the pre-existing [133] potential | level of the membrane is kept raised or lowered by applying a polarizing current, either with microelectrodes or external electrodes, the current pulse needed to make the fiber contract is quite predictable from the difference between the actual potential and 57 millivolts. Dr. Jenerick has raised the potential to 100 or 110 millivolts and lowered it to 70, and the threshold always varies nicely with the change in membrane potential. Decreased excitability thus goes with a positive potential shift and increased excitability with a negative one; and inhibition again appears as the inverse of excitation, not as its absence. (When the membrane potential falls below 57 millivolts under the action of chemicals or currents, propagated responses drop out and only local shortening occurs. This is a separate phenomenon.)

How these changes might be brought about we do not know. There is always the possibility, which I am not inclined to take very seriously, of two different kinds of synaptic endings on a neuron. Another possibility, which I have taken seriously, is of different patterns of synpatic[!] endings on the neuron. I tossed out the notion, in the early thirties, that inhibitory endings might be gathered at the end of the soma relative to the axon hillock and excitatory ones at the other end. This spatial asymmetry, especially associated with an electrical gradient in the soma (urged also by Gesell), would account for much of the interaction of excitation and inhibition. It also has certain difficulties. When Eccles first reported his Golgi cell hypothesis of inhibition, which he has now discarded, during a symposium at Canberra, Australia, in 1948, I said he was proposing the same explanation except that he had cut a dendrite off and made it into another cell. I must not pursue this problem further now. Note, also, that I am leaving out peripheral inhibition. Here, perhaps most clearly of all cases, it is helpful to think of qualitative differences, including chemical ones, between excitatory and inhibitory endings.

Let us look at the other side of the picture. That inhibition may not be inverse excitation but merely absent excitation has been suggested many times. The first theory that I know of (shortly after Howell's chemical theory of inhibition, due to potassium liberation, of the heart by the vagus) is Lucas's suggestion based on Wedensky inhibition. Inhibition, in this view, results from piling up impulses at such a frequency that each travels in the relatively refractory period of its predecessor, is thus kept at a subnormal intensity, and so cannot get through a region of difficult conduction. Gasser's captured interneuron is another instance of interference. If one reflex arc is active and an interneuron which is necessary to this arc as well as to a second one is thereby engaged, the second reflex cannot come through because this link in its path is not [134] available. Still another variation, perhaps best | linked to Renshaw's early work but given attention by Lorente de Nó and recently emphasized by Lettvin and by Frank, places the block in the afferent fiber as it approaches the motoneuron. The incoming axon divides into ever finer twigs and presumably some are too fine to carry an impulse. The critical diameter at which conduction blocks is, in turn, a function of the chemicals about the twig, of eddy currents from contiguous fibers, and similar environmental factors.

The experiments that I now present bear on the question: Is inhibition due to an action on the motoneuron or to an interference with the arrival of excitatory impulses at the motoneuron? In the latter case, if a block is simply produced somewhere along the input by the inhibitory process, the properties of the motoneuron, on direct test, should not be altered during inhibition. The motoneuron can be tested, excited, directly in at least two ways. One is to place a microelectrode or semi-microelectrode in the cord near the neuron and to excite it directly by an electrical stimulus; the other is to stimulate its efferent axon and send an antidromic impulse back into the cell body. The antidromic impulse can induce the motor cell to fire back again, a one-cell reflex, so to speak. There have been conflicting reports about the effect of reflex inhibition on the antidromic events. Eccles reported a decrease in the response of a motoneuron to an antidromic impulse while the neuron was under the influence of an inhibitory afferent; Renshaw reported no change. I don't know what the facts are.

The experiments I have to offer (and the work I shall mention from now on has been done mainly by Dr. Karl Frank in my laboratory) use focal stimulation with a fine electrode in the cord, instead of antidromic stimulation. This is following Renshaw's technique. The needle is pushed into the cord, more or less by trial and error, until it is in such a position (rather dorsal to the motor nucleus) that a current flowing into it, of appropriate strength, excites about equal numbers of motoneurons directly and via presynaptic fibers. The evidence that this has been achieved is a response in the ventral root with two waves of about equal amplitude. The first is the response of some motoneurons to direct stimulation of their cell bodies; the second is the response of other motoneurons, one synaptic delay later, to impulses set up in presynaptic fibers close to their synaptic terminals. This dual response should alter, of course, under the action of an inhibitory afferent. If the motoneuron itself is involved in the inhibitory process, the first wave should be increased as well as the second. If the motoneuron's irritability has not been altered but there has been a decrease in the presynaptic barrage to that motoneuron, the second wave should be diminished during inhibition but the first show no change. Experimentally, that is the fact; there is no change in the first wave. How|ever, one must be sure, for example, that the direct stimulus is to the cell [135] body and not to its axon, but I won't go into the secondary problems involved.

Pitts: Excuse me, but this is during Lloyd's inhibition, is it?

Gerard: Yes, the ordinary spinal reflex inhibition.

The second line of evidence is obtained with a needle in the same region, but used to pick up electrical events instead of to induce physiological ones. After a dorsal root is stimulated, such an electrode registers a complicated electrical pattern. In this, the first waves represent the arrival of the presynaptic impulse, later comes a wave that has been called a synaptic potential, and, finally, a large wave signals discharge of the motoneuron. Under the action of an inhibitory conditioning stimulus, this last wave is diminished, since fewer cells discharge to give the inhibited reflex response. This obviously must be so, but there is a report by Eccles that the synaptic potential under these conditions is not changed.

Pitts: I believe he thinks the decrease is very slight.

McCulloch: The decrease is very slight, that is what he said.

Gerard: Good. What we seem to have found is that the still earlier waves, and also the synaptic potential, decrease under the action of an inhibitory conditioning volley.

If that is correct, whether or not there is any direct inhibitory influence on the motoneuron, there must certainly be a cutting down of the input to that motoneuron. These first waves are too early to be anything other than the incoming presynaptic impulses.

A third line of evidence is from the action of polarization on the motoneurons in the cord. If the irritability of motoneurons in a pool is raised or lowered by appropriate polarization, through their axons in the ventral root and an indifferent lead from the cord, more or fewer neurons will respond to a given afferent volley and the size of the reflex potential will be increased or decreased. This is all perfectly straightforward. But what should be the effect of such polarization on the amount of inhibition of this reflex response produced by a conditioning inhibitory afferent volley. If these are changes in the properties of the neuron itself, increased or decreased irritability, they should hardly be symmetric for excitation and inhibition. It would be most unlikely that the percentage inhibition of reflexes of widely differing absolute magnitudes would remain unchanged. But if the inhibition is acting presynaptically, so that only the amount of activation reaching the motoneuron is decreased, polarizing the neuron should change its absolute response but leave the percentage inhibition unaltered. In other words, on that basis, the curves of inhibition of the reflex, as a function of interval between conditioning and test impulses, should be [136] | superimposable when plotted as percent of the uninhibited response whether the absolute value of the uninhibited response has been multiplied severalfold or cut down severalfold by the polarization. This is, indeed, the case; the complex curves are fully superimposable.

The last bit of evidence I shall mention is, to me, the least convincing but I give it for completeness. Attention of workers in this field has been focused overwhelmingly on the two-neuron part of the reflex response evoked by stimulating a dorsal root or an afferent nerve, and not much attention has been given the multineuron responses. I am sure that those of you who have worked with these preparations are aware of the great variation from animal to animal in the relative sizes of the two- and multi-neuron elevations. If preparations are used in which the multineuron response is relatively large instead of relatively negligible, so that the two elevations can be easily compared, it is our experience that a dorsal-root conditioning volley that gives profound inhibition of the two-neuron response does not affect the multineuron response at all. Of course, it might be argued that different motoneurons are involved in the two responses and that only those of the two-neuron arc are reached by the conditioning inhibition, but this is rather a *post hoc* assumption.

The other interpretation would be based on the fact that, in the two-neuron response, the input afferents go directly to the motoneuron, while in the other response interneurons are interposed. It does not seem an arbitrary assumption that the two different sets of connections reaching the motoneuron are not morphologically identical. Primary afferents might arborize quite differently from interneuron fibers. If the incoming fibers split into finer and finer branches until some of the branches become so small that the impulse dies out in them before reaching the motoneuron, changes in the environment of those branches can alter the critical diameter – not necessarily the diameter of the branch but that diameter which it must have to permit conduction – so that more or fewer of the impulses will die out on the way to the motoneuron. Such changes could be produced, as I indicated earlier, by electrical fields set up around them as a result of activity in other fibers, or they could be produced by chemicals liberated from other units nearby or entering intercellular spaces due to altered blood-brain barrier permeability, or they could come about in still other ways; but this takes us to a second level of hypothesis and so is a good place to stop.

KUBIE: May I ask two questions, Ralph: what role, if any, does mere synchronization of the activity of groups of cells play in this whole process, with relation to corrective action, for one thing, and in relation to the EEG, for another? My second question has to do with | the matter of the refractory phase. To what extent does the effect of excitation or inhibition depend upon the relative number of grouped neurons which are in a refractory phase at the same time, irrespective of whether the refractory phase involves the cell body, the fiber, or the synapse? [137]

GERARD: The first question, I think I can answer. I am not sure about the second.

KUBIE: The first, of course, refers back to Adrian's paper (1).

GERARD: Yes. You are asking about the area which I explicitly omitted from discussion, namely, the field effects. What this must amount to is as follows: Since there are potential changes around a neuron and its processes when it becomes active as compared with the preceding situation; since these must lead to a flow of currents in the territory about it; since this will happen for each cell; since the more cells there are undergoing changes in the same direction at the same time, the more will these currents (if the geometry is right, at least) tend to add up at a given point in space, that is, at an appropriate part of some other cell; it follows that the more synchronous activity exists, the greater is going to be the tendency for a group of beating cells to recruit additional ones and bring them into phase with it. This is a mechanism that has had much study, and one can even measure it quantitatively in terms of how hard it is to produce desynchronization. It is true for the brain that the more regular the electrical beat (the alpha rhythm, if you want to call it that, although this has been done on the frog brain) the more difficult it is to break it up. It has also been shown in model experiments on Nitella by Fessard, who saw individual cells, scattered around in a dish and connected only by a conducting medium, recruit to an active center, so that eventually all were giving their electrical discharges in synchrony.

What the state of refractoriness of a given fraction of this population of a pool would do to relative excitatory and inhibitory results, I should like to think about a little before I try to answer.

BATESON: I feel very much challenged by this paper, and I think it is a challenge to the whole group to discuss what Ralph has said, not at the level of asking about the domestic details of the particular servo that he is investigating, but at the level of asking how his handling of these problems and his presentation of them and conceptualization of them bear upon how those of us who are not neurophysiologists should think about problems in personal communication, in evolution, in all fields to which the cybernetic approach is applicable. We are not here as a neurophysiological technical group, trying to solve neurophysiological problems, but we are profoundly interested in how those who solve neurophysiological problems do it. We are interested in the | nature of such problems, because similar problems occur in a vast number of other fields. [138]

I have been racking my brains to consider what order of question I can put which will, at least implicitly, focus discussion upon general philosophical and epistemological problems, even though it be verbally a domestic neurophysiological question. I think of two such questions. Ralph has handled the neuron as if it were not itself of servo complexity, as though it were a relay within a servo but not itself of the order of complexity of a circular causal chain. One question would be, how would the expected characteristics of the total servo, of the brain, shift if the neuron is alive, if it is itself a self-correcting element?

YOUNG: He excluded that consideration by his initial remarks.

BIGELOW: We are having such difficulty in understanding how a system will work, making very simple assumptions about the building bricks of which it is composed, that if you try to complicate the assumptions, we will be in even more of a morass.

BATESON: Well, I see a second type of question which one might explore. We have, from the input-output characteristics of the larger system, arrived at the word »inhibition.« Does it make very much difference to the major characteristics of the system how this inhibition is produced at the relay level ? Or have we here a case of the sort that Pitts was referring to, in which, whatever the specifications, the machine can always be built?

MCCULLOCH: May I, in part, come back to your first question. Von Neumann pointed out at the Hixon Symposium that the total number of particles in a brain is fabulously great. It is of the order of 10^{27}, 10^{28}, ultimate physical particles. If such a problem can be split into two parts, one, the size of a neuron, which is something of the order of 10^{16} ultimate physical parts, and second, the number of neurons in the order of 10^{10}, you have a very nice simplification of it. You are then able to consider the behavior of the unit composed of 10^{16} parts as a totality, or, since they have relatively simple functions as components in the more elaborate circuit you are able to think about the more elaborate circuit, taking properties for granted for the individual neurons.

GERARD: Let me make an additional comment. Your question would resolve itself, I think, into taking into account changes in the threshold of the neuron, whether produced during the recovery cycle as a result of stimulation or appearing independently of any deliberate manipulation. This latter is the spontaneous play of threshold, which has been recognized and studied and found to be essentially Gaussian. Perhaps the thermal movement of ions in the membrane is enough to give changes in threshold, [139] even to the extent of actual discharge. I think | that is really what your question would resolve itself into.

BATESON: I remember that two or three years ago (2) we had a report of something that von Neumann had said, which he did not actually say in this room but Dr. McCulloch reported on it. It was that the order of complexity of the brain, the number of relays there, was much too low, and that what was needed was not 10^{10} relays, but something of the order of the square of that number, in order to perform the informational and computing functions that the brain seems to perform.

PITTS: That was by no means demonstrated.

BATESON: It was not demonstrated, but the figures were fantastic.

MCCULLOCH: It was not demonstrated, but there were too few neurons to account for human memory if memories were based on interlocking memories. He wanted something smaller than the neurons. He wanted information stored *within* individual neurons.

WIESNER: But this is based on some very simple assumptions of what memory is.

PITTS: Simple combinatorial assumptions that I think were not convincing at all. As a matter of fact, I don't believe it.

MCCULLOCH: I don't think he was right, but that is what he said.

YOUNG: It assumes the independence of different bits of memory, too.

WIESNER: That's right. That is the most serious fault.

MCCULLOCH: Even without that, it always leads to a theory of simple location, which is very bad in physiology. We don't seem to find that kind of location.

PITTS: I should like to say, giving a more concrete answer to Gregory Bateson's question, if the present line of attack, in which Ralph Gerard and Lorente and ourselves are included, succeeds as well as it promises, we may be in a position at the end of some number of years to be able to say something about the qualitative nature of the kind of relaying that occurs whenever the afferents to a given neuron are in a certain geometric relation with its cell body and its axons and its dendrites. If the geometrical relations are of a certain kind, we should expect the result to be inhibitory or excitatory depending upon the flow of current generated in the medium, the directionality in relation to the cell, and so forth, by the presynaptic elements as they are arranged.

As soon as this can be done, there is a wide field of additional possibilities. We know a rather fair amount about the anatomy of the brain, although certainly not as much as we should like to know about its fine details of construction, but most of this, until now, has been almost completely uninterpretable except on the simple level of there being a possible path between thus and such a place and thus and such a place. We can say that thus and such a nucleus generates a track | which leads to thus and such [140] another nucleus; we can stimulate A and record from B, so we say there is an interaction at least from A to B and possibly in the reverse direction. We can say that information is communicated from one to the other. But our picture of the nervous system as derived from anatomy and physiological experimentation has largely divided it into subsystems, that is, particular regions in which something or other is computed, we don't know what, and the result of that computation is transmitted to somewhere else, where it is combined with some other information, where something else is computed, and the ultimate result is, of course, the output. But we have had absolutely no way of interpreting what we know about the details of the anatomy, of the particular nuclei, the particular stations where things are computed, to be able to make even a plausible guess as to what is computed. If we have a physiological theory which allows us to make physiological inferences from knowledge of relative geometrical positions of the elements which come in contact, we can begin to say what is done in the nucleus and then, of course, we can go immensely further.

BATESON: I am not criticizing the beauty or the relevance of the attack on the problem, nor the choice of problem, nor am I criticizing the exposition, which was excellent. I am saying that there is a challenge to this group to meet these data in general terms as well as in terms of the specific problems at which the research is aimed.

PITTS: Very often, we can say that a given sort of computation is localized to a given place, but if we could, for example, say that whenever the afferent inputs to a neuron end in a certain place and in a certain direction in relation to the cell body, the result must be that the afferents inhibit the neuron or that they excite it, then we can begin to say how the function is carried out, from knowing the anatomy.

BATESON: I agree.

PITTS: And that would be an enormous step beyond anything we have ever had before.

KUBIE: I still think that there is one important missing link in our thinking about this. I had this in mind with my initial question. There is some danger that we may be using a single word for two quite different phenomena, namely, inhibition or excitation in terms of the individual cell, or in terms of the summation of activity which includes the brain as a whole or large parts of the brain, coordinated with the body as a whole, and manifesting itself in terms of total activity. Adrian used as his figure of speech the difference between giving an order to an unruly mob or giving an order to a well-trained regiment (1). Better still, let us think of a democracy consisting of many factions, individuals and groups, churning around in an enormous amount of internal activity. These extremely active groups of individual units would | cancel each other [141] out so that, in terms of external action, very little would happen. For the body as a whole, there would be a paralysis of external action. Then along comes a war, producing a synchronization of everyone's effort, a patterning of purposes, so that a form of action occurs which is effective externally, yet requiring a great deal of inhibition as far as the individual units are concerned. What I asked about synchronization, or at least what I had in mind, was whether activation or inhibition with respect to total function necessarily parallels the state of activation or of inhibition of individual neurons.

BIGELOW: There is a strong chance that it does.

GERARD: As you know, what you are talking about now I have played with for two decades while what I presented is something I have only been thinking about in detail for the last year or two. What has always bothered me is how the final effector elements, by which I mean the motoneurons, are engaged after all these processes which go bouncing around inside the nervous system have done their bouncing. I am convinced, in fact, I made an issue of it at an earlier meeting of this group, that these field effects or mass effects are extremely important in that synchronization by nonsynaptic mechanisms is extremely important. All I am saying now is that whatever the mechanisms may be, eventually they must engage discrete units, and the kind of excitation and inhibition I am talking about appears at those neurons. The serious problem is how these holistic processes, whether they are discrete impulses running around geometrically identifiable nets or are profiles of potential patterns in an area or a volume, finally settle down at the specific motor unit. This requires exploring the point of junction between the whole and the unit.

YOUNG: I am a little worried at your separating synaptic events from field effects. Aren't there all transitions in between?

GERARD: As I indicated, both mechanisms come into play; but finally you've got to account for neuron A firing or not firing at time T.

KLÜVER: At one of these meetings you emphasized the semiliquid nature of the nervous system. Apparently you did this under the impact of a conference on genetic neurology (3) held at that time. Some neuroanatomists forty or fifty years ago said that the dendrites are moving.

YOUNG: I say it now.

KLÜVER: I don't know whether this quivering or moving is really taken seriously by electrophysiologists. I have seen no evidence of it in their publications.

You started your discussion by talking about behavior. You mentioned, for instance, the fact that the same behavioral objective may be achieved by a variety of bodily movements. You even mentioned the | transposition of melodies. I had hoped you would come back to this at the end, but you did not. [142]

GERARD: One of my favorite tricks is to quote the closing paragraph of an article which I wrote in 1933, comparing the brain to an orchestra, but I manfully avoided that today. The only point of speaking of the transposition of melodies was that a shift of pattern could occur along several dimensions and yet there is still a sequence of individual units. It was not the recognition of melody that I was getting at. On the other question you raise, I think I will defer to John, for he has done much recent thinking about it. I think you are wrong, Heinrich, in saying that physiologists have not paid attention to it.

KLÜVER: I was merely raising a question.

GERARD: They don't often come out in print with their thoughts because they haven't felt that they were too helpful. But Speidel has tried to explain the effect of electroshock in curing schizophrenia in terms of retraction and regrowth of axons and dendrites. Eccles is now trying to explain inactivity of a reflex following functional denervation, and of the recovery with activity, by a shrinkage of the terminal knob and a reswelling when nerve impulses arrive over the nerve fiber. There are changes in the diameter of the nerve fiber with the conduction of the impulse, as both David Hill in England and Julian Tobias at Chicago have shown. Hebb, in his monograph, has come out for memory in terms of changes in the geometry. Go ahead, John.

YOUNG: Terzuolo (4) showed the development of the dendrites in the ciliary ganglia of chickens with age. The preganglionic fiber in a young chicken is a relatively simple basket, but as the chicken ages, it develops a tremendous complexity of the dendritic apparatus. The great complexity of the dendritic apparatus develops in relation to the input fiber. As the pecking reflex gets to work, the ciliary ganglion is presumably working. There is no experimental evidence to show whether that is due to function, but that is a possible inference.

GERARD: It has been reported, by whom I have forgotten, that, as a result of stimulation of afferent nerves, one can show an increase in glia around the active neurons in the dorsal root ganglion. I don't know how valid that may be.

BIGELOW: How many terminals do you get in the case of the chicken? Are there 100 or 500?

YOUNG: There is one axon coming up, branching out into a tremendous area of contact. That is one of the things I wanted to ask Ralph; actually, his whole conception of an ending is one that the histologist does not find.

BIGELOW: Suppose you take the first point of contact where any one fiber touches and count them, would you have 50 points or 100 points? | [143]

YOUNG: Of that order, yes.

WIESNER: Does that all apply to one receptor?

YOUNG: Yes, one receptor.

PITTS: These are all axonal branches?

YOUNG: All axonal branches, and one preganglionic fiber in contact with dendritic branches.

PITTS: Do the dendrites also increase?

YOUNG: Yes, tremendously.

BIGELOW: Is there any information on the number of states that this thing can achieve?

YOUNG: Well, I suppose it is only two. Ralph, isn't that right?

GERARD: Yes.

BIGELOW: Well, I understand this controls the eye?

YOUNG: It controls the accommodation muscles, presumably.

BIGELOW: This one ganglion does the whole job, or is it one of a family?

YOUNG: It is one of a family of cells and they presumably have a variety of jobs. They operate the various parts of the accommodation mechanism.

BIGELOW: Is there any information about how many states of accommodation the eye has?

GERARD: That would be just a matter of how many separate muscle cells are involved and whether they can respond in all-or-none fashion.

BIGELOW: Is this known at all? Can you guess at it?

GERARD: I would say it is a large enough number to be great compared to the number of nerve cells here.

BIGELOW: Thousands?

MCCULLOCH: What is more, in the case of muscle as in the case of the lens itself, there is a lag in the effector, so that a change in the rate of impulses over any given fiber will cause a change in the adaptation of that eye.

TEUBER: But as I understand it, Dr. Young, there is no evidence as yet that such processes are not strictly embryological.

YOUNG: Due to maturation.

TEUBER: I had the impression that there is a certain amount of evidence rather supporting the notion that dendritic-outgrowth is sheer growth, independent of function.

YOUNG: Oh, no. I would be prepared to support the opposite view. Axons and dendrites tend to atrophy when cut off from their input or when cut off from their effective means of operating.

PITTS: Anything more positive?

ASHBY: There is another piece of evidence given by Carey (5), who found that the [144] motor end plate is ameboid, that it will retract after injury. |

YOUNG: There is some evidence that any motor nerve cell gets smaller in diameter if its output is reduced or if its input is reduced.

TEUBER: But is that generally true? I know we have particularly striking trans-synaptic changes in the visual system, when the visual input is cut off, but that is a special situation. Is it at all likely to occur in other neuronal systems?

YOUNG: Yes, I think so.

TEUBER: I was not aware of that.

YOUNG: I am prepared to say that it is much more generally true than is suspected.

BIGELOW: Can you do things like anesthetizing the organ for a long time and see if its growth is pathologic?

YOUNG: There is the famous experiment of animals kept under anesthesia from the time of hatching; when let out of anesthesia, they behaved as normal.

BIGELOW: Their nervous systems were perfectly normal as far as could be told by examination?

YOUNG: Histologically, it has not been done very completely, but a lot of maturation can go on without receiving outside information. The animal develops with the information received by heredity.

GERARD: I don't know of any evidence of spatial changes associated with inhibition. Of course, the minute you start postulating these changes for long-lasting recording of past events, for memory, in contradistinction to the fairly immediate response and after-discharge and so on, then you get into difficulty. As Warren already mentioned, this leads to localization of memories and there is strong (but not unanimous) evidence against that.

YOUNG: Let's not get into that problem.

BIGELOW: But you made one other statement, which seemed to me very conclusive, a statement about the energy consumption as a function of whether you inhibit or whether you excite. Was that for a single neuron?

GERARD: No, that is a ganglion, a group.

BIGELOW: That is for a ganglion? You inhibit, and the energy consumption definitely decreases. Is that correct?

GERARD: Yes.

YOUNG: And there is no muscle in that.

GERARD: No, this is just nerves.

BIGELOW: I think those experiments are quite remarkable.

GERARD: Yes, I think so. Actually, comparable results have been obtained independently by another worker.

BIGELOW: When you inhibit, you are inhibiting the entire ganglion? Exactly how was [145] the inhibition produced? |

GERARD: The ganglion was placed in a microrespirometer, with its ordinary efferent nerves to the heart cut, of course, but with the preganglionic connections attached.

Stimulation of the preganglionic inhibitory nerve, which stops or slows the heart when it is attached, cuts the oxygen consumption of the isolated ganglion to one-fifth.

PITTS: Can you tell me anything more about the microscopic findings on that ganglion? Is there any geometric relation to the cells of the two afferents?

McCULLOCH: Which ganglion is it?

GERARD: It is the Limulus heart ganglion. The geometric relation may be known, but I doubt if it is.

VON BONIN: The endings on the second cervical nerve in the cat have been investigated, I forget by whom. That nerve sends some fibers across to the other side, and it controls the small muscles that turn the head. There is here, as well as somewhere down in the tall region, an inhibition of the muscles of the other side. The person who did this work investigated the degenerating end buttons after cutting the posterior roots, and they are distributed helter-skelter all over the cell. There seems to be no particular pattern.

McCULLOCH: That is the inhibitory group?

VON BONIN: Yes. They are on both dendrites and bodies.

YOUNG: That is not conclusive, is it?

GERARD: And in the Mauthner cell there is evidence of localized group endings.

VON BONIN: Yes.

YOUNG: But surely the separation of the inhibitory and excitatory afferents there is rather speculative, to put it kindly.

VON BONIN: I am talking about the nerves that innervate the obliquus capitis superior between the epistropheus and the occiput. The one side contracts and the other side relaxes. After cutting one posterior root, the author investigated the degenerating fibers which cross over to the other side of the spinal cord.

YOUNG: Do you know that there are no excitatory fibers, going across the cord? How can you identify the cell bodies of those neurons which are only being inhibited from the opposite side of the cord? I would say that was a speculative thing.

VON BONIN: I would like better evidence, I grant you.

McCULLOCH: The cells in the tail region of the cat, that Lloyd has investigated fairly carefully, are only inhibitory.

VON BONIN: But the same situation applies, so far as my information goes, to the cervical end. Again, it is only inhibitory.

McCULLOCH: But I don't think the cervical region has been as fully investigated as the caudal region. | [146]

VON BONIN: That is true, too.

McCULLOCH: May I come back once more to Bateson's question? I think detailed knowledge of the housekeeping, so to speak, of the individual neurons is important in a way that we don't take into account fully at first glance. It would be possible to construct a nervous system out of neurons, each of which only fire spontaneously, and to bring to it axons whose impulses inhibited, and only inhibited, those neurons. It would then be capable of full combinatorial handling of information, but the number of neurons would have to be vastly increased and the times required would be much longer. What would be necessary would be to construct a Sheffer stroke function, and a neuron which was inhibited and only inhibited would then have to go to another neuron which was inhibited and only inhibited to get the equivalent of an assertion. If you take almost any line in the *Principia Mathematica* and convert it into Sheffer stroke functions, you will find that for a relatively short expression in Principia nomenclature, you have some 270-odd strokes to worry about. Now, if you have neurons that can be excited and inhibited, you can, in a single synaptic delay, achieve combinatorial power far greater than you can otherwise.

PITTS: That is also true of Gasser's type of inhibition (6). The difficulty lies in the long time such inhibition would require.

MCCULLOCH: Yes. Gasser's type of inhibition is too time-consuming to be used throughout. The nervous system apparently has at least two kinds of inhibition that it utilizes besides the Gasserian: The one Ralph Gerard has been talking about is clearly demonstrable, I believe. The second kind is far more lasting and does result in a rise in threshold of the cells; it is brought about in spinal cord by excitation of the bulboreticular (so-called inhibitory) mechanism, which is very important in the overall inverse feedback necessary when hosts of closed-loop control mechanisms are combined. There must be an adder of some kind to secure stability. The fact that there are these kinds of control over the next relay makes for much greater overall economy in the construction and for much greater rapidity in the action of the nervous system in question.

BATESON: This is the sort of thing I am asking about.

GERARD: I interpreted your opening comment, Gregory, as saying you would like us not to get into a neurophysiological argument about how much current there is or which way the dendrites go or something like that, which I was very glad to have you say.

BATESON: Unless that argument illustrated the wider approaches.

GERARD: That's right. And Warren has exhibited one concrete instance of it. I would [147] like to answer your question at the level of | greatest generality, or at least at one of greater generality. Many of the group here are working fairly seriously to create models, whether mathematical or mechanical or other, that will do what the nervous system does. Each time we are able to specify a little more precisely what those models must do, the way they must act if they are to be like the nervous system, the easier that job becomes. Let me point that up by reminding you that we all talk and think in terms of the brain being made up of neurons, and the fact that the brain is made up of neurons is just this kind of a housekeeping detail that some poor laboring neuroanatomists worked out in the last century. We are simply feeding to the theoreticians some limiting conditions for their theories.

MCCULLOCH: May I put it in Dave Rioch's phrase? The best model for the behavior of the brain is the behavior of the brain.

BATESON: The moral which I, working in another field, should draw from what goes on is: Divide the problems in the sort of way that Warren has been talking about and get the limiting specifications, if you can, at the various levels of that division.

ASHBY: In a sense, I agree with what Dr. Gerard has said. Restricting a model too closely may nullify the whole work. Suppose, for instance, that a modelmaker found a new way of preparing proteins, and suppose he made miniature neurons, put them together, and built a brain exactly like the human brain, and found that it behaved exactly like a human brain, he would learn nothing from it. We learn things only by building the model within a wide conceptual field, so that we can see what is the relation of the brain as it is to what the brain would be if it were built differently. We can understand the brain only in relation to a wider scheme of things.

GERARD: I would agree with that, that is, if you are asking for further conceptual generalizations from the particular effects. On the other hand, I would disagree with you that, if you could handle the protein molecules and had built a brain, you would not have learned anything, because you could not have done it without learning all the necessary things to put it together properly.

FREMONT-SMITH: We might also point out that such a brain would be without a history and therefore you would learn a lot from what it couldn't do as compared to the brain that had grown up from a baby.

ASHBY: I have learned that already.

PITTS: I think modelmakers in general really have two functions. First they want to demonstrate that thus and such a function, which various people suppose can only be done by the mind's substance or some other nonphysiological entity, not by any mechanism, can in fact be done by some mechanism. This is an extremely important educational task, since certainly the vast majority of the world would refuse | to accept for a [148] moment the assumption that what they regard as specifically psychological functions can be done by any mechanism whatsoever; but, so to speak, from our point of view, we can regard that as a decided question. As soon as the psychological function can be specified strictly enough, as soon as the psychologists know exactly what it is they want the brain or the mind in question to do, we can certainly build a machine to do it. That is the first thing, and I should say that for us it is superfluous, but it is extremely important with respect to the rest of the world.

The second function of modelmakers is to find models that throw a light, either directly in the sense of making a mechanism out of components as much like what we know about the neurons as possible, so that we can perhaps form direct suggestions as to how in fact the brain does something, or else indirectly, by means of elements which have certain of the formal properties, although by no means necessarily all. We can see what sort of mechanism would efficiently and reasonably perform functions that we know occur psychologically, so that we can get some sort of insight in the possible regions in the brain that actually do them. It is very important to distinguish between these two functions of the modelmaker. What the neurophysiologist can tell us in the way of limiting conditions is important for the second function of the modelmaker, although, of course, not at all for the first.

KLÜVER: As a psychologist, I have made up my mind about a few things that the »brain« ought to be able to do. Dr. Gerard, for instance, referred to an equivalence of motor reactions when he pointed out that the same end can be achieved by diverse movements. Similarly, on the sensory side, we find an equivalence of stimuli when the organism is handling a heterogenous sensory input. Confronted with widely different stimuli having hundreds of properties, the organism may fall back on its ability to pick out, isolate, identify, attend, or respond to one aspect only or to one common property. There is no doubt that neuroanatomists and neurophysiologists have come forward with respectable suggestions as to brain mechanisms involved in discriminating stimuli of the external world; they have been strangely silent as to the possible mechanisms involved in identifying or isolating one or several aspects in heterogenous stimulus situations.

PITTS: That is where the modelmaker, of course, can have his greatest utility.

VON BONIN: May I come back to something that Dr. Kubie said a while ago? He said that Adrian referred to »an unruly mob« in one place and »a well-trained regiment« in another place. I think what Adrian had in mind at that time was the cerebral cortex and the spinal cord. I wonder whether the rather precise synaptic organization of the | [149] spinal cord can be postulated for the cerebral cortex. Are there boutons terminaux in the cerebral cortex? I don't think anybody would like to stick his neck out and assert it. The organization in the cerebral cortex may really be on an entirely different principle.

GERARD: This came out once before, when I spoke about digital and analogical mechanisms at the synapse.

KLÜVER: The ability to pick out and respond to a common factor in widely different stimuli is found, of course, even in animals without a cerebral cortex.

VON BONIN: What I wanted to keep open was the question as to how the neurons are put together. We know that in the mammalian spinal cord, they are put together in a certain way, but we are not at all sure how they are put together in, well, take the octopus.

GERARD: I will, foolishly, react to that; to say anything is unwise. It would be awfully nice if there were so simple an explanation as mere differences in morphology to account for the vast differences between the regimented behavior of the spinal cord and the imaginative, creative behavior of the brain. I don't think it is going to come out that easily.

VON BONIN: It probably won't.

YOUNG: There are differences in probability at the interneuronal connections. There are some synapses with a high degree of contact, where there is, very high probability of transmission, and the series goes down through a cord synapse, where the whole system is less highly determined than that, but still fairly highly determined, until it comes to the cortex, where effects at considerable distances may be more important and the probability of transmission along any one line is low.

GERARD: I think that is right.

VON BONIN: We approach a random net.

FRANK: Gregory's earlier statement has implications for interdisciplinary meetings of this kind. Here we come with different bodies of knowledge, experience, assumptions, and conceptions. I take it that what he was asking – and it seems to me this is pretty important for all our future meetings – is that the presentations made, of whatever degree of particularity, be looked at by people of different levels, who say, »Look, are the assumptions that you make such that we can take those when we deal with the social sciences?« and, conversely, those who are dealing with neurology have to come back and say »Are the assumptions you are making in social science compatible with the kind of assumptions and knowledge we have about the nervous system?« That is the point I think Gregory is after, and it seems to me, unless we do that, we merely resolve ourselves into a series of small, discrete disciplinary groups who take in each [150] other's washing. |

BATESON: Which is to derive the analogic lesson in how to tackle problems of a certain sort.

TEUBER: But on what side of the synapse should it happen? Does it devolve upon the speaker or the listener?

BATESON: Both.

MEAD: It was a very interesting thing that happened. In the course of that discussion, Gregory used »domestic« to mean a discipline or a small group of disciplines. It was transformed into a denigrating word: »housekeeping.«

MECHANICAL CHESS PLAYER

W. ROSS ASHBY

The question I want to discuss is whether a mechanical chess player can outplay its designer (1). I don't say »beat« its designer; I say »outplay.« I want to set aside all mechanical brains that beat their designer by sheer brute power of analysis. If the designer is a mediocre player, who can see only three moves ahead, let the machine be restricted until it, too, can see only three moves ahead. I want to consider the machine that wins by developing a deeper strategy than its designer can provide. Let us assume that the machine cannot analyze the position right out and that it must make judgments. The problem, then, becomes that the machine must form its own criteria for judgment, and, if it is to beat its designer, it must form better judgments than the designer can put into it. Is this possible? Can we build such a machine? The problem that faces the designer is the same as that of the father who is not a good chess player and who wants his son to become world champion. Obviously, he must be very careful about what he teaches the boy. If he teaches him rules like: »Always get your queen in the middle of the board as quickly as you can,« he may do permanent injury to the child's chess-playing powers.

The problem, then, is how is the machine to develop better criteria of judgment than the designer himself can produce. We can get a line of argument by considering a chess position – I haven't one available, but probably you all can supply one out of your memories – that looks fairly ordinary and yet, in fact, has a powerful move possible. How are we to get the machine to play that move? Suppose the move is so subtle that not even our best players can see it; how are we to get that move played at all? I say that there are only two classes of players that are capable of making that move. One is the beginner, who is so bad that he can make any silly move, and the other is a random player that just draws its moves out of a hat. The one player who can never make that move is the mediocre player: he has his rules of thumb for playing and they are not good enough.

Wiesner: I don't think that necessarily follows. He may be so mediocre that he is playing a random game, too.

Ashby: In that case, he is in the other class.

Wiesner: The fact that a man is a mediocre player doesn't necessarily mean that his rules exclude his making a superhuman move.

Ashby: Not rigorously, but the tendency for the mediocre player is to reject the very good moves and go for the mediocre moves, not | necessarily, of course, but usually our first deduction, then, is that if the designer wants to get moves that are better than he can provide himself, he must go for them to a random source of moves.

The next principle is that the machine, as it faces a position, must form a great variety of transformations from the position, must form a valuation from each transformation, must follow the line of action that these valuations suggest, and then, when the game is over, must go back to modify the transformations and valuations simply according to whether the game was won or lost. The transformations and valuations can be formed, at first, entirely at random. A Geiger counter suitably worked up will provide variety and may lead to such rules as: always keep rooks as far apart as possible, or, always keep one bishop on a white square and one on a black.

Pitts: You can't do otherwise.

Bateson: They stay that way anyway.

ASHBY: That was a good random one, then. The result is, of course, that the first games will be silly. But if there is corrective feedback that is operated by results, such a machine, breaking up the transformations when the game is lost and holding them when it is won, will inevitably move its population of transformations from the completely random toward those transformations that are the right ones for winning the game.

BATESON: The noisegenerator on your machine is now in a different relationship to the organism and the environment from what it was when you were using the machine to illustrate homeostasis.

ASHBY: Yes. It is wanted simply to provide variety. It could be a Geiger counter, providing a stream of irregularly varying numbers. If this machine is a determinate machine, it cannot make randomness out of nothing; it will demand a specific instruction. If you say »Watch the Geiger counter, take the last three numbers on it, and form your transformation in that way,« you are giving it specific instructions that it can follow; but because it is getting its instructions off a Geiger counter, you are getting possibilities which are not limited by the limitations of the designer. If the designer said, »I have provided you with a great number of good transformations; select the best,« the machine is restricted by the very best that he can produce. But if the transformations take Brownian movements as their source, theoretically, they have no limitations at all. However complicated and subtle the really good transformations are, Brownian movement can provide them. I suggest that something like Brownian movement is the only place where they can be found. They can't be provided by hypothesis.

BIGELOW: Why not?

[153] ASHBY: That is my basic hypothesis, that our intelligence goes so | far and then stops, and that we cannot provide these transformations. We are trying to find something better than what we can provide from our own skill.

WIESNER: You can provide all the elements of the transformation.

ASHBY: Yes, but then we have still got to provide some instructions for their combination. If we try to construct it in detail as a determinate machine, it will be limited by our ideas. With Brownian movement, we can just let it go.

BIGELOW: It is not at all clear that the addition of the Brownian movement adds one iota of information to the system.

WIESNER: Suppose you tell the machine to try all the possible combinations. One main difficulty is that it would come out no better, as we said previously, than the Brownian movement.

BIGELOW: Exactly.

WIESNER: If you have a stack of cards and you shuffle through to find something, without knowing anything about the order, it doesn't matter much whether you do it in a systematic way, if there are a fair number of operations to perform, or do it randomly, provided you examine each thing only once. If you inject the Brownian motion, you run the possibility of sometimes taking longer because you do certain operations more often. Perhaps occasionally you will come out better, but on the average, I think you will come out exactly the same.

BIGELOW: Exactly so. Furthermore, I see no possible way to distinguish between the analysis of the situation and the formulation of a strategy in a game like this. I think if you put any limitations on the ability of either machine or human operator to analyze, you put an exact equivalent limitation on the ability to form strategy. To that extent, the problem is closed as soon as you state it. It has little further interest. If you limit the ability of the person or the machine to analyzing three moves ahead, then you put an absolute limit on the variety of strategies they can choose.

ASHBY: I am suggesting just the opposite, that this random method can get past the limit.

BIGELOW: I thought you started out by saying that you chose to discuss those games on which this limitation is imposed.

HUTCHINSON: Your idea is that some games would be played in which the play could not be known more than three moves ahead to be good and therefore could not be chosen on any reasonable criterion. Such good moves could, however, come in from the noise-producing mechanism and so will happen to be played. There is then a selector device so that when this happens, memory comes in, and the process is of some use in the future?

ASHBY: Yes. | [154]

HUTCHINSON: So the real criterion is what the opponent can learn.

ASHBY: Yes.

BIGELOW: It is impossible to separate the function of analysis – and by analysis, I mean computing the possible things that can happen – from the operation of formulating the best strategy for a move at any one point. They are one and the same.

ASHBY: I deny that.

BIGELOW: Well, let's take it up.

ASHBY: I went through Capablanca's *Chess Fundamentals* (2) the other day, and found many sentences each of which gave clear advice in a general way without making any specific analysis on specific squares. One example I will quote because it did in fact go beyond my standard of play. When the game has reached a point at which there are only pawns and bishops left, the beginner always takes his pawns off the squares that the enemy bishop covers. Capablanca, however, advises that the pawns should be moved on to the squares the enemy bishop covers to restrict the enemy bishop's movements, regarding that as more important than the mere safety of the pawns. His statement has nothing to do with analysis in the sense of following out in detail the exact position on the board.

BIGELOW: But this is a statistical observation about what beginners do. This is an analysis.

ASHBY: Let me define what I mean by »analysis.« I mean the actual working out on 64 squares that if this bishop moves to that square, it will attack that knight, let this piece in, and so on. By »analysis,« I understand specific reference of the actual pieces to the actual squares on the board.

BIGELOW: But precisely the statement you made is the summation of experience with such restricted types of analysis as that. It is a stastistic[!] of how people behave in playing this game. The question as to whether or not a computing machine can outplay its maker really contains the question as to whether or not it can gather statistics on the probabilities of the situation at each move more rapidly than its maker can, so as to ever exceed the amount of information about the probabilities that its maker has. This is the question, in my language, that you are trying to discuss.

ASHBY: Yes, I agree.

HUTCHINSON: May I make my few remarks before I have to leave, Mr. Chairman?

MCCULLOCH: Please do.

[155]

TURBULENCE AS RANDOM STIMULATION OF SENSE ORGANS

G. EVELYN HUTCHINSON

I WANTED to say a few words about something from an exceedingly different field that I think is rather exciting in this connection because it seems to show that organisms actually do need randomness in the environment and can't get on properly without it. We have some fairly precise evidence about this from one group of animals. They are all of them rather small, marine and fresh-water plankton, not more than 2 or 3 millimeters long, with correspondingly small nervous systems. Phylogenetically, they are probably all rather simplified from larger more elaborate ancestors, having lost a certain number of circuits in the evolutionary process.

The thing that set me thinking about this is that in aquaria, in marine biological stations, there is a very common device consisting of a tincan hung at one end of a beam. The can fills up slowly with water from a faucet. When the water reaches a critical level, it siphons out, and this is used to move a counterpoise, which consists of a glass plate, up and down in the aquarium tank. It was always believed that this was a method of oxygenating the water. It is, however, almost certain that that is not the case. What it does, I think, is to produce turbulence in the water and provide randomness for the organisms, a continual random stimulation of the sense organs. Some quite interesting and critical work was done in our laboratory on this by a former student and present colleague of mine, Dr. J. L. Brooks.

In the common fresh water crustacean, *Daphnia*[1], there are some species which, in the summertime, grow an enormous helmet (Figure 5). There are several processes involved in producing this helmet. One of the most important factors is the temperature at which the egg develops, but in still water in the laboratory, these helmets never grow as large as they do in nature. In ordinary laboratory vessels, the animal shows a sort of microencephaly.

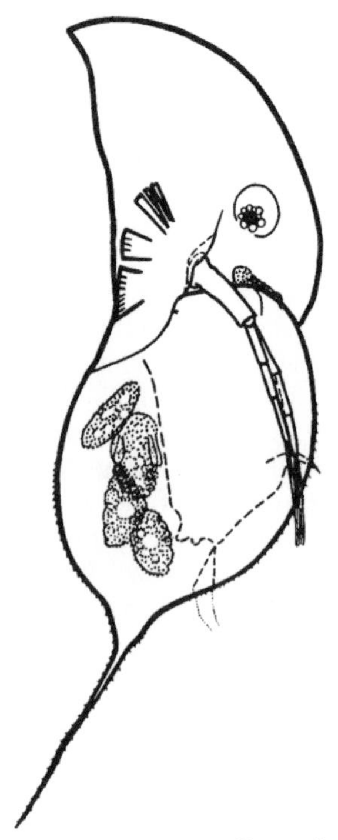

FIGURE 5.
By courtesy of Dr. J. L. Brooks.

It has been proved quite clearly that the reason for this failure of helmet growth is that in the laboratory vessel, there is not turbulent water, while in nature there is a continual randomness in the water, a whole spectrum of turbulence. The antennae are used in swimming and are stretched out laterally. In turbulent water, they must receive a whole series of random proprioceptive stimuli. We suspect that the reception of an adequate number of such stimuli is responsible for the normal development of the head in some species. There are other species which probably cannot grow at all without turbulent disturb|ance. [156] It would seem, therefore, that, although there are an enormous number of interconnecting, internal things that are also obviously required, here is a case that bears some resemblance to the mechanical chess player.

KLÜVER: Your system reminds me of the Chemostat devised by Novick and Szilard for growing bacterial populations (1). In this device, the contents of the growth tube are continuously stirred by bubbling air through so that the bacteria are kept homogeneously dispersed at all times.

HUTCHINSON: Yes, but I don't quite know how you get the randomness.

QUASTLER: Then nutrient is dropped in, so there really is some turbulence.

HUTCHINSON: Is the turbulence essential ?

QUASTLER: I don't know whether it is.

HUTCHINSON: There are lots of places where turbulence makes a difference but I don't think they are quite analogous to this, because here it looks to me as though this is turbulence acting on a receptor. It is providing a level of activity.

YOUNG: It isn't a survival factor at all. These large hoods tend to | appear on races living in lakes, don't they, and small hoods in small ponds? [157]

HUTCHINSON: That story is fantastically complicated. Everything that has been written about it has proved to be wrong. That is almost axiomatic now. Woltereck has written more on this than all the other people put together, and it is all wrong.

YOUNG: Does the stimulus produce the growth of a different part? I am not quite clear whether the stimulus falls on the antennae.

HUTCHINSON: Presumably, the stimulus produces a distortion of the antennae. I don't think it is likely to have anything to do with the actual development of the creature.

QUASTLER: Is there any functional significance to the size of of[!] the hood?

HUTCHINSON: The functional significance is unknown. There have been two theories: one, that it is a steering device to keep the thing horizontal; the other is that it reduced the sinking rate. There is a good deal of evidence to show that both are wrong.

BOWMAN: There might be some stabilizing power at the muscles that move the antennae.

HUTCHINSON: No.

BOWMAN: Where do the muscles attach?

HUTCHINSON: The attachment of the muscle defines a line beyond which it is customary to measure the height of the helmet. Harvey at Plymouth, who has had a great deal of experience, perhaps more than anybody else, tells me that he feels that a continual movement of the water is essential to the plankton.

GERARD: Are you sure that the element of randomness or turbulence is involved in this, or is it merely an adequate amount of stimulation? That is, could they get it, presumably, by a steady current flowing on them?

HUTCHINSON: There is no evidence that it wouldn't do just as well. I can see very great technical difficulties in trying to prove it. I don't quite know what kind of an aquarium would have to be used to put a directional stimulus on the thing, because to get a directional response, the aquarium would have to be enormously long to enable the animal to keep moving in that direction throughout the fifteen days of the experiment.

PITTS: It could be annular.

HUTCHINSON: An annular one might be possible, but it would be a horribly difficult thing to set up.

BOWMAN: How about a centrifuge with the load of the water superimposed radially inward?

1 Brooks, J. L.: Turbulence as an environmental determinant of relative growth in Daphnia. *Proc. Nat. Acad. Sc.* 33, *141 (1947).*

HUTCHINSON: The density of the egg is considerably greater than that of the organism. It would centrifuge through it. | [158]

BATESON: I thought that some of these organisms had fluffy appendages of one kind or another. It would seem to me that wherever you do find appendages of that kind, there is some justification for saying they are like the seed of a tree which depends upon a wind scatter, and, in fact, that the organism is likely to sink if turbulence is not provided. The turbulence is a randomizing circumstance which keeps the organism up, rather than provides stimulation.

HUTCHINSON: I think that may sometimes be the case. With some discussion, which would take too long now, I think I could convince you that that was not true in this particular case. There is quite all elaborate theory which has been worked out by Riley, Stommel, and Bumpus and there is no doubt whatever that it is exceeding pertinent, but I don't think it is involved in this particular case.

VON BONIN: Don't you help the oxygen concentration by stirring the water around the animal?

HUTCHINSON: The animals are moving the entire time, and I don't think that reduced oxygen concentration in the actual neighborhood of the animal can be involved. The appendages are moving sufficiently fast for a feeding current to go through. You can always tell whether they are getting anoxemic because they turn red, developing hemoglobin. Brooks never got anything of that sort, and the overall oxygen tension is precisely the same in the medium of the control and of the experimental groups. About the question of providing adequate stimulation, that seems to me to be merely a rephrasing of the noisemaker in the mechanical chess player. What is the point of having adequate stimulation just as a general thing except to provide noise?

GERARD: Well, your leg muscles will atrophy when you don't walk, and you don't have to walk around in circles.

HUTCHIN[S]ON: There is plenty of activity in these organisms, but I can't think of any place where there isn't activity in the still water that isn't present in the turbulent water, except in a continual attempt to correct the position of the antennae, which must be more or less random. I just wanted to interject this as an interesting case, which is very far away from the kind of things we have been talking about.

YOUNG: What it really amounts to is that Ashby's postulation of the environmental stimulus, if you like, is a necessary approach. It is another way of saying the same thing, isn't it?

HUTCHINSON: Yes, I think so.

YOUNG: I quite agree. Ralph says you get no organism if you don't stimulate it. It atrophies.

HUTCHINSON: The stimulus has to be direct in this particular case.

YOUNG: But it has a gravity component, hasn't it?

HUTCHINSON: That will make little difference in the turbulence pattern in the water.

INVESTIGATIONS ON SYNAPTIC TRANSMISSION [159]

WALTER PITTS[1]

The central nervous system, like the heart, produces electrical signs of its action at body boundaries; but that these signs can be interpreted in terms of function tells more of the glory of nature than the ingenuity of biologists, for not by polarity height, frequency, or distribution is it possible to relate the potentials recorded on the surface to events within. No more useful are the records from single electrodes thrust into the nervous felt, except where they show the impulses of blocked fibers or butted cells. We can tell the firing of single cells or fibers and measure their frequencies, but the chaos of synapsis is less neatly handled, and it is this chaos to which most of the pen-wiggles are attributed.

There are three ways to get electrical information about the central nervous system. The first two are indirect, and have been largely exploited. One is to examine the input-output relations of the whole structure: the other is to use emergent fibers as electrotonic probes of events affecting their intramedullary extensions. The results of the former method are ambiguous, without detailed knowledge of the anatomy and physical properties of junctions. Morever, it is only rarely that the tracts entering or leaving a nucleus can be isolated well enough for separate stimulation and recording. The data provided by the latter method are susceptible of quantitative analysis only under most controversial assumptions, such as that the properties of the fiber do not change within the cord, even before their geometry is altered by bifurcation, demyelination, expansion, or coarctation. Accordingly, we have devised a third technique to examine activity in the central nervous system at the place where it is.

As you know, an impulse along a neuron is a region of depolarization, receiving current from an adjacent polarized membrane. This carries the impulse farther. If the neuron lies in a nonconductor, an electrode on one region pitted against a far-distant electrode perceives the potential of the local membrane with respect to the other; the passing impulse is seen as the familiar negative monophasic spike. But if the insulating medium is replaced by a conductor, such as electro | lyte, the potential of the electrode [160] is proportional to the current across the membrane, which is very nearly the second derivative along the neuron of the monophasic spike which would have been recorded if the neuron had been in oil. But it is attenuated because the electrode computes a weighted average over a volume of a function whose own average is zero. To this external medium, the nerve appears to produce, absorb, then produce current when a spike passes; from this probe's standpoint of view, the impulse is a source, followed by sink, followed by source. Their algebraic sum in time is zero, unless the impulse stops either at the electrode or short of it. In the former case, the preceding source is averaged with the initial part of the sink, but the stationary decaying sink is recorded as a large negative potential with respect to a remote electrode. In the latter case, the electrode perceives only a source, near it, rising as the impulse approaches the block, then decaying as the impulse fades; it therefore turns monophasically positive. For this reason, it is possible to record single fiber activity in the central nervous system, when the micro-electrode blocks individual fibers. The recorded spikes may be either purely

1 This work was done in collaboration with Drs. Lettvin, Wall, and McCulloch. We are grateful to the Office of Naval Research and to the Department of Public Welfare of the State of Illinois for financing this research and to Dr. A. P. Bay for his assistance in its prosecution at the Manteno State Hospital, Manteno, Illinois. The work is being continued and expanded at the Massachusetts Institute of Technology.

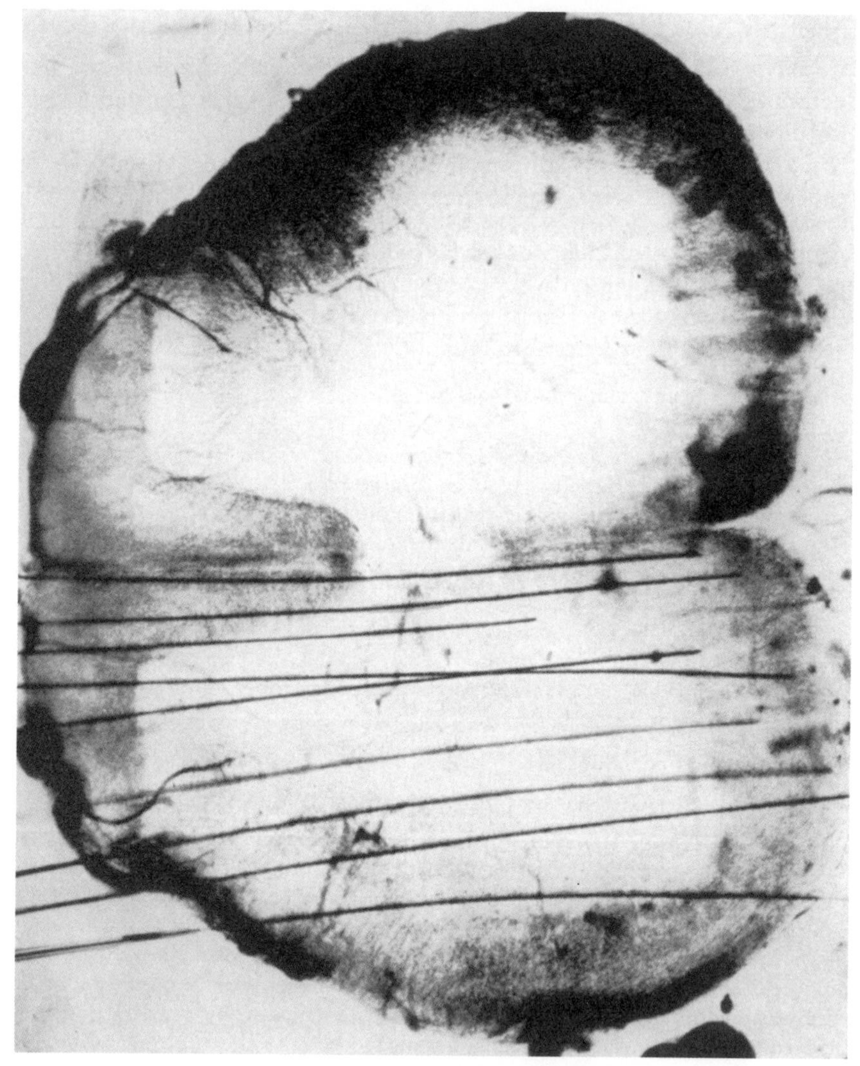

FIGURE 6

positive or positive and then negative. It is unnecessary to suppose that spikes are only recorded from cell bodies, since such records can be obtained from peripheral nerve with micro-electrodes. It is likely that some of the large potential swings recorded in synaptic regions are records of the impulses having reached the ends of presynaptic arbors.

Let us consider for a moment a parallel bunch of axons entering a bounded conductor in which it is blocked. The passage of the impulse is small to any boundary electrode: only the entry of the impulse into the volume and the dying dipole at the block are conspicuous. Mapping the potentials all over the boundary would hardly determine the point of entry and the place of block, since any distribution of potentials on the surface can be realized by an infinite number of widely different configurations of sources and sinks within. Unless we can narrow the choice enormously from additional knowledge we hardly ever possess, we can only speculate, as people do freely, apparently not realizing their temerity, where the disturbance ties. It is, in a word, impossible to infer the pattern of central nervous system activity from the most detailed surface measurements on cord or cortex, or from records at single points in the substance.

Sources and sinks in a volume conductor are strictly additive in their effects on the potential in the external medium. The potential at each point is a sum of contributions from all of the sources and sinks present at that time anywhere. Its magnitude and sign are consequently no direct indicators of elements active in the immediate neighborhood of a recording electrode. In principle, however, if the potential V is | known in [161] enough detail, the source density ρ can be inferred from it by the well-known equation of Poisson:

$$\rho(x, y, z) = -\frac{\delta}{\delta x}\frac{1}{R}\frac{\delta V}{\delta x} - \frac{\delta}{\delta y}\frac{1}{R}\frac{\delta V}{\delta y} - \frac{\delta}{\delta z}\frac{1}{R}\frac{\delta V}{\delta z}$$

if R is bulk resistance of the fluid medium. We therefore decided to [Figure 6] | repeat [162] Berry Campbell's mapping of the potentials in the spinal cord after dorsal root stimulation, but in enough detail and under such conditions that it should be legitimate to employ Poisson's equation, or an approximation to it, to compute ρ.

To examine the activity of the cord in its substance, it is necessary to introduce a grid of electrodes arranged densely enough at accurately known stations to make out the progress of activity in fine fibers and to take simultaneous records from this grid. We cannot do this, if only because the volume of electrodes would seriously disturb the cord if they were large enough to record the signals above noise level. But we have found that an isolated cord, whose degree of anesthesia, temperature, circulation, and oxygenation stay reasonably constant, keeps its input-output curves and micro-electrode records tolerably the same for a long time, suggesting that we may treat succes-

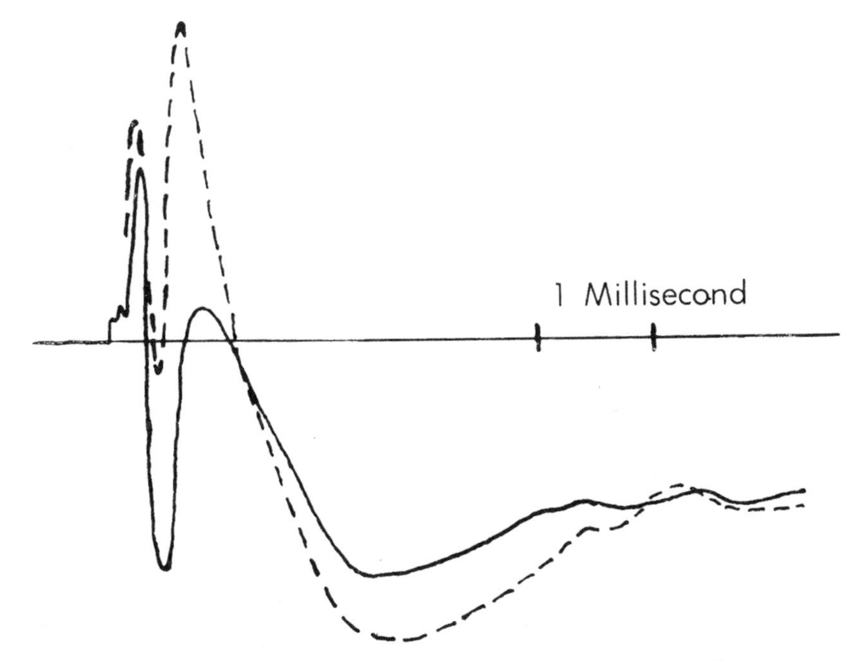

FIGURE 7. Solid line is record taken just touching surface of cord following a dorsal root volley. Broken line is record taken pressing electrode down to dimple the surface. The same stimulus has been used.

sive records for repeated constant stimuli as if they were simultaneous. Supposing this, we introduced micro-electrodes serially, penetrating by measured steps to a final station, recording at each step, where we cut them loose from the apparatus and left them [163] in situ for later verification of position. [Figure 7] |

These micro-electrodes were spaced as closely as would not interfere with the monitored response in the ventral root, or about three per millimeter. Such a group is seen in Figure 6.

A detail of micro-electrode use is important enough to be mentioned here. It must penetrate easily and without dimpling the surface either on entry or further translation, for the cord is bounded by oil in these experiments and dimpling introduces serious errors by distorted surface reflections, as well as by pressure on neurons (as shown in Figure 7). This means it must have a microscopically smooth, sharp point, with no shoulder at junction with insulation and no change in shank diameter. These requirements are very necessary since we do not remove pia for fear of local damage. One such micro-electrode tip appears in Figure 8. Further, it must be small enough to minimize tissue damage, but not so small as to obscure the record with electrode noise. This critical diameter lies near 10 micra, the smallest wire it is possible to draw bare in most metals. A more extended discussion of electrode problems and properties will appear in a later publication. One further point deserves comment, and that is that withdrawing such an electrode through its own path introduces a distortion because of the reduced resistance of the track of destruction, an additional reason for considering only records taken on the way in.

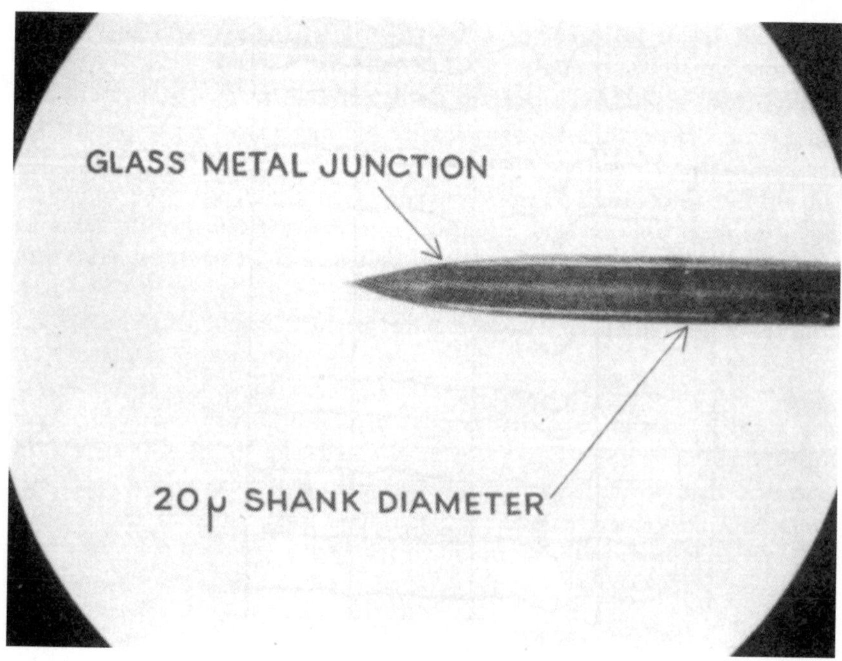

FIGURE 8

[164] | [Figure 9] | In the experiment shown, we have used nine electrodes, giving a grid of [165] about 300 points in 9mm^2 of ½ of a transverse section of a segment adjoining that lumbar segment whose dorsal root we stimulated. Previous studies suggested that no great longitudinal currents flowed in the times we meant to consider, thus allowing us

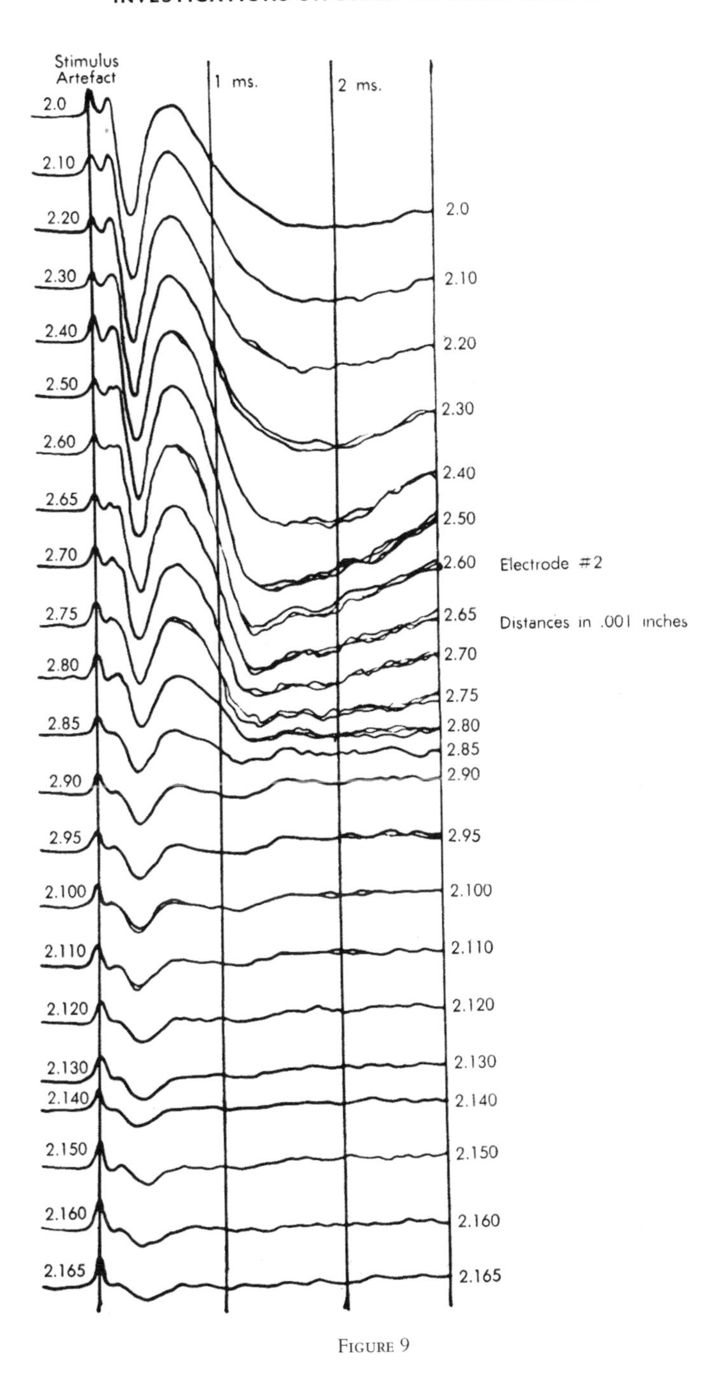

FIGURE 9

to record across a single plane. Figure 9 exhibits a sample of the records from one electrode track, three traces per position. The scatter for any one position is not excessive for computation. The spatial distribution of potential at one instant following the dorsal root stimulus is interpolated to a square lattice of about the same density, whence the LaPlacin can be approximated simply with a precision sufficient for the noise level.

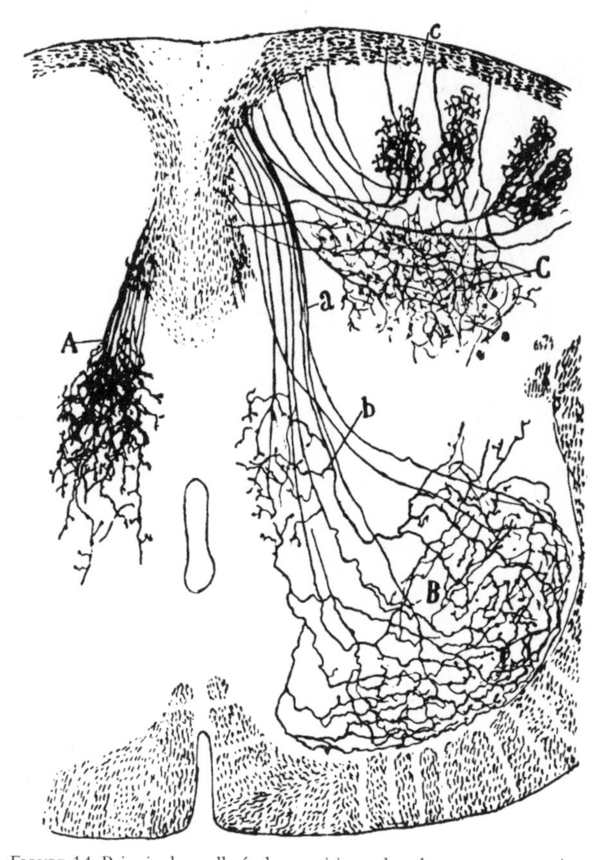

FIGURE 14. Principales collatérales sensitives, cheq le rat nouveau-né. Méthode de Golgi. A, collatérales du noyau gris intermédiaire; B, arborisations embrassant les noyaux moteurs; C, ramifications étendues dans la tête de la corne postérieure; *a*, faisceau sensitivo moteur, *b*, collatérales p fondes de la substance de Rolando. Reprinted by permission, from RAMON Y CAJAL, S.: *Histologie du Systéme Nerveux*. Paris, Maloine, 1909.

Naturally, since we interpret only differences (second differences, in fact) of potential, the position of the reference electrode is quite irrelevant. It might quite well be taken on the stimulated dorsal root. It does not matter whether it is »indifferent.« We are thereby also freed from specifying the position of the reference electrode. To present these records clearly, we filled a square around each computed value with the number of dots proportional to the value in arbitrary units, green for source and red for sink. In Figures 10 through 13, we present the isopotential map on the left and the source-sink map on the right for the successive times after dorsal root stimulus given in milliseconds in the upper left-hand corners. It is clear that the sources and sinks cannot be seen by simple inspection of the isopotential map.

Figure 14 is a drawing by Ramon y Cajal of the fiber distribution from dorsal root in the lumbar spinal cord of a newborn rat, whose white matter is thinner than appears in our section. While it is premature to push the anatomical interpretation of the records, let me remark on the set of fibers entering medially and curving back into the substantia gelatinosa. The series of maps shows a sink followed by source proceeding from the place where the fibers enter the gray matter at the base of the dorsal column, and moving laterally and then up into the substantia gelatinosa.

While the degree of resolution is coarse, and the noise of experiment, necessarily amplified by this kind of computation, is high, we think that this set of pictures shows more accurately than heretofore the places where activity occurs, how large it is, and in which direction it moves, in the cord of a spinalized cat anesthetized with Dial-cum-urethane. We mean to apply this technique to the problem of synaptic transmission. | [Figure 14] [166]

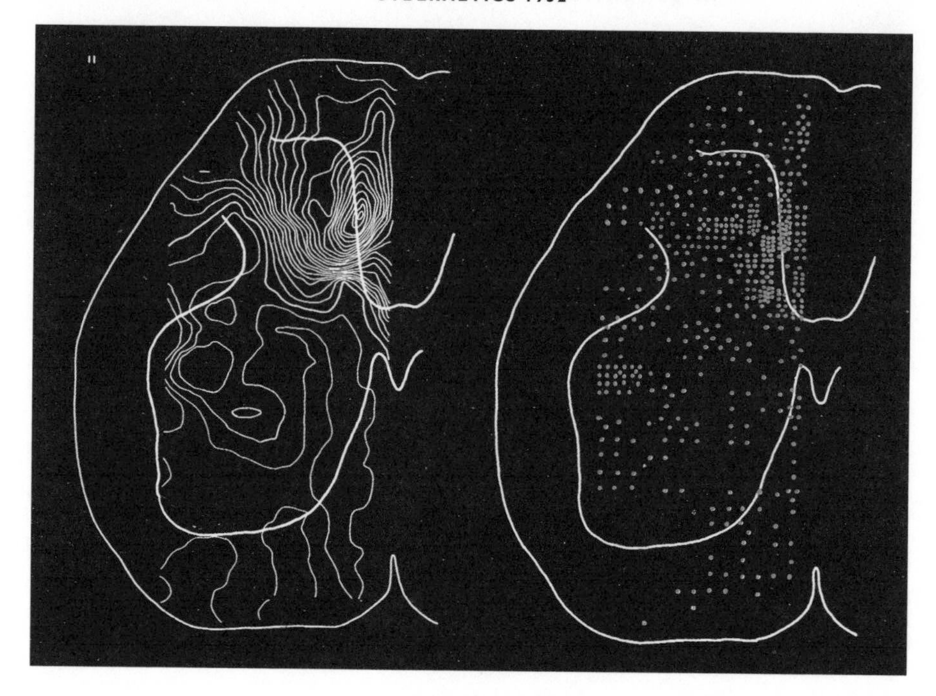
11

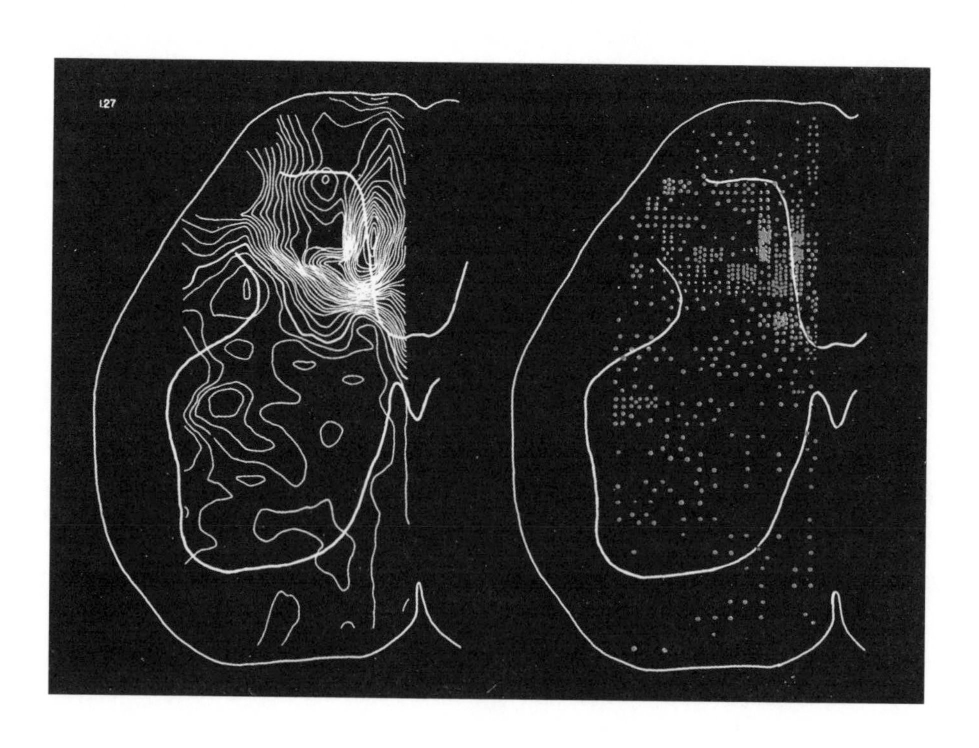
127

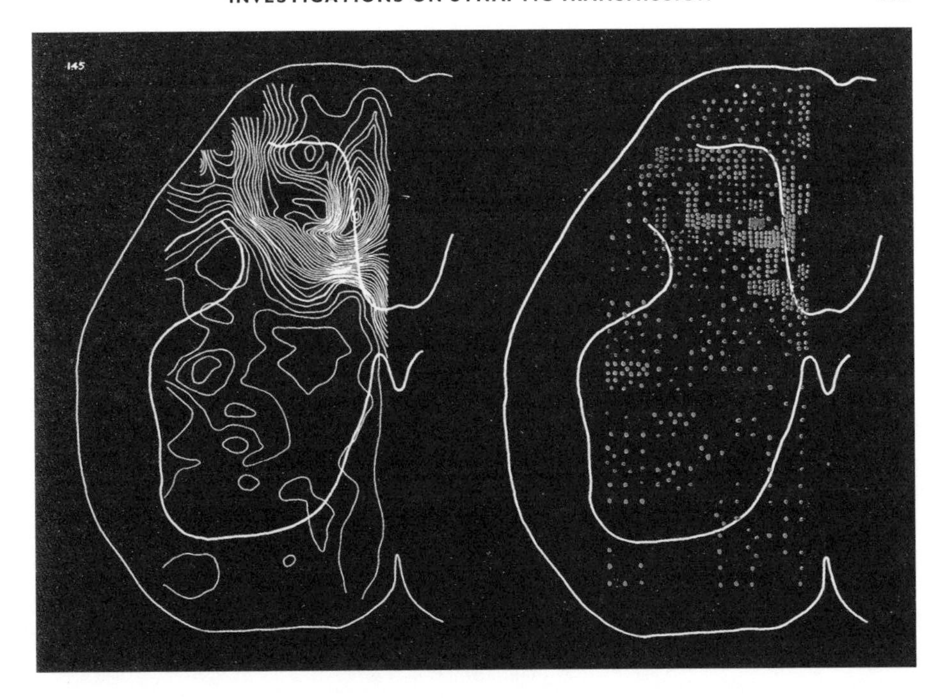
145

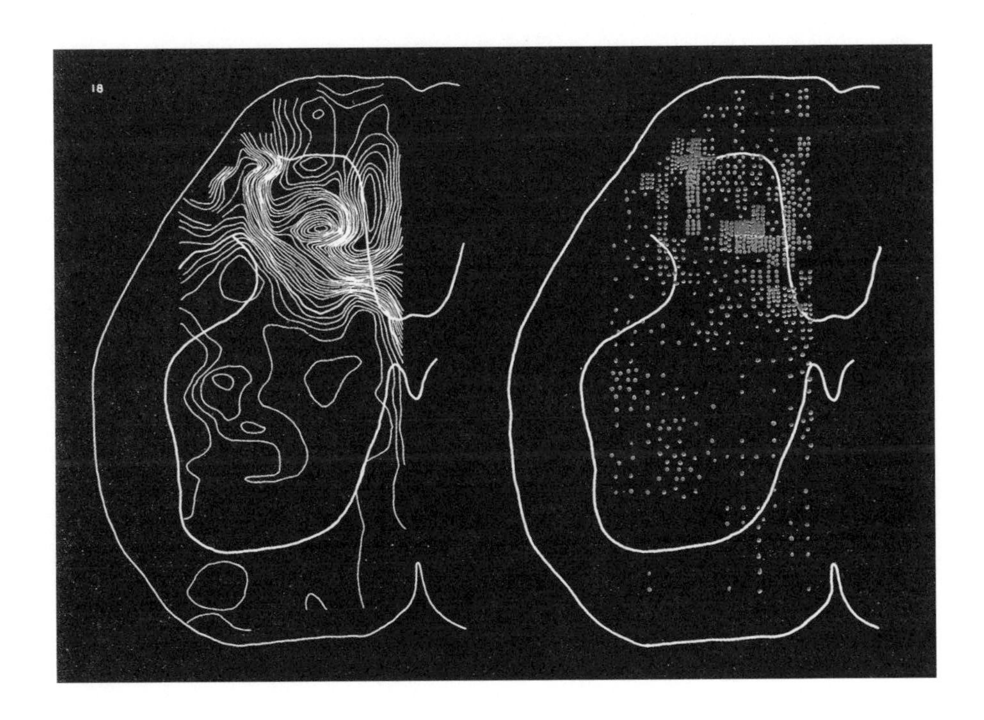
18

[167] # FEEDBACK MECHANISM IN CELLULAR BIOLOGY

HENRY QUASTLER

FEEDBACK REGULATION SIMPLE ENZYME SYSTEMS

FEEDBACK mechanisms are readily recognizable in even the simplest of biological systems, enzyme catalysis. The scheme in Figure 15 brings out some of the feedback characteristics of a simple reaction.

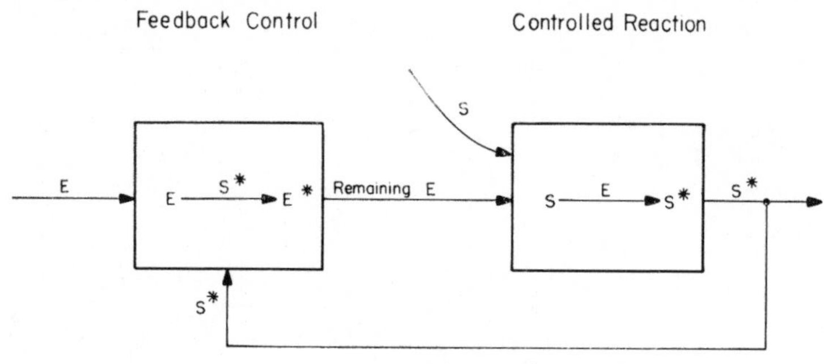

FIGURE 15

In a reaction labeled »Controlled Reaction,« a substrate, S, is changed into a product, S⋆, with the help of enzyme, E. As a rule, an enzyme in inhibited by the products of its activity. This feature is indicated by feeding enzyme and some of the reaction product into a second reactor labeled »Feedback Control« in which the enzyme, E, is changed into some inactive form, E⋆, by the reaction product, S⋆. This feedback loop provides a unilateral control; it cannot accelerate the reaction but can only slow it down. Thus, it can protect the system against accumulation of the reaction product, S⋆. It is often complemented by a second feedback system which can be represented as follows:

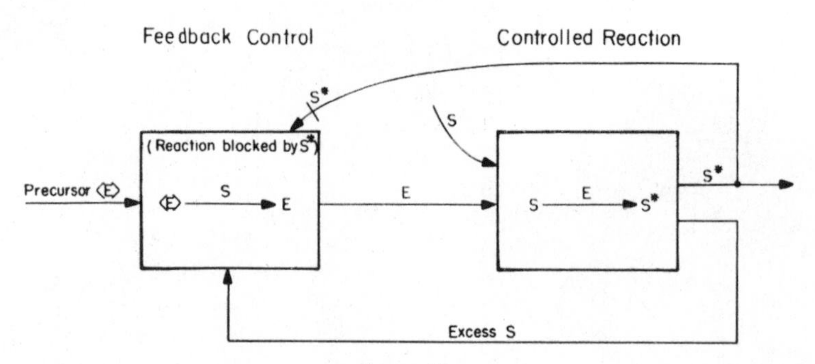

FIGURE 16

[168] | Figure 16 represents the increased enzyme formation in response to excess substrate, or the process of enzymatic adaptation. In the reactor labeled »Feedback Control,« the enzyme, E, is produced from a precursor, <E>, under the influence of excess substrate, S. By increasing the amount of enzyme, the loop can protect against accumulation of the substrate. It appears that the enzyme formation can be blocked by excess product; this feedback loop operates in the same direction as the inactivation of

enzyme but with a different time factor. Thus, a triple feedback mechanism, controlling the concentrations of enzyme, substrate, and product, can be diagramed.

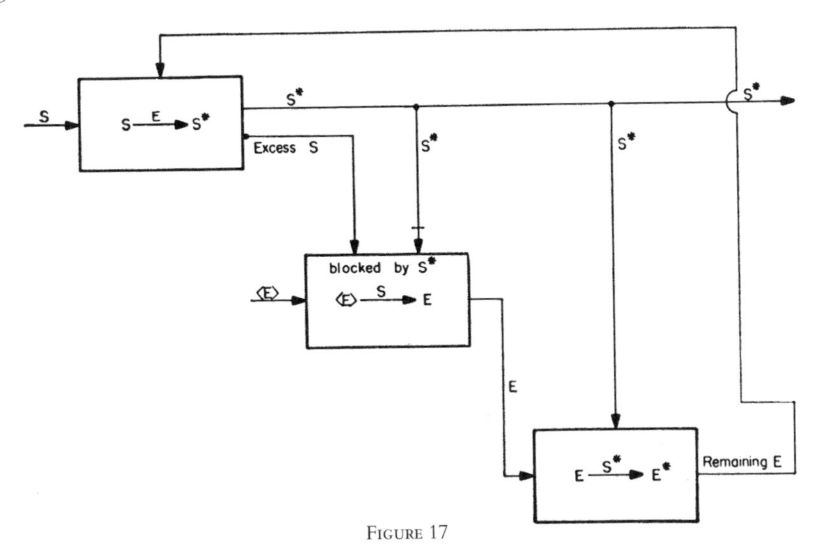

FIGURE 17

Such a scheme (Figure 17) has no other purpose than that of emphasizing the feedback mechanisms in enzyme systems; in general, the traditional approach of chemical kinetics is preferable. Certain features are brought out very clearly by a scheme due to Mr. Robert Baer.

In this diagram, the various substances are represented by circles, reactions by boxes, changes by horizontal arrows, and facilitation by vertical arrows. Thus, the substrate, S, is brought into the system, changed into S⋆, and carried off. During this process, S promotes the change from the precursor <E> into E; this enzyme catalyzes the change $S \rightarrow S^{\star}$, and is, itself, drawn off by the reaction product.

The enzyme product tends to inhibit the enzyme in various ways, such as combining with the enzyme in the same way as the substrate, or changing the pH to one not favorable for the enzyme action. The diagram only indicates that the amount of enzyme is, in effect, reduced by the product of its activity. | [169]

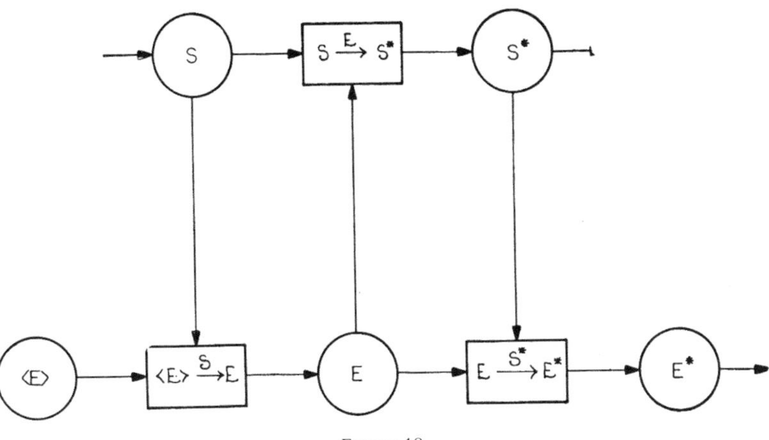

FIGURE 18

FEEDBACK CONTROL OF GROWTH AND MULTIPLICATION

We now consider the enzyme, E, as part of an enzyme-bearing structure which is self-reproducing (which might be a cell constituent, a whole cell, or a whole organism). In this case the precursor <E>, becomes E itself, and the adaptive reaction is written:

$$E \xrightarrow{S} \text{more E},$$

i.e., E duplicates if furnished with a suitable substrate. A slight rearrangement of the enzyme diagrams yields a graphic representation of the feedback loops adjusting rates of growth and multiplication:

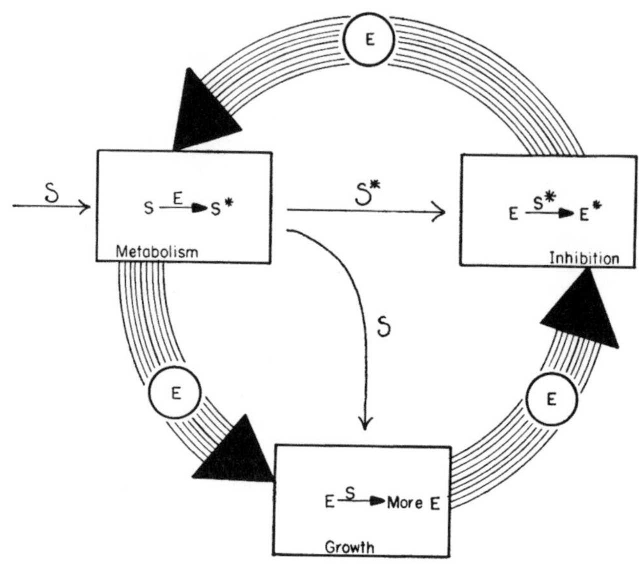

FIGURE 19

[170] | The structure, E, grows by duplication in the presence of a favorable substrate, S. The amount of growth is regulated by a double feedback; passively, by the substrate S being removed by E, and actively, by the inhibitory effect of S* resulting from the activity of E.

NUCLEOCYTOPLASMIC FEEDBACK SYSTEMS IN PARAMECIUM AURELIA

I have presented, very sketchily, some highly schematized feedback diagrams which apply to biological Structures very generally. I will now discuss some interesting, specific cases. They were all found in *Paramecium aurelia* (1). This is a unicellular organism, or better, an acellular structure; i.e., its cytoplasm is not divided into cells, and it has the equivalent of about 40 somatic nuclei lumped into a single formation called a macronucleus. It also has a pair of micronuclei which are analogous to the germinal nuclei in multicellular organisms. During vegetative growth, all three entities, cytoplasm, micronucleus, and macronucleus, are self-duplicating and interdependent, as indicated in Figure 20. The vegetative cycle is interrupted by a process of fertilization and reorganization, during which the old macronucleus disappears, and a new one is formed by the micronucleus; this process can be accompanied by profound changes in the cytoplasm. Thus, we have a complicated system of interaction and feedback:

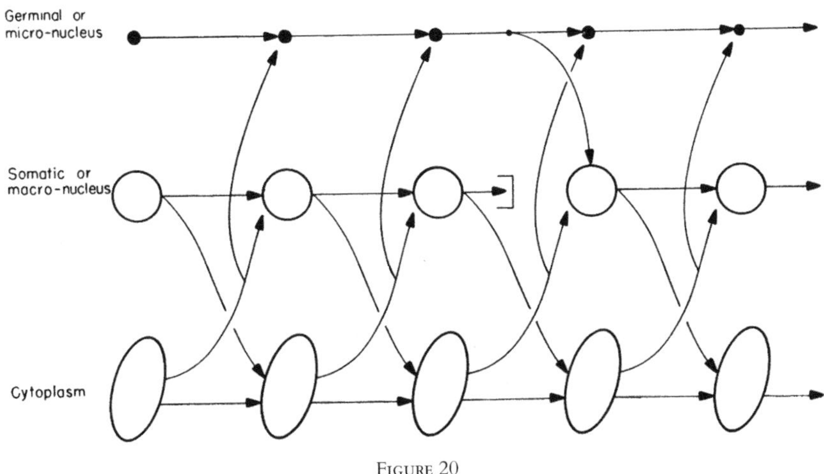

FIGURE 20

The control of three traits in *Paramecium aurelia*[1] has been examined in considerable detail (1). The one with the longest history is the | so-called »killer« trait. Some paramecia have the faculty of secreting a substance (called paramecin) which will kill other paramecia, known as sensitives. The killer trait is known to depend on the presence of a self-duplicating cytoplasmic particle called kappa. A cell which has lost its kappa-particles cannot form them *de novo*, and becomes a sensitive. On the other hand, the ability of a cell to maintain its kappa concentration depends on the presence of a nuclear gene called K. This gene, while it does not enable an animal to form kappa and paramecin, is needed to sustain kappa. Thus, we have the following control mechanism: [171]

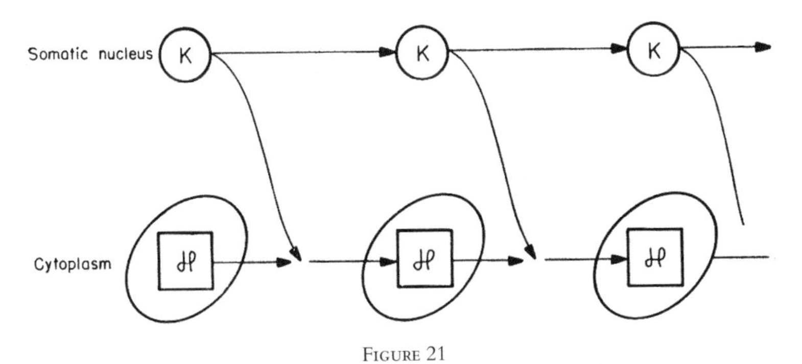

FIGURE 21

There is good reason to believe that this type of control system occurs rather commonly; however, it does not hold for the other two well-investigated traits in the same organism. The second trait is sex. *Paramecium aurelia* has two sexes, and there is good evidence that one of them has twice as many macronuclear units (or twice as high a degree of polyploidy) as the other. It also seems fairly certain that the macronucleus secretes a polyploidizing substance into the cytoplasm, which, if present in sufficient amount, will cause the macronucleus of the next generation to go through one extra division. Thus, the degree of polyploidy is maintained by a feedback through nucleo-cytoplasmic interaction:

1 Nanney, D. C.: The control of hereditary traits in *Paramecium aurelia* (unpublished article).

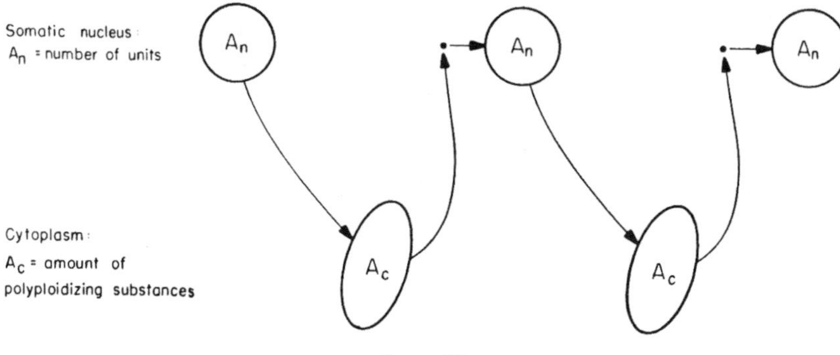

FIGURE 22

[172] | The third trait is the antigenic property. The antigenic type is definitely a cytoplasmic property; it is controlled by the nucleus insofar as a given genome will allow a certain spectrum of alternative antigenic types. Which of these possibilities is realized is decided at the time of reorganization, at which time it can be affected by external influences. Apparently, in this control system, the cytoplasm does not affect the nucleus:

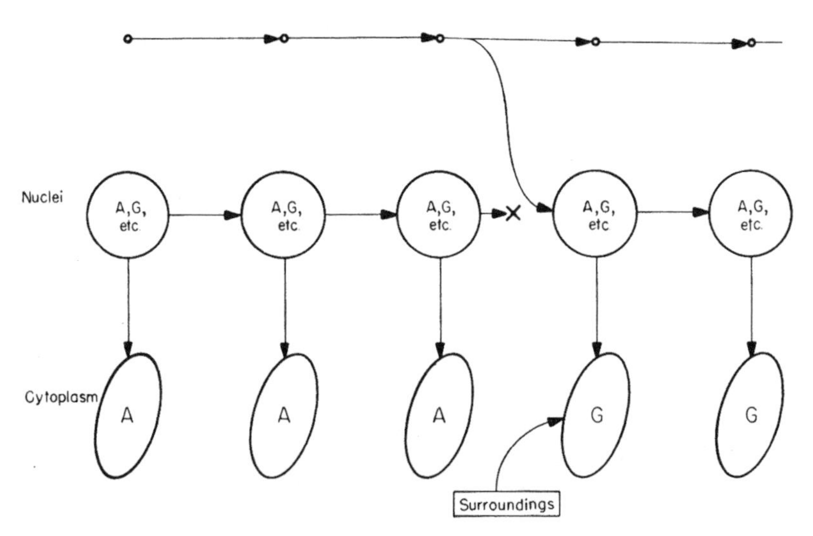

FIGURE 23

Thus, there are three traits in *Paramecium aurelia* in which the control mechanisms, or at least the distribution of controlling power between nuclei and cytoplasm, are reasonably well known. I have given a superficial and incomplete description, but I hope I have made it clear that for those three traits, there are as many different methods of control. This is an exceedingly sad fact. The analysis of the three traits represents the results of years of work by one of the greatest contemporary biologists, Tracy M. Sonneborn, with excellent collaboration and equipment. There are easily some thousand traits in *Paramecia,* and there just aren't enough Sonneborn-years available to analyze them one by one. I give this consideration in an attempt to justify, possibly, the approach I am going to take next, namely, to try to estimate the complexity of various living structures without actually analyzing the components in detail.

THE COMPLEXITY OF LIVING STRUCTURES

a. Complexity and Information Content

We have considered feedback-regulated systems on various levels of | organization. The [173] performance of these systems depends on successful coordination of the various parts of a functional unit, or on the ability of the whole structure to carry out certain tasks of ordering. We may ask: How complex must a structure be in order to be capable of embodying a certain number of functionally connected systems of the types described?

The first approximation in trying to estimate the total degree of complexity is a count of structural elements. Thus, the famous count of 10^{10} neurons is thought to be indicative of the complexity of the central nervous system. However, the count of elements is not a very satisfactory measure; by this yardstick, a large cornfield is much more complex than a small one, and the liver of a whale much more complex than that of a mouse. While this is probably true to some degree, our intuitive feeling for measure of complexity demands that this measure should not increase as fast as that of the size. I propose to show that the measure of information content, or negative entropy, approximates more closely than the count of elements the requirements for a satisfactory measure of complexity. This approach is due to the late Dr. Sidney Dancoff.[1]

It must be stated right now that there is one facet of complexity which is not considered in the measurement of information content, namely, the richness of different kinds of interrelations between the details of the structure. An additional step will be needed to introduce this feature into the measure of complexity.

One often runs into semantic difficulties if information theory is used in fields beyond that for which it was originally designed. Therefore, with due apologies to the experts here, I shall begin by rewording the definition of the quantity H; I stand ready to be corrected.

We can call H the negative entropy, or the information content, or degree of organization, or amount of design, or amount of significant detail, or degree of specificity, and probably quite a few other terms. Actually, all these concepts are such that they can be measured properly by the same yardstick. The measuring procedure is as follows: We consider some item (an object, a state, an action, etc.); this item is investigated within the context of some well-defined class of items to which it belongs. We wish to estimate in how many ways items belonging to this class can be different, or, to be more accurate, differ to a degree which we consider significant. We shall count the ways of being different on a binary scale, that is, we will reduce all differences to dichotomies. Furthermore, we make all dichotomies most efficient, | that is, we [174] arrange them in such a way that the a priori probabilities for our item to be on either side of the dichotomy are as nearly as possible equal. Then, the expectation of the number of dichotomies needed to specify an item among the other items in the same class, specifying it with what can be an arbitrary degree of resolution, is the measure of H (2, 3). This is a long and elaborate way of saying:

$$H = -\sum_{i} \mathrm{p_i} \log_2 \mathrm{p_i}$$

The estimate of the information content of a structure depends critically on the degree of resolution, or on the size of the smallest significant detail, in other words, on the size of what is considered a building stone[2] [3] (4). Living structures have many dis-

1 Dancoff, S. M., and Quastler, H.: Information rates and error rates of living things. (Unpublished article.)

2 Quastler, H.: Size and duration of living structures. (Unpublished article.)

3 Dancoff, S. M., and Quastler, H.: Information rates and error rates of living things. (Unpublished article.)

tinct levels of organization, and have well-defined elements which can be considered as building stones of the next larger units: elementary particles, atoms, simple molecules, large compound molecules, macromolecules, organelles, cells, and so on. There is a sharp boundary line between macromolecules and compound molecules. The former (proteins and related substances) have distinct personalities and can be identified, at least in principle, as belonging to a certain organ of a certain species, or even, to a certain individual. Organic molecules (and smaller units) are impersonal; thus, some large molecule, for example, some prophyrine is exactly the same whether it comes from a plant or animal or out of a test tube. We conclude that molecules (up to large organic molecules) can be properly considered as interchangeable building stones, without intrinsic complexity for any biological objects, whereas macromolecules themselves are the smallest and simplest structures we must consider in the study of the complexity of living things.

b. Information Content of an Enzyme

The best-known class of macromolecules are the enzymes. All enzymes are proteins, and it is not unlikely that all proteins have enzymatic activity. Every living unit contains many enzymes, which act in series and in parallel and interact in the sense of both facilitation and inhibition. The proper functioning of each system depends on the interaction of substrate, enzymes, and reaction products. Each component reaction is regulated by the results of othe[!] other reactions. This implies that information, in some form, must be transmitted between the places where the component reactions occur. A compound system will be traversed, at any given moment, by numerous such »communications.« As far as each single system is concerned, the communications [175] relating to other systems are »noise.« This implies | that each system must have some selective mechanism to separate its own »signals« from the general »noise.« The separation can be achieved by separation of communication channels (as in some parts of the nervous system) or by coding the »address« into some structural feature of the molecule carrying the message. Whatever method is used, it is clear that a certain amount of information is needed to guarantee the proper ordering of coexisting systems.

The minimum amount of information needed to achieve functional separation of enzyme systems is the amount of information needed to select any enzyme from all other enzymes, in other words, the information content of an enzyme. This could be easily estimated if we knew how many different kinds of enzymes there are and what their relative frequencies are. I found biochemists extremely reluctant to make even rough guesses. However, one can get some agreement about possible ranges. There are certainly more than 100 different enzymes and probably not as many as 10^6; their relative frequencies are certainly unequal, with a few enzymes undoubtedly making up a large fraction of the total bulk. For a number of reasonable distributions within this range, H has been found not to deviate much from 10.0. We shall, therefore, use this number as a crude estimate of the »specificity« of an enzyme, or as that minimum amount of information which, acted upon in some way or another (we don't have to specify how), will suffice to produce a suitable association between an enzyme and its substrate.

An enzyme might be specified with 10 bits with respect to the biological unit to which it belongs. But it is also a protein and is distinguishable from all or most other proteins. I do not know how many proteins can be differentiated by serological and other methods. The number is certainly large, but probably not of a larger order of magnitude than the number of existing species because we know of several proteins coming from very different sources which are, immunologically, closely related. It is likely that some 25 bits are sufficient to specify any given protein among all other proteins.

c. Information Content of a Workable Complement of Enzyme Systems

We consider now the aggregate information content of a compound system consisting of a complete, workable set of simple enzyme systems. I mentioned before that we do not know how many enzymes constitute a minimum working complement. As a crude and highly preliminary guess, we might adopt the number 1000. Each of these must be specified with not less than 10 bits. Therefore, a total of not less than 10^4 bits of information must be processed in the manufacture of a workable complement of enzymes. | [176]

The instructions on how to manufacture enzymes are transmitted from generation to generation. The structure carrying this information is called the genome. In principle, it would not be necessary to provide separately the instructions for each single enzyme, but there is good evidence that this is the case, at least to a considerable degree. The manufacture of single enzymes can be modified by certain perturbations in restricted parts of the genome without apparent effect on the manufacture of other enzymes. Therefore, we assume that the specifying information (not less than 10 bits per enzyme) for each of 1000 enzymes is coded into the genome separately. Then, a genome, in order to sponsor the reproduction of a visable enzyme complement, must contain not less than 10^4 bits of information. In the simplest organisms, the complexity of the adult should be not much larger than that of the genome.

We arrive at the same figure by a different approach. The smallest object capable of independent life is a bacterial spore of about 10^8 atoms. Of those, about nine-tenths are comparatively featureless, thick-shell substances that go into solution the moment germination starts, food, etc. This leaves 10^7 atoms for essential structures, or about enough for 10^3 enzyme molecules. Defining each of these with 10 bits yields also a total of 10^4 bits. This, therefore, might be an extremely crude estimate of the minimum information content needed for independent life.

KLÜVER: Does this mean that you would exclude virus particles from the category of living things, such as those of the tobacco mosaic virus which have been reported to contain 7 million atoms?

QUASTLER: Tobacco mosaic virus is not capable of independent life, On the other hand, it has a good deal of excess information, because one can treat tobacco mosaic virus with X-rays and find that each »hit« has only a small chance of scoring a kill; yet, we know that an X-ray hit destroys information within the neighborhood.

d. An Estimate of the Complexity of Man

I have made some wild guesses, but I should like to proceed to a still more hazardous one and estimate the information content of a very highly developed living thing, such as a human being. The features which distinguish a certain man from other men, or from all other living things, depend on surroundings and heredity. The relative importance of these two factor groups are not the same for various traits, but, in general, there is agreement that not less than 10 per cent (on the average) of all relevant information concerning an adult is genetically determined. This 10 per cent, then, must be coded into the genome. Thus, if we obtain an estimate of the information content of the genome, the result should come to within an order of magnitude of the information content of a whole organism. | [177]

Genomes can differ from each other in a large number of ways; each independent source of distinctness is called a gene. The number of genes in a higher organism is not known, but most geneticists agree that it is of the order of tens of thousands. Each gene can assume a number of distinguishable states called alleles. Once more, there are no good estimates of numbers of allelic states. From the old idea of a very limited number of alleles of each gene, we have come to a point where it appears that the number of allelic states is limited by the patience of the investigator to isolate and distinguish them. Still, the number 1000 would seem to be a high estimate. It takes 10 bits to specify any one allele out of 1000; therefore, the state of a genotype with some 10^4 genes would be defined by some 10^3 bits of information.

I shall attempt to obtain an estimate of the possible information content of the genotype in a very different manner. We know that all the information carried in the genome must be coded, in some fashion, in the chromosomes of the gametes. The chromosomes carry a good deal of structural and protective material. It probably is safe to assume that the essential genetic information is contained in not more than one cubic micron, or an amount of matter consisting of some 10^{11} atoms.

The number of different configurations which can be produced with 10^{11} atoms is large but not infinite. We can estimate the amount of information needed to specify any given configuration as follows.[1] There are some 60 different kinds of atoms in living matter. They occur at very different frequencies. With efficient coding (2), it will take about 1.5 bits, on the average, to determine the nature of any atom. Next, we wish to specify the position of the nucleus; in order to locate it within a cell of dimensions commensurate to the amplitude of thermal vibrations, about 23 bits are needed. It appears that once the nature and position of atomic nuclei are given, additional information concerning electrons is made unnecessary by the constraints of molecular physics. Therefore, 23.5 bits will suffice to specify any atom in living matter and, consequently, not more than 3×10^{12} bits will specify the particular cubic micron of living matter which carries the genotype,

In arriving at this estimate, we have used only knowledge of atoms and some very general principles of chemistry. We know from biochemistry that only a limited number of the physically possible configurations of atoms are actually found in living matter; in other words, a large fraction of the information space previously considered is empty. If we perform the coding in terms of biochemistry, we find | that we need, on the average, 7 bits to specify molecular aggregates of about 100 atoms. On this basis, not more than $7 \times (1/100) \times 10^{11}$ or about 10^{10} bits will specify any genotype. This number still neglects all present and future knowledge of specific chromosome chemistry, and is, therefore, an upper bound rather than an estimate.

[178]

Pitts: Have you used the permutations of the atoms? This is assuming that an individual atom is distinguishable from all others of the same kind, which, of course, it is not.

Quastler: That is right.

Pitts: That is an enormous reduction.

Quastler: The calculation is based on the assumption that each building stone, for a given location, must be separately described. Interchange of identical building stones, within structural units, won't change the information content.

If we assume that the information content of a genotype is between 10^5 and 10^{10} bits, then the information content of an organism such as man should be between 10^6 and 10^{11} bits; of course, if an adult organism is analyzed in the same way in which we treated the material carrying the genotype, a very much larger figure is obtained, but this just proves that a large proportion of information so estimated is redundant. It appears quite certain that it is not necessary to specify in detail the state of every one of the millions of cells which constitute the liver. It is a wide open question whether such

1 Dancoff, S. M., and Quastler, H.: Information rates and error rates of living things. (Unpublished article.)

a separate treatment is indicated in the case of the 10^{10} neurons. That I don't know at all, and there I shall leave this discussion of complexity.

KLÜVER: It is known that isomers may differ in physiologic as well as in chemical and physical properties. Life is supposedly associated with the existence of extremely complex macromolecules. Since the number of possible isomers of such complex macromolecular substances as the proteins staggers the imagination and figures such as 10^{1278} have been quoted, one is apparently faced with enormous possibilities for variation. just what is the relation between the genes and this immense range of possible variations?

QUASTLER: Yes, macromolecules seem to be quite sensitive to structural details. If you feed a cell which is building enzymes some amino acid analog, something which looks enough like an amino acid to be accepted by the cell, the result is a nonfunctional enzyme.

The number I arrived at is small, but this smallness is a bit misleading. If the information content of a genotype is, say 10^7, that implies that there are:

$$2^{10^7}$$

or about $10^{3,000,000}$ different genotypes possible. This number is large, but it is not infinite; it is a number one can deal with.

TEUBER: How do these quantities tie in with the empirical results | that come from [179] the X-ray bombardment data on mutation?

QUASTLER: I tried to work that out. However, right now, we seem to know very little about the mechanism of X-ray induced genetic changes. A few years ago, the mechanism seemed to be beautifully clear. Thus, we do not have a solid basis for an evaluation of the data.

ASHBY: I should like to stress the analogy between these enzymes systems and the cerebral processes of adaptation, because I think knowledge of either side may shed light on the other. For instance, these complex interlocking enzyme systems are obviously under the same restriction that neuronic systems are, for both can persist only if they are stable. Again, each must contain something autocatalytic if it is not to be totally inert, and with it comes, to each of them, a chance of instability. In fact, when these possibilities are coupled with the fact that the genes can change from generation to generation, and that evolution will reject or accept certain gene combinations, the result is a process which is, I think, homologous with what would occur in a random neuronic network and in the homeostat.

If the enzymes fit badly so that the combination is unstable and develops a runaway, that enzyme system, and the organism that bears it, will be eliminated by natural selection. The selection will go on until the genes produce a pattern whose enzyme systems form a balance so that the systems are immune to small disturbances and are, of course, properly related to the other requirements of the organism. The homology is now clear: the enzyme system with its small fluctuations, the substrates going a little up and a little down according to the needs of the moment, corresponds to the small fluctuations of the current variables in the homeostat; the gene-changes, which are of step-function type from generation to generation, with a back effect from the environment acting selectively, give the process which, in the homeostat, corresponds to the finding of a proper set of values for the step-functions.

QUASTLER: I very much agree. I might make an additional point. Adaptation in a cell or in a species corresponds formally to what goes on in the homeostat; also, as in the homeostat, adaptation will, on the average, not increase complexity (in the sense of information content). In this way, adaptation is not a method of evolution in the sense of development from something simple to something more complex; if one state is replaced by another state, the total amount of detail remains the same, although, as

Nelson Wax has pointed out to me, one state may be replaced by a less likely one, and this would increase information content.

FREMONT-SMITH: Isn't evolution always at the cost of something? That is, isn't the [180] adaptation which makes it possible to function better | in relationship to A inevitably leading to a lesser functioning with respect to B, C, or D?

QUASTLER: Yes, and that is why adaptation is not a complete model of evolution, because in evolution some gain must be made without concomitant loss in order to reach higher complexity. Incidentally, if you accept my figures, the total gain in complexity achieved in biological evolution is not so very high: say, from 10^4 to, maybe, 10^{10}. It took two billion years to achieve that.

FREMONT-SMITH: There are some biological examples of that kind of adaptation, although it is not always easy to see what the loss is. There is some work by MacNider on kidney and liver cells which are adapted to survive what would otherwise be a completely lethal dose of a toxin; there is a change in the size and shape of the cells, a recession toward a more transitional type of primitive cell, which then is able to resist completely lethal doses of a series of nonspecific poisons.

QUASTLER: One conceivable mechanism of evolution is that not infrequently living structures, after duplication, fail to separate. If one component of the twin structure changes, then new complications are introduced.

ASHBY: The moment an enzymic system has become stable and established, it provides a fixed point for other systems to work on, to take as a basis for their own adaptation. The achievement of stability in one place immediately makes easier the achievement of stability in the neighborhood. In that way, evolution at one stage much facilitates evolution at the next stage.

KLÜVER: In connection with evolution, it is perhaps of interest to consider not only the smallest number of atoms associated with life but also the number of atoms in the largest living things. Some of the big trees, for example, *Sequoia gigantea* may contain 10^{32} atoms. Paleontology seems to show that the giants have frequently died out. Are there any chemical reasons why there should be no plants or animals with more than 10^{32} atoms?

QUASTLER: I wouldn't know.

ASHBY: One disadvantage that the big animals suffer from is that they are slow in growing, so they are left behind in the evolutionary race. While the big animal has grown one generation, some little animal has passed through ten generations and has tried out ten sets of mutations.

MEAD: Why should human beings survive?

KLÜVER: In a very interesting, if not startling, book on *Malignancy and Evolution* (5) Roberts has tried to explain the destruction of whole species and genera in the course of evolution on the basis of his theory of »hostile symbiosis.« In his opinion, the body [181] is composed of differ|entiated cell colonies living in a state of hostile symbiosis with the result that the precarious and uneasy balance between organs and tissues may be easily lost, with disastrous consequences as regards survival of the species. He even looks at the brain of man as a morbid over-growth or tumor and considers the possibility that the tremendous increase in brain size may not only be one of the causes of malignancy but, ultimately, the cause of the extinction of the human race.

ASHBY: The argument is a very approximate one.

MCCULLOCH: Gentlemen, I hate to break it up but the hour is waxing late. I think we had better say goodbye to each other for another year.

APPENDIX I
REFERENCES

[182]

THE POSITION OF HUMOR IN HUMAN COMMUNICATIONS

1. RUESCH, J., and BATESON, G.: *Communication. The Social Matrix of Psychiatry.* New York, Norton, 1951 (See: The Conventions Of Communication, p. 212).
2. RUSKIN, J.: *The Stones of Venice.* London, Smith, 1853, and New York, Dutton, 1907 (Vol. III).
3. CARRINGTON, J. F.: *The Talking Drums of Africa.* London, Carey, Kingsgate, 1949.
4. Simmel G.: *Soziologie. Untersuchungen über die Formen der Vergesellschaftung.* Leipzig, Duncker & Humblot, 1908.
5. COBB, S., and LINDEMANN, E.: Neuropsychiatric observations. *Management of the Cocoanut Grove Burns at the Massachusetts General Hospital. Philadelphia,* Lippincott, 1943 (p. 14-34).
6. MONNIER, M.: Experimental work on sleep and other variations in consciousness. *Problems of Consciousness.* Abramson, H. A., Editor. Third Conf. New York, Josiah Macy, Jr. Foundation, 1952 (p. 107).
7. BARACH, A. L.: Continuous immobilization of the lungs by residence in the equalizing pressure chamber in the treatment of pulmonary tuberculosis. *Dis. of Chest* **12,** *3 (1946).*
8. BARACH, A. L., EASTLAKE, C., JR., and BECK, G. J.: Clinical results and physiological effects of immobilizing lung chamber therapy in chronic pulmonary T. B. *Dis. of Chest* **20,** *148 (1951).*
9. KUBIE, L. S., and MARGOLIN, S.: A physiological method for the induction of states of partial sleep and securing free associations and early memories in such states. *Transactions of the American Neurological Association* Richmond, Va., Byrd, 1942.
10. —: An acoustic respirograph. A method for the study of respiration through the graphic recording of the breath sounds. *J. Clin. Investigation* **22,** *221 (1943).*
11. MAGOUN, H. W.: An ascending reticular activating system in the brain stem. (cf. Bibliog.) *Arch. Neurol. & Psychiat.* **67,** *145 (1952).*

THE PLACE OF EMOTIONS IN THE FEEDBACK CONCEPT

1. KUBIE, L. S.: Problems and techniques of psychoanalytic validation and progress. *Psychoanalysis as Science.* The Hixon Lectures, California Institute of Technology. Stanford Univ. Press, 1952 (p. 46).
2. MARGOLIN, S.: The behavior of the stomach during psychoanalysis. *Psychoanalyt. Quart.* **20,** *349 (1951).*
3. HEDIGER, H.: *Wild Animals in Captivity.* London, Butterworths Sci. Publ., and New York, Academic Press, 1950.
4. RICHTER, C. P.: Domestication of the Norway rat and its implications for the problem of stress. *Res. Publ. Assoc. Res, Nerv. & Ment. Dis.* **29,** *79 (1950).* | [183]
5. PAPEZ, J. W.: A proposed mechanism of emotion. *Arch. Neurol. & Psychiat.* **38,** *725 (1937).*
6. KLÜVER, H., and BUCY, P. C.: Preliminary analysis of functions of the temporal lobes in monkeys. *Arch. Neurol. & Psychiat.* **42,** *979 (1939).*

HOMEOSTASIS

1. Ashby, W. R.: *Design for a Brain.* New York, Wiley; London, Chapman & Hall, 1952.
2. —: The nervous system as physical machine, with special reference to the origin of adaptive behaviour. *Mind* **56,** *44 (1947).*
3. —: The homeostat. *Electronic Engineering* **20,** *380 (1948).*
4. —: The cerebral mechanisms of intelligent behaviour. *Perspectives in Neuropsychiatry.* Richter, D., Editor. London, Lewis, 1950.
5. The stability of a randomly assembled nerve-network. *Electroencephalog. & Clin. Neurophysiol.* **2,** *471 (1950).*
6. Pomerat, C. M.: Pulsatile activity of cells from the human brain in tissue culture. *J. Nerv. & Ment. Dis.* **114,** *430 (1951).*
7. Schleich, C. L.: *Vom Schaltwerk der Gedanken. Neue Einsichten und Betrachtungen über die Seele.* Berlin, Fischer, 1925.
8. Fisher, R. A., and Yates, F.: *Statistical Tables for Biological, Medical and Agricultural Research.* Edinburgh, Oliver & Boyd, 3rd ed., 1943.
9. Harlow, H. F.: The formation of learning sets. *Psychol. Rev.* **56,** *51 (1949).*
10. Störring, G. E.: *Gedächtnisverlust durch Gasvergiflung, Ein Mensch ohne Zeitgedächtnis,* Leipzig, Akad. Verlagsgesellschaft, 1936.

DISCRIMINATION AND LEARNING IN OCTOPUS

1. Boycott, B. B., and Young, J. Z.: The comparative study of learning. *Symp. Soc. Exper. Biol.* **4,** *432 (1950).*
2. Young, J. Z.: Growth and plasticity in the nervous system. *Proc. Roy. Soc., Lond.* **B. 139,** *18 (1957).*

GENERAL EXCITATION AND INHIBITION

1. Adrian, E. [?].: The mental and physical origins of behavior. *Internat.* 1. *Psycho-Analysis* **27,** *1 (1947).*
2. Von Foerster, H., Editor: *Cybernetics.* Trans. Sixth Conf. New York, Josiah Macy, Jr. Foundation, 1949 (p. 12).
3. Weiss, P.: *Genetic Neurology. International Conference on the Development, Growth, and Regeneration of the Nervous System.* Chicago, Univ. Chicago Press, 1950.
4. Terzuolo, C.: Ricerche sul ganglio ciliare degli uccelli. Connessioni, mutamenti in relazione all'eta e dopo recisione delle fibre pregangliari. *Ztschr. f. Zellforsch. u. mikr. Anal.* **36,** *255 (1951).* | [184]
5. Carey, E. J., *et al.:* Studies of ameboid motion and secretion of motor end plates; experimental pathologic effects of traumatic shock on motor end plates in skeletal muscle. *J. Neuropath. & Exper. Neurol.* **4,** *134 (1945).*
6. Gasser, H. S.: The control of excitation in the nervous system. *Harvey Lect.* **32,** *169 (1937); Bull. New York Acad. Med.* **13,** *324 (1937).*

MECHANICAL CHESS PLAYER

1. Ashby, W. R.: Can a mechanical chessplayer outplay its designer? *Brit. J. Philos. Sci.* **3,** *44 (1952).*
2. Capablanca, J. R.: *Chess Fundamentals.* London, Bell, 7th ed., 1942.

TURBULENCE AS RANDOM STIMULATION OF SENSE ORGANS

1. NOVICK, A., and SZILARD, L.: Description of the Chemostat. *Science* **112,** *715 (1950).*

FEEDBACK MECHANISMS IN CELLULAR BIOLOGY

1. SONNEBORN, T. M.: Recent advances in the genetics of Paramecium and Euplotes. *Advances Genet.* **1,** *263 (1947).*
2. SHANNON, C. E., and WEAVER, W.: *The Mathematical Theory of Information.* Univ. of Illinois Press, 1949.
3. WIENER, N.: *Cybernetics.* New York, Wiley, 1948.
4. VON NEUMANN, J.: The general and logical theory of automata. *Cerebral Mechanisms in Behavior. (The Hixon Symposium, 1948.)* New York, Wiley, 1951 (p. 1).
5. ROBERTS, M.: *Malignancy and Evolution. A Biological Inquiry into the Nature and Causes of Cancer,* London, Nash & Grayson, 1926.

CYBERNETICS

CIRCULAR CAUSAL AND FEEDBACK MECHANISMS IN BIOLOGICAL AND SOCIAL SYSTEMS

Transactions of the Tenth Conference
April 22, 23, and 24, 1953, Princeton, N. J.

Edited by

HEINZ VON FOERSTER
DEPARTMENT OF ELECTRICAL ENGINEERING
UNIVERSITY OF ILLINOIS
CHAMPAIGN, ILL.

Assistant Editors

MARGARET MEAD
AMERICAN MUSEUM OF NATURAL HISTORY
NEW YORK, N. Y.

HANS LUKAS TEUBER
DEPARTMENT OF PSYCHIATRY AND NEUROLOGY
NEW YORK UNIVERSITY COLLEGE OF MEDICINE
NEW YORK, N. Y.

Sponsored by the
JOSIAH MACY, JR. FOUNDATION
NEW YORK, N. Y.

PARTICIPANTS

Tenth Conference on Cybernetics[1]

MEMBERS

WARREN S. McCULLOCH, *Chairman*
Research Laboratory of Electronics, Massachusetts Institute of Technology; Cambridge, Mass.

HEINZ VON FOERSTER, *Secretary*
Department of Electrical Engineering, University of Illinois; Champaign, Ill.

GREGORY BATESON
Veterans Administration Hospital; Palo Alto, Cal.

ALEX BAVELAS[2]
Department of Economics and Social Science, Massachusetts Institute of Technology; Cambridge, Mass.

JULIAN H. BIGELOW
Department of Mathematics, Institute for Advanced Study; Princeton, N. J.

HENRY W. BROSIN
Department of Psychiatry, University of Pittsburgh School of Medicine; Pittsburgh, Pa.

LAWRENCE K. FRANK
72 Perry St., New York, N.Y.

RALPH W. GERARD[2]
Departments of Psychiatry and Physiology, University of Illinois College of Medicine; Chicago, Ill.

GEORGE EVELYN HUTCHINSON
Department of Zoology, Yale University; New Haven, Conn.

HEINRICH KLÜVER
Division of the Biological Sciences, University of Chicago; Chicago, Ill.

LAWRENCE S. KUBIE
Department of Psychiatry and Mental Hygiene, Yale University School of Medicine; New Haven, Conn.

RAFAEL LORENTE de NO[2]
Rockefeller Institute for Medical Research; New York, N.Y.

DONALD G. MARQUIS
Department of Psychology, University of Michigan; Ann Arbor, Mich.

MARGARET MEAD
American Museum of Natural History; New York, N.Y.

F. S. C. NORTHROP
Department of Philosophy, Yale University; New Haven, Conn.

WALTER PITTS
Research Laboratory of Electronics, Massachusetts Institute of Technology; Cambridge, Mass.

ARTURO S. ROSENBLUETH[2]
Department of Physiology, Instituto Nacional de Cardiologia; Mexico City, D. F., Mexico

LEONARD J. SAVAGE
Committee on Statistics, University of Chicago; Chicago, Ill.

T. C. SCHNEIRLA
American Museum of Natural History; New York, N.Y.

HANS LUKAS TEUBER
Department of Psychiatry and Neurology, New York University College of Medicine; New York, N. Y.

GERHARDT VON BONIN
Department of Anatomy, University of Illinois College of Medicine; Chicago, Ill.

GUESTS

VAHE E. AMASSIAN
Department of Physiology and Biophysics, University of Washington; School of Medicine; Seattle, Wash.

Y. BAR-HILLEL
Department of Philosophy, Hebrew University; Jerusalem, Israel

JOHN R. BOWMAN
Department of Physical Chemistry; Mellon Institute of Industrial Research, University of Pittsburgh; Pittsburgh, Pa.

YUEN REN CHAO
Department of Oriental Languages, University of California; Berkeley, Cal.

JAN DROOGLEEVER-FORTUYN
Department of Neurology, University of Groningen; Groningen, Holland

W. GREY-WALTER
Burden Neurological Institute; Stapleton, Bristol, England

HENRY QUASTLER
Control Systems Laboratory, University of Illinois; Urbana, Ill.

CLAUDE SHANNON
Bell Telephone Laboratories, Inc., Murray Hill Laboratory; Murray Hill, N. J.

THE JOSIAH MACY, JR. FOUNDATION

FRANK FREMONT-SMITH, *Medical Director*

JANET FREED LYNCH, *Assistant for the Conference Program*

1 This is the final conference.
2 Absent.

FOREWORD

The three presentations published herewith were presented and discussed at the Tenth and last Conference on Cybernetics (Circular Causal and Feedback Mechanisms in Biological and Social Systems) in the usual informal style as represented in the previous four publications of this conference series. On review of the verbatim transcript of the discussion, it became evident to the Editors that in this instance the presentations repeatedly interrupted by discussion would not produce an effective publication. Accordingly each of the authors has been asked to pull his material together into a single consecutive statement, and the discussion has been omitted.

FRANK FREMONT-SMITH, M.D.
Medical Director

THE JOSIAH MACY, JR. FOUNDATION CONFERENCE PROGRAM

[11]

CONTINUOUS ADVANCE IN the field of medicine requires not only new discoveries at the frontiers of knowledge but also effective communication among investigators. New »break-throughs« at any one spot may depend upon the integration of insights and technical skills derived from widely disparate areas of scientific investigation.

The growing volume of scientific publications in itself imposes a heavy burden upon the investigator to keep abreast of advances in his own field and discoveries in other branches of science pertinent to his particular interest. But there is another aspect of communication, a more personal one which has to do with the tendency of scientists, at their meetings, to accept the lecture or other formal, and hence uninterrupted, presentation as the major means of communication. There are a number of obstructions to this method of communication, some obvious such as problems of national and technical language. Others, less evident and therefore more difficult to cope with, are psychological and cultural. These have to do with unrecognized blind spots, prejudices, and over-attachment to or dependence upon an »authority« or upon too narrowly conceived criteria of credibility. Such hidden obstructions to communication form a major source of misunderstanding and even hostility among scientists and threaten to delay the acceptance or proper evaluation of new data and particularly to prevent genuine cross-discipline understanding and multidiscipline team work.

The Conference Program of the Josiah Macy, Jr. Foundation has gradually evolved in an effort to deal more effectively with these hidden obstructions to communication. The participants are limited to twenty-five – fifteen to twenty regular members who attend the five annual meetings of each group and the remainder guests, invited to one or more meetings. The group lives together for 2 days at a small inn, away from the distractions of a large city, where the informality of arrangements contributes to the development of a warm and friendly atmosphere. | [12]

The emphasis upon discussion, provided for by limiting the presentations to one, or at the most two, per day and by encouraging interruptions at any time during the presentation, and the tradition, now well established, that »authority« carries little weight in evaluating the credibility of ideas, concepts, and data, help to make the conference a forum for searching examination of differences of opinion and of the reasons for contradictory experimental results. Overgeneralizations are quickly met with the question »with respect to what?«

In the atmosphere of such a meeting it is often possible to discover the basis for contradictory findings. Thus the »failure to confirm« is frequently found to be due to previously unrecognized differences in experimental procedure. At such a meeting a scientist may discover that his hostility to the concepts or data of another is engendered by the cross-cutting of two contradictory but overgeneralized conclusions. Once the overgeneralizations are »cut down to size,« i.e., made to conform to the limits within which there are supporting data, the threat to his position disappears and hostility is often replaced by acceptance or constructive criticism.

Members of the conference group become friends; spontaneous collaboration follows naturally. With the growth of mutual confidence, the members bring unpublished data and plans for experiments to the conference in order to obtain critical judgment and suggestions from the group.

Finally, as an atmosphere of free-floating security is established, the group becomes increasingly creative. New suggestions for research and working hypotheses are freely put forward, to be discarded, amended or subsequently tested by experiment. Often the most constructive suggestion comes from a participant not immediately concerned with the problem under discussion and able, therefore, to see the issue with fresh perspective. A genuine partnership in the growth of ideas is the goal.

The transactions of the conferences are published in order to share the experience of the meetings with a larger audience. Although verbatim reporting is impractical, every effort is made to preserve the spirit of the conference. Each participant is given an [13] opportunity to edit his own contributions. The Medical Editor | in co-operation with the Foundation staff reserves, however, the liberty to make some rearrangement of material to provide better continuity for the reader.

The Foundation looks upon the Conference Program as an experiment in communication, still in progress.

FRANK FREMONT-SMITH, M.D.
Medical Director

INTRODUCTORY REMARKS[1]

WARREN S. McCULLOCH
Chairman

I believe this group has been guilty of a certain irreverence with respect to the subconscious or the unconscious. Therefore, I should like to review how the idea of unconscious mental mechanisms came into the field of psychology. No less a figure than Leibnitz[!] (1) was responsible for initiating it. He was interested only in perception, but be regarded it as a sort of integral of our awareness of the world. His »Petite Perception« was not something of which we could say, »I know this, I know that, I know the other.« We could only say such things of the integral. Out of these *petites perceptibles,* which served, if you will, as infinitesimals in his calculus, he supposed that we formed our notions. He found himself coping with a kind of froth, under which there was a kind of sea, without being able to come to grips with the controlling variables.

That is not new, either in biology or in physics. In physics, it led first to the notion of potential energy, which certainly is not energy in the sense in which energy does anything. It is only that it may, when it gets going. In biology, it led to a corresponding observation of significant variables: the conservation of species which in Aristotle (2) became »Entelechy.« Since we are not aware of all the significant variables, many of them, in every panpsychism, have to be supposed to operate *sub rosa.*

If we wish to follow the notion from then on, it will be found next appearing in Hume (3) in two forms: one as an instinct, or »habit of mind,« which leads us to group events into what we thereafter call time and space; this we shall recognize later in Kant (4) as the forms of sensation, and its organization is the synthetic *a priori.* The other, which underlies the notion of causality, becomes in Kant the category of reason. I did my best to persuade Professor Jean Piaget to be with us, because I think he has a clearer | concept of this subject, and certainly better data on how the idea of causality arises in children, than any of the rest of us. It is certain that if what we call cause and consequence are separated sufficiently in time, then the consequence appears as a spontaneous act. Think for a moment that we have a ball rolling up to another ball; the first ball arrives at the second ball, and the second ball takes off. If it happens promptly, we have a notion that the first ball kicked the second. If it happens after ten minutes, the second ball did it on its own. What the mechanism is, and how it operates in us, I do not know, but it is fairly clear that it does, and I should have liked to hear an excellent observer of human beings, such as Professor Piaget, tell us how it arises in children, because I think that is the only way we are ever going to make sense out of it. I think Hume would have agreed.

Causality became, in Immanuel Kant's category of reason, a little bit twisted. The forms of sensation and the category of reason underwent a Hegelian twist and led to the basis of Marxism, on the one hand, and to the dynamic ego and much of our so-called dynamic psychology, on the other. I have studied rather carefully the beginnings of these notions and their spread in Europe. The German school is blatant. Its best protagonist was von Hartmann, who has written and published repeatedly on this subject, so that at the time Freud began to write, he had sold some nine editions, running to ten thousand or more items each. The best exemplar in Scotland was Laycock. (5) He was Professor of Neuropsychiatry in Edinburgh; and in 1869, the second edition of his book, *Mind and Brain,* appeared. You will find there a brief history of the growth of this idea on the Continent. It became in France the normal way of understanding hys-

1 This was distributed to the participants in advance of the Tenth Conference.

teria in this period. You probably have not read a most entertaining volume which is entitled *Unconscious Memory* and written by Samuel Butler (6), author of *Erewhon* (7). It deals with unconscious memory, and is charming in its self-revelation, and in its horror at von Hartmann's notions.

I am inclined to believe that in all our discussions, the unconscious has suffered a gratuitous insult. I was a psychologist before I went into physiology, and I went into physiology because I was convinced that I had been dealing with a froth and that there [17] were significant variables lying below any level which any flow, I don't | care how relaxed it was, could ever reach. I believe that our willingness to participate in this conference indicates that we are not too much worried about how we appear to our neighbors or to ourselves and are quite willing to make fools of ourselves provided we can propose some mechanism which may be at the basis of what is going on. As every scientist knows, that is a hazardous affair, and I should mention that my own notions of how the brain ever gets an idea have had several holes poked in them.

Norbert Wiener, not I, first proposed that there was some scanning mechanism in vision (8). I tied it up with a particular structure and I was probably wrong; Lashley (9) is quite sure I was wrong. I think his arguments are not entirely conclusive, but they are persuasive. An erstwhile member of this group, Donald MacKay, while working in my laboratory, constructed a square which could be made to balloon out, shrink down at any preassigned rate, or could be coupled back from one's own brain waves. According to all my notions, that square should have upset our ability to perceive form. It failed completely to do so, and gave us peculiar distortions of color vision. I think MacKay has thrown the biggest rock that can be thrown through my hypothesis. I am rather upset by a paper which Sholl and Uttley (10) sent me not long ago, in which they said that there is no theory of perception which is subject to a test. I can only answer, »I did propose one, and it is probably wrong.« To insist on being wrong is to insist on there being something which can be checked. It is my notion that every scientific hypothesis has a reasonable expectation of being disproved; certainly none can be proved, and it is my great woe, with most of my friends who are interested in psychodynamics (above all, those who are particularly interested in the subconscious), that I fail to find hypotheses set up by them which are capable of experimental disproof. The best of psychoanalysts, I am quite sure, are as much troubled by this as I am.

Ten years ago I tried an experiment. There was a meeting of the American Psychiatric Association in Detroit. I was then intrigued by the Lear Komplex, which is the subject of a paper coming out of Vienna concerning the play of King Lear, and it proves that Mrs. Lear is the all-important person, because she is not once mentioned. [18] My experiment was the following: In the hotel in Detroit | there was a rather large and comely bear. Drs. Frank Fremont-Smith and Molly Harrower cottoned on to the bear and they paraded it through the hotel. They gave me the idea of what I think is the nicest of all yams I have ever invented. The story is the following: They brought the bear to the conference room. One psychiatrist after another would look at the bear sitting among them, and then snap his head back to the front. You could count ten, and each one would take a second look and snap his head back to the front. Several years later I asked Drs. Fremont-Smith and Harrower whether any of the psychiatrists had said anything to either of them about seeing a bear at the meeting. They said, »No.« I have told this story wherever psychiatrists were gathered together, and shall continue to tell it. My esteemed friend, Dr. Alexander Forbes, heard me tell the story and at once went to Dr. Harrower and asked, »Did you really have a bear at the meeting?« She replied, »No.«

STUDIES ON ACTIVITY OF THE BRAIN

[19]

W. GREY-WALTER
Burden Neurological Institute
Stapleton Bristol, England

Cybernetics is a term that means all things to some men and nothing to many. To me, as a British neurophysiologist, it means, on the one hand, the mechanical apotheosis of reflexive action, on the other, the incarnation of information. If these conferences have done nothing else, they have encouraged some of us who would otherwise have remained isolated in the laboratory to risk an occasional sortie out into the universe in order to experience unity. Not Unity with a mystic capital, but the simple unity of principle in communication and causality that Norbert Wiener was the first to emphasize so clearly. Cybernetics may change its maiden name and breed a score of changelings, but we shall never again be targets for the gibe of knowing more and more about less and less.

I was brought up in a fairly rigorous school of physiology in Cambridge, England and spent some of my earlier years studying the nerve impulses in peripheral systems. Then for a while I was engaged in the study of conditioned reflexes. It was rather a nice juxtaposition I reached, passing from the purely academic study of peripheral neurophysiology to my first direct acquaintance with the Pavlovian school of conditioned reflexology, under the wing of a pupil of Pavlov's. At this time I met Pavlov himself and discussed with him the lines they were following.

The first thing I learned, in branching out from axonology to conditioned reflexes, was that not all animals are the same. In physiology, it has always been a sort of axiom that one sort of nerve from an animal will be like the same sort of nerve from another sort of animal. You could speak of *the* action potential of *the* nerve. What became clear in studying conditioned reflexes was that often you cannot speak of *the* dog at all; usually you can speak with assurance only of this particular dog at this particular time. | [20]

That introduced me to the general notion that individuality and personality – and not only consistent personality factors but personality as a function of time – would be an important thing in investigating how brains work and how we get ideas. From that, I went on to the study of the brain itself in application to clinical problems.

A few years later I arrived at the more elegant problem of normal brain functions in relation to exactly those matters that Dr. McCulloch has outlined for us – this question of how the brain does get ideas and why it is that the mechanism of idea formation seems to be so inaccessible and yet to obsess people so much. My aim during the last 5 or 10 years has been to try to identify and specify the elements of »awareness,« if you like, in terms of physiological mechanism, not trying to creep up furtively to the brain through the spinal cord, but to examine the brain as a whole organ and see whether information coming out of it was going to tell us anything about how it works, and what it prefers to work on.

The proposition I am going to bring up for discussion is that, in fact, quite apparently superficial study of brain mechanisms can tell us an extraordinary amount about how the brain works, much more than we ever expected a few years ago. You have all seen the sort of work that has been done in brain physiology in humans, mainly for clinical purposes. The ordinary electroencephalograms and pictures of that sort are very tantalizing and disappointing. If you work out the amount of information you can obtain from an ordinary brainwave record in terms of what is there, the results are very discouraging. The redundancy is absolutely fantastic. I think a first class research team of workers dealing with ordinary brain records would be very lucky to obtain one part in a hundred thousand of the information from their brain records into their own brains, to understand it and appreciate it, and be able to see what it means; and that, I think, is why we all have been so discouraged until recently about the possibility of examining the human brain in its normal, conscious state, from outside and without interfering with the individual too much – simply because of this vast redundancy.

There is nothing more discouraging than to have to read a language in which you [21] understand, for example, only one sign, not | even one word but one sign, in a hundred thousand. Most of us in trying to read Chinese might recognize, after a while, the simple sign for »man« or something like that; but we would not be content with so little. It is a very discouraging experience to have a cipher constantly slipping through one's fingers. That is just about the situation of the average lay physiologist dealing with the brain.

We have spent the last few years in developing all sorts of ways of displaying the activity of the brain, mainly in conventional terms, to see how it relates to the psychological events that go on during growth and the establishment of a series of private notions. Parallel with that sort of work, we have taken the other obvious step of taking our results as they have come out of the machinery, gradually, and building models of them, not just for fun, but to see whether, if we take the mechanisms as they seem to be working in the flesh and transpose them from the paper to the metal, the models will work the way the mechanisms seem to work when they are in the brain. This is a perfectly reasonable and logical step. It is simply a question of putting on to the table, in the form of mechanical devices, the algebraic expressions which our records, in effect, are.

One field of endeavor in studying the brain or mind has been to make pictures of how it looks, of its structure. I think probably many of us here have suffered for some years, in our various schools, the embarrassment of looking at pictures of the brain or the spinal cord, stained, drawn and photographed in various ways, and trying to discover what in this was related to what had actually been happening there. I myself have always found particular difficulty in understanding these things.

We have made some pictures to give an idea of the amount of information we can deal with. The method used to obtain these pictures is derived from a system used during the war for radar purposes. The signals which we get from the head, from electrodes attached in the usual way, are amplified by a factor of a million or so, and are made to modulate the brilliance of small cathode-ray oscilloscopes which are assembled in a pattern, as though you were looking down on the head of a patient from above. You can see 22 tubes displaying in their correct spatial positions the outputs from the various channels. There is, then, a picture of rhythmic, flashing lights, which [22] gives a very good simulation of what Sir Charles | Sherrington called »an enchanted loom where flashing shuttles weave a dissolving pattern.« Unfortunately, it *is* a dissolving pattern; in other words, it is constantly changing, and the eye follows the changes but it does not extract the invariants; to extract the invariants from this assembly of patterns, we have to make use of photography which we do not consider a satisfactory method of recording, but the photographic emulsion is a good integrator and builds up the integral of the invariant components very well.

We distinguish the invariant components from the irregular and noisy ones by having on each tube a scanning vector. What is seen on the tubes, if nothing is happening, is a line of light, and if there is no signal coming in at all, it is motionless. But this line of light can be rotated around the tubes like the hands of a clock in various ways. It can be rotated at a controlled speed, so that it turns, for example, ten times per second, in which case something appears that looks rather like a slice of pineapple or a cartwheel with luminous spokes; or, more interestingly, the rate at which this line is scanned round the tubes can be made dependent on the brain itself. Any one component or part of the display system can be used to drive the scanning machinery inside the apparatus. Now, the thing we hoped for from this device, which seems to be a reasonably satisfactory one, is that in doing this, we are making the activity in the brain drive the machine, in such a way that the different parts of the display system are reacting to whatever the brain is doing at any moment. The time ratios are thus local time. There need be no absolute time scale. We are not trying to constrain the brain activity to display itself on an arbitrary, absolute time scale of seconds or milliseconds. We are allowing time to vary; the speed at which this vector goes around may be quite erratic; but the speed it goes at is determined by what is happening in the person's brain at that time and that place; so whatever else is happening elsewhere in the brain at that time and at that rate will be synchronized in rotation on the clock faces. What you have, in fact, are 24 channels comprising 2 monitors, and 22 clock faces. The bands of these 22 clocks, all driven by a master generator, are, as it were, lit up by the brain activity at that particular time; so a picture results only of those events in the brain which are synchronized with, or driven by, or are driving, the activity you have | chosen to act as a [23] synchronizer. In this manner, a direct impression of coincidence, congruence, and contingency can be acquired. I think causality is the root problem of the whole brain function and hope to be able to go into that question in more detail at a later time. What are the causal relations between these various events?

The technical virtues of the display system are that there is almost no redundancy. Furthermore, noise – whether physical noise in the sense of thermal agitation, or brain noise, i.e., the sort of noise in the background which has to do with the administrative work of the brain in keeping itself alive and fed, and so on – is not seen at all. This machine might almost be defined as a device with which it is impossible to see noise, because noise, by definition, produces no pattern. You cannot synchronize the machine on noise from any one channel and have the noise of any other channel synchronized, by definition, because the noise in one region of the brain has no phase relation or regular time relation with the noise in any other channel. Therefore, those events which are significantly related to one another in some time sense are chosen.

The device can also be arranged not only to be driven by the brain activity, but also to drive the brain activity, to administer to the individual under study a series of stimuli – light signals, sound signals, or touch signals – which in themselves may give rise to a pattern in time or space; and this pattern of input signals can be delivered at any desired rate. We use this equipment regularly in both these ways; that is, we use it, first of all, with the brain driving the machine, and then the reverse process, with the machine driving the brain.

The results are rather surprising, first of all, in relation to this notion of scansion which Dr. McCulloch has mentioned and which has been attributed to various authors. We have observed, unequivocally, some type of time dissection, occurring in the brain. I coined a word for this which I am using now for only the second time in public; the word is »abscission.« I have deliberately tried to avoid using the word »scanning,« because it has so many associations in ordinary speech and it has become rather tainted, perhaps, with too many simple-minded ideas. But I think there certainly is | [24] some type of abscission, of cutting off, of signals in the brain and their reprojection on a local time base.

We see such abscission particularly clearly as a separating out of the parts of a visual time pattern, in the visual association area of the back of the head. The appearance is easily recognized; one part of the pattern appears in one place and another part of the pattern appears later in another place. Abscission certainly happens; it is there that the incoming ensemble is broken down into parts. But the exciting thing to me when I saw it the first time was the fact that as long as the pattern which the person is seeing is novel or surprising or interesting, the abscission process has its converse in other parts of the brain. The abscission process is turned around and the pattern is abstracted; it reappears again in parts of the brain not generally regarded as being attached very directly to peripheral mechanisms, in the association or silent areas. In those regions, the pattern appears not only in its original form, in whatever time pattern has been introduced, but in all possible transforms of its elements. If a threefold stimulus is put in, it may come out as a three-in-one pattern, or the various signals, when obtained, may be closed or opened up, or two of them together and the third nearby; this continues for a matter of a few minutes. If you leave a person subjected to this type of stimulation for a minute or two, possibly 20, 50, or 100 repetitions of this situation may take place. Then in time the appearance fades away. If the stimulation is stopped before its novelty has worn off, the pattern is preserved for several seconds in the temporal association areas, and then it slowly melts away.

The fading is not a question of fatigue. The time is too short for fatigue to mean anything in relation to the tiredness of the brain. The excitement fades away, as far as we can see, because it becomes boring. The repetition of this pattern, which is a perfectly real signal, a real physiological stimulus, begins to lose meaning to the person. When it is not followed by anything, when it does not seem to signify or imply anything, this remote, diffusely projected and abstracted pattern fades away and subsides. If you then change the pattern at all, if you introduce into it even a slight alteration of [25] the rhythm or shape or brilliance, then immediately the distant syn | thesis shows again; the whole pattern is re-established; it re-forms in the association areas.

We are now going to take an important step further – to associate the arbitrary, novel, but mental signal pattern with an unconditioned stimulus. We have not yet got very far with this. All we have done so far is to establish a defensive reflex which, as we all know from personal experience and training, is much more easily established than an appetitive one. All I need say at present is that when an unpleasant sensation is given in association with certain patterns of this type, the abscission-abstraction-preservation process is enormously enhanced, building up to something quite literally sensational.

The size of photographs taken to record this phenomenon is small compared with ordinary recording, and the patterns are minute by ordinary recording standards. But this elaborate extraction of information goes on in a perfectly regular, automatic way and the results can be recognized and identified quite well. It is satisfying to note that they are of a very personal, very individual character, and are enormously dependent on the subject's experience.

Information about the way these things happen must be handled at the present stage in terms of a rigorous hypothesis. There are two auxiliary methods of study which suggest themselves: One is the making of models which do this sort of thing; the other one is the study of people who have widespread and deep-seated lesions in their brains. In the last few years, neurosurgeons have begun to remove not little bits of brains but half-brains. It is quite commonplace now in neurological clinics to see people who have had this so-called hemispherectomy operation. It is not really hemispheric, in the physiological sense, as the whole hemisphere is not taken out. Quite a large amount of the thalamus and the base of the brain is left, but the cortex is removed almost completely.

We have studied a few of these people, one boy in particular, to see how much the cortex is involved in the abscission process. It is obviously important, in seeking to relate this process to the information from anatomy and physiology, to see whether we are dealing with something which is whole-brain function, or whether the different [26] parts contribute different things. |

The results are quite straightforward in these cases. In people who lose a hemisphere to this extent, with nearly complete loss of cortex, the mutilated side is by no means dead. Quite large voltages can be observed to arise in the side where no cerebral cortex remains, but the spontaneous activity, the unstimulated rhythmic activity, is always considerably attenuated on that side. The activity produced by stimulation in the mutilated side arises mainly in the temporal and transverse regions. The equivalent geometry of the generator is a transverse lozenge, about 4 or 5 centimeters long, pointing out into the central region rather diffusely. In the normal side of the brain, the pattern stimulus is translated as a multidimensional integral complex of activity; on the decorticate side, there is only one component, only this transverse radially oriented component of the pattern.

When the stimulus is cut off in order to test the preservation system of the brain, it flicks off at once on the decorticated side. The moment the stimulus goes, the activity vanishes. But on the intact side, the signaling continues – always, of course, assuming that the signal has been given some significance or that it retains some novelty. Yet, on the decorticate side, although the pattern, when showing, may be nearly as large as on the intact side, there is no preservation at all. The cortex, therefore, does seem to contribute to this process in a rather interesting way. It receives information from below in a much more elaborate and diffuse way than was indicated by the classical physiological data, and it is on the cortex that the preservation process seems to depend.

One other aspect of the matter is clear, namely, that the geometry, or topology, if you like, of these processes is a very intricate one, literally intricate because these changes are interwoven or interlaced. These abstracted components come in great profusion. Some of them come from below; others we have seen projected radially; others again come transversely in the profusely projected activity. I think the best illustration I can give of this is to compare these processes with the action of a sewing machine. I am not certain that I know how a sewing machine works, as it is a quite complicated piece of machinery, but I know there are two threads. This machine does not operate like a mechanical seamstress who simply puts her needle in and out of the fabric on both sides so | that while she is sewing there is one thread in the top needle and [27] another in the shuttle underneath. The needle goes down; the other thread is pushed through it and knotted; this continues until a seam is made. However, if a sewing machine is wrongly adjusted, it will handle the material, the needle will rise and fall, and there will be a seam, but if the thread is pulled at the end, the seam will pull out straightaway because it is not knotted underneath. As far as I can see, the brain works very much like the machine. It sews a seam in which one set of events is one thread and another set of events is another, and the two are interwoven; and so, from the threads of circumstance, a knot of probability results.

In the majority of cases, evidently there is not simply a knot, but something more nearly comparable with the enormous complexity that we find in brain records, a very tangled skein. But we can now begin to disentangle it, to separate it out.

The problem is how the brain decides whether or not it is worth weaving a particular fabric. So many things are happening to us continuously. How do we decide that some things matter? How does the brain decide that this or that particular set of events is really important and has to be woven into this tough, durable fabric of ideas and memory? How does it get an idea about things that occur, which makes certain sets of events more important than others?

I have described my electronic models elsewhere in some detail, but the hypothesis about this measure of significance is this: It would seem that inside the brain there are structures which are phylogenetically very ancient and ontogenetically very variable, which separate the events, as they come in, into two coded sets of information streams, quite separate from the specific sensory projection volleys. Only two are necessary for the simple hypothesis. They will be found quite complex enough for the time being. These two streams have the following characteristics: One stream is essentially a set of time derivatives of events; their counterpart in physiological records is a series of spikes, by far the most prominent feature of most physiological records – responses which appear only or mainly at the beginning of stimulation periods.

The other set I see as almost the converse or complement of the first. It is a stretched [28] series of events, extending the duration of | signals that come in. What comes out from this second system is a long stretched output which, from the physiological standpoint, is an after-discharge. This, it will be recognized, is a perfectly standard or classical physiological process.

In such hypotheses it is obviously desirable not to postulate that the brain is doing anything magical. If a sense organ or the spinal cord will account for an experience of the brain, that is a little more satisfactory than some quite new thing. Moreover, I take it for granted that the chemical substances, of which we hear so much and know so little, may be the prime movers. Acetylcholine and other mechanisms must be involved. I disclaim any notion that the brain is a purely electrical machine.

The second set of operative signals, then, is a stretched series, which long outlasts the duration of the stimulus. It preserves the information that a particular signal or set of signals has occurred. After it has occurred, these two signals have to be combined in some way, not with one another again, but each with the complementary stream of signals from some other stimulus source. This is the important point; a signal comes in, for example, through the eyes, and is divided into these two streams which go forward to the brain as coded information to nearly all areas. I have dubbed these streams of inoperative but still significant signals, »propaganda.« They are to be propagated only, not acted on; they are not to be remembered because they *mean* nothing. There are going into our nervous system all the time innumerable streams of information, 99 out of 100 of which are insignificant to us. If we paid conscious attention to them, we should go crazy. The brain is dealing with them summarily. I mean such facts as that the lights in the room are on, a fact which is not important to us in itself but only for the light provided for our work. It might be important. For example, if one of the lights started to flicker or go out, it might be important. We are prepared to pay attention to such facts, but they are not memorable.

Such things become memorable only if two streams of signals are repeated often enough to overlap in their complementary forms more often than would be expected by chance. This is where the seam in our fabric comes in. One of these threads, let us [29] say, from the needle, is the derivative. The shuttle which moves across catch|ing the other thread and fixing it is the stretched, extended form of signal. So we have this constant interweaving of the clipped form, from one source, with the stretched form, from the other, which forms are then interwoven into a permanent and durable seam of congruence and contingency, provided that their coincidence is greater than chance expectation. That is why I call this a knot of probability, because what the making of the seam depends upon is precisely the measure of probability that these two series of events from different sources are connected, that one may imply the other.

In order to learn more about this measure of probability, about the odds of association, I have devised an electronic model, an instrument which can receive two streams of signals. Some criterion of the likelihood that two series of events will be regularly and frequently associated with one another has thus been provided.

Other questions that come to mind are how the model works, and more cogently, how the brain or an animal builds up a notion of probability, how it begins to reckon the odds, and how it learns what the odds are in favor of significance, in favor of two events being significantly connected.

Consider an experience that is common but that is happening for the first time. We are aware of a flash of light outside; a few seconds later, we hear thunder. The first time it happens we may think that someone turned on a light and someone hit a drum. If the two things happen regularly, however, we begin to build up a significant idea of a thunderstorm. We then think that when there is lightning, there is thunder and we may generalize in this way. From a statistic one may conjecture, but should not assert, a cause. Yet we may be tempted to go a step further, and say, that lightning causes thunder, unless we are metaphysical or mystical and prefer to say that lightning and thunder have a common cause, which we may call Zeus or electricity, according to the age we live in. We may well ask again what the criterion of significance incorporated in the brain is. But we know the answer already: for tentative action, the same as you would ask for in a preliminary scientific observation, about 20 to one against chance; for definitive action, about 100 to one; for »certainty,« 1,000 to one, and so on. | [30]

Here we have some very simple evidence of the way human brains may build up notions of significance, from which they can derive hypotheses of causality, and may even generate some kind of notion even of causal unity. That leads back in a circle to what Dr. McCulloch started with, the notion of how causality and creativeness and imagination are based on physiological mechanisms.

During the years that these conferences have been held, partly because of them and certainly concurrently with them, notions about brain function have been evolving in a highly significant way. One of these evolutionary trends I have just outlined; it may be summarized by saying that many of us are beginning to consider what, for want of a better phrase, is called »higher nervous activity,« as the behavior of an error-operated analog statistical computor. In this company I need not point out how much such a development owes to the basic thinking of Norbert Wiener, Warren McCulloch, Claude Shannon, and their colleagues and pupils. In England the essays and models of Ross Ashby and Albert Uttley also have encouraged this sort of approach. Mention of Ashby and Uttley leads to the question with which I should like to close my contribution: the fundamental problem of organization *versus* randomness in brains and machines. Almost by definition, a statistical computor[!] need have no program and may not have any fixed circuit diagram. To be sure, whatever components it uses, it must have some arrangement to take care of their metabolism, to keep the cells or transistors polarized if you like, but the interconnections between the elements might be as random as you please. Both Ashby and Uttley have actually made convincing models which assume very little about what may happen or not happen, and I have explicitly postulated this condition in my hypothesis of learning. The rigorous application of Occam's razor seems to shave off most of the whiskers around this problem and to leave something very like randomness as a jumping-off state for a docile machine, whether flesh or metal. But, and this is why I put the question, we know that in the primate brain there is in fact a great deal of elegant and intricate circuitry. There is nothing like random interconnection in any vertebrate and we must plunge into the sea and study coelenterates before we find anything like it. Even apart from the tract and ganglion neuroanatomy in the brain, even in the cortex, there is a highly systematic | stratification and some considerable topographic differentiation. You may say, of [31] course, that this proves the rule; the coelenterate nerve-net was the jumping-off state phylogenetically. It is questionable whether or not this is a just view. It is certainly attractive and the fact that jellyfish are not very clever need not dismay us, for Ashby's Homeostat is not very clever either in that sense – in fact I have classified it as *Machina Sopora* because it seems to be able to stay asleep however rudely it is disturbed. Evolution of animal brains seems to have operated according to the principle »no specialization without delegation,« so that at each growing point of our family tree there is a sort of bud of atavistic randomness. This vestigial network exists in the brain, it seems, and it will be one of our most difficult and absorbing tasks in the near future to find out how much this random excipient contributes to the properties of the statistical learning filters which I have suggested should provide us with trustworthy evidence about the world we live in and ourselves.

Finally, it is legitimate in this context to speculate about the relations between living brains and artificial ones. In the evolution of computing engines, the process seems quite unlike the organic case. The Babbage machine, the modern electronic binary calculator, and all of that breed have very systematic and determined anatomy as well as extravagant metabolism. Obviously they represent only molecules, as it were, of the whole universe of intelligent machinery, and it is interesting to observe that only now are engineers beginning to think about machines that are competent to answer »perhaps« as well as »yes« or »no« or belch out strings of numbers. If I were asked – as indeed I often am – what will be the prosthetic brains of the future, I should not suggest anything like the instruments of the present day; rather I should envisage some sort of »hunch generator« that would deal speculatively with events much as I think our brains do, but would begin life as metallic medusas and would acquire human wisdom without the shocks and disappointments that, in us, breed prejudice and despair. Whether, like the legendary Medusa, they would petrify their observers, will depend on the reflective cunning of our descendants.

SEMANTIC INFORMATION AND ITS MEASURES [33]

YEHOSHUA BAR-HILLEL
Department of Philosophy
Hebrew University
Jerusalem, Israel

The following presentation will be confined mainly to the theory of semantic information as it was developed recently by Dr. Carnap and myself (1). Also it will be concerned with the relation of this concept to the concept of amount of information, as developed by C. E. Shannon (2), and an attempt will be made to show that the two are not in competition. The presentation will occasionally touch upon the relationship between thermodynamical entropy and amount of information, emphasizing both the positive connections and the nonidentity; and it will also touch upon the distinction between statistical probability and inductive probability.

Information Theory, in the sense in which this expression is used in the United States, namely, Theory of Information Transmission, is not generally spoken of as dealing with the content of the messages transmitted, and hence it is regarded as a discipline having no connection with semantics (3). (At a previous conference, Dr. MacKay (4) pointed out to this group that, in British usage, Information Theory has a different connotation.)

However, few could resist falling into the semantical trap laid by the use of the word »information« and not assume sooner or later that this field of interest is applicable to semantics and does reveal certain aspects of meaning. Claude Shannon is, to my knowledge, the only major information theoretician who consistently refrained from drawing illicit inferences.

The theory, an outline of which is presented here, is meant overtly and exclusively to deal with the concept of semantic information conveyed by a statement, or the semantic content of a statement, and various measures for this concept. | [34]

Many would not consider this a worthwhile aim. They would insist that it is sensible to talk about information in the communicational sense and, as a next step, about information in the fullblooded pragmatical sense, i.e., about how much information a certain statement carries for someone in a given state of knowledge. This latter concept is obviously of greatest concern to psychologists, teachers, and other people involved in transmission of information or in the study of this transmission process. But they would deny that the intermediate concept of semantical information – which is pragmatics-free abstracts from the users of language and deals only with the relationships between linguistic entities and what they stand for, or designate, or denote – is of any importance. I believe that this view is mistaken. However, instead of starting a polemic, I shall present merely the outline of the theory and allow you to decide whether or not it can be fruitfully applied.

To repeat, the following is a systematic *explicatum* of what is meant by the content of a statement, or the information carried by a statement, or when it is said that the information conveyed by statement *j* is greater than that conveyed by statement *i*. Only statements that convey or carry information are being used as arguments. This is done deliberately. Not that I want to deny that there are other entities that are said, in ordinary discourse, to carry information. Communication theoreticians assign amounts of information to all kinds of signs or symbols, and scientists talk about the information conveyed by the outcome of an observation or an experiment, or by any other kind of event, in general. I shall not at this time go into the interesting technical question of to what degree these usages are reducible to information carried by statements. What regards assignment of information to events, i.e., entities expressible by statements, I shall adopt this usage as an alternative to my regular usage, whenever this proves profitable. Though there are some interesting philosophico-logical questions involved in this, I shall not discuss them here.

Let me then give you immediately what seems to Dr. Carnap and me to be a reasonable *explicatum,* for the presystematic concept of information, in its semantical sense. We call it *content* and denote it by »*Cont,*« with a capital »*C.*« According to the [35] scholastic dictum, *omnis determinatio est negatio,* we take *the content of a* | *statement* to be *the class of those possible states of the universe which are excluded by this statement,* that is, the class of those states whose being the case is incompatible with the truth of the statement.

For technical reasons, however, we prefer to talk about *state-descriptions* instead of states. The final definition for the content of a statement is, consequently, *the class of all state-descriptions excluded by this statement.* One of the advantages of the shift to state-descriptions is that this forces us to relativize our treatment with respect to given languages. Unfortunately, however, we are able so far to present reasonably adequate definitions for rather restricted language systems only, namely for what is known in the profession as *applied first-order language-systems with identity.* These languages are certainly too restricted to serve as languages adequate for all science. Their major drawback is the absence of functors for quantities such as mass, temperature, etc. This drawback is admittedly serious, and attempts are being made to extend the impact of our theory to richer languages (5). For the time being, however, I beg you to keep this restriction of scope in mind.

To illustrate the concepts I am going to introduce, I shall use a specimen-language, and this will help you to visualize their applicability. The specimen language has the following structure. It contains, in addition to the customary connectives, »~,« »∨,« »&,« »⊃« and »≡,« being, respectively, the signs for negation, disjunction, conjunction, (material) implication, and (material) equivalence, three individual-signs, »*a,*« »*b,*« and »*c,*« and two primitive predicates, »*M,*« and »*Y.*« The interpretation is as follows: A census has to be taken in a small hamlet of three inhabitants. The census is of a very restricted scope. The census taker is required to find out only whether the inhabitants are male or female and young or old (defined, respectively, as being under 35 years of age, or otherwise). The three individual-signs denote, then, the inhabitants of the hamlet, in a certain fixed order, and the two primitive predicates the properties Male and Young. For »~*M*« and »~*Y,*« denoting, respectively, the properties Not-Male, or Female, and Not-Young or Old, we shall use the abbreviations »*F*« and »*O.*«

Any sequence of one of the letters »*M,*«»*F,*« »*Y,*« and »*O*« and one of the letters »*a,*« [36] »*b,*« and »*c*« is called a *basic statement.* |

»*Ma,*« for instance, is a basic statement signifying that *a* is male. All the statements of our language consist of the twelve basic statements and the statements formed out of them with the help of the connectives.

Our specimen language is extremely poor. What it loses thereby will, I hope, be outbalanced by the perspicuity of the applicability of the rather abstract concepts to be introduced presently.

An example of a state-description in our language is as follows: Since such a state-description denotes a possible state of the universe dealt with in this language, it tells us, for each individual, whether this individual is male or female, young or old. It consists, consequently, altogether of a conjunction of six basic statements. »Ma & Ya & Fb & Yb & Mc & Oc,« for instance, is a state-description telling that *a* is male and young, *b* female and young, *c* male and old. This conjunction describes our universe completely. As soon as I add a basic statement which is not identical with any of the six components of this conjunction, this basic statement can only be the negation of one of these components so that the resulting conjunction will be self-contradictory. We conclude, then, that any statement which logically implies, or is stronger than, a state-description is self-contradictory. In this sense, a state-description is a strongest synthetic statement in its language.

We take as the *explicatum* for the information conveyed by a given statement the class of the state-descriptions excluded by that statement. Each statement in our specimen language excludes, then, none, one, or more (up to 2^6, or 64, which is the number of all state-descriptions) of the state-descriptions. An analytic, or logically true, statement, such as »*Ma* ∨ *Fa*,« excludes none; a self-contradictory, or logically false, statement, such as »*Ma* & *Fa*,« excludes all; a synthetic, or logically indeterminate, statement excludes some, but not all, state-descriptions. An analytic statement has minimum content, and a self-contradictory statement maximum content. This is not surprising. A self-contradictory statement tells too much, it excludes too much, and is incompatible with any state of the universe, whereas an analytic statement excludes nothing whatsoever and is compatible with everything.

A state-description itself, as a strongest synthetic statement, excludes all other state-descriptions. We shall deal somewhat later | with its counterpart, a weakest synthetic statement, that excludes just one state-description. [37]

Please notice that we have so far explicated only the notion of information itself. In a prior meeting, Shannon told this group that he did not define the concept of information itself and was not interested in it. He wanted to define only the concept of amount-of-information, with undeletable hyphens. To my knowledge he never uses the notion of information as such in any essential way, nor is there any reason why a communication engineer should do so. We offer an explication for information itself, in its semantical sense, for whatever it is worth. Whether *Cont* is a good *explicatum*, you may decide for yourselves.

Our next problem is now to define *measures of content* to serve as *explicata* for amount-of-information, in the semantical sense. So long as we are talking only about content itself, the most we can say is that a certain statement has a larger content than another one, and this in case that the class of state-descriptions excluded by the first statement includes the class of state-descriptions excluded by the second one as a proper part. Thus we would say that

$$\text{Cont(»Ma«)} \supset \text{Cont(»Ma} \vee \text{Yb«)},$$

since the class of state-descriptions excluded by »Ma« (there are 32 of them) contains the class of state-descriptions excluded by »Ma ∨ Yb« (there are 16 of them) as a proper part. But if two contents are exclusive, i.e., have no member in common, or overlapping, i.e., have some but not all members in common, we can say no more about the statements whose contents they are, although we certainly would like to be able to say, and justify, that »*Ma*« has a larger content, conveys more information, than »*Mb* ∨ *Yb*.«

In order to do just this, we must go over to the stage of talking about measures of content, to define – as the mathematicians would say – a measure-function ranging over the set of contents. Fortunately, we do not have to start from the beginning. Carnap (6) has developed a rather extensive theory of measure-functions ranging over, not, to wit, contents, but something very similar to them, namely, what he calls *ranges*. (The range of a statement is the class of those state-descriptions that logically imply that statement.) Those measure-functions, which he calls *m-functions*, are | meant to explicate the presystematic concept of *logical* or *inductive probability*. [38]

Now, since a measure-theory of ranges has already been developed, we can clearly make full use of it and define our content measure-functions on the basis of Carnap's *m*-functions. For each such *m*-function, a corresponding function can be defined in some way that will measure the content of any given statement.

Let *m* be a measure-function of ranges. What kind of function of *m* shall we take as a measure for contents? We have a multiplicity of choices. But all choices will have to fulfill the following condition that seems clearly indicated for any adequate explication of amount of information: *the greater the logical probability of a statement, the smaller its content measure.* The fulfilment of this inverse relationship will be the guiding requirement for our choices.

The mathematically simplest relationship that fulfills this requirement is the complement to 1. Let $m(i)$ be the logical probability of the statement *i*. Then $1 - m(i)$ can be taken as a plausible measure for the content of *i*. We call this measure simply the *content measure* of *i* and denote it by »*cont*(*i*)« with a lower case »*c*«, in distinction to the uppercase »*Cont*« standing for content. The formal definition is, then,

$$\text{cont}(i) =_{\text{df}} 1 - \text{m}(i).$$

It can easily be seen that *cont* fulfills, in addition to the aforementioned condition of adequacy, other requirements of an adequate explication of amount of information, *but not all.* It fails to fulfill a certain requirement of additivity, the counterpart of which plays a great role in Shannon's theory. This requirement, in our case, would be that the content measure of two statements that are *inductively independent* – meaning thereby roughly that the logical probability of either statement should not be changed by being given the other statement – should be equal to the sum of the content measures of each of them taken separately. But, it is not the case, in general, that if *i* and *j* are inductively independent, then $cont(i\&j) = cont(i) + \text{cont}(j)$. There is, indeed, an additivity theorem for *cont,* but the condition under which it holds, is not that *i* and *j* should be inductively independent but rather that they should be *content-exclusive,* i.e., that [39] there should be no | state-description excluded by both *i* and *j*. This additivity condition makes sense though it is at odds with the more customary one of inductive independence. It makes sense to say that the content measure of the conjunction of two statements should be equal to the sum of their content measures if, and only if, they are content-exclusive. Then *cont* has certain plausible properties though it lacks certain other properties which are equally plausible. Since, however, no concept can have both these plausible properties simultaneously, we are led to the idea – and there are many other arguments pointing in the same direction – that we do not have in our mind *one* clearcut, unique, presystematic concept of amount of information but at least two of them (and both still in the semantic dimension). This is not so strange. On the contrary, it is a rather common phenomenon that two related but different concepts are regarded as being identical although contradictory properties are required for their *explicata.*

Though *cont* seems to be a very natural and simple systematic correlate for the presystematic concept of amount of semantical information, neither it nor, what is perhaps somewhat more surprising, its statistical counterpart have been discussed much so far. On second thought, however, it is perhaps not difficult to account for this neglect. It seems that in communication engineering, the requirement of additivity under the statistical counterpart of inductive independence is much more important and practical than such a requirement under the counterpart of content exclusiveness, and it is doubtful whether this condition makes sense there at all. Another reason will be given further on in the presentation.

At this point, it will probably be of some help to present to you a pair of content-exclusive statements in our specimen-language:

»Ma« and »Ma ⊃ Fb.«

You can easily see that the contents of these two statements do not overlap if you transform the second statement, according to the rules of the ordinary propositional calculus, into »Fa ∨ Fb.« More generally, any two statements, one of which is an implication statement having the other as its antecedent, are content-exclusive.

I already mentioned that we chose the roundabout way of introducing the concept of content measure *via* the *m*-functions only | because an extensive theory of these [40] functions stood at our disposal. For those unacquainted with that theory, it would probably have been pedagogically wiser to base the whole thing not on state-descriptions but on their »dual,« which we call *content-elements*. You get the »dual« of a state-description if you replace the »&«-signs by »∨«-signs and each capital letter by its complement. The dual of the state-description, for instance, I mentioned before for illustrative purposes, is:

$$Fa \vee Oa \vee Mb \vee Ob \vee Fc \vee Yc.$$

Just as the state-descriptions are the strongest synthetic statements, so the content-elements are the weakest synthetic statements. I could then have defined the content of *i* as the class of all content-elements logically implied by *i*. I could then finally have defined *cont*(*i*) as a measure-function of contents dually analogous to Carnap's definition of *m*-function.

Since *cont* is not additive under inductive independence, we need another *explicatum* for amount of information that will have this property and will assign to »*Ma* & *Yb*,« for instance, an information measure that is equal to the sum of the information measures of »*Ma*« and »*Yb*« since these two statements are inductively independent for any adequate *m*-function I can think of. Being given the information that *a* is male, the logical probability of *b*'s being young should not be affected. This seems rather obvious. Under a certain normalization, the information-measure of each of these two statements turns out to be *one* (bit). We would then like the information measure of their conjunction to be 2. But this does not bold for *cont*. The *cont*-value of the conjunction is smaller than the *cont*-value of the components, since these two statements are certainly not content-exclusive: They both imply, for instance, the statement »*Ma* ∨ *Yb*.« Since a synthetic statement is logically implied by both statements, their contents cannot be exclusive.

Insisting on additiveness on condition of inductive independence, we obtain another set of measures for amount of information which we call this time *information measures* and denote by »*inf*.« We can define »*inf*« either with the help of »*cont*« as

$$\text{inf (i)} =_{df} \log_2 \frac{1}{\text{cont(i)}}$$

| or else directly on the basis of *m* as [41]

$$\text{inf (i)} =_{df} \log_2 \frac{1}{\text{m(i)}} \ (\equiv \log_2 \text{m(i)})\,,$$

which immediately recalls the standard definition given by many people, though not by Shannon himself, for the amount of information carried by a single signal *i*. We have only to replace »*m*(*i*)« by »*p*(*i*)« for this purpose.

It can easily be shown that *inf*, as defined by either of these definitions, fulfills the above-mentioned requirement.

There is another requirement which causes dissension among people looking for an explication of amount of information. Some would insist that the amount of information of *any* two statements should always be at most equal to the sum of the amounts of information of these statements; others would not want to commit themselves on this point. It can be shown that *cont* fulfills this requirement, i.e., that for any *i* and *j*,

$$\text{cont(i\&j)} \leq \text{cont(i)} + \text{cont(j)},$$

whereas between *inf(i&j)* and the sum of *inf(i) and inf(j)* all three possible relationships of magnitude may subsist.

One standard objection, to be found among those who believe that an explication of the semantical sense of information is of little or no use, is that the concept of information, as ordinarily used, is essentially subjective. The statement, »Johnny is hungry,« carries no information for someone who already knows that Johnny is hungry, a moderate amount of information for someone who is ignorant about Johnny's state in this respect, and a very large amount of information for him who believes that Johnny is not hungry.

In analogy to the situation with respect to probability – and this analogy is, of course, no accident in view of the close relation between probability and amount of information – this objection must be split up into two quite different parts, only one of which points to an essential subjectiveness. This part is only the relatively trivial one in which the same statement can be said to carry different amounts of information even for two people with the same beliefs. Admitting this, we admit no more than we [42] were ready to | do long before, namely, that there is a concept of pragmatical information which is badly in need of explication. But the other part, based upon the fact that different people may have different sets of beliefs, need not necessarily be interpreted as pointing to *subjectiveness* but can better be interpreted as pointing toward *relativity,* toward the fact that the same statement might carry different informations, *objectively* different semantic informations, *relative* to other statements, taken as objective evidence. With regard to our illustrations, we would prefer to formulate the situation by saying that the statement, »Johnny is hungry,« carries different informations when taken relative to the statement, »Johnny is hungry,« as evidence, when taken absolutely, and when taken relative to the statement, »Johnny is not hungry,« as evidence. This points only to the necessity of distinguishing clearly between the information carried by a statement absolutely, taken by itself, and the information carried by it relative to other statements. This we do, of course, and distinguish *Cont (i/j)* – the content of *i* relative to, or given, or on the evidence of, or simply on, *j* – from *Cont(i), cont(i/j)* from *cont(i), inf(i/j)* from *inf(i)*. The definitions look alike, that of *inf (i/j)* is, for instance, as you would probably predict,

$$\mathrm{inf}\,(i/j) =_{df} \mathrm{inf}\,(i\&j) - \mathrm{inf}\,(j).$$

Denoting any analytic statement by »*t*,« we find that

$$\mathrm{inf}\,(i/t) = \mathrm{inf}\,(i),$$

an equation that would have allowed us to define *inf(i)* in terms of *inf (i/j)*, i.e., the absolute information measure in terms of the relative information measure, instead of the other way round.

Here the question arises: What are the relations of the relative content and information measures to the relative, or conditional, probability measures? Defining *m(i/j)*, i.e., the inductive probability of i given j, in the customary fashion as

$$m(i/j) =_{df} \frac{m(i\&j)}{m(j)}$$

we arrive at the remarkable result that

$$\mathrm{inf}(i/j) = \log_2 m(i/j),$$

in complete analogy to

$$\mathrm{inf}(i) = -\log_2 m(i).$$

[43] | There is no corresponding theorem for *cont,* i.e., it is not the case that

$$\mathrm{cont}(i/j) = 1 - m(i/j).$$

This is then the other reason, intimated previously, for the preference given in communication theory, and everywhere else, to the correlate of *inf.* It seems that many authors take it for granted, more or less, that the amount of information of *i* relative to *j* should be the same function of the probability of *i* given *j* as the absolute amount of information of *i* of the absolute probability of *i*. The fulfillment of this requirement leads indeed to a log type of function.

I claimed previously that the fact that both *cont* and *inf* exhibit plausible properties indicates that the presystematic concept of amount of semantic information is ambiguous and apparently an amalgam of at least two different concepts, one of which is explicated by *cont,* the other by *inf.* (In addition to this, the vital distinction between the absolute and the relative concepts is often not made, thereby still increasing the confusion.) To prove this once more: A moment ago, a property of *inf* was pointed out which many regard as essential for an amount of information function. I am going to point now at a property that characterizes *cont* but not *inf,* which is considered by many as a *desideratum* for an amount of information function. When I asked people what they regard as the appropriate relation between the absolute amount of information of a given statement *i* and its amount of information relative to any *j,* most of them were very positive that no increase in the evidence should increase the amount of information, though it might not necessarily decrease it. It can, however, easily be shown that whereas indeed

$$\text{cont}(i/j) \leq \text{cont}(i),$$

the corresponding statement for *inf* does not hold.

I shall show this for our specimen language. For any adequate *m*-function, *m*(»*Ma*«/*Fb*«) will be less than *m*(»*Ma*«), since, according to the *principle of instantial relevance,* the instance »*Fb*« is negatively relevant to »*Ma*.« Therefore $-\log_2$ *m*(»*Ma*«/»*Fb*«) will be greater than $-\log_2$ *m*(»*Ma*«), hence

$$\text{inf}(»\text{Ma}«/»\text{Fb}«) > \text{inf}(»\text{Ma}«), \text{ q.e.d.}$$

| It is perhaps not too far from the point if we regard *cont* as a measure of the *substantial aspect* of a piece of information, and *inf* as a measure of its *surprise value.* When the census taker learns that *Ma,* when he first comes to the hamlet, he learns that the universe he is interested in is not in any of certain 32 states out of the 64 states it could possibly have been in. If this is the second thing he learns, having learned first thing that *Fb,* the substantial increase of his knowledge about that universe is less, since »*Ma*« tells him now only that the universe is not in any of 16 states out of the 32, it could still have been in, these 16 states forming a subclass of the class of 32 states, in the first case. But though his knowledge increases less substantially, he is (or should be) now more surprised than he was in the first case. Knowing nothing, he expects *a* to be *M* as much as *F,* but having observed that *b* is *F* first, he then expects *a* to be *F* rather than *M* – it is important to remember that this is *all* he knows, in our fictitious situations; he knows nothing about attraction of sexes and all the other many things we know and which are relevant to the census – and is therefore rightfully surprised when *a* turns out to be *M* after all. [44]

By illustrating in psychologistic terms, I am afraid I created more puzzles than enlightenment. In order not to fall into the trap laid thereby, it might perhaps be preferable to speak of *inf* as a measure of (objective) *unexpectedness* rather than of surprise.

Though *inf* is a monotonic transform of *cont,* it is not a linear transform. Consequently, not only are the *cont*-values and *inf*-values in general different, but so are the *cont* and *inf* ratios. It will, therefore, in general make a difference whether, in a given practical situation, one is going to use the *cont* or *inf* measure.

This can be illustrated by means of the following story, inspired by the title of a book by Agatha Christie entitled *Bridge Murder Case.* There was a bridge party in A's villa, with B, C, D, and E participating; A was the host and only kibitzed. When the last rubber was finished and the guests were looking for A to take leave of him, they found him murdered in the garden. Everyone of the four players had been the dummy at one time or another and had left the room for refreshments. Each one had, on the available [45] evidence, an equal opportunity for murdering A. A reward was | promised to those who could forward information leading to the identification of the murderer.

A day later, X came and produced evidence sufficient to prove that B could not have been the murderer. The next day, Y showed, to the district attorney's satisfaction, that C was innocent. The following day, Z did the same for D. Whereupon E was duly convicted and electrocuted.

The problem now for the district attorney was how the reward should be distributed; he had to adopt some numerical proportion. He could evaluate the information given by the three informants according to the absolute *cont* value of their statements, or according to their absolute *inf* value, or according to their measures, relative to the information he received, or according to no explicit function whatsoever. I believe that American law does not require him to justify his method of distribution.

I asked six friends at MIT how they would have handled the situation. I received six different answers. One, a newcomer to MIT, with little previous contact with Information Theory, would have distributed the reward equally between X, Y, and Z. Another claimed that the information supplied by Y was worth more than that supplied by X, since X's testimony excluded one suspect out of four, whereas Y's testimony eliminated one out of three, and similarly for Z. He was in favor of distributing the reward according to the proportion ¼ : ⅓ : ½. A third agreed with the second's evaluation of the situation but argued for a distribution of the reward according to a logarithmic scale. A fourth wanted to give all of it to Z, since he alone achieved the identification of the murderer. A fifth, an Iranian, was sure that if the story had happened in his country some years ago, the attorney would have kept the reward for himself, which is probably exaggerated; and I have forgotten what the sixth had to say.

The story could be usefully elaborated in many different directions. I shall not do this here, but I hope I have made one point clear: A numerical measure of information content is of interest not only for science but even for everyday situations.

The next concept we introduce is the *estimate of the amount of information* conveyed [46] by a statement. Consider, for instance, that | we are about to perform an experiment with so many possible outcomes and that these outcomes form what we call an *exhaustive system,* so that one and only one outcome must occur. Under these conditions, it makes sense to ask: What is the amount of information which the outcome of this experiment can be expected to carry?

Let H be the exhaustive system, and $h_1, h_2, \ldots, h_n$ the n possible outcomes. Let $p_1, p_2, \ldots, p_n$ be their respective logical probabilities with $\sum_i p_i = 1$. We define now *the estimate of the amount of information carried by* (the members of) *H* as the weighted average of the information carried by each h_i, with the probabilities serving as the weights, In symbols:

$$\text{est(in,H)} =_{\text{Df}} \sum_i p_i \times \text{in(i)}$$

(where »in« stands for any amount-of-information function). For the specific *inf* function, we have

$$\text{est(inf,H)} = -\sum_i p_i \times \log_2 p_i \,.$$

If there are, for instance, 16 possible outcomes and these outcomes are equiprobable, then each of these outcomes will carry four units of information according to *inf,* and the expectation-value will then, of course, be also four. But if the probabilities are not equal, the estimate will be less than four, with the difference depending upon the specific distribution.

We then go on to define even more complex notions like *amount of specification and estimate of amount of specification*. However, I shall not go into a discussion of them at this time. May I point out, however, that the whole theory now takes on an aspect which should be very familiar to those who are acquainted with current (communicational) Information Theory. The task of thoroughly comparing these two theories is still before us, though many preliminary attempts have been made. Pointing out that they deal with entirely different subject matters in a manner that shows very far reaching formal analogies is a good start but certainly not all that could and must be said. But no more will be said at this time.

Many of you are acquainted with the considerations, not to say speculations, that accompany another formal analogy subsisting | between the basic formulas of Infor- [47] mation Theory and mechanical statistics, and the concepts of entropy and negentropy have popped up in the discussions of this group more than once. Some of our best thinkers have expressed the view that this analogy is much more than just any analogy, and statements identifying thermodynamics and communicational information theory, requiring a revision of the second principle of thermodynamics, and even statements identifying thermodynamics with logic have been made recently. I believe that these declarations were more than an attempt to explode old-fashioned ways of thinking and to force people to go deeper into the foundations of thermodynamics than they did so far. But, perhaps for the first time in my life, I find myself on the side of the scientific conservatives. I find it utterly unacceptable that the concept of physical entropy, hence an empirical concept, should be identified with the concept of amount of semantic information, which is a logical concept, and this in spite of the recent attacks by excellent logicians on the logical-empirical dichotomy (7).

I think that my initial attitude will carry more weight if it is shown what might have brought about this identification, in addition to the formal analogy. One source of this mistake lies in the fact that many physicists are accustomed to say that entropy – physical, thermodynamical entropy – is a concept that depends upon the state of knowledge of the observer, so that if someone is going to compute the entropy of a given system, for example, a container of gas, he will have to do this relative to his state of knowledge. If this were so, then it turns out that the entropy concept in physics is becoming a psychological concept, which is at least as disturbing as saying that it is becoming a logical concept (and probably behind this latter statement). This is, of course, not the first time that physical functions have been described as being dependent on the state of the observer, including his state of knowledge. It has even become a kind of fashion, probably originating with certain expositions of relativity theory, to introduce everywhere such dependencies. However, this talk about a physical function being dependent upon the state of knowledge of the observer can hardly be taken seriously. It is probably no more than what Carnap has termed »qualified psychologism,« i.e., formula | tions from which the psychological terminology can be eliminated with- [48] out loss. It may be of some didactical importance to present certain arguments in terms of the observer, but one can do without. The introduction of the observer is only a *façon de parler.*

Not being a physicist, what I am about to say now in conclusion may well be pure nonsense, but I feel that it is necessary to say this in order to provoke clarification of the prevailing situation by those who are more qualified to do so than I.

As I see it, the entropy of a system is a determinate quantity. However, being fallible human beings, we are unable to determine this quantity, at least in general. If the outcome of some action of ours is a function of the entropy of a system, then we would like to act on our knowledge of the *true* value of this quantity. However, all we can do is to act on our *estimate* of this value. This is utterly trivial and no different from, for example, the case of the length of a table. This quantity is, of course, uniquely determinate, though in order to act, we have to rely on estimates of this length derived, for instance, from certain measurements through averaging operations. Now, of course, estimates are relative to the available evidence, hence, in a sense, to the state of knowledge of the estimator. Someone's estimate of the entropy of a given system depends upon his state of knowledge. I would urge not to formulate this situation by saying that the entropy of the system depends upon his state of knowledge. The treatment of estimate functions of entropies, of lengths, or of any other quantities, belongs to (inductive) logic, but this does not mean, of course, that the treatment of entropies belongs to logic.

MEANING IN LANGUAGE AND HOW IT IS ACQUIRED

YUEN REN CHAO
Department of Oriental Languages
University of California

I COME HERE as a linguist, and by a circular route, from a long distance away. I was born and brought up in an old country, with very early linguistic experiences of various sorts. As a child, I had to move from place to place because my family traveled a great deal. As most of you know, the dialects in China sometimes vary so much that they are really entirely different languages. Before I came in contact with foreign languages, I had to learn to read and write in the ancient style, as was the custom, and also in different dialects, so in that way I became very conscious of language. At the turn of the century children were still required to read all the classics, even though they did not know what they meant. However, later on, suddenly everything began to take on meaning. That gave me the experience of »understanding« without knowing the exact meaning of the language. It took on an autonomous meaning, rather like music. That is one of the points I hope to develop later in this presentation.

Then I came into contact with foreign languages, and as soon as I had the opportunity I went abroad to the New World, planning to study electrical engineering; however, on the way, I met a friend who had been in the New World, and he explained to me the difference between pure science and applied science. This influenced me to study pure science, so by the time I arrived in the New World, I had changed to mathematics and physics. I found physics too specialized a subject to satisfy me, so in my graduate work I went into technical philosophy. However, because the use of words was so important in my work, I decided that I should resume the study of language. Therefore, by a circuitous route, I had arrived back where I started. | [50]

I returned to the Old World and spent 20 or 30 years in studying and recording the languages and dialects there, sometimes even trying to change the language and writing of the country. That is how I got into the study of language, and I have remained there ever since. I have very little to do with physiological or medical studies, but I feel at home with those of you here who work in those fields, because, as Dr. Norbert Wiener said in his autobiography, »Chao married a charming Chinese woman doctor« (1).

What I have to say may be, in part, a repetition of your previous discussions. We have already had a very rigorous and eloquent presentation of the meaning, from the point of view of strict semantics, of language. I shall start by saying that I am speaking of the general problems of meaning in natural languages; then, later, I shall take up the pragmatic aspect of that meaning. Perhaps the various approaches to language, and to meaning of language, are mutually complementary. If we wish to be rigorous and clear, and be certain of each step we take, then we can say very little. On the other hand, if we wish to speak of what we are really interested in, taking all factors into account, then what we say will be not only less certain as to truth, but perhaps even less clear as to content.

I shall try to take an intermediate point of view. Perhaps the everyday work of a linguist is somewhere between the very rigorous methodology of semantics on the one hand, and the full-blooded concrete study of the speaking human on the other. The speaking organism is studied by psychology and psychiatry, but among linguists there is a comparatively recent tendency to lean toward the easier, neater, cleaner, more formal aspect of language, and to leave out that which is more interesting and concrete. I am probably representative of the majority of linguists in that 95 per cent of my work has to do with formal linguistics, the dry description of the stuff of language; only 5 per cent has to do with the more meaty, interesting, and human side. I will say also that, at least as far as I am concerned, the 5 per cent of my professional work is somewhat amateurish, and what I have to say about the meaning and pragmatic aspects of natural [51] languages will have to be judged on its merits. |

I shall divide the discussion into three parts. First, I shall say something about the acquisition of language, especially with regard to meaning. Secondly, I wish to say something about the continuity of form and meaning: the gradual shifting line between the stuff of language itself, and the meaning, in whatever sense you take it, of the form. Thirdly, I should like to present to you an idea which I call the nonplasticity of form, and I shall tell you what that means. I make that contradiction in terms intentionally.

1.

There are already two excellent papers on the acquisition of language in the Transactions of the Seventh Conference: »On the Development of Word Meanings« by Heinz Werner (2) and »The Development of Language in Early Childhood« by John Stroud (3). Therefore, I shall not say very much on what has been dealt with in those papers.

As to the acquisition of meaning of linguistic forms: When adults, who have forgotten how they learned meaning in their own language, learn a foreign language, they usually remember some of that process. If we wish to know what »chat« means in French, we are told it means »cat.« But the other day my granddaughter asked me, »How does an English-speaking child learn what ›cat‹ means?« That is a difficult thing for me to answer in the vocabulary that she has.

That example, perhaps, is not so very difficult; in things like cats and dogs it is easy enough to tell by what is known as »ostensive definition.« They may be pointed out. However, as Dr. Mead has said, if we point at a cat and ask, »What is that?« we may be told that it is a »finger.« In fact, that is how the etymology for the classical form of the Chinese word for »self« came about. The graph for that is a picture of the nose, by ostensive definition.

Now, as meanings become less and less clearly related to easily delineated parts of the child's world, it becomes more and more difficult to locate and point at them, so the child learns very slowly and uncertainly during this process, and makes mistakes constantly. He gets into habits that diverge from the general usage of the people around [52] him, and so has to be constantly corrected. We tell | a child, »You don't write pictures, you draw them; no, you don't sing a story, you tell a story.« Or you say, »Yes, that is a tomato, too; it's a yellow tomato.«

Jesperson (4) told a story of a child who wanted to have some peace in the house. The adults around him could not understand what he wanted until, after a while, it began to occur to the father that a few days earlier there had been company in the house and they had been served beer. The little child had insisted on having some and had been refused; so after a while, the father had said, »All right; let's have some peace in the house!« That was an ostensive definition, which was given without enough varying circumstances to narrow it down so as to agree with the usage of the adult language. So, out of this »big, buzzing, blooming confusion,« and out of »things« and »not things« – in Stroud's paper these latter terms were used – in a child's early life, it takes a lot of trial, error, and focusing before the child can isolate and recognize approximately recurring features and items of his world, and then relate them as to form and meaning in ways that agree with those of the people around him.

Now, from the point of view of complete system building, a rather different point of view from the rigorous language system as described earlier by Dr. Bar-Hillel, one might like to imagine that all linguistic forms could be related in some vast systems of multilingual, bilingual, and monolingual dictionaries, equating one thing to another. That would leave only a relatively small list of basic forms with primary meanings, and these could be merely defined ostensively, perhaps in a central museum of meanings. I imagine such a museum would have a Hall of Colors, a Hall of Gestalten, a Hall of Smells, and then, in a central court, it would have a Hall of Time where it would be shown what it is to be before, after, or simultaneous. Yes, right and left – there's something that will tax the ingenuity of the curator or even the director of the museum. How can one define that if the visitor insists that he cannot understand what that means? Martin Gardner (5) has recently written on this subject.

I started this museum of ostensive definition idea as something incidental; a kind of side show. What I wished to do was to set up a straw man and knock him down. In the actual process of | learning the meaning of language, we do not have such superhuman power as to know all the primitive meanings embodied in these fundamental ostensive definitions, or all the combinations and deductions so that we can know the rest of the language or languages involved. Actually, ostensive definition goes on all the time, whether in agreement with previous experience of an individual, or in disagreement, and what a child does in acquiring meaning, so as to agree in the main with adult usage, is constant correction. This means constant denial, in some respect, of previous ostensive definitions. [53]

So, from this point of view, perhaps we could comment on one of the points brought up earlier as to whether, if we say 17 times 19 is 323, it gives any information or not. From the point of view of the actual learning of the meaning of numbers, it does give a lot of information, because we have a lot of experience with 17 and 19, but perhaps not so much with 323; still, it does not come exclusively from the basic postulates for arithmetic. We certainly have many other ways of contact with the number 323. Likewise, in the example of the statement that the three medians of a triangle meet at one point, if we stick to the initial Euclidean postulates as all that are necessary, it seems to give us no new information. But in practice, if we draw a triangle and three medians with ruler and compass, they often do not quite meet in the same point, and by revising the drawing, we try to make them meet in one point. The result is that we make better medians. Of course in geometry, as it is written, we do not define medians that way. It is the common practice to treat certain parts as belonging to the postulated parts, and other parts as theorems, although the division is quite flexible. But in practice, one is just as meaningful, or just as useful, as the other.

That is typical of the way one learns meaning in language. One learns it without regard to independence of elements, independence of initial ostensive definitions, and without regard to consistency. The actual meaning one finally arrives at is a composite photograph – and composite photographs are always blurred – of the various associations one has met in life, so that ostensive definition is really going on all the time. The acquirement of meaning is a process of change in the individual, and that change, in the case | of normal growth, will stabilize in approximate conformity with the society around one. [54]

The society around one is, however, also subject to change, if at a much slower rate. The acquirement and change of meaning in the child's language parallels to some extent the same processes in history and may be compared with ontogenetic recapitulation. The earliest uses of language by a child or mankind tend to lump together emotional outlet, influencing of action, and factual observation, all three in an undifferentiated way. This undifferentiated use of language seems to be the state of affairs as found among the so-called primitive peoples and in early literatures.

Usually one thinks of a child as uttering sounds which are an expression of some emotional state; then the expressing of wants comes rather earlier than factual comment, or reporting on observations of the world. But sometimes the opposite seems to be the case. In my granddaughter's speech, the word »water« was used in commenting on the presence of water months before she discovered, to her great surprise and satisfaction, that the mentioning of the word could bring about the thing itself. In Helen Keller's account (6) of her memory of the first word she learned as a word, the word »water« was also understood in a somewhat cognitive sense, though very much charged with feeling.

I do not need to say much about the shift of scope in meaning which occurs in the child's language, as it does in the history of the language. I shall only note one factor in the change of meaning which may be of relevance to our discussion in that it has to do with quantitative information theory, namely, the matter of frequency of occurrence. In consonance with the idea in information theory that the occurrence of a frequent item out of a list of possible items gives less information than that of a rarer one, a very common word or phrase or any linguistic form »means« less than an unusual one, whether it is for the purpose of simple information, literary appreciation, or for influencing action. That is why basic English, no matter how skilfully composed. always seems to taste so insipid and that is why poets search for less hackneyed words and expressions in order to obtain fresher effects. For words charged with meaning and significance fade with use and have to be replaced by newer ones until they in turn have [55] to be replaced. |

2.

My next topic is the continuity of form and meaning, or to use Mr. Bateson's terminology, the continuity of communication and metacommunication.

In my very unmathematical treatment of this subject, I should say that two points make a dichotomy and three points a *continuum*. Most descriptive linguists today, counting myself as one, would agree, whether they are in psychological theory behaviorists or not, that language is a type of social behavior. Linguists as a class are less ready to regard meaning also as behavior or something reducible to behavior, but I suppose everyone will agree that meaning is context in some sense; that is an expression used as early as 1909 in E. B. Titchener's (7) work on the psychology of thought processes. And he was a typical introspective psychologist.

The word »context« can have various graded senses. If it is linguistic context, such as context in the use of pronouns, then of course it is already form, but in most cases it is the social context, or at least the context of nonlinguistic items in the speaker's experience.

These things called nonlinguistic are not simply nonlinguistic. I think of them as having various gradations of linguistic status, and have classified them as follows: Zero grade, or I, is form, apparently without meaning. II is some arbitrary association of form and meaning. III is form, and some kind of behavior which is not usually regarded as linguistic form, but which I believe can be treated in the same way as we treat linguistic forms. IV is my own particular interest, i.e., stylistic elements; roughly speaking, it includes gesture, intonation, voice quality, and so on. V is a certain subtle, less tangible situation, which seems to be the meaning of the form, but often not verbalized. Finally, VI is the literal meaning, the core of the meaning of the form.

I shall take up the first instance of form, which apparently has no meaning. We have all probably had the experience of reading through a page while daydreaming about something else, and then, suddenly realizing that we did not understand what we had read, we have gone back for the meaning. Here is an experience of what seems to be pure form in the use of language. The example I men|tioned earlier, of learning by [56] heart all the classics without knowing what they mean, seems to me to have at least the emphasis on the learning of it; that is, one begins to learn the form. In such a case, language is an experience rather like listening to music; you enjoy it, but it does not necessarily have a further reference. There is a famous essayist (8) who was proud that when he read books he did not attempt to understand them too thoroughly because he was enjoying just the form of the language.

Secondly, we have the form, and also some arbitrary association, which seems to be its meaning but really has nothing to do with the use of that language, so far as the socially sanctioned usage is concerned. There is probably a medical or psychological term for the experience we have just before going to sleep, when various things pass through the mind, and somehow one thing seems to mean something else. And yet afterward, in recalling it, we see that they have actually nothing to do with each other. A fine example of that was mentioned by Dr. Klüver (9): his dream about a sack of Idaho potatoes was a perfect illustration of »the synthetical unity of the manifold in all possible intuitions.«

I now come to the third item: the continuity of form and overt behavior. Piaget (10), in reporting his observations of the behavior of children, reports that children use language as behavior in the classroom, for instance, along with other behavior. The two interchange or alternate, and the children do not necessarily use them for the purpose of communication. In one of the experiments, a child, whom be calls the »explainer,« is told a story, or is given an explanation of some mechanism or physical object. The child is then asked to explain it to another child. Even though it is often not successful as an explanation, and the listening child may not really understand, yet very often it works out as a game consisting of gestures and speech, or one alternating with the other, and in that way the second child learns.

Another example would be the comparison between the early days of the sound movie and the later form. When it first became possible to put speech into a movie, there was speech going on all the time, as though a continuous stream throughout the story were considered necessary. But it was soon realized that a movie is just another kind of mixture, and that if the artistic purpose of the film | is to be taken into [57] account, the two media should be integrated into one articulate whole.

In language teaching, the conventional method is to give the forms in the foreign language, and then equate them to words, phrases, and sentences in the language of the learner. Contrasted with this is the so-called direct method, which tries to articulate the speech with what is going on; it has less to do with ostensive definitions of terms than with the actual using of the language in various situations, very much as a child learns language. One of the forms of this method, called the Coué Method, consists of demonstrations performed by the teacher, such as this: »I am taking this piece of chalk; I am writing on the blackboard; I turn around; I sit down,« and so forth. It gets a little monotonous, with the limited situations in the classroom, but the reason why it is effective is because it exemplifies the actual condition under which language and behavior merge into each other when language is actually used.

Under this heading, I might mention the meaning of certain things, half remembered, in the form of some behavior attitude not clearly verbalized. I wish to make a telephone call, and the telephone is around the comer to my right, but something interrupts me. Then, after the interruption is over, I remember that I have to do something on this side, but the verbalization, or whatever other linguistic form it takes, of the making of that call is gone, and what remains is not specific enough to direct me to do exactly what I wished to do. I still remember the direction, and I have some tension in my muscles, but that seems to be all that is left of the meaning of that intended telephone call.

Now I come to the fourth category, that of »stylistic elements.« This is not primarily in the sense of literary style, but the sense of those elements of language other than the distinctive units usually represented in writing in letters, or in other forms of writing. They are intonation, dynamics, over-all loudness, rhythm, voice quality, gesture, and, last but not least, diction, i.e., the frequency distribution of the kinds of words that we choose. The last mentioned is in fact the primary, though not the only, element of literary style. There are two approaches to the treatment of such elements: One is to [58] regard them as just elements of language. One studies them, | lists them, finds out what they mean, and perhaps symbolizes and teaches them.

For they have recognizable recurring patterns, with either conventionalized or physiologically natural meanings.

The other approach is to put such elements on a separate level, because they are more complicated. They are usually outside the normal list of sounds of the language; sometimes we do not have to deal with them, and, in reading a book, we do not find them spelled out. Henry Lee Smith[1] and George Trager (11) have regarded the study of at least some of these aspects as part of metalinguistics. As Dr. Fremont-Smith has said, here is an example where there is a rather serious conflict in the use of terms, because metalinguistics in this sense does not mean the study of any metalanguage. But we shall not discuss the merits and disadvantages of either usage.

In the stylistic sense, metalinguistics is sometimes concerned with the language being used, and comments upon it; but most of the time it has nothing to do with the language. It has really more to do with the behavior of the speaker or the speaking situation. One says:

»Good (mid rising) night (low-to-mid rising)!«

on parting; but one says to the hostess:

»Good (low) night (mid-high-low-mid double circumflex)!« meaning »Please listen to me; excuse me for interrupting, but I am leaving.«

It is a historical accident that we have one part of language – to be sure, a very important part – recorded in writing, whereas most of the other parts are not recorded; however, they can be. A while ago, I was talking with Dr. Savage about the recording of expression in the dialogue of a play. That has actually been done. H. H. Davies' play, *The Mollusc* (12), has been transcribed by Dorothée Palmer in the International Phonetic Alphabet – but that is not the point, it could have been in ordinary orthography – with all the expression marks, mostly intonation marks, according to the system of Harold E. Palmer. That gives one specific interpretation in which the [59] play could be given. Of course actors may not like it; they | might say, »I should rather read it some other way.« On the other hand, consider a similar kind of recording, the matter of musical notation. Formerly, a composer wrote out the figured bass and let the organist do what he liked; there were few or no expression marks. But the modern practice is to write out even the exact time values of grace notes, and make very detailed expression marks, dynamic and otherwise, for all that is part of the music. The question is: Is the composer supposed to do all this? How much is the composer expected to compose, and how much is the playwright expected to write? In the case of the transactions of this conference, how much are we expected to include? It is a

1 Smith, H. L., Jr.: *An Outline of Metalinguistics.* (Unpublished data)

relevant question, because in some instances, one cannot understand the meaning for lack of the necessary significant or distinctive linguistic elements.

I am inclined to agree that the two usages of the term »metalanguage,« on the one hand, and »metalinguistics« on the other, overlap in certain situations, but I do not believe that they are completely of the same scope. I can also cite examples where there would be shifting from one to another. If we say, »*I* do not believe it,« we have the ordinary sense, plus the metalinguistic element in the Smith and Trager sense, of the voice qualifier, with a contrasting stress on »I.« But if we are translating into French, we cannot say, »*Je* ne le crois pas,« because we would be changing it to something that is not French; we would have to say, »Moi, je ne le crois pas,« which consists of changing only the ordinary items of the phonemes and morphemes of the sentence, without using any special element that would be called metalinguistic. There are other cases of that kind. Take the intonation which H. E. Palmer calls »the swan,« because the intonation curve looks like a swan, giving a concessive implication. »It's good (mid-high-low-mid)« means that it is good, but there are some other objections to it. That, of course, is metalinguistic in the sense that it is put on top of the predicate »good.« But if I try to translate that into Chinese, it would take the form, »Good is good,« meaning, »As for being good, it is good (but as for something else, it may not be)«; so that if I were building that sort of equivalence into a translating machine, I should have to build in it: »Input, swan intonation: output, predicates, verb ›to be,‹ repeat predicate,« etc., to respond to that situation. Thus, on the one hand, we have the ordinary as well as | metalinguistic elements, but in the other language it is just rephrasing of the material. [60]

I should like to go on to the fifth heading of less tangible situations, as shown in the example Mr. Bateson gave of the hand stretching out from the frame,[1] and as shown still more vividly in Dr. McCulloch's example[2] of the ground crew calling in the pilot. The fact that the pilot is talking continually is behavior which may be interpreted, in addition to what he is saying (which may be nonsense or unimportant things). The pilot knows that he is talking, but he does not think of this as a coded message, but rather as routine aviation procedure. The fact that he does that would place it under metalinguistics in the Smith and Trager sense. Because it is about the language, it would also be a metalanguage, but perhaps not in Mr. Bateson's sense, because what the pilot says may be nonsense or unimportant. He is not talking *about* it; he knows

1 Let us consider a picture showing a man holding a glass of whiskey. There is a frame around him, and the frame is there as part of the advertiser's message, to attract and focus the attention on that which is to be the figure as opposed to the ground in the message. The frame, then, appears to be, »Oyeh, oyeh!« a listening exclamation, a command; »attend to my message about this whiskey.« Then it becomes apparent that the frame is being used in other ways; this device is very common in liquor advertisements. For example, the hand that is holding the whiskey will be projecting from the picture outside the frame, in the style of trumboleic pictures, in which the frame, which suggested the unreality and limitation of the picture, is now used as a counterstatement, so that an unreal reality can be attributed to the picture by transcending the frame, and so on. The messages of these various components – the picture, the hand, the frame – are all we have of messages about each other.

2 When a plane is coming into an airport, the pilot is »talked in,« with a steady stream of conversation. It is the business of that conversation to say, »I am in contact with you, I am ready to pick up whatever you want,« and so on and so forth, so that the man knows his line of communication is open. The statement that the line of communication is currently open is not the message that is conveyed by the speech as speech. That speech as speech is saying, »You're on the beam, pull a little more to the right, stay with it,« that sort of thing. That is the official concept of it. But the steadiness of the stream of this is the assertion. »This line of communication is open.« In a sense, this is a remark concerning the channel or concerning the language.

that he is on the beam for the next second or two, so it may not be relevant; it may not be metalinguistic in the other sense.

I shall give another example, which I have observed in more than one child. There is a scene; a mother scolds or punishes a child, and the child cries. Then, after a while, the mother comments on something else, quite unrelated. From the behavioral situation, the child learns that that incident is closed. After some experience with that sort of thing, the child will begin to test the situation by asking questions, or asking for [61] things, meanwhile watching how the mother | reacts. Then the mother, knowing from experience that this is a test, will or will not change the subject, as she sees fit. This intentional manipulation, which hitherto has been part of the total behavioral situation, will become part of the language, so it would be a metalinguistic element which is a little more subtle than intonation or stress patterns.

Another example which is similar to the ground-to-plane talk is this: When I back my car, I say to my wife or daughter, »Is there anyone behind?« or, »Is there a car behind?« and they will say, »No,« and stop. Then I will say, »Keep saying ›No car, no people; no car, no people,‹ until I have backed out.« That is quite parallel to the talk between the ground and the pilot. It is not only the message itself, but the fact that it is continually being sent that is an important message. In developing this technic for backing a car, I did not know about the aviation technic; it began merely as behavior between people. Neither I, nor those who took part in this behavior, treated it as a language symbol, but after a while it developed as a symbol for those who co-operated in using it.

I think my item VI, the literal meaning of linguistic forms, is something of an anticlimax. Since we have already discussed it under the general idea of ostensive definition, I shall not spend any more time on it. I shall go on to the third aspect of the pragmatic language, which is here more formal than pragmatic, because I am more concerned with forms; namely, the nonplasticity of forms. If you do not like that apparent contradiction, you may speak of the autonomy of forms, or perhaps even the primacy of forms, because by »form« I mean the stuff of language itself.

3.

Forms have a way of being themselves and going their own ways. They are made by speakers according to the nature of their speech and hearing organs. They have been adapted by communities through the activity of individuals who have exercised their vocal organs in ways that are recognized and accepted by others of the same community. They are forms which have been used more or less effectively in their original application or have been adapted to other activities; there is no reason why they should [62] not be well adapted to the purpose of the interpersonal and intrapersonal com | munication. However, it is highly improbable that the habits of formation of sound sequences should have any simple or systematic relationship to the rest of the life of the individual in the community.

As I said, the processes of natural formation, conventional design, learning and transmission, and perhaps also of forgetting of linguistic forms have their own special ways. They have their physical and physiological conditions. They have a strong resistance to any attempt to mold or change them at will. They are not wholly plastic in the ordinary sense. There are strong cultural traditions for keeping a language as it is and allowing only very slow changes, especially in the larger structure of words and sentences, and there seem to be also strong physiological, noncultural conditions of the human organism, which put rather narrow limits on the possibilities of linguistic elements and their manner of combining.

Most languages in the world can conveniently be analyzed into phonemes and morphemes. The morpheme is the smallest unit of one or more phonemes that has a meaning, and the majority of morphemes have one or two syllables.

Many linguists have tried to put both phonemics and morphology under one uniform method of analysis by treating both on a distributional basis, that is, according to the manner in which elements of various sizes (phonemes and morphemes) typically occur and recur. While a morpheme defined in distributional terms is a very different sort of thing, intensionally, from a morpheme defined as the minimum unit with meaning, there is no reason why it should be incompatible with it, extensionally, in actual application to any given language. In practice the old definition is still the only one used by practicing linguists, including those who develop the new approach. Harris (13) has made an analysis of the morpheme on a purely distributional basis. For a different but more rigorous than the traditional treatment of the morpheme see Hockett's manual of phonology (14).

In Hawaiian, there is a fish called a *homohomonukunukuapua,* and you cannot analyze that into smaller meaningful units. But at the same time, in that same language, there is another fish which | is called *ø*.[1] In inflected languages, however, accidences or elements of inflection are morphemes, even though they are often less than a syllable. Expressive or stylistic elements will usually spread over more than one syllable, and over a phrase or a whole utterance if they are included under morphemes, but on the whole they are of medium length. In some languages we have morphemes which do not consist of continuous rows of elements, but groups of, for example, three consonants, with spaces filled in by vowels which do not belong to that morpheme and which may indicate something else. I would not call such forms rare, but it is certain that they are to be found in only a minority of the well-known languages. [63]

By and large, the size of the morpheme is less than one, or two syllables, or, roughly, four or five phonemes. As to the number of phonemes in a language, most languages have a moderately small number. The list rarely falls below ten, and rarely even approaches 80, so that, speaking roughly again, the number of phonemes in a language is of the order of two to the fifth power, or five bits of information. This would give a total of over 20 bits to each morpheme. If any phoneme could combine with any other in any order, it would yield a possible vocabulary of the order of millions. Actually, however, the succession of phonemes is so limited to characteristic patterns that the total number is always only a very small fraction of the possible number. In other words there is great redundancy, in the informational sense, in the actual use of phonemes in a language. This has been clearly demonstrated in detail by Cherry, Halle, and Jakobson (15), who have analyzed Russian, with a typical repertory of 42 phonemes, not by just counting the phonemes, but by going into the subpbonemic distinctive features.

If we consider the vocabulary of basic English, it is supposed to have a list of 850 words, but in this case it is necessary to put »words« in quotes, because the system of counting is quite special; for example, we count »what« as the neuter gender of »who.« That would be an example of one word. It is by that method of scoring that we arrive at the number of 850. It is really a good deal more than one thousand by one of the more usual ways of counting words. | [64]

Again, if we count the number of syllables in Mandarin Chinese, it comes to 1,279 syllables, counting the difference in tone, because tone is a part of the phonemic con-

1 This short-named fish was mentioned at a public lecture that I attended but I have not as yet been able to find further reference to it in any publication.

stituents. This is apart from the question of intonation, because intonation in Mandarin is something else than tone.

This in no way suggests that linguistic forms are quite flexible and plastic, because it shows that languages tend on the whole to take meaningful units of a certain size and shape, no matter how we make use of them. Wherever there are marked deviations from this average condition, leaving such exceptional cases as *homohomonukunukuapua* and *ø*, there is usually a tendency to bring it back to this approximate average, shall we say, by negative feedback of some sort. A morpheme must not be too large, and if there are too many phonemes in an unanalyzed whole, the speaker tends to analyze it anyway, and that is one of the causes of folk etymology. There is too long a string of sounds, and we wish to put meaning into parts of them. If they are similar to some morphemes already in the vocabulary, we break it up accordingly.

On the other hand, if the natural historical wear and tear of distinctive sounds have lowered the informational capacity of any linguistic item, resulting in coalescence of words originally distinct, then there is another compensatory change. When the romance word for »bee« was worn down to a single vowel »*e,*« it dropped out of the French language and a modified word was used in its place. Of course such things are not consciously done on the part of the speakers in any way, but rather indirectly through the choice or preference for certain forms to others. We have a situation like that in the change from ancient to modern Chinese. In the ancient Chinese of 601 A.D., there were 3877 distinguishable syllables, each of which was, on the whole, the size of single morphemes. When the distinctions had been worn down and reduced to modern Mandarin, of 1279 syllables, what happened at the same time of the change was that we had a great number of dissyllables, which, although analyzable historically and meaningfully on the part of the educated, are spoken by the illiterate as unanalyzed wholes, so that we come back to a number much larger than 1,000.

I have been speaking of the size of morphemes as one of the most interesting of the [65] cases of the nonplasticity of forms, but that | is not the only aspect of language that is nonplastic. Conditions of transmission and of reception of signals and codification will affect the morphemes too, and a change of method of written records will also change the style. The style of pen strokes and drawing, in fact the whole system of writing, can be affected according to whether one uses a stylus, brush, pen, pencil, or stenotype. As I was putting down these notes in Chinese, I quite unconsciously changed the order of the items, which is rather symptomatic of this phenomenon. I had at first the order: »lead writing-instrument, steel writing-instrument« and so on. That is the usual order. But when I came to typing the list in English, it became »pen, pencil,« because »pen« is the simpler unit and »pencil« is »pen« with something added to it, and just because of the mechanics of the thing itself, it changed the very order of the terms without my knowing what I had done.

A more extreme example of the autonomy of forms going their own ways is the divergence of literary Chinese, which leans heavily on visual differentiation of homophonous characters, from the spoken language, which must satisfy the requirement of auditory intelligibility. I have constructed a story in literary Chinese, consisting of only the syllable *shih* (in four tones, to be sure) pronounced 106 times and therefore not auditorily intelligible, and yet, in characters:

施氏食獅史

石室詩士施氏嗜獅誓食十獅氏時時適市
視獅十時氏適市適十碩獅適市是時氏視是
十獅恃十石矢勢使是十獅逝世氏拾是十獅
屍適石室石室濕氏使侍拭石室石室拭氏
始試食是十獅屍食時始識是十碩獅屍實十
碩石獅屍是時氏始識是實事實試釋是事

it is as clear and idiomatic as any other prose telling the story of Mr. Shih eating lions:

> | »Stone house poet Mr. Shih was fond of lions and resolved to eat ten lions. The gentleman from time to time went to the market to look for lions. When, at ten o'clock, he went to the market, it happened that ten big[1] lions went to the market. Thereupon, the gentleman looked at the ten lions and, relying on the *momenta* of ten stone arrows, caused the ten lions to depart from this world. The gentleman picked up the lions' bodies and went to the stone house. The stone house was wet and he made the servant try and wipe the stone house. The stone house having been wiped, the gentleman began to try to eat the ten lions' bodies. When he ate them, he began to realize that those ten big lions' bodies were really ten big stone lions' bodies. Now he began to understand that this was really the fact of the case. Try and explain this matter.« [66]

The nonplasticity of forms makes it easy for communication between persons with the same or similar forms of schemas, and it makes it difficult between persons who have different forms, such as classical and modern Chinese. In the Transactions of the Eighth Conference on Cybernetics, Donald MacKay (16), in his paper on symbols, made a distinction between communication by what he called prefabricated representations, on the one hand, and the theory of scientific information, the designing and learning of new representations, on the other. When we come to the meaning of natural languages, it is largely in the form of prefabricated representation, at least so far as communication between adults is concerned. But even in the case of communication between children, we find that at a certain age in acquiring prefabrications, they tend to follow a recalcitrant tendency of keeping to what they already have.

I should like to quote from Piaget (17) again on that point. He says, in *The Language and Thought of the Child*, »It is not paradox to say that at this level« – that is, between the ages of 7 and 8 – »understanding between children occurs only in so far as there is contact between two identical mental schemas already existing in each child. In other words, when the explainer and his listener have had at the time of the experiment common preoccupations and ideas, then each word of the explainer is understood, because | it fits into a schema already existing and well defined within the listener's mind. In all other cases, the explainer talks to the empty air. He has not, like the adult, the art of seeking and finding in the other's mind some basis on which to build anew.« That is MacKay's second problem. But it is interesting to see that Piaget adds a foot- [67]

1 The word for »big« has an alternate pronunciation *shuo*^{4}, if it is of any help.

note to this discussion, which seems somewhat germane to our discussion here. He says that Nicolas Roubakine (18) came to an analogous conclusion in his studies on adult understanding in reading. He showed that when reading each other's writings, adults of different mental types do not understand each other. But we adults of different mental types do make honest efforts, as we have been doing, in the building of new representations, and in the explication of prefabricated representations. So, in order to conclude my remarks on a positive note, I will make a correction in what I said at the beginning about the complementarity of rigor and content: If you wish to be precise and accurate, you cannot say much, and if you wish to say a great deal, you cannot be clear and precise; as though there were a parameter there that had to be constant, and which we could not do anything about. I feel that, with our efforts to approach from both sides and the efforts to break or interpret our code, we have slightly increased the value of that parameter. For a parameter is a constant which is variable. Even if the product of quantity and accuracy must remain less than or equal to an tipper limit, I visualize it as a parameter which had one value yesterday, and has another value nearer to the upper limit today. Dare one hope that that tipper limit is perfect mutual understanding among all the disciplines?

APPENDIX I
SUMMARY OF THE POINTS OF AGREEMENT REACHED IN THE PREVIOUS NINE CONFERENCES ON CYBERNETICS[1]

WARREN S. McCULLOCH
Research Laboratory of Electronics Massachusetts Institute of Technology Cambridge, Mass.

EINSTEIN ONCE DEFINED truth as an agreement obtained by taking into account observations, their relations, and the relations of the observers. In his case, the observations were the coincidences of signals at points in frames of reference; their relations were matters of space and time in those frames; his observers were reduced to what Helmholtz called a *locus observandi,* devoid of prejudices and imagination; and the only relations he had to consider among them were their relative positions, motions, and accelerations. The truth he had in mind is a picture of the world upon which all observers can agree, because it is expressed in a manner invariant under the transformations required to represent the relations of the observers. It is a paradigm for what »scientific agreement« may mean.

Unfortunately for us, our data could not be so simply defined. It has been gathered by extremely dissimilar methods, by observers biased by disparate endowment and training, and related to one another only through a babel of laboratory slangs and technical jargons. Our most notable agreement is that we have learned to know one another a bit better, and to fight fair in our shirt sleeves. That sounds democratic, or better, anarchistic, as you have twice reminded me. Aside from the tautologies of theory, and the authority of unique access by personal observation of a fact in question, our consensus has never been unanimous. Even had it been so, I see no reason why God should have agreed with us. For we have | been very ambitious in seeking those notions which prevade all purposive behavior and all understanding of our world: I mean the mechanistic basis of teleology and the flow of information through machines and men. In our own eyes we stand convicted of gross ignorance and worse, of theoretical incompetence.

Our meetings began chiefly because Norbert Wiener and his friends in mathematics, communication engineering, and physiology, had shown the applicability of the notions of inverse feedback to all problems of regulation, homeostasis, and goal-directed activity from steam engines to human societies. Our early sessions were largely devoted to getting these notions clear in our heads, and to discovering how to employ them in our dissimilar fields. Between sessions many of us made observations and experiments inspired by them, but we generally found it difficult to collect sufficient appropriate data in the 6 months between meetings. At the end of the first five sessions, of which there are no published transactions, we elected to meet but once a year, keeping our group together as nearly as possible, replacing a few who were lost to us, and inviting a few speakers to help us where help was needed most.

By the time we made this change, we had already discovered that what was crucial in all problems of negative feedback in any servo system was not the energy returned but the information about the outcome of the action to date. Our theme shifted slowly and inevitably to a field where Norbert Wiener and his friends still were the

1 This material was distributed to the participants in advance of the Tenth Conference on Cybernetics.

presiding genii. It became clear that every signal had two aspects: one physical, the other mental, formal, or logical. This turned our attention to computing machinery, to the storage of information as negative entropy. Here belong questions of coding, of languages and their structures, of how they are learned and how they are understood, including the theme of this, our last meeting, in which we expect to range from the most formal aspects of semantics, to its most contental contact with the world about us. For all our sakes I wish Wiener were still with us, but I understand that he is at present happily immersed in the clear and serene domain of relativity.

To refresh our memories and inform our guests, let me recapitulate, in logical rather than chronological order, the topics we have considered, and on which I believe the [71] majority of us have been | of one mind, to the limit of our ability to understand the evidence or the theory. You may find the consensus more frequently in my statement than in our published transactions. I am compelled to watch your faces, and to guess, before I let you have the floor, whether you will speak to the point or not, and from which side of the fence. With malice aforethought I have given the malcontent the floor, because he disagreed, or doubted, however unreasonably. Before I knew you so well this happened by accident, but as time went on, and we learned one another's languages, I learned that it was the best way to keep our wits on their toes. Our guests have been remarkably good sports, but the transcriptions of our discussions inevitably sometimes result in misunderstandings and altercations instead of agreement. Of those who understood and agreed the transaction reveals nary a trace.

Feedback was defined as an alteration of input by output; gain was defined as ratio of output to input; feedback was said to be negative or inverse if the return decreased the output, say by subtracting from the input. The same term, inverse or negative feedback, was used for a similar effect but dissimilar mechanism, wherein the return decreased the gain. The transmission of signals requires time, and gain depends on frequency; consequently, circuits inverse for some frequencies may be regenerative for others. All become regenerative when gain exceeds one. Regeneration tends to extreme deviation or to schizogenic oscillation, unless gain decreases as the amplitude of the signal increases. Inverse feedback determines some state to be sought by the system, for it returns the system to that state by an amount which increases with the deviation from that state. Servomechanisms are devices in which the state to be sought by the system is determined by signals sent to that system from some other source. These notions were applied to machines, including the steam engine and its governor, to the steering engines of ships, to well-regulated power packs, telephonic repeaters, self-tuning radios, automatic gun-pointing machinery, etc., and thereafter to living systems. Homeostasis was first considered in terms of reflexive mechanisms, in which change initiated in some part of the body caused disturbances, including nervous impulses, which were reflected eventually to that part of the body where they arose, [72] and there stopped or reversed the processes that | had given rise to them. Similar regulatory circuits entirely within the central nervous system were found to resemble the automatic volume control of commercial radios. Appetitive behavior was described as inverse feedback over a loop, part of which lay within the organism, part in the environment. When a target or a goal could be indicated, a description of appetitive behavior was found to be couched in the same terms as that for self-steering torpedos and self-training guns, whether these devices emitted signals reflected by their targets, or merely depended upon signals emitted by the target to readjust subsequent behavior to the outcome of previous behavior so as to minimize its error. Wiener drew a most illuminating comparison between the cerebellum and the control devices of gun turrets, modern winches, and cranes. The function of the cerebellum and of the controls of those machines is, in each case, to precompute the orders necessary for servomechanisms, and to bring to rest, at a preassigned position, a mass that has been put in motion which otherwise, for inertial reasons, would fall short of, or overshoot, the mark. These notions have served to guide subsequent neurophysiological research in the functional organization of the nervous system for the control of position and motion, some carried out in my laboratory in Chicago, and others by Wiener, Pitts, and Rosenblueth in the Institute of Cardiology in Mexico City, as well as by our friends in other laboratories. The general organization was found to consist of multiple closed loops of control, but the circuit action was extremely nonlinear, and consequently not amenable to any general simple mathematical analysis in terms of the Fourier Theory. Generally, multiple loops, severally stable by inverse feedback, may be unstable in conjunction, but the system can be stabilized by adding a portion of each of the returns and subtracting the sum from one or more of the servos. Such a system was found in the central nervous system by Setchenow in 1865, and rediscovered by Magoun. A group of us is studying the detail of its multiple afferents and its mode of affecting all reflexive activity; we shall use destructive lesion, and shall stimulate and map sources and sinks in various parts of the nervous system by methods presented at the last meeting. With failure of inhibitory signals or increased gain, the stretch reflex becomes regenerative, producing a rise in tone and a series of contradictions known as *clonus*. This has been elegantly analyzed, quantitatively, by Rosen|blueth, Pitts, and [73] Wiener, as described at our conference. Moreover they were able to demonstrate that the pool of relays of the so-called monosynaptic are showed two numerous groups of relays, and a third less numerous, as judged by the random distribution of thresholds around three maxima. It will be years before we have fully exploited these notions.

Closed loops within the central nervous system – first suggested by Kubie as a substitute for undiscoverable motor activity proposed by the behaviorists to explain thinking in terms of reflexes, and by Ranson to account for homeostatic processes within the central nervous system, and independently discovered and demonstrated in the case of nystagmus by Dr. Rafael Lorente de Nó – were mentioned as possibly accounting for transitory memories by McCulloch and Pitts, who indicated that they were logically sufficient, but physiologically improbable, as an explanation for all forms of memory. Livingston has suggested that such mechanisms might account for causalgic symptoms after blocking or removal of perverted peripheral circuits which had been rendered regenerative by some trauma resulting in streams of impulses over small afferent neurons appreciated as burning pain. Kubie had proposed that the core of every neurosis was a reiterative process in some closed loop.

I have summarized and presented to the Royal Society of Medicine evidence along all of these lines, with much more obtained from many varieties of intervention in causalgia. It is clear that the notions of feedback are the appropriate ones for the understanding of the normal function and diseases of the structures in question. Since that time, Dr. Galarvardin of Lyons, studying patients with auditory hallucinosis accompanied by muscular activity of month, tongue, and larynx, has had removed, bilaterally, the post-central somesthetic area for the face. The consequent disappearance of the hallucinosis had lasted 18 months when I last heard from him. This brings one symptom of a clearly organic psychosis into line with the findings on those obsessive compulsives who have been at least temporarily helped by frontal lobotomy, in that the central pathways of some reverberative process within the brain have been partially interrupted. | [74]

Again in terms of these notions, we have been able to make sense out of some aspects of what the psychologists have called goal-directed activity, and our attention has been duly called to the asymmetry of advance and escape, for in the former, the object sought is kept near the center of the receptive field of the sense organs, and behavior duly modified to approach it, whereas, in escape, learning along these lines cannot occur, and the behavior may easily become stereotyped. The most complex situations we have heard discussed are the stabilities engendered by inverse feedback in social structures of isolated communities reported principally by social anthropologists. Their devices have been extremely elaborate, depending, in some cases, on many interwoven loops. They seem to have utilized elaborate forms of distinctions and rules with respect to kinship, forms of address, hazing, bullying, praise, blame, and even rituals with respect to eating. Examples from ecology and from the behavior of anthills have extended these notions of inverse feedback.

Our members interested in economics and the polling of public opinion made use of these notions to explain fluctuations of the market, the banter leading to fight in roosters and boys, and the armament races initiating wars. In such circular systems, it becomes difficult to detect the causal relations. Wiener handled this by pointing out that it was possible to detect causality in the statistical sense by auto- and intercorrelations with lag, in those situations in which correlation was not perfect between the time series of the related component events, and explained how, with such devices, optimum predictions could be obtained. He doubted the applicability of this method to social problems, because of the shortness of our runs of the time series of information concerning human behavior. In these terms we discussed how a fielder catches a ball and a cat a mouse.

The question of conflict between motives was then raised by the psychiatrists, who, like psychologists, would like to have some common measure of value among human desires, comparable to what economists believe they have in the doctrine of marginal utilities and price in an open market. Kubie raised the question of the urgency of dissimilar ends, beginning with the need for moderate temperature, air, drink, food, [75] sleep, and sex, the most urgent | need resulting in the simplest response, and the least urgent allowing elaborate play. I indicated that an organism endowed with six neurons, constituting three chains of inverse feedback, and interrelated either by the requirement of summation or inhibitory links, was sufficiently complicated to exhibit the value anamoly, and if organization were left to chance would do so half the time. That is, given A and B it would prefer A; given B and C it would prefer B; but given C and A it would prefer C. A similar question was raised concerning dominance in the pecking order of chickens, but there was no adequate data as to the number of circles in coops of given numbers to settle the question. By this time we had become so weary of far-flung uses of the notion of feedback that we agreed to try to drop the subject for the rest of the conference.

Two interesting digressions appeared at this point: The first concerned cardiac flutter, which appears as a propagated disturbance running around the periphery of an area it cannot cross and it therefore cannot stop itself, whereas fibrillation appears as a disturbance which wanders over changing paths determined from moment to moment by shifts of threshold produced by previous activities at those points. Its mathematical analysis was indicated but not presented to the group. Second, Pitts presented a theory of disturbances in random nets, such perhaps as the cerebral cortex, in which it was possible to find a value around which to perturb the activity; namely, that probability of a signal in a neuron is equal to the probability of a signal in the neurons that are afferent to it.

Moreover, we had all come to realize that for problems of feedback, energy was the wrong thing to consider. The crucial variable was clearly information.

We began by considering computers as »analogue,« if the magnitude of some continuous variable like voltage, pressure, or length were made proportional to a number entering into a computation; and as »digital« if they were a set of stable values (at least two) separated by regions of instability, and the number was represented by the configuration of the stable state of one or more components. Analogue devices showed tendencies for errors to appear in the least significant place, but were limited by precision of manufacture and could not be combined to secure additional places. Digital | [76] devices might show errors in any place (a limitation inherent in all positional nomenclatures), never required extreme accuracy, and could always be combined to secure another place, at the same price per place as previously. When components are relays, the digital devices sharpen the signal at every repetition. We considered Turing's universal machine as a »model« for brains, employing Pitts' and McCulloch's calculus for activity in nervous nets. It uses the calculus of propositions of the *Principia Mathematica*, subscripted for the time of occurrence of an impulse of a given neuron. We demonstrated the equivalence of all general Turing machines, and how they could be designed to answer any nonparadoxical question which could be put to them in an unambiguous manner. We considered the far-flung conclusions that followed here from Goedel's arithmetizing logic. It became clear that having ideas required circuits capable of computing invariants under the necessary groups of transformation, that is, reverberant activity preserving the form of its afferent, initiated at one time, or inverse feedback leading some figure of input by some path to a canonical presentation out of its many legitimate ones. Gestalt notions led only to multiplications of particulars with distortions attributed to »cortical fields« in which currents are conserved, though in nature there are sources and sinks, and although the areas of cortex they are said to pervade are anatomically discontinuous. The discrete action of nervous components was considered the only way in which they could normally function to handle the amount of information transmitted through them. Gross disturbances of function (epilepsy, etc.) were seen to be accompanied by gross fluctuations affecting most of the neurons in a given area in much the same way, thus producing a loss of information. Emotions were considered as expressions of some overrunning of parts of the computer, producing somewhat fixed responses to diffuse and variable inputs, as if in a Turing machine the computed value of an operand ceased to affect subsequent operations. Wiener proposed that by glandular means emotions might broadcast a »to whom it may concern« message, causing items to be locked in, or remembered. It was suggested that the best way to find out what an unknown machine did was to feed it a random input; clearly it had to be random in tems of the aspects of the input that the machine could discriminate. This was likened to the Rorschach Test, and its auditory equivalent, | and it was noted that the gibberish produced by free association was apt [77] to cause the psychiatrist to project his own difficulties on his patient.

Three kinds of the storage called memory were discussed at length: The first, active reverberation, such as in the acoustic tank, was recognized as responsible for nystagmus and the only storage left in presbyophrenia. J. Z. Young made use of the same notion to describe the residual memory after the destruction of the main memory organ of the octopus. The second type of storage has been found only in the octopus, where it occupies a separate structure with well defined and separate access and egress. The organ itself is composed of a host of small cells; the nature of its synapses is not yet well known. This is the storage that has excited theoretical physicists because of the immense number of bits retained by it. Von Foerster computed its size from access-time times access-channels against mean half-life of the trace, and Stroud from the number of snapshots one-tenth second each at, for example, a thousand bits per frame. Figures are in rough agreement that it lies between 10^{13} and 10^{15}. Instead of declining asymptotically to zero, a few per cent of the items are retained forever. Von Foerster has proposed mechanisms to account for this, requiring *circa* 0.02 watt; the brain is a 24-watt organ. Access to this store is probably not by simple addresses sought seriatim.

Recall seems to rest on a process locating items by their contents. This was discussed by Von Neumann concerning similarities, and by Klüver concerning stimulus equivalences, but both had apparently noted retention of eidetic fragments, a topic which should be gone into much more thoroughly hereafter. Peculiarities of this kind of storage in man seem to be that the contents are a series of snapshots, each devoid of motion; they are accessible in the order of filing, not in the reverse order; there is a delay of about a minute between the making of the trace, and the time when it is first accessible; and finally, a snapshot too similar to the one before it may upset the process. These traces cannot be simply localized; each bit is an alteration of synapses effective somewhere in a net, and the alteration is not confined to some one junction. The third type of storage seems to behave more like the growth-with-use characteristic of muscle, and shows fatigue on too frequent testing. Shurrager ap|parently has evidence that it can occur at a monosynaptic reflex level, and changes with use have been seen where the vagus contacts ganglionic cells in the frog's auricle, but there is no evidence that such a change persists anywhere in the central nervous system, and there is no anatomical evidence that it ever happens there. Some change of organization with use does occur perhaps as suggested by Ashby. The organization of the visual cortex with use is a case in point. When congenital cataracts were removed, the difficulty with vision was in part found to be attributable to antithetical organization of these mechanisms by impulses from elsewhere in the central nervous system.

[78]

»Traffic jams« of brains become increasingly probable with increase in volume, for the number of long distance connections cannot be expanded to keep pace with the number of relays to be connected, except by increasing cable-space disproportionately. It was suggested that potentiation, described by Lloyd, may serve to lock in lines temporarily on a basis of their previous use; this would resemble a scheme proposed in Holland for more efficient use of limited telephonic facilities. Repetitive firing of cerebrospinal neurons leaving the cell body and dendrites largely depolarized when axons were hyperpolarized (P2 after-potential) would thus account for that component of facilitation marked by surface negativity and depth positivity of the cerebral cortex, as a matter of lowered threshold, with increased voltage of volley delivered to cord. The same mechanism would account for the stiffening up of the Parkinsonian patient.

Considerable time was spent discussing the way in which the actual flow of information determined the structure of groups, and discussing the way in which command moved from moment to moment to that place in the net where most information necessary for action was concentrated. In parallel computing machines, including brains, when one part is busy or damaged, another will serve for the same computation. This requires that the whole machine be tended by some part of the machine which can switch the problems to them. Such a machine might give correct answers when most of it was out of commission. What appears on one side as the problem of redundancy of neurons and channels composed of them, appears on the other as the problem of securing infallible perform|ance from fallible components. Von Neumann's last work on this score, delivered at a conference on the West Coast, is titled »*Probabilistic Logic.*«

[79]

With respect to language, as second only to vision as a source of information to brains, and all important in human communication, not to mention psychoanalysis, it was generally admitted that we almost never say anything unless we wish someone to do something about it. Apart from this general hortatory aspect, language contains a few signs such as »hum,« »um hum,« »unh unh,« and »huh,« that are specifically so, but otherwise contentless. The question arises whether the logical particle comes from the signs used by dogs and small children. It was generally agreed, as stated by Dr. Mead, that when all definitions must be ostensive, as in learning a language where no possibility of translation exists, e.g., as a child, or as a newcomer to an island of an alien tongue, it is best to learn from children because they will repeat indefinitely. Learning was first defined as an alteration of transition probabilities. Speech, broken into phonemes, distinguished, according to Jakobson, by a few decisions between opposites, poorly represented at best in the conventional spelling of English, and studied by Licklider's method of chopping and distorting it to an incredible extent, retains its intelligibility when little is left beyond an indication of when pressure waves cross the axis, and even this is enormously redundant. One point, returning to ten snapshots per second, is the peculiarity of speech to remain intelligible when each tenth of a second is half speech and half noise many decibels louder. Total amount of information conveyed by speech is probably not more than ten bits per second, though it takes a thousand bits per second to produce a sound indistinguishable from it. Shannon's work on redundancy of English in reducing amount of information conveyed per symbol was studied from the position he shares with Wiener, that information is negative entropy. The recipient has a set of entities to match the signals be is intended to receive, and the signal causes him to make the selection. This selective information was found to be comparable to MacKay's logon information but not to his metron information, the point being that the entropic cost of a metron of information goes up as the square of the number of metrons, rather than as the numbers. | [80]

We have considered Zipf's law – that the number of kinds of any given rarity is proportional to the square of the rarity – but I do not think we are satisfied either as to the validity of the law, the basis of the exceptions, or the universe it presupposes. Finally, we have proposed to look into the amount of information conferred upon us by our genes, and have tried to straighten out for ourselves those difficulties that have arisen because of confusion of the level of discourse. It is my hope that by the time this session is over, we shall have agreed to use very sparingly the terms »quantity of information« and »negentropy.«

APPENDIX II: REFERENCES

INTRODUCTORY REMARKS

1. Leibnitz[!], G. W.: *The Philosophical Wrintings of Leibnitz.* Translated by C. R. Morris. New York, Dutton (Everyman's Library), 1934.
2. Aristotle: *De Anima. The Works of Aristotle.* Vol. III. W. D. Ross, Editor. New York, Oxford (Clarendon Press), 1931.
3. Hume, D.: *Treatise of Human Nature.* New York, Dutton (Everyman's Library), 1911.
4. Kant, I.: *Critique of Pure Reason.* Translated by J. Meiklejohn. New York, Dutton (Everyman's Library), 1934.
5. Laycock, T.: *Mind and Brain.* 2nd ed., 2 vols. New York, Appleton, 1869.
6. Butler, S.: *Unconscious Memory.* rev. ed. New York, Dutton, 1910.
7. —: *Erewhon.* Press, Canterbury, N. Z. 1863.
8. Pitts, W., and McCulloch, W. S.: How we know universals. The perception of auditory and visual forms. *Bull. Math. Biophys.* **9**, *127 (1947).*
9. Lashley, K. S.: Functional interpretation of anatomic patterns. *A Research Nerv. & Ment. Dis. Proc.* **30**, *529 (1952).*
10. Sholl, D. A., and Uttley, A. M.: Pattern discrimination and the visual cortex. *Nature* **171**, *387 (1953).*

SEMANTIC INFORMATION AND ITS MEASURES

1. Carnap, R., and Bar-Hillel, Y.: *An Outline of a Theory of Semantic Information.* Technical Report No. 247, Research Laboratory of Electronics, Cambridge, Mass., Massachusetts Institute of Technology, 1952.
2. Shannon, C. E., and Weaver, W.: *Mathematical Theory of Communication.* Urbana, Univ. Illinois Press, 1949.
3. Cherry, E. C.: A history of the theory of information. *Proc. Inst, Elect. Engnrs.* **98**, *383 (1951).*
4. MacKay, D.: In search of basic symbols. *Cybernetics.* H. Von Foerster, Editor, Trans. Eighth Conf. New York, Josiah Macy, Jr. Foundation, 1952 (p. 181). | [82]
5. Kemeny, J. G.: A logical measure function. *J. Symb. Logic* **18**, *289 (1953).*
6. Carnap, R.: *Logical Foundations of Probability.* Chicago, Univ. Chicago Press, 1950.
7. Quine, W. V. O.: *From a Logical Point of View,* Cambridge, Mass., Harvard Univ. Press, 1953.

MEANING IN LANGUAGE AND HOW IT IS ACQUIRED

1. Wiener, N.: *Ex-Prodigy.* New York, Simon & Schuster, 1953 (p. 299).
2. Werner, H.: On the development of word meanings. *Cybernetics.* H. Von Foerster, Editor. Trans. Seventh Conf. New York, Josiah Macy, Jr. Foundation, 1951 (p. 187).
3. Stroud, J.: The development of language in early childhood. *Cybernetics.* H. Von Foerster, Editor. Trans. Seventh Conf. New York, Josiah Macy, Jr. Foundation, 1951 (p. 205).
4. Jesperson, O.: *Language; Its Nature, Development, and Origin.* London, Allen & Unwin, and New York, Holt, 1922 (p. 116).

5. GARDNER, M.: Is nature ambidextrous? *Philosophy & Phenoinenological Res.* **13**, *200 (1952).*
6. KELLER, H. A.: *The Story of My Life.* New York, Doubleday, 1903 (p. 23).
7. TITCHENER, E. B.: *Lectures on the Experimental Psychology of the Thought Processes.* New York, Macmillan, 1909 (p. 175).
8. T'AO, CH'IEN (365-427 A. D.): Wu-liu-Hsien-sheng Chuan (Biography of the Gentlemen of the Five Willows) *Collected Works,* 1910.
9. KUBIE, L. S.: The relationship of symbolic function in language formation and in neurosis. *Cybernetics.* H. Von Foerster, Editor. Trans. Seventh Conf. New York, Josiah Macy, Jr. Foundation, 1951 (p. 233).
10. PIAGET, J.: Types and stages in the conversation of children between the ages of four and seven. *The Language and Thought of the Child.* 2nd ed. New York, Harcourt, 1932 (p. 50).
11. TRAGER, G. L.: The field of linguistics. *Studies in Linguistics: Occasional Papers No. 1.* Norman, Okla., Battenburg Press, 1949.
12. DAVIES, H. H.: *The Mullusc,* Annotated phonetic edition with tone-marks by Dorothee Palmer. Cambridge, Heffer, and New York, Appleton, 1929.
[83] 13. HARRIS, Z. S.: From phoneme to morpheme. *Language* 31, *190 (1955).* |
14. HOCKETT, C. F.: *A Manual of Phonology.* Baltimore, 1955 (p. 15).
15. CHERRY, E. C., HALLIE [!], M., and JAKOBSON, R.: Toward the logical description of languages in their phonetic aspect. *Language* **29**, *34 (1953).*
16. MACKAY, D. M.: In search of basic symbols. *Cybernetics.* H. Von Foerster, Editor. Trans. Eighth Conf. New York, Josiah Macy, Jr. Foundation, 1952 (p. 182).
17. PIAGET, J.: Understanding and verbal explanation between children of the same age between the years of six and seven. *The Language and Thought of the Child.* 2nd ed., New York, Harcourt, 1932 (p. 120).
18. FERRIÉRE, M. A.: La psychologie bibliologique. D'après les documents et les travaux de Nicolas Roubakine. *Arch. de Psych.* **16**, *101 (1917).*

INDEX

N

O

P

Q

R

Tobias Harks, Sebastian Vehlken (eds.)
Neighborhood Technologies
Media and Mathematics of Dynamic Networks

240 p. ▪ Softcover
ISBN 978-3-03734-523-8 ▪ US$ 45,00 ▪ € 39,95

Neighborhood Technologies expands upon sociologist Thomas Schelling's wellknown study of segregation in major American cities, using this classic work as the basis for a new way of researching social networks across disciplines. Up to now, research has focused on macrolevel behaviors that, together, form rigid systems of neighborhood relations. But can neighborhoods, conversely, affect larger, global dynamics? This volume introduces the concept of "neighborhood technologies" as a model for intermediate, or meso-level, research into the links between local agents and neighborhood relations. Bridging the sciences and humanities, Tobias Harks and Sebastian Vehlken have assembled a group of contributors who are either natural scientists with an interest in interdisciplinary research or tech-savvy humanists. With insights into computer science, mathematics, sociology, media and cultural studies, theater studies, and architecture, the book will inform new research.